The Forest of the Lacandon Maya

Suzanne Cook

The Forest of the Lacandon Maya

An Ethnobotanical Guide

Springer

Suzanne Cook
Department of Linguistics
University of Victoria
Victoria, BC, Canada

Additional material to this book can be downloaded from http://extras.springer.com.

Videos can also be accessed at http://link.springer.com/book/10.1007/978-1-4614-9111-8

ISBN 978-1-4614-9110-1 ISBN 978-1-4614-9111-8 (eBook)
DOI 10.1007/978-1-4614-9111-8

Library of Congress Control Number: 2015950469

Springer New York Heidelberg Dordrecht London

Printed on acid-free paper

Springer Science+Business Media LLC New York is part of Springer Science+Business Media (www.springer.com)

Preface

When I first drove into Naha', in 1990, all the Lacandon men of the community of Naha' were at work in the middle of a logging road erecting electric power poles. Although electricity failed to arrive for another decade, it was clear that a dramatic shift in the Lacandones' way of life was underway.

I was inspired at that time by a book which described the Lacandones as "The Last Lords of Palenque".[1] It was a romantic portrayal of a Mayan people supposedly forgotten by time. The book has since been debunked for its premise by most scholars: today it is confirmed that the Ch'ol Mayas, rather than the Lacandones, were the builders of Palenque. Furthermore, at the time of its writing, the traditional culture has largely disintegrated, and the forest had already shrunk to a few protected reserves. When I arrived, the government-owned logging company was still extracting timber, and in its wake, immigrants were clamouring to establish homesteads.

While conservationists have made some progress towards curbing the destruction, only a few of them have considered the attrition of Lacandon traditional ecological knowledge. Gertrude "Trudy" Blom, wife of archaeologist Franz Blom and proprietor of Na Balom, a research station in San Cristobal, was a powerful voice among those wanting to save the Lacandon forest. As the trees were being felled, she alone expressed great concern for the Lacandones' emotional, spiritual, and psychological well-being, believing that without the forest they would lose their culture and be forced to join the ranks of the disenfranchised, indigenous people of Mexico. Others were Nations and Nigh, who were the first to show how Lacandon traditional knowledge sustained the ecological diversity of the forest, in their paper "The Evolutionary Potential of Lacandon Maya Sustained-Yield Tropical Forest Agriculture" (1980). Their thesis was that the Lacandones' traditional agricultural practices not only preserved the forest but facilitated its regeneration, and that without the forest the Lacandones' ecological knowledge would be lost. Thus, the fates of the forest and the Lacandones were intertwined, each depending on the other to survive.

This book is an attempt to record as much of that knowledge as possible, with the hope that it will aid conservationists in their ongoing efforts to save the Lacandon forest, and that it will also interest Mayanists, Mayan enthusiasts, and most of all, the Lacandones.

The videos for this book can be accessed at:
http://www.springer.com/us/book/9781461491101

Victoria, BC — Suzanne Cook

[1] V. Perera and R. D. Bruce, *The Last Lords of Palenque: the Lacandon Mayas of the Mexican Rainforest* (Berkeley and Los Angeles: University of California Press, 1986).

Acknowledgements

I am indebted to the multitude of people who helped me with taxonomic identification. They are: A. C. Aceby, Estación de Biología Tropical "Los Tuxtlas" Instituto de Biología, UNAM; Alfonso Delgado-Salinas, Departamento de Botánica Instituto de Biología, UNAM; Andrew Henderson, MOBOT; Andrew MacDonald, UTPA; Axel Dalberg Poulsen, Oslo Botanical Garden, Natural History Museum University of Oslo; Billie L. Turner, University of Texas at Austin; Carlos R. Beutelspacher Baights, Herbario Eizi Matuda, Universidad de Ciencias y Artes de Chiapas; Daniel Austin, Arizona-Sonora Desert Museum; Daniel L. Nickrent, Southern Illinois University; Donald Hodel, University at Davis, U. Cal; Duncan Golicher, El Cologio de la Frontera Sur (ECOSUR), San Cristobal; Eduardo Gomes Goncalves, Depto. de Botânica, ICB, UFMG, Brazil; Eleazar Carranza, Instituto de Ecología, A. C., Centro Regional del Bajío, Pátzcuaro, Mich.; Erin Tripp, Rancho Santa Ana Botanic Garden; Filiberto Malagón, Laboratorio de Malariología, Facultad de Medicina, Universidad Nacional Autónoma de México; Francisco Morales, Instituto Nacional de Biodiversidad (INBio); George Staples, Singapore Botanic Garden, National Parks Board; Gerald D. Carr, Department of Botany and Plant Pathology Oregon State University; Gib Cooper, Bamboo of the Americas Organization (BOTA); Gilberto Cortes, BOTA; Harold Robinson, Smithsonian Institution; Heike Vibrans Lindemann, Colegio de Postgraduados en Ciencias Agrícolas, Mexico DF; Jens G. Rohwer, Biozentrum Klein Flottbek und Botanischer Garten, Hamburg, Germany; Jerome M. Levi, Carleton College, MN; Jim Conrad, author, naturalist, botanist, Chichen Itza, YUC., Mexico; Job Kuijt, University of Victoria, BC; John Gaskin, Botanist/Research Leader PMRU, Acting Research Leader ASRU,USDA ARS NPARL; John Longino, University of Utah; John M. MacDougal, Harris-Stowe State University; John R. Paul, Colorado State University; Jorge E. Arriagada, St. Cloud State University, Minnesota; Julio Enrique Morales, Herbario USCG, Guatemala; Kenneth A Langeland, Center for Aquatic and Invasive Plants, University of Florida (IFAS); Kent D. Perkins, University of Florida Herbarium (FLAS); Lauren Raz, Universidad Nacional de Colombia; M. C. Alvaro Campos Villanueva, Estación de Biología Tropical "Los Tuxtlas", Instituto de Biología, UNAM; Mark E Olson, Universidad Nacional Autónoma de México; Michael Grayum, MOBOT; Nee, Michael, New York Botanical Garden; Nelson Zamora, Instituto Nacional de Biodiversidad (INBio); Oscar Farrera Sarmiento, Flora Jardin Botanico F. Miranda, Chiapas; Paul Maas; Peter Jorgensen, MOBOT; Peter W. Fritsch, California Academy of Sciences; Reinaldo Aguilar, Director of the Flowering Plants of the Osa Peninsula, Costa Rica; Robbin Moran, NY Botanical Garden; Rosamond I. Coates, Estacion de Biologia Tropical "Los Tuxtlas", Instituto de Biología, UNAM; Roy Erkens, Institute of Environmental Biology, Utrecht University; Servando Carvajal, Herbario del Instituto de Botanica de la Universidad de Guadalajara, Mexico; Sohrab Kheradmandan, International Investigation Centre for Natural Sciences, I2CNS GmbH; Susan G. Letcher, Purchase College; Thomas Croat, MOBOT; Yero R. Kuethe, James Cook University, Australia.

Many thanks go to Robert M. Laughlin, Department of Anthropology, National Museum of Natural History, Smithsonian Institution, and Nancy J. Turner, School of Environmental Studies, University of Victoria, for reading the book proposal and giving their nod of approval.

Finally, I am indebted to Barry F. Carlson, for his financial and emotional support. Without him this book would not have been possible.

Contents

List of Videos xxv

1 Introduction 1
1.1 The Lacandones 1
1.2 History of Research 2
1.3 Collection Methods 2
1.4 Lacandon Consultants 2
1.5 How to Read This Book 5
1.5.1 Inventory Format 5
1.6 Abbreviations 6
References 7

2 Language 9
2.1 The Language System 9
2.2 Previous Research 9
2.3 Transcription Conventions in This Work 10
References 10

3 The Lacandon Rain Forest 13
3.1 Geography 13
3.2 Climate 14
3.3 Vegetation 14
References 24

4 The Role of Plants in Traditional Lacandon Culture 25
4.1 Overview 25
4.2 Food 25
4.2.1 Maize 25
4.2.2 Beans 26
4.2.3 Fruits and Vegetables 27
4.2.4 Fungi 27
4.2.5 Ritual Offerings 28
4.3 Agriculture 37
4.3.1 Overview 37
4.3.2 *Kol 'milpa'* 37
4.3.3 *Kolil nah* 'House Garden' 39
4.3.4 *Paakche' kol* 39
4.3.5 Subsistence Activities 41
4.3.6 Current Situation 44
4.4 Housing 45
4.4.1 Layout 45
4.5 Artefacts 61
4.5.1 Furniture 61
4.5.2 Containers 61

4.5.3 Fibre and Cordage ... 64
4.5.4 Religious Objects ... 65
4.5.5 Weapons and Tools for Hunting, Trapping, and Fishing ... 73
4.6 Crafts ... 79
4.6.1 Jewellery ... 79
4.6.2 Bows and Arrows ... 80
4.6.3 Pottery ... 80
4.6.4 Musical Instruments ... 81
4.6.5 Spinning ... 82
4.6.6 Weaving ... 83
4.6.7 Dyes ... 86
4.7 Clothing ... 88
4.7.1 Men's Clothing ... 88
4.7.2 Women's Clothing ... 88
4.7.3 Children's Clothing ... 90
4.8 Medicine and Curing ... 90
4.9 Lacandon Plant Classification ... 97
4.9.1 Lacandon System of Botanical Classification and Nomenclature ... 97
4.9.2 Correspondence Between Folk Taxa and Botanical Taxa ... 99
References ... 100

5 Botanical Inventory ... 103
Botanical-Lacandon Index ... 281
English-Lacandon Index ... 292
References ... 302

6 Ethnographic Inventory ... 309
6.1 Agriculture ... 309
6.2 Artefacts ... 311
6.2.1 Accessories ... 311
6.2.2 Containers ... 311
6.2.3 Cordage ... 312
6.2.4 Craft ... 313
6.2.5 Furniture ... 314
6.2.6 Hunting, Fishing, Trapping Items ... 315
6.2.7 Musical Instruments ... 316
6.2.8 Religious Objects ... 317
6.2.9 Tools ... 318
6.3 Clothing ... 320
6.4 Deities ... 321
6.5 Food ... 322
6.6 Geography ... 325
6.7 Housing ... 327
6.8 Medical Conditions ... 330
References ... 333

Appendix 1 Lacandon Plants Unidentified Botanically ... 335

Appendix 2 General Botanical Terms in Lacandon ... 343

Appendix 3 Seasons ... 353

Appendix 4 Weather ... 355

Appendix 5 Lacandon Texts 357
Take Care of the Forest, by CK, Naha', 1991 357
Secret of the *Balche'*, by AM, Naha', 2003 358
Song of the Cacao, by JK, Naha', 2003 360
How to Build a House, by BM, Naha', 2011 361
Flint Song, by AM, Naha', 2003 362
Song for Spinning Thread, by JK, Naha', 2003 363
Song for Grinding, by JK, Naha', 2003 365

Index 371

List of Figures

Fig. 1.1 AM (Lt) and BM (Rt) ... 3
Fig. 1.2 AM offers balche' to the gods ... 3
Fig. 1.3 KP with unripe poochin fruit ... 4
Fig. 1.4 KM stripping vine for baskets ... 4
Fig. 1.5 Säkho'ol ... 5
Fig. 1.6 Photographer, Chan K'in Daniel ... 5

Fig. 3.1 Chiapas physical map ... 15
Fig. 3.2 Lacandon communities, northern and southern ... 16
Fig. 3.3 Lacandon area ... 17
Fig. 3.4 Northern Lacandon territory ... 18
Fig. 3.5 Fallow milpa ... 18
Fig. 3.6 Ek' lu'um 'black soil' ... 19
Fig. 3.7 Soil structure ... 19
Fig. 3.8 Canoe by rock cliff ... 20
Fig. 3.9 Waterfall in the southern Lacandon territory ... 20
Fig. 3.10 Forest, 10 year fallow ... 20
Fig. 3.11 Lagoon ... 21
Fig. 3.12 Lagoon ... 21
Fig. 3.13 Cattle in pasture ... 22
Fig. 3.14 Naha', 2010 ... 22
Fig. 3.16 Lacandon homestead ... 23
Fig. 3.15 Schoolyard ... 23

Fig. 4.1 Making tortillas ... 35
Fig. 4.2 Making tortillas ... 35
Fig. 4.3 K'u'umbil wah ... 36
Fig. 4.4 Straining sa' ... 36
Fig. 4.5 Balche' bark ... 37
Fig. 4.6 Fallow milpa first year ... 38
Fig. 4.7 Fallow milpa second year ... 38
Fig. 4.8 Sketch of household and gardens ... 40
Fig. 4.9 House garden ... 40
Fig. 4.10 Slashed milpa ... 42
Fig. 4.11 Milpa, 2010 ... 43
Fig. 4.12 Doubled-over maize ... 43
Fig. 4.13 Flooded milpa, 2010 ... 45
Fig. 4.14 Two extended families' compounds ... 46
Fig. 4.15 Typical homestead ... 46
Fig. 4.16 House garden, compost ... 47
Fig. 4.17 House garden of KP ... 47
Fig. 4.18 Koh's backyard ... 48
Fig. 4.19 Homestead ... 48
Fig. 4.20 Outdoor sink ... 49

Fig. 4.21 Storage shed 49
Fig. 4.22 Chicken coops 50
Fig. 4.23 Typical Lacandon house 50
Fig. 4.24 Woman on hammock 51
Fig. 4.25 Cooking hearth 51
Fig. 4.26 Cooking range 52
Fig. 4.27 The Kohs' kitchen, 2000 52
Fig. 4.28 The old Kohs' kitchen 53
Fig. 4.29 The old Kohs' kitchen 53
Fig. 4.30 The new Kohs' kitchen, 2011 54
Fig. 4.31 Stove 54
Fig. 4.32 Cooking on a traditional stove 55
Fig. 4.33 Kun thatch 55
Fig. 4.34 Battens and shelf for god pots 56
Fig. 4.35 Parts of a god house 56
Fig. 4.36 Lashing 57
Fig. 4.37 Smoke vent 57
Fig. 4.38 Battens in a god house 58
Fig. 4.39 God house interior 58
Fig. 4.40 Ceremonial kitchen 59
Fig. 4.41 Vertical pole house walls 59
Fig. 4.42 Concrete house 60
Fig. 4.43 Jailhouse 60
Fig. 4.44 AT and his gourd beehive for meliponines, stingless bees 61
Fig. 4.45 Gourd beehive hangs from house rafters 62
Fig. 4.46 Gourd beehive interior 62
Fig. 4.47 KM opening lek gourd, to make a lekil wah 'tortilla plate' 62
Fig. 4.48 K'u'umbil wah tortillas in a lekil wah 63
Fig. 4.49 Hama' balche'cups 63
Fig. 4.50 Gourd bowls used for preparing cocoa 64
Fig. 4.51 Me'et 'trivet' used to stabilize gourd cups 64
Fig. 4.52 Hol drying on line 65
Fig. 4.53 Cord made from hol 66
Fig. 4.54 Rolling fibres of hol 66
Fig. 4.55 Rolling fibres of hol 67
Fig. 4.56 Hol hanging over rafters in bedroom 67
Fig. 4.57 Pounded bark fibre of bit'skal hu'un (*Ficus* sp.) 68
Fig. 4.58 Wooden fibre beater 68
Fig. 4.59 Aerial root of k'i'ilix holop' (*Philodendron radiatum*) 68
Fig. 4.60 God house with balche' chem in foreground 69
Fig. 4.61 Mashing pom in the chemil pom 69
Fig. 4.62 K'ayum drum 70
Fig. 4.63 God pot 70
Fig. 4.64 Kiln 71
Fig. 4.65 Pak, the balche' serving urn 71
Fig. 4.66 Men's pak (Lt) and women's pak (Rt) 72
Fig. 4.67 Xikal 72
Fig. 4.68 Rubber 72
Fig. 4.69 Rubber several weeks after it was made 73
Fig. 4.70 Rubber figurines 73
Fig. 4.71 Cigars 73
Fig. 4.72 Ka' grindstone 74
Fig. 4.73 Earliest depiction of chocolate, from the Princeton Vase Late Classic period AD. 750 74
Fig. 4.74 Net full of gourd cups 75

Fig. 4.75 Ceremonial kitchen table, utensils, and bowls 75
Fig. 4.76 Cooking utensils 76
Fig. 4.77 Xämäch clay griddle 76
Fig. 4.78 Arrow set 77
Fig. 4.79 Arrow fletching 77
Fig. 4.80 Arrow heads 78
Fig. 4.81 Arrow binding 78
Fig. 4.82 Bamboo fish spear 79
Fig. 4.83 Seed necklace of *Ormosia schippii* (*red and black*), *Hymenaea courbaril* L. (*large brown*), and *Cois lacryma-jobi* (*grey*) 79
Fig. 4.84 Seed necklace of *Sapindus saponaria* (*black*) and *Ormosia isthmensis* (*red*) 79
Fig. 4.85 Necklace with 'hamburger beans' (*Mucuna sloanei*) 80
Fig. 4.86 Toucan beak 80
Fig. 4.87 Bow and arrow set (ca. 1950s) 81
Fig. 4.88 Female doll 81
Fig. 4.89 Male doll 82
Fig. 4.90 Clay animals 82
Fig. 4.91 Spinning top (*Quercus corrugate*) 83
Fig. 4.92 Ceremonial rattle 83
Fig. 4.93 Backstrap loom 84
Fig. 4.94 Chan K'in wearing the hand spun cotton xikul 84
Fig. 4.95 Basket 85
Fig. 4.96 Basket rim 85
Fig. 4.97 Basket bottom 86
Fig. 4.98 Hanging basket, weave 87
Fig. 4.99 Hanging basket 87
Fig. 4.100 Ba'ay 'net bag' 88
Fig. 4.101 Woven bag, loop 88
Fig. 4.102 Woven purse 88
Fig. 4.103 Woven purse, weave 89
Fig. 4.104 Traditional hammock 89
Fig. 4.105 Hammock, knots 90
Fig. 4.106 Cotton xikul 91
Fig. 4.107 Bark cloth xikul 91
Fig. 4.108 Leather purse-front 91
Fig. 4.109 Leather purse-back 92
Fig. 4.110 Traditional women's dress 92
Fig. 4.111 (**a**) Feather hair ornament (**b**) *Clusia flava* blossom 93
Fig. 4.112 (**a**) Lacandon girls, 2011 (**b**) Lacandon boys on hammock 93

Fig. 5.1 'Aak'älyoom *Cestrum nocturnum* 104
Fig. 5.2 (**a**) 'Akunte' *Acacia mayana* (**b**) 'Akunte' *Acacia mayana* thorn 105
Fig. 5.3 (**a**) 'Ak su'uk *Bothriochloa laguroides* (**b**) 'Ak su'uk *Bothriochloa laguroides* tiller (**c**) 'Ak su'uk *Bothriochloa laguroides* 106
Fig. 5.4 (**a**) 'Anis *Abelmoschus moschatus* unopened pod (**b**) 'Anis *Abelmoschus moschatus* flower (**c**) 'Anis *Abelmoschus moschatus* seed pod 107
Fig. 5.5 Axux ak' *Cydista aequinoctilis* 108
Fig. 5.6 (**a**) 'Äm ch'uplal *Ormosia isthmensis* (**b**) 'Äm ton *Ormosia schippii* 109

Fig. 5.7 (**a**) Babah 'oonte' (kakache') *Ocotea cernua* (**b**) Babah 'oonte' *Ocotea cernua* seeds 110
Fig. 5.8 Bahun che' *Cordia alliodora* 111
Fig. 5.9 Baknikte' *Polianthes tuberosa* 111
Fig. 5.10 Balumte' *Theobroma bicolor* 113
Fig. 5.11 Bamax *Pseudolmedia* sp. 113
Fig. 5.12 (**a**) Bambu *Bambusa vulgaris* 'Vittata' (**b**) Bambu *Bambusa vulgaris* 'Vittata' 114
Fig. 5.13 (**a**) Bey ule' tähte' *Podocarpus matudai* (**b**) Bey ule' tähte' *Podocarpus matudai* 115
Fig. 5.14 (**a**) Bilim suli' *Dioscorea bulbifera* (**b**) Bilim suli' *Dioscorea bulbifera* 115
Fig. 5.15 Booxa'an *Acoelorraphe wrightii* 117
Fig. 5.16 (**a**) Bu'ul box *Vigna unguiculata* (**b**) Bu'ul box *Vigna unguiculata* seeds (**c**) Bu'ul box *Vigna unguiculata* flower 118
Fig. 5.17 (**a**) Bubuhchi' (yo'och ahpek') Canavalia flower (**b**) Bubuhchi' *Canavalia ensiformis* (**c**) Bubuhchi' (bu'ul boox) *Canavalia ensiformis* (**d**) Bubuhchi' *Canavalia ensiformis* (**e**) Bubuhchi' *Canavalia ensiformis* (**f**) Bubuhchi' *Canavalia ensiformis* (**g**) Bubuhchi' *Mucuna puriens* (**h**) Bubuhchi' *Mucuna puriens* 120
Fig. 5.18 (**a**) Buk'luch' *Vanilla planifolia.* (**b**) Buk'luch' *Vanilla planifolia* flower 121
Fig. 5.19 (**a**) Chank'ala' *Canna indica* (**b**) Chank'ala', bamboo, and petsk'in necklace (**c**) Chank'ala' flower 121
Fig. 5.20 Chawah ik *Capsicum frutescens* 122
Fig. 5.21 (**a**) Chäkchob *Lantana trifolia* (**b**) Chäkchob *Lantana trifolia* 122
Fig. 5.22 Chäklah *Bursera simaruba* 123
Fig. 5.23 (**a**) Chäklanpix *Clibadium arboreum* (**b**) Chäklanpix *Schistocarpha eupatorioides* (**c**) Chäklanpix *Schistocarpha eupatorioides* 124
Fig. 5.24 (**a**) Chäk halol *Heliocarpus appendiculatus* (**b**) Chäk halol *Heliocarpus appendiculatus* 125
Fig. 5.25 Chäk k'o'och *Ricinus communis* 126
Fig. 5.26 Chäk mäkäl *Xanthosoma mafaffa* 126
Fig. 5.27 Chäk me'exk'in *Calliandra calothyrsus* 127
Fig. 5.28 (**a**) Chäk nikte' *Plumeria rubra f. rubra* (**b**) Chäk nikte' *Plumeria rubra f. rubra* blossom 127
Fig. 5.29 (**a**) Chäk pach *Ternstroemia tepezapote* (**b**) Chäk pach *Ternstroemia tepezapote* 128
Fig. 5.30 (**a**) Chäk ts'in *Manihot esculenta* (**b**) Chäk ts'in *Manihot esculenta* roots 129
Fig. 5.31 Chäk ts'ulha' *Crinum amabile* 130
Fig. 5.32 (**a**) Chechem *Metopium brownei* (**b**) Chechem *Metopium brownei* 131
Fig. 5.33 Chiah *Dieffenbachia* sp. 132
Fig. 5.34 Choochel ak' *Rhynchosia erythrinoides* 133
Fig. 5.35 Chukuch le' uxiiw *Heliconia librata* inflorescence 134
Fig. 5.36 Chukum che' *Croton draco* 134
Fig. 5.37 Chunup *Clusia lundellii* 135
Fig. 5.38 (**a**) Ch'ayok' *Lycianthes heteroclite* (**b**) Ch'ayok' *Solanum americanum* (**c**) Ch'ayok' *Solanum rovirosanum* 136
Fig. 5.39 (**a**) Ch'äm *Bromelia pinguin* (**b**) Ch'äm *Bromelia pinguin* 136
Fig. 5.40 Ch'ibix bo'oy (**a**) *Chamaedorea elegans* (**b**) Ch'ibix uk'um *Chamaedorea cataractarum* 137
Fig. 5.41 Ch'it bo'oy *Chamaedorea* sp. 139
Fig. 5.42 Ch'obenche' *Trichilia* sp. 139
Fig. 5.43 Ch'om mäkäl (hach ch'om) *Xanthosoma robustum* 140
Fig. 5.44 (**a**) Ch'un ak' *Passiflora aff. ambigua* interior (**b**) Juliana and Omar with ch'un ak' fruit 141

Fig. 5.45 Ek' bahche' *Guatteria anomala* 141
Fig. 5.46 (**a**) Ha'as *Pouteria mammosa* (**b**) Ha'as *Pouteria mammosa* 143
Fig. 5.47 Haban uts'u' *Podachaenium eminens* 143
Fig. 5.48 Habonsiyo/habón *Sapindus saponaria* 144
Fig. 5.49 (**a**) Hach 'akte' *Astrocaryum mexicanum* thorns (**b**) Hach 'akte' *Astrocaryum mexicanum* 144
Fig. 5.50 (**a**) Hach balche' *Lonchocarpus* sp. (**b**) Hach balche' seeds (**c**) Hach balche' seed pod (**d**) Hach balche' trunk 146
Fig. 5.51 (**a**) Hach boox *Musa acuminata* ×*Musa balbisiana* (**b**) Hach boox *Musa acuminata* × *Musa balbisiana* (**c**) Hach boox *Musa acuminata* × *Musa balbisiana* 147
Fig. 5.52 (**a**) Hach bo'oy *Chamaedorea pinnatifrons* (**b**) Hach bo'oy *Chamaedorea pinnatifrons* 148
Fig. 5.53 (**a**) Hach chiah (chiah k'aax) *Philodendron inaequilaterum* leaves (**b**) Hach chiah *Philodendron inaequilaterum* 149
Fig. 5.54 Hach halol *Heliocarpus donnell-smithii* 150
Fig. 5.55 (**a**) Hach hänan *Desmoncus orthocanthus* (**b**) Hach hänan *Desmoncus orthocanthus* (**c**) Hach hänan *Desmoncus orthocanthus* leaves and stem 151
Fig. 5.56 (**a**) Hach hoben *Piper auritum* (**b**) Hach hoben *Piper auritum* 152
Fig. 5.57 (**a**) Hach hut'kih *Ipomoea alba* flower (**b**) Hach hut'kih *Ipomoea alba* stem (**c**) Hut'kih *Ipomoea alba* latex 152
Fig. 5.58 (**a**) Hach hu'un *Ficus* sp. (**b**) Hach hu'un inner bark 153
Fig. 5.59 Hach 'ilon *Monstera* sp. 153
Fig. 5.60 Hach luuch *Crescentia cujete* 154
Fig. 5.61 (**a**) Hach mäkuuläm *Piper aduncum* (**b**) Hach mäkuuläm *Piper aduncum* (**c**) Hach mäkuuläm *Piper aduncum* leaves 155
Fig. 5.62 (**a**) Hach 'oochin *Philodendron donnell-smithii* (**b**) Hach 'oochin *Philodendron donnell-smithii* leaf 156
Fig. 5.63 Hach 'oox *Brosimum* sp. leaves 156
Fig. 5.64 Hach pätan *Musa acuminata* x*Musa balbisiana* 157
Fig. 5.65 (**a**) Hach sukal *Saccharum officinarum* (**b**) Hach sukal *Saccharum officinarum* 157
Fig. 5.66 (**a**) Hach suli' *Dioscorea alata* (**b**) Hach suli' *Dioscorea alata* 158
Fig. 5.67 Hach su'uk *Digitaria setigera* 158
Fig. 5.68 Hach taw *Belotia mexicana* 159
Fig. 5.69 (**a**) Hach ts'in *Manihot esculenta* leaves (**b**) Hach ts'in *Manihot esculenta* (**c**) Hach ts'in *Manihot esculenta* roots 160
Fig. 5.70 (**a**) Hach xiiw (chukuch le uxiiw) *Heliconia* sp. (**b**) Hach xiiw *Heliconia* sp. 161
Fig. 5.71 Halo'och k'i'ix *Rubus* sp. 162
Fig. 5.72 Holop' *Philodendron smithii* 163
Fig. 5.73 (**a**) Huhup *Spondias* sp. (**b**) Huhup *Spondias* sp. (**c**) Huhu' (huhup) bark 164
Fig. 5.74 (**a**) Hunbuhk'ä'winik pätan *Musa acuminata* × *Musa balbisiana* (**b**) Hunbuhk'ä'winik pätan *Musa acuminata* × *Musa balbisiana* fruit 165
Fig. 5.75 'Ibil k'aax *Oxyrhynchus trinervius* seed 166
Fig. 5.76 (**a**) Iich'ak tolok (nukuch tak'lan che') *Struthanthus orbicularis* vine (**b**) Iich'ak tolok (nukuch tak'lan che') *Struthanthus orbicularis* vine (**c**) Iich'ak tolok (nukuch tak'lan che') *Struthanthus orbicularis* fruit 166
Fig. 5.77 'Is *Ipomoea batatas* 167
Fig. 5.78 (**a**) Its'in säk wawal *Calathea crotalifera* flowers (**b**) Its'in säk wawal *Calathea crotalifera* leaf 168
Fig. 5.79 Kahanche' *Clusia flava* 169

Fig. 5.80 Kawe' *Coffea* L 169
Fig. 5.81 Kih *Agave sisalana* 171
Fig. 5.82 Kimil sukun Asclepiadaceae 171
Fig. 5.83 (**a**) Koke' ak' *Smilax luculenta* (**b**) Koke' ak' *Smilax luculenta* 172
Fig. 5.84 Komo' (te' 'usin) *Renealmia* sp. 172
Fig. 5.85 (**a**) Kopo' *Ficus* sp. (**b**) Kopo' wits *Ficus* sp. 173
Fig. 5.86 Korason de mono *Entada gigas* 174
Fig. 5.87 Kubuh *Pachira aquatica* leaves 174
Fig. 5.88 Kulantro *Eryngium foetidum* 174
Fig. 5.89 (**a**) Kun *Cryosophila stauracantha* plants (**b**) Kun *Cryosophila stauracantha* fruit 175
Fig. 5.90 Kunkunche' *Oecopetalum mexicanum* 176
Fig. 5.91 Kunkunche' kab *Clethra* sp. 176
Fig. 5.92 Kuti' *Talauma mexicana* fruit 176
Fig. 5.93 Ku'uchnook' Bidens odorata 177
Fig. 5.94 Kuut' *Calathea macrosepala* 177
Fig. 5.95 (**a**) *Ganoderma applanatum* (**b**) Kuxumche' (kib lu'um) *Cookeina tricholoma* 178
Fig. 5.96 (**a**) K'ak'che' *Aegiphila monstrosa* flowers (**b**) K'ak'che' *Aegiphila monstrosa* leaves (**c**) K'ak'che' *Aegiphila monstrosa* growing tip (**d**) K'ak'che' *Bauhinia rubeleriziana* leaves (**e**) K'ak'che' *Bauhinia rubeleriziana* seed pods 179
Fig. 5.97 (**a**) K'alok' che' *Billia colombiana* seed (**b**) K'alok' che' *Billia colombiana* seed 180
Fig. 5.98 (**a**) K'ante' *Erythrina berteroana* seeds (**b**) K'ante' *Erythrina berteroana* flowers (**c**) K'ante' *Erythrina berteroana* plant 180
Fig. 5.99 (**a**) K'äb ta'k'in *Oreopanax obtusifolius* plant (**b**) K'äb ta'k'in *Oreopanax obtusifolius* inflorescence 181
Fig. 5.100 (**a**) K'än 'abäl *Spondias purpurea forma lutea* ripe fruit (**b**) K'än 'abäl *Spondias purpurea f. lutea* leaves (**c**) K'än 'abäl *Spondias purpurea f. lutea* tree 182
Fig. 5.101 (**a**) K'än bo'oy *Chamaedorea neurochlamys* leaves (**b**) K'än bo'oy *Chamaedorea neurochlamys* fruit 183
Fig. 5.102 (**a**) K'än chi' *Byrsonima crassifolia* ripe fruit (**b**) K'än chi' *Byrsonima crassifolia* unripe fruit (**c**) K'än chi' *Byrsonima crassifolia* tree (**d**) K'än chi' *Byrsonima crassifolia* leaves 184
Fig. 5.103 K'eben *Chamaedorea ernesti-augusti* 185
Fig. 5.104 (**a**) K'i'ilix holop' *Philodendron radiatum* leaves (**b**) K'i'ilix holop' *Philodendron radiatum* aerial roots (**c**) K'i'ilix holop' *Philodendron radiatum* inflorescence (**d**) K'i'ilix holop' *Philodendron radiatum tuch'ul* (**e**) K'i'ilix holop' *Philodendron radiatum* plant 186
Fig. 5.105 (**a**) K'i'ix 'uk'unche' *Solanum torvum* leaves and unripe fruit (**b**) K'i'ix 'uk'unche' *Solanum torvum* plant (**c**) K'i'ix 'uk'unche' *Solanum torvum* flowers 187
Fig. 5.106 (**a**) K'ik' *Castilla elastica* sapling (**b**) K'ik' *Castilla elastica* latex (**c**) K'ik' latex from bark slash (**d**) K'ik' fruit 188
Fig. 5.107 (**a**) K'ik'che' *Virola guatemalensis* fruit (**b**) K'ik'che' *Virola guatemalensis* sapling (**c**) K'ik'che' *Virola guatemalensis* leaves 188
Fig. 5.108 K'ik'ni' balum *Ardisia compressa* 189
Fig. 5.109 K'o'och *Cecropia* spp. 190

Fig. 5.110 (**a**) K'o'och ba'ats' *Louteridium donnell-smithii* plant
(**b**) K'o'och ba'ats' *Louteridium donnell-smithii* leaves
(**c**) K'o'och ba'ats' *Louteridium donnell-smithii* flower
(**d**) K'o'och ba'ats' *Stanhopea tigrina* ... 191
Fig. 5.111 K'uche' *Cedrela odorata* ... 192
Fig. 5.112 (**a**) K'ul ak' *Dioscorea bartlettii* burl
(**b**) K'ul ak' *Dioscorea bartlettii* leaves ... 192
Fig. 5.113 (**a**) K'um *Cucurbita moschata* (**b**) K'um *squash blossoms* ... 193
Fig. 5.114 (**a**) K'um sek *Passiflora biflora* fruit
(**b**) K'um sek *Passiflora biflora* fruit interior
(**c**) K'um sek *Passiflora biflora* leaves and immature fruit on vine
(**d**) K'um sek *Passiflora biflora* leaf ... 193
Fig. 5.115 K'unche' pahok *Geonoma oxycarpa* ... 194
Fig. 5.116 (**a**) K'uxub *Bixa orellana* flowers (**b**) K'uxub *Bixa orellana* flowers
(**c**) K'uxub *Bixa orellana* plant ... 194
Fig. 5.117 K'u' kan *Selaginella silvestris* ... 195
Fig. 5.118 Läkte' *Zanthoxylum* sp. ... 195
Fig. 5.119 (**a**) Lek *Lagenaria siceraria*
(**b**) Lek *Lagenaria siceraria* seeds ... 196
Fig. 5.120 (**a**) Le'mo' *Epiphyllum* sp. (**b**) Le'mo' *Epiphyllum* sp. ... 196
Fig. 5.121 (**a**) Lo'k'in *Parathesis* sp. plant
(**b**) Lo'k'in *Parathesis* sp. leaves and fruit ... 197
Fig. 5.122 Maasan kitan *Esquisetum hyemale* ... 198
Fig. 5.123 Mahaas *Quararibea funebris* ... 199
Fig. 5.124 Mehen ch'alol ... 200
Fig. 5.125 Mehen ch'eel koox *Chione chiapasensis* ... 201
Fig. 5.126 (**a**) Mehen kopo' *Ficus pertusa* fruit
(**b**) Mehen kopo' *Ficus pertusa* cultivated tree ... 201
Fig. 5.127 Mehen k'än'ol *Astronium graveolens* ... 202
Fig. 5.128 (**a**) Mehen mäkuuläm plant
(**b**) Mehen mäkuuläm leaves and fruit
(**c**) Mehen mäkuuläm leaves ... 203
Fig. 5.129 (**a**) Mehen 'oochin *Philodendron hederaceum*
(**b**) Mehen 'oochin *Syngonium podophyllum* ... 203
Fig. 5.130 Mehen sukal *Saccharum officinarum* ... 204
Fig. 5.131 Mehen tak'lan che' *Phoradendron crassifolium* ... 204
Fig. 5.132 Mehen tuch *Thevetia ahouai* ... 204
Fig. 5.133 Mehen ts'us *Vitis tiliifolia* ... 205
Fig. 5.134 Miis *Carludovica labela* ... 205
Fig. 5.135 (**a**) Mo'ak' *Rhynchosia pyramidalis* seeds
(**b**) Mo'ak' *Rhynchosia pyramidalis* vine ... 206
Fig. 5.136 (**a**) Mumun che' *Telanthophora grandifolia*
(**b**) Mumun che' *Saurauia yasicae*
(**c**) Mumun che' (muxan che') *Gonzalagunia thyrsoidea* ... 207
Fig. 5.137 Muxan che' *Alchornea latifolia* ... 208
Fig. 5.138 Nikte' kisin *Bouvardia longiflora* ... 210
Fig. 5.139 Ni' ah'och *Anthurium* sp. ... 210
Fig. 5.140 Ni' ts'ul *Mucuna argyrophylla* ... 211
Fig. 5.141 (**a**) Nukuch bits' *Inga punctata* pod
(**b**) Nukuch bits' *Inga punctata* sweet aril
(**c**) Nukuch bits' *Inga punctata* seeds ... 211
Fig. 5.142 Nukuch koyoh *Persea schiedeana* ... 212
Fig. 5.143 Nukuch k'än'ol *Mosquitoxylum jamaicense* ... 212

Fig. 5.144 (**a**) Nukuch 'oochin *Philodendron sagittifolium* (**b**) Nukuch 'oochin *Philodendron sagittifolium* (**c**) Nukuch 'oochin *Syngonium chiapense* (**d**) Nukuch 'oochin *Syngonium podophyllum* 214
Fig. 5.145 (**a**) 'Ooh *Gynerium sagittatum* plants (**b**) 'Ooh *Gynerium sagittatum* leaves (**c**) 'Ooh Gynerium sagittatum flower stalks 216
Fig. 5.146 'Oopil k'aax *Annona scleroderma* 217
Fig. 5.147 Pahsa' ak' *Costus pulverulentus* 219
Fig. 5.148 (**a**) Pakyon *Chamaedorea tepejilote* plant (**b**) Pakyon *Chamaedorea tepejilote* ripe berries (**c**) Pakyon *Chamaedorea tepejilote* male inflorescence (**d**) Pakyon *Chamaedorea tepejilote* male inflorescence 220
Fig. 5.149 Patoh ak' *Caesalpinia major* 220
Fig. 5.150 (**a**) Pähk'ol *Spathiphyllum phryniifolium* plants (**b**) Pähk'ol *Spathiphyllum phryniifolium* inflorescence 221
Fig. 5.151 (**a**) Pähpähche' *Chione chiapasensis* tree (**b**) Pähpähche' *Chione chiapasensis* fruit and leaves 221
Fig. 5.152 (**a**) Pähpäh xibil *Ardisia* sp. fruit and leaves (**b**) Pähpäh xibil *Ardisia* sp. fruit and leaves 222
Fig. 5.153 (**a**) Pähpox *Annona muricata* tree (**b**) Pähpox *Annona muricata* immature fruit (**c**) Pähpox *Annona muricata* inflorescence (**d**) Pähpox *Annona muricata* leaves 223
Fig. 5.154 (**a**) Päh ak' *Disciphania calocarpa* vine with fruit (**b**) Päh ak' *Disciphania calocarpa* fruit 223
Fig. 5.155 (**a**) Päh pätan *Musa acuminata* ×*Musa balbisiana* fruit (**b**) Päh pätan *Musa acuminata* × *Musa balbisiana* plant with fruit 224
Fig. 5.156 Peesache' *Pimenta dioica* 225
Fig. 5.157 (**a**) Pets'k'in *Schizolobium parahybum* seed (**b**) Pets'k'in *Schizolobium parahybum* trunk 225
Fig. 5.158 Pichik' (chäk, säk) *Psidium guajava* red and white forms 226
Fig. 5.159 Pixan k'ambul *Quercus skinneri* 226
Fig. 5.160 Pooch'in *Passiflora serratifolia* 227
Fig. 5.161 (**a**) Puhan *Muntingia calabura* tree (**b**) Puhan *Muntingia calabura* flowers and leaves 228
Fig. 5.162 (**a**) Pukte' *Terminalia amazonia* tree (**b**) Pukte' *Terminalia amazonia* seeds 228
Fig. 5.163 (**a**) Put *Carica papaya* tree (**b**) Put *Carica papaya* flowers (**c**) Put *Carica papaya* fruit 229
Fig. 5.164 (**a**) Puuna' *Swietenia macrophylla* tree (**b**) Puuna' *Swietenia macrophylla* seeds (**c**) Puuna' *Swietenia macrophylla* sapling 230
Fig. 5.165 P'akan *Physalis pruinosa* 230
Fig. 5.166 (**a**) P'ix *Sechium edule* fruit (**b**) P'ix *Sechium edule* leaves (**c**) P'ix *Sechium edule* vines 231
Fig. 5.167 (**a**) Sanyah ahch'o' *Melothria pendula* vine with leaves and immature fruit (**b**) Sanyah ahch'o' *Melothria pendula* flower and fruit (**c**) Sanyah ahch'o' *Melothria pendula* flower, tendril and stem 232
Fig. 5.168 San Lorenso (ts'ak se'em) *Kalanchoe pinnata* 232
Fig. 5.169 (**a**) Sasakche' *Dendropanax arboreus* fruit (**b**) Sasakche' *Dendropanax arboreus* inflorescence and fruit 233
Fig. 5.170 Säha' (deleted text) *Doliocarpus dentatus* 233
Fig. 5.171 (**a**) Säkpahen *Coix lacryma-jobi* plants (**b**) Säkche' *Coix lacryma-jobi* flowers 234

Fig. 5.172 (**a**) Säkpähk'ak' che' *Psychotria poeppigiana*
(**b**) Säkpähk'ak' che' *Psychotria poeppigiana* 234
Fig. 5.173 Säk bahche' *Guarea glabra* leaves 235
Fig. 5.174 (**a**) Säk bo'oy *Chamaedorea arenbergiana* plant
(**b**) Säk bo'oy *Chamaedorea arenbergiana* leaves 236
Fig. 5.175 (**a**) Säk chulul *Platymiscium dimorphandrum* trunk
(**b**) Säk chulul *Platymiscium dimorphandrum* seeds 237
Fig. 5.176 (**a**) Säk halol *Heliocarpus donnell-smithii* tree
(**b**) Säk halol *Heliocarpus donnell-smithii* leaves
(**c**) Säk halol *Heliocarpus donnell-smithii* trunk 238
Fig. 5.177 (**a**) Säk mäkäl *Xanthosoma mafaffa*
(**b**) Säk mäkäl *Xanthosoma mafaffa* 238
Fig. 5.178 (**a**) Säk muxän *Heliconia librata* inflorescence
(**b**) Säk muxän (chukuch ule' xiiw) *Heliconia* sp. leaves 239
Fig. 5.179 (**a**) Säk 'oop *Annona cherimoya* immature fruit
(**b**) Säk 'oop *Annona cherimoya* fruit interior
(**c**) Säk 'oop *Annona cherimoya* cultivated tree 240
Fig. 5.180 (**a**) Säk wawal *Calathea lutea* (**b**) Säk wawal *Calathea lutea* 241
Fig. 5.181 (**a**) Säk ya' *Chrysophyllum mexicanum* leaves
(**b**) Säk ya' *Chrysophyllum mexicanum* young trees 242
Fig. 5.182 (**a**) Semet' *Rhipidocladum bartlettii*
(**b**) Semet' *Rhipidocladum bartlettii* 242
Fig. 5.183 Siib *Cladium jamaicense* 243
Fig. 5.184 (**a**) Sikilte' che' *Jatropha curcas* fruit and seeds
(**b**) Sikilte' che' *Jatropha curcas* leaves 244
Fig. 5.185 (**a**) *Sapindus saponaria* fruit and seed
(**b**) *Sapindus saponaria* blossom (**c**) *Sapindus saponaria* suds 245
Fig. 5.186 Subul *Dipholis* sp. 246
Fig. 5.187 (**a**) Suumil bits' *Inga oerstediana*
(**b**) Suumil bits' *Inga oerstediana* 247
Fig. 5.188 (**a**) Taw leaves (**b**) Taw inner bark
(**c**) Chäk taw *Trema micrantha* tree 248
Fig. 5.189 (**a**) Tähte' *Pinus* sp. bark slash
(**b**) Tähte' *Pinus* sp. wood chips 248
Fig. 5.190 Tämän che' *Malvaviscus arboreus* flower 249
Fig. 5.191 Te' usin *Alpinia purpurata* inflorescence 250
Fig. 5.192 Tu'xikin *Ipomoea* sp. flowers 251
Fig. 5.193 T'elen bits' *Inga pavoniana* fruit and leaves 252
Fig. 5.194 (**a**) Tsaah *Cnidoscolus multilobus* leaves
(**b**) Tsaah *Cnidoscolus multilobus* leaf, top side
(**c**) Tsaah *Cnidoscolus multilobus* unopened flowers
(**d**) Tsaah *Cnidoscolus multilobus* plant
(**e**) Tsaah *Cnidoscolus multilobus* stinging hairs 253
Fig. 5.195 (**a**) Tsaayentsaay *Vigna umbellata*
(**b**) Tsaayentsaay *Desmodium tortuosum* 254
Fig. 5.196 (**a**) Tsatsel pom *Protium copal* leaves
(**b**) Tsatsel pom *Protium copal* fruit 255
Fig. 5.197 (**a**) Tsäk'ats *Licania platypus* cultivated tree
(**b**) Tsäk'ats *Licania platypus* fruit (**c**) Tsäk'ats *Licania platypus* leaves (**d**) Tsäk'ats *Licania platypus* small tree 256
Fig. 5.198 (**a**) Tsäläm *Leucaena leucocephala* trees
(**b**) Tsäläm *Leucaena leucocephala* leaves, seeds, and flowers
(**c**) Tsäläm *Leucaena leucocephala* seed pods
(**d**) Tsäläm delonix regia (**e**) Tsäläm delonix regia 257
Fig. 5.199 Tsits *Justicia* sp. leaves 258

Fig. 5.200 (**a**) Tso'ots *Eriobotrya japonica* inflorescence (**b**) Tso'ots *Eriobotrya japonica* (**c**) Tso'ots *Eriobotrya japonica* ripe fruit on tree (**d**) Tso'ots *Eriobotrya japonica* fruit interior (**e**) Tso'ots *Eriobotrya japonica* leaves 259
Fig. 5.201 (**a**) Tso'ots bamax *Pseudolmedia oxyphyllaria* leaves (**b**) Tso'ots bamax *Pseudolmedia oxyphyllaria* sap 260
Fig. 5.202 (**a**) Tsuk'in ak' *Smilax* sp. vine with leaves (**b**) Tsuk'in ak' *Smilax* sp. stem and leaves 260
Fig. 5.203 (**a**) Ts'ak kan ak' snake medicine vine *Piper amalago* leaves (**b**) Ts'ak kan ak' snake medicine stems (undetermined species) (**c**) Ts'ak kan ak' snake medicine leaves (**d**) Ts'ak kan ak' snake medicine vine 261
Fig. 5.204 (**a**) Ts'oy *Clusia lundellii* plant (**b**) Ts'oy *Clusia lundellii* flowers 262
Fig. 5.205 (**a**) Ts'ubtok *Hampea stipitata* flowers (**b**) Ts'ubtok *Hampea stipitata* leaves, stem, and flowers (**c**) Ts'ubtok *Hampea stipitata* leaf, underside (**d**) Ts'ubtok fruit 263
Fig. 5.206 (**a**) Ts'ulha'il petha' *Hymenocallis littoralis* flower (**b**) Ts'ulha'il petha' *Hymenocallis littoralis* flower (**c**) Ts'ulha'il *Hymenocallis littoralis* 264
Fig. 5.207 (**a**) Ts'us *Vitis* sp. unripe fruit (**b**) Ts'us *Vitis* sp. leaves and fruit 265
Fig. 5.208 (**a**) 'Ukunte' *Sapium lateriflorum* tree (**b**) 'Ukunte' *Sapium lateriflorum* tree (**c**) 'Ukunte' *Sapium lateriflorum* growing tip (**d**) 'Ukunte' *Sapium lateriflorum* latex 266
Fig. 5.209 (**a**) 'Uk'unche' *Solanum torvum* (**b**) 'Uk'unche' *Solanum erianthum* flowers (**c**) 'Uk'unche' *Solanum erianthum* leaves 267
Fig. 5.210 (**a, b**) 'Uuch *Diospyros digyna* 268
Fig. 5.211 (**a**) Wakäx su'uk *Pennisetum purpureum* leaves, stems (**b**) Wakäx su'uk *Pennisetum purpureum* flower stalks 269
Fig. 5.212 (**a**) Wich ahyuk *Mucuna sloanei hilum* (**b**) Wich ahyuk *Mucuna sloanei* seed 269
Fig. 5.213 (**a**) Winik su'uk *Andropogon glomeratus* seed head (**b**) Winik su'uk *Andropogon glomeratus* plant 269
Fig. 5.214 (**a**) Wolche' *Cordia dodecandra* fruit (**b**) Wolche' *Cordia dodecandra* leaves 270
Fig. 5.215 Xa'an *Sabal mexicana* 270
Fig. 5.216 Xänä' mook *Chamaesyce hirta* 270
Fig. 5.217 Xikin *Schizophyllum commune* 271
Fig. 5.218 (**a**) Xinich (xilich) *Parathesis* sp. fruit (**b**) Xinich *Parathesis* sp. leaves 271
Fig. 5.219 (**a**) Xoyok *Morinda panamensis* trunk (**b**) Xoyok *Morinda yucatanensis* leaves (**c**) Xoyok *Morinda panamensis* inner bark (**d**) Xoyok *Morinda yucatanensis* leaves (**e**) Xoyok young tree 272
Fig. 5.220 (**a**) Ya'axche' *Ceiba pentandra* (**b**) Ya'ax balche' *Lonchocarpus* sp. (**c**) Ya'ax balche' *Lonchocarpus rugosus* flowers (**d**) Ya'ax balche' *Lonchocarpus* sp. leaves, top side (**e**) Ya'ax balche' *Lonchocarpus* sp. leaves, underside (**f**) Ya'ax balche' *Lonchocarpus rugosus* tree (**g**) Ya'ax balche' bark (**h**) Ya'ax balche' leaves 274

Fig. 5.221 (**a**) Ya'ax ch'ib muxan xiiw *Calathea macrosepala*
(**b**) Ya'ax ch'ib muxän xiiw *Calathea macrosepala* 276
Fig. 5.222 (**a**) Ya'ax hänan *Chamaedorea elatior*
(**b**) Ya'ax hänan *Chamaedorea elatior*
(**c**) Ya'ax hänan *Chamaedorea elatior* 276
Fig. 5.223 Ya'ax'oon 277
Fig. 5.224 Ya'ax 'oox *Brosimum alicastrum* ssp. *Alicastrum* 277
Fig. 5.225 Yahche' ts'in *Manihot esculenta* 278
Fig. 5.226 (**a**) Yo'och ähbaach' *Siparuna thecaphora*
(**b**) Yo'och ähbaach' *Siparuna thecaphora* 279
Fig. 5.227 (**a**) Yo'och ik mehen *Psychotria* sp.
(**b**) Yo'och ik mehen *Psychotria* sp. 279
Fig. 5.228 Yo'och ik nukuch *Psychotria panamensis var. panamensis* 280
Fig. 5.229 (**a**) Yuyul holop' *Philodendron tripartitum*
(**b**) Yuyul holop' *Philodendron smithii* 280

Fig. 6.1 Hach chaachib 'authentic colander' 319
Fig. 6.2 Häxal k'ak' 'firedrill' 319

List of Tables

Table 2.1 Lacandon Orthographic Conventions 10
Table 2.2 Pronunciation Guide 10

Table 4.1 Food Plants 29
Table 4.2 Plant Materials (in alphabetical order of scientific name within delimited categories) 31
Table 4.3 *Hulup' che'* Species 39
Table 4.4 Birds and Mammals in the *Paakche' Kol* 41
Table 4.5 Agricultural Round 42
Table 4.6 Dyes 90
Table 4.7 Plant Medicines 94

List of Videos

The videos for this book can be accessed at:
http://www.springer.com/us/book/9781461491101

Video 1.1	Take Care of the Forest
Video 4.1	*Balche'* Preparation 1
Video 4.2	*Balche'* Preparation 2
Video 4.3	*Ut'anil Balche'* 'Secret of the Balche'
Video 4.4	Cooking Traditional Foods
Video 4.5	Cooking *Pakyon*
Video 4.6	*Uk'aay Käkow* 'Song of the Cacao'
Video 4.7	*Uk'aay Huuch'* 'Song for Grinding'
Video 4.8	Clearing a New *Milpa*
Video 4.9	*Milpa* Harvest
Video 4.10	How to Build a House
Video 4.11	Tour of the God House
Video 4.12	*Tulis K'ik'*
Video 4.13(a)	Copal 1
Video 4.13(b)	Copal 2
Video 4.14	Tour of the Ceremonial Kitchen
Video 4.15	*Uk'aay Tok'* 'Song of the Flint'
Video 4.16	Making Arrows
Video 4.17	Seed Necklace
Video 4.18	AM Describes the Ceremonial Rattle
Video 4.19	*Uk'aay K'uch* 'Song for Spinning Thread'
Video 4.20	Weaving on the Backstrap Loom
Video 4.21	Hammock
Video 4.22	Making Natural Dyes
Video 4.23	KM Makes Annatto Paste
Video 4.24	Curing Strings
Video 4.25	Cooking hearth
Video 6.1	*Hach chaachib* 'authentic colander'

Introduction 1

1.1 The Lacandones

The Lacandones comprise two ethnically distinct groups whose ancestors were Yucatec-speaking Mayas that fled into the Chiapas forest from Guatemala, Campeche, and the Yucatan peninsula between the seventeenth and eighteenth centuries to escape assimilation or extermination. They moved into the vacuum left by the Ch'ol Lacandones, whom the Spanish had expunged from the forest a century earlier. Arriving in separate waves, the northern and southern Lacandones established their homesteads in areas adjacent to their original homelands.

They, along with other apostate Mayas, were called Lacandones, which meant 'wild Indians, heathens'. According to Thompson (1970), the word did not designate any particular group of people; rather it was a name of a place in the Maya lowlands that was occupied by a group of Ch'ol Mayas. The origin of the word has been analyzed differently by some scholars: Bruce suggests it derives from *ah-akan-tun-oob* 'they who set up stone' (Perera & Bruce 1986: 8), whereas Tozzer believes it derives from *acan-tun* 'to groan/thunder stone' (1907: 4). The Spaniards transformed it according to the rules of their language, which resulted in *El Acantún*. The word underwent further analysis, becoming *El Lacantún* and finally *El Lacandón*. The Lacandones, however, refer to themselves as the *hach winik* 'true people'.

Although they share a common history and cultural background, the modern Lacandones comprise two distinct groups: northern and southern. These names reflect their geographical location, more than anything else. The northern Lacandones occupy the north-western corner of the Lacandon forest, while the southern Lacandones, the south-east corner near the Usumacinta River. Historical accounts, and archaeological and linguistic evidence suggest that the northern Lacandones are closely affiliated with the Itzáj Maya (in the Guatemala Petén), whereas the southern Lacandones are closer to the Yucatec Maya. More research is needed.

Traditionally endogamous, the two groups are connected through the intermarriage of some of their members. There are a number of northern Lacandones living in the southern community and a few southerners that have moved into the northern community. As such, the division that exists between the two has become somewhat blurred. Nevertheless, each group views the other as being different, which is reflected in their terms for one another. The northern Lacandones refer to their southern neighbours as the *chukuch nook'* 'long tunics', while the southern Lacandones call the northerners *naachil winik* 'far away people' or *huntul winik* 'other people' (Boremanse 1998: 8). Moreover, although they speak a mutually intelligible variety of Lacandon, each group considers the other's speech to be deficient and, at times, unintelligible (R. Bruce, personal communication, 1992).

The Lacandones were never a large group. Even today they number approximately 600 men, women, and children. There are roughly 300 northern Lacandones, approximately 250 of which live in Naha', and the rest live in Mensäbäk. The southern Lacandones constitute the other half of the Lacandon population. Although the majority live in their jungle settlements and restrict their movements to travelling back and forth among the three villages, their population numbers change during peak tourist seasons, when ten percent of the population moves to the nearby city of Palenque to peddle their souvenirs (J. McGee, personal communication, 2000).

The northern Lacandones of Naha' are the focus of this book, primarily because they retained their traditional culture longer than the southern Lacandones. Under the leadership of Chan K'in Viejo, their civic and religious leader of more than four decades, they continued to worship their pantheon of deities. But after his death in 1996, the people lost their faith, with many converting to Christianity. This, along with deforestation and increasing numbers of colonist settlements in the forest, has affected the Lacandones traditional beliefs, knowledge, and most of all, their connection with the forest.

Electronic supplementary material: The online version of this chapter (doi:10.1007/978-1-4614-9111-8_1) contains supplementary material, which is available to authorized users. Videos can also be accessed at http://link.springer.com/chapter/10.1007/978-1-4614-9111-8_1.

S. Cook, *The Forest of the Lacandon Maya*,
DOI 10.1007/978-1-4614-9111-8_1, © Springer Science+Business Media New York 2016

1.2 History of Research

A detailed botanical inventory of the southern Lacandon territory of Lacanha' Chan Sayab is reported by Levy et al. (2006). This adds to extensive research carried out on this community's agro-forestry methods (Levy-Tacher et al. 2000; Levy-Tacher and Aguirre-Rivera 2005; Diemont et al. 2006). Less botanical work has been conducted in the northern Lacandon territory. Nations and Nigh (1980) provide a comprehensive list of the wild and cultivated plants in both northern and southern areas along with their Lacandon names. Durán (1999) provides an extensive inventory of the plants in Naha', that forms the basis for the botanical index published by the Comisión Nacional de Áreas Naturales Protegidas (2006).

1.3 Collection Methods

Data were gathered following a qualitative approach to answer how the Lacandones perceived their natural environment, which plants they deemed important, and whether this was reflected in a conceptual framework akin to a botanical taxonomy. Attention was given to how plants were named, as a guide to uncovering their system of plant classification. For the most part, consultants provided information through informal interviews. To a lesser extent, they participated in formal elicitation sessions that involved picture recognition tasks and tasks that involved sorting plants into categories. Most plants were identified and discussed on walks through the forest with Lacandon consultants. Some plant parts were removed and taken back to the community for verification and further elaboration by other Lacandones, or for use in demonstrations.

Lacandon names were cross-checked with those recorded in the Lacandon botanical literature and published inventories, specifically, Breedlove (1986), Comisión Nacional de Áreas Naturales Protegidas (2006), Durán (1999), Levy et al. (2006), and Nations and Nigh (1980). At times, the same language consultant would give two or more names for the same plant; when this was the case other speakers of the community were asked to select the common name. Consultants who were unfamiliar with the common Lacandon name for a species provided descriptive phrases, a common practice in all linguistic communities. When species were unavailable to inspect, the Lacandon names or descriptions were cross-referenced with botanical data collected by other researchers in the area and compared with images of specimens in various electronic herbaria. Consultations were also carried out with botanists familiar with the area.

Botanical names and taxonomic information on most of the species in this work were obtained from electronic botanical databases, botanical literature, and botanists at various institutions, including the Missouri Botanical Gardens (MOBOT), Facultad de Ciencias Biológicas, Universidad de Ciencias y Artes de Chiapas, Dept. of Ecology, El Cologio de la Frontera Sur (ECOSUR), Estación de Biología Tropical "Los Tuxtlas", Instituto de Biología (UNAM), and the New York Botanical Garden (NYBG). Botanists at these institutions helped identify species from photographs provided to them.

Numerous recordings and associated texts of Lacandon narratives, folklore, songs, rituals, demonstrations, and interviews are included, to illustrate the role of plants in Lacandon traditional culture. Most of the texts were transcribed in the field with trained consultants.

A number of computer programs were used to compile the inventories and analyze the texts. First, the recordings were digitized and then converted into wav and mpeg files using Cool-Edit.[1] Then, Transcriber[2] was used to segment and transcribe the recorded discourse. These transcriptions, and Lacandon botanical and ethnographic word-lists were then exported to Shoe Box 5, an integrated data management and analysis computer program developed at the Summer Institute of Linguistics.[3] This program was used to create the inventories. The transcriptions and associated recordings were then synchronized using computer programs developed at the Max Planck Institute for Psycholinguistics.[4] The English translations of the texts were then added to the video files.

1.4 Lacandon Consultants

None of this work would have been possible without the support and enthusiasm of the Lacandon community. Twenty-two Lacandon consultants provided the information for this study. Chan K'in Antonio Martinez (AM) and Bol Ma'ax Garcia (BM) were field guides and teachers, providing identifications and descriptions for a majority of the plants, as well as giving demonstrations, singing songs, and narrating stories (Fig. 1.1).

A dedicated practitioner of traditional religion, AM is one of the few remaining *hach winik* 'true men' left. Born in *Sa'am* (Monte Líbano) around 1925, he was raised by his mother and grandparents. He was raised in a traditional Lacandon setting at a time before the Lacandones still lived in isolated homesteads. He can still recall the song his mother used to sing every time she ground corn on a traditional millstone, watching his grandmother weave on the backstrap

[1]Originally Syntrillium Software and now Adobe® Audition® software 2003 http://www.adobe.com/special/products/audition/syntrillium.html/

[2]Transcriber—Copyright 1998–2008, DGA http://trans.sourceforge.net/

[3]Copyright 1989, 1990, 1991, 1993, 1996, 2000 by SIL International http://www-01.sil.org/computing/shoebox/Overview.html

[4]http://dobes.mpi.nl/

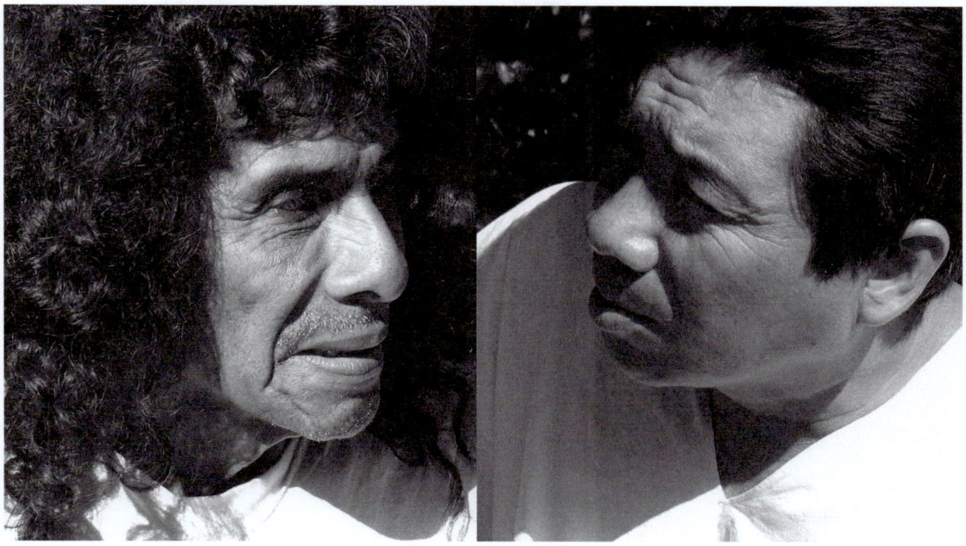

Fig. 1.1 AM (Lt) and BM (Rt). Photo Credit: Suzanne Cook

Fig. 1.2 AM offers balche' to the gods. Photo Credit: Suzanne Cook

loom, and helping his grandfather tap trees and prepare copal to be used in ceremonial offerings to the gods. At 89 years old, he is still energetic, rising early in the morning to tend his three milpas. He spends the hottest hours of the day in the cool of the forest or in his god house (Fig. 1.2).

BM is AM's son-in-law. He lives next door, following the customary uxorilocal residence pattern in Naha'. One of Chan K'in Viejo's (CK) eldest sons, BM benefited from his father's knowledge, learning how to do things in the traditional manner. He also remembers the myths, folktales, and ritual protocols. He has an intimate knowledge of the forest and continues to cultivate his milpas using traditional methods. As well as being one of the main guides for this book, he helped with the transcription and translation of many of the audio-video recordings and provided a number of traditional narratives.

Other individuals provided additional information or confirmation of names and uses for the species collected from the main consultants. Three, in particular, deserve mention: Koh Maria (KM), Koh Paniagua (KP), and Chan K'in "Sakho'ol" Garcia (SK) (Figs. 1.3 and 1.4).

KM and KP are Chan K'in's widows. They live together in the family household where they raised 20 children between them. Exemplars of the traditional Lacandon woman, they wear the traditional shirt tunic over a colourful skirt and wear their long hair in a single braid tied with bird pelts and feathers. They conduct their lives as did their mothers before them, rising early to grind maize, patting out several

Fig. 1.3 KP with unripe poochin fruit. Photo Credit: Suzanne Cook

Fig. 1.4 KM stripping vine for baskets. Photo Credit: Suzanne Cook

dozen tortillas for the day's meals, cultivating their milpas, and harvesting bushels of produce before noon. Afternoons are spent washing cloths, shelling maize, chopping firewood, and processing a variety of foods harvested from their *milpas* 'cornfields' and collected in the forest. They also demonstrated methods of cooking traditional foods, and weaving fabric and baskets.

SK is the son of KP and Chan K'in Viejo. Born in the late 1980s, he is one of the youngest of their children. He was one of the few people of his generation to have retained an interest in the traditional stories and ceremonial rites, and traditional methods of cultivation. He still recalls much of what he learned from his father, and is conversant about forest species and their traditional uses. He devoted countless hours to identifying plants and explaining their uses, besides helping to transcribe the texts. He also sang songs, narrated stories, and demonstrated the method of making natural dyes (Fig. 1.5).

The following lists the other consultants who contributed one way or another to the research for this work: Atanasio (AT), Bol Pedro Martinez Garcia (BMjr), Chanuk Garcia Martinez (BMw), Chan K'in Obregón (CKO), Chanuk Garcia (CNK), Chaxnuk Garcia (CHX), Juana Koh (JK), K'ayum Arturo Garcia (KA), K'ayum Segundo Garcia, K'ayum Garcia (KyP), Mateo Viejo Garcia (M), Irma Nuk Garcia (NI), Nuxi' Solorzeno (NX), Nuxi's wife (NXw), and Miguel Garcia (MCG). Chan K'in Daniel Garcia (CKD) took many of the photos (Fig. 1.6).

Fig. 1.5 Säkho'ol. Photo Credit: Suzanne Cook

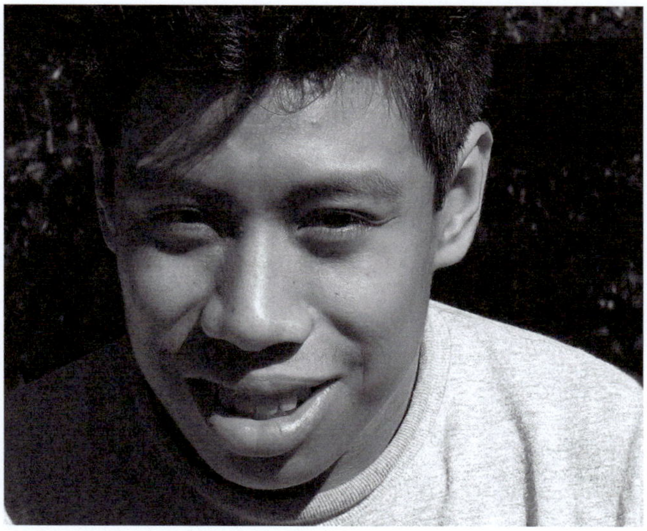

Fig. 1.6 Photographer, Chan K'in Daniel. Photo Credit: Suzanne Cook

The author is indebted to all of these people for their patience and good humour, even during trying times.

1.5 How to Read This Book

This work serves as a comprehensive guide to the botanical heritage of the northern Lacandon (Maya), providing, in both the Lacandon language and in English translation, over 450 Lacandon names for plants with full descriptions of the plants and their uses, as well as audio-video recordings that demonstrate their relevance in cultural contexts.

Technical jargon is kept to a minimum; where it does occur, an explanation in non-technical terms follows. The names of plants from Yucateco, Itzáj, Mopan, and the southern Lacandon dialect have been included, since words for the same plants can vary across these languages, between the dialects of the same language, and between people in the same linguistic community.

The work is presented in a format similar to most other ethnobotanies. It approaches the data from an ethnographic and botanical perspective, by providing both a botanical inventory and an ethnographic inventory. These inventories constitute the main body of the work. Chapters addressing the socio-historical context, the geography of the study area, and the role of plants in Lacandon culture precede the inventories. Ethnographic photographs and cross-references to audio-video files and Lacandon texts are included in these chapters. The book provides reverse English/Lacandon and Latin/Lacandon indices, and four glossaries that cover general Lacandon plant terms, geographical terms, and seasons and weather terms. An orthographic table, a pronunciation chart, and a note on transcription conventions used in this work follow a brief overview of the Lacandon language.

1.5.1 Inventory Format

1.5.1.1 Botanical Inventory

1) Both inventories are presented in dictionary format, with main entries headed by Lacandon names in alphabetical order:
   a, ä, b, ch, ch', e, (g), h, i, k, k', l, m, n, o, p, p', (r), s, t, t', ts, ts', u, w, x, y
2) The main entry word for each dictionary entry is a Lacandon lexeme or headword. It is indicated in bold type.
3) Phonetic transcriptions in [square brackets] follow the headword, indicating alternate pronunciations.
4) English common names follow. Where more than one name exists, they are also listed within parentheses. When the common name is unknown, this is indicated as (common name unknown).
5) A literal translation is included, whenever possible.
6) Botanical species are italicized. Although every effort has been made to identify the exact species, where there was insufficient evidence to determine a species, the botanical designation contains the genus name followed by sp., e.g., *Solanum* sp. Where the Lacandon plant refers to a number of species within the same genus, the plural form is used, spp., e.g., *Solanum* spp. The family name precedes the genus and species names, to help those readers unfamiliar with the flora of the region, e.g., Solanaceae *Solanum nigrum*.
7) Descriptions of plants and their uses are given in the Lacandon language and are followed by an English translation. The initials of the Lacandon consultants who provided the description are enclosed in parentheses and placed after the English translation. Additional

information on the species follows, where further clarification is required.

8) Each entry includes information about its taxonomic status according to the Lacandon classification system. Gen. (generic) and Spec. (specific) labels indicate the next higher and next lower level taxa, respectively. When a generic taxon is the headword (the main entry), e.g.,***'abäl*** 'mombin', all known specifics are listed in that entry, e.g., *k'än'abäl* 'yellow mombin', *chäk'abäl'* 'red mombin', *ya'ax'abäl* 'green mombin'.
9) Generics are cross-referenced with their specifics. For example, in the entry for ***k'än'abäl*** 'yellow mombin',*'abäl* is listed as the generic. Conversely, in the entry for***'abäl*** 'mombin' *k'än'abäl* 'yellow mombin' is listed as the specific (See note 8, above). Taxa at the same level, namely "companions" (the Lacandon designation for group members), are marked as Sim. For example, in the entry ***k'än'abäl*** 'yellow mombin', *ya'ax'abäl* 'green mombin' and *chäk'abal* 'red mombin' are listed as Sim.
10) Synonyms or alternative names, and borrowed names are included, if available.
11) Media associated with the inventory entries are specified by file name and the type of file, e.g., *uk'aay huch'*.mp4, *uk'aay huuch'*.doc.
12) Additional information is provided in notes, including Lacandon words collected by previous researchers of Lacandon and cognate words in the other Yucatecan languages—Mopan, Itzáj, Yucatec, and the southern Lacandon dialect. Except for northern Lacandon, the language is indicated within square brackets following the word in abbreviated form, i.e., [Yuc.], [Itz.], [Mop.], [S. Lac.]. The sources for these words follow in parentheses.

1.5.1.2 Enthnographic Inventory

The ethnographic inventory focuses on the cultural significance of plants. It is divided into semantic, or cultural, domains. These domains are provided in English and organized alphabetically. Within each domain, the entries are listed alphabetically according to the Lacandon language. The domains include:

6.1 Agriculture
6.2 Artefacts
6.2.1 Accessories
6.2.2 Containers
6.2.3 Cordage
6.2.4 Craft
6.2.5 Furniture
6.2.6 Hunting, Fishing, Trapping
6.2.7 Musical Instruments
6.2.8 Religious Objects
6.2.9 Tools
6.3 Clothing
6.4 Deities
6.5 Food
6.6 Geography
6.7 Housing
6.8 Medical Conditions

Fields in each entry of the ethnographic inventory are cross-referenced with the relevant species in the botanical inventory. In the ethnographic inventory, the field label *Mat* (Material) displays the species used; in the botanical inventory, the field label *Use* displays what the species is used for and how.

1.6 Abbreviations

Field markers used in the entries

Activity	Pertains to the type of agricultural activity undertaken
Ant	Antonym
Female	Pertains to the sex of the species
From	Borrowed word
Gen	Generic. A broader term which subsumes the headword. Generic-specific relationships are reciprocally cross-referenced
Indicator	Refers to the indicator species that signal the time to commence an agricultural activity
Lit	Literally
Loc	Location associated with the entry
Male	Pertains to the sex of the species
Mat	Material used to make headword, or material of which it is composed
Morph	Morphology
Note	Notes
Phase	Phase in the agricultural cycle or milpa
Prep	Preparation
Ref	Reference to audio-video recordings and written texts
SD	Semantic domain
sd2	Semantic domain, second level, e.g., cordage, food.
See	Cross-reference
Sim	Similar. Near synonyms or other terms at the same level of the taxonomy and subsumed under the entry
Sit	Situation
Source	Source of information, apart from cited sources
Spec	Specific. A term subsumed under the entry, i.e., types of a kind. Generic-specific relationships are reciprocally cross-referenced

Syn	Synonym
Thes	Thesaurus. A term designating the highest level of the Lacandon botanical taxonomy, i.e., Life Form
Time	Pertains to the timing of an event in the agricultural cycle, e.g., sowing
Use	Uses associated with the plant or object
Variant	Variant forms, including synonyms

Other abbreviations

Aff.	Affinis. Used when the identity of a species is unknown but which is very similar to a known species.
AGT.	Agent of an action
forma (plural formae).	Refers to a "secondary" taxonomic rank, below that of variety.
Itz.	Itzáj
Mop.	Mopan
S. Lac.	Southern Lacandon dialect
sp.	Species
spp.	More than one species
ssp.	Variety (botanical)
Tzo.	Tzotzil
var.	Variety (botanical).
Yuc.	Yucatec

Morphological symbols

-	Dashes are used to segment morphemes.
/ /	Slashes indicate the underlying representation of words.
[]	Square brackets indicate the surface representation of words.
<	Etymological derivation
=	Equal sign segments a compound word.

References

Boremanse, D. (1998). *Hach Winik: The Lacandon Maya of southern Chiapas, Mexico*. Institute for Mesoamerican Studies Monograph 11. Albany, NY: University of Albany.

Breedlove, D. E. (1986). *Listados Florísticos de México IV. Flora de Chiapas*. México: Instituto de Biología, UNAM.

Comisión Nacional de Áreas Naturales Protegidas. (2006). *Programa de Conservación y Manejo Área de Flora y Fauna Nahá*. 1a edición. Tlalpan C.P., México, DF: Dirección General de Manejo para la Conservación, CONANP. http://www2.ine.gob.mx/publicaciones/consultaPublicacion.html?id_pub=558&id_tema=12&dir=Consultas ISBN: 968-817-814-4

Diemont, S. A. W., Martin, J. F., Levy-Tacher, S. I., Nigh, R. B., et al. (2006). Lacandon Maya forest management: Restoration of soil fertility using native tree species. *Ecological Engineering, 28*(3), 205–212.

Durán, F. A. (1999). *Estructura y etnobotánica de la selva alta perennifolia de Nahá Chiapas*. M.Sc. Thesis. Universidad Nacional Autónoma de México, Mexico.

Levy-Tacher, S. I., & Aguirre-Rivera, J. R. (2005). Successional pathways derived from different vegetation use patterns by Lacandon Mayan Indians. *Journal of Sustainable Agriculture, 26*, 49–82.

Levy-Tacher, S. I., Aguirre-Rivera, J. R., García-Perez, J. D., & Martínez-Romero, M. M. (2006). Aspectos florísticos de Lacanhá Chansayab, Selva Lacandona, Chiapas. *Acta Botanica Mexicana, 77*, 69–98.

Levy-Tacher, S. I., Durán-Fernández, A., & Sánchez-Carrillo, D. (2000). Contribución al conocimiento de la flora útil de la selva Lacandona. *Informe final del ProyectoM002*. http://www.conabio.gob.mx/institucion/proyectos/resultados/InfM002.pdf

Nations, J. D., & Nigh, R. B. (1980). The evolutionary potential of Lacandon Maya sustained-yield tropical forest agriculture. *Journal of Anthropological Research, 36*(1), 1–30.

Perera, V., & Bruce, R. D. (1986). *The Last Lords of Palenque: the Lacandon Mayas of the Mexican Rainforest* (Berkeley and Los Angeles: University of California Press).

Thompson, J. E. S. (1970). *Maya history and religion*. Norman: University of Oklahoma Press.

Tozzer, A. M. (1907). *A comparative study of the Mayas and the Lacandones*. London: Macmillan.

2 Language

2.1 The Language System

Lacandon is one of the least known of the Middle American languages (Campbell 1979: 928; Andy Hofling personal communication; Nora England personal communication). It belongs to the Yucatecan Maya family and is divided into a northern and southern variety. Because the Lacandones did not experience centuries of political, cultural, religious, and linguistic domination by the Colonial Spaniards and the Mexican State, the language displays less Spanish influences than its sister languages, Itzáj, Yucatec, and Mopan.

Lacandon exhibits a consonant inventory similar to other Maya languages, having stops and affricates that occur in pulmonic and glottalic series. The system includes only one voiced obstruent, /b/. There are six vowel qualities and vowel length is distinctive.

The clause structure is morphologically ergative. Verbs take suffixes to indicate valency (transitivity, causation, and reflexivity), tense, aspect, mood, and person. Nouns take possessive suffixes. Both nouns and verbs are preceded by clitics of personal reference (possessor and subject). The verb complex is introduced by tense, aspect, and mood markers. These markers co-occur with corresponding suffixes on the verb.

For a more detailed account of the language, please refer to the Lacandon Cultural Heritage website at http://web.uvic.ca/lacandon/Language.htm

All Lacandon children still grow up speaking Lacandon. Toddlers are usually reared by their mothers and grandmothers, who invariably speak Lacandon to them. Until very recently, grandmothers were monolingual, and so the children were only exposed to Lacandon during their formative years. Even today, in bilingual Lacandon and Spanish households, mothers still begin speaking to their infants in their native language.

In the past, children began to acquire Spanish during late childhood or early adolescence. But today, with the introduction of a state-run primary school and an increase in their interaction with tourists, government officials, and other Mayan groups, Lacandon children are learning Spanish at an earlier age. With the expansion of colonization in the Lacandon forest, Tzotzil, Tzeltal, Ch'ol, and Tojolobal Mayas now reside in close proximity to Lacandon communities. Of these groups, the Tzeltal is the largest and most ascendant. Intermarriage between the Tzeltales and the Lacandones is not uncommon, and in these households, Spanish is the main language spoken. All of these factors will have an impact on the survival of traditional Lacandon knowledge, and potentially, the Lacandon language itself.

2.2 Previous Research

A grammar (Bruce 1968) is available for northern Lacandon. Grammatical sketches of southern Lacandon can be found in Baer and Baer (1952) and in Baer and Merrifield (1971, 1972). There is also an unpublished dictionary of southern Lacandon from the San Quintín area (Canger 1970). Thompson (1977) provides a comparative study of Lacandon and Yucatec, while Swadesh (1961) gives a survey of Mayan languages. Comparative studies of Lacandon and other peninsular Mayan languages are provided in Fisher (1973), Romero-Castillo (1977), and Tozzer (1907). Northern and southern Lacandon oral performances are examined in McGee (1987, 1997a, b) and Boremanse (1981a, b, c), respectively. More abundant texts from the northern Lacandones are published in Bruce (1974, 1975–1979 vol. 2, 1976).

S. Cook, *The Forest of the Lacandon Maya*,
DOI 10.1007/978-1-4614-9111-8_2, © Springer Science+Business Media New York 2016

2.3 Transcription Conventions in This Work

Lacandon is a polysynthetic language, which means that words, phrases, and clauses are formed by affixation and compounding, e.g., *t-in-lah-han-t-ah* 'I ate it all up'. As word structure devices, the hyphens separating the morphemes are omitted in this work. Hence, the example above would be written as tinlahantah.

Lacandon texts are written phonemically, which means that allophones, or non-distinctive phonetic differences, are disregarded.

Spanish grammatical and orthographic conventions, i.e., syllable stress and the letter j, are ubiquitous in the extant Lacandon literature. Because these conventions are ungrammatical in Lacandon, they are not used in this work (see Tables 2.1 and 2.2).

Table 2.1 Lacandon Orthographic Conventions

Lacandon (Bruce)	Lacandon (Baer)	Yucatec (Bricker)	Yucatec (Tozzer)	Itzaj (Hofling)	Colonial Maya	IPA
'	'	ʔ		'		ʔ
b	**b**	**b'**	**b**	**b'**	**b**	**b**
ch	č	č	**tš**	**ch**	**ch**	ʧ
ch'	č'	č'	**tš'**	**ch'**	**cħ**	ʧ'
h	**j**	**h**	**h**	**j**	**h**	**h**
k	**c, qu**	**k**	**k**	**k**	**c**	**k**
k'	**c', q'u**	**k'**	**q**	**k'**	**k**	**k'**
l(r)	**l(r)**	**l**	**l**	**l(r)**	**l**	**l (r)**
m	**m**	**m**	**m**	**m**	**m**	**m**
n	**n**	**n**	**n**	**n**	**n**	**n**
p	**p**	**p**	**p**	**p**	**p**	**p**
p'	**p'**	**p'**	**p'**	**p'**	**p, pp**	**p'**
s	**s**	**s**	**s**	**s**	**ç, z**	**s**
t	**t**	**t**	**t**	**t**	**t**	**t**
t'	**t'**	**t'**	**t'**	**t'**	**tħ, th**	**t'**
ts	**ts**	**¢**	**c**	**tz**	**tz**	**ts**
ts'	**ts'**	**¢'**	**c'**	**tz'**	**dz**	**ts'**
w	**w**	**w**	**w**	**w**	**u, v**	**w**
x	**x**	**š**	**š**	**x**	**x**	ʃ
y	**y**	**y**	**y**	**y**	**i, y**	**j**
a	**a**	**a**	**a**	**a**	**a**	**a**
ä	ʌ			**ä**	ə	ə
e	**e**	**e**	**e**	**e**	**e**	**e**
i	**i**	**i**	**i**	**i**	**i**	**i**
o	**o**	**o**	**o**	**o**	**o**	**o**
u	**u**	**u**	**u**	**u**	**u**	**u**

Table 2.2 Pronunciation Guide

Consonants	
'	Glottal stop--brief closure of the vocal cords restricting the flow of air.
b	Voiced bilabial stop, **b**aby
ch	Alveolar affricate, **ch**urch
ch'	Glottalized alveolar affricate, catch it [kæ**ch'**It]
h	Glottal fricative, **h**ouse
k	Velar stop, **k**ey
k'	Glottalized velar stop, take it [te**k'**It]
l	Lateral approximant, **l**ate
m	Bilabial nasal, **m**other
n	Alveolar nasal, **n**iece
p	Voiceless bilabial stop, **p**u**pp**et
p'	Glottalized voiceless bilabial stop, stop it [sta**p'**It]
s	Alveolar voiceless fricative, **s**un
t	Voiceless alveolar stop, **t**une
t'	Glottalized alveolar voiceless stop, kitten [kI**t'**n]
ts	Alveolar affricate, bi**ts**
ts'	Glottalized alveolar affricate, **itsy**-bi**tsy** [**Its'**i bI**ts'**i]
w	Bilabial glide, rounded velar, **w**o**w**
x	Alveopalatal voiceless fricative, **sh**ine
y	Palatal glide, **y**ellow
Vowels (descriptions refer to the position and height of the tongue in the vocal tract.)	
a	Central low vowel, sounds like f**a**ther, but shorter
ä	Unstressed central mid vowel, like a relaxed /a/ sound, **a**bout
e	Mid-front lax, t**e**ll [tɛl]; mid-front tense, l**a**bour [lebɚ]
i	High front, p**ee**p
o	Mid-back, **o**pen
u	High back, sh**oo**t
Long vowels are pronounced the same as their short counterparts, except that the sound is sustained for a longer period of time.	

Note: In this work, Yucatec, Itzáj, and Mopan words are written in the Lacandon orthography.

References

Baer, M. & Baer, P. (1952). Materials on Lacandon culture of the pethá (pelhá) region. *Microfilm Collection of Manuscripts on Middle American Cultural Anthropology* 34. Chicago: University of Chicago Library.

Baer, P., & Merrifield, W. R. (1971). *Two studies on the Lacandones of Mexico*. Norman, OK: Summer Institute of Linguistics, University of Oklahoma.

Baer, P., & Merrifield, W. R. (1972). *Los Lacandones de México: Dos estudios*. I.N.I. México. [trans. C. Viqueira].

Boremanse, D. (1981a). Una forma de clasificación simbólica: Los encantamientos al balche' entre los Lacandones. *Journal of Latin American Lore, 7*(2), 191–214.

Boremanse, D. (1981b). A southern Lacandón Maya account of the moon eclipse. *Latin American Indian Literatures, 5*(1), 1–6.

Boremanse, D. (1981c). Tomorrow: The day of the jaguar. *Latin American Indian Literatures, 6*, 45–53.

Boremanse, D. (1998). *Hach Winik: The Lacandon Maya of southern Chiapas, Mexico*. Institute for Mesoamerican Studies Monograph 11. Albany, NY: University of Albany.

Bruce, S. R. D. (1968). *Gramática del Lacandón*. Departamento de Investigaciones Antropológicas (Vol. 21). México: Instituto Nacional de Antropología e Historia.

Bruce, S. R. D. (1974). *El libro de Chan K'in*. Colección Científica, Lingüística, 12. México: Instituto Nacional de Antropología e Historia.

Bruce, S. R. D. (1975). *Lacandon dream symbolism 1: Dream symbolism and interpretation*. México: Ediciones Euroamericanas Klaus Thiele.

Bruce, S. R. D. (1976). *Textos y dibujos Lacandones de Najá*. (Trilingual Edition: Lacandón-Spanish-English). Departamento de Lingüística Colección Científica, Lingüística, Núm 45. México: Instituto Nacional de Antropología e Historia.

Campbell, L. (1979). Middle American languages. In L. Campbell & M. Mithun (Eds.), *The languages of Native America: Historical and comparative assessment* (pp. 902–1000). Austin and London: University of Texas Press.

Canger, Una. (1970). *Lacandón de San Quintín, Vocabulary of San Quintín*. Unpublished manuscript.

Fisher, W. M. (1973). *Towards the reconstruction of Proto-Yucatec* (Doctoral dissertation). University of Chicago, IL.

McGee, R. J. (1987). Metaphorical substitution in a Lacandón Maya ritual song. *Anthropological Linguistics, 29*(1), 105–118.

McGee, R. J. (1997a). Narrative structure of Lacandón creation mythology. *Latin American Indian Literatures, 13*(1), 1–22.

McGee, R. J. (1997b). Natural modeling in Lacandón Maya mythology. In F. A. Salamore & W. R. Adams (Eds.), *Explorations in anthropology and theology* (pp. 175–190). Oxford: University Press of America, Inc.

Romero-Castillo, M. (1977). La unidad lingüística del Maya peninsular. *Anales del I.N.A.H. 8a*(1): 83–108.

Swadesh, M. (1961). Interrelationes de las lenguas mayenses. *Anales del I.N.A.H., 13*, 231–267.

Thompson, J. E. S. (1977). A proposal for constituting a Maya subgroup, cultural and linguistic, in the Petén and adjacent regions. In G. D. Jones (Ed.), *Anthropology and history in Yucatán* (pp. 3–42). Austin, TX: University of Texas Press.

Tozzer, A. M. (1907). *A comparative study of the Mayas and the Lacandones*. London: Macmillan.

The Lacandon Rain Forest

3

3.1 Geography

The Lacandon Forest is the largest unbroken tropical rain forest north of the Amazon. It covers 1,800,000 ha (Levy-Tacher et al. 2002: 513) of the Eastern Highlands[1] of Chiapas, Mexico. The region is classified as part of the Maya Lowlands, but this is an historical-political designation that belies the physiographic reality: the terrain is essentially a series of steep, mountain ridges and narrow gullies, ranging in elevation from 400 to 1500 m (1312–4921 ft.) (Breedlove 1981: 2).

The mountain ranges traverse the landscape in a northwest to southeast direction, gradually descending into the rich, alluvial basin of the upper Usumacinta River on the Chiapas and Guatemala borders. The watershed formed by the confluence of the Lacanjá, Pasión, Chixoy (Salinas), Lacantún, and Tzaconeja Rivers, which drain into the Usumacinta River, the seventh largest river in the world, encompasses 106,000 km^2 (65,720 mile^2). The perennially moist soils and humidity support the only tract of true Tropical Rain Forest in the Lacandon Forest (Breedlove 1981: 7), and a diversity of freshwater habitats. The land formation is predominantly marine limestone, producing sinkholes, springs, complex underground drainage systems, and caves.

Variations in elevation, topography, and climate give rise to a number of different vegetation formations in the Lacandon Forest. The major ones have been classified with various names, depending on the approach taken by researchers and authors. Breedlove (1981) broadly delineates them into "optimal" and "seasonal" formations. Optimal formations include Tropical Rain Forest, Lower Montane Rain Forest, Montane Rain Forest, and Evergreen Cloud Forest. They occur throughout a wide range of elevations where conditions are almost constantly humid and seasonal variation is imperceptible. Seasonal formations include Evergreen Seasonal Forest, Tropical Deciduous Forest, Thorn Woodland/Short Tree Savanna, Pine-Oak-Liquidambar Forest, and Pine-Oak Forest. They occur in diverse landscapes that experience distinct wet and dry seasons. In addition to these formations, bands of transitional forest occur between adjacent formations.

The diverse vegetation formations make the Lacandon Forest the richest, most biologically diverse ecosystem north of the Amazon (De la Maza 1997). Although it occupies 0.4 % of the landmass of Mexico, it supports 40,000 species of flora and fauna. An estimated 4000 species of vascular plants have been recorded, including 1500 species of trees (cf. Dirzo and Miranda 1991). One hectare of forest may support up to 160 species of plants, and a single tree, 70 species of orchids. The forest is also home to 25 % of the country's mammals, 48 % of its birds, 33 % of its bats, 11 % of its reptiles, and 500 species of diurnal butterflies (Vásquez and Ramos 1992). Although species have been reduced by deforestation and human activities, much of the original wild flora and fauna persist in protected areas. Threatened species include: *Guatteria anomala*; *Cryosophila argentea*; *Chamaedorea arenbergiana*; *C. pinnatifrons*; *C. ernesti-augusti*; *C. metallica*; *Calophyllum brasiliense*; *Magnolia grandiflora*; *M. schiedeana*; *Talauma mexicana*; *Astronium graveolens*; *Podocarpus matuda*; *Rheinhardtia gracilis*; *Tillandsia festucoides*; *Legeophila clavigera*; and, *Litsea glaucescens* (Hernández-Nava 2003).

Study Area. Naha' and Mensäbäk form the northern limit of the Lacandon Forest, in a region referred to as the Subregíon Zona Norte (Northern sub-region). The area covers approximately 87,867 ha, or 9.2 %, of the forest (INE 2000a: 47). The terrain is a succession of deeply folded plateaus and mountain ranges. A layer of marine limestone covers the high ridges. Siltstone and shale derived from marine sediments cover the valley floors. An assortment of soils accumulates in niches, but the predominant soil types are the reddish-brown rendzina, rich in lime formed from the underlying chalk rocks, and the black, dust-like skeletal lithosols.

[1] One of the seven physiographic regions outlined by Müllerried (1957).

S. Cook, *The Forest of the Lacandon Maya*,
DOI 10.1007/978-1-4614-9111-8_3, © Springer Science+Business Media New York 2016

The structure of both types is nominally complex and the fertility is relatively low, save for a thin, 50 cm (19″) top layer of organic matter (Durán 1999: 30).

Naha' is located between 16° 56′41″ and 17° 00′42″ latitude north and between 91° 32′52″ and 91° 37′43″ longitude west. Elevation ranges between 910 and 1100 m (2985–3609′). It covers an area of 3847.41 ha. Mensäbäk is located between 17° 08′36″ and 17° 04′53″ latitude north and between 91° 34′42″ and 91° 40′09″ longitude west. Elevation ranges between 470 and 920 m (1542–3018′). It covers an area of 3368.35 ha (Hernández-Nava 2003: 1).[2]

The region represents the most important lacustrine area in the Lacandon Forest. Nine lakes are located in Naha', occupying 127 ha. The largest of these are Naha' and Ocotalito. Mensäbäk has 21 lakes, the largest being Ts'ibatnah and Mensäbäk. Most of the lakes are interconnected by underground streams and the Naha' River. This wetland network plays an integral role in the complex hydrologic system of the Usumacinta watershed, recharging the aquifers in the sub-basin of the Lacanjá River (Hernández-Nava 2003: 2). The wetlands also support a unique ecosystem of Haematoxylon campechianum, numerous species of orchids, bromeliads, high forest evergreens, and remnants of pine-oak forest.

3.2 Climate

The Northern sub-region is classified as a warm and humid thermal tropical zone, or Aw2 (w) (i') g,[3] under the Köppen climatic classification system. Temperatures and precipitation are largely influenced by humid winds coming off the Gulf of Mexico. The monthly average temperature is 23.6 °C (74 °F) with an annual fluctuation of 5.6 °C (42 °F). The coldest month is January, with an average daytime temperature of 20.9 °C (69 °F) and cooler nights. May and June are the warmest months, with an average temperature of 25.6 °C (78 °F).

Total annual rainfall is 1862 mm (73″), distributed over two well-defined wet and dry periods. Annual rainfall is greatest at the lower altitudes, decreasing somewhat at the higher altitudes. The rainy season extends from May to December. As January approaches, the weeks become progressively drier and warmer, punctuated by xämäns, or northers. In March, the "dog days" descend and the sun scorches the fields and exposed forest floors, and streams evaporate. This period is brief, for toward the end of April, the rains set in again (Fig. 3.1, 3.2, 3.3, and 3.4).

[2]Domínguez-Vázquez and Islebe (2008: 327) say that the altitudinal range of Naha' is between 800 and 1200 m.

[3]Comisión Nacional de Áreas Naturales Protegidas (2006: 19).

3.3 Vegetation

Lower montane rain forest is the predominant formation, while montane rain forest occurs on the moist, cool upland slopes. Riparian zones in the low-lying areas support a unique ecological niche where pines intermingle with lowland, rainforest species. Large areas have been extensively transformed by settlers, and timber and oil companies, creating a patchwork of secondary forests, cultivated fields, and *acahuales* (fallow fields) in various stages of secondary growth.

Once continuous with the Lacandon rainforest to the south, the northern territory has been reduced to two forest reserves in just over 50 years; yet, the region still exhibits the greatest biologically diverse transitional zone between the Nearctic and Neotropical ecozones (Hernández-Nava 2003: 3); the high evergreen forest of Naha' that occurs in the upper margins of the Lacandon forest is composed of a complex plant community structurally similar to the general pattern of other high forests and some distinctive species communities of its own. Botanical species in Naha' and the surrounding area have been well documented (Durán 1999) and indexed (Comisión Nacional de Áreas Naturales Protegidas 2006). Records indicate 779 vascular plant species, 452 genera, and 116 families, with over half of the species belonging to Araceae, Arecaceae, Bromeliaceae, Euphorbiaceae, Fabaceae, Lauraceae, Melastomataceae, Meliaceae, Moraceae, Orchidaceae, and Rubiaceae families.

Forest Formations. in the northern territory. Among the numerous formations of vegetation in the state of Chiapas, six occur in the study area (Hernández-Nava 2003).[4] Vegetation formations follow Breedlove's (1981) classifications.

1. Lower Montane Rain Forest.[5] This formation covers most of the Eastern Highlands (Breedlove 1981). It spans a broad range of elevations, from 350 to 1100 m and displays the greatest diversity of vegetation. While physically similar to the Tropical Rain Forest, it consists of only two strata of trees (Breedlove 1981: 8). The upper canopy is characterized by trees that range between 25 and 45 m (82–147.6′) in height, their branches adorned with epiphytes and interlaced with lianas. Many of the species are also found in the intermediate canopy of the Tropical Rain Forest. These include: *Billia colombiana*; *Dipholis stevensonii*; *Aspidosperma megalocarpon*;

[4]Durán describes four main formations: (1) Selva Alta Perennifolia; (2) Selva Mediana perennifolia; (3) secondary forests; and, (4) stands of *Pinus chiapensis* (around the lakes of Naha' Ocotalito.)

[5]Corresponds to Bosque Tropical Perennifolio, in part (Rzedowski 1978) and Selva Alta Perennifolio, in part (Miranda and Hernández 1963; Gómez Pompa 1965; Pennington and Sarukhán 1968; Flores et al. 1971).

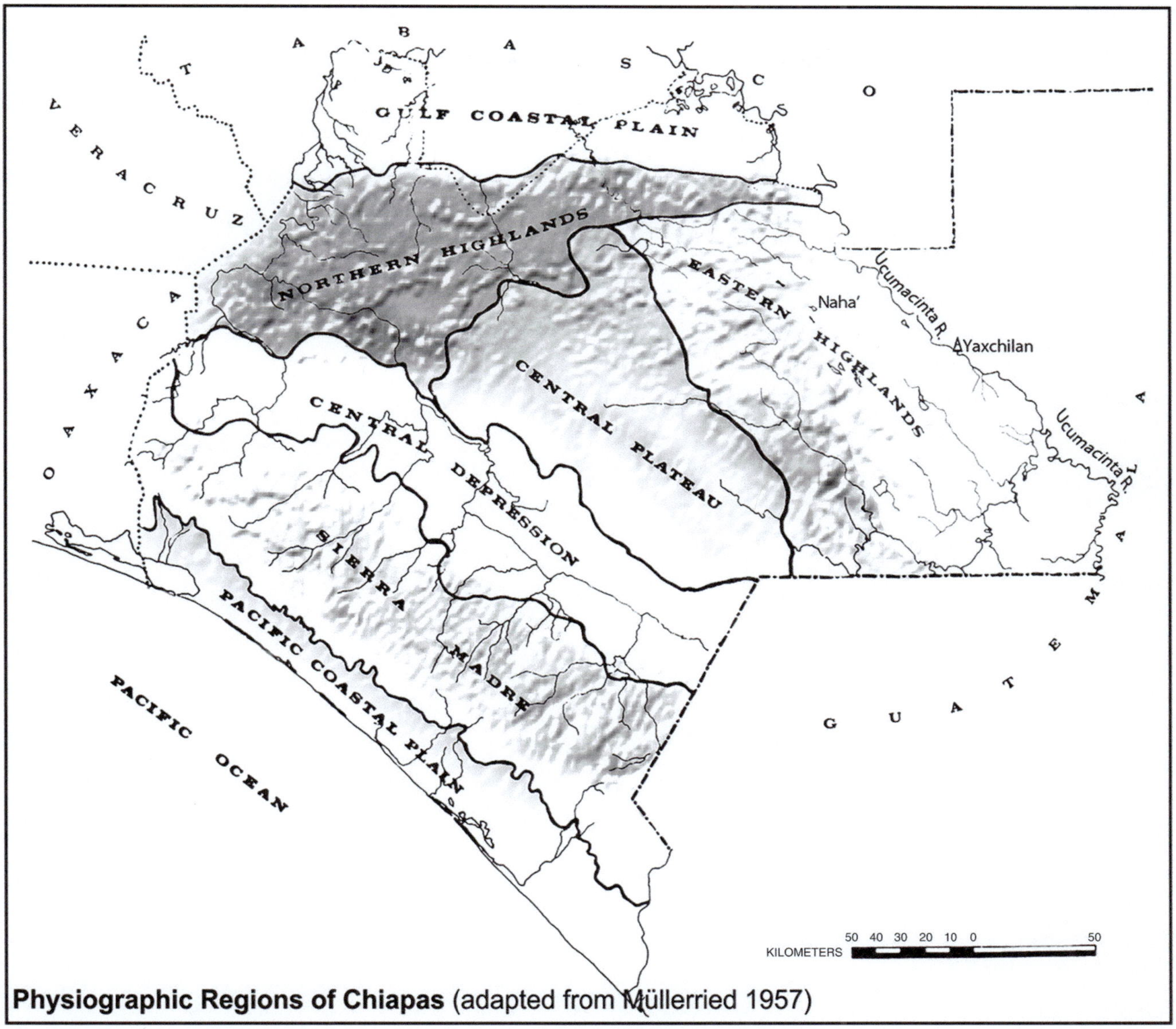

Physiographic Regions of Chiapas (adapted from Müllerried 1957)

Fig. 3.1 Chiapas physical map. Photo Credit: Suzanne Cook (adapted from Müllerried 1957)

Pseudomedia oxiphyllaria; *Manilkara achras*; *Terminalia amazonia*; *Coccoloba hondurensis*; *Talauma mexicana*; *Pimenta dioica*; *Oecopetalum mexicanum*; *Calophyllum brasiliense*; *Pimenta dioica* (Comisión Nacional de Áreas Naturales Protegidas 2006: 20–21).

2. Montane Rain Forest.[6] This formation occurs on the steep slopes and mountain ridges. It is a three- or sometimes two-storied formation. The upper canopy is irregular, with trees reaching up to 35 m (115′), occasionally interspersed with taller trees (Breedlove 1981: 10). Herbaceous shrubs and small trees are abundant in the intermediate and lower strata, in addition to a profusion of aroids, bromeliads, orchids, ferns, and club mosses.
3. Pine-Oak-*Liquidambar*.[7] This formation occurs above 1000 m (3281′) in the northern and north-eastern regions of Naha' (Hernández-Nava 2003: 6). Here, a diversity of deciduous and semi-deciduous trees forms a continuous canopy. Species include: *Oecopetalum mexicanum*; *Podocarpus matudai*; *Calatola laevigata*; *Billia colombiana*; *Quercus corrugate*; *Q. skinneri* (Comisión

[6]Corresponds to Bosque Tropical Perennifolio, in part (Rzedowski 1978) and Selva Mediana or Baja Perennifolia, in part (Miranda and Hernández 1963; Gómez-Pompa 1965; Pennington and Sarukhán 1968); Bosque Mesófilo de Montaña, in part, and Selva Alta Perennifolia (INEGI 1980).

[7]Corresponds to Bosque Mesófilo de Montaña (Rzedowski 1978; Rzedowski and McVaugh 1966), and Cloud Forest, in part (Leopold 1950, Martin 1958); Pine-Oak-Liquidambar (Carlson 1954).

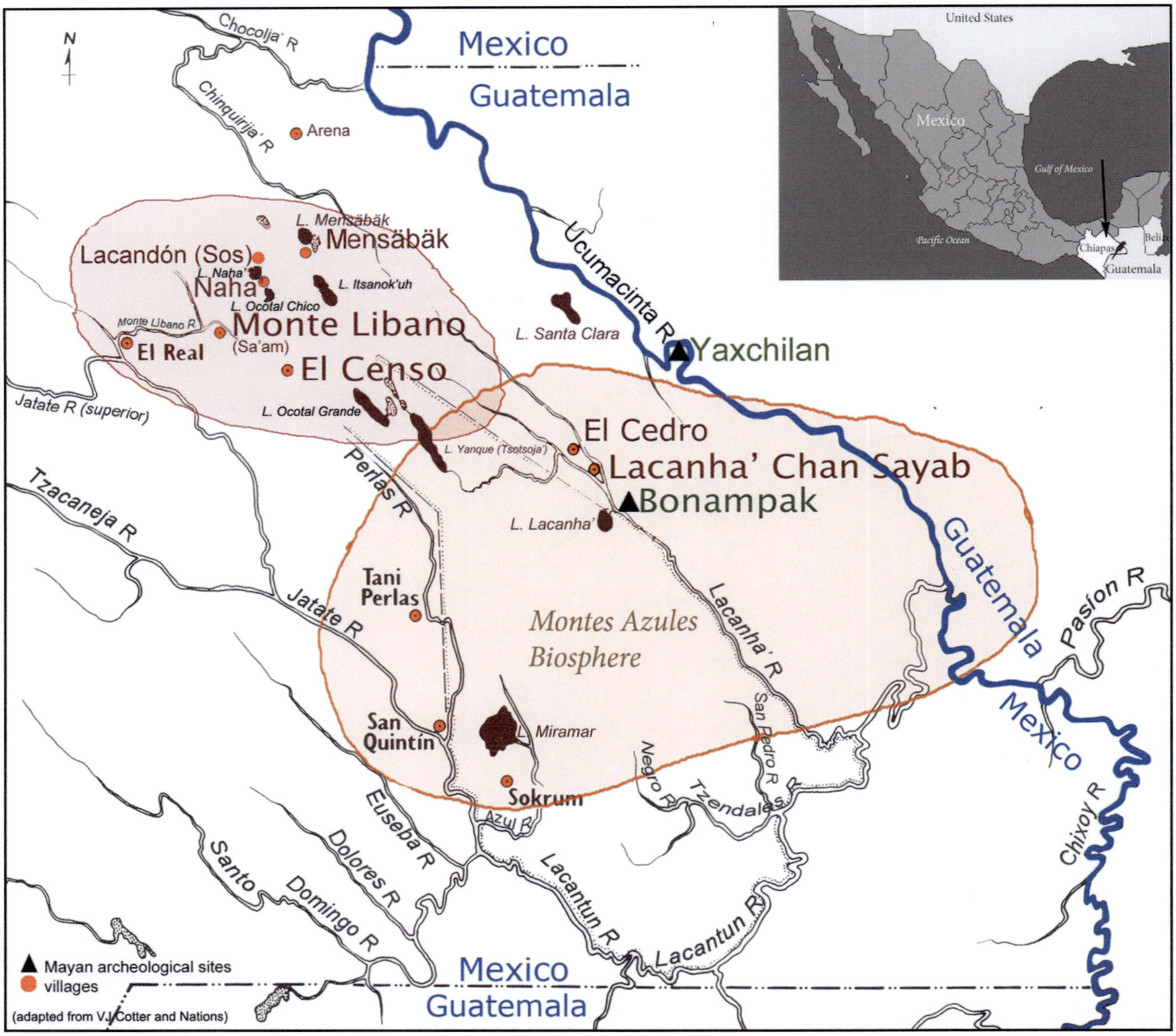

Fig. 3.2 Lacandon communities, northern and southern. Photo Credit: Suzanne Cook (adapted from V. J. Colter and Nations, in Blom, Gertrude 1984)

Nacional de Áreas Naturales Protegidas 2006: 21); *Q. Anglohondurensis*; *Q. Pedunculari*; Photiniamicrocarpa; *Clusia rosea*; *Pseudolmedia oxyphyllaria*; *Eugenia* spp.; *Sapindus* sp.; *Pithecellobium matudai*; *Hedyosmum mexicanum*; *Lonchocarpus* spp.; and, various Lauraceae (Hernández-Nava 2003: 6). The understory is a dense complex of shrubs, vines, mosses, ferns, and Chamaedorea palms, including: *Chamaedorea elatior*; *C. erenbergiana*; *C. oblongata*; *C. elegans*; *C. pinnatifrons*; and, *Cryosophila argentea* (Comisión Nacional de Áreas Naturales Protegidas 2006: 21).

4. Pine Forest.[8] This formation occurs on the upper slopes and ridges above 1000 m in elevation, where the air is cool and dry, and the soil, thin. The forest is populated by a few species of pines and oaks ranging between 15 and 40 m (49–131′) in height (Breedlove 1981: 19). The dom-

[8]Corresponds to Bosque de Pino y Encino in Rzedowski and McVaugh (1966).

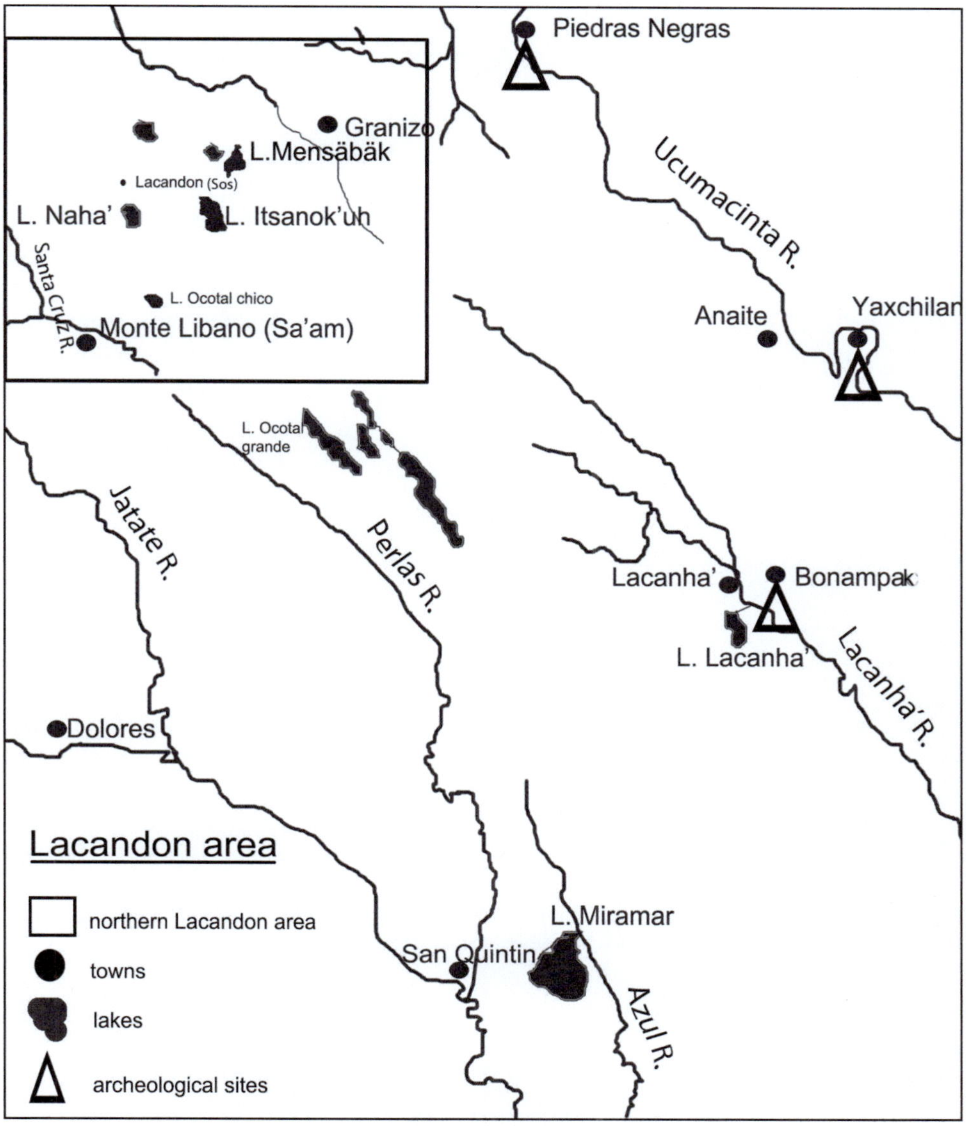

Fig. 3.3 Lacandon area. Photo Credit: Suzanne Cook

inant species are *Pinus maximinoi*; *P. oocarpa*, *P. pseudostrobus*, and *P. chiapensis*. Pines also occur around the lagoons of Naha' and Ocotalito, notably *P. chiapensis* (Durán 1999), where they intermingle with tropical rain forest species, shrub thicket, and low trees armed with thorns. Other species recorded in this formation include: *Pinus tenuifolia*; *Podocarpus matudai*; *Eugenia* sp.; *Rondeletia* sp.; *Ardisia* sp.; *Cryosophila argentea*; *Anthurium chiapasense* var. *chiapasense*; *Rapanea myricoides*; *Myricacerifera*; *Hauyaheydeana*; *Litsea glaucescens*; and, *Oreopanax* sp. (Comisión Nacional de Áreas Naturales Protegidas 2006: 22).

5. Flooded Thorn Woodland.[9] Uncommon in other parts of the Lacandon Forest, this riparian forest is unique to Naha' and Mensäbäk. Dense populations of *Haematoxylum brassileto* are found here along with epiphytes, *Pachira aquatica* and other arboreal species. Together they form a narrow band around the lakes (Hernández-Nava 2003: 6).
6. Secondary Vegetation. A patchwork of secondary forest in various stages of recovery is distributed in small areas to the east and northeast of Naha'. Also called second-

[9]Corresponds to Bosque espinoso inundable (CIEDAC 1991).

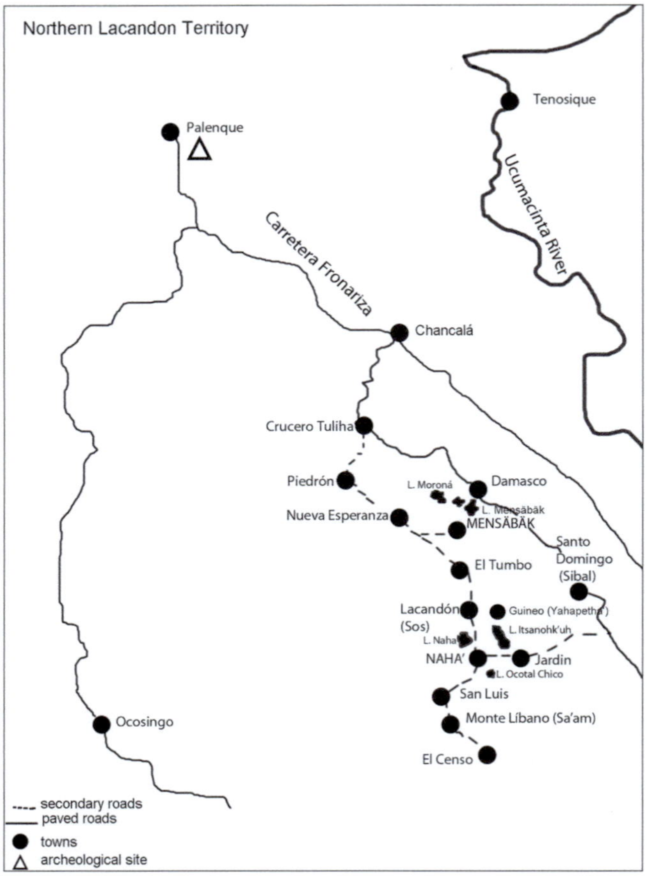

Fig. 3.4 Northern Lacandon territory. Photo Credit: Suzanne Cook (adapted from C. Morales, http://www.ecotourism-adventure.com/index.htm)

growth and successional forest, this vegetation type refers to primary forest that has been altered by natural causes and human intervention, and displays various stages of re-growth (successions) (Breedlove 1981: 24).

Immediately after the felling of the original forest, and from one to two years thereafter, a number of large plants develop, including annual species of Chenopodiaceas, Amaranthaceas, and Compositae. Two or three years later, shrubby vegetation develops. Common species include: *Acacia* spp.; *Mimosa* spp.; *Calliandra* spp.; *Opuntia* spp.; *Arctostaphylos pungens*; and *Dodonaea viscose*. As the forest matures, trees between 15 and 20 m (49–65′) in height dominate the landscape. They include: *Belotia mexicana*; *Heliocarpus appendiculatus*; *H. donnell-smithii*; *Hampea stipitata*; *Cecropia peltata*; *C. obtusifolia*; and *Quercus* spp. In the shade of the understory flourish species from the ginger family, such as maranta, calathea and heliconia, *Xanthosoma robustum*, species of piper, ferns, vines, and grasses.

Secondary vegetation in more open areas, especially along roadsides, comprises species such as: *Saurauia scabrida*; *Dendropanax arboreus*; *Clethra suaveolens*; *Clusia* sp.; and *Verbesina myriocephala* (Comisión Nacional de Áreas Naturales Protegidas 2006: 22) Figures 3.5–3.16 provide a snapshot of the the study area.

Fig. 3.5 Fallow milpa. Photo Credit: Suzanne Cook

Fig. 3.6 Ek' lu'um 'black soil'. Photo Credit: Suzanne Cook

Fig. 3.7 Soil structure. Photo Credit: Suzanne Cook

Fig. 3.8 Canoe by rock cliff. Photo Credit: Suzanne Cook

Fig. 3.9 Waterfall in the southern Lacandon territory. Photo Credit: Suzanne Cook

Fig. 3.10 Forest, 10 year fallow. Photo Credit: Suzanne Cook

Fig. 3.11 Lagoon. Photo Credit: Suzanne Cook

Fig. 3.12 Lagoon. Photo Credit: Suzanne Cook

Fig. 3.13 Cattle in pasture. Photo Credit: Suzanne Cook

Fig. 3.14 Naha', 2010. Photo Credit: Suzanne Cook

Fig. 3.16 Lacandon homestead. Photo Credit: Suzanne Cook

Fig. 3.15 Schoolyard. Photo Credit: Suzanne Cook

References

Breedlove, D. E. (1981). *Flora of Chiapas Part 1: Introduction to the flora of Chiapas*. San Francisco, CA: California Academy of Sciences.

Carlson, M. C. (1954). Floral elements of the pine-oak-*Liquidambar* forests of Montebello, Chiapas, Mexico. *Bulletin of the Torrey Botanical Club, 81*(5), 387–399.

Centro de Investigaciones de Energía, Desarrollo y Conservación. (1991). *Naturaleza histórica en la Selva Lacandona*. México: CIEDAC.

Comisión Nacional de Áreas Naturales Protegidas. (2006). *Programa de conservación y manejo área de flora y fauna Nahá*. 1a edición. Tlalpan C.P., México, DF: Dirección General de Manejo para la Conservación, CONANP. http://www2.ine.gob.mx/publicaciones/consultaPublicacion.html?id_pub=558&id_tema=12&dir=Consultas ISBN: 968-817-814-4

De la Maza, C. L. (1997). Beneficios monetarios en áreas silvestres protegidas. Aplicaciónde los métodos de valorización contingente y costo de viaje. CONAF/UNDP.

Dirzo, R., & Miranda, A. (1991). Altered patterns of herbivory and diversity in the forest understory: A case study of the possible consequences of contemporary defaunation. In P. W. Price, T. M. Lewinsohn, G. W. Fernandes, & W. W. Bensen (Eds.), *Plant-animal interactions: Evolutionary ecology in tropical and temperate regions*. New York, NY: John Wiley and Sons.

Domínguéz-Válzquez, G., & Islebe, G. A. (2008). Protracted drought during the late Holocene in the Lacandon rain forest, Mexico. *Vegetation History and Archaeobotany, 17*, 327–333.

Durán, F. A. (1999). *Estructura y etnobotánica de la selva alta perennifolia de Nahá Chiapas* (Unpublished master's thesis). Universidad Nacional Autónoma de México, Mexico.

Flores, M., Jiménez L., Madrigal, S., Moncayo, R., Takaki, F. (1971). *Mapa y descripción de los tipos de vegetación de la república mexicana*. México, DF: Dirección de Agrología, Secretaría de Recursos Hidráulicos.

Gómez-Pompa, A. (1965). La vegetación de México. *Boletín de la Sociedad Botánica de México, 29*, 76–120.

Hernández-Nava, J. (2003). *Areas de protección de flora y fauna de Nahá y Metzabok*. Chiapas, México: Ficha Informativa de los Humedales de Ramsar.

Instituto Nacional de Ecología (INE). (2000a). *Programa de manejo de la reserva de la biosfera Montes Azules*. México, DF: Instituto Nacional de Ecología.

Instituto Nacional de Ecología (INE). (2000b). *Zonificación de la reserva de la biosfera Montes Azules, Chiapas*. México, DF: Instituto Nacional de Ecología.

Leopold, A. S. (1950). Vegetation zones of Mexico. *Ecology, 31*(4), 507–518.

Levy-Tacher, S. I., Aguirre Rivera, J. R., Martinez-Romero, M. M., & Durán- Fernández, A. (2002). Caracterización del uso tradicional de la flora espontánea en la comunidad Lacandona de Lacanhá, Chiapas, México. *INCI, 27*(10), 512–520. http://redalyc.uaemex.mx/pdf/339/33907302.pdf

Martin, P. S. (1958). A biography of reptiles and amphibians in the Gomez Farias region, Tamaulipas, Mexico. *Misc. Publ. Mus. Zool. Univ. Michigan* 101.

Miranda, F., & Hernández, E. (1963). Los tipos de vegetación de México y su clasificación. *Boletín de la Sociedad Botánica de México, 28*, 29–179.

Müllerried, F. K. G. (1957). *La geología Chiapas. Gobierno del Estado de Chiapas*. Chiapas, México: Tuxtla Gutiérrez.

Pennington, T. D., & Sarukhán, J. (1968). *Manual para la identificación de Campo de los principales árboles tropicales de México*. México: INIF/FAO.

Rzedowski, J. (1978). *Vegetación de México*. México DF: Limusa.

Rzedowski, J., & McVaugh, R. (1966). La vegetación de Nueva Galicia. *Contributions from the University of Michigan Herbarium IX, 9*, 1–123.

Vásquez, S. M. A., & Ramos, M. A. (1992). Reserva de la biosfera Montes Azules, selva Lacandona: Investigación para su conservación. *Publicaciones especiales Ecosfera* 1. México.

4 The Role of Plants in Traditional Lacandon Culture

4.1 Overview

Plants played a vital role in the Lacandon traditional culture. Almost everything was made wild or cultivated species. Some manufactured items, such as metal, firearms, and woven cotton, were introduced through trade with chicleros (chicle tappers) and lumbermen when the forest was opened up to outside industries. But for the most part, the Lacandones relied on their environment to provide them with the food and materials they needed to survive. At the time of contact, the Lacandones had already developed an extensive knowledge of their natural world, drawing on it for spiritual inspiration and incorporating the plants and creatures into their mythology and folklore.

Today, the Lacandones rely less on the forest, as the new generation acquires most of their material goods and food from city markets or travelling merchants. Although young families still cultivate *milpas* (corn fields), the number of crops are drastically reduced compared to what the typical Lacandon farmer cultivated in the past. Nevertheless, there are still a few of the older generation who still obtain most of their material and nutritional requirements from the forest, *milpas*, and fallow lands. What follows is a description of the subsistence and material culture of such traditionalists.

Food Plants. At least 100 species of food plants are still exploited. (See Table 4.1 Food Plants for a list). Of the introduced species, one of the most important is *Musa acuminata* ×*Musa balbisiana* 'bananas and plantains'. In the past, these fruits were added to corn dough when stocks of maize were low. Other cultivated food plants include citrus fruits, mangos, pineapples, and, until recently, sugarcane.

Wild species are also cultivated. The most important include: species of palms (*Chamaedorea tepejilote*), which are harvested for their immature male inflorescences and hearts; custard apples (*Annona reticulata* and *A. cherimoya*); elephant ears (*Xanthosoma mafaffa*); and, monkey hand piper (*Piper umbellatum*).

Material Plants. Over 133 plants are used in Lacandon technology. (See Table 4.2 Plant Materials for a list). Approximately 50 species of trees provide materials for tools, utensils, furniture, weapons, and house construction. Twenty-two of these are used for fuel. Nineteen species of trees, vines, palms, and herbaceous plants are used for cordage, clothing, and mats. Seventeen species are exploited for dyes, paints, and resins. Some 23 species are used for adornment, mainly jewellery, while a number of other species serve a variety of purposes, such as wrappers, toys, cleaning agents, and medicines. The following sections provide details on the cultural domain in which certain plants are used.

4.2 Food

The Lacandones enjoy a wide variety of both wild and cultivated fruits, vegetables, and until recently, game, fish, and species of ants and grubs. Most foods are cooked thoroughly by boiling, roasting, baking, frying, and steaming, but almost all fruits are eaten raw.

4.2.1 Maize

Corn is the most important staple in the Lacandon diet. Dried corn, *chich näl* 'hard corn', is shucked and boiled in a solution of water and *k'ulta'an* 'quicklime', to soften the kernels and loosen the hulls. This process is known as nixtamalization. Not only does this process make corn easier to grind, it also boosts the nutritional value of the grain. The processed form is known as hominy, or *k'u'um*. The dough prepared from it is called *k'u'umbil*, which forms the basis for a variety of corn-based foods, including tortillas and different kinds of

Electronic supplementary material: The online version of this chapter (doi:10.1007/978-1-4614-9111-8_4) contains supplementary material, which is available to authorized users. Videos can also be accessed at http://link.springer.com/chapter/10.1007/978-1-4614-9111-8_4.

S. Cook, *The Forest of the Lacandon Maya*,
DOI 10.1007/978-1-4614-9111-8_4, © Springer Science+Business Media New York 2016

gruels. The common varieties of tortillas and gruels are described below:

Tortillas. Tortilla dough is pressed out by hand on a banana leaf or plastic sheet. It is laid carefully onto the xämäch 'griddle', turned twice, and then pressed with the palm of the hand to make it puff up. At this point it is either eaten immediately or stored (Figs. 4.1, 4.2, and 4.3).

Extra ingredients may be added to tortilla dough, such as mashed plantain and bananas, ground breadnut, partially cooked and ground beans, the seeds of *ha'as* 'mammee sapote' and *k'uxub* 'annatto'. Of the number of variations on the basic tortilla, the typical ones are the following:

'Is wah is a sweet tortilla made from the dough of fresh, ripe maize. The raw kernels are pared from the cob, ground into dough and then patted out. The fresh corn makes a slightly sweet, soft tortilla.

Ch'uhuk wah 'sweet tortilla' is prepared with sugar, salt, and lard. This is a Tzeltal recipe introduced into the Lacandon kitchen. It is rarely made.

K'u'umbil wah is a large, 25 cm (10") disc about 6 mm (0.25") thick. They are made in great quantities every day and eaten at every meal. Those not eaten are stored until later, when they are reheated over an open fire. Unused dough is stored overnight and made into delicious, sour-dough tortillas, the next morning. These are considered the most nutritious of all the tortilla varieties.

Säkpet is a small, 15 cm (6") crisp tortilla, similar to the Mexican tostada. It is made from fresh dough or from the inflated side of prepared tortillas and then grilled over an open fire. As they remain fresh for a long time, they are the perfect food for lunches in the *milpa*.

Still another kind of tortilla is prepared with raw *ts'in* 'manioc' and maize.

Tamales. Tamales are small packages of corn dough wrapped in banana leaves or corn husks. They are usually filled with meat or beans and are either steamed or baked in hot coals. There are at least five main kinds of tamales that Lacandon women prepare:

Bu'ulil wah 'bean tamale' is made with mashed beans spread between layers of corn dough. Dozens of them are steamed in large pots over an open fire for several hours.

Chäkbil wah 'cooked tortilla' is made of cooked or raw meats placed between layers of corn dough. These are steamed.

Hach wah 'authentic tortilla' is a 10–15 cm (4–6") round tortilla that is usually filled with beans. It is toasted on the *xämäch* 'griddle'.

Huybil wah 'stirred tortilla' consists of scrambled eggs and pieces of tortilla wrapped up in a large leaf. Ground chicken and spices are sometimes added. It is baked on hot coals.

Mukbil wah 'buried tortilla' is a large, oblong tamale filled with beans or chicken. It is baked on hot coals.

Gruels. Gruels constitute the bulk of the Lacandon diet. Typically, they are made of corn dough that has been dissolved in water and are drunk, rather than eaten with a spoon. The following typical gruels are listed from the most to the least frequently consumed.

Ma'ats' is the most common gruel, made with *k'u'um*. Sometimes it is left to sour, which enhances the flavour and nutritional value. On special occasions, particularly during ceremonies, ground cacao is added with a dollop of the foam spooned on top.

K'ayem is similar to *ma'ats'* only cooked.

Sa' is prepared with the kernels of fresh corn that have been ground and left overnight to sour. The liquid is strained before it is boiled to a thick consistency. It is usually sweetened (Fig. 4.4).[1]

K'äh is ground maize that is toasted and then boiled in water. It is sweetened with honey or sugar. Although somewhat gritty, it has a pleasant nutty taste.

U'ul is a hot gruel made with fresh corn kernels that have been left to sour before being boiled. In religious contexts, it is called *säkha'* 'white water' (Baer and Baer 1952).

'Ulisa' is a gruel made with fresh corn (Bruce 1975). This may be the same as *u'ul.*

Fresh corn gruels are only available for a short period, July-August, when the ripe ears of corn are still tender.

K'ool is a porridge thickened with either *k'u'um* or softened pieces of tortilla. The basic *k'ool* is called *k'oolbil wah*. More elaborate versions include meats, fungi, and vegetables. The dishes are named according to these additional ingredients, e.g., *k'oolbil kax* 'chicken porridge', *k'oolbil'ulum* 'turkey porridge', *k'oolbil k'ayoch'* 'oyster mushroom porridge', and *k'oolbil mäkäl* 'elephant ears porridge'.

4.2.2 Beans

In a country that is based on the three sisters—corn, beans, and squash—beans are further down on the list of Lacandon dietary preferences. Nevertheless, they add substantial nutrients to the Lacandones' otherwise corn-based diet. Beans are dried and shelled; green pods are rarely, if ever, eaten. *Ek' bu'ul* 'black beans' are preferred over *chäk bu'ul* 'red beans',*'ip'* 'lima beans', and *box bu'ul* 'cow peas'. Lesser beans, such as *tsaayentsaay* 'rice bean' (*Vigna umbellate*) grow rampant in house gardens and are often pulled out and disposed of with the rest of the weeds. Beans are boiled and eaten as is, with little or no seasoning, or boiled, mashed, and then fried with onions in oil.

[1] Baer and Baer (1952) say that the southern Lacandones do not strain it.

4.2.3 Fruits and Vegetables

Over 66 species of wild plants, trees, vines, and palms augment the basic diet of corn and beans. Despite this, fruits and vegetables constitute a small percentage of the daily diet. Most berries, fruits, seeds, leaves, and tubers are gathered from the forest and mature *acahuales* (fallow corn fields). Highly esteemed species, such as some species of Chamaedorea and Xanthosoma, are transplanted from the wild and cultivated in house gardens, providing a close-at-hand supply of hearts of palms and immature inflorescences. Some species considered weeds are carefully tended, such as *ch'ayok'* (*Solanum amercianum*).

Fruits. A variety of wild fruits are gathered from the forest and cultivated in *milpa*s and house gardens. Among the most varied and numerous fruits are bananas and plantains. Next to maize, these are the most valued food.

Pätan 'plantains' are boiled, roasted, or fried, as they are too coarse and mealy to be eaten raw, except if they have ripened to black. According to Baer and Baer (1952), green plantains are boiled with meats.

Boox 'bananas' are always eaten raw. Entire stalks are harvested when the fruits are still green and hung up to ripen indoors. Both plantains and bananas are mashed and incorporated into corn dough, especially when maize is scarce.

Other common fruits include: *put* 'papaya'; *pach'* 'pineapple' and *ch'äm* 'pinguin';*'oop* 'custard apple', *pähpox* 'soursop'; *ha'as* 'zapote'; *'oon* 'avocadoes'; *chinah* 'oranges'; and, *chäk chinah* 'mandarins'. A complete list of the fruits is provided in Table 4.1 Plant Foods.

Vegetables. More intensively cultivated species include*'is* 'sweet potato', *k'um* 'calabasa, pumpkin', *mäkäl* 'elephant ears', *p'ak* 'tomatoes', *p'ix* 'vegetable pear', and *ts'in* 'manioc'.*'Is* and *ts'in* are eaten baked in either the white coals of a cooking fire or in pits.

K'um (*Cucurbita moschata*) is a more versatile vegetable, providing both flesh and seeds. It is cured in the field or on top of hot tin roofs until the skin is thoroughly hardened. At this point, it is chopped up into chunks, seeded, and then boiled with sugar or roasted plain. The protein-rich seeds are toasted and eaten out of hand or made into a confection similar to brittle. The toasted seeds are also ground and added to *ma'ats'* and ceremonial tamales.

Mäkäl (*Xanthosoma mafaffa*) is cultivated for its starchy tubers and tender leaves. The tubers are peeled and then boiled, while the very young, slightly unfurled leaves are boiled and added to soups.

Of the tomato species, the one most widely cultivated is a red, cherry tomato, *p'ak* (*Lycopersicon esculentum* var. *cerasiforme*). The fruits are tart and acidic, and so they are always cooked before eating. They are boiled or fried with onions and then added to stews and soups.

P'ix is a pear shaped, green squash that grows on a robust vine. Three varieties are cultivated: *säk p'ix* 'white vegetable pear', *ya'ax p'ix* 'green vegetable pear', and *k'i'ilix p'ix* 'spiny vegetable pear'. There is little difference in flavour or consistency among them. They are slightly watery and taste like potatoes. The soft seed is also eaten. The leaves are boiled and added to soups and stews.

Several introduced species are cultivated. Of these, *axux* 'garlic', *k'äxeex* 'epazote', *ts'ak kax* 'mint', and *ts'ak k'ek'en (ts'ak kitam)* 'wild scallions' are the most popular. Most families also cultivate *kulix* 'cabbage', *lechuga* 'lettuce', and *rábano* 'radishes' in small, raised beds near the house.

Lacandon cuisine is plain by Mexican culinary standards. Of the few spices and seasonings that are used, most are fresh. Among the most common, are *perehir* 'cilantro' and *kurantro* (*Eryngium foetidum*).*'Ik* 'chile' varieties number in the dozens, yet they are seldom, if ever, used to spice dishes. Rather, they are eaten raw as a condiment. *Axux ak'* 'garlic vine' is a native plant whose flowers, stems and leaves taste like garlic, and was used extensively before the introduction of axux 'garlic'. *Ts'akax* 'mint' is used mainly to season chicken stew. *K'uxub (k'uxu')* 'annatto' seeds are rendered down to a paste, then reconstituted for use as food colouring and mild seasoning for meats and tamales.*'Itsänte'* 'glory-bower' (*Clerodendrum* sp.)[2] is cultivated for its leaves, which are used to season fish. The oily seeds of *sikilte' che'* 'physic nut' (*Jatropha curcas*) are dried, roasted, and ground. The nut powder is cooked with annatto and then used to season meats.[3] Before the introduction of commercial cooking oil, the ground seeds were boiled to extract the oil.

4.2.4 Fungi

Although eaten infrequently, a small variety of fungi is gathered from the wild and in *milpas*. As the men are more often in the forest, they collect the bulk of the fungi. Women collect the species that grow in the *milpa* or near the home-steads. The varieties eaten by the northern Lacandones include: *k'ayoch'* (*Pleurotus djamor*); *lo'loh* (*Auricularia* spp.); *säk muluch'* (*Oudemansiella canarii*); and, *xikin* (*Schizophyllum commune*).

The Lacandones they always cook fungi, believing that eating raw fungi will rot their insides. They usually prepare it *k'oolbil* style.

[2] ahtsunte', in Baer and Baer (1952).

[3] Baer and Baer (1952).

4.2.5 Ritual Offerings

Collectively, offerings of every description are called *si(hi)l* 'gift' in Lacandon.[4] Offerings fall into one of two categories: inedible and edible. Inedible offerings transform into food for the gods when they are burned, whereas edible offerings are consumed by ceremonial participants. Edible offerings are described below. Inedible offerings are described in Sect. 4.5.4.

Edible Offerings. *Balche'* (*Lonchocarpus* spp.) is the most important of the food offerings. It plays a central role in every ceremony, and is sometimes the only offering. The *balche'* tree was one of the first trees created by the Lacandones' supreme deity *Hachäkyum*, as a means to venerate him. It is the main food offering in ceremonies, as it is believed that *balche'* makes the gods happy. Ritual intoxication is obligatory, because the sight of men drunk and in good humour pleases the gods. The men are also obligated to consume the entire log of *balche'* in one sitting. Tozzer (1907: 136) describes ritual participants drinking gourd after gourd and only stopping to vomit before continuing. Ritual intoxication among the Mayas is an ancient practice. Frey Diego de Landa observed the practice while he was in the Yucatan, more than five centuries ago.

Balche' is brewed in the *chemil balche'*, a large canoe made from the trunk of *puuna'* 'mahogany'. To prepare the brew, first the inner bark of the *balche'* tree is removed from lengths of cut branches. The bark is pounded until pliable and then hung up to dry. Strips of the dried bark are then submersed in numerous litres of water and several kilos of sugar (honey was used in the past). These are reused until the bark extracts have been exhausted. At this point, a few strips of fresh inner bark are added (Fig. 4.5).

Two videos showing AM preparing the *balche'* for an upcoming ceremony are presented in Video 4.1 Balche' Preparation 1, and Video 4.2 Balche' Preparation 2.

To ensure the *balche'* ferments, one must then recite *ut'anil balche'*, a charm that summons species of plants and animals with hot or venomous qualities or the ability to agitate the brew, i.e. snakes, frogs, fish, turtles, and the *chäk xok* 'supernatural sirens'. If the charm is delivered incorrectly or incompletely then the brew will either fail to ferment or cause severe diarrhea. Tozzer (1907) provides a version of this chant (1907: 177), which differs from the one presented in Video 4.3.

After the *balche'* has been prepared, the canoe is covered with layers of leaves of *pahok* or *k'unche' pahok* (*Geonoma oxycarpa*) and left to ferment for 3 days.

Before the *balche'* is served, several small rituals are performed, including the consecration of the balche', presented in Video 4.13a. *Balche'* ceremonies may be large or small, depending on their purpose. Tozzer provides a detailed account of a *balche'* ceremony performed during the changing of the god pots, together with a number of chants, while Davis (1978) includes comprehensive descriptions of the offerings and the ritual protocol she observed in Naha' in the 1970s.

Today, *balche'* ceremonies are rare. AM is the only one in the community of Naha' (and possibly of all Lacandon communities) who still performs the ceremony, and usually only at the request of tourists or researchers (Fig. 4.5).

Käkow 'cacao' is the most sacred food in Mesoamerica, after maize, and is thus consumed in ceremonial contexts. The beverage is rarely, if ever, drunk on its own; it is normally mixed with corn gruels. The froth that is raised while it is beaten is spooned on top. Frey Diego de Landa describes the concoction that he witnessed being made in the Yucatan at the turn of the sixteenth century: "They make of ground maize and cacao a kind of foaming drink which is very savoury and with which they celebrate their feasts. And they get from the cacao grease which resembles butter and from this and maize they make another beverage which is very savoury and highly thought of." ([1566]1941: 90)

The process of preparing cacao is so arduous that it constitutes a ritual in itself. In the Florentine Codex, Sahagún (1590) describes a process that mirrors the Lacandon method: "She grinds the beans; she crushes, breaks, and pulverizes them. She adds water, sparingly, conservatively, and then aerates it, making it form a head. She then removes the head, makes it thicken, makes it dry, and stirs water into it." (1950–1982 [1590], cited in Green 2009: 327)

Additionally, the Lacandones roast the cacao before grinding the beans and then beat it, adding the resin of sarsaparilla vine, which makes the beverage foam. After this they stir it into *ma'ats'*. They call this preparation *käkow näl* 'cacao maize', according the KP.

Other edible offerings are b asically the same foods prepared for daily consumption. The only difference is that the corn-based offerings are prepared with maize that has not been processed in quicklime. The main food offerings include the following:

Ch'ulha' 'wet water'[5] is the same as ma'ats 'posol';

Hach huybil ma'ats 'true stirred posol' is the same as *huybil ma'ats*;

Hach ma'ats 'true posol' is the same as *ma'ats'*;

Nahwah/nahil wah 'great tortilla' are of two kinds: *Chäkbil wah* 'cooked bread/tortilla' is a tamale filled with meat, and *bu'ulil wah* is a tamale filled with beans;

Säkha' 'atole' 'white water' is the same as *sa'*;

K'uxub 'annatto' is a red paste rendered from the fruit of the 'lipstick tree' (*Bixa orellana*). It is patted into small cakes, dried in the sun, and then stored. In ritual contexts it

---

[4] *sih-il* <*sih* 'give a gift' *-il* NOM. Cf. siih [Yuc.] 'donate' (Bricker et al. 1998), siy [Itz.] 'give' (Hofling and Tesucún 1997).

[5] Bruce (1968,1975) and McGee (2002) gloss this word as 'sacred water'.

represents blood. Up until the middle of the twentieth century, the Lacandones sacrificed their own blood in ritual blood-letting ceremonies, believing that the gods relished the sight and smell of blood. When the religion began to disintegrate, blood-letting ceased and human blood was replaced by annatto (McGee 2002: 142). *K'uxub* is added to tamales and used to paint *chäk hu'un*, strips of bark cloth worn by ceremonial participants.

The preparation of most of the edible food offerings is done by women (see Video 4.4 Cooking Traditional Foods, Video 4.5 Cooking Pakyon). It is gruelling work, owing to the number of different dishes and the quantities of food that ceremonies require. To ease their burden and speed up the process, they recite magical charms to invoke certain creatures and the life forces of the plants. Two of these work songs, *uk'aay käkow* 'The Cacao Song' and *uk'aay huch'* 'The Grinding Song', are included in this book. Davis (1978: 216–219) presents a version of the song for frothing cacao. It is similar to the one presented here. Transcriptions and translations of both songs are provided in Appendix 5.

First Fruits. In addition to these prepared offerings, the Lacandones also give the first fruits of their crops in small rituals called *uts'ahik ti' ho'ol* 'give it to the head'. There is no special method of preparation for the fruits and vegetables; most are just boiled.

The first fruit rituals had already begun to decline in the 1970s along with the disintegration of the traditional religion (Davis 1978, McGee 2002). Today, few people, apart from AM, still perform them.

Table 4.1 Food Plants

Species	Parts Used							
	Fruit, seeds	Roots, tubers	Leaves	Flowers	Stems, hearts, branches	Bark	Resin	Non-edible uses
'abäl *Spondias* spp.	x							
hach'akte' *Astrocaryum mexicanum*				x	x			
'akunte' *Acacia mayana*	x							
'axux *Allium sativa*		x						
'axux ak' *Cydista aequinoctialis*			x	x				
bamax *Pseudomelia* spp.	x							
bits' *Inga* spp.	x							
boox *Musa acuminata* ×*Musa balbisiana*	x		x					
box *Cucurbita* sp.	x							
bu'ul *Phaeolus vulgaris*	x							
chäkchop' *Lantana trifolia*	x							
ch'äm *Bromelia pinguin*	x							
ch'ayok' *Solanum amercianum*			x					
chi' *Byrsonima crassifolia*	x							
chiah *Dieffenbachia* sp.				x				
ch'ibix uk'um *Chamaedorea cataractarum*				x				
ch'ibix wits *Chamaedorea woodsoniana*				x	x(?)			
ch'i'ib *Chamaedorea tepejilote*				x				
chi'kam *Pachyrhizus erosus*		x						
chinah *Citrus sinensis* **chäk chinah** *Citrus reticulata*	x							
ch'ink'ak'il *Passiflora cookii*	x							
ch'um *Passiflora* aff. *ambigua*	x							
ha'as *Pouteria mammosa*	x							
hach balche' *Lonchocarpus* sp.						x		Fermented
hach hänan *Desmoncus orthocanthus*					x			
hach hoben *Piper auritum* **säk hoben** *Piper umbellatum*			x					
hach k'uuts *Nicotiana tabacum*			x					Smoking
huhup *Spondias* sp.	x							
ik *Capsicum* spp.	x							

(continued)

Table 4.1 (continued)

Species	Parts Used							
	Fruit, seeds	Roots, tubers	Leaves	Flowers	Stems, hearts, branches	Bark	Resin	Non-edible uses
'ib *Phaseolus lunatus*	x							
'is *Ipomoea batatas*		x						
'itsänte' *Clerodendrum* sp.			x					
k'äb ta'k'in *Portulaca oleracea*			x					
k'äxeex *Chenopodium ambrosiodes*			x					
k'ayoch' *Pleurotus djamor*								
hach k'ik'ni' balum *Ardisia compressa*	x							
komo' *Alpinia purpurata*			x					
kopo' *Ficus* spp. **nukuch kopo'** undetermined			x					
koyoh (nukuch) *Persea schiedeana*	x							
kubuh *Pachira aquatica*	x							
kulantro *Eryngium foetidum*			x					
kulix *Brassica oleracea* var. *alba*			x					
k'um *Cucurbita moschata, Cucurbita pepo*	x							
k'unche' pahok *Geonoma oxycarpa*					x			
k'uxub *Bixa orellana*	x							
k'uyuch ak' undetermined	x							
le'mo' *Epiphyllum* sp.	x							
lo'loh *Auricularia* spp.								Fruiting body
mahaas *Quararibea funebris*					x			Grill
mäkäl *Xanthosoma mafaffa*		x	x					
mäkuuläm *Piper* spp.			x		x			Grill
mankoh *Mangifera indica*	x							
mäya' *Musa acuminata ×Musa balbisiana*	x		x					Wrapper
muluch' *Oudemansiella* spp., *Pluteus harrisii*								Fruiting body
muxan *Heliconia librata, Calathea* sp.			x					
naba' *Myroxylon balsamum*					x	x		Grill, smoking
näl *Zea mays*	x		x					Husks used for tamale wrappers
nukuch'oochin *Philodendron sagittifolium, Syngonium chiapense*	x			x				
'oon *Persea americana*	x							
'oop *Annona* spp.	x							
o'ot' ya' *Pouteria campechiana*	x							
'oox *Brosimum alicastrum*	x							
pach' *Ananas comosus*	x							
päh äk' *Disciphania calocarpa*	x							
pähk'ol *Spathiphyllum phryniifolium*				x				
pähpäh xibil *Ardisia* sp.	x							
pähpox *Annona muricata*	x							
p'ak *Lycopersicon esculentum*	x							
p'akan *Physalis* sp.	x							
pakyon *Chamaedorea tepejilote*				x				
pätan *Musa* spp.	x		x					Wrapper
peesahche' *Pimenta dioica*	x		x					
perehir *Coriandrum sativum*			x					
pichik' *Psidium guajava*	x							
p'ix *Sechium edule*	x							

(continued)

Table 4.1 (continued)

Species	Parts Used							
	Fruit, seeds	Roots, tubers	Leaves	Flowers	Stems, hearts, branches	Bark	Resin	Non-edible uses
pooch'in *Passiflora serratifolia*	x							
puhan *Muntingia calabura*	x							
put *Carica papaya*	x							
pu'uhte' *Ardisia* sp.	x							
sanyah *Citrullus lanatus*	x							
seboya *Allium cepa*		x						
sensi' *Zingiber officinale*		x						
suli' (bilim) *Dioscorea bulbifera* **(hach)** *Dioscorea alata*		x						
ts'ak kax *Dysphania ambrosioides*			x					
ts'ak k'ek'en *Allium glandulosum*		x	x					
tsäk'ats *Licania platypus*	x							
ts'in *Manihot esculenta*		x						
tso'ots *Eriobotrya japonica*	x							
tsuk'in ak' *Smilax domingensis*								To froth cocoa
ts'us *Vitis* spp.	x							
tuch *Thevetia ahouai*	x							
tuk' *Acrocomia aculeata*	x							
tuts *Attalea butyracea*					x			
'uuch *Diospyros digyna*	x							
wolche' *Casimiroa* sp.	x							
xikin *Schizophyllum commune*								
xinich *Parathesis* sp.	x							
ya' *Manilkara* spp.	x						x	Chewing gum

Table 4.2 Plant Materials (in alphabetical order of scientific name within delimited categories)

Materials for Construction, Implements, Containers		
Lacandon	Scientific Name	Use
tuk'	*Acrocomia aculeata*	Hard shell of the nut used for the *tukil* 'spindle whorl'
pähpäh xibil	*Ardisia* sp.	Branches used for *xuul* 'planting stick'
sayok'	*Aspidosperma megalocarpon*	Wood used for *säkche'* 'rafters', *taan* 'girt, horizontal beams', *hiil* 'battens', *waan* 'girt (side)'
säkche'	*Bauhinia* sp.	Wood used for *akam* 'foundation posts'
hach taw	*Belotia mexicana*	Wood used for *hi'che' ti' k'anche' näl* 'walls for the corncrib', *chemil balche'* 'balche' canoe'
ponche'	*Bernoullia flammea*	Wood used for *hi'che'* 'vertical wall boards', *ch'aak* 'beds', *k'anche'* 'chairs', *ho'olan che'il* 'ridgepole'
k'alok' che'	*Billia colombiana*	Wood used for horizontal wall boards
säkyuuhche'	*Blepharidium mexicanum*	Wood used for *ho'olan che'il* 'ridgepole'
babah	*Calophyllum brasiliense*	Wood used for *hiil* 'battens', *säkche'* 'rafters', *taan* 'girt', *ho'olan che'il, waan* 'girt (side)', *uleh'il ch'iich'* 'bird traps', *chem* 'canoes'
k'uche'	*Cedrela odorata*	Wood used for *hi'che'* 'vertical wall boards', *chemil balche'* 'balche' canoe'
säk ya'	*Chrysophyllum mexicanum*	Wood used for *chuste'* 'arrow foreshafts', *chulul* 'bows', *akam* 'foundation posts'
bahun che'	*Cordia alliodora*	Wood used for *hiil* 'battens'
hach luuch	*Crescentia cujete*	Rinds of fruits used for *luuch* 'cups', *hama'* 'ceremonial drinking cups', *cha'achil* 'colander'
kun	*Cryosophila stauracantha*	Leaves used for *nahil* 'roof thatching'
ton k'uk'	*Cymbopetalum* sp.	Wood used for *pakat* 'bird bolt'
k'ul ak'	*Dioscorea bartlettii*	Basal burl for *lekil k'uh* 'god pots'
pakah che'	undetermined (Flacourtiaceae)	Saplings used for *hiil* 'battens'

(continued)

Table 4.2 (continued)

Materials for Construction, Implements, Containers		
Lacandon	Scientific Name	Use
pahok	*Geonoma Geonoma oxycarpa*	Leaves used as *nahil* 'roof thatch', *pixil chem* 'cover for the *balche'* canoe', *xuul* 'planting stick'
ek' bahche'	*Guatteria anomala*	Wood used for *hi'che' nah* 'vertical wall boards for the house', *hiil* 'battens', *pohche'* 'tables'
'ooh	*Gynerium sagittatum*	Stalks used for *häläl* 'arrows', *hiil* 'battens'
ek'	*Haematoxylon* spp.	Wood used for *okman* 'pillars'
halol	*Heliocarpus* spp.	Wood used for *hi'che'* 'vertical wall boards', *ka'che' näl* 'corncrib'
chintok'	*Krugiodendron ferreum*	Wood used for *chulul* 'bows'
chuh, lek	*Lagenaria siceraria*	Rinds of fruit used for *hama'* 'ceremonial drinking cups', *ha' chuh* 'water jug', *habänche' kab* 'beehive', *lekil wah* 'tortilla plate'
tsäläm	*Leucaena leucocephala*	Trunk hollowed out for *chem* 'canoes'
ya'ax balche'	*Lonchocarpus* spp.	*pa'te' kax* 'chicken coop'
chäk ya'	*Manilkara achras*	Wood used for *chulul* 'bows', *akam* 'foundation posts', *okman* 'pillars', *bahbil* 'horizontal wall boards', *ho'olan che'il* 'ridgepole', *chuste'* 'arrow foreshaft', *kekche'* 'cocoa beater', *hälälte'* 'loom sticks', *but'u'o'och* 'weaving shuttle', *xuunche'* 'end bars of the loom'
hach ya'	*Manilkara zapote*	*akam* 'foundation posts'
kakache'	*Ocotea cernua*	Wood used for *okman* 'pillars'
säk chulul	*Platymiscium dimorphandrum*	Wood used for children's practice bows
haban uts'u'	*Podachaenium eminens*	Stalks used for *k'anche' näl* 'corncrib', *hiil* 'battens'
hach häläl	*Phragmites australis*	Stalks used for *häläl* 'arrows'
mäkuuläm	*Piper* spp.	Branches used for *pa'ate' kax* 'chicken coop', *ch'am* 'grill'
tso'ots bamax	*Pseudolmedia oxyphyllaria*	*säkche'* 'rafters'
bamax	*Pseudolmedia* spp.	Saplings used for *hiil* 'battens' *ho'olan che'il* 'ridgepole'
mahaas	*Quararibea funebris*	Wood used for *pakat* 'bird bolt', *ti' uch'uyik bäk', nook'* 'hangers for meat and clothes', *tukil* 'spindle whorl', *pecheech* 'spindle', *hi'che' ahpek'* 'dog fence', *ch'am* 'grill'
hach ch'alol	*Quercus corrugata*	*ch'alol* 'spinning top'
semet'	*Rhipidocladum bartlettii*	Ramets used for *lama' käy* 'fish spear'
xa'an	*Sabal mexicana*	Leaves used for *nahil* 'roof'
che' chäk'an	savanna tree, e.g., *Vernonanthura patens*	Branches used for *hiil* 'battens', *hulu'che' uho'ol* (two poles perpendicular to the ridgepole used for securing the thatch to the roof)
puuna'	*Swietenia macrophylla*	Wood used for *hi'che'* 'vertical wall boards', *hool* 'door', *pohche'* 'tables', *k'anche'* 'stools', *chemil balche'* 'balche' canoe', *chemil pom* 'canoe for mashing incense', *xikal* 'incense board', *ba'ab* 'paddles'
hach chulul	*Tabebuia guayacan*	Wood used for *chulul* 'bows', *häläl* 'arrows'
kuti'	*Talauma mexicana*	Wood used for *hopil k'ak'* 'coal pan'
pukte'	*Terminalia amazonia*	Wood used for *puche'* 'bridges', *hi'che'* 'vertical wall boards', *pohche'* 'tables'
tuch	*Thevetia ahouai*	Wood used for *okman* 'pillars', *ka'che' näl* 'corncrib'
ch'obenche'	*Trichilia* sp.	Wood used for *okman* 'pillars'
k'ik' che'	*Virola guatemalensis*	Wood used for *hi'che'* 'vertical wall boards'
k'än chululche'	*Wimmeria bartlettii*	Wood used for *okman* 'pillars'
yache' kab	undetermined	Wood used for *tan* 'girt', *säkche'* 'rafters'
k'isis che'	undetermined	Saplings and small branches used for *hiil* 'battens', *säkche'* 'rafters'

Materials for Fuel		
Lacandon	**Scientific**	**Comments**
säkyuuhche'	*Blepharidium mexicanum*	Choice
chäk'oop che'	*Clarisia biflora*	Wood is used in kiln
bu'ulche'	*Cojoba* spp.	
wolche'	*Casimiroa* sp.	Choice
kun	*Cryosophila stauracantha*	Burning coals of leaf stems placed under hammocks at night for warmth
wäch'	*Dialium guianense*	
hach pahok	*Geonoma* sp.	
säk bahche'	*Guarea glabra*	

(continued)

Table 4.2 (continued)

Materials for Construction, Implements, Containers		
Lacandon	Scientific Name	Use
ek'	*Haematoxylon* spp.	
bits'	*Inga* spp.	
chechem	*Metopium brownie*	
ek''oonte'	*Nectandra globosa*	
'äm	*Ormosia* spp.	
tähte'	*Pinus* spp.	Large sticks used as torches and to start fires
bamax	*Pseudomelia* spp.	
chäk'ax	*Simira salvadorensis*	Shavings used as kindling while using the fire drill
yo'och ähbach'	*Siparuna thecaphora*	
ton tsimin	*Tabernaemontana amygdalifolia*	
chäk ya'	*Manilkara achras*	Choice
ch'obenche'	*Trichilia* sp.	
k'än chululche'	*Wimmeria bartlettii*	Choice
ak' tsup	undetermined	Dried sections of the vine used for a drill fire

Fibres and Vines for Ropes, Binding, Baskets, Mats, Clothing

Lacandon	**Scientific Name**	**Use**
kih	*Agave sisalana*	Fiber twisted to make *häxan* 'string, fine cord', used for bow strings, *ch'im* 'shoulder bag', *leh* 'traps'
säk ak'	*Arrabidaea verrucosa*	Small vines used to tie together house poles, rafters to beams, and the girt to the house posts
ya'ax ak'	*Arrabidaea* sp.	Vines used for tying the girt to the pillars
taw	*Belotia mexicana*	Fibre from inner bark used for *bilich* 'mattress'
nikte' ak'	undetermined	Vines used for tying posts to rafters and cross-beams, and as frame for *ch'uyu'* 'hanging baskets'
ya'ax hänan	*Chamaedorea elatior*	Ramets used for making the rings for *ch'uyu'* 'hanging baskets'
hänan	*Desmoncus orthocanthus*	Vines woven into rings called *me'et* upon which the *balche'* cups are placed, and for headband for *pak*, the ceremonial *balche'* pot (a.k.a. *Bol* 'God of Wine')
ak' hu'un	*Ficus* sp.	Inner bark used for *hu'un* 'bark cloth'
bits'kal hu'un	*Ficus* sp.	Inner bark used for *hu'un* 'bark cloth'
hach hu'un	*Ficus* sp.	Inner bark used for *hu'un* 'bark cloth'
tämän	*Gossypium hirsutum*	Fibre from seed heads spun into *k'uch* 'cotton thread'
ts'ubtok	*Hampea stipitata*	Fibre from stalks used to make *ba'ay* 'rough net bags', twine for tying house poles and thatch to *hiil* 'battens', bow strings, hanging baskets, and *k'an* 'hammocks'
halol	*Heliocarpus* spp.	Fibre used for *ba'ay* 'net bags', *k'an* 'hammocks', *taabil ba'ay* 'tumpline', and *ch'uyu'* 'hanging baskets'
'ilon	*Monstera* spp.	Vines used for *xaak* 'baskets', *me'et* 'trivet', and *xaakil chukkäy* 'fish trap'
ch'ink'ak'il	*Passiflora cookii*	Vines used to tie beams and rafters
k'i'ilix holop'	*Philodendron radiatum*	Vines used for tying together battens, rafters, and thatch
yuyul holop'	*Philodendron smithii* *P. tripartitum*	Vines used for tying roof thatch to battens
säk hoben	*Piper umbellatum*	Leaves used to cover *k'ät* 'clay'
ak' hu'un	*Poulsenia armata*	Inner bark used for *hu'un* 'bark cloth'

Dyes and Resins

Lacandon	**Scientific Name**	**Use**
k'uxub	*Bixa orellana*	Fruits boiled down to a red paste ceremonial paint
sik'	*Carthamus tinctorius*	Flowers processed to produce a yellow-orange dye
k'ik'	*Castilla elastica*	Latex used to make *k'ik'* 'rubber'
chäk'oopche'	*Clarisia biflora*	Fruits used like a stamp to stain clothes, the resin from the bark is mixed with *pom* 'incense' and *kibil kab* 'wax' to seal the binding on arrow shafts
k'ante'	*Erythrina berteroana*	Roots are scraped and soaked to produce a yellow dye

(continued)

Table 4.2 (continued)

Materials for Construction, Implements, Containers		
Lacandon	Scientific Name	Use
hut'kih	*Ipomoea albans*	Resin from stems used to vulcanize the latex from *Castilla elastica*
tsits	*Justicia* sp.	Leaves seared and soaked to produce a blue-purple dye
ts'ak k'ik'	*Merremia tuberose*	Resin used to vulcanize the latex of the *Castilla elastica* to make *k'ik'* 'rubber'
xoyok	*Morinda panamensis*	Roots scraped and soaked to produce a yellow dye
ni' ts'ul	*Mucuna argyrophylla*	Resin from the leaves used to seal *k'anil luts* 'fishing lines'
tähte'	*Pinus* spp.	Sap used as *pom* 'incense' for offerings
tsatsel pom	*Protium copal*	Sap used as *pom*
chäk'ax	*Simira salvadorensis*	Bark is used to extracted a red dye for painting the *chäk hu'un* 'ceremonial bark headbands'
puuna'	*Swietenia macrophylla*	Bark soaked to produce a red dye
chäk pach	*Ternstroemia tepezapote*	Berries pressed onto fabric to stain it reddish-orange
läkte'	*Zanthoxylum* sp.	Resin is used to prevent *tu' bok* 'cockroaches' from entering houses
ch'ox	undetermined	Leaves crushed to produce a dark blue dye

Materials Used in Crafts (excluding dyes and fibres)		
Lacandon	**Scientific Name**	**Use**
anis	*Abelmoschus moschatus*	Seeds for jewellery
tuts	*Attalea butyracea*	Seeds for jewellery
chank'ala'	*Canna indica*	Seeds for jewellery
kahanche'	*Clusia flava*	Flowers strung on jewellery
säkpahen	*Coix lacryma-jobi*	Seeds for jewellery
hach hänan	*Desmoncus orthocanthus*	Seeds for jewellery
corazón de mono	*Entada gigas*	Seeds for jewellery
pets'k'in	*Enterolobium cyclocarpum*	Seeds for jewellery
k'änte' ak'	*Erythrina* sp.	Seeds for jewellery
pakay	*Hymenaea courbaril*	Seeds for jewellery
chuh, lek	*Lagenaria siceraria*	Rinds used for toys, *so'ot* 'rattle'
tsäläm	*Leucaena leucocephala*	Seeds for jewellery
gorila	*Merremia discoidesperma*	Seeds for jewellery
ni' ts'ul	*Mucuna argyrophylla*	Seeds for jewellery
wich yuk	*Mucuna sloanei*	Seeds for jewellery
'äm	*Ormosia* spp.	Seeds for jewellery
mehen ch'alol	*Quercus* sp.	Seeds for jewellery
mo'ak'	*Rhynchosia pyramidalis*	Seeds for jewellery
choochel ak'	*Rhynchosia erythrinoides*	Seeds for jewellery
xa'an	*Sabal mexicana*	Seeds for jewellery
habonsiyo, subul	*Sapindus saponaria*	Seeds for jewellery
pets'k'in	*Schizolobium parahybum*	Seeds for jewellery
buk'luch'	*Vanilla planifolia*	Dried pieces of whole beans are strung in necklaces

Materials for Other Purposes		
Lacandon	**Scientific Name**	**Use**
k'ul ak'	*Dioscorea bartlettii*	The basal burl used for *lekil k'uh* 'god pots'
säha'	*Doliocarpus dentatus*	Leaves used to sand *chulul* 'bows'
ya'ax balche'	*Lochchocarpus* spp.	Pulverized bark used as fish poison
chechem	*Metopium brownei*	Pulverized bark used as fish poison
xiiw	Zingiberales	Leaves used as covers for food and other small items

Fig. 4.1 Making tortillas. Photo Credit: Suzanne Cook

Fig. 4.2 Making tortillas. Photo Credit: Suzanne Cook

Fig. 4.3 K'u'umbil wah. Photo Credit: Suzanne Cook

Fig. 4.4 Straining sa'. Photo Credit: Suzanne Cook

Fig. 4.5 Balche' bark. Photo Credit: Suzanne Cook

4.3 Agriculture

4.3.1 Overview

Nations and Nigh (1980) were the first to describe in detail the Lacandones' intensive agricultural method. For these researchers, the Lacandon system answered the question of how the ancient, lowland Maya managed to feed several million citizens on crops produced from thin, forest topsoil. The Lacandones practice a form of tropical swidden agriculture, a system whereby the farmer slashes and burns a small site, cultivates it for a number of years, and then carves out and cultivates another site, while the abandoned field is left fallow for long periods to recover its fertility. This rotation of sites not only assists the natural regeneration of a mature forest, it also ensures an abundance of food all year round. Although untended, fallow fields still provide fruits, leafy greens, and roots. Moreover, they attract game animals, which the Lacandones hunt to supplement their mainly vegetarian diet with high quality protein.

The Lacandon household rotates a number of plots, each measuring 1 ha, or 10,000 square meters. The active *kol* is cultivated for 2–5 years and then laid fallow for a number of years. During this time, it goes through stages of secondary succession, or transformations.

The northern Lacandones differentiate four stages: (1) *lo'obil*, characterized by grasses and other herbaceous plants; (2) *mehen che'*, characterized by small bushes and trees; (3) *nukuch che'*, characterized by mature, fast growing evergreen trees.; and, (4) *tämän che'* 'dense forest', comprised of large deciduous trees (Figs. 4.6 and 4.7).

An additional phase, *jurup che*, has been identified in the extant Lacandon agro-forestry literature. It is described as a transition from *mehen che'* to *jurup che*, wherein the short-lived, dominant canopy trees are replaced by intermediate to long-lived secondary species (Nigh 2008: 237). Only one Lacandon whom I interviewed was familiar with this term. He described it as a fallow *milpa* with small trees clear of undergrowth where the soil beneath is soft and fertile.[6] Among the species found in this succession are: *bits'* (*Inga* spp.); *hach taw* (*Belotia mexicana*); *che'il chäk'an* 'savannah tree', e.g., *Vernonanthura patens*; *tsaah* (*Cnidoscolus multilobus*); and, *halol* (*Heliocarpus* spp.). Table 4.3 lists all the dominant *jurup che* species reported in Nigh (2008: 239).

4.3.2 *Kol* '*milpa*'

The *milpa* provides the bulk of the crops used for local consumption and thus demands most of the Lacandones' time and effort. Between 40 and 50 different crops are grown in

[6] K'ayum Arturo Segundo (personal communication 2010).

Fig. 4.6 Fallow milpa first year. Photo Credit: Suzanne Cook

Fig. 4.7 Fallow milpa second year. Photo Credit: Suzanne Cook

Table 4.3 *Hulup' che'* Species

Lacandon name	English	Botanical name
chäklah	Gumbo limbo	*Bursera simaruba*
k'o'och	Pumpwood	*Cecropia* spp.
säk halol	White majagua	*Heliocarpus donnell-smithii*
bits'	Ice cream bean	*Inga* spp.
ya'ax balche'	Lancepod	*Lonchocarpus* spp.
(chuhum)	Balsa tree	*Ochroma pyramidale*
mäkuuläm (mehen)	Jamaican pepper, matico	*Piper aduncum, P. hispidum*
hoben	Mexican pepperleaf	*Piper* spp.
nukuch le' si'si'k'uuts	American burnweed	*Erechtites hieracifolia*
huhup	Hog plum	*Spondias* sp.

both active and fallow *milpas* at any one time. Unlike monocropped fields, crops in Lacandon *milpas* are dispersed throughout the *milpa*, avoiding large plantings of single species. Below the surface, root crops are planted at varying depths, while above-ground growth includes papayas, bananas, and other forest species. This strategy of dispersing a variety of plants in time and space is a characteristic of traditional, tropical, swidden systems, which efficiently use available space, water, and soil nutrients. Such a system circumvents widespread crop loss caused by insects, herbivores, and disease (Nations and Nigh 1980: 11).[7]

4.3.3 *Kolil nah* 'House Garden'

The majority of the Lacandones in Naha' are consummate horticulturalists and are constantly transplanting forest plants and trees into their home gardens. The cultivation of these species ensures a reliable and nearby supply of wild foods, decorative seeds, medicines, seasonings, gums and resins, and bouquets of fresh flowers. Typical species include: *buk'luch'* 'vanilla (Vanilla planifolia)'; *peesache'* 'Jamaican allspice (Pimenta dioica)', *pakyon* and *ch'i'ib*, which provide edible flower buds; *axux ak'* 'garlic vine (*Cydista aequinoctilis*)'; *balche'* 'lancepod (*Lonchocarpus* sp.)', whose bark is used to produce balche' liquor; *'akte'* 'bullhorn acacia (Acacia mayana)';*'aak'älyoom* 'night-blooming jasmine (*Cestrum nocturnum*)'; *bits'* 'ice cream bean (*Inga* spp.)'; *chank'ala'* 'canna lily (*Canna indica*)'; and, *k'ante'* 'coral bean tree (*Erythrina* spp.)'. These are in addition to a veritable cornucopia of citrus trees, bananas of various kinds, edible palms, and flowering shrubs and herbs.

A small section of the property is reserved for a kitchen garden, with demarcated beds of cabbages, radishes, carrots, and lettuce. Clumps of mint, garlic, and green onions spring up throughout. Flowering lianas, beans, and the ubiquitous chayote squash drape over fences, and pepo and moschata squash vines form an impenetrable network underfoot. Figure 4.8 gives an example of the landscape of Kohs' property. As can be seen, lush vegetation, orchards, palms, herbaceous plants, and flowers surround the buildings. On the bottom left of the image a small corner has been cleared for a vegetable patch (Fig. 4.9).

4.3.4 *Paakche'kol*

Paakche' kol is a *milpa* without trees. AM describes it as *uchbenchal*, a fallow *milpa* that is has become old. Previous researchers in the area frequently write "pak che kol" and gloss it as 'planted tree *milpa*' and claim that the Lacandones encourage the growth of certain tree species to facilitate forest regeneration (Diemont et al. 2006; Levy Tacher and Golicher 2004, in Nigh 2008: 239). While this may be true for the southern Lacandon community, where most of the agro-forestry research has been conducted, the transcription erroneously assumes that the word is *pak'* 'to plant'. A better interpretation of *paakche' kol*[8] would be "an old *milpa* that is undergoing or has attained second-growth regeneration."

Although it appears to be an abandoned field, on closer inspection the *paakche' kol* is a stable larder of over 50 species of fruiting trees and vines, root and leafy crops, and fibre-bearing plants (Nations and Nigh 1980: 15). Not only does this abundance supply a sustained yield of food and materials, but it also attracts a diverse assortment of wild animals and birds that provide meat, pelts, and feathers. Table 4.4 presents an example of the diversity of fauna that visits the *paakche' kol*.

Lacandon farmers influence the development of forest regeneration by constant weeding (Nigh 2008: 235). This reduces the weed seeds and facilitates the establishment of native seedlings from the surrounding forest. After 4 years in fallow, AM says the soil will have regained enough fertility to yield good harvests. He attributes this to certain fast growing trees: *bits'* (*Inga* spp.), *halol* (*Heliocarpus* spp.), *hach taw* (*Belotia mexicana*), *ponche'* (*Bernoullia flammea*), *ek' bahche'* (*Guatteria anomala*), and *pukte'* (*Terminalia amazonia*). He says that all of these trees shed an abundance of leaves, which decompose quickly creating a soil that is very black.[9]

Another tree that increases soil fertility is *chuhum* (*Ochroma pyramidale*), according to researchers in the

[7]Most of the Lacandon agro-forestry research has been carried out in the southern Lacandon community of Lacanha' Chan Sayab. Presumably, *"jurup che"* is a southern Lacandon term, although the word is not found in Canger's (1970) dictionary of San Quintín (a southern Lacandon community). The word *hú'ub'*, in Yucatec, refers to a fallow field less than three years old (Bricker et al. 1998: 113).

[8]The long vowel shortens in this compound.

[9]Other trees include *Hampea stipitata, Sapium lateriflorum, Cecropia obtusifolia*, and *Piper auritum* (Diemont 2006: 4–5).

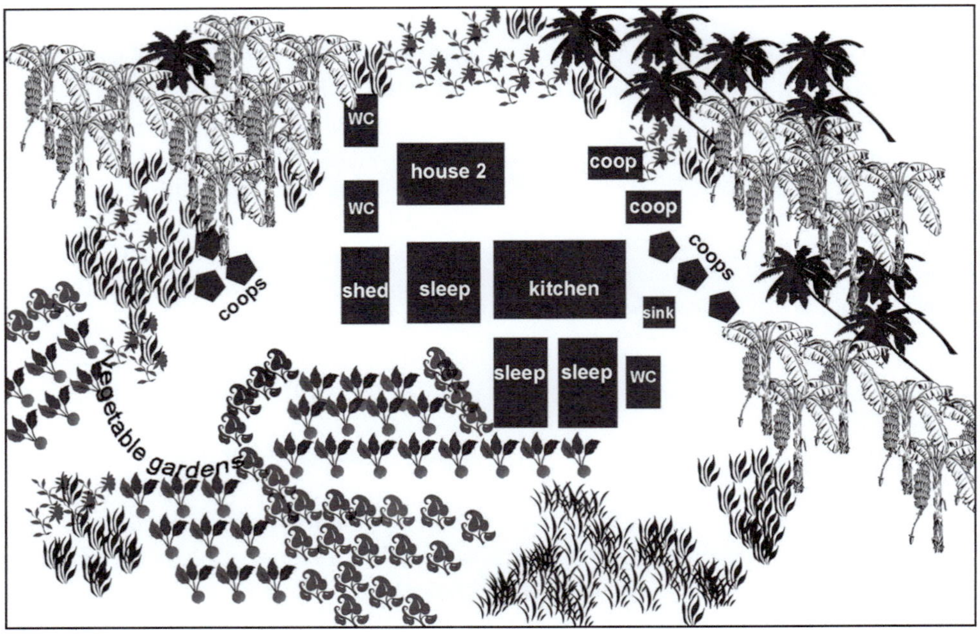

Fig. 4.8 Sketch of household and gardens. Photo Credit: Suzanne Cook

Fig. 4.9 House garden. Photo Credit: Suzanne Cook

Table 4.4 Birds and Mammals in the *Paakche' Kol*

Birds		
täkay	Myiarchus and kiskadee-like flycatchers	
kachok'	Mealy (blue-crowned) parrot	*Amazona farinosa*
mo'	Scarlet macaw	*Ara macao*
k'ili'	Parakeet	*Aratinga canicularis*
ts'aapin	Saltators	*Atlapetes gutturalis*
chuhum	Zebra-backed woodpeckers	*Centurus* spp., *Dendrocopus* sp.
ts'ul	Dove	Columbidae
tuch	Pigeon	Columbidae
ts'iil	Blue-black grosbeak	*Cyanocompsa cyanoides*
sensi'	Green jay	*Cyanocorax yncas*
ixkit	Turquiose-browed motmot	*Eumomota superciliosa*
xet'et'	Red-throated ant-tanager	*Habia fuscicauda*
yuyum ch'iich'	Yellow-backed oriole	*Icterus* spp.
baach	West Mexican chachalaca	*Ortalis vetula*
koox	Crested guan	*Penelope purpurascens*
t'ut'	Small parrots, white-crowned parrot	*Pionus senilis*
paaya	Brown jay	*Psilorhinus morio*
k'ok'tah	Robin	Psittacidae
pichik'	Collard araçari toucan	*Pteroglossus torquatus*
pän	Keel-billed toucan	*Ramphastos sulfuratus*
pupuch'	Thrashers	*Toxostoma longirostre*
Mammals		
ba'ats'	Howler monkey	*Alouatta palliata*
ma'ax	Spider monkey	*Ateles geoffroyi*
sek	Bat	Chiroptera
hale'	Spotted cavy	*Cuniculus paca*
yuk	Brocket deer	*Mazama americana*
keh	Mule deer	*Odocoileus virginianus*
ch'ämäk	Gray fox	*Urocyon cinereoargenteus*

southern Lacandon area (Diemont 2006; Levy et al. 2006; Nigh 2008), where the southern Lacandones deliberately plant it to improve the soil. The northern Lacandones whom I interviewed were unfamiliar with the name of this species, conceivably because it does not grow in their area. Moreover, they said that they do not intervene in the regeneration of the fallow, but leave this task to nature.

4.3.5 Subsistence Activities

The Lacandones' diversified agricultural system incorporates activities to produce not only food and materials for local use, but also for outside markets. Thus they must balance a number of various activities throughout the year. Besides farming, other activities include the commercial production of materials for crafts, notably a giant cane they call *'ooh* (*Gynerium sagittatum*), which provides arrow shafts, the harvesting of *xate* (*Chamaedorea* spp.) leaves, which they sell to the global florist industry.

4.3.5.1 Farming

Ninety percent of the Lacandon farmers slash and burn the fallow *milpas* and secondary vegetation. Although many hire Tzeltal (Maya) labourers to clear the land and, on occasion, cultivate the maize crop, every family member participates in the various farming activities. Women shoulder a significant share of the load, weeding, clearing debris, and cultivating the herbaceous crops. Men undertake the heavy work, felling trees, slashing underbrush, and burning the fields. They also sow the maize. The entire family takes part in the harvest, while women alone process it.

Uyanchunil kol 'the first *milpa*' is made in March before the rain begins and the ground is still dry.

Originally, *uyanchunil kol* referred to a *milpa* cut from virgin forest. But since the area was decreed a protected site in the 1990s, this practice is infrequent. Instead, fallow *milpas* are rotated and reverted back into cultivation. So today, the *uyanchunil kol* refers to a *milpa* cut from secondary forest.

The best sites are those where the soil is *siis* 'cool' and *hup* 'friable, deep soil'. These are places where *'oox* (*Trophis mexicana*), *ya'axche'* (*Ceiba pentandra*), and *bo'oy* (*Chamaedorea* spp.) flourish. The presence of these species is taken into account when selecting a site. The brush is cut and the small trees are felled with axes and machetes (see Video 4.8 Clearing a New *Milpa*). The debris is then left to dry for approximately *huntu(u)lwinik* 'twenty days'. When it has dried sufficiently, it is burned. Before burning, firebreaks are cut on the windward side of the field, to contain the flames. After the flames diminish, the field is left to cool for 3 days or more. After this, cultivation begins (Fig. 4.10). Burning the fields benefits the farmer in two ways: it drives out rats and deposits a heavy layer of ash, which fertilizes the soil.[10]

The main felling and cultivation period is called *nah kolil*. Farmers observe the flowering of certain forest trees to guide their agricultural activities. These "indicator" species signal the foot of the agricultural cycle (Nations and Nigh 1980: 11). When the *halol* (*Heliocarpus* spp.) and *pet ak'* (unidentified) bloom, Lacandones clear the brush and unwanted trees. They sow the first crop of maize when *puuna'* (*Swietenia macrophylla*) flowers and then sow all the rest of the crops when the tree sheds its blossoms. The blossoms of *pukte'* (*Bucida buceras*) also signal the time to sow maize, while those of *ek' bahche'* (*Guatteria anomala*) signal the time to sow black beans and tobacco, and *wäch* (*Dialum guianese*) blossoms signal the time to plant sweet potatoes and tobacco. Lacandones also plant sweet potatoes when the *ek' 'oonte'*

[10] Nations and Nigh (1980) say it is one month.

Fig. 4.10 Slashed milpa. Photo Credit: Suzanne Cook

Table 4.5 Agricultural Round

Month	Activity and Crops	Indicator species
Jan–Feb	*Halik loobil, ch'äkik uche'* 'Pull out the weeds, fell the trees.'[a]	*halol*
March	*Kuch'uktik utihil, kutookik* 'Wait for (the brush) to dry and then burn it.'	*pukte'*
April	*Uta'anil, kuch'uktik, kupiktik kol, upäk'ik näl (es lo mejor maize). Päk'ik ak'il bu'ul, p'ak, k'um, ik, tuwolol* 'Wait for the ashes (to cool off), then sow the main crop of maize. When the maize is about 20 cm (8") high, plant pole beans, tomatoes, squash, chile peppers, and then everything else.'[b]	*pukte'*
May	*Upäk'ik che'il bu'ul. Utalha' chichin* 'Plant bush beans. It starts to rain a bit.'[c]	*puuna'*
June	*Tan uk'äntal (näl), tan utihil. Halik uloobil. Puede upäk'ik näl, también.* 'The maize is turning yellow, it is drying. Pull out weeds. One may plant more maize.'[d]	*puuna'*
July	*Mixba'al* 'Nothing happens.'	*ek' bahche'*
August	*Uwuts'ik näl* 'Bend over the maize stalks.'[e]	*ek' bahche'*
Sept–Oct	*Päybil näl* 'Clear weeds, pull out spent maize stalks, and then plant maize in the empty spaces'.[f]	*ek' 'oonte', wäch*
November	*Mäna' meyah* 'There isn't any work.'[g]	*ma'ax ak', halol*

[a]Maize and watermelon are sown at this time (Nations and Nigh 1980: 12).
[b]Maize is sown in late April (Nations and Nigh 1980: 12).
[c]Corn is sown in late May (Nations and Nigh 1980: 12).
[d]Maize, rice, and peanuts are sown in early June (Nations and Nigh 1980: 12).
[e]Tobacco and black climbing beans are sown at this time (Nations and Nigh 1980: 12); KyP says that beans are planted at this time.
[f]Nations and Nigh (1980: 12) indicate that tobacco is sown at this time. Baer and Baer (1952, fn. 10) note that corn is doubled over in September and removed in November.
[g]Corn may also be planted in late November to late December (Nations and Nigh 1980: 12).

(*Nectandra* sp.) blooms. Towards the end of the season, when the *ma'ax ak'* (*Paullinia* sp.) blooms again, the Lacandones sow a second crop of white maize and tobacco.

In the past, this traditional system of planting according to the flowering cycle of indicator species roughly coincided with the months of the year; and, the Lacandones may still refer to the Gregorian calendar when explaining the agricultural cycle to foreigners. But due to changing weather patterns, the traditional system is a more reliable method. Table 4.5 presents SK's description of the agricultural round according to the months of the year.

Crops. The major crop is maize, followed by beans. Bananas were also considered a major crop, providing vital sustenance

Fig. 4.11 Milpa, 2010. Photo Credit: Suzanne Cook

when maize yields were low. Despite their nutritive value, beans and bananas are of secondary importance in the Lacandon diet. According to Mateo Viejo and Chan K'in Viejo, two of the most respected elders of the Naha' community, these foods are more like snacks, whereas maize is the the Lacandones' foremost sustenance (1990). Of the other crops, the most widely cultivated include: *'is* 'sweet potato'; *ts'in*'manioc' (*Manihot esculenta*); *k'um* 'squash (*Cucurbita pepo*, *C. moschata*)'; and, *p'ix* 'vegetable pear' (*Sechium edule*). *K'uuts* 'tobacco' (*Nicotiana tabacum*) was an important crop for the older generation, but most of the younger generation does not smoke traditional cigars and so it is not widely cultivated anymore. Much of the tobacco that is grown is sold to outsiders.

Wild plants and weedy greens are also semi-cultivated or protected, a practice that has coexisted with agriculture for thousands of years (Alcalde and Hernández 1972; Casas et al. 1987; Basurto et al. 1998; Bye 2000). Popular species include: *ch'i'ib* and *pakyon* 'pacaya palm' (*Chamaedorea alternans, C. tepejilote*); *ch'ayok'* 'nightshade' (*Solanum* spp.); and, *mäkäl* 'elephant ears' (*Xanthosoma mafaffa*).

Harvest. Men and women take part in harvesting the crops, particularly the maize. Maize is harvested in August and September. Before the harvest, the stalks are doubled over, to allow the ears to dry and prevent the rain from rotting the kernels (Figs 4.11 and 4.12).

When sufficiently dry, the ears are harvested. According to KyP (personal communication), this is done when the

Fig. 4.12 Doubled-over maize. Photo Credit: Suzanne Cook

moon is full; if they were removed while the moon was still "tender", the kernels would rot, turn into dust, and quickly dissolve. The ears are pulled from their stalks and stacked in neat rows in the *k'anche'* 'corncrib'. The best are reserved for the next year's seed. Meanwhile, the stalks are cleared, to give room for the tobacco to develop.

Tobacco is cut in March, when it is about one meter (5.5′) tall (Baer and Baer 1952). Pole beans are collected from the vines and bush beans pulled up by their roots. The vines are laid out in the sun to dry. before being shelled and stored. Squash is left to dry on the vine before it is stored. Other vegetables and fruits are picked as they mature, except for bananas, which are cut while still green and the stalk hung up in the house or a shed to gradually ripen. In Video 4.9 *Milpa* Harvest, KM, KP and their children are shown harvesting their *milpa*.

Preservation. Maize, beans, and cacao seed are dried and stored in corncribs and net bags. Meats are barbequed until black, or it is dried, salted, or smoked. Maize that has not fully dried on the stalks is hung by the shucks over the hearth. Tortillas, dough, and perishable foods are stored in baskets in the rafters. Today many perishables are also stored in refrigerators.

Wrapping foods also preserves their freshness. Banana, plantain, and heliconia leaves serve as the main wrapping material for dough, and corn boiled in their shuck and tamales may stored as is for several days (Baer and Baer 1952: 28). Fruit is often stored green in net bags while bananas and plantains are hung by their stalks to ripen indoors.

4.3.5.2 Commercial Enterprises

While "making *milpa*" is the main subsistence activity, commercial pursuits have become important as Lacandones have become more reliant on processed foods and commercial products. The most important cash-crop is *'ooh* 'giant cane' (*Gynerium sagittatum*), which is used for arrows, because the production of bows and arrows for the outside market has intensified in recent years. The cultivation of *'ooh* is a male activity, since the making of weapons was a male activity. Today women help their husbands help by preparing and applying the binding around the arrow shafts. Because the production of arrows requires a great amount of effort and time, it has become the principal commercial activity for more than 90% of Lacandon families (Durán 1999: 81).

Another cash crop is *xate*, certain species of chamaedorea palms whose fronds are used as filler in floral bouquets sold throughout North America. The income derived from this activity is meagre by western standards, Lacandones receiving on average the equivalent of less than 40 US cents per frond. This piecework enterprise compels Lacandones and *xateros* (men who harvest *xate* palm) to cut as many fronds as possible in the shortest amount of time, to make their efforts worthwhile. Concern over the ecological impact of careless harvesting has prompted the establishment of *xate* nurseries in local communities throughout lowland Chiapas, Guatemala, and Belize. The Lacandones are cooperating with biologists and conservationists to develop and manage nurseries of other palms and trees as well.

4.3.6 Current Situation

When Nations and Nigh published their 1980 article, 20% of the Lacandones were still using their traditional farming methods. Since then, the number of traditional *milpas* has declined. Today, traditional farmers are progressively reducing the number of species they cultivate, while the younger generation has shifted to monocropping maize. In addition, burgeoning colonization and an increase in forest reserves compels the Lacandones to rotate their existing *milpas* more often.

The reduction of crop species, monocropping, and curtailed fallow periods will undoubtedly influence soil characteristics that once supported the succession of ideal vegetation associations and ultimately reduce subsequent agricultural productivity. Adding to the problem is the lack of motivation among the younger generation to farm and the effects of climate change. The younger generation is having fewer children, and so there are fewer hands to work in the fields and process the harvest. As a result, many young families are investing less of their time in farming and devoting more energy to making and selling their crafts.

Climate change has also affected the agriculture round by altering the cultivation cycle and jeopardizing crops. Maize, normally planted in March, must now be planted in May, or even June. BM stated that, as a result, the roots are frequently underdeveloped and that the plants often succumb to *k'än k'up'*, a disease that swiftly yellows and withers the stalks and leaves. He says that, whereas 20 years ago the rains were more regular but not heavy and the climate cooler, now there is often severe flooding, which destroys the bean crops (Fig. 4.13). Today, the Lacandones' reliance on government emergency supplies of maize has become the norm rather than the exception.

Of all the threats to traditional practices the most egregious is the unbridled expansion of cattle ranching, which decimates large tracts of tropical forest and turns it into African grass lands. The permanent loss of the forest ecosystem can only be avoided if current government policies change, and Lacandon youth and immigrants are encouraged to incorporate traditional Lacandon farming methods into land management practices.

Fig. 4.13 Flooded milpa, 2010. Photo Credit: Suzanne Cook

4.4 Housing

4.4.1 Layout

The traditional Lacandon home consists of a large property with a number of buildings that surround an open courtyard. Vegetable gardens together with a small orchard form the perimeter.

Larger families have a greater number of buildings. The main building is the kitchen and the gathering place. Most of the daily activities take place there. Hammocks may occupy a corner, but sleeping quarters are located in separated buildings, with boys in one cabin, and girls and their parents, in another. Other buildings include a water closet, shower stall, storage shed, and several chicken coops. An outdoor concrete wash table with a sink is located just outside of either the front or back entrance.

Married children, their spouses, and children live in their own, self-contained compound on the family property (Figs. 4.14, 4.15, 4.16, 4.17, 4.18, 4.19, 4.20, 4.21, 4.22, 4.23, and 4.24).

The kitchen is divided into at least three areas: (1) a grinding area, (2) two or more cooking hearths, and (3) an eating area. The *k'oben* 'hearth' is located just inside the back door. It consists of three stones arranged in a circle on the ground with a metal grill and a *xämäch* 'grilling plate' placed on top. Fire brands are fed in between the stones. This kind of hearth is used primarily for boiling large pots of maize and tamales.

The main hearth is a raised platform approximately 1 m (3′) high, and made of *puuna'* 'mahogany' and concrete. It is covered with a thin piece of sheet metal with a coating of quicklime. Firebrands are inserted through the gap between the floor of the hearth and the metal top. A short tube extends up and out of the back of the oven, to vent the smoke (Figs 4.25, 4.26, 4.27, 4.28, 4.29, 4.30, 4.31, and 4.32).

4.4.1.1 Construction

Materials from the forest are used in the construction of traditional houses, kitchen areas, god houses, and corncribs. The foundation of a typical structure is a rectangular perimeter of four *akam* 'foundation posts'. The posts have a *xaay* (side- or v-notch) carved into their upper ends, to cradle the *tan nah/ tanah* 'girt, horizontal beams' and the *kuch säkche' 'load-*bearing beams'. As the name implies kuch säkche' support the load of the säkche' 'rafters'. Two taller posts, *kelil nah*,[11] stand between the *akam* at both ends of the house, to support the *ho'olan che'il* 'ridgepole'. Inside, additional poles span the distance between the walls, resting on top of the *kuch säkche'*. They provide a base for shelves upon which food baskets, brooms, and other items are stored.

[11] Term from Baer and Baer (1952).

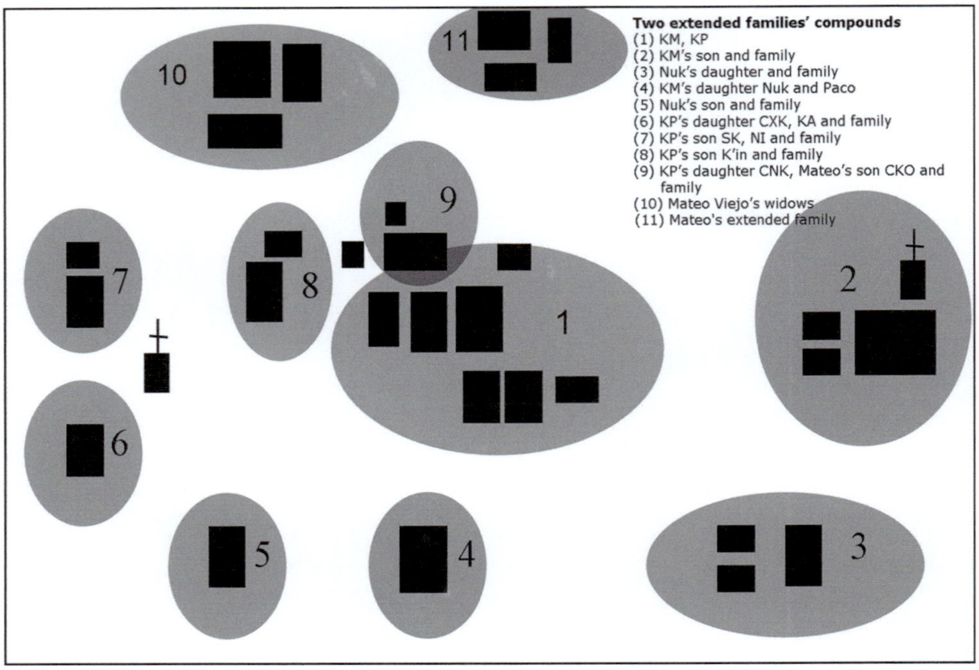

Fig. 4.14 Two extended families' compounds. Photo Credit: Suzanne Cook

Fig. 4.15 Typical homestead. Photo Credit: Suzanne Cook

Rot-resistant timber, such as *chäk ya'* (*Manilkara achras*) or *sak ya'* (*Chrysophyllum mexicanum*), provide the vertical supports for the beams and the ridgepole. *Bamax* (*Pseudomelia* sp.) is sometimes used for the rafters. *Ya'ax ak'* (*Arrabidaea* sp.) is used to lash interior posts and beams, where it will not be dried out by the sun. All other ties are made from the more durable *nikte' ak'*, possibly a species of Bigoniaceae.

The roof is constructed with four sides that slope down from the ridgepole to create a hip roof. Rafters extend from the load-bearing beams up to the ridgepole, forming the slope of the roof. The rafters are notched and tied on each

Fig. 4.16 House garden, compost. Photo Credit: Suzanne Cook

Fig. 4.17 House garden of KP. Photo Credit: Suzanne Cook

Fig. 4.18 Koh's backyard. Photo Credit: Suzanne Cook

Fig. 4.19 Homestead. Photo Credit: Suzanne Cook

Fig. 4.20 Outdoor sink. Photo Credit: Suzanne Cook

Fig. 4.21 Storage shed. Photo Credit: Suzanne Cook

end, to keep them from sliding. Thick layers of broad palm leaves of *k'unche' pahok* (*Geonoma oxycarpa*) or *kun* (*Cryosophila stauracantha*) cover the roof. The fronds are tied to the roof poles in sets of three, with the stems facing upward and secured to *hiil*, rows of slender poles laid across the rafters. A roof cap is made by bending fronds over the peak, creating a water tight seal.

Pole walls, called *hi'che'* (or, *hiilche'*), are made with the timber from *halol* (*Heliocarpus* spp.). It is either split into narrow planks or left whole if the branches are straight. These are tied to slender, horizontal poles that run between the foundation posts.

The *hool* 'door' is a stable door, which is split in two halves. They are assembled with vertical planks of *puuna'*

Fig. 4.22 Chicken coops. Photo Credit: Suzanne Cook

Fig. 4.23 Typical Lacandon house. Photo Credit: Suzanne Cook

'mahogany' and are hung in a frame. The threshold is a heavy piece of timber that has been sunk a few inches into the ground.

A *mooy* may be added. It is a canopy that extends beyond the main frame of the house, supported by four posts set in a semi-circle. If added, the rafters of the main house extend beyond the main frame, sloping down, to form the framework for the roof. BM describes how to build a house, in the Video 4.10 How to Build a House.

The god house*, yatoch kuh*, is built in a similar fashion but without walls. Instead, the thatch is extended closer to the ground, to provide privacy and protection from the elements.

Fig. 4.24 Woman on hammock. Photo Credit: Suzanne Cook

Fig. 4.25 Cooking hearth. Photo Credit: Suzanne Cook

A sheet of pounded bark from the *hach taw* (*Belotia mexicana*), is stretched across the back of the hut for added privacy, for this is where the shelf of *lekil k'uh* 'god pots' are located. Additionally, the tops of the four corner posts have a square hole cut through them, instead of being notched, and the beams are inserted through these holes (Figs. 4.33, 4.34, 4.35, 4.36, 4.37, 4.38, and 4.39).

A ceremonial kitchen, called *uchi' pohche'*, adjoins the god house. It is a simple, fragile hut built without walls. This is where women prepare the food offerings. The utensils and equipment are made of natural materials and are reserved only for this purpose (Fig 4.40).

A description of the utensils is provided in the subsection on ceremonial kitchen utensils in 4.5.2.

Fig. 4.26 Cooking range. Photo Credit: Suzanne Cook

Fig. 4.27 The Kohs' kitchen, 2000. Photo Credit: Suzanne Cook

Traditional Lacandon houses have been modified in recent years. The vertical pole walls are being replaced by wide, horizontal planks, called *bahbil* 'nailed', and the thatch roofs are now tar paper, corrugated tin, or, more recently, laminated sheets. New houses are being built with metal framing, concrete block walls, metal doors and shutters, and poured concrete floors. They are divided into several rooms. The concrete houses are not normally preferred and usually end up as storerooms or accommodations for tourists (Figs. 4.41, 4.42, and 4.43).

Fig. 4.28 The old Kohs' kitchen. Photo Credit: Suzanne Cook

Fig. 4.29 The old Kohs' kitchen. Photo Credit: Suzanne Cook

Fig. 4.30 The new Kohs' kitchen, 2011. Photo Credit: Suzanne Cook

Fig. 4.31 Stove. Photo Credit: Suzanne Cook

Fig. 4.32 Cooking on a traditional stove. Photo Credit: Suzanne Cook

Fig. 4.33 Kun thatch. Photo Credit: Suzanne Cook

Fig. 4.34 Battens and shelf for god pots. Photo Credit: Suzanne Cook

Fig. 4.35 Parts of a god house. Photo Credit: Suzanne Cook

Fig. 4.36 Lashing. Photo Credit: Suzanne Cook

Fig. 4.37 Smoke vent. Photo Credit: Suzanne Cook

Fig. 4.38 Battens in a god house. Photo Credit: Suzanne Cook

Fig. 4.39 God house interior. Photo Credit: Suzanne Cook

Fig. 4.40 Ceremonial kitchen. Photo Credit: Suzanne Cook

Fig. 4.41 Vertical pole house walls. Photo Credit: Suzanne Cook

Fig. 4.42 Concrete house. Photo Credit: Suzanne Cook

Fig. 4.43 Jailhouse. Photo Credit: Suzanne Cook

4.5 Artefacts

4.5.1 Furniture

The typical Lacandon house is sparsely furnished, with a dining table, a kitchen work table, wooden benches and chairs, and one or more hammocks. Shelves, made of raised wooden planks that straddle the rafters, are used to stow bowls, brooms, and other assorted implements. All of the furniture is made by the men using either homemade tools or purchased ones that have been modified. Each item is described below.

Ch'aak 'beds' are used for sleeping, and were in use well before the introduction of the hammock. The bed consists of a wooden platform supported by four heavy corner posts driven into the ground. The wood may be *pukte'* (*Terminalia amazonia*), *ek' bahche'* (*Guatteria anomala*), or *puuna'* (*Swietenia macrophylla*). Originally, the sleeping platform was made of cord woven in a lattice design. This was covered with a *bilich*, a heavy pad made from the inner bark of *hach taw* (*Belotia mexicana*) and blankets. Today, almost everyone sleeps on store-bought mattresses.

K'an 'hammocks' are used for lounging. Originally they were made of knotted cord from the inner fibre of the halol tree (*Heliocarpus* spp.); today, however, the hammocks are those ubiquitous, nylon ones sold in Mexican markets.

K'anche' 'stool, chair' were stools made of a single piece of *puuna'* 'mahogany', approximately 30 cm (12") long by 22.8 cm (9") wide, and 30 cm (12") high. Although these have been replaced by the typical, Mexican, vertical-backed, wooden chair, they are still used in the god house. Another kind of seat is a long, crude, wooden bench, which is located just inside the door for visitors.

Pohche' 'tables' are made of *pukte'* (*Terminalia amazonia*), *ek' bahche'* (*Guatteria anomala*), or *puuna'* 'mahogany' are constructed like the bedstead, with four corner posts driven into the ground, and topped with a rectangular slab of wood.

Additional furnishings include purchased chests for storing clothing and other personal items and, in some households, a refrigerator and a gas stove.

4.5.2 Containers

In the past, containers were made from natural materials. The most common sources were the gourds of the annual vine *Lagenaria siceraria* and evergreen calabash trees (*Crescentia cujete, C. alata*). These were used for bowls, cups, water bottles, and beehives (Figs. 4.44, 4.45, and 4.46).

Clay was another source, used primarily for cooking vessels and god pots. Baskets and bags were made with woven vines from species of *Monstera*, *Ficus*, and *Heliocarpus*.

Fig. 4.44 AT and his gourd beehive for meliponines, stingless bees. Photo Credit: Suzanne Cook

Fig. 4.45 Gourd beehive hangs from house rafters. Photo Credit: Suzanne Cook

Fig. 4.47 KM opening lek gourd, to make a lekil wah 'tortilla plate'. Photo Credit: Suzanne Cook

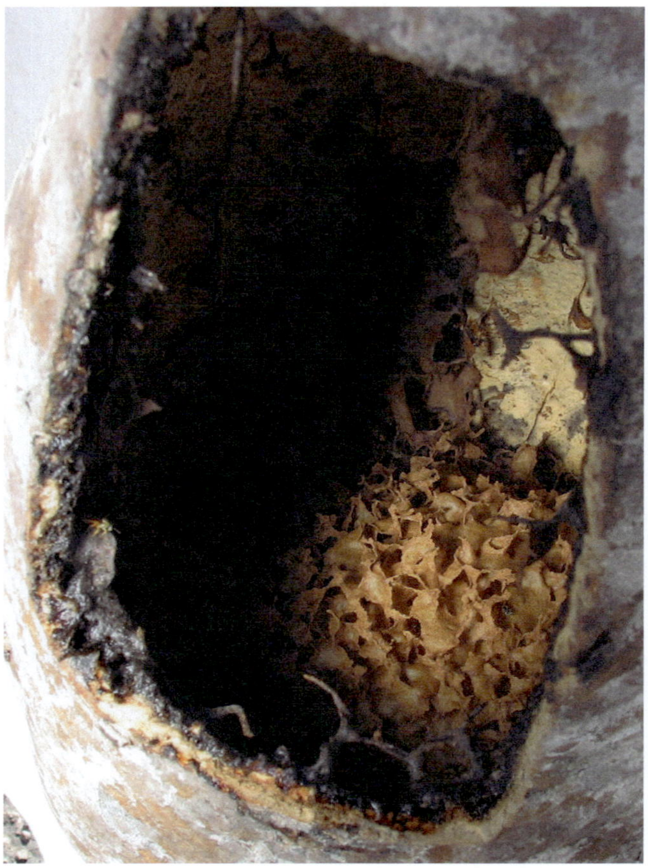

Fig. 4.46 Gourd beehive interior. Photo Credit: Suzanne Cook

4.5.2.1 Gourd Containers

Chuh 'bottle gourd' range in shape from an hour glass to an elongated pear. Ripe fruits are harvested and a small hole is punched in the stem end, to allow the pulp drain. They are then submerged in a river, and left there until the pulp loosens and rots. The gourd is then removed and the rotted pulp, removed. The shells are left in the sun to dry out. The gourds serve as containers for liquids and as bee hives.

Lek 'calabash gourd' is a round, squat, moderately ribbed gourd that grows on the *Lagenaria siceraria* vine. A mature gourd is cut horizontally through the centre and scraped clean of the pulp. The shell is then left to dry in the sun. It is primarily used as a plate for storing tortillas (Figs. 4.47 and 4.48).

Box is a large, almost spherical or slightly elongated, gourd. When ripe, it is sliced across the top. It is cured in the same way as the other calabashes. The rinds serve as drinking vessels and minnow traps (Baer and Merrifield 1971: 147).

Luuch containers are made from the fruits of the perennial *Crescentia* 'calabash tree', tropical evergreens that bear smooth, bright green, round gourds. *Hach luuch* 'authentic luuch' are fashioned from the cannon-ball sized fruits of *Crescentia cujete. Sina'an luuch* 'stretched out luuch' are smaller, 5–7.6 cm (2–3") elongated fruits of *Crescentia alata.* The fruit is harvested when ripe, cut in half, and placed near a fire to loosen the pulp. The empty shell is then placed in the sun to dry. *Hach luuch* are used as *hama'*, a large drinking cup for

Fig. 4.48 K'u'umbil wah tortillas in a lekil wah. Photo Credit: Suzanne Cook

Fig. 4.49 Hama' balche'cups. Photo Credit: Suzanne Cook

balche', gruels, and chocolate beverages. *Sina'an luuch* are used as ladles and small scoops. The variety of sizes and shapes of gourd cups used in ceremonial contexts are listed below:

Chan pak is a small gourd cup used to dip out *balche'* from the *chemil balche'* and pour it into the *pak* '*balche'* urn'.

Etsel is a small, shallow cup, similar to a ladle but without a handle. It is used during the ritual to ferment *balche'*. Four portions are scooped out of the freshly prepared *balche'* and held over the *balche'* canoe while the host of the ceremony chants the *ut'anil balche'* 'Secret of the *Balche'* to activate fermentation.

Hama' is a large drinking gourd incised with images and used for drinking corn gruels (Fig. 4.49).

K'ä' luuch (or, *k'äb luuch*) 'hand cup' is smaller than the *hama'* and used for drinking *balche'*.

Suulil lek 'steeping bowl' is a shallow cup used in the preparation of *käkow* 'cacao' (Fig. 4.50).

Fig. 4.50 Gourd bowls used for preparing cocoa. Photo Credit: Suzanne Cook

Me'et 'trivets' are woven from the stems of *Desmoncus orthacanthus* and *Chamaedorea elatior* and are used to stabilize the gourd cups (Fig. 4.51).

4.5.2.2 Pottery Containers

Pottery bowls and cooking cauldrons are made of red clay mixed with sand. They are hand-moulded and plain. The large vessels are used for cooking ceremonial foods, whereas smaller bowls were used for daily use (Baer and Baer 1952). Both men and women made pottery, but all religious artefacts were made by the men. Women made pottery vessels used in the kitchen, such as the *xämäch* 'griddle' and *kum* 'large pot'. *Kumil nahwah* 'pot for ceremonial tamales' and *kumil puk'*, 'pot for corn gruels', were used specifically for cooking large amounts of food offerings.

4.5.3 Fibre and Cordage

A number of raw materials are used in making textiles, baskets, and cordage. These include *kih* 'agave'; *halol* 'balsa'(*Heliocarpus* spp.); *tämän* 'cotton', *ts'ubtok* (*Hampea stipitata*) and *ak'* 'vines'. The following describes the most commonly used materials and how they are made.

4.5.3.1 Fibre

Kih is the inner fibre of the henequen, or agave plant (*Agave sisalana*). The exterior of the leaves is removed by rasping

Fig. 4.51 Me'et 'trivet' used to stabilize gourd cups. Photo Credit: Suzanne Cook

Fig. 4.52 Hol drying on line. Photo Credit: Suzanne Cook

them on a *pakche'il kih* 'mahogany board' with the dull side of a machete. The exposed long, yellowish fibres are washed and then laid out in the sun to bleach. The dried fibres are then twisted together to form a double-stranded twine, which is used for small net bags called *ch'im*, bow strings, and fishing lines.

Hol is the name for another fibre made from the *halol* 'balsa' (*Heliocarpus* spp.). The fibre is separated from the outer bark by soaking the bark in a river for one month. Then it is hung up to bleach in the sun. After it has dried, it is stripped into long ribbons about 6.35 mm (0.25") wide and then stored for future use (Figs. 4.52, 4.53, 4.54, 4.55, and 4.56).

Bark from the *ts'ubtok* is processed the same way as *hol*. Because it breaks easily, it is a merely substitute for the more desirable *hol* twine.

Bark is also pounded flat for making pads and bark cloth. The inner bark from taw (*Belotia mexicana*) is pounded into thick, spongy pads for mattresses, whereas thinner sheets of bark cloth are obtained by pounding the inner bark from several species of ficus, including *hach hu'un*, *bits'kal hu'un*, and *ak' hu'un* (Figs. 4.57 and 4.58).

Tämän 'cotton' (*Gossypium hirsutum*) is cultivated in *milpas* to make cotton thread. After it is harvested, the bolls are removed, laid out to dry, and then the seeds are removed. The bolls are spun with a drop spindle, called *pecheech*. A demonstration of spinning is presented in Video 4.19 Uk'aay K'uch 'Song for Spinning Thread'.

4.5.3.2 Vines

Vines are used for a variety of purposes, obtained from numerous species. *'Ilon* (*Monstera* spp.), *holop'* (*Philodendron* spp.), and *ya'ax hänan* (*Chamaedorea elatior*) are stripped of their green epidermis and immediately woven into baskets and the fish trap. Frames for hanging baskets are made from the vines of *nikte' ak'*. Trivets for gourd cups are made from this vine or the stems of *ya'ax hänan* (*Chamaedorea elatior*). *K'i'ilix holop'* (*Philodendron radiatum*), *yuyul holop'* (*Philodendron smithii*), *ch'ink'ak'il* (*Passiflora cookie*), *nikte' ak'*, *säk ak'* (*Arrabidaea verrucosa*), and *ya'ax ak'* (*Arrabidaea* sp.) are primarily used for lashing (Fig. 4.59).

4.5.4 Religious Objects

Religion and ritual have played a central role in the traditional culture of the Lacandones. Because the men conduct the ceremonies, they alone made the sacred artefacts. Women made the clay vessels used in the ceremonial kitchen, while the men carved the wooden implements, and built the table and the millstone. These ritual objects are described below and in Video 4.11 Tour of the God House.

4.5.4.1 God House Artefacts

Chaachibil balche'/nok'il balche' 'strainer' is a red, cotton bandana used to strain debris from the *balche'* as it is poured into the *pak*, a large clay urn (described below). Before commercial fabric was available, the strainer was made of bark cloth that had been dyed red (Davis 1978: 113).

Chemil balche' 'fermentation canoe' is made from the trunk of *puuna'* 'mahogany'. It is similar to the dug-out canoe used for navigation (Fig. 4.60).

Chemil pom is a small canoe used for mashing copal. It is usually made from an old *chemil balche'*. A long pole made from a tree branch is used to mash copal in the *chemil pom*. Two short paddles, *ba'ab*, are used for mixing and removing the mashed copal (Fig. 4.61).

Huub 'horn' is the shell of the queen conch (*Strombus gigas*), a large, edible mollusc found in the Caribbean. It is used to summon ceremonial participants.

Hu'un 'bark cloth' made of *hach hu'un* (*Ficus* sp.), *bits'kal hu'un* (*Ficus* sp.), and *ak' hu'un* (*Poulsenia armata*) is used to make ceremonial headbands, traditional tunics, cushions, and the drape that is hung behind the shelf for the god pots.

Huyub k'uh is a small wooden spoon used for feeding the god pots.

Huyubil chulha' is a larger wooden spoon used for eating *chulha'*, a kind of ceremonial gruel. Both are made of cedar (*Cedrela odorata*). Other spoons are made from the rolled up

Fig. 4.53 Cord made from hol. Photo Credit: Suzanne Cook

Fig. 4.54 Rolling fibres of hol. Photo Credit: Suzanne Cook

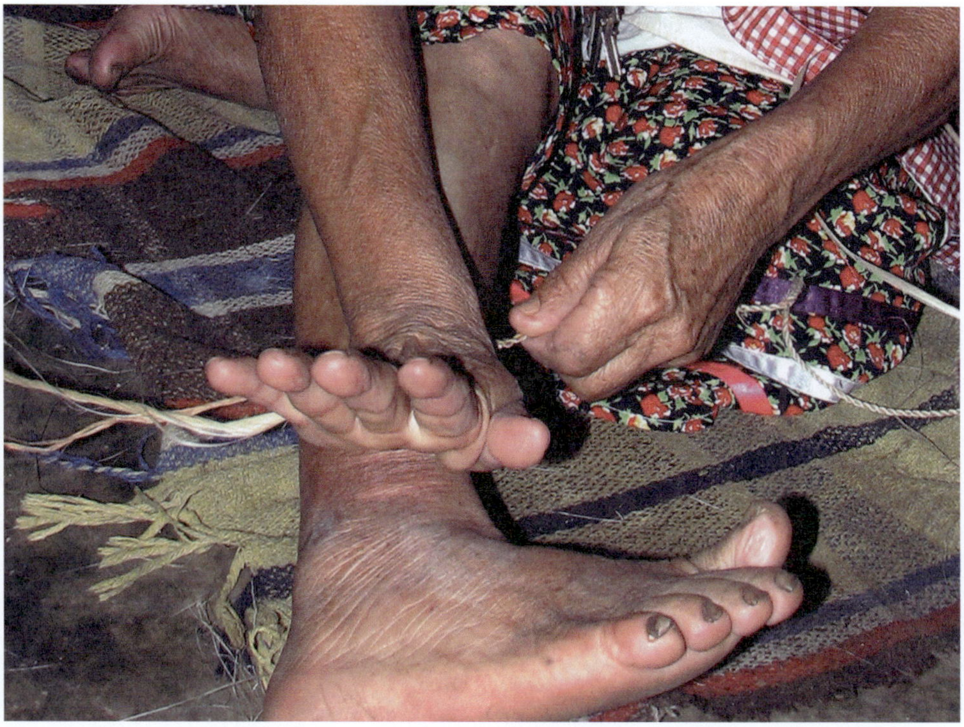

Fig. 4.55 Rolling fibres of hol. Photo Credit: Suzanne Cook

Fig. 4.56 Hol hanging over rafters in bedroom. Photo Credit: Suzanne Cook

Fig. 4.57 Pounded bark fibre of bit'skal hu'un (*Ficus* sp.). Photo Credit: Suzanne Cook

Fig. 4.59 Aerial root of k'i'ilix holop' (*Philodendron radiatum*). Photo Credit: Suzanne Cook

Fig. 4.58 Wooden fibre beater. Photo Credit: Jay Levi

leaves of *Chamaedorea oblongata* and *C. pinnatifrons* and used to dribble *balche'* into the mouths of the god pots.

K'anche' 'stools' are made of a single piece of mahogany.

K'ayum is the ceremonial clay drum named after the Lord of Song. It is relatively small, about 25 cm (10") tall, wide at the base, and tapering to a narrow neck. The opening at the top is covered with an animal skin, which is tightened by turning small sticks inserted through a sisal cord around the circumference of the neck. A human face is formed in clay on the front of the drum (Fig. 4.62).

Lekil k'uh 'god pot' is a brazier in which offerings for the gods are placed. They function as a means of communication to address the gods and receive assistance or protection from them. Each deity has its own clay brazier with a face that has a protruding bottom lip on which edible offerings and *balche'* are placed. Incense is burned inside the bowls. Once the pots have been formed, they are baked, given a white wash of quicklime, and then painted with red and black designs: vertical stripes represent male pots and a crosshatch pattern represents female pots (Figs. 4.63 and 4.64).

Luuch 'cup' is similar to the typical gourd cup made from the calabash tree (*Crescentia cujete*), except that it comes in a variety of sizes, the largest ones include images inscribed on the exterior. See Sect. 4.5.2.[12]

Pak is a large clay serving urn for *balche'*. It is also referred to as *Bol* 'Lord of Wine, the deity responsible for making the *balche'*. Similar to the other god pots, it has a human face on the front of the pot, and is inscribed with simple designs of suns, waves, arcs, feathers, and a frog. These images may be purely decorative, but the frog plays a role in churning the *balche'* (Figs. 4.65 and 4.66).

Pop 'mats' of chamaedorea palm leaves are placed under all sacred objects, to prevent them from touching the ground.

P'ätälche' is a hand-hewn board of mahogany on which the god pots are placed during ceremonies.

[12]See also Davis (1978: 114).

Fig. 4.60 God house with balche' chem in foreground. Photo Credit: Suzanne Cook

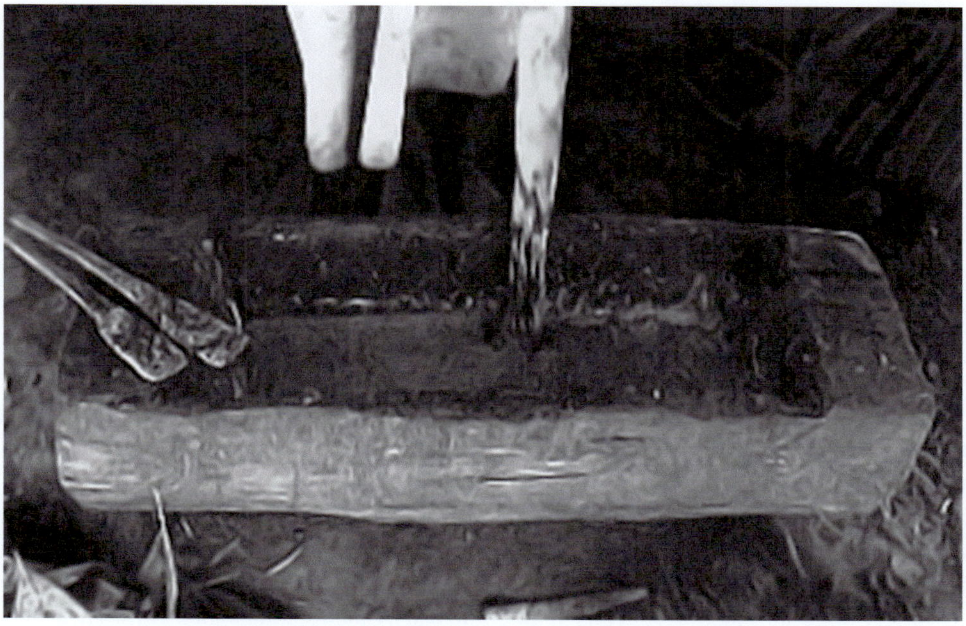

Fig. 4.61 Mashing pom in the chemil pom. Photo Credit: Suzanne Cook

So'ot 'rattle' is described in Sect. 4.6.4 Musical instruments. AM describes the ceremonial rattle in Video 4.18.

Xikal 'copal board' is a large mahogany paddle with a long handle on which balls of *päk'änil pom,* knobs of copal resin, are placed before they are burned as offerings (Fig. 4.67).

4.5.4.2 Inedible Offerings

Chäk hu'un 'bark cloth headbands' are ceremonial headbands made of *hach hu'un* (*Ficus* sp.) and *ak' hu'un* (*Poulsenia armata*). They are painted with geometrical designs in *k'uxub* 'annatto' dye; the ends are cut into different shapes. They are

Fig. 4.62 K'ayum drum. Photo Credit: Suzanne Cook

Fig. 4.63 God pot. Photo Credit: Jay Levi

worn by ceremonial participants and offered to the gods during certain ceremonies, such as the changing of the god pots.

K'ik' 'rubber' is made from the latex of the *Castilla elastica* mixed with the resin of certain kinds of morning glory, e.g. *Ipomoea alba*, which give rubber its bounce. AM demonstrates how this is done in Video 4.12 Tulis K'ik'. According to Tozzer (1907), the offerings of raw rubber frequently made by the Lacandones were often combined with *pom* 'copal'. This appears to be an ancient ritual, as remnants of rubber encrusted with incense have been excavated from Mayan ruins (Tozzer 1907: 127).

Tulis k'ik' 'whole rubber' are miniature, 5–7 cm (2–3"), anthropomorphic figurines of men and women that are offered to the gods to carry out gender-specific tasks on behalf of the supplicant. They are offered only in ceremonies of the greatest importance (Bruce et al. 1971: 52). Before burning them, the supplicant awakens them by wafting burning incense over their bodies while chanting instructions. A version of this chant is provided in Davis (1978: 137–141). After they have been awakened their souls are carried up to the gods on the rising smoke (Figs. 4.68, 4.69, and 4.70).

K'uuts 'home-grown tobacco' was offered in the form of cigars (Tozzer 1907), but by 1980, this practice had been abandoned along with many other activities associated with the traditional religion (McGee 2002: 45). Tobacco is still grown as a minor cash crop, providing a small source of extra income (Fig. 4.71).

Pom 'incense' is as important as maize in traditional Mayan cultures, and the two are symbolically related. Whereas maize is the sustenance of the people, *pom* transforms into maize to feed the gods. The *pom*-maize reversal extends into Lacandon dream interpretation, where mashed *pom* represents corn dough and corn dough represents *pom* (Bruce 1975: 210). *Pom* is gathered from a number of different resin-rich species of *Bursera* and *Pinus*. The trunk is slashed then a leaf is placed in the wedge and left for several days, to allow the resin to collect. When the leaf is full, it is carried back to the god house and mashed in the *chemil pom*, a small canoe used only for this purpose. The mashed *pom* is then stored in leaves or corn husks until ready to use. On the day of the ceremony, it is rolled in balls, called *päk'änil pom*. These are then lined up in rows on a large paddle of *puuna'* 'mahogany' called a *xikal*. They are consecrated and then placed in the god pots of certain gods that were selected through divination, to receive the offerings. Like *k'ik'* offerings, the nodules of *pom* are designated by gender (Tozzer 1907: 128), the male representations arranged in cones and the female representations, as flattened disks stacked in threes (Stross 1997).[13] Before burning them, the host of the ceremony awakens them by wafting the smoke of burning

[13] Copal nodules in the Yucatan display a similar arrangement (Stross 1997).

Fig. 4.64 Kiln. Photo Credit: Suzanne Cook

Fig. 4.65 Pak, the balche' serving urn. Photo Credit: Suzanne Cook

incense over them while chanting instructions to them. A version of this chant is provided in Tozzer (1907: 178–179). In smaller ceremonies, *pom* is put directly into the incensories without being formed. Videos 4.13a and 4.13b demonstrate how *pom* is made and offered in a *balche'* ceremony.

4.5.4.3 Ceremonial Kitchen Utensils

The ceremonial kitchen is a small shelter annexed to the god house. Ceremonial food offerings are prepared here using equipment that never leaves the site. Most, if not all, the equipment is hand made from natural materials. Because they are never used daily, ceremonial pots, tools, and utensils become well-preserved antiques. In Video 4.14, KyP describes the equipment and the tasks involved in the preparation of ceremonial food offerings.

The common utensils and tools include the following:

Chaachib 'colander' is made from the fruit of the calabash tree (*Crescentia cujete*). It is used for straining the hulls from corn kernels during the preparation of corn gruels.

Huts'o' che'[14] is a small paddle on a 30 cm (12") slender stick used for stirring pots of food. This spoon was also used in secular contexts, before the introduction of metal flatware. The wooden spoon continued to be used for ceremonial cooking.

Ka' 'millstone' is a slab of rough stone, elevated on three short legs; the back leg is higher than the two front legs to facilitate grinding as the pestle is pushed down and away from the person who is grinding. The pestle is a long, stone bar that tapers at either end (Fig. 4.72).

Kum are large cooking pots made of clay. Two pots are used: *kumil nahwah* is for bean and meat tamales and *kumil puk'* is for boiling maize and beans.

Mulinuh 'cocoa beater'[15] is made from the soft wood of the *mahaas* (*Quararibea funebris*), or *chäk ya'* 'red chi-

[14]Meaning is uncertain. Compare huts' [Yuc.]. 'to approach, place near' (Bricker et al. 1998: 113).

[15]Mulinuh, from molino (Spanish), also designates the hand-cranked and electric corn grinders.

Fig. 4.66 Men's pak (Lt) and women's pak (Rt). Photo Credit: Suzanne Cook

Fig. 4.67 Xikal. Photo Credit: Suzanne Cook

Fig. 4.68 Rubber. Photo Credit: Suzanne Cook

cle' (*Manilkara achras*).[16] The beater is a 30 cm (12") long stick with paddles at the base. The handle is rapidly twisted back and forth between the palms, to blend and froth the beverage. The device was introduced to the Mayas by the Spanish, along with the name, *molinillo*

[16]Baer and Baer (1952: 15) report that the beater is made from the wood of luwin (*Ampelocera hottlei*) in the southern Lacandon community.

Fig. 4.69 Rubber several weeks after it was made. Photo Credit: Suzanne Cook

Fig. 4.70 Rubber figurines. Photo Credit: Suzanne Cook

'hand mill'. Prior to the Conquest, Mayan women frothed cocoa by pouring the beverage from one vessel to another (Coe 1994: 51) (Fig. 4.73).

Pohche' is a work table for dough preparation made from a single piece of wood set on four legs that have been driven into the ground, with one end slightly elevated. The edges of the table top are slightly curved up to retain the ground maize and to create a run-off channel for the water used to form the dough (Figs. 4.74, 4.75, and 4.76).

Xämäch 'griddle' is a large, round, shallow dish made of clay. It is used for toasting cacao beans, ground corn, and tortillas (Fig. 4.77).

Today the ceremonial kitchen is seldom used. AM's wife may be the only Lacandon woman who still prepares ceremonial foods there. All of the other kitchens, along with the god houses, have deteriorated or have been dismantled.

4.5.5 Weapons and Tools for Hunting, Trapping, and Fishing

4.5.5.1 Hunting

In the past, Lacandones hunted with the bow and arrow, which was introduced to the Mayan lowlands by Putún mercenaries from the Valley of Mexico, during the Postclassic period. Before this, the Maya used spears and clubs (Thompson 1970:178). At the turn of the twentieth century, the Lacandones discovered a market for their bows and arrows, and began producing and selling them in exchange for foreign goods, including Spanish muzzle-loaders, and later, .22 caliber rifles (Tozzer 1907: 54). While many Lacandones continued to use the bow and arrow for hunting

Fig. 4.71 Cigars. Photo Credit: Jay Levi

Fig. 4.72 Ka' grindstone. Photo Credit: Suzanne Cook

Fig. 4.73 Earliest depiction of chocolate, from the Princeton Vase Late Classic period AD. 750. Drawing by Diane Griffiths Peck, in Coe and Coe (2010:48)

and defence until the mid-twentieth century, the bow and arrow had begun to lose its ritual significance.[17]

Today, the Lacandones only produce bows and arrows for tourists. This is the main source of income for the community in Naha' and is the principal activity for 90% of the families. Less care, however, goes into the making of these souvenirs; compared to the traditional set, the tourist replicas resemble toys. The bow and arrow are still crafted in the traditional way, although softer wood, chicken feathers, and cotton binding are used instead of hard woods, parrot feathers, and sisal cord. Both Nations (1989) and Tozzer (1907) provide detailed descriptions of the bow and arrow. The following summarizes these descriptions.[18]

Bows. The bows are made from a number of different woods, including *säk ya'* 'wild star apple' (*Chrysophyllum mexicanum*), *chäk ya'* 'red chicle' (*Manilkara achras*), and, according to Durán (1999), *Dipholis salicifolia*. Nations (1989) reports that *hach chulul* 'authentic bow' was originally made of *Lignum vitae* (*Guaiacum sanctum*), one of the hardest, most resilient woods. The wood from this tree may have been imported, because it does not occur in the Lacandon forest, or at least, not now. Conceivably, *hach chulul* refers to *Tabebuia guayacan* (Seem.) Hemsl., a tree known for its hard and resilient wood (Hernandez-Solano 1999: xix; Miranda 1953).

Long, square pieces of timber are carved into a narrow, elliptical shape that tapers at both ends. The bow is the straightened by holding it over an open fire for about half an hour (Nations 1989: 450) and then sanded smooth with the leaf of the jungle vine, *säha'* 'caskfruit' (*Doliocarpus dentatus*), or a whetstone.

Arrows. Arrows are predominantly made of the hollow flower stalks of *'ooh* (*Gynerium sagittatum*). The entire stalk is cut down when it reaches its full height, about 1.8 m (6′), and then laid out in the sun to dry. There are four kinds of arrows, each made to kill specific game. Most are fitted with a *chuste'* 'foreshaft' made of *k'uche'* 'cedar' (*Cedrela mexicana*), *säk ya'* 'sapodilla' (*Chrysophyllum mexicanum*), *ek'* 'logwood' (*Haematoxylum* sp.), or *puuna'* 'mahogany' (*Swietenia macrophylla*).

[17] Tozzer (1907: 61) and Nations (1989: 456) remark that the bows and arrows were presented to the deities during a boy's puberty rite. However, this rite seems to have been abandoned.

[18] The arrows made by the northern Lacandones are longer than those made by their southern counterparts (Nations 1989: 456, fn. 1).

Fig. 4.74 Net full of gourd cups. Photo Credit: Suzanne Cook

The *tok'* 'arrow head' is either flint or *tah* 'obsidian'. These points are used to kill large mammals. Arrows used for monkeys are fitted with a *k'ek' che'* 'barbed wood', which prevents them from pulling them out. The arrow used for fish is a pointed length of wood. Arrows for birds lack the foreshaft; this is replaced by a *pakat* (*p'äkat?*), a wide, blunt-pointed knob carved from the soft wood of *mahaas* (*Quararibea funebris*). When knapping flint with a new bone hammer, usually a piece of deer antler, the Lacandones sing the 'Song of the Flint', which invokes birds with straight, sturdy wing bones and hard skulls, and specific plants with straight, durable midribs to empower the hammer to flake high-quality blades. A version of this song by AM is presented in Video 4.15 Uk'aay Tok' 'Song of the Flint'.

A *tsuy* 'nock' is carved into the butt-end of the arrow and filled with a wedge of *puuna'* or another durable wood. The nock serves to keep the arrow in place on the bowstring as it is drawn back.

Fletching. Feathers are attached to the shaft with *hol*, made from the *halol* 'balsa' (*Heliocarpus* spp.). If *halol* is unavailable, then *kih* 'henequen' (*Agave sisalana*) is used (KyP, personal communication). Fletching is made from the tail and wing feathers of parrots, motmots, hawks, and falcons.

Binding. Fibre from *kih* (*Agave sisalana*) or *halol* (*Heliocarpus* spp.) is sealed with a mixture of *kibil kab* 'bees wax' from stingless bees (Meliponinae), mixed with the soot of burnt

Fig. 4.75 Ceremonial kitchen table, utensils, and bowls. Photo Credit: Suzanne Cook

Fig. 4.76 Cooking utensils. Photo Credit: Suzanne Cook

Fig. 4.77 Xämäch clay griddle. Photo Credit: Suzanne Cook

incense and the resin of either *ni' ts'ul* (*Mucuna argyrophylla*) or *chäk'oopche'* (*Clarisia biflora*). The sticky strands are then rolled together down the shin or thigh into a long thread. The thread is wound around the foreshaft, the arrowhead, and at intervals along the shaft to stabilize the arrow in flight.

The processes described above are demonstrated in the video, Video 4.16 Making Arrows.

Quiver. Lacandones carry 9–15 arrows in a quiver[19] made from a single sheet of bark from *halol* (*Heliocarpus* spp.). To remove the bark, a length of the trunk is heated over coals until the bark softens. Then the bark is slit down the length of the trunk, pounded all over, to loosen it, and then slipped off. It is fastened shut with strips of *halol* (Baer and Baer 1952: 45) (Figs. 4.78, 4.79, 4.80, and 4.81).

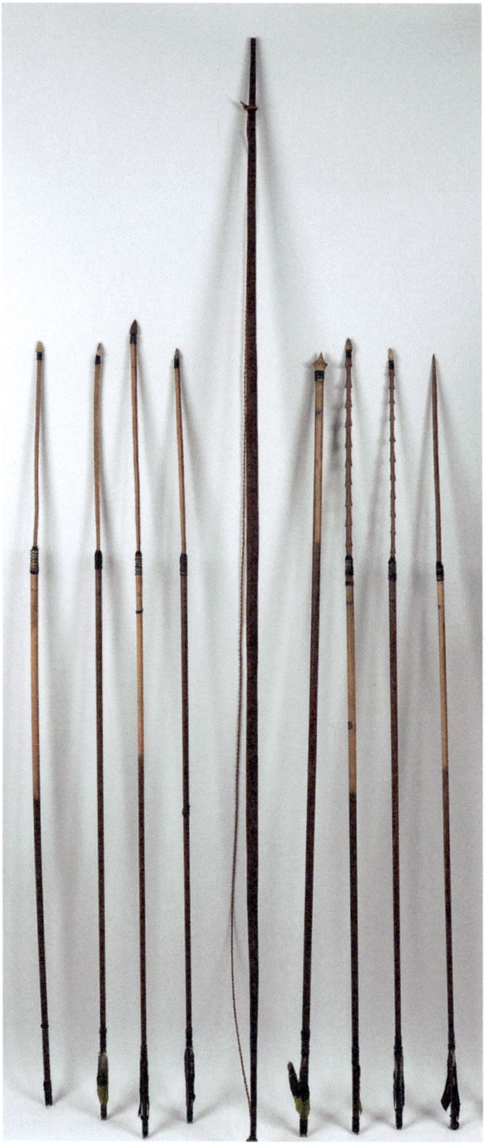

Fig. 4.78 Arrow set. Photo Credit: Jay Levi

[19] KyP says there are nine and Tozzer (1907) says there are fifteen.

Fig. 4.79 Arrow fletching. Photo Credit: Jay Levi

4.5.5.2 Trapping

Traps were made primarily to protect crops from marauding creatures, such as the spotted cavy, squirrels, and weasels. Because spotty cavy and squirrels are also eaten, trapping also supplemented the Lacandones' diet with high quality protein. The traditional trap is a snare consisting of a noose connected to a pole-under-tension system. The trap is triggered when an animal tries to remove the bait, which releases the tension and quickly tightens the noose around the animal's hind quarters and yanks it into the air. A simpler type of trap consists of a baited, upturned basket or metal tub propped up with a stick.

4.5.5.3 Fishing

Fishing was done with a variety of implements: bow and arrow, spears, fish traps, hand lines, and poison. Today only nylon hand lines with metal hooks are used. The following describes the traditional fishing gear.

Lama', lomo' 'spears' were made of the stems of palm *Rhipidocladum bartlettii* and other species (FIg. 4.82).

Fig. 4.80 Arrow heads. Photo Credit: Suzanne Cook

Fig. 4.81 Arrow binding. Photo Credit: Suzanne Cook

Luts käy 'fish hook' was probably made from a thorn of some species, before the introduction of steel hooks.

Xaakil chukäy 'fish trap' is an oblong box about 91 cm (3′) long that widens at one end. A flap door hinged at the top is held open by the current. When full of fish, the trap is turned against the current, to close the door. It is made of *'ilon* (*Monstera* spp.).

Yäk ba'al 'poisonous thing' is made from the pulverized bark of *chechem* (*Metopium brownie*) and a species of wild yam, e.g., *Dioscorea bartlettii* (Nations 2006: 96). Pieces of bark or root are scattered on the surface of the water, which deprives the fish of oxygen. The fish float to the surface, where they are easily caught.

Fig. 4.82 Bamboo fish spear. Photo Credit: Gib Cooper

Fig. 4.83 Seed necklace of *Ormosia schippii* (*red and black*), *Hymenaea courbaril* L. (*large brown*), and *Cois lacryma-jobi* (*grey*). Photo Credit: Suzanne Cook

4.6 Crafts

The term is applied here to the small-scale production of goods. Since the Lacandones at one time made everything by hand, these goods were, for the most part, utilitarian. They include containers, utensils, tools, weapons, and furniture. The artistic skill and attention to detail displayed in the ancient Mayan artefacts is lacking in Lacandon crafts. Nevertheless, the simplified forms etched on gourd cups and god pots add a charming effect to these otherwise basic objects.

The inventory of crafts includes only those artefacts that involve some skill and are decorative, rather than utilitarian. Fabric weaving is included, given that a considerable amount of experience and time are required to produce two rectangular pieces of cloth to make the *xikul*, the traditional male garment.

4.6.1 Jewellery

Women contribute to family income by making and selling seed necklaces, earrings, and bracelets to tourist outlets in San Cristobal (Figs. 4.83, 4.84, 4.85, and 4.86).

Fig. 4.84 Seed necklace of *Sapindus saponaria* (*black*) and *Ormosia isthmensis* (*red*). Photo Credit: Suzanne Cook

Strings of coloured plastic beads worn only by older women are no longer in style, while the old glass beads acquired through trade long ago have all but disappeared. Natural materials other than seeds were also used. These included shells, toucan beaks, alligator teeth, vanilla beans, and flowers. In Video 4.17, KP uses an Archimedes screw drill for drilling small holes in seeds. In the past, the women would have used an awl of bone or hard wood.

4.6.2 Bows and Arrows

Bows and arrows are the principal items made exclusively for the tourist market. Almost every family member contributes to the production of bows and arrows, as these generate the bulk of their income. These are scaled-down versions of the original bow and arrow, made in haste from soft woods and turkey feathers (Fig. 4.87). A description of the traditional bow and arrow is provided in Sect. 4.5.5.1 Hunting.

Fig. 4.85 Necklace with 'hamburger beans' (*Mucuna sloanei*). Photo Credit: Suzanne Cook

4.6.3 Pottery

Mayan ceramics date back to the Middle Preclassic Period, although the use of clay among the Lacandones seems to be a relatively recent development. One Lacandon consultant said that before they discovered clay, the Lacandon ancestors used the burl of the barbasco vine (*Dioscorea bartlettii*) for their god pots.

Baer and Baer (1952) recall when clay food bowls and large cooking vessels were used. The clay vessels were hand-moulded, possibly using the coil technique. They were unslipped, undecorated, and unglazed, similar to the utilitarian ware of the Mayan proletariat. Only the sacred *balche'* urn had a rounded base raised on four, short legs, resembling the tetrapod mammiform supports developed during the Late Preclassic period. Many of these artefacts had a ritual function.

The Lacandon god pot comes closest to resembling the style of the antique ceramics, featuring a human face applied on one side and decorated with simple black and red strips on a white background made from quicklime. The ceremonial drum, *K'ayum*, dedicated to Lord of Song, also displays an anthropomorphic face. Images of the drum are provided in Sect. 4.5.4 Religious Objects.

Fig. 4.86 Toucan beak. Photo Credit: Jay Levi

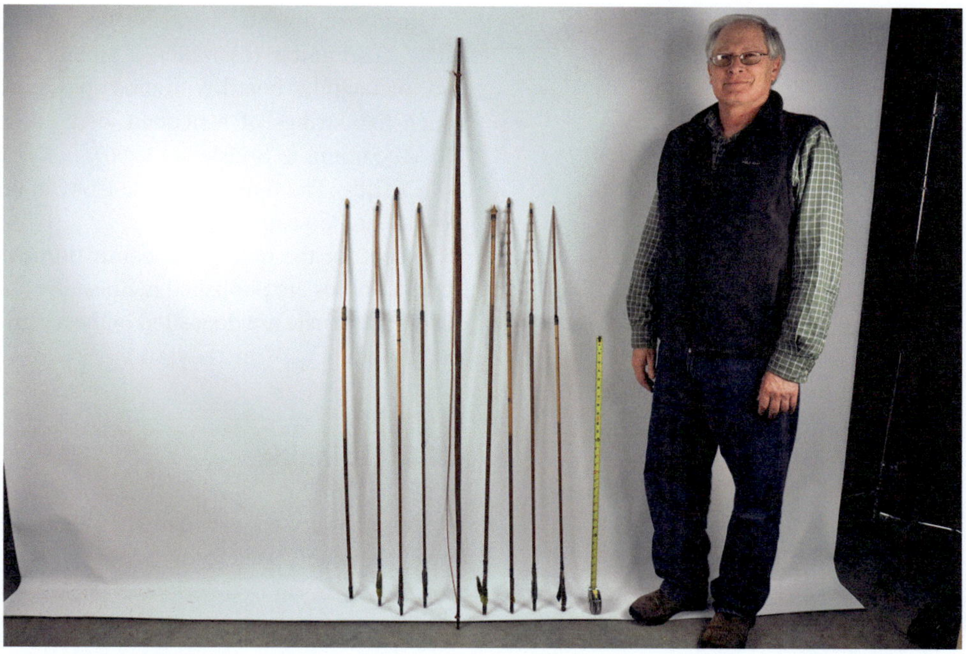

Fig. 4.87 Bow and arrow set (ca. 1950s). Photo Credit: Jay Levi

Although metal pots, pans, and plastic dishes have replaced the traditional earthenware, the large, clay cooking vessels continued to be used to prepare ceremonial foods until the traditional religion was abandoned. Both women and men still make clay dolls and figurines of animals, birds, turtles, and alligators. While they were toys for children in the past, they are made primarily for tourists now (Figs. 4.88, 4.89, and 4.90).

Another old toy was the spinning top. It was made out of an acorn from the *hach ch'alol* tree (*Quercus corrugate*). A hole was bored into the nut and an agave fibre cord was passed through it and attached to a thin rectangular piece of wood. These have long been replaced by commercial toys (Fig. 4.91).

4.6.4 Musical Instruments

Music played a small role in the lives of the Lacandones. A few instruments were played during ceremonies. The drum, *K'ayum*, represented the Lord of Song, and thus occupies a special place in the god house along with the god pots and *Bol*, Lord of Wine. Another sacred instrument is *so'ot* 'rattle', which consists of a single round gourd from the *Lagenaria siceraria* on a long handle. The rattle is painted with red and black designs. The handle is wrapped with bark cloth and tied at the end with a tassel of bark cloth. AM explains that it is only played in accompaniment to the 'Song of the Copal', which is sung during certain, large ceremonies (Fig. 4.92). AM describes the ceremonial rattle, in Video 4.18.

Fig. 4.88 Female doll. Photo Credit: Jay Levi

Fig. 4.89 Male doll. Photo Credit: Jay Levi

Other instruments were the *pax* 'guitar' and the *chul* 'flute'. The guitar resembled a zither, composed of a flat, thin, rectangular board with cross-pieces at either end over which five strings of henequen were strung (Tozzer 1907: 75). According to Soustelle (1966), it was likely introduced from Africa. The flute was made from an undetermined species of bamboo called *'ämäy*.

Except for the rattle, musical instruments have been discarded. Photos are published in Soustelle (1966). Recordings of the flute music are deposited in the archives of the Walters Art Museum.

4.6.5 Spinning

Women spin cotton they grow in their *milpas*. The cotton thread is not only used for weaving, but is worn as amulets to protect sick or injured people, especially pregnant women, and new mothers and their babies.

Cotton is spun on a supported drop spindle, called *pecheech*. It is a slender, light weight shaft made of *mahaas* (*Quararibea funebris*) with a *tukil*[20] 'spindle whorl' of *mahaas*, turtle shell, or the nut shell of *pakay* (*Hymenaea courbaril*), attached near the bottom. The *tukil* is set in a gourd or cocoa bean shell, to reduce friction and help maintain momentum. The shaft is held between the knees, or placed before the spinner on the ground. She stretches and twists the fibre with one hand as she spins with the other. As

Fig. 4.90 Clay animals. Photo Credit: Suzanne Cook

[20]Called *tuchul* in Baer and Baer (1952).

Fig. 4.91 Spinning top (*Quercus corrugate*). Photo Credit: Jay Levi

Fig. 4.92 Ceremonial rattle. Photo Credit: Suzanne Cook

she spins, the thread collects around the shaft directly above the whorl. While she spins, she recites a magical charm, to speed up the process and increase the amount of thread. This charm is recorded in Video 4.19.

4.6.6 Weaving

Weaving was strictly women's work. Weaving was only done to create material for personal garments, as it was time-consuming work. The practice started to decline once commercially produced cotton became available. Women still weave baskets and net bags for their own use, and occasionally sell them to tourists for extra income.

The weaving techniques are similar to those employed in other Mayan communities. Fabric is woven in a simple checkerboard pattern that involves crossing the weft threads over and under the warp threads. The basket weave is similar, except that the weft crosses over and under two warp strands instead of one.

Säkälbil nook' 'woven fabric' was woven from homespun cotton on a backstrap loom. The simplest of all looms, it consists of *xuunche'*, the two end bars used to hold the warp (vertical threads) to the upper and lower ends of the loom. One end is lashed to a pole or tree and the other to the weaver's waist by means of a backstrap. The end bars and the various weaving sticks are made from *chäk ya'* 'red chicle' (*Manilkara achras*).

Although simple, the backstrap loom is difficult to master, given the large number of moving parts involved. There are several wooden rods called *hälälte'*: one *hälälte'* 'shed rod' separates the upper and lower warp yarns, to create *ts'aap* 'shed' (the space between the warp threads); another *hälälte'* 'heddle stick' lifts and separates alternate threads of the warp to allow insertion of the *mämäk* 'heddle' (the thread cords used to separate the warp threads, and to make a path for the shuttle); the *but'u'o'och* 'shuttle' is passed back and forth through the shed between the yarn threads of the warp, to weave in the weft; a third *hälälte'* serves as the beater, to tighten the weft (horizontal threads) as they are woven in (Fig. 4.93).

It takes the experienced weaver one full month, working 2–3 hours a day, every day, to complete the front and back panels of the *xikul* 'traditional woven tunic'. To speed up the process, the weaver recites a magical charm, called *uk'aay ti' säkälbil nook'* 'Song of the Woven Cloth'. All of the elderly people who would have seen their mothers weave or would have done it themselves, i.e., KM, KP, AM, AM's wife, claimed not to remember the charm. Perhaps this is because weaving on the backstrap has become a long lost art in Naha' (Fig. 4.94).

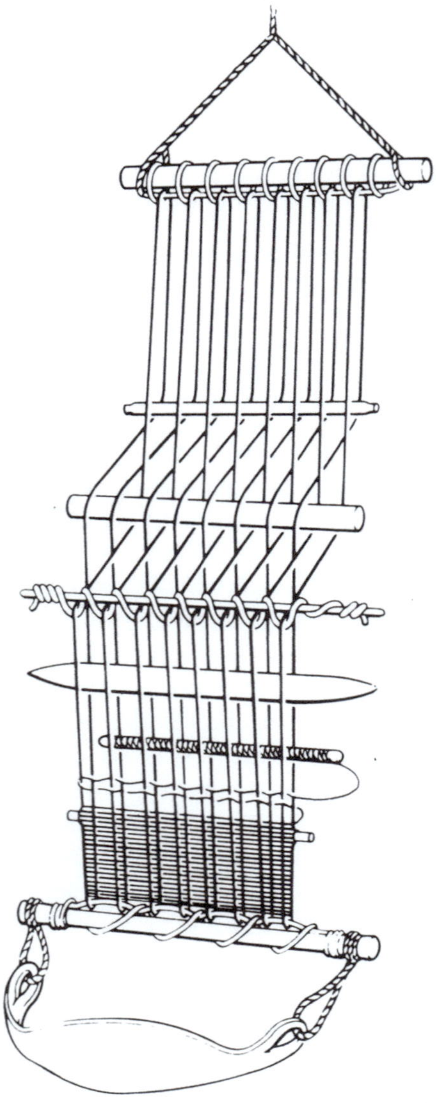

Fig. 4.93 Backstrap loom. Photo Credit: Krystyna Deuss, 1993

Fig. 4.94 Chan K'in wearing the hand spun cotton xikul. Photo Credit: Vic Sarin

KM demonstrates how to weave on an antique loom that belonged to her late husband's grandmother, in Video 4.20.

Xaak 'basket'. The basket is a simple bowl with a flattened base and sides that gradually curve up, flare out in the middle, and then pull in toward the top. The sturdy, flexible vines of *'ilon* (*Monstera* spp.) are first stripped of their epidermis. Then numerous strands are laid out to form the spokes that serve as the foundation of the basket and the vertical structure, or ribs. The horizontal weave is passed over and under the spokes, progressing up the sides of the basket. The loose ends of the spokes are twisted around each other, folded over, and then tucked back into the weave of the basked, creating a simple, looped rim. Sizes vary, but the average basket is about 34.3 cm (13.5") wide at the top and 19 cm (7.5") deep. Women also produce smaller baskets for the tourists, to earn extra income (Figs. 4.95, 4.96, and 4.97).

Ba'ay 'net bags'. These are unstructured, sling-type bags made from the fibre of *halol* (*Heliocarpus* spp.) Two techniques are employed. One is knotless netting, which produces a simple loop pattern similar to that used in crocheting. This

Fig. 4.95 Basket. Photo Credit: Suzanne Cook

Fig. 4.96 Basket rim. Photo Credit: Suzanne Cook

Fig. 4.97 Basket bottom. Photo Credit: Suzanne Cook

is the most widespread type in the New World.[21] The other technique is knotted netting, which entails knotting strands of cord using three fingers to measure the distance between each knot and the size of the space in between. The roll of cord is used as a shuttle. The knots are called *uwich* 'the eyes' and the space between the knots, the net, is called *ule'* 'the leaf'. The spokes radiating from the centre, the warp, are called *uyaal* 'the rows'. Both techniques are used for bags, whereas knotted netting is used for hanging baskets and hammocks (Figs. 4.98, 4.99, 4.100, 4.101, 4.102, and 4.103).

K'an 'hammock'. There are two types of woven hammocks. One serves as the bed springs for the traditional bedstead. The webbing is woven on the bedstead, in a simple basket weave pattern. The other type is the typical sling hammock, which is woven on two upright, parallel poles set up at either end of the warp. The weaving method entails knotted netting, as described above. Both types are made from *halol* fibre (Figs 4.104 and 4.105). A Lacandon woman demonstrates the technique of weaving the sling hammock, in Video 4.21.

4.6.7 Dyes

Lacandones used a variety of vegetative materials and one mineral to produce dyes, stains, and paint. The basic colours produced were blue, black, white, yellow, and red. Dyes, stains and paints were used to colour cotton thread used for the red and blue stripes down the front of the male *xikul* and stripes along the hem of women's skirts. Black, white, and red was the palette used to paint the god pots. Red was used primarily for ceremonial artefacts, such as the *chäk hu'un* 'bark headbands' and the clothes of participants in important ceremonies, such as the changing of the god pots.

The colours were extracted from the stems, leaves, flowers, fruits, bark, and roots of trees, shrubs, and vines. These were scraped, mashed, boiled, or seared, and then the resins diluted with water. Exceptions include hematite and soot, which were used to produce red and black, respectively. White was obtained by diluting quicklime to a thin consistency. It was used as a whitewash for the god pots. SK and KM demonstrate how to make dyes from these various sources, in Video 4.22.

Dyes are no longer made, the practice having been abandoned when commercial ribbons and threads were introduced. Before the decline of the traditional religion, however, natural dyes were still being made for the ceremonial objects; and the red paste made from the annatto bush is still prepared for food colouring and seasoning. The time-consuming process of making annatto paste is demonstrated in Video 4.23. A list of the dyes and the materials used is found in Table 4.6 Dyes.

[21] See Davidson (1935).

Fig. 4.98 Hanging basket, weave. Photo Credit: Suzanne Cook

Fig. 4.99 Hanging basket. Photo Credit: Jay Levi

Fig. 4.100 Ba'ay 'net bag'. Photo Credit: Jay Levi

Fig. 4.101 Woven bag, loop. Photo Credit: Jay Levi

4.7 Clothing

4.7.1 Men's Clothing

The traditional garment for Lacandon men is the *xikul*, a white tunic made of two panels sewn up the sides (Fig. 4.106). In the past it was made of hand-woven cotton,

Fig. 4.102 Woven purse. Photo Credit: Jay Levi

but today it is made of commercial, bleached cotton. Tunics were also made of bark cloth, but it is said that these were rarely worn (Fig. 4.107). Another style of tunic was made for ceremonial use. It too was made of hand-woven cotton, but a seam was sown up the middle of the garment, rather than up the sides. Two stripes of red and blue cotton were woven into the seam.

Because the tunic lacks pockets, men carry a small pouch of skin or hide with a closing flap, called *pooxal* in which they carry cigars, matches, and ammunition. A *ch'im*, a net bag from agave is used to carry larger items (Figs. 4.108 and 4.109). This attire is common among the older men; the new generation prefer to wear Western clothing.

In the past, the men wore *k'äxnäk*, cotton loincloths, underneath their tunics. The loincloth common attire among the men of the Yucatan peninsula as well. Today, shorts or trousers are worn instead.

All men today wear the ubiquitous, clear plastic sandals. When they go into the fields or forest, they often don black rubber boots. In the past, both men and women went barefoot.

4.7.2 Women's Clothing

The traditional attire of the northern Lacandon women consists of a printed, cotton *pik* 'skirt' under a short cotton *xikul*

Fig. 4.103 Woven purse, weave. Photo Credit: Jay Levi

Fig. 4.104 Traditional hammock. Photo Credit: Jay Levi

blouse. While the *xikul* is plain white, the hem of the skirt is decorated with a few lengths of coloured ribbons (Fig. 4.110). In their hair married women wear *k'uk'man*, a tassel of bird pelts from toucans, woodpeckers, and chicken hawks (Fig. 4.111a). Like the men, women went barefoot, but now most wear plastic slippers.

Today, modern women wear long skirts and blouses made of brightly coloured and patterned cotton, polyester, and rayon fabrics. Almost all women wear jewellery, typically necklaces made of seeds and plastic coloured beads. In the past the necklaces included alligators' teeth, pieces of vanilla bean, and the waxy, fragrant blossoms of *kahanche'* (*Clusia flava*) (Fig. 4.111b).

Fig. 4.105 Hammock, knots. Photo Credit: Jay Levi

Table 4.6 Dyes

Hue	Plant	Preparation and Objects
Blue	***ch'ox*** (undetermined)	Leaves crushed
	tsits *Justicia* sp.	Leaves seared and soaked
Red	***chäk'oopche'*** *Clarisia biflora*	Fruit cut in half and pressed on fabric
	chäk'ax *Simira salvadorensis*	Bark soaked
	k'uxub *Bixa orellana*	Fruit boiled and rendered to a paste
	puuna' *Swietenia macrophylla*	Bark soaked
Red-orange	***chäk pach*** *Ternstroemia tepezapote*	Berries crushed onto fabric
Yellow	***k'ante'*** *Erythrina berteroana*	Roots scraped and soaked
	xoyok *Morinda yucatanensis*	Roots scraped and soaked
	sik' *Carthamus tinctorius*	Flowers crushed
Black	***ek'*** *Haematoxylon* spp. ***tsits*** *Justicia* sp.	Bark shavings from *ek'* mixed with resin from *tsits* and added to soot

4.7.3 Children's Clothing

Traditionally, children wore the same attire as adults. Today, many of them are dressed in Western clothes, although on special occasions boys dress in clean, white *xikuls*. Most children go barefoot (Figs. 4.112a, b).

4.8 Medicine and Curing

The most frequent medical complaints reported by the Lacandones include: respiratory infections (*se'em* 'common cold', *säk se'em* 'catarrh'); gastrointestinal complaints (*chibal hämnen* 'stomach ache', *hubil hämnen* 'diarrhea', *xeh* 'vomiting'); fever (*chäkwilil* 'fever'); aches and pains (*yah uho'ol* 'sore head', *yah ubakel* 'sore bones'); oral conditions (*yah ukoh* 'sore teeth', *yah uchi'* 'sore mouth'); dermatological conditions (*k'ak'il* 'boils, measles, chicken pox, small pox'); and, pregnancy (*ma'uts* 'unwell') (Table 4.7).

Illnesses are believed to be caused by spiritual forces. In the past, all Lacandones believed that illnesses were caused by a vengeful god in retaliation for being neglected (Davis 1978: 37). Averting or curing illnesses was accomplished by appeasing the gods through prayer and payment (AM, personal communication). In addition to petitions and supplications, therapeutic incantations, or curing spells, were aimed at alleviating symptoms, such as headaches, stomach aches and nausea, easing childbirth, and preventing or

Fig. 4.106 Cotton xikul. Photo Credit: Suzanne Cook

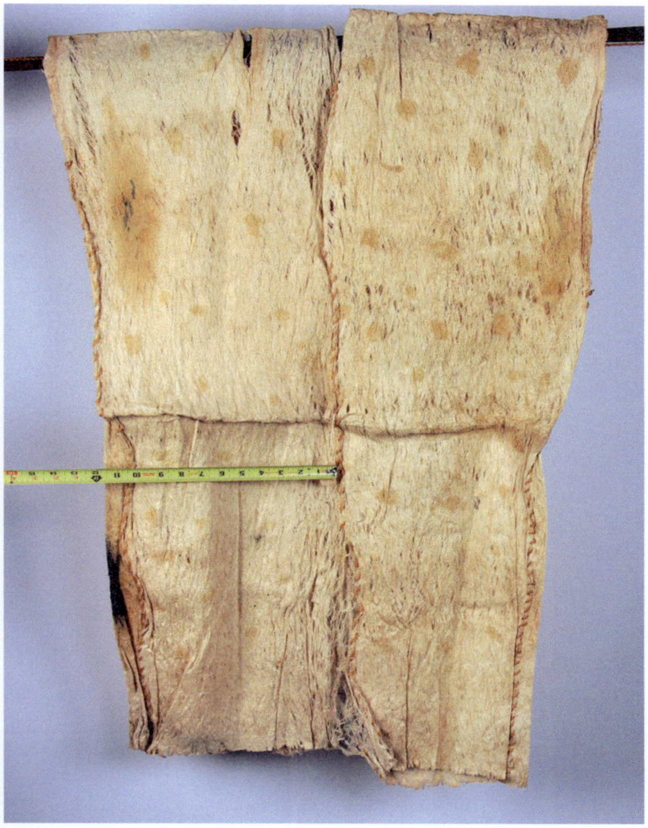

Fig. 4.107 Bark cloth xikul. Photo Credit: Jay Levi

Fig. 4.108 Leather purse-front. Photo Credit: Jay Levi

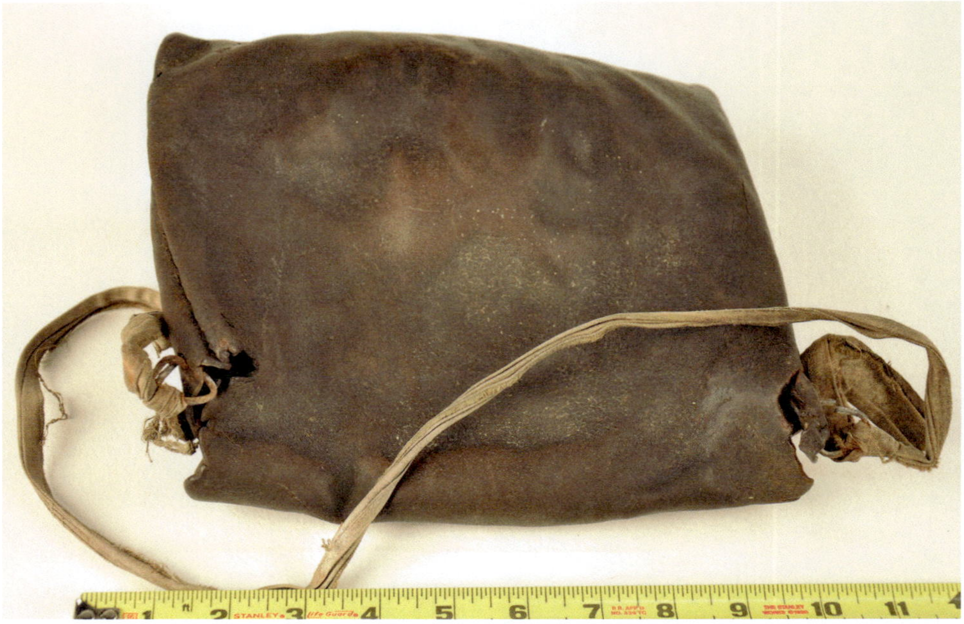

Fig. 4.109 Leather purse-back. Photo Credit: Jay Levi

Fig. 4.110 Traditional women's dress. Photo Credit: Suzanne Cook

staunching bleeding. These spells originated with the principal deity, *Hachäkyum*, who taught them to his son-in-law, *Äk'inchob*, who, in turn, taught them to the Lacandones (Davis 1978: 44).

Although the root of most illness is believed to be supernatural, Lacandones also recognize contagion as a cause of illness. Having contracted foreign diseases, such as mumps and measles, through contact with outsiders and foreigners, Lacandones regard these illnesses as the creation of *Äkyanto'*, the God of Foreigners (Bruce 1968, 1975; Davis 1978; Boremanse 1998; McGee 2002). Because these diseases were not created by *Hachäkyum*, prayer and petitions are futile; the only remedy is Western medicine, also created by *Äkyantho'* (Bruce 1968; Bruce 1975; McGee 2002).

Fig. 4.111 (**a**) Feather hair ornament. Photo Credit: Suzanne Cook (**b**) *Clusia flava* blossom. Photo Credit: Suzanne Cook

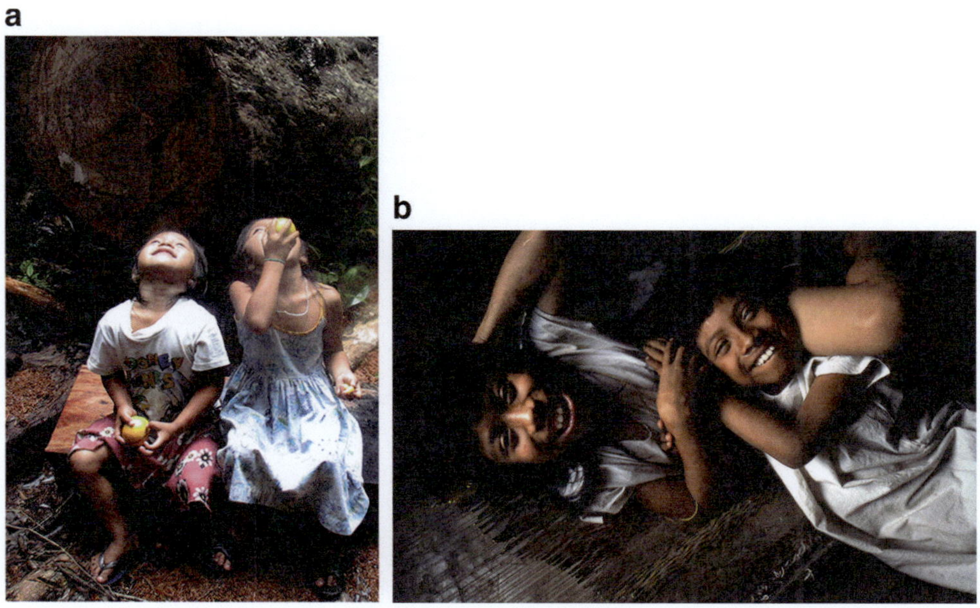

Fig. 4.112 (**a**) Lacandon girls, 2011. Photo Credit: Suzanne Cook (**b**) Lacandon boys on hammock. Photo Credit: Suzanne Cook

Pregnancy is viewed an illness, reflected in the phrase *ma' uts* 'unwell'. Pregnancy is a dangerous condition, as it was the main cause of death for women and their newborns. The introduction of Western medicine and access to clinics has considerably reduced obstetric complications, miscarriages, and injury to newborns. Nevertheless, they do still occur, and when they do, they are believed to be caused by the "evil eye". According to Blom and Blom (1969), when a woman in the community becomes pregnant, all the men spend a great deal of time praying to *äkna'il Äk'inchob*, the wife of *Äk'inchob* and the goddess of childbirth, weaving, and everything feminine (Bruce 1975). But pregnant women are also believed to have great healing ability (Blom and Blom 1969). Thus, women who are pregnant and those who are still fertile spin cotton curing strings to give to other pregnant women and their newborns, to protect them. This practice is said to have been passed down from the *haawo'* [haaβo'], messengers of *Hachäkyum* (Davis 1978: 47, 53). In Video 4.24, CHX and her newborn receive curing strings from a number of female relatives.

Lacandon women believe it is dangerous to even discuss their pregnancies for fear of miscarrying or losing the infant soon after childbirth (Boremanse 1998: 77; McGee 2002: 69). This fear is reflected in naming practices: For the first year of its life a child is simply referred to as *'ooch* 'possum'. Conceivably, this is done so that family members will not get attached to the child should it die. Another reason a woman does not discuss her pregnancy is because she is afraid that should she not be given curing strings, she, her baby, or both would die (Boremanse 1998: 77). Therefore she keeps her pregnancy a secret to avoid this risk.

Table 4.7 Plant Medicines

Medical Categories	Condition	Lacandon and Scientific Names	Parts Used and Preparation	Collector
Aches	Knee pain	chäk wawal *Calathea* sp.	Leaves applied to area	SC
	Bone ache	chek ah (undetermined)	Tea from leaves	K & Mc
	Muscle ache	chekoch *Chione chiapasensis*[a]	Tea from growing tips	D
	Headache	ches ke' (undetermined)	Tea from leaves	K & Mc
	Headache	chi ke' *Chrysophyllum mexicanum*	5–10 leaves ingested	K & Mc
	Leg pain	ch'obenche' *Trichilia* sp.	Leaves applied to area	SC
	Knee pain	hach k'uuts *Nicotiana tabacum*	Roasted leaves applied to area	SC
	Rheumatism	koch *Cecropia peltata*[b]	Tea from leaves	D
	Rheumatism	kopo *Clusia salvinii*[c]	Tea from branches	D
	Bone pain	maasankitan *Esquisetum hyemale*	No data	SC
	Headache	maka che *Rinorea hummelii*	Tea from whole plant	D
	Headache	nikte' ak' *Clematis* sp.[d]	Tea from bark	K & Mc
Colds, Coughs, Catarrh	Chest cold	hach 'ak *Cymbopogon citratus*	Tea from leaves	SC
	Cough	jai patan ak *Senna racemosa*[e]	Tea from sap	D
	Asthma	mulix *Citrus aurantifolia*	Drink from fruit and leaves	K & Mc
	Cold and flu	mulix *Citrus aurantifolia*	Juice and tea from leaves	SC
	Cough	*Tillandsia festucoided*	Tea from leaves	D
	Cold and flu	San Lorenzo *Kalanchoe pinnata*	Tea from leaves	SC
	Coughing phlegm	ah sensi *Zingiber officinale*	Tea from roots	K & Mc
	Coughing blood	ah thus *Vitis* spp.[f]	Tea from roots	K & Mc
Skin conditions	Inflammation	bäbäh	Sap applied to affected area	K & Mc
	Boils	babah *Calophyllum brasiliense* Camb. var. *rekoi*	Resin applied to draw out puss	SC
	Insect bites	chäk ak' (undetermined)[g]	Bark applied to area	K & Mc
	Irritation	chechem *Metopium brownei*[h]	Sap used as eyewash	K & Mc
		chäklah *Bursera simaruba*	Sap used to counteract the effects of chechem	N (2006)
	Mange	ch'ulkeh (undetermined)	Infusion of leaves scrubbed on dogs' skin	SC
	Insect bites	kuch nok (undetermined)[i]	Leaves wrapped around area	K & Mc
	Insect bites, irritated skin	k'uh che' *Cedrela odorata*[j]	Chewed leaves applied to area	K & Mc
	Insect bites	kul ak' *Magnolia mexicana*[k]	Sap applied to area	K & Mc
	Counteract effects of *chechem*	pahsa' ak' *Costus pulverulentus*	Resin from stalks applied to area	SC
	Boils, pimples,	suki ak *Gouania lupuloides*[l]	Crushed branches made into a poultice and applied to affected areas	D
	Warts	ts'ak ah'ax (undetermined)	Resin applied to area	SC
	Scabies	'ukunte' *Sapium lateriflorum*	Resin applied to area	SC
	Rash, baby	xo yok che' *Psychotria panamensis*	Bath infused with leaves	D
	Rash, baby	yo'och ik mehen *Psychotria* sp.	Bath infused with leaves	SC
	Rash, adult	yo'och ik nukuch *Psychotria panamensis* var. *panamensis*	Bath infused with leaves	SC
Fever		ma'ax ak' *Paulina pinnata* [sic.]	Tea from stalk and leaves	K & Mc
		mulix *Citrus aurantifolia*	Juice applied to forehead, head, and chest, to reduce sweating	SC
		puna *Swietenia macrophylla*	A decoction made from the juice of the fruit is drunk	K & Mc
		Spigelia humboldtiana	Bath made with infusion of leaves	D
Gastrointestinal conditions	Stomach ache	bitz *Inga* spp.[m]	Leaves eaten	K & Mc
	Stomach ache	chäkla *Bursera simaruba*[n]	Tea from roots	K & Mc
	Stomach ache, diarrhea	hach mäkuuläm *Piper aduncum*	Tea from leaves	SC
	Diarrhea	hayan le che *Hamelia longipes*	Tea from flowers	D
	Diarrhea	kach che ak/A kan ak *Securidaca diverifolia*[o]	Tea from fruit	D
	Stomach ache	kakawat *Arachis hypogaea*	Tea from leaves	K & Mc, SC
	Diarrhea	kan che, ukam che *Aphananthe monoica*	Fruits eaten raw	D
	Diarrhea	kansak' *Unacaria tomentosa*	Sap ingested	K & Mc
	Diarrhea	kanti ak' *Erythrina coralloides*[p]	Bark and sap ingested	K & Mc
	Stomach ache	kusunche *Garcia nutans*	Warm tea from whole plant	D

(continued)

Table 4.7 (continued)

Medical Categories	Condition	Lacandon and Scientific Names	Parts Used and Preparation	Collector
Gastrointestinal conditions	diarrhea	kuti' *Talauma mexicana*	Tea from bark	SC
	Dysentery	kutik *Magnolia* sp.	Tea from branches	D
	Diarrhea	lol k'in (undetermined)[q]	Tea from bark	K & Mc
	Stomach ache	maka che' *Rinorea hummelii*	Tea from whole plant	D
	Gastritis	makurum *Piper hispidum*	Tea from leaves	D
	Stomach ache	mäk' ulami k'ax (undetermined)	Tea from leaves	K & Mc
	Stomach ache	nikte' ak' (undetermined)	Drink from resin	SC
	Intestinal discomfort	pakam che *Clidemia* aff. *petiolaris*	Tea from whole plant	D
	Diarrhea	pesa *Eugenia caryophyllata*	Tea from leaves	K & Mc[r]
	Colic	pesa che *Pimenta dioica*[s]	Tea from leaves	D
	Stomach ache	pimienta *Pimenta dioica*	Tea from leaves	K & Mc
	Stomach ache	puham *Muntingia calabura*	Leaves eaten	K & Mc
	Stomach ache	säk'atz *Licania platypus*[t]	Tea from leaves	K & Mc
	Diarrhea	sayuk *Aspidosperma* spp.	Tea from bark	K & Mc
	Stomach ache	sensi *Zingiber officinale*	Tea from roots	K & Mc, SC
	Stomach ache	tsak ti tsem *Artemisia* spp.[u]	Leaves eaten	K & Mc
	Stomach ache	tsaka cho che' (undetermined)[v]	Tea from leaves	K & Mc
	Diarrhea	tsäk'ats *Licania platypus*	Tea from seeds	SC
	Diarrhea	xu'u *Allium sativa*[w]	Tea from stalk	K & Mc
	Dysentery	yo och uk *Chysophyllum mexicanum*[x]	Tea from branches	D
Inflammation	*chupul* 'swelling'	k'uxub *Bixa orellana*	Name invoked in *kunyah*	SC
	chupul	naba' *Myroxylon balsamum*	Name invoked in *kunyah*	SC
	Pain Swelling	säk hoben *Piper umbellatum*	Leaves blistered over flame and applied to affected area	K & Mc
	chupul uwich 'swelling of the face'	ya', hach ya' *Manilkara achras*	Name invoked in *kunyah*	SC
Nervousness	Bedwetting, nervousness	chäk chop *Lantana* spp.	Raw leaves eaten before bed time	K & Mc
	mal de orín en niños, and nervousness in adults	koch *Cecropia obtusifolia*	Infusion from the growing tips	D
Oral conditions	*ts'uk chi'* 'mouth rot' *k'ik' chi'* 'bleeding mouth'	chukum che' *Croton draco*	Sap from the bark applied to lips, tongue, and gums	SC
	ts'uk chi', *säkpähk'ak'* (stomatitis?) and *chibal koh* 'toothache'	ik *Capsicum annuum*	Fruit chewed in context of *kunyah*	SC
	Cankers, tooth aches	k'änse ak' (undetermined)	Resin applied to area	SC
	Toothache	peesache' *Pimenta dioica*	Leaves chewed	D, SC
	Toothache	puna *Swietenia macrophylla*	Decoction from fruit drunk; fruit applied to gums	K & Mc
	Gum inflammation, toothache	säk chawah ik, ya'ax chawah ik *Capsicum annuum*	Fruit chewed	SC
	säkpähk'ak' stomatitis (?)	säk yolte' ik *Capsicum* sp.	Fruit chewed	SC
	säkpähk'ak'	säkpähk'ak' che' *Psychotria poeppigiana*	Name invoked in a *kunyah*	SC
	Toothache	sensi' *Zingiber officinale*	Raw root applied to area	SC
	Ulcer	suki ak *Gouania lupuloides*	Branches crushed and applied to area	D
	Toothache	tuch *Ficus* sp. 'wild dogbane'	No data	N & N
	Ulcer	tuch *Thevtia ahouai*	Infusion of the resin and leaves used as mouthwash	D
	Toothache	ya'ax'oox *Brosimum alicastrum* var. *alicastrum*	Cotton soaked in resin applied to tooth	SC
Other	Kidney problems	chäklah *Bursera simaruba*	Tea from the bark	SC
	Counteract the venom of the *ch'up*, a type of furry caterpillar	ts'ak ahch'up *Cissus biformafolia*	Resin applied to area	SC
	Dogs that have ingested *chok* (*Rourea glabra*)	yo'och ahpek' *Fabaceae Canavalia* spp.	Beans toasted and then fed to dogs	SC
Parasites	*ts'ikin nok'ol* 'botfly'	hach k'uuts *Nicotiana tabacum*	Nicotine juice applied to draw out the worm	SC
	Helminths	k'äxeex *Chenopodium ambrosiodes*	Tea from roots	SC
	Leishmaniasis	k'o'och *Cecropia obtusifolia, C. peltata*	Tea from leaves and tender shoots applied to area	SC

(continued)

Table 4.7 (continued)

Medical Categories	Condition	Lacandon and Scientific Names	Parts Used and Preparation	Collector
Snakebite		amäk wäm *Ficus maxima*	Moistened leaves applied to bite	K & Mc
	Any venomous snake	chak pach *Ternstroemia tepezapote*	Decoction made from bark applied to area	D
	Any venomous snake	makurum *Piper hispidum*	Tea from leaves	D
	Any venomous snake	ts'ak kan ak' (undetermined)	Leaves applied immediately, and tea made from the root used later	SC
	Any venomous snake	ts'ak kan ak' *Piper amalago*	No data	SC
	Bite from the nauyaca *Bothrops asper*	uzan kan *Strychnos tabascana*[y]	Poultice made from crushed branches; tea made from branches and other plants	D
Wounds, burns	Burns	chak cra che *Bursera simaruba*	Bark crushed and applied to area	D
	Skin abrasions	pa chac che *Casearia* aff. *aculeata*	Poultice from crushed leaves	D
	Skin wounds	suki ak *Gouania lupuloides*	Poultice from crushed branches	D
	Wounds	tuch *Thevia ahouai*	Poultice from the resin and leaves	D

Collectors indicated by initials: Cook (SC), Durán (1999) (D), Kashanipour and McGee (2004) (K & Mc), Nations (2006) (N), and Nations and Nigh (1980) (N & N). Lacandon and scientific names are those of the collectors, and remain unaltered. Underlying forms of Lacandon words are presented between slashes in the footnotes along with other comments.

[a]/ch'eel koox/ named after the crested guan.

[b]/k'o'och/.

[c]/kopo'/.

[d]*Nikte' ak'* corresponds to a species of Bignoniaceae, possibly *Callichlamys latifolia*, according to the descriptions provided by my Lacandon consultants.

[e]*hay patan ak'*.

[f]/ts'us/.

[g]Possibly a species of *Smilax*.

[h]This is an unusual remedy. The sap from the *chechem* tree is highly caustic, causing blindness if it gets into the eyes.

[i]/ku'uchnook"/ *Bidens odorata* Cav.

[j]Always pronounced [k'uche'], although the underlying form is likely k'uhche' 'god's tree, holy tree".

[K]/k'ul ak'/

[l]/tsuk'in ak'/

[m]/bits'/

[n]Three forms of this word have been registered: *chäklah*, *chäkalah*, and *chäkah*.

[o]/ahkan ak'/

[p]/k'ante' ak'/

[q]According to my Lacandon consultants, *lo'k'in* corresponds to a species of *Parathesis*.

[r]*Eugenia caryophyllata* does not occur in Chiapas; *peesache'* refers exclusively to *Pimenta dioica* (Jamaican allspice).

[s]/peesache'/

[t]/tsäk'ats/

[u]/ts'ak ti' se'em/, lit. 'cold medicine'.

[v]/ts'ak ahchoochel/, lit. 'medicine for the gut, intestines'.

[w]Probably *axux*.

[x]/yo'och' uuk/, lit. 'dove's food'.

[y]Probably /uts'ak kan/ 'snake medicine'.

Air and wind are prominent sources of illness, especially among the Mayas of the Yucatan. These may be natural breezes, which cause imbalances in the body, or metaphorical winds that harbour substances that enter the body and cause disease. Although the Lacandones originated in the Yucatan and adjacent regions, these beliefs are not apparent in their curing rituals.

Lacandones acknowledge the link between the properties of hot and cold, and states of health and illness: siis 'cold' correlates with *se'em* and *siisil* 'chills', and *chäkal* 'hot' correlates with *chäkwilil* 'fever'. Certain plants and animals are regarded as being hot or cold. During curing rituals their names are invoked to produce the opposite effect in the patient. For example, chiles are invoked during rituals to reduce fevers and the throb of toothaches. During the changing of the god pots ceremony, there is a ritual, *k'ämkih*, that involves three of the hottest chiles offered to cool the tempers of irate gods who are angry that their god pots have been neglected for so long.

The hot/cold opposition also figures in prophesies of or explanations for illnesses and mishaps. To dream of creatures and foods with either hot or cold properties foretells an illness with the corresponding properties. For example, birds in general foretell *chäkwilil* 'fever', but parrots foretell *siisil* 'chills' and *se'em* 'catarrh', as do lizards, snakes, alligators, and crocodiles. To dream of cool things, such as water or green sprouts, when one has a fever foretells recovery (Bruce 1975).

Unlike highland Maya groups, the Lacandones have no herbalists, diviners, pulse-readers, bone-setters, masseuses, or mid-wives. Family health is a private concern and the responsibility of the male head of the household. Also absent is the complex pharmacopoeia of plant-based products commonly found in other Mayan societies, particularly the Tzeltales (Berlin et al. 2003: 833). Although the Lacandones use some natural remedies, their herbal pharmacopoeia was never very extensive. According to AM, who is one of the most traditional Lacandones in Naha', the only traditional recourse to avert or cure illness is to petition the gods. Kashanipour and McGee (2004) identified some individuals with special knowledge of medicinal plants and their uses, but the majority of these individuals said they acquired their knowledge from Tzeltales and other outsiders. Durán (1999), and Nations and Nigh (1980) also collected data on medicinal plant use among the Lacandones, but they neglected to mention whether this knowledge was borrowed or traditional.

4.9 Lacandon Plant Classification

4.9.1 Lacandon System of Botanical Classification and Nomenclature

One of the main ways people organize their natural world is by sorting living kinds into groups on the premise of shared attributes, i.e., x is a kind of y. Determining which living things belong in the same category, or rank, begins with sorting them according to their basic traits and then repeating the processes on subsequent groups.[22] This recursive process yields a taxonomy of mutually exclusive taxa (subgroups) headed by a taxon (group). These levels are referred to as ranks. Diagram 4.1 illustrates the hierarchical structure.

The highest rank is the unique beginner (UB). It is the most inclusive, subsuming all other taxa below it (Berlin et al. 1968: 290). At this level, the natural world is partitioned into kingdoms. Most familiar to the botanist are the kingdoms **Plantae** and **Animalia**. English uses the terms "plants" and "animals" for these domains.[23] But like many other languages, Lacandon does not recognize this dichotomy in the same way. Although the Lacandones recognize the difference between animals and plants, their language does not encode an animal and a plant kingdom, per se. Plants are divided into two categories: forest plants, referred to as *tuwolol k'aax* and cultivated plants, referred to as *pak'al*.[24] Taken together, *tuwolol k'aax* and *pak'al* constitute a discrete, albeit unnamed, or covert, category. When asked to differentiate plants from animals, Lacandon consultants note the morphological features that all vegetative forms share in common, viz. roots, stalk, stems, leaves, flowers, and seeds. Animals, humans included, lack these features. Moreover, their language contains numeral classifiers that suffix to the number of objects being counted. If they are sentient beings, the suffix *-tuul* is used, e.g., *hun-tuul winik* 'one person', *hun-tuul mis* 'one cat'; if they are plants, the general suffix *-p'el* 'thing' is used, e.g., *hum-p'el che'* 'one tree', *hun-p'el nah* 'one house'. Of course, this system does not distinguish plants from other inanimate objects.

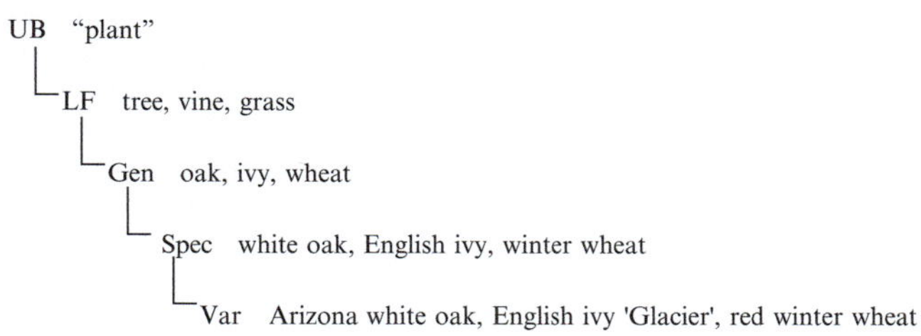

Diagram 4.1 Taxonomic ranks

4.9.1.1 Life Forms

The next level below the unique beginner is the life form. The number of life forms is small, as plants are classified according to gross morphological configurations, such as stem habit. Life forms include many subgroups, or generics. The Lacandones divide vascular plants (those that conduct water and minerals via lignified tissues) into three main categories: *che'*, which encompasses plants with woody trunks, stems and branches; *ak'*, herbaceous vines or woody lianas; and, *su'uk*, fast-growing plants with small, leafy shoots and rhizomes.

Atypical vascular plants, such as epiphytes, ferns and fern allies, reeds, most palms, and Zingiberales, e.g., bananas and elephant ears, are unaffiliated with any life form, as are non-vascular plants, such as fungi and lichens. The latter, together with mosses, are usually referred to as *kuxum* 'mould', because they appear to live on other, usually decayed, plants. Mosses are also called *tso'ots*, because of their furry appearance.

Although apparently taxonomically irrelevant, a number of unaffiliated plants are economically important to the Lacandones. A few examples are: edible species of fungi (*Oudemansiella* spp., *Pluteus harrisii*, *Schizophyllum commune*, *Pleurotus djamor*, *Auricularia* spp.); a species of tall reed, *Gynerium sag-*

[22] Either by means of attribute reduction and "configurational recoding". The first entails selecting a few of the most obvious features and ignoring the rest. The second entails chunking a number of features together to form a single attribute (Bruner, Goodnow, and Austin 1956: 46, in Andrade 1995: 93).

[23] Plant in is broadest sense, viz. vegetation. Ordinarily, plant has a restricted meaning in English, typically denoting small forms (houseplants) and excluding trees. For more insight into this problematical category, see (Wierzbicka 1992).

[24] The same is true for "animal". Lacandones have no name for animal. The word *bäk'* comes closest, however it refers almost exclusively to game animals. It also means meat, specifically red meat. Moreover, when asked to draw their idea of "animal", Lacandon consultants produced an image with four legs, a tail, fur, and spots. This all suggests that *bäk'* is a functional category, not a taxonomic one; and that the concept of animal is not a unique beginner but a life form on par with bird, fish, and snake.

ittatum, whose rigid flower stalks are used for arrows; certain species of *Chamaedorea* that provide food and materials for thatch, mats, and ritual implements; and, maize.

4.9.1.2 Generics

Generics are the next level below life form. Generics have an inherent psychological reality or an "essence" that cannot be reduced to a set of observable properties (Berlin 1972: 54–55). Because of their uniqueness, they are given a "proper name" (Berlin 1990: 64).[25] The names comprise a primary lexeme, i.e., a simple or compound word, e.g., oak, maple, robin, a compound, e.g., blackbird (a species of bird opposed to a black bird), or a noun phrase that cannot be analyzed into smaller units of meaning without changing the sense of the term, e.g., poison oak (which is not a kind of oak tree). This level comprises the largest number of taxa than any other. Of the 638 plants recorded in this work, almost 300 are generics.

Generics are elicited by exploring natural kinds within the frame, 'X is a kind of (generic)', e.g., white oak is a kind of oak (*Quercus* sp.). In Lacandon, *k'än bo'oy* is a kind of *bo'oy* 'chamaedorea palm' (*Chamaedorea* sp.), *säk 'oop* is a kind of *'oop* 'custard apple' (*Annona* sp.), and *mehen bits'*, *nukuch bits'*, and *t'elen bits'* are kinds of *bits'* 'ice cream bean' (*Inga* spp.). Often, generics are obvious, since many names include the name of the life form to which they belong.

Just as often, however, the name of a generic refers to certain features of the plant, such as:

form or habit (*'aak'älyoom* 'night foam' *Cestrum nocturnum*, *muluch'* 'mound' *Oudemansiella* spp., *tsaayentsaay* 'spliced together here and there' *Vigna umbellate* and *Desmodium* spp.);

economy of use (*häläl* 'arrow' *Gynerium sagittatum*, *hach hu'un* 'true bark cloth' *Ficus* sp., *luuch* 'cup' *Crescentia cujete*, *ts'ak k'eken* 'peccary spice' *Allium glandulosum*);

innate properties (*säkpähk'ak' che'* name is invoked during a curing chant to treat *säkpähk'ak'* [stomatitis?], *ku'uchnook'* 'sticks on clothes' *Bidens odorata*);

resemblance to another plant (*axux ak'* 'garlic vine' *Cydista aequinoctialis* < *axux* 'garlic', *tämän che'* 'cotton tree' *Malvaviscus arboreus* < *tämän* 'cotton' *Gossypium hirsutum*);

resemblance to an animal, animal part, relationship to an animal (*neeh k'ambul* 'curassow tail' *Anthurium schlechtendalii*, *ni' ah'ooch* 'opossum's nose' *Anthurium huixtlense*, *'oochin* (hemi)epiphytic Araceae species <*'ooch* 'opossum', with reference to the way the young hug on to their mothers, or, according to one consultant, that opossums like to eat the fruit of these plants); and,

traditional beliefs (*ni' ts'ul* 'foreigner's nose' *Mucuna argyrophylla*).

Names may also be borrowed from other indigenous languages spoken in the area, historically or recently, or from the language of colonizers. A small number of Spanish borrowings occur in Lacandon, but these mainly correspond to introduced species, most are food plants. Examples include: *bambu* (<Spanish <Malay); *ch'ikam* 'jicama' <xicamatl (Nahuatl); *sanya* 'watermelon' <sandía (Sp.); *sukal* 'sugar' <azúcar (Sp.); and, *su'uk* <so'k (Mix-Zoq) 'grass'.

4.9.1.3 Specifics and Varietals

Specifics. Generics can be either monogeneric (one of a kind) or polytypic (a generic that subsumes several kinds). Whereas monogenerics constitute single species that do not include different kinds, polytypic generics are further divided into specifics, a level immediately below that of generic and one that further distinguishes different kinds of a generic, e.g., white oak, garry oak, red oak.

In this work, 212 monotypic generics and 209 polytypic generics are recorded. They are typically labelled by secondary lexemes, a binomial term in which one lexeme includes the name of the immediately superordinate class plus an attribute, usually colour, texture, and size. A secondary lexeme is one that contrasts with other members in the set, e.g., white oak, chicarro oak, garry oak, but not poison oak. Specifics can be elicited by asking: "Are there kinds of X?" or "Does X have companions?"

Specifics are common in tropical areas, where there is great biological diversity, and in agricultural communities, where farmers are particularly concerned about the effect that plants have on of plants on each other. At this level, the distinctions are made on the basis of cultural importance and thus influences the naming and categorization of taxa. See Sect. 4.9.2 Correspondence between Folk Taxa and Botanical Taxa.

Varietals. Specifics may further divide into varietals. When they do, they include the name of the specific plus an attribute, e.g., Arizona white oak. Varietals are rare in Lacandon, with only three recorded in this work: *tso'ots bamax* 'hairy bamax', *chäk tso'ots bamax* 'red hairy bamax', *k'än tso'ots bamax* 'yellow hairy bamax'; *hach chulul* 'authentic chulul', *chäk hach chulul* 'red authentic chulul', *ek' hach chulul* 'black authentic chulul'; and, *chawah'ik* 'dangling chile', *säk chawah'ik* 'white dangling chile', *ya'ax chawah'ik* 'green dangling chile'. An example of a generic and its specifics and varietals is provided in Diagram 4.2:

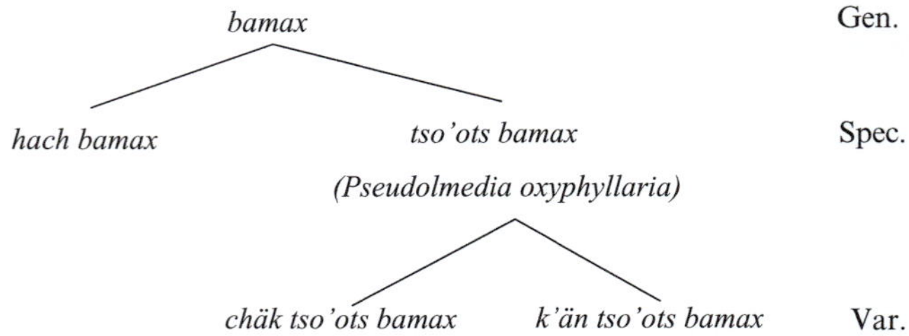

Diagram 4.2 Generic, specifics, varietals

[25] See Atran (1987a, 1987b, 1987c), Keil (1989), Kripke (1972), Putnam (1975, 1977), and Schwartz (1978).

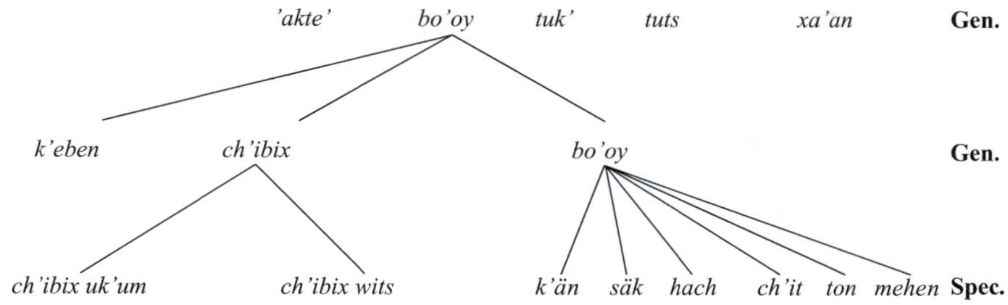

Diagram 4.3 Taxonomic hierarchy

4.9.1.4 Intermediate Generics

There are other cases where generics include other generics. *Bo'oy* (*Chamaedorea* spp.) is an example of a named intermediate generic. It is subsumed under the covert (unnamed) category "palms" (Arecaceae). *K'eben* and *ch'ibix* are kinds of *bo'oy*, although speakers typically omit the word *bo'oy* in everyday discourse. That *k'eben*, *ch'ibix*, and *bo'oy* share the same taxonomic level (generic) is verified by consultants who describe them as *bäho'* 'companions'. Thus, *k'eben*, *ch'ibix* and *bo'oy* form a generic subgroup of *bo'oy*, while *bo'oy* shares the same taxonomic level as other Arecaceae, i.e., *tuts, tuk', akte',* and *xa'an*. Diagram 4.3 illustrates the taxonomic hierarchy.

4.9.2 Correspondence Between Folk Taxa and Botanical Taxa

It is generally the case that "folk generics" and botanical "species" correspond fairly well, since this level are determined by both folk taxonomists and botanists according to easily recognized, gross morphological features. Above and below this taxonomic level, however, correspondence between folk and botanical taxonomies begins to diverge, because other factors are taken into account. Folk taxonomists typically base their observations on perceptual as well as cultural and functional uses, whereas botanists consider evolutionary history and molecular composition. Frequently, correspondences between folk generics and botanical species correspond one-to-many (over differentiation), and many-to-one (under differentiation) (Berlin 1987). Diagram 4.4 shows these three kinds of correspondence between Lacandon generic and specifics, and botanical species.

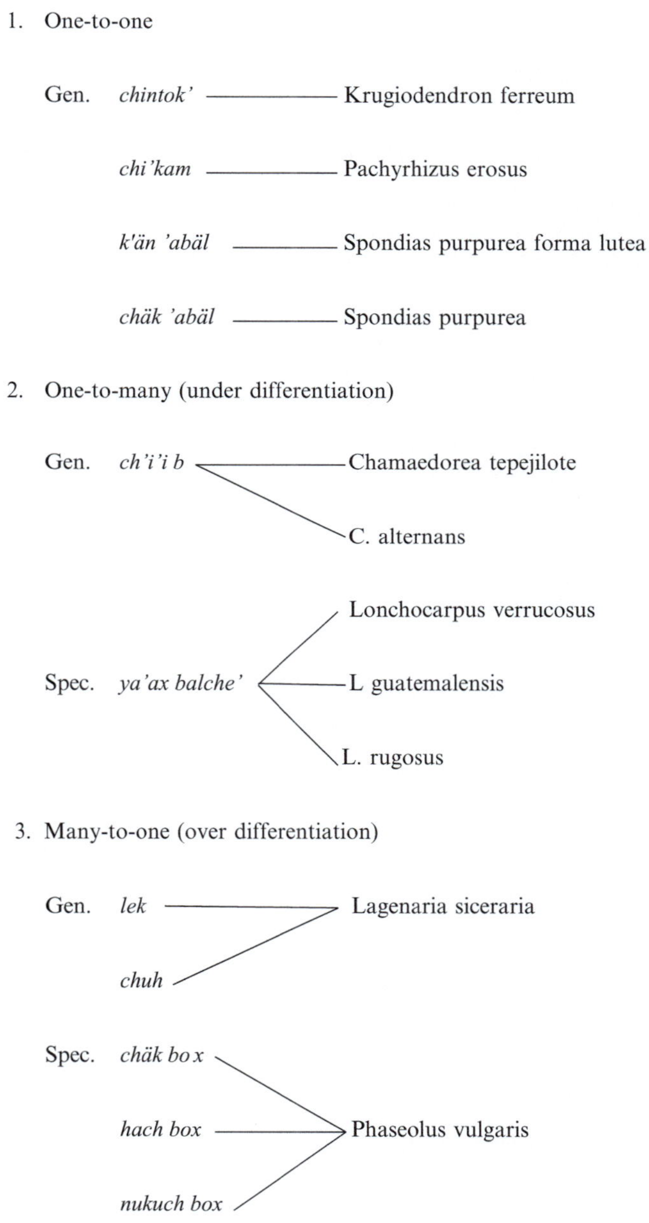

Diagram 4.4 Three kinds of correspondences

References

Alcalde, B. S., & Hernández, E. X. (1972). Estudio preliminar sobre la competencia nutrimental entre arvenses y el maíz, y sus efectos sobre el rendimiento del cultivo. In *Resúmenes* (pp. 94–96). Sociedad Botánica de México: Primer Congreso Latinoamericano.

Atran, S. (1987a). Origins of the species and genus concepts. *Journal of the History of Biology, 20*, 195–279.

Atran, S. (1987b). Ordinary constraints on the semantics of living kinds: A commonsense alternative to recent treatments of natural-object terms. *Mind and Language, 2*, 27–63.

Baer, M. & Baer, P. (1952). Materials on Lacandon culture of the pethá (pelhá) region. *Microfilm Collection of Manuscripts on Middle American Cultural Anthropology* 34. Chicago: University of Chicago Library.

Baer, P., & Merrifield, W. R. (1971). *Two studies on the Lacandones of Mexico*. Norman, OK: Summer Institute of Linguistics, University of Oklahoma.

Basurto, P. F., Martínez, A. M. A., & Villalobos, G. C. (1998). Los quelites de la Sierra Norte de Puebla, México: Inventario y formas de preparación. *Boletín de la Sociedad Botánica de México, 62*, 49–62.

Berlin, B. (1972). Speculations on the growth of ethnobotanical nomenclature. *Language in Society, 1*, 51–86.

Berlin, B. (1987). Folk systematics in relation to biological classification and nomenclature. *Annual Review of Ecology and Systematics, 4*, 259–271.

Berlin, B. (1990). The chicken and the egg-head revisited: Further evidence for the intellectualist bases of ethnobiological classification. In D. A. Posey, W. L. Overal, C. R. Clement, M. J. Plotkin, et al., (Eds.), *Ethnobiology: Implications and applications* (Vol. 1). Proceedings of the First International Congress of Ethnobiology (Belem, 1988) (pp. 19–33). Belem, Brazil: Museu Paraense Emilio Goeldi.

Berlin, E. A., Berlin, B., & Stepp, J. R. (2003). Maya of Highland Mexico. *Ember, 2*, 30–52.

Berlin, B., Breedlove, D., & Raven, P. (1968). Covert categories and folk taxonomies. *American Anthropologist, 70*, 290–299.

Blom, G., & Blom, F. (1969). The Lacandones. Ethnology, Part One. In Evon Z. Vogt (Ed.), Robert Wauchope, (Ser. Ed.), *Handbook of Middle American Indians* (Ch. 13, Vol. 7, pp. 276–97). Austin: University of Texas Press.

Boremanse, D. (1998). *Hach Winik: The Lacandon Maya of southern Chiapas, Mexico*. Institute for Mesoamerican Studies Monograph 11. Albany, NY: University of Albany.

Bricker, V., Po'ot, E., & de Po'ot, O. D. (1998). *Dictionary of the Maya Language: As spoken in Hocaba Yucatan*. Salt Lake City: University of Utah Press.

Bruce, S. R. D. (1968). *Gramática del Lacandón*. Departamento de Investigaciones Antropológicas (Vol. 21). México: Instituto Nacional de Antropología e Historia.

Bruce, S. R. D. (1975). *Lacandon Dream Symbolism 1: Dream symbolism and interpretation*. México: Ediciones Euroamericanas Klaus Thiele.

Bruce, R., Robles, C., & Ramos-Chao, E. (1971). *Los Lacandones: Cosmovision Maya*. Mexico City: Instituto Nacional de Antropologia e Historia.

Bye, R. A., Jr. (2000). Quelites—Ethnoecology of edible greens—Past, present and future. In P. Minnis (Ed.), *Ethnobotany* (pp. 197–213). Norman, OK: University of Oklahoma Press.

Canger, Una. *Lacandón de San Quintín Vocabulary. of San Quintín*. Unpublished manuscript.

Casas, A., Viveros, J. L., Katz, E., & Caballero, J. (1987). Las plantas en la alimentación mixteca: Una aproximación etnobotánica. *America Indigena, 47*(2), 317–343.

Coe, S. D. (1994). *America's first cuisines*. Austin, TX: University of Texas Press.

D'Andrade, R. G. (1995). *The development of cognitive anthropology*. Cambridge, UK: Cambridge University Press.

Davidson, D. S. (1935). Knotless netting in America and Oceania. *American Anthropologist, 35*, 117–134.

Davis, V. D. (1978). *Ritual of the Northern Lacandon Maya*. (Unpublished doctoral dissertation). Tulane University, New Orleans, LA.

Diemont, S. A. W. (2006). *Ecosystem management and restoration as practiced by the indigenous Lacandon Maya of Chiapas, Mexico* (Unpublished doctoral dissertation). Ohio State University, OH.

Diemont, S. A. W., Martin, J. F., Levy-Tacher, S. I., Nigh, R. B., et al. (2006). Lacandon Maya forest management: Restoration of soil fertility using native tree species. *Ecological Engineering, 28*(3), 205–212.

Durán, F. A. (1999). *Estructura y etnobotánica de la selva alta perennifolia de Nahá Chiapas* (Unpublished master's thesis). Universidad Nacional Autónoma de México, Mexico.

Green, J. S. (2009). Feasting with foam: Ceremonial drinks of cacao, maize, and pataxte cacao. In J. E. Staller, & M. D. Carrasco (Eds.), *Pre-Columbian foodways: Interdisciplinary approaches to food, culture, and markets in Ancient Mesoamerica* (pp. 315–343). New York: Springer.

Hernandez-Solano, J. (1999). *Estudio integral de la selva Lacandona, Chiapas* (Unpublished tesis de licenciatura). Universidad Autónoma Chapingo, Texcoco, Estado de México.

Hofling, C. A., & Tesucún, F. F. (1997). *Itzaj Maya-Spanish-English Dictionary-Diccionario Maya Itzaj-Español-Ingles*. Salt Lake City, Utah: The University of Utah Press.

Kashanipour, R. A., & McGee, R. J. (2004). Northern Lacandon Maya medicinal plant use in the communities of Lacanja Chan Sayab and Naha', Chiapas, Mexico. *Journal of Ecological Anthropology, 8*, 47–66.

Keil, F. (1989). *Concepts, kinds, and cognitive development*. Cambridge, MA: MIT Press.

Kripke, S. (1972). Naming and necessity. In D. Davidson & G. Harman (Eds.), *Semantics of natural language* (pp. 253–355). Dordrecht: Reidel. 763–769.

Levy-Tacher, S. I., Aguirre-Rivera, J. R., García-Perez, J. D., & Martínez-Romero, M. M. (2006). Aspectos florísticos de Lacanhá Chansayab, Selva Lacandona, Chiapas. *Acta Botanica Mexicana, 77*, 69–98.

Levy-Tacher, S. I., & Golicher, J. D. (2004). How predictive is traditional ecological knowledge: The case of the Lacandon Maya fallow enrichment system. *Interciencia, 29*, 496–503.

McGee, R. J. (2002). *Watching Lacandon Maya*. Boston: Allyn and Bacon.

Miranda, F. (1953). *La vegetación de Chiapas (segunada parte)*. Tuxtla Gutiérrez, México: Gobierno del Estado de Chiapas.

Nations, J. D. (1989). Lacandon Maya bow and arrow: An ethnoarchaeological example of Postclassic lowland Maya weapon manufacture. In M. Gaxiola & J. E. Clark (Eds.), *La Obsidiana en Mesoamerica* (pp. 449–457). Mexico: Instituto Nacional de Antropología e Historia.

Nations, J. D. (2006). *The Maya tropical forest: People, parks and ancient cities*. Austin: University of Austin Press.

Nations, J. D., & Nigh, R. B. (1980). The evolutionary potential of Lacandon Maya sustained-yield tropical forest agriculture. *Journal of Anthropological Research, 36*(1), 1–30.

Nigh, R. (2008). Trees, fire, and farmers: Making woods and soil in the Maya Forest. *Journal of Ethnobiology, 28*(2), 231–243.

Putnam, H. (1975). *Mind, language and reality. Philosophical papers* (Vol. 2). Cambridge: Cambridge University Press.

Putnam, H. (1977). Realism and reason. *Proceedings and Addresses of the American Philosophical Association, 50*, 483–498. Repr. in Meaning and the Moral Sciences (1978), pp. 123–138.

Sahagún, F. B. de (1950-1982 [1590]). *General history of the things of New Spain: The Florentine Codex*. Trans. A. J. O. Anderson and C. E. Dibble. Salt Lake City: University of Utah Press.

Schwartz, S. (1978). Putnam on artifacts. *Philosophical Review, 87*, 566–74.

Soustelle, G. (1966). *Collections Lacandons*. Paris: Museum National D'Histoire Naturelle.

Stross, B. (1997). Mesoamerican copal resins. *U-Mut Maya, 6*, 177–186. http://www.utexas.edu/courses/stross/papers/copal.htm

Thompson, J. E. S. (1970). *Maya history and religion*. Norman: University of Oklahoma Press.

Tozzer, A. M. (1907). *A comparative study of the Mayas and the Lacandones*. London: Macmillan.

Wierzbicka, A. (1992). What is a life form? Conceptual issues in ethnobiology. *Journal of Linguistic Anthropology, 2*(1), 3–29.

5 Botanical Inventory

A–a

'aak'älyoom night-blooming cestrum (night-blooming jasmine, night jessamine). *Lit:* 'night foam'. Solanaceae *Cestrum nocturnum* L. **Hachäkyum tumeenta Kisin. Tumeenta yok'ol toop' aak'älyoom Kisin. Hachäkyum tumeentah humbuk'ä'k'in Kisin.** 'Hachäkyum (our True Lord) created Kisin (Lord of Death). He created him in the flower of the *'aak'älyoom*. He created Kisin in 5 days.' (Chan K'in, in Bruce 1974: 99) *Cestrum nocturnum* is an evergreen, woody shrub with simple, lanceolate, smooth, and glossy leaves. Greenish-white, tubular flowers are born in cymose inflorescences. They are followed by small, purple berries, 10 mm (0.39″) long and 5 mm (0.20″) in diameter. All parts of the plant are toxic. It is an invasive species throughout its range—Mexico, Belize, Costa Rica, El Salvador, Guatemala, Honduras, Nicaragua, Panama, and Cuba (USDA-ARS, 2013). In Naha', it is cultivated as an ornamental. In the southern Lacandon community of Lacanha' Chan Sayab, the wood is used for fuel and the leaves, for fodder (Levy et al. 2006: 94). *Use:* **tsoy uwich** 'ornamental'. *Thes:* **che'** (Fig. 5.1). *[Note:* aak'älyoom, 'aak'älyoom. "espuma de la noche" (Bruce 1974); makab-yom, akab-nicte [Yuc.]. *Cestrum diurnum* L. (Roys 1931); ak'ab'jom [Yuc.]. *Cestrum nocturnum* (in, Atran et al. 2004); ix jaway, ix 'akäjaway [Itz.]. *Cestrum nocturnum* (Atran et al. 2004).*] [Source:* AM; BM; CK*] [*\sd2 folklore_and_mythology*]*

'abäl ['abäl, 'abäh, 'abah, 'aboh, 'abon] mombin (hog plum). Anacardiaceae *Spondias* spp. Mombins are deciduous or semi-evergreen trees between 25 and 30 m (82–98.4′) tall. They bear drupes 4–10 cm (1.57–3.9″) long with yellow, orange, red, or green flesh surrounding one or more large stones. The Lacandones identify three kinds based on the colour and texture of the fruit. *K'än 'abäl* 'yellow hog plum' and *ya'ax 'abäl* 'green hog plum' begin to fruit in September. The yellow variety matures earlier than the green, and is sweeter and softer. The green variety is a little larger, somewhat elongated, with firm, tart flesh (BM). *K'än 'abäl* is *Spondias purpurea* var. *lutea*. *Ya'ax 'abäl* may refer to *Spondias mombin* or *S. radlkoferi*. The third kind is *chäk 'abäl* 'red 'abäl' (Baer and Baer 1952), which likely corresponds to the red-fruited form of *Spondias purpurea*. Nations and Nigh's (1980) inventory from the Lacandon region includes *"k'inim"*, which corresponds to *Spondias mombin*. *k'inim* corresponds to *ju'ju'* (*Spondias mombin*), in Itzaj, a language closely related to Lacandon (Atran et al. 2004). *Use:* **hanal** 'food'; *Part:* **wich** 'fruit'; *Prep:* **che'che'** 'raw'; *Spec:* **chäk 'abäl** 'red plum (*Spondias purpurea*)'; *Spec:* **k'än 'abäl** 'yellow plum (*Spondias purpurea* forma *lutea*)'; *Spec:* **ya'ax 'abäl** 'green plum (*Spondias* sp.)'. *See:* **huhup**. *Thes:* **che'**. *[Note:* 'ab'äl, ix joob'oj 'ab'äl (aj chäk, aj k'än) [Itz.]. *Spondias purpurea* (Atran et al. 2004); ju'ju', k'inim [Itz.]. *Spondias mombin* (Atran et al 2004); abal [Yuc.] a general term applied to the various forms of the *Spondias* (Roys 1931: 213).*] [Source:* BM; CKD; CNK; NI; OM; SK*] [*\sd2 food*]*

'ak$_1$ grass. Poaceae. **'Ak es igual que su'uk. No tiene che'il. No es che', ma' che'; es suave, ma' chich. Ne mihin uche'il, ne mihin. Yan ukanil, retoño, ukanil uyak'il.** 'Ak is the same as grass. It doesn't have a trunk. It isn't a tree; it's soft, not hard. The stem is small, tiny. It has shoots, a stolon.' (BM) In Yucatec, a closely related language, *'àak* is the folk generic for grass (Bricker 1998, Roys 1931). This word is rare in Lacandon, occurring in only three contexts: *ah-'ak* (= *hach 'ak*), *'ak su'uk*, and *'ak lo'obil* 'weed grass'. The common Lacandon name for grass is *su'uk*, of which the Lacandones identify numerous kinds. According to BM, *'ak* is one kind of *su'uk*. *Spec:* **'ak su'uk** 'silver beardgrass'; *Spec:* **'ak (ah)** 'lemongrass'. *Thes:* **su'uk**. *[Note:* 'àak (1) [Yuc.]. grass (Bricker et al. 1998); ac [Yuc.]. *Andropogon* sp. Barbon (Roys 1931)*] [Source:* AM; BM; KM*]*

'akte' palm, chocho (choco palm, spiny palm). Arecaceae *Astrocaryum mexicanum* Liebm. ex Mart., syn. *Bactris cohune* S. Watson . **Yan uche'il como pahok. Yan uk'i'ixel i chukuch ule'. Ule' yan uk'i'ixel tu' yalam k'ä'. Uhantik uyol i uyaal. Ne ki' uyaal. Yan uwich, uwich yan uk'i'ixel, ek'. Ne**

S. Cook, *The Forest of the Lacandon Maya*,
DOI 10.1007/978-1-4614-9111-8_5, © Springer Science+Business Media New York 2016

Fig. 5.1 'Aak'älyoom *Cestrum nocturnum*. Photo Credit: Suzanne Cook

yaab uwich. Wolis uwich, batak 6.35 cm (2.5″). P'iis ubok koko. Uchunik uyaal en noviembre i mayo. 'It has a trunk like that of *pahok*. It has spines and long leaves. The leaves have spines underneath the branches. One eats the hearts and the tasty, immature inflorescences. It produces numerous black, spiny fruits, about 6.35 cm (2.5″) in diameter. They smell like coconut. The inflorescences develop in November and May.' (BM) *Hach 'akte'* and *'akte'* are names that are usually interchangeable, but, according to AM, *'akte'* is the "older brother" of *hach 'akte'*, which suggests that *'akte'* and *hach 'akte'* are different in some way. According to BM, there are two kinds of *hach 'akte'*—one is red and the other, green. Conceivably, this is the distinction AM is referring to. If there is a hach *'akte'*, then there must be at least one other kind, apart from *'akte'*, that has yet to be identified. The leaves are used for thatch in the northern Lacandon communities of Mensäbäk and Monte Líbano (Baer and Baer 1952: 3); people in Naha' prefer to use *kun* (*Cryosophila stauracantha*) and *kunche' pahok* (*Geonoma oxycarpa*) for this purpose. The fruits of *Astrocaryum mexicanum* are also eaten fried or baked in the hearth coals (Baer and Baer 1952; Nations and Nigh 1980). According to KA, the *chuk'il kan* 'fer-de-lance' (*Bothrops atrox*) lives among the branches. *Use:* **hanal** 'food'; *Part:* **uyol** 'heart'; *Part:* **uyaal** 'immature inflorescences'; *Prep:* **Päytan apokik, pachil ahoytik uhabanil, pachil atsähik i ahuuytik yetel ye'el kax, wa amäk'ik yetel ch'ooch' yetel mulix** 'First you roast it, then you scoop out the insides, fry them, and then stir them in with eggs, or you eat them with salt and lime juice'. *Variant:* **hach akte'**. *[Note:* ak té. *Astrocaryum mexicanum* (Nations and Nigh 1980); akte (chapay). *Astrocaryum mexicanum* (Durán 1999); chapay [Itz.]. *Astrocaryum mexicanum* (Atran et al. 2004); chapay [Mop.]. *Astrocaryum mexicanum* (in Atran et al. 2004).*] [Source:* AM; BM; KA; KM*]* /\ sd2 folklore_and_mythology \sd2 food*]*

'akunte' acacia, bull horn. Fabaceae *Acacia mayana* Lund., syn. *Vachellia mayana* (Lund.) Seigler & Ebinger. **'Akunte' es che', chen ma' ne nuk. Uche'il, ne yan uki'ixel. Ne nuk uk'i'ixel. Pechapech uk'i'ixel. Tuwolol uche'il, wolol k'i'ix. Ule', chan mihin ule'. Uwich, chukuch. Ne chäk usool uwich. Unoy, ne säk. Unek' p'iis sanya: ne yaab unek', chan mihin wolis. Wa amäk'ik, tuwolol alähhantik. Alähmäk'ik yetel unoy yetel nek'. Mix uk'anik, porque ne k'i'ix. Uwich, uxupul ti' noviember. Yan usinik. Ne ts'ik. Ne yan ch'ik pach ha'. Subinte' ubäho'. Se parece ket yetel, pero mas chäk uk'i'ixel.** "'Akunte' is a tree, but it's not very large. The trunk has a lot of thorns. The thorns are very large. The thorns stick out all over the trunk. The leaves are tiny. The fruit is long and very red. The pulp is very white. The seeds are like those of watermelon: numerous, tiny, and round. You eat the whole fruit, seeds and all. No one really bothers with it though, because the tree is so thorny. The fruit is finished by November. It has very fierce ants. Many trees grow on the other side of the lagoon. *Subinte'* is its brother. It looks the same, but the thorns of *subinte'* are redder.' (BM) *Acacia mayana* is a species that grows in lowland wet forests and forest margins in Chiapas, Oaxaca, Tabasco, and Veracruz, Mexico, and the departments of Alta Verapaz and El Petén, Guatemala. It is the most easily recognized of the "swollen acacias" by the pair of blade-like flanges that extend from the thorn base to the tip on either side of the spine. The inflorescences taper to a point. There are rachis glands (secretory structures) between each pair of leaflets. The fruits are borne on a long, thin stalk before swelling out around the seeds (Janzen 1974: 93). *Loc:* **ch'ik pach ha'** 'on the far side of the lagoon'; *Use:* **hanal** 'food'; *Part:* **wich** 'fruit'; *Prep:* **che'che'** 'raw'; *Sim:* **subinte' (ah)** '*Acacia collinsii*'. *Thes:* **che'** (Fig. 5.2a, b). *[Source:* AM; BM*]* /\sd2 food*]*

'ak (ah)$_2$ lemongrass. Poaceae *Cymbopogon citratus* (DC.) Stapf. This is a clump-forming, perennial grass with numerous, stiff stems arising from a short, rhizomatous rootstock. It is a native of tropical and semi-tropical areas of Asia where it is used as a food. It is also cultivated in South and Central America, and Africa. It may have been introduced into Mesoamerica and the Caribbean by African slaves. The plant is also known for its antioxidant, anti-inflammatory, and anti-fungal properties (Bansod and Rai 2008; Lee et al. 2008; Tiwari et al. 2010). According to AM, lemongrass is a traditional Lacandon medicinal for treating colds and flu. *Loc:* **kol** 'milpa'; *Use:* **ts'ak ti' se'em** 'medicine for colds'; *Part:* **le'** 'leaves'; *Prep:* **Achäkik ule' ich ha', pachil awuk'ik** 'You boil the leaves in water and then you drink it'. *Variant:* **hach 'ak**. *Thes:* **su'uk**. *[Note:* ak. *Cymbopogon citratus* (Nations and Nigh 1980); aj/ix su'uklimoon, ix su'ukkastil [Itz.]. *Cymbopogon citratus* (Atran et al. 2004); sakate limon [Mop.]. *Cymbopogon citratus* (in, Atran et al. 2004); sakate limon [Yuc.]. *Cymbopogon citratus* (in, Atran et al. 2004).

Fig. 5.2 (**a**) 'Akunte' *Acacia mayana*. Photo Credit: Suzanne Cook (**b**) 'Akunte' *Acacia mayana* thorn. Photo Credit: Suzanne Cook

Scientific name from Germplasm Resources Information Network (GRIN). http://www.ars-grin.gov/cgi-bin/npgs/html/taxon.pl?12797. Retrieved August 2011.*]* *[Source:* AM; BM; SK*]* *[*\sd2 beverage \sd2 medicinal*]*

'ak lo'obil (äh) beardgrass, silver. *Lit:* 'grass weed'. Poaceae *Bothriochloa laguroides* (DC.). *Loc:* **kol** 'milpa'; *Gen:* **'ak** 'grass'. *See:* **lo'obil**. *Variant:* **'ak su'uk**. *[Source:* BM*]*

'ak su'uk beardgrass, silver. *Lit:* 'grass grass'. Poaceae *Bothriochloa laguroides* (DC.) Herter. **'Ak su'uk, lati' ne ma' tsoy. Lati' su'uk, es su'uk. Yan umots kubin yok'ol lu'um. Yan uyak'ilil. Uyak'il ukanil kubin. Ukanil ne p'iis k'i'ix, yan uni'. Ubin yok'ol lu'um. Yaab. Ukanil es raiz. Way otro tichil, way uyaal, way uyaal, i ubin uyak'il otros ubin, uyak'il. Ese es, se dice, ukanil. Todo eso kubin ukanil yak'il. Es como pätan. Pätan yan utichil ukanil, tich ukanil, pero pätan no tiene ak'. Ne ma' tsoy. Ubin yok'ol lu'um. Ne ma' tsoy 'ak su'uk.** *'Ak su'uk* is very bad. It is grass. It has roots that go over the ground. They have vines [stolons]. The vines of the tillers go out. The tillers are really like thorns, they have a point. They travel on the ground. There are a lot. The tiller is a root. Here another comes up, here is its offspring, and then their tillers go out. These are called *ukanil*. All of these go out. It's the same with bananas. Bananas have [suckers] that sprout up, but bananas don't have vines. *'Ak su'uk* is really bad, because it travels along the ground.' (BM) Silver beardgrass is an invasive grass that ranges from the central United States to South America (Hanan-Alipi & Mondragón-Pichardo 2009a). Each plant consists of numerous stems growing from a single root and produces multiple stems (tillers). The inflorescence is a plume panicle up to 20 cm (7.8″) long. The shaft of the inflorescence is covered with abundant, long, white, silky hairs. Each floret contains a single seed. It is found growing in milpas and disturbed areas. The plant is considered one of the worst weeds in the *milpa*, because it is difficult to pull out and impossible to control. Not only do the rhizomes spread rapidly, producing numerous young offshoots, but the seeds are carried on the wind far and wide. *Sim:* **wakäx su'uk** 'cow grass (*Pennisetum purpureum*)'; *Sim:* **winik su'uk** 'man grass (*Andropogon glomeratus*)'; *Sim:* **hach su'uk** 'authentic grass (*Digitaria setigera*)'. *Variant:* **'ak lo'obil**. *Thes:* **su'uk** (Fig. 5.3a, b, c). *[Source:* AM; BM*]*

ak'il bu'ul (ah) beans, climbing (pole beans). *Lit:* 'vine of beans'. Fabaceae *Phaseolus vulgaris* L. There are two varieties, *ek'* 'black' and *chäk* 'red'. They have bush counterparts, called *che'il bu'ul* 'bush beans'. Beans are rarely eaten green, but are left to dry on the plants. At harvest time, the beans are picked off the vines, thrashed in a sack to split the shells, and then stored for later use. *Use:* **hanal** 'food'; *Part:* **nek'** 'seeds'; *Prep:* **chäkbil, wa mäxik i tsähbil yetel seboya, nahwah, mukbil wah** 'boiled, or mashed and fried with onions, ceremonial tamales, or baked in a large tamale'; *Spec:* **chäk ak'il bu'ul** 'red climbing bean'; *Spec:* **ek' ak'il bu'ul** 'black climbing bean'; *Ant:* **che'il bu'ul (ah)** 'bush beans'; *Gen:* **bu'ul** 'beans'. *[Source:* AM; BM; KM*]* *[*\sd2 food*]*

ak' hu'un mastate. *Lit:* 'vine bark cloth'. Moraceae *Poulsenia armata* (Miq.) Standl. **Ak' hu'un es che'. No pega yetel che'. Kuch'ihil tuhunal. Uch'ihil ti' uyäkanil o ich as wits, wits äka'an también el crece. Ese es k'aax, no es paakche' kol, ich k'aax. Nuk ne ka'anan ubakel. Tint'an yan viente metro ubakel, uka'anan ubakel. Usool, pim usool, pim usool. Yan**

a b c

Fig. 5.3 (**a**) 'Ak su'uk *Bothriochloa laguroides*. Photo Credit: Chan K'in Daniel (**b**) 'Ak su'uk *Bothriochloa laguroides* tiller. Photo Credit: Suzanne Cook (**c**) 'Ak su'uk *Bothriochloa laguroides*. Photo Credit: Suzanne Cook

uyits. Uyits, ne säk uyits. Uyits, ne ts'ak. Yan ne ts'ak. Wa tälik, uyits ne ts'ak. Mix medicina, ha'li' ts'ak uyits. Pero uts'oolik, yan mak uts'oolik uyoot'el usool, pachil uka'ts'ilik, yan uyaalal usool. Uyaalal usool es dos doble cascaras. Encima es mero cascaras, en el medio es como carnes. Pero ese que va a hacer hu'un bäxik. Ese es lo que va servir. Eh, hach uyoot'el no sirve. Es el uts'ilik asi delgalito uhook'ol. Ma' ne pim. O wa se puede ts'oolik asi tamaño tanto, lo mas grande. Ese es hu'un, ak' hu'un. Ule', ma' ne nuk ule', batak. Kooch utan, batak, as chukuch ule'. Yan uyits ule'. Cuando ne yi'h ule', k'änk'än uhutul. Läh yan uyits. Yan utop'. Säk utop'. Ma' ne nuk utop'. Uwich, as nunuk uwich, batak uwich. Cuando täk'än, este, k'änk'än uwich, täk'än. Ti' noviembre yan uwich, täk'än ti' noviembre. Winik ma' umäk'ik, ti' bäk'. Bäk' umäk'ik, por ejemplo, hale' i tsup. Wa ch'iich' también poco comen. Uch'iktal mas kabal que bits'kal hu'un. *'Ak' hu'un* is a tree. It doesn't stick to trees. It grows on its own. It grows in the low-lying flats or on the lower slopes of the hills. It also grows on level areas in the hills. This is a forest (tree), it isn't in the acahuales (fallow cornfields); it's in the forest. It's very tall. I think it's about twenty meters (65.6′). Its bark is thick. It has resin that is very white and strong. If you strike it, the resin gushes out. It's not medicine, just strong resin. One peels off the bark, peels away the inner bark, and then tears it into strips. There are two layers of bark. The outside layer is the true bark and the middle layer is like its flesh. This is what you pound into bark cloth. This is what it is used for. The outer bark doesn't work. One peels off the inner bark in thin strips. It's not very thick. You also can peel off larger strips. This is called *hu'un, ak' hu'un*. Its leaves are not very large. They are wide and somewhat long. The leaves have resin. When the leaves are mature and yellow they drop off. They're full of resin. It has white flowers. They aren't very large. The fruit is somewhat large. The ripe fruit is yellow. The fruit ripens in November. People don't eat it, it's for the animals. The animals eat it, such as the spotted cavy and the agouti. Birds also eat it a little bit. It is shorter than *bits'kal hu'un*.' (BM) The tree BM describes is likely *Poulsenia armata*. It grows to 27 m (88.5′) tall and 90 cm (35″) in diameter, with low buttresses. Stems, petioles, stipules, and the lower midrib of leaves are armed with 1–3 mm (0.04–0.12″) long spines (Croat 1978: 360). Branches are horizontal or ascending. It has a rounded or irregular, dense crown. The bark is finely fissured, dark brown or grey. The inner bark is white, changing to drab pink and fibrous (Pennington and Sarukhán 2005: 150). When slashed, the bark exudes copious whitish, rose-coloured, or yellowish-brown latex. The fibre was traditionally pounded into bark cloth for clothing and ceremonial headbands. Although my consultants mentioned no other uses for the tree, Nations and Nigh (1980: 22) report that the Lacandones eat the bark. According to Levy et al. (2006: 90), the southern Lacandones use the tree for food, but these authors fail to mention which part is eaten. This group also makes instruments from the timber (Levy et al. 2006). According to Durán's (1999) northern Lacandon consultants, *ak' hu'un* corresponds to *Ficus petenensis*, which was also used for making bark cloth. *Loc:* **Yaha'nah** 'La Cueva'; *Loc:* **Mensäbäk**; *Loc:* **Yahapetha'** 'Guineo'; *Use:* **hu'un, chäk hu'un, hach nook'** 'bark cloth, bark cloth headbands, authentic clothing'; *Part:* **uyoot'el usool** 'its inner bark'; *Prep:* **Uts'oolik uyoot'el usool, pachil uka'ts'ilik, pachil ubäxik** 'One peels off the inner bark, then tears it into strips, and then pounds it'; *Sim:* **hach hu'un** 'authentic hu'un'; *Sim:* **bits'kal hu'un** 'strangle hu'un'; *Gen:* **hu'un**. *Thes:* **che'**. *[Note:* ak' hun. *Poulsenia armata* (Nations and Nigh 1980); ak ju'un. *Poulsenia armata* (Nigh 2008); akun (Pennington and Sarukhán 2005); chimo, hun, ak hun che.

Fig. 5.4 (**a**) 'Anis *Abelmoschus moschatus* unopened pod. Photo Credit: Suzanne Cook (**b**) 'Anis *Abelmoschus moschatus* flower. Photo Credit: Suzanne Cook (**c**) 'Anis *Abelmoschus moschatus* seed pod. Photo Credit: Suzanne Cook

Ficus petenensis (Durán 1999). English common name from The Wood Explorer database, http://www.thewoodexplorer.com/maindata/we970.html.*]* *[Source:* AM; BM*]* *[*\sd2 clothing*]*

'anis₁ candle tree (cow okra, cucumber tree). Bignoniaceae *Parmentiera edulis* DC., syn. *P. aculeata* (Kunth) Seem. This is a medium-sized tree that grows to 15 m (49.2′) high with a trunk girth of 30 cm (11.8″). Branches are armed with pairs of short, stout spines at each joint. It is normally much branched from the base, where fissures produce suckers. It bears large, greenish-white, trumpet-shaped blossoms 5–6.5 cm (1.9–2.56″) long, singly or in clusters in the leaf axils, on the trunk, and at the tips of branches. Fruits are tapering pods with numerous, vertical ridges. Mature fruits are yellowish green. (Pennington and Sarukhán 2005: 474) Although my consultants failed to mention any uses for the tree, the Lacandones reportedly eat the fruit (Levy et al. 2006; Nations and Nigh 1980). Atran et al. (2004: 94) say that the Itzaj eat the fruit to treat coughs, flu symptoms, and gallstones. According to Levy et al. (2006), southern Lacandones regard the fruit as a medicinal as well. They also use the wood for fuel. *Loc:* **k'aax, Yahapetha'** 'forest, Guineo'; *Use:* **hanal** 'food'; *Part:* **wich** 'fruit'; *Prep:* **che'che'** 'raw'. *Variant:* **kat**. *Thes:* **che'**. *[Note:* kat, änis. Guajilote. *Parmentiera edulis* (Nations and Nigh 1980).*]* *[Source:* BM*]* *[*\sd2 food*]*

'anis₂ musk mallow (musk okra, musky-seeded hibiscus, ornamental okra, tropical jewel hibiscus). Malvaceae *Abelmoschus moschatus* Medik, syn. *Hibiscus abelmoschus* L. *Abelmoschus moschatus* is an herbaceous plant, 2.5 m (8′) tall with soft hairy stems, alternate, palmately lobed leaves with coarsely toothed margins, and nodding, hibiscus-like yellow flowers with a red splotch at the base of each petal. The fruit is a capsule up to 18 cm (7″) long and contains numerous, hard, black seeds. The dehiscent capsule resembles the tapering fruits of okra, only larger. The plant is an introduced weed usually found in open and disturbed areas (Espinosa-García 2000). In its native India the seeds are mainly used as an antiseptic, stimulant, and aphrodisiac. Despite these benefits, the Lacandones whom I consulted did not attribute any medicinal value to the seeds or any other organ of the plant. *Loc:* **paakche' kol, kol, kolil nah** 'fallow milpa, milpa, house garden'; *Use:* **uuh** 'necklaces'; *Part:* **nek'** 'seeds' (Fig. 5.4a, b, c). *[Note:* English common names from M.M.P.N.D. http://www.plantnames.unimelb.edu.au/Sorting/Abelmoschus.html#moschatus. Retrieved February 2013.*]* *[Source:* NI; SK*]* *[*\sd2 craft*]*

'axux ['axux, 'axox] garlic. Amaryllidaceae *Allium sativa* L. This is an introduced, soft neck variety of garlic. Apart from its culinary use, Kashanipour and McGee (2004) report that the Lacandones also make a tea from the stalk to treat diarrhea, but the Lacandones whom I consulted said that they had never heard of this. *Loc:* **kol, kolil nah** 'milpa, house garden'; *Use:* **hanal** 'food'; *Part:* **wi'** 'tuber'; *Prep:* **chäkbil** 'cooked'; *Use:* **hubul hämnen** 'diarrhea'; *Part:* **che'il** 'stalk'. *[Note:* xu'u. *Allium sativa* (Kashanipour and McGee 2004); ahshush. *Allium sativa* (Nations and Nigh 1980); 'aajoj [Itz.]. *Allium sativa* (Atran et al. 2004). Garlic remedy from Kashanipour and McGee (2004).*]* *[Source:* BM; CNK; KM; KP; SK*]* *[*\sd2 food*]*

'axux ak' garlic vine. Bignoniaceae *Cydista aequinoctialis* (L.) Miers, syn. *Bignonia aequinoctialis* L. **Lo comimos 'axux ak' antes era axux. P'iis ts'ak, p'iis 'axux.'** 'Before there was garlic, we ate this. It's the same as food, the same as garlic.' (KP) A non-invasive, evergreen liana found primarily in tropical moist forests throughout the Neotropics, the garlic vine is distinguished by its lavender-blue, trumpet-shaped flowers and garlic-scented foliage. Leaves are mostly smooth and bifoliolate (having two leaflets). Leaflets are ovate to elliptic, 9–16 cm (3.5 6″) long, 4–11 cm (1.5–4″) wide. Flowers are produced in racemes or panicles in terminal or axillary nodes. Each blossom measures between 5 and 8 cm (1–3″) long, and ranges in colour from pale lavender to white with yellow, pubescent throats and lavender nectar guides on the 3-lobed lower lip of the corolla. Fruits are smooth, shiny,

Fig. 5.5 Axux ak' *Cydista aequinoctilis*. Photo Credit: Jayesh Patil (Flickr: Garlic Vine) [CC-BY-2.0 (http://creativecommons.org/licenses/by/2.0)]

flattened capsules, 30–56 cm (11–22″) long and 2–2.5 cm (<1″) wide. (Croat 1978: 763–764). *Loc:* **kol, kolil nah** 'milpas and house gardens'; *Use:* **ts'ak** 'food'; *Part:* **le', top'** 'leaves and flowers'; *Prep:* **chäkbil yetel sopa, bu'ul** 'cooked in soups and with beans'. *Thes:* **ak'** (Fig. 5.5). *[Note:* ashush ak'. garlic vine. *Cydista aequinoctialis* (Nations and Nigh 1980); 'aajoj 'ak' = u-'ak'il 'aajoj [Itz.]. *Cydista aequinoctialis* (Atran et al. 2004); ak' xuux, xuux ak' [Yuc.]. *Cydista aequinoctialis* (in, Atran et al. 2004).*] [Source:* KP; SK*]* *[*\sd2 food*]*

Ä–ä

'äm bead tree (horse-eye beans). Fabaceae *Ormosia* Jacks. **Che', nukuch che'. Ne ka'anal ubakel. Chukuch ule' batak 10 cm (4″). Utop', as k'änk'än utop', wolis utop'. Mäna' ubok. Uchunik utop' ti' mayo. Uwich yan usool, chich usool. Uwaalkal ti' nah ya'axk'in, marzo.** 'This is a large, very tall tree with leaves that are about 10 cm (4″) long and yellowish, round flowers without any scent. The flowers begin to bloom in May. The seeds are encased in a hard shell, which splits open in the spring (March).' (BM) BM adds that the tree prefers a warm climate. According to other consultants, there is a male and a female tree. The male tree is called *'äm ton* and the female tree, *'äm ch'uplal*. They are distinguished by the kind of seed they produce: the male tree produces bi-coloured black and red seeds, whereas the female tree produces somewhat smaller, solid red seeds. The seeds of both kinds are elliptic, while the female seed is considerably compressed. Two *Ormosia* species—*O. isthmensis* and *O. schippii*—are recorded from the Lacandon forest (Comisión Nacional de Áreas Naturales Protegidas 2006; Levy et al. 2006). The former is the female tree and the latter, the male tree. Like the Lacandones, indigenous peoples of the Amazon distinguish male and female *huayruro* (*Ormosia*). The male (*O. amazonicas*) produces red seeds with a black splotch, whereas the female (*O. macrocalyx*) produces solid red seeds (Rudd 1965: 279). *Loc:* **k'aax** 'forest'; *Use:* **uuh** 'necklaces'; *Part:* **nek'** 'seeds'; *Use:* **che'il k'ak'** 'firewood'; *Male:* '**äm ton** (*Ormosia schippii*)'; *Female:* **'äm ch'uplal** (*Ormosia isthmensis*). *Thes:* **che'**. *[Note:* 'am [Itz.]. colorín. coral bean tree, coama wood, tiger wood (Hofling and Tesucún 1997); 'am, pitoche', chäk'am [Itz.]. *Erythrina standleyana* (Fabac.) (Atran et al. 2004).*]* *[Source:* AM; BM; CNK*]* *[*\sd2 craft \sd2 fuel*]*

'äm ch'uplal Isthmus ormosia. *Lit:* 'female 'äm'. Fabaceae *Ormosia isthmensis* Standl. *Ormosia isthmensis* belongs to the American Section Unicolores (single colour) Series isthmensis (Rudd 1965). It is a tree about 50 m (164′) tall, with pairs of leaflets 3–5 cm (1–1.9″) apart, coriacious, ovate, oblong, or obovate-oblong, 3–35 cm (1–13.7″) long and 2–10 cm (0.79–3.9″) wide, acute to acuminate at the tip and truncate at the base. The surface is smooth and underside, finely pubescent. Flowers are 10 mm (3.9″) long with white and pinkish-purple petals. Fruits are black or dark brown, smooth, dehiscent capsules, 3–7 cm (1–2.7″) long by 2–3 cm (0.79–1.18″) wide, generally containing one solid red seed, 10–13 mm long (0.39–0.5″),10–11 mm (0.39–0.43″) wide, and 6–8 mm (0.23–0.31″) thick. The species is found growing in rain forests from southern Mexico to northern Colombia, 5–800 m (16–2625′) in elevation (Rudd 1965: 319). The Lacandon women whom I consulted refer to the similar-looking, solid red seeds of *k'ante'* (*Erythrina berteroana*) as *'op' ch'uplal*, which is another name for the seed of *Ormosia isthmensis*. *Use:* **uuh** 'necklaces'; *Part:* **nek'** 'seeds'; *Male:* **'äm ton, 'op' ton** '*Ormosia schippii*'; *Gen:* **'äm** 'Ormosia'. *Variant:* **'op' ch'uplal**. *Thes:* **che'** (Fig. 5.6a). English names from Grandtner (2005). *[Source:* CNK; NI*]* *[*\sd2 craft*]*

'äm ton Schipp ormosia (John Crow bead [Bel.]). Fabaceae *Ormosia schippii* Pierce ex Standl. & Steyerm. *Ormosia schippii* is a medium sized 50′ tall and 12″ in diameter, with yellow wood, and reddish-purple flowers (Stirton 1987). It belongs to the American Section Ormosia, series Amazonicae (Rudd 1965: 308). It produces bi-coloured red and black seeds. The species is usually found in swampy areas in rain forests below 150 m in elevation (Rudd 1965: 326), but it has also been recorded from the southern Lacandon territory, at 350 m (Levy et al. 2006: 71). *Use:* **uuh** 'necklaces'; *Part:* **nek'** 'seeds'; *Female:* **'äm ch'uplal, 'op' ch'uplal** (*Ormosia isthmensis*); *Gen:* **'äm** (Ormosia). *Variant:* **'op' ton**. *Thes:* **che'** (Fig. 5.6b). English name from Grandtner (2005). *[Source:* CNK*]* *[*\sd2 craft*]*

B–b

baba' soapberry. *Lit:* 'soap'. Sapindaceae *Sapindus saponaria* L. Also called the soapnut tree, native to both the Old

Fig. 5.6 (**a**) 'Äm ch'uplal *Ormosia isthmensis*. Photo Credit: Suzanne Cook (**b**) 'Äm ton *Ormosia schippii*. Photo Credit: Suzanne Cook

and New Worlds, this tree bears fruit whose pulp was used as soap. In addition, the Lacandones use the sap to relieve inflammation (Kashanipour and McGee 2004: 64) and as a medicine for dogs (Nations and Nigh 1980: 17), and the seeds for jewellery. *Use:* **uuh** 'necklaces'; *Part:* **nek'** 'seeds'; *Use:* **ts'ak ti' chupul** 'medicine for swelling'; *Part:* **uyits** 'sap'; *Prep:* **Ha'li' apulik chichin yok'ol uyah** 'You just put a little bit on the sore area'. *See:* **subul$_1$**. *Variant:* **subul, habon, habonsiyo**. *Thes:* **che'**. *[Note:* baba', habon. soap (Bruce 1975); babah. *Sapindus saponaria* (Nations 2006); bäbä. *Sapindus saponaria* (Kashanipour and McGee 2004).*]* *[Source:* Bruce (1975)*]* *[*\sd2 craft \sd2 medicinal*]*

babah false mamery (Brazil beauty-leaf, Santa Maria). Clusiaceae *Calophyllum brasiliense* Camb. var. *rekoi* Standl. **Nukuch che'. Usool, pim usool. K'än usool. Uche'il as k'änk'än. Yan uyits, as k'änk'än uyits. Ne tsaay uyits, komo chicle. Ne ka'anan ubakel. Ule', ma ne nuk ule', chan mihin ule', pim ule'. Yan utop', chan mihin utop'o', säk utop'. Uwich, yan uwich, wolis uwich, ya'ax uwich. Yan unoy. Yan unek' ich unoy. Hump'eli' unek'. El tronco se usa para hacer los cayukos. Cuando está muy joven, se usa para hacer las vigas y las trampas de los pájaros.** 'This is a large tree with thick bark. The bark is yellow. The wood is somewhat yellow. It has resin, which is also somewhat yellow. It is very sticky, like chicle. The tree is very tall. The leaves are small and thick. The flowers are very small and white. The fruit is round and green. The fruit has pulp and contains one seed. The trunk is used to make canoes. When it is very young, it is used to make battens and bird traps.' (BM) *Calophyllum brasiliense* is an evergreen tree found in low and flooded areas of primary and old secondary forests, near rivers, and in canyons, below 1200 m (3937′) elevation. It ranges from southern Mexico to the northern region of South America. It grows 20–45 m (65.6–147.6′) tall with a trunk girth of up to 1.3 m (4.26′). It has a dense, rounded crown. The external wood is yellowish and the heartwood, reddish. It exudes a yellow latex. Leaves are opposite, entire, elliptic to oblong or obovate, leathery, smooth, glossy-green above and pale green below. Flowers grow in panicles 2–5 cm (0.8–2″) long. Each blossom is 10–13 mm (0.39–0.51″) in diameter, with four white sepals and 1–4 white petals. Fruit is a fleshy drupe, 2.5–3 cm (1–1.2″) long, greenish yellow when ripe, with a single, large seed. (Vázquez-Yanes et al. 1999) Fish, monkeys, and bats disperse the seeds. According to AM, the fruit is also relished by the great curassow (*Crax rubra*). In addition to canoes and construction, the straight trunks and saplings are used to make corncribs and paddles (Durán 1999). The latex is used to draw out puss from boils (AM) and to seal bark paper (Nations and Nigh 1980). *Use:* **k'ak'il** 'boils, skin eruptions'; *Part:* **uyits usool** 'latex from the bark'; *Prep:* **Ha'li' ats'ahik uyits yok'ol uk'ak'il** 'You just apply the latex to the boils'; *Use:* **hiil** 'battens'; *Use:* **säkche'** 'rafters'; *Use:* **taan** 'girt (front)'; *Use:* **akam tanil** 'front pillars'; *Use:* **waan** 'girt (side)'; *Use:* **ho'olan che'il** 'ridgepole'; *Use:* **uwaan** 'beams running the length of the roof frame'; *Part:* **mehen che', k'äbche', chunche'** 'saplings, branches, trunk'; *Use:* **ulehil ch'iich'** 'bird trap'; *Part:* **k'äbche', mehen che'** 'saplings'; *Use:* **chem** 'canoe'; *Part:* **chunche'** 'trunk'. *Thes:* **che'**. *[Note:* bahbah, barí. *Calophyllum brasiliense* (Nations and Nigh 1980); babaj. *Calophyllum brasiliense* (Nigh 2008); babah. barril. *Calophyllum brasiliense* Camb. var. *rekoi* (Diemont 2006: 37); baba, bari, kisik che. *Calophyllum brasiliense* (Durán 1999); b'ariiyoj [Itz.]. *Calophyllum brasiliense* (Atran et al. 2004). Common names from the AgroforestryTree database (http://www.worldagroforestrycentre.org/sea/products/afdbases/af/asp/SpeciesInfo.asp?SpID=1737)*]* *[Source:* AM; BM; K'ayum Segundo*]* *[*\

a

b

Fig. 5.7 (a) Babah 'oonte' (kakache') *Ocotea cernua*. Photo Credit: Suzanne Cook (b) Babah 'oonte' *Ocotea cernua* seeds. Photo Credit: Suzanne Cook

sd2 housing \sd2 hunting_and_fishing \sd2 transportation\sd2 medicinal*]*

babah mehen 'oonte' ocotea, drooping (timber sweet, cayenne rosewood). Lauraceae *Ocotea cernua* (Nees) Mez. *See:* **babah 'oonte'**.

babah 'oonte' ocotea, drooping (timber sweet, cayenne rosewood). Lauraceae *Ocotea cernua* (Nees) Mez. **Ket yetel mehen 'oonte', ha'li' ule' mas mihin. Uch'ihil paakche' kol. Uwich ket yetel.** 'It's exactly the same as *mehen 'oonte'*, only its leaves are smaller. The fruit is the same.' (AM) No uses were reported. *Loc:* **k'aax** 'forest', paakche' kol 'fallow milpa'; *Sim:* **säk 'oonte'** 'white 'oonte' (*Licaria alata*)'; *Sim:* **ek' 'oonte'** 'black 'oonte' (*Nectandra globosa*)'; *Sim:* **mehen 'oonte'** 'small 'oonte' (*Nectandra* sp.)'; *Sim:* **nukuch 'oonte'** 'large 'oonte' (*Nectandra* sp.)'; *Gen:* **'oonte'**. *Variant:* **kakache'**. *Thes:* **che'** (Fig. 5.7a, b). *[Note:* Botanical identification provided by Jens G. Rohwer (pers. comm.). This tree was called *kakache'* by KP. kaki che, mejen onte. *Nectandra coriacea* (Durán 1999); mejen onte. *Ocotea cernua* (Durán 1999). English names from Grandtner (1997–2010).*] [Source:* AM; BM; KP*]*

bahun che' [bahun che', bahum che'] onion cordia (Spanish elm). Boraginaceae *Cordia alliodora* (Ruiz & Pav.) Oken. **Ne tah uche'il. Nuk uche'il. Chukuch chan le' i as wolis. Mäna' uyits. Yan utop', säk utop', p'iis puuna'. Uwaakal ti' marzo. Chan mehen uwich, ek'. Yan ähweex. Chan mehen ähweex, ek'. Uchibal. Kuximbal tupach.** 'It's a very straight tree with a large trunk. The leaves are small, long, and somewhat round. There isn't any resin. It bears white flowers, the same as mahogany. They bloom in March. The fruit is very small and black. There is a kind of small, black, biting ant that walks along the bark of the tree.' (BM) *Cordia alliodora* is a tall, straight tree, up to 35 m (115′) high, with open branches and a small crown. Stinging ants (*Azteca longiceos*) live inside of the swollen nodes at the ends of branches, guarding the tree against the incursion of insects and other herbivores. Panicles of white flowers develop in the dry season, followed by small, oval drupes. It ranges from Mexico to Panama in secondary forests at low to mid elevations in a wide range of environmental conditions and soils. The timber is used in house construction. *Use:* **hiil** 'battens'; *Part:* **mehen che', k'äbche'** 'saplings, branches'. *Variant:* **bahun**. *Thes:* **che'** (Fig. 5.8). *[Note:* bajum. *Cordia alliodora* (Diemont 2006); bajom. *Cordia alliodora* (Nations 1992, in Atran et al. 2004); b'ojom [Itz.]. *Cordia alliodora* (Atran et al. 2004). English common name from USDA (http://plants.usda.gov/java/profile?symbol=COAL)*] [Source:* AM; BM*] [*\sd2 housing*]*

baknikte' [baknikte', bäknikte', boknikte'] tuberose. *Lit:* 'bone blossom tree'. Asparagaceae *Polianthes tuberosa* L. **P'iis ts'ulha'. Chukuch ule', uwaan batak 20.32 (8″), ma' nuk utan, batak 1.27 cm (0.5″). Uch'ihil yok'ol hach lu'um (no es xul petha', tikin). Yan utop', chan säk ule' utop', much'much' utop'. Yan ubok, ne ki'. Casi ubäho' ahts'ulha', pero as nuk ule', as nuk ubakel. Ne tak bekil utop'. Säk. Yan ubok.** 'It's similar to ts'ulha' (*Hymenocallis* sp.). Its leaves are long and narrow, about 20.32 (8″) long by

Fig. 5.8 Bahun che' *Cordia alliodora*. Photo Credit: Suzanne Cook

1.27 cm (0.5″) wide. It grows in true soil (not at the edge of the lagoons). It has small, white clusters of flowers. They're very fragrant. *Baknikte'* is almost the brother of *ts'ulha'*, except that it has somewhat larger leaves and it is a bit taller. The flowers are almost the same: they're white and fragrant.' (BM) A native of Mexico, *Polianthes tuberosa* is a night-blooming plant with tubular, white, waxy blossoms borne in long, terminal racemes. They are extremely fragrant, especially at night. In Nahuatl, the language of the Aztecs, the tuberose is called *omixochitl* 'bone flower' (Trueblood 1973: 157). Similar species include *Polianthes gracilis, P. longiflora*, and *P. pringlei*. All were sold as "nardo" in the Mexican markets during the 16th century (Trueblood 1973: 158). In Lacandon mythology, this flower is the birthplace of the Lacandon principal deities, *Hachäkyum, Äkyanto'*, and *Sukunkyum*. *Loc:* **kolil nah** 'house gardens'; *Use:* **tsoy uwich** 'ornamental'; *Myth:* **utop' ti' ahk'uh** 'flower of the gods'. *Morph:* **baak=nik-te'** (Fig. 5.9). *[Note:* bäknikte'. flor (el nardo) (Bruce 1968).*] [Source:* AM; BM*] [*\sd2 folklore_and_mythology \sd2 other*]*

Fig. 5.9 Baknikte' *Polianthes tuberosa*. Photo Courtesy: Forest and Kim Starr, Creative Commons Attribution 3.0 License

balche'$_1$ [balche', bache'] lancepod. Fabaceae *Lonchocarpus* spp. Lonchocarpus is a large genus with approximately 270 species. Twenty-three are listed for the state of Chiapas (Breedlove 1986). Of these, twelve have been recorded from the northern Lacandon territory: *L. amarus; L. castilloi; L. cruentus; L. guatemalensis; L. hondurensis; L. latifolius; L. lineatus; L. luteomaculatus; L. punctatus; L. robustus; L. rugosus;* and, *L. verrucosus* (Comisión Nacional de Áreas Naturales Protegidas 2006). *Lonchocarpus castilloi, L. guatemalensis, L. punctatus,* and *L. rugosus* occur in the southern Lacandon territory (Levy et al. 2006: 86). Lacandones lump all of these species into two groups: *hach balche'* 'authentic balche'' and *ya'ax balche'* 'green balche''. They differentiate the two on the basis of leaf shape and colour, flower colour, bark texture, and the degree of toxicity of the resin in the bark. *Hach balche'* provides the bark which the Lacandones use to make *balche'*, their fermented mead. They avoid *ya'ax balche'*, saying that it causes severe stomach aches and diarrhea. Legend has it that *Hachäkyum's* assistants brewed the first batch of *balche'* from this tree, drank it, and died. Conceivably, the cause of their demise, and the gastrointestinal discomfort described by the Lacandones, was rotenone. This is the active ingredient extracted from the bark and roots that is used to stun fish in the Neotropics (Higbee 1947: 427) and used throughout the world as an insecticide. *Use:* **balche'** 'fermented beverage'; *Spec:* **hach balche'** 'authentic balche' (*Lonchocarpus* sp.)'; *Spec:* **ya'ax balche'** 'green balche' (*Lonchocarpus* spp.)'. *Morph:* **ba'al-che'**. *Thes:* **che'**. *[Note:* b'aal che' [Yuc.]. *Lonchocarpus violaceus* (in, Atran et al. 2004).*] [Source:* AM; BM; SK*] [*\sd2 religion \sd2 food*]*

balumchan$_1$ [balumchan, balunchan] cacao, wild. Sterculiaceae (Malvaceae) *Theobroma* sp. This may be a synonym for *balumte'* or it may refer to another wild species of cacao. According to my consultants, it does not occur around Naha'. *Use:* **uk'ul, hanal** 'beverage, food'; *Part:* **nek'** 'seeds'; *Prep:* **k'elbil, huch'bil** 'toasted and ground'; *Sim:* **käkow** '*Theobroma cacao*'; *Sim:* **balumte'** '*Theobroma bicolor*'. *Thes:* **che'**. *[Source:* BM; CKO*] [*\sd2 food*]*

balumte' [balumte', balunte', barunte'] cacao, mocambo. *Lit:* 'jaguar tree'. Sterculiaceae (Malvaceae) *Theobroma bicolor* Humb. & Bonpl., syns. *Theobroma ovatifolia* Moc & Sessé ex DC., *Cacao bicolor* (Humb. & Bonpl.) Poir, *Tribroma bicolor* (Humb. & Bonpl.) Cook, *Theobroma cordata* Rufz & Pavón. This is a wild species of cacao that grows in fallow *milpas*. Although its botanical identification has not been confirmed, Nations (2006) believes that it is a feral form of *käkow* (*Theobroma cacao*). The cultural importance of cacao was evident during Classic Maya era: Many of the Maya paintings, particularly on vases and pottery, depict the complex association between cacao and maize, the former representing death, and the latter, rebirth. The God of Maize, the Lord of Death, and the Jaguar God of the Underworld were depicted as a cacao tree (Martin 2006). If the interpretation is correct, then the association between *käkow* 'cacao' and *balamte'* 'jaguar tree' becomes significant. *Balumte'* corresponds to cacao in a number of Maya languages, e.g., *b'ajrum=tye', b'ajlum=tye'* [Ch'ol], *balumte, bahlamte* [Ch'olti'], and *balamté* [Yuc.]. In Q'eqchi' (Kekchi), the connection between the jaguar and cacao is made clear in the phrases *b'a:lam kakaw, kakaw b'a:lam* (Haeserijn 1979: 53, 79). Given the pan-Mayan significance of *balamte'*, it is not surprising to encounter the word in the Lacandon lexicon. My Lacandon consultant, KP, said that *balumte'* is similar to *käkow*, but the flavour is far superior. If *käkow* corresponds to *Theobroma cacao*, then *balumte'* could refer to *Theobroma bicolor* (Nicaragua chocolate tree). Another common name for this species in Mexico and Guatemala is *pataxte'*. It differs from *T. cacao* mainly in size and the exterior texture of the fruit: the tree is much larger, although wild forms of *T. cacao* can grow 23 feet or more (Rosenblum 2006: 7), and the texture of the skin is coarsely netted or honeycomb-like. *Theobroma bicolor* is cultivated throughout the Neotropics for its sweet pulp and seeds. *Pataxte'* and *käkow* are mentioned in the Popul Vuh, the Quiché Maya book of creation, in which both are listed among the first foods that the gods created for their mortals. Residue in excavated Mayan vessels confirms that both *T. cacao* and *T. bicolor* were prepared by the ancient Maya (Hurst 2006). Although *T. bicolor* was enjoyed by the Mayas and the Aztecs, the Spanish considered it inferior to *T. cacao* (Hellmuth 2010a). This and another wild cacao, *T. angustifolium*, were, and still are, used as substitutes for or adulterants of *T. cacao* in Mexico. Although the fresh fruit of *T. bicolor* is delicious, as far as can be determined, the Lacandones in Naha' only eat the beans. *Use:* **uk'ul, hanal** 'beverage, food'; *Part:* **nek'** 'seeds'; *Prep:* **Uk'elik unek', uhuch'ik yok'ol ka', ubut'ik yok'ol ha' yetel ahtsuk'in, uhäxik yetel ukekche', pachil ubut'ik yok'ol ma'ats'** 'One toasts the seeds, grinds them on the metate (millstone), puts (the powder) in water with tsuk'in (Sarsaparilla), beats it with the cacoa beater, and then puts (the mixture) into ma'ats' (type of corn gruel)'; *Sim:* **käkow** '*Theobroma cacao*'; *Sim:* **balumchan** '*Theobroma* sp.'. *Thes:* **che'** (Fig. 5.10). *[Note:* balamté [Yuc.]. *Theobroma bicolor* Humb. & Bonpl. (Standl.). A variety of cacao (Motul.) (Roys 1931: 216, 326).*] [Source:* BM; KM; KP*] [*\sd2 food*]*

bamax (äh) [bamax, bamäx] goat's milk tree. Moraceae *Pseudolmedia* spp. **Es che'. Yan utop', chan nuk, batak 2.45 cm (1″) i wolis. Säk utop'. Uwich ti' hach bamax as chukuch, batak 1.27 cm (0.5″). Ne yul uwich. Ne chäk. Hump'eli' unek'. Ch'ihil ich k'aax, yok'ol wits. Tso'ots bamax yan uwich, wolis uwich, batak 2.45 cm (1″). Chan tso'otsel upach. Yan ka'p'eli': k'än i chäk. Abril utahal. Uch'ihil ich k'aax yok'ol äkan lu'um. Ba'ats', k'ambul, tuwolol ch'iich' umäk'ik uwich. Hale' i tsub umäk'ik unek'.(BM) Ne ki' uche'il. Ma' utupul tulahsastal (AM).** 'Bamax is a tree. It has flowers that are a little bit large and round, about 2.45 cm (1″) in diameter. The flowers are white. The fruit of the *hach bamax* 'authentic bamax' is somewhat long, about 0.63–1.27 cm (1/4–1/2″). The fruit is very smooth and very red. It has one seed. It grows in the hills of the forest. *Tso'ots bamax* 'hairy bamax' grows in the forest in the low lying flat areas. It has round fruits, about 1.27–2.45 (1/2″-1″) in diameter. The skin has small hairs. There are two kinds of *tso'ots bamax*: a yellow kind and a red kind. They ripen in April. Howler monkeys, curassows, and all the birds eat the fruit. The spotted cavy and the agouti eat the seeds.' (BM) 'The wood provides excellent firewood, because it burns all night.' (AM). In addition to fruit size and texture, leaf shape is

Fig. 5.10 Balumte' *Theobroma bicolor*. Photo courtesy of Nicholas Hellmuth, FLAAR

another feature that Lacandones generally use to distinguish the types of *bamax*. According to BM, the leaves of *tso'ots bamax* are slightly hairy, whereas those of *hach bamax* are smooth. However, the leaf he identified as *tso'ots bamax* was smooth, leathery, and somewhat wrinkled; its venation was arcuate (secondary veins diverge from the main vein in an arc) and reticulate (tertiary veins are netted); the leaf margin was wavy. This is in contrast to the leaf he identified as belonging to *hach bamax*, which displayed more than 18 pairs of sub-horizontal secondary veins, smooth, glossy blades with slightly wavy leaf margins. Despite these distinctions, both correspond to *Pseudolmedia oxyphyllaria* (Servando Carvajal, pers. comm.). Although *bamax* grows wild, Lacandones also transplant seedlings into their house gardens and *milpas*. *Loc:* **k'aax** 'forest'; *Use:* **hanal** 'food'; *Part:* **wich** 'fruit'; *Prep:* **che'che'** 'raw'; *Use:* **che'il k'ak'** 'firewood'; *Use:* **hiil** 'battens'; *Use:* **ho'olan che'il** 'ridge-pole'; *Part:* **mehen che'** 'saplings'; *Spec:* **hach bamax** '*Pseudolmedia oxyphyllaria*'; *Spec:* **tso'ots bamax** '*Pseudolmedia oxyphyllaria*'. *Thes:* **che'** (Fig. 5.11). *[Note:* Botanical identification provided by Servando Carvajal (pers. comm.). bamäsh. *Brosimum* sp. (Nations and Nigh 1980); ba am bax. *Brosimum* sp. (Diemont 2006); bamax, tzot bamax, has bamax. *Pseudolmedia oxyphyllaria* (Durán 1999); u-xib'-al mä'na'ax [Itz.]. *Pseudolmedia oxyphyllaria* (Atran et al. 2004); u-ch'up-al mä'na'ax [Itz.]. *Pseudolmedia spuria* (Atran et al. 2004); mänax [Mop.]. *Pseudolmedia oxyphyllaria* (in, Atran et al. 2004); manax [Yuc.]. *Pseudolmedia oxyphyllaria* (in, Atran et al. 2004).*] [Source:* AM; BM; CKO; SK*] [*\sd2 food*]*

Fig. 5.11 Bamax *Pseudolmedia* sp. Photo Credit: Suzanne Cook

bambu bamboo, golden (striped bamboo, painted bamboo). Poaceae *Bambusa vulgaris* 'Vittata' Schrad. ex J.C. Wendl. This is an introduced Asian variety of a sturdy, 10–20 m (30–70′) tall bamboo that forms moderately loose clumps of thick-walled, inflexible, unarmed culms. Horizontal branches bear sword-shaped leaves on stalked blades. Lacandones cultivate the species primarily for its ornamental appeal. Chiapas has the greatest number of bamboo species in Mexico, with four herbaceous species and 25 woody species (Cortés and Inés-Aquilar 2002: 91, 94). Despite this, only six species of Bambuseae have been recorded in the Lacandon forest: *Guadua amplexifolia, Lasiacis divaricata, L. procerrima, Merostrachys* sp., *Olyra latifolia*, and *Rhipidocladum bartlettii* (Durán 1999; Comisión Nacional de Áreas Naturales Protegidas 2006; Levy et al. 2006; Nations and Nigh 1980). *Use:* **tsoy uwich** 'beauty'; *Sim:* **semet'** '*Rhipidocladum bartlettii*'; *Sim:* **'ämäy** '*Guadua amplexifolia* (?)'. *From:* bambu (Malay) (Fig. 5.12a, b). *[Note:* Taxonomic identification provided by Gib Cooper (pers. comm.). chum jalal (Nations 1992, in Atran et al. 2004); b'amb'u' [Itz.]. *Bambusa vulgaris* (Atran et al. 2004); b'amb'u [Mop.]. *Bambusa vulgaris* (in, Atran et al. 2004); b'amb'u [Yuc.]. *Bambusa vulgaris* (in, Atran et al. 2004).*] [*\sd2 ornamental*]*

bä'ch'ich' coccoloba. Polygonaceae *Coccoloba* spp. A native to the Neotropics, Coccoloba is an evergreen tree with alternately arranged, large, pinnately-veined, elliptic leaves with a nearly heart-shaped base. The trunks of young trees display stipular scars where the leaves have detached, and

Fig. 5.12 (a) Bambu *Bambusa vulgaris* 'Vittata'. Photo Credit: Suzanne Cook (b) Bambu *Bambusa vulgaris* 'Vittata'. Photo Credit: Suzanne Cook

enlarged rings. Flowers are produced in spikes followed by a dry, three-angled, one-seeded fruit surrounded by brightly coloured flesh. Four species have been recorded in Naha': *C. cozumelensis; C. tuerckeimii; C. barbadensis;* and, *C. hondurensis*. In the southern Lacandon territory, *C. barbadensis* and *C.* aff. *liebmannii* occur (Levy et al. 2006). The fruits of these last two species are edible. *Loc:* **k'aax** 'forest'; *Use:* **hanal** 'food'; *Part:* **wich** 'fruit'; *Spec:* **mehen bä'ch'ich'** 'small bä'ch'ich' (*Coccoloba hondurensis*)'; *Spec:* **nukuch bä'ch'ich' (ah)** 'large bä'ch'ich' (*Coccoloba barbadensis*)'. *Thes:* **che'**. *[Source:* BM*]* *[*\sd2 food*]*

beyanchan tomato, husk. Solanaceae *Physalis* sp. **Yan uyak'il. Wolis uwich, batak 5 cm (2″). Ya'ax u ts'ibal usool. Säk unoy. Ch'uhuk. Säk utop', p'iis p'ak. Lo'obil.** 'It has vines and 5 cm (2″) round fruit with green-striped skin. The flesh is white and sweet. The flowers are white, like those of tomato plants. It's a weed.' (AM) Very few Lacandones were familiar with this word, so I was unable to locate the plant. However, AM and KP describe the fruit as a husk tomato. *Beyanchan* is mentioned in Tozzer (1907: 21) and Soustelle (1966: 105), both authors glossing it as 'tomato' (*Lycopersicum esculentum*). The species that is generally grown in Lacandon *milpas* is a cherry tomato (*Solanum lycopersicum* var. *cerasiforme*); but, KP says that *beyanchan* is larger than this. *Use:* **hanal** 'food'; *Part:* **wich** 'fruit'; *Prep:* **che'che'** 'raw'; *Prep:* **chäkbil yetel sopa, bäk', uye'el kax** 'cooked with soups, meats, and eggs'; *Sim:* **p'akan** 'ground cherry'. *[Note:* beyanchan. *Lycopersicum esculentum* (Tozzer 1907); beyanchan. tomate (Soustelle 1966).*]* *[Source:* AM*]* *[*\sd2 food*]*

bey ule' tähte' (ah) podocarp, Jalisco. *Lit:* 'like the leaves of *tähte*''. Podocarpaceae *Podocarpus matudae* Lund., syns. *P. matudae* Lund. var. *reichei* (J. Buchholz & N. E. Gray) de Laub. & Silba, *P. matudae* Lund. var. *macrocarpus* J. Buchholz & N. E. Gray, *P. matudae* Lund. var. *jaliscanus* de Laub. & Silba. This is a descriptive term that AM provided for a sapling. It may be his pronunciation of *bedetate* (Spanish). BM identified the tree by its fruits, which were on the ground at the base of another tree. He called the tree ch'ibil wits. The fruits are eaten by the *kacho'* (mealy blue-crowned parrot). *Podocarpus matudae* is a species of conifer native to Mexico and Guatemala. It grows to 20 m (65.5′) tall with a trunk diameter of 1.5 m (5′) or more (Standley and Steyermark 1958). Leaves are narrowly lanceolate, leathery, and ridged above the midvein. Fruit is a cone that is borne on a peduncle 4–6 mm (0.16–0.24″) long, with a receptacle of 2 thick, fused scales of up to 6 mm (0.24″) in length enclosing a seed approximately 10×8 mm (0.4×0.3″) with a blunt crest. (Dallimore et al. 1967, cited in Earle 2013b) The tree is usually found in pine-oak-liquidambar or cloud forests, 1900–2600 m (6233–8530′) in elevation, but it also occurs in mountainous areas around Naha', up to1300 m (4265′) (Comisión Nacional de Áreas Naturales Protegidas 2006: 21). Although my Lacandon consultants reported no uses for the tree, Durán's Lacandon consultants report using the timber to make rifle butts, axe handles, and other tools (Durán 1999: 147). *Morph:* **bey u-le' täh-te'**. *Variant:* **ch'ibil wits**. *From:* bedetate (Spanish)? *Thes:* **che'** (Fig. 5.13a, b). *[Note:* chibix xi wits, bedetate. *Podocarpus matudai* (Durán 1999). English and Spanish common names from Multilingual Multiscript

Fig. 5.13 (**a**) Bey ule' tähte' *Podocarpus matudai*. Photo Credit: Suzanne Cook (**b**) Bey ule' tähte' *Podocarpus matudai*. Photo Credit: Suzanne Cook

Fig. 5.14 (**a**) Bilim suli' *Dioscorea bulbifera*. Photo Credit: Suzanne Cook (**b**) Bilim suli' *Dioscorea bulbifera*. Photo Credit: Suzanne Cook

Plant Name Database (M.M.P.N.D.) http://www.plantnames.unimelb.edu.au/Sorting/Frontpage.html. Retrieved February 2013. Botanical synonyms from the gymnosperm database http://www.conifers.org/po/Podocarpus_matudae.php. Retrieved March 2013.*] [Source:* AM; BM*]*

bilim suli' [bilim suli', bilin suli'] potato, air. Dioscoreaceae *Dioscorea bulbifera* L. BM described the plant as being like a potato that grows on a vine. The aerial tubers are hard and, according to BM, not very tasty. SK added that *bilim suli'* is a very old variety they used to cultivate, but rarely now. You can still find them hung up in trees and rambling over bushes in abandoned house gardens and old *milpas*. *Dioscorea bulbifera* is a perennial vine that can reach as high 45.7 m (150′). It forms edible bulbils in the leaf axils of twining stems and tubers similar to potatoes. Native to South Asia and cultivated to a limited extent in tropical and subtropical areas of East Africa to Polynesia, it is an invasive species in the Neotropics and Florida (National Tropical Botanical Garden 2013). It is superficially similar in appearance to *hach suli'* (*Dioscorea alata*). *D. alata* differs by having square stems with wide ridges, or wings, opposite leaves, and enormous tubers (Langeland and Burks 1998). *Loc:* **paakche' kol** 'fallow milpas'; *Use:* **hanal** 'food'; *Part:* **wich** 'fruit'; *Prep:* **chäkbil** 'boiled'; *Sim:* **hach suli'** 'yam (*Dioscorea alata*)'; *Gen:* **suli' (äh)** (Fig. 5.14a, b). *[Note:* bilim suli'. potatoes or batata (Bruce 1975).*] [Source:* BM; BM's wife; KM*] [*\ sd2 food*]*

bits' [bits', biits'] inga (shimbillo). *Lit:* 'tighten'. Fabaceae *Inga* spp. **Es che'. Yan humbuhk'äb—suumil bits', t'elen bits', bits'il k'aax, y mehen bits'. Hay un otro—nukuch**

bits'. Ti' kah. Se vende por los campesinos. T'elen bits' i suumil bits' uch'ihil paakche' kol. Mehen bits' uch'ihil ich k'aax wa paakche' kol. Crecen solito. Uwich, kupäk'ik ähch'iich'. Nukuch bits' päk'bil. Säk utop'. Mäna' ubok. Bits'il k'aax uch'ihil ich k'aax. Kooch uwich, batak 2.54 (1″). Uwaan batak 7.6 cm (3″). Mayo utahal. Octubre utahal uwich ti' mehen bits'. Suumil bits' yetel nukuch bits' utahal uwich ti' Agosto. 'It's a tree. There are five kinds: 'rope inga', 'ridged inga', 'small inga', 'large inga', sold by the villagers, and 'wild inga'. 'Ridged inga' and 'rope inga' grow in the fallow milpas. 'Small inga' grows in the forest or in the acahuales. It grows on its own [uncultivated]. Its seeds are dispersed by the birds. The 'large inga' is cultivated. Its flowers are white. They don't have any scent. *Bits'il k'aax* 'wild inga' grows in the forest. Its fruit is wide, about 2.54 cm (1″) wide and about 7.6 cm (3″) long. It ripens in May. *Mehen bits'* 'small inga' ripens in October. The fruit of the 'rope inga' and the 'large inga' ripen in August.' (BM) *Inga* is a genus of tropical trees and shrubs in the legume family. They have pinnate leaves, white, pompom flowers, and large, glossy, sometimes curved, pods. In some species, the seeds are surrounded by a white, sweet, fibrous aril that resembles cotton-candy. Seventeen species occur in Chiapas (Breedlove 1986). Eleven of these are recorded from Naha': *I. belizensis, I. leptoloba, I. vera, I. acrocephala, I. laurina, I. pavoniana, I. pinetorum, I. punctata, I. rodrigueziana, Inga*. sp., and *I. vera* (Comisión Nacional de Áreas Naturales Protegidas 2006). In addition to the five species described by BM, Nations and Nigh (1980) recorded *bits'il petha'* 'inga of the lagoon' (*I. spuria*). Durán (1999) recorded "bits wits", which corresponds to *I. belizensis* (Durán 1999). The latter is a synonym of *bits'il k'aax* (BM). According to AM, *Inga* is a good tree to have in the *milpa*, because its leaves amend the soil. This fact has been confirmed by a number of agro-forestry researchers in the Lacandon forest (Nations and Nigh 1980; Diemont 2006; Levy et al. 2006; Nigh 2008). The fruit is eaten out of hand and the timber is used for fuel. In the southern Lacandon territory, the leaves of some species (e.g., *I. pavoniana*) are eaten to relieve stomach aches (Diemont 2006; Kashanipour and McGee 2004). The southern Lacandones use the timber of *I. punctata* for firewood (Levy et al. 2006). *Use:* **hanal** 'food'; *Part:* **wich** 'fruit'; *Prep:* **che'che'** 'raw'; *Use:* **chibal hämnen** 'stomach ache'; *Part:* **le'** 'leaves'; *Prep:* **Ha'li' acha'achtik** 'You just chew (the leaves)'; *Use:* **che'il k'ak'** 'firewood'; *Spec:* **suumil bits'** 'rope bits' (*Inga oerstediana*)'; *Spec:* **mehen bits'** 'small bits' (*Inga* sp.)'; *Spec:* **t'elen bits'** 'ridged bits' (*Inga pavoniana*)'; *Spec:* **bits'il k'aax** 'forest bits' (*Inga belizensis*)'; *Spec:* **nukuch bits'** 'large bits' (*Inga punctata*)'; *Spec:* **bits' petha'** 'lagoon bits' (*Inga spuria*)'. *Thes:* **che'**. *[Note:* bitz. *Inga pavoniana* (Diemont 2006); bitz. *Inga* spp. (Kashanipour and McGee 2004); bitz. *Inga pavoniana* (Nigh 2008); bitz'. jinicuil. *Inga* spp. (Nations and Nigh 1980); *Inga pavoniana* and *Inga punctata* (Levy et al. 2006); (aj/ix) b'itz' [Itz.]. *Inga vera, I. leptoloba* (Atran et al. 2004); b'iitz' [Yuc.]. (in, Atran et al. 2004). b'itz [Mop.] (in, Atran et al. 2004).*] [Source:* AM; BM; KA; SK*] [*\sd2 food*]*

bits'il k'aax inga, Belize. *Lit:* 'forest bits''. Fabaceae *Inga belizensis* Standl. This species is a synonym for "bits wits", which corresponds to *Inga belizensis* (Durán 1999). *Wits* 'hills, mountains' and *k'aax* 'forest' are commonly used interchangeably to refer to high forest. According to SK, the fruits are small and not as tasty as the fruits of the other kinds of *bits'*. *Loc:* **k'aax** 'forest'; *Use:* **hanal** 'food'; *Part:* **wich** 'fruit'; *Prep:* **che'che'** 'raw'; *Sim:* **suumil bits'** 'rope bits' (*Inga oerstediana*)'; *Sim:* **mehen bits'** 'small bits' (*Inga* sp.)'; *Sim:* **t'elen bits'** 'ridged bits' (*Inga sapindoides*)'; *Sim:* **nukuch bits'** 'large bits' (*Inga punctata*)'; *Sim:* **bits' petha'** 'lagoon bits' (*Inga spuria*)'; *Gen:* **bits'**. *Thes:* **che'**. *[Note:* bits wits. *Inga belizensis* (Durán 1999).*] [Source:* BM; SK*] [*\sd2 food*]*

bits'kal hu'un fig, strangler. *Lit:* 'tighten neck bark cloth'. Moraceae *Ficus* sp. **Casi lo mismo que ak' hu'un. Uch'ihil ich k'aax. Ne yäk uyits. Yan ubok ne yäk. Ne k'än uwich, p'iis 'abäl. Uk'axik ubäk'ik uche'.** 'It's almost the same as *ak' hu'un* (*Poulsenia armata*). It grows in the forest. Its resin is very strong: it has a very strong odour. The fruits are very yellow, like hog plums (*Spondias* spp.). It wraps around trees.' (BM) According to AM, the resin burns the hands, making it difficult to work with. *Use:* **hu'un, hach nook'** 'bark cloth, authentic clothing'; *Part:* **uyoot'el sool** 'inner bark'; *Prep:* **ubäxik uyoot'el yetel che'** 'One pounds the inner bark with a stick'; *Sim:* **ak' hu'un** '*Poulsenia armata*'; *Sim:* **hach hu'un** '*Ficus* sp.'; *Gen:* **hu'un**. *Morph:* **bits'=kaal**. *Thes:* **che'**. *[Note:* b'its' [Yuc.]. tighten (Bricker et al. 1998).*] [Source:* AM; BM; KM; KP*] [*\sd2 clothing*]*

bits' petha' guaba_inga (guaba, pan chock [Jam. Eng.], river-icolco [Puerto Rico], river-koko [Jam. Eng.], Spanish ash). Fabaceae *Inga spuria* Humb. & Bonpl. ex Willd, syn. *Inga vera* ssp. *spuria* (Humb. & Bonpl. ex Willd.) J. León. This is a tree 10 m (33′) tall, with a rounded crown and straight, cylindrical trunk. The brown bark is smooth or slightly furrowed. Leaves are alternate, with 5–9 pairs of leaflets arranged opposite to each other on the rachis. They are 3–17 cm × 2–4 cm (1.2–6.7″ × 0.8–1.6″), oblong or elliptic, acuminate at the apex, rounded at the base, and have entire margins. Petioles and rachises are winged. The slightly fragrant, greenish-white, pompon-like flowers are borne in racemes. Fruit is a cylindrical pod, 15 cm (5.9″) long with four ribs. It is yellowish when ripe. The tree grows in sunny locations, primarily in riparian vegetation (Gutiérrez-Carvajal and Dorantes-López 2003–2004). *Loc:* **petha'** 'lagoon'; *Use:* **hanal** 'food'; *Part:* **wich** 'fruit'; *Prep:* **che'che'** 'raw'; *Sim:* **suumil bits'** 'rope bits' (*Inga oerstediana*)'; *Sim:* **mehen bits'**

'small bits' (*Inga* sp.)'; *Sim:* **t'elen bits'** 'ridged bits' (*Inga sapindoides*)'; *Sim:* **bits'il k'aax** 'forest bits' (*Inga belizensis*)'; *Sim:* **nukuch bits'** 'large bits' (*Inga punctata*)'; *Gen:* **bits'**. *Thes:* **che'**. *[Note:* u bitz' petha. calahuite. *Inga spuria* (Nations and Nigh 1980). English names from Grandtner 2005.*] [Source:* BM*] [*\sd2 food*]*

boox [boox, box] banana. Musaceae *Musa acuminata* ×*Musa balbisiana* Coll. Boox refers to bananas with relatively thin skins and soft, sweet flesh. Almost all modern, edible bananas come from two wild species, *Musa acuminata* and *Musa balbisiana*. Both originated in Southeast Asia and Oceania. In addition to "Cavendish", which comprises over 40% of the edible bananas produced worldwide, cultivars include 'Manzana', 'Lady Finger', 'Red', 'Red Spanish' (a.k.a. Red Cuban or Colorado), and 'Orinoco' (a.k.a. Horse, Hog, or Burro) (Morton 1987: 29–46). A number of these are cultivated in Naha'. Next to maize, bananas and plantains are the most important food in the Lacandones' diet, according to AM. The leaves are also indispensable as food wrappers. And before cooking oil was introduced, they were used as non-stick cooking surfaces, according to KP. Today, they are still used as surfaces to pat out tortillas. According to AM, the consumption of too many dessert bananas gives children worms. He attributes this to the fruit's high sugar content. Whereas other Mayan cultures use bananas and plantains for medicinal purposes as well as food, the Lacandones only mention its food value. Indirectly, however, bananas and plantains carry a connotation of illness: to dream of bananas foretells anemia (Bruce 1975: 192). *Use:* **hanal** 'food'; *Part:* **wich** 'fruit'; *Prep:* **che'che'** 'raw'; *Sim:* **pätan** 'plantain'; *Spec:* **hach boox** 'authentic banana'; *Spec:* **chäk boox** 'red banana'; *Spec:* **maasam (boox) (äh)** 'Lady Finger (*Musa acuminata*)'. *Thes:* **xiiw**. *[Note:* bosh, patan. banana, plantain (Nations and Nigh 1980); ix gineeyoj [Itz.]. banano (Atran et al. 2004). *Guineo* appears to be the Spanish word for dessert banana. *Ha'as* was the original name for *Pouteria mammosa*, but now refers to plantains and bananas in all the Yucatecan languages except Lacandon.*] [Source:* BM; CKD; KP*] [*\sd2 food*]*

booxa'an [booxa'an, bonxa'an] palm, Everglades. Arecaceae *Acoelorraphe wrightii* (Griseb. & H. Wendl.) H. Wendl. ex Becc., syn. *Paurotis wrightii* (Griseb. & H. Wendl) Brittonex Britton & Schafer. **Uch'ihil xok'ol campamento. Yan uk'uch ti' ule'. Ma' ak'anik ti' ho'olnah.** 'It grows near the camp (near the lagoon). There are threads on the leaves. We don't use them for roof thatch.' (BM) This is a large palm with stiff, fan-shaped leaves. The leaflets narrow to a sharp point. Thin, white fibres dangle from leaf margins. Petioles are armed. It is native to Central America, south-eastern Mexico, the Caribbean, and the extreme end of southern Florida. It grows in swamps and periodically flooded forests, low-lying areas inundated with fresh or brackish water, thickets in savannas, and, occasionally, in pine woods near sea level and low elevations (Henderson et al. 1995: 53). According to BM, the leaves are not used for thatch, because the sections that comprise the leaf blade separate and then stick together, creating large gaps. In the language of the Itzaj, *b'onxa'an* corresponds to *Sabal mexicana*. These people eat the fruit, make hats and baskets from the leaf fibres, and obtain salt from the leaves and heart (Atran 1999: 135). The Lacandones call *Sabal mexicana*, *xa'an*, and use it for the same purposes (Fig. 5.15). *[Note:* b'onxa'an [Itz.]. hat palm (Hofling and Tesucún 1997); aj b'onxa'an [Itz.]. cabbage palm. *Sabal mexicana* (Atran 1999); tasiistej [Itz.]. *Acoelorraphe wrightii* (Atran et al. 2004); b'onxá'an [Yuc.]. *Sabal mexicana* (Bricker et al. 1998); bom [Yuc.]. *Sabal* sp. (Roys 1931); xaan [Maya] *Sabal jajapa* (Standl. 1920–26: 71, in Roys 1931). Synonyms from the M.M.P.N.D., http://www.plantnames.unimelb.edu.au/Sorting/Frontpage.html. Retrieved December 2011.*] [Source:* BM*]*

Fig. 5.15 Booxa'an *Acoelorraphe wrightii*. Photo Credit: Suzanne Cook

bo'oy (äh)$_2$ [booy] chamaedorea. *Lit:* 'shade'. Arecaceae *Chamaedorea* spp. **Bo'oy yan uche'il. Ne mihin uche'il. Yan umots. Bo'oy. Che'il, ne chich. Ma' ne nuk uche'il. Ya'ax uche'il. Yan ule', yan uch'ibil uche' ule'. Yan uchan mehen che' ule'. Eh, uk'äb, ma' ne tah uk'äb ule', i chukuch uwaan. K'äb ule' es como gajos de las hojas. Yaab uk'äb, yaab ule', su' – way hump'el, way hump'el, way hump'el. Tatahan, tatahan yaab ule'. Chukuch uwaan ule' bo'oy. P'iis ch'i'ib. Yan uwich, yan uwich. Uwich, wa nen ya'ax, muun. Wa nen täk'än, ek'. Ele, ne täk'än umäk'ik ch'iich'. Ha'li' pän umäk'ik. Tuwolol ch'iich' ma' hach ma' hach umäk'ik. Ne yaab unek', ne yaab. Tint'an yan cien, o más que cien, uwich. Yan säk bo'oy, yan k'än bo'oy, i yan hach bo'oy. Hach bo'oy lati' tintsikbaltah. Ya'ax uwich utahal, ek', ek' uwich. Asi lati' hach bo'oy.** '*Bo'oy* has a trunk. It's a small trunk. It has roots. The trunk is very hard, but it's not a large trunk. The trunk is green. The leaflets have a main vein. The stems have leaflets. The leaf stems are not very straight, and they are long. There are many branches thick with leaves–here is one, here is another one, and here is another one. There are a lot of leaves, one after the other. The leaves of *bo'oy* are long, like those of *ch'i'ib* (*Chamaedorea tejilote, C. alternans*). Its fruit is green and turns black when ripe. The birds eat them when they are very ripe. Well, only the toucan eats them; all the other birds don't really care for them. There are a lot of seeds—100 or more, I think. There is *säk bo'oy* 'white bo'oy', *k'än bo'oy* 'yellow bo'oy', and *hach bo'oy* 'authentic bo'oy'. *Hach bo'oy* is what I was talking about. It has green fruits that ripen to black. That's *hach bo'oy*.' (BM) Chamaedoreas are pinnate (feathery-leaved) palms with green stems that are ringed, similar to bamboo. Most are under 3 m (9′) but some may reach 6 m (20′) or more. Some species are multi-stemmed, while others are solitary-stemmed. The Lacandones distinguish at least eleven species. *Hach bo'oy*, as its name signifies, is the 'true bo'oy', or prototypical species. It is primarily used in ceremonial contexts, serving as mats, spoons, and fans. Several of the other species are sold in great quantities to the floral and textile industries in the United States. The species include *C. elegans, C. oblongada*, and *C. ernesti-augusti*. Collectively, they are called *xate*. The *xate* industry has become so large in lowland Chiapas, Belize, and Guatemala that illegal harvesting of the leaves from the wild now threatens the survival of these species. Efforts are underway to establish nurseries in a number of forest communities. One has already been established in Naha'. *Loc:* **k'aax** 'forest'; *Spec:* **hach bo'oy** 'authentic bo'oy'; *Spec:* **ch'ibix bo'oy** '*Chamaedorea* spp.'; *Spec:* **säk bo'oy** 'white bo'oy'; *Spec:* **k'än bo'oy** 'yellow bo'oy'; *Spec:* **chan mehen bo'oy** 'very small bo'oy'; *Spec:* **ch'it bo'oy** '*Chamaedorea* sp.'; *Spec:* **ton bo'oy** '*Chamaedorea* sp.'; *Sim:* **k'eben** '*Chamaedorea ernesti-augusti*'. *[Note:* bo'oy. chamaedora (species of small palm tree) (Bruce 1975); xate. *Chamaedorea concolor* (Durán 1999); ix yat. *Chamaedorea* spp. xate (Nations 2006); boi. *Chamaedorea elegans* (Atran 1999); b'ó'oy N. shadow, shade (Bricker et al. 1998). K'ayum Segundo also called this species "cola de pescado" (fish tail). He may be referring to *k'eben* (*Chamaedorea ernesti-augustii*), which is also called *bo'oy.] [Source:* AM; BM; K'ayum Segundo*]* [\sd2 industry\sd2 religion*]*

box₁ bean, shelling. Fabaceae Phaseoleae. According to AM, all beans are called box, i.e., *Phaseolus vulgaris, Vigna* spp., and *Canavalia* spp. An exception is *'ib* 'lima bean' (*Phaseolus lunatus* cv. 'Sieva'). *Use:* **hanal** 'food'; *Part:* **wich** 'fruit'; *Prep:* **chäkbil, tsähbil, nahwah, bu'ulil wah** 'boiled, fried, ceremonial tamales, bean tamales'; *Spec:* **hach box** 'authentic box'; *Spec:* **chäk box** 'red box'. *Variant:* **box bu'ul**. *Thes:* **bu'ul**. *[Source:* AM*]* [\sd2 food*]*

box₃ calabash. *Lit:* 'shell'. Cucurbitaceae *Cucurbita* sp. This is a vine that produces inedible squashes used only for containers (Baer and Merrifield 1971; Bruce 1975). It likely gets its name from its hard rind. Containers made from the fruits are also called *box*. *Use:* **box** 'kind of container'; *Use:* **luuch** 'cup, bowl'; *Part:* **wich** 'fruit'. *[Note:* box. calabash, gourd (a large type); drinking vessel made of the same (Bruce 1975).*]* [\sd2 containers*]*

box bu'ul₁ cow pea (black eyed pea). *Lit:* 'box bean'. Fabaceae *Vigna unguiculata* (L.) Walp. *Vigna unguiculata* is a vigorous climbing annual vine from Asia that was introduced to tropical America in the 17th century by the Spanish (Madamba et al. 2006). It bears lilac-coloured blossoms followed by long, straight or slightly curved, pencil-thin pods that hang in pairs. Seeds are oblong, beige or black, although the species produces more colours (Cook et al. 2005). The Lacandones eat it as a dry bean and, occassionally, as a fresh string bean. Levy et al. (2006) report that this species is used medicinally in the southern Lacandon territory, although these authors provide no details. *Use:* **hanal** 'food'; *Part:* **nek'** 'seeds'; *Prep:* **chäkbil, mäxbil i tsähbil** 'boiled, mashed and fried'; *Part:* **wich** 'fruit'; *Prep:* **chäkbil** 'boiled'; *Sim:* **ek' bu'ul** 'black bean'; *Sim:* **ek'el bu'ul** 'coloured/spotted bean'; *Sim:* **chäk bu'ul** 'red bean'; *Sim:* **hach bu'ul** 'authentic bean'; *Gen:* **bu'ul**. *See:* **box₁**. *Thes:* **bu'ul** (Fig. 5.16a, b, c). *[Note:* ix pelon [Itz.]. *Vigna unguiculata* (Atran et al. 2004); ixpeelon [Yuc.]. *Vigna unguiculata*

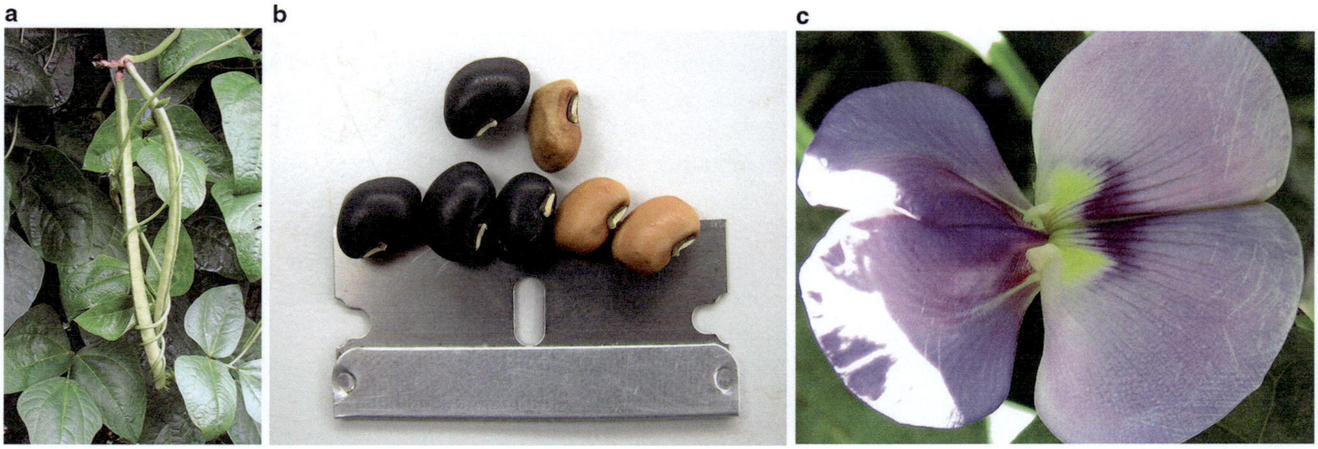

Fig. 5.16 (**a**) Bu'ul box *Vigna unguiculata*. Photo Credit: Suzanne Cook (**b**) Bu'ul box *Vigna unguiculata* seeds. Photo Credit: Suzanne Cook (**c**) Bu'ul box *Vigna unguiculata* flower. Photo Credit: Suzanne Cook

(in, Atran et al. 2004); b'ox-b'ú'ul [Yuc.]. black bean. *Phaseolus vulgaris* (Bricker et al. 1998).*] [Source:* AM; BM; CKO; KM*]* *[*\sd2 food*]*

box bu'ul$_2$ bean, shelling. Fabaceae Phaseoleae *Phaseolus vulgaris, Canavalia* spp. *See main entry:* **box$_1$**. *Thes:* **bu'ul**. *[Source:* AM; KM; KP*]*

bubuhchi' (ah)$_1$ bean, Jack (Chickasaw lima bean). *Lit:* 'split mouth'. Fabaceae *Canavalia* spp. *Bubuhchi'* likely refers to the exotic blossoms, which are comprised of a bell-mouth chalice with two lips, the lower lip broad and shallowly cleft, and the upper lip, curved down with the sides folded up and flared out on either side. This structure produces the effect of a beak and two lower teeth. *Canavalia* species are also referred to as *yo'och ahpek'* 'dog food', because the toasted seeds are fed to dogs that have eaten the berries of *chok* (*Rourea glabra*). The beans are believed to purge the toxins lethal to dogs. *Loc:* **kolil nah** 'house garden'; *Use:* **ts'ak ahpek'** 'dog medicine'; *Part:* **nek'** 'seeds'; *Prep:* **k'elbil** 'toasted'. *See:* **yo'och ahpek'**. *Thes:* **ak'** (Fig. 5.17a, b, c, d, e, f). *[Source:* Nuk; AM; KM*]* *[*\sd2 medicinal*]*

bubuhchi' (ah)$_2$ velvet bean (cowhage, cowitch). *Lit:* 'split mouth'. Fabaceae *Mucuna pruriens* (L.) DC., syn. *Stizolobium pruriens* (L.) Medik. This is a vigourous, herbaceous vine 3 m-15 m (9.8′–49′) in length that bears racemes of waxy, pea-like blossoms. The Lacandon name likely refers to the split corolla, as it is also applied to the somewhat similar blossom of *Canavalia*. Although normally purple, the flower colour of *Mucuna pruriens* can vary (Heike Vibrans, pers. comm.). The fruit is a brown pod covered with numereous, brittle, barbed hairs. Leaves are tripinnate, ovate to rhombus-shaped. Considered *lo'obil* 'a weed'. No uses were reported, but the southern Lacandones are said to eat it (Levy et al. 2006: 86). *Loc:* **paakche' kol** 'fallow milpa'. *Morph:* **bu-buh=chi'**. *See:* **ni' ts'ul**. *Thes:* **ak'** (Fig. 5.17g, h). *[Source:* KM*]*

buk'luch' (äh) vanilla. Orchidaceae *Vanilla planifolia* Jacks. ex Andrews, syn. *Vanilla fragrans* Ames. A native of Mexico, vanilla is a vine that produces clusters of large, smooth, thick-skinned pods approximately 12.7 cm to 15.24 (5–6″) long that contain numerous, miniscule black seeds resembling coarse dust. According to KM and KP, the resin in the pods severely irritates the eyes, so caution must be taken when picking and processing them. Dried pieces are tied onto necklaces, to add fragrance. *Use:* **uuh** 'necklaces'; *Part:* **sool uwich** 'shell of the bean'; *Prep:* **Ahayk'intik uwich, pachil axet'ik uxot'** 'You spread out the fruits in the sun to dry and then break them into small pieces'. *Thes:* **ak'** (Fig. 5.18a, b). *[Note:* buclux. *Vanilla* sp. (Durán 1999); siis-lik', siisb'ik' [Itz.]. *Vanilla fragrans* (Atran et al. 2004); siis b'ik [Yuc]. *Vanilla fragrans* (in, Atran et al. 2004).*] [Source:* BM; KM; KP*]* *[*\sd2 craft*]*

bu'ul [bu'ul, buul, buur] bean. Fabaceae *Phaseolus vulgaris* L. *Use:* **hanal** 'food'; *Part:* **nek'** 'seeds'; *Prep:* **chäkbil, mäxbil i tsähbil** 'boiled, mashed and fried'; *Spec:* **box bu'ul** 'shelling bean'; *Spec:* **ek' bu'ul** 'black bean'; *Spec:* **ek'el bu'ul** 'coloured/spotted bean'; *Spec:* **chäk bu'ul** 'red bean'; *Spec:* **hach bu'ul** 'authentic bean'. *See:* **ak'il bu'ul (ah)**; **che'il bu'ul (ah)**. *[Note:* bu'ul. *Phaseolus vulgaris* (Nations and Nigh 1980): b'u'ul [Itz.]. *Phaseolus vulgaris* (Atran et al. 2004); b'ú'ul [Yuc.]. beans (Bricker et al. 1998: 38); b'u'ul [Mop.]. *Phaseolus vulgaris* (in, Atran et al. 2004).*]* *[*\sd2 food*]*

bu'ulche' [bu'ulche', buulche'] cojoba. *Lit:* 'bean tree'. Fabaceae *Cojoba* spp. **Hay dos clases: nukuch bu'ulche' y mehen bu'ulche'. Nukuch bu'ulche' uche'il 20–30 metros (65.6–98′). Ne mihin ule'. Utop', säk. Uwich, p'iis bu'ul. Ne tah uwich. Yan 6–7 unek'. Chäk usool unek'. Uch'ihil xok'ol ya'ax petha'. Mehen bu'ulche' uch'ihil 15 metros (49′). Unek', ek'. Säsäk utop'. Uch'ihil xok'ol ya'ax petha'.** 'There are two kinds: *nukuch bu'ulche'* 'large bu'ulche' and *mehen bu'ulche'* 'small bu'ulche''. The large *bu'ulche'* grows to 20–30 m (65.6–98′). The leaves are very small, about 0.6 mm (1/4″). The flowers are white. The fruit is like a bean. It is very straight. There are 6–7 seeds. The skin of the seed is red. It grows near the green lagoon. The small *bu'ulche'* grows to 15 m (49′). The seeds are black. The flower is somewhat white.' (BM) *Mehen bu'ulche'* may be *Cojoba arborea* L. (Britton & Rose) or *C. graciliflora* (S. F. Blake) Britton & Rose. *C. arborea* is used for house construction and fuel in the southern Lacandon community (Levy et al. 2006). My consultants reported only using the wood for fuel. The identity of *nukuch bu'ulche'* is undetermined. *Loc:* **ya'ax petha'** 'green lagoon'; *Use:* **che'il k'ak'** 'firewood'; *Spec:* **nukuch bu'ulche'** 'large bu'ulche'; *Spec:* **mehen bu'ulche'** 'small bu'ulche'. *Thes:* **che'**. *[Note:* buche. *Cojoba arborea* (Durán 1999); buul-che [Yuc.]. Lit. bean-tree (Roys 1931: 219); b'ulu(l)che' [Itz.]. *Platymiscium yucatanum* (Atran et al. 2004); b'uluche' [Itz.]. *Swartzia lundellii* (Hofling and Tesucún 1997: 188).*] [Source:* BM*]* *[*\sd2 fuel*]*

Ch–ch

chank'ala' lily, canna (Indian shot, false ginger). Cannaceae *Canna indica* L., syns. *C. edulis* Ker Gawl, *C. lutea* Larrañaga (nom. illeg.). Native to tropical America and the West Indies, *Canna indica* is a medium-sized, spreading shrub 22–30 cm (9–12″) tall and 90 cm-2.4 m (3–8′) wide. Its large, ribbed, boat-shaped leaves are 20.4 cm (8″) wide and 46 cm (18″) long. Scarlet flowers are born in spikes on slender stems, 1 m (3.5′) tall. Flowers are followed by large 3 cm (1″) oval, bristly pods containing numerous small, hard, black, round seeds. Flowers and seed pods occur simultaneously. Plants

Fig. 5.17 (**a**) Bubuhchi' (yo'och ahpek') Canavalia flower. Photo Credit: Suzanne Cook (**b**) Bubuhchi' *Canavalia ensiformis*. Photo Credit: Suzanne Cook (**c**) Bubuhchi' (bu'ul boox) *Canavalia ensiformis*. Photo Credit: Suzanne Cook (**d**) Bubuhchi' *Canavalia ensiformis*. Photo Credit: Suzanne Cook (**e**) Bubuhchi' *Canavalia ensiformis*. Photo Credit: Suzanne Cook (**f**) Bubuhchi' *Canavalia ensiformis*. Photo Credit: Suzanne Cook (**g**) Bubuhchi' *Mucuna puriens*. Photo Credit: Suzanne Cook (**h**) Bubuhchi' *Mucuna puriens*. Photo Credit: Suzanne Cook

Fig. 5.18 (**a**) Buk'luch' *Vanilla planifolia*. Photo Credit: Fpalli (Own work) [CC-BY-SA-3.0 (http://creativecommons.org/licenses/by-sa/3.0)] (**b**) Buk'luch' *Vanilla planifolia* flower. Photo Credit: H. Zell (Own work) [CC-BY-SA-3.0 (http://creativecommons.org/licenses/by-sa/3.0)]

Fig. 5.19 (**a**) Chank'ala' *Canna indica*. Photo Credit: Suzanne Cook (**b**) Chank'ala', bamboo, and petsk'in necklace. Photo Credit: Suzanne Cook (**c**) Chank'ala' flower. Photo Credit: Suzanne Cook

are cultivated for their ornamental appeal and for their seeds. *Loc:* **kolil nah** 'cultivated in house garden'; *Use:* **uuh** 'necklaces'; *Part:* **nek'** 'seeds' (Fig. 5.19a, b, c). *[Note:* ×chank'ala' [Yuc.]. false ginger. *Canna indica* (Bricker et al. 1998); chankalá [Yuc.]. lengua de dragón. *Canna edulis* (Roys 1931); ix chikila' [Itz]. *Canna lutea* Larrañaga (Atran et al. 2004); chiki-lab' [Mop.]. *Canna lutea* Larrañaga (in, Atran et al. 2004).*] [Source:* CKO; KM; KP*]* *[*\sd2 craft*]*

chan mehen bo'oy chamaedorea. *Lit:* 'very small bo'oy'. Arecaceae *Chamaedorea* sp. It is unclear if this is actually a species or simply a small, generic *bo'oy*. No uses were reported. *Loc:* **k'aax** 'forest'; *Sim:* **hach bo'oy** 'authentic bo'oy'; *Sim:* **ch'ibix** (*Chamaedorea* spp.); *Sim:* **säk bo'oy** 'white bo'oy'; *Sim:* **k'än bo'oy** 'yellow bo'oy'; *Sim:* **ch'it bo'oy** (Chamaedorea sp.); *Sim:* **ton bo'oy** (*Chamaedorea* sp.); *Gen:* **bo'oy**. *[Source:* AM*]*

chawah ik (ah) [chawah, chawa'] pepper, chile. *Lit:* 'dangling chile pepper'. Solanaceae *Capsicum frutescens* L. This is a variety of hot chile plant with ellipsoid-conical shaped fruits. There is a white and a green variety. *Loc:* **kolil nah, kol** 'house gardens, milpas'; *Use:* **hanal, ts'ak** 'food, spice'; *Part:* **wich** 'fruit'; *Prep:* **che'che'** 'raw'; *Spec:* **säk chawah ik** 'white chawah'; *Spec:* **ya'ax chawah ik** 'green chawah'; *Gen:* **ik** (Fig. 5.20). *[Note:* chawah. ornamental pepper (Bruce 1976); chaua ic [Yuc.]. *Capsicum* sp. (Bolles 2001); ×chawa'ìik [Yuc.]. dangling chili. *Capsicum annuum* (Bricker et al. 1998). This name was unfamiliar to CKO, BM, BMjr and his mother.*] [Source:* AM; Bruce (1975); CK; KP*]* *[*\sd2 food*]*

chäkche' (äh) poinsettia (Christmas star, Christmas flower). *Lit:* 'red tree'. Euphorbiaceae *Euphorbia pulcherrima* Willd. ex Klotzsch. Although this is the only species that was unanimously called by this name,; other woody plants with red flowers may also be called *chäk che'* 'red tree'. Native to southern Mexico and Guatemala, it is a large shrub or small tree noted for its large, red bracts. It is commonly sold at Christmas and cultivated as a house plant throughout the world. Although it grows wild in the area, the Lacandones also cultivate it as a garden ornamental. *Loc:* **k'aax** 'forest, house garden'; *Loc:* **kolil nah** 'house garden'; *Use:* **tsoy uwich** 'ornamental'. *Morph:* **chäk=che'**. *Thes:* **che'**. *[Note:* chac top che. *Rondeletia capitellata* (Durán 1999); ix paaskuaj [Lac.]. *Euphorbia pulcherrima* (Nations 1992, cited in Atran et al. 2004); u-/ix top'il paaskwaj [Itz.]. *Euphorbia pulcherrima* (Atran et al. 2004).*] [Source:* BM; CKD; KM; KP*] [*\sd2 ornamental*]*

Fig. 5.20 Chawah ik *Capsicum frutescens*. Photo Credit: Chan K'in Daniel

chäkchob [chäkchob, chäkchop'] lantana, threeleaf (lavender popcorn). Verbenaceae *Lantana trifolia* L., syn. *Lantana maxima* Hayek. **Es lo'obil. Da flores como azul. Chan wolis utop', ne mihin. Su fruta chan mihin, como ubas. Come las frutas.** 'It's a weed. It produces bluish-flowers. The flowers are a little bit round and tiny. Its fruit is very small, like grapes. You eat the fruit.' (BM) *Lantana trifolia* is one of 150 species in the genus. Native to the Neotropics and Africa, it is a perennial shrub growing to 2 m (6′) or more in height. Leaves are ovate-lanceolate, with serrated margins and dark green blades with the texture of sandpaper. Three leaves occur at each node. Umbels of small, tubular florets are produced in combinations of pink-lavender with a circle of yellow, blue, or white in the centre. Fruits are clusters of fleshy, pinkish-purple, mildly sweet berries approximately 3.5 mm (0.14″) in diameter. Each contains a single seed. (Hanan Alipi & Mondragón-Pichardo 2009e; MacDonald et al. 2008) The plants grow freely in *acahuales* (fallow cornfields) and house gardens. Fruits are occasionally eaten out of hand. Kashanipour and McGee (2004: 64) report that the leaves are eaten to prevent or cure nervousness and bedwetting. *Use:* **hanal** 'food'; *Part:* **wich** 'fruit'; *Prep:* **che'che'** 'raw'; *Use:* **mal de orín en niños** 'bedwetting (and nervousness)'; *Part:* **le'** 'leaves'; *Prep:* **Ha'li' ahantik ule' che'che'** 'You just eat the leaves raw' (Fig. 5.21a, b). *[Note:* The meaning of *chob* is unclear, but it could refer to the cob-like appearance of the densely-packed berries. *[Note:* x-chak-chòob' [Yuc.]. red corn used for whooping cough (Bricker et al. 1998: 59). chäk chop. *Lantana* spp. (Kashanipour and McGee 2004); chäk chop. *Lantana trifolia* (Nations and Nigh 1980).*] [Source:* AM; BM; KP*] [*\sd2 food*]*

a b

Fig. 5.21 (**a**) Chäkchob *Lantana trifolia*. Photo Credit: Suzanne Cook (**b**) Chäkchob *Lantana trifolia*. Photo Credit: Suzanne Cook

chäklah [chäklah, chäkrah, chäkarah, chäkah] gumbo limbo (red gumbo limbo, tourist tree, West Indian birch). Burseraceae *Bursera simaruba* (L.) Sarg. This is a large tree with peeling red bark and smooth inner bark. Bruce (1975: 138) calls it "madrona tree". Madrona refers to many species of Bursera, e.g., *B. simaruba, B. excelsa,* and *B. bipinnata*, and to *Euphorbia fulva*, a type of spurge (Schoenhals 1988: 84, 205). Medicinal uses have been attributed to the bark to treat illnesses such as stomach aches (Kashanipour and McGee 2004: 61), diabetes (Diemont 2006: 83), boils and burns (Durán 1999), and kidney problems (AM). According to AM, this is not traditional Lacandon medicine, but belongs to the Tzeltales. The Itzaj use the bark to treat a number of ailments, from foot fungus to kidney problems (Atran et al. 2004: 98). Additionally, the Lacandones eat the seeds toasted (Durán 1999) and burn the resin in their incensories (Durán 1999; Nations and Nigh 1980). In Lacandon mythology, the tree was created by *Hachäkyum*, the Lacandones' supreme deity, to help the ancestors escape from the *xutneh ayim*, a kind of mythical, tail-less crocodile that ran about on dry land chasing people to eat them. The slippery bark made it impossible for the crocodiles to climb. But when *Hachäkyum* saw that the women were unable to climb the tree, he gave the crocodiles tails to slow them down (AM). Eventually, *Hachäkyum* interred them at the bottom of a deep cave. *Use:* **yah uch'aay** 'kidney pain'; *Part:* **soolche'** 'bark'; *Prep:* **Achäkik ich ha', pachil awuk'ik 'oxwäts' k'in ti' hunts'it na'** 'You boil it in water and then you drink it three times a day for one month'; *Use:* **chuhul** 'burn'; *Part:* **sool** 'bark'; *Prep:* **Amuxik usool pachil le pulik yok'ol uchuhul** 'You crush the bark and then put it on the burn'; *Use:* **chechem** '*Metopium brownei*'; *Part:* **uyits** 'sap'; *Prep:* **Ha'li' apulik yok'ol uyahil** 'You just put it on the wound'; *Use:* **chibal hämnen** 'stomach ache'; *Part:* **mots** 'roots'; *Prep:* **te'** 'tea'. *Thes:* **che'** (Fig. 5.22). *[Note:* ch'acaj. *Bursera simaruba* (Nigh 2008); *Bursera simaruba* (Nations 2006); chäklah, huhup, tasi pom. *Bursera simaruba* (Nations and Nigh 1980); chac cra che. *Bursera simaruba* (Durán 1999); chäk chäkaj [Itz.]. *Bursera simaruba* (Atran et al. 2004); chacah [Yuc.]. *Bursera simaruba* (Roys 1931). Chechem remedy from Nations (2006). Lacandon translation by author.*]* *[Source:* AM; BM; K'ayum Segundo*]* */*\sd2 medicinal\sd2 folklore_and_mythology*]*

chäklanpix (ah)₁ daisy family. Asteraceae. Refers to both *Clibadium arboreum* and *Schistocarpha eupatorioides*. Each are visited by the butterfly called *ahchäklanpix* (*Heliconius ismenius*). *Loc:* **paakche' kol** 'fallow milpa'. *See:* **ki' bok (äh)**. *[Source:* AM; KM*]*

chäklanpix (ah)₂ *Clibadium arboreum* (common name unknown). Asteraceae *Clibadium arboreum* Donn. Sm., syns. *C. donnell-smithii* J. M. Coult., *C. oligandrum* S. F. Blake, *C. pueblanum* S. F. Blake. This is a tall, highly

Fig. 5.22 Chäklah *Bursera simaruba*. Photo Credit: Suzanne Cook

branched woody shrub with an erect main stem and erect to lax branches. Leaves are ovate, 10–20 cm, attenuate at base, membranaceous, veins keeled underneath; mature leaves are light grey-green on upper surface, pilose, with serrated margins. Petioles are long, sometimes partially winged. Flower heads are cymose (flat-topped cluster) with small, white tubular corollas and peripheral florets. Numerous black anthers extend above the blossoms. *Loc:* **paakche' kol** 'fallow milpa'. *Variant:* **ki' bok (äh)** (Fig. 5.23a). *[Note:* Synonyms from the Catalogue of Life, 2010 Annual Checklist. http://www.catalogueoflife.org/annual-checklist/2010/details/species/id/7095767. Retrieved March 2013.*]* *[Source:* KM*]*

chäklanpix (ah)₃ catalina. Asteraceae *Schistocarpha eupatorioides* (Fenzl) Kuntze. **Es lo'obil. Uyo'och ähchäklanpix.** 'It's a weed. It's food for the *chäklanpix* (*Heliconius ismenius*).' (AM) *Schistocarpha* is distinguished by its pappus, a modified calyx made up of a ring of fine, capillary bristles that aid the wind dispersal of the fruit. *Schistocarpha bicolor* and *Schistocarpha longeligula* have also been recorded from Naha' (Comisión Nacional de Áreas Naturales Protegidas 2006). *Loc:* **paakche' kol** 'fallow milpa'. *See:* **ki' bok (äh)** (Fig. 5.23b, c). *[Note:* Common name from the

Fig. 5.23 (**a**) Chäklanpix *Clibadium arboreum*. Photo Credit: Suzanne Cook (**b**) Chäklanpix *Schistocarpha eupatorioides*. Photo Credit: Suzanne Cook (**c**) Chäklanpix *Schistocarpha eupatorioides*. Photo Credit: Suzanne Cook

Smithsonian Tropical Research Institute http://biogeodb.stri.si.edu/biodiversity/species/22762/] *[Source:* AM; KM; KP*]*

chäk 'abäl mombin, red (hog plum). Anacardiaceae *Spondias purpurea* L. Although the species was unavailable to inspect, it likely refers to *Spondias purpurea*, a shrub or small, low-branched tree, up to 12 m (39.4′), that bears pinkish-red flowers and bright purplish-red or yellow ovoid fruits on short stalks along the branches. The sapwood is whitish and the heartwood is soft and brittle, rendering the timber unsuitable for most construction purposes (Orwa et al. 2009). *Spondias purpurea* ranges from sea-level to 1700–1800 m (5500–6000′) in Mexico and Central America (Morton 1987). It is a component of the dominant stratum of low deciduous forest, growing in a wide variety of soils (Pennington and Sarukhán 2005: 330). Cultivated. *Use:* **hanal** 'food'; *Sim:* **k'än 'abäl** 'yellow mombin (*Spondias purpurea* forma *lutea*)'; *Sim:* **ya'ax 'abäl** 'green mombin (*Spondias* sp.)'; *Gen:* **'abäl** 'hog plum (*Spondias* spp.)'. *See:* **k'än 'abäl**. *[Note:* ajchäk 'ab'äl [Itz.]. red hogplum. *Spondias* sp. (Hofling and Tesucún 1997); ix joob'oj abäl (aj chäk, aj k'än) [Itz.]. *Spondias purpurea* (Atran et al. 2004).*]* *[Source:* Baer and Baer (1952)*]* *[*\sd2 food*]*

chäk ak' *Lit:* 'red vine'. Smilacaceae *Smilax* sp. (?). This is a vigorous climber with thin, reddish stems armed with sharp, straight thorns. Kashanipour and McGee (2004) report that the bark is used to relieve insect bites. *Loc:* **paakche' kol** 'fallow milpa'; *Use:* **Uyoomtik käkow yetel** 'One froths cacao with it'; *Part:* **yits uyol** 'resin from the growing tips';

Prep: **Ubäxik uyol, yäts'ik uyits, pachil uhäxik yetel käkow** 'One pounds the growing tips, squeezes out the resin, and then whips it with cacao'; *Use:* **chibal** 'insect bites'; *Part:* **sool** 'bark'; *Prep:* **Mäna'** 'None'; *Sim:* **koke' ak'** 'sarsaparilla (*Smilax* sp.)'. *Variant:* **chäk nikte' ak'**. *Thes:* **ak'**. *[Source:* AM; BM*]* *[*\sd2 food*]*

chäk boox$_2$ banana, Cuban red (Jamaican red, colorado, red dacca). *Lit:* 'red banana'. Musaceae *Musa acuminata* ×*Musa balbisiana* Coll. This is a tall variety with fat, red-skinned fruits and sweet, cream-coloured to yellow-orange flesh. It is cultivated throughout the West Indies, Mexico, and Central America. *Use:* **hanal** 'fruit'; *Part:* **wich** 'fruit'; *Prep:* **che'che'** 'raw'; *Sim:* **hach boox** 'authentic boox'; *Gen:* **boox**. *[Note:* May correspond to *aj chäk b'oxja'as* [Itz.]. red guineo morado (Atran et al. 2004). KM referred to a stubby, yellow-skinned variety as *chäk boox*, saying that it was called *chäk* because the leaves were reddish. The variety she identified had yellow-orange flesh.*]* *[Source:* BM; CNK; KM; KP*]* *[*\sd2 food*]*

chäk box *Lit:* 'red bean'. Fabaceae *Phaseolus vulgaris* L. *Use:* **hanal** 'food'; *Part:* **nek'** 'seeds'; *Prep:* **chäkbil, tsähbil** 'boiled, fried'; *Sim:* **hach box** 'authentic box'; *Gen:* **box**. *Thes:* **bu'ul**. *[Source:* AM*]* *[*\sd2 food*]*

chäk bu'ul *Lit:* 'red bean'. Fabaceae *Phaseolus vulgaris* L. *Use:* **hanal** 'food'; *Part:* **wich** 'fruit'; *Prep:* **chäkbil, tsähbil** 'boiled, fried'; *Spec:* **box bu'ul** 'black bean'; *Spec:* **ek' bu'ul** 'black bean'; *Spec:* **ek'el bu'ul** 'coloured/spotted bean'; *Spec:* **hach bu'ul** 'authentic bean'; *Gen:* **bu'ul**. *[Note:* AM refers to the beans of *Vigna umbellata* as *chäk bu'ul* also.*]* *[*\sd2 food*]*

chäk che'il bu'ul bush beans. *Lit:* 'red bush beans'. Fabaceae *Phaseolus vulgaris* L. *Use:* **hanal** 'food'; *Part:* **wich** 'fruit'; *Prep:* **chäkbil, tsähbil yetel säk seboya yetel axux** 'boiled, fried with onions and garlic'; *Sim:* **ek' che'il bu'ul** 'red bush beans'; *Gen:* **che'il bu'ul** 'bush beans'; *Ant:* **ak'il bu'ul (ah)** 'pole beans'. *Thes:* **bu'ul**. *[*\sd2 food*]*

chäk chinah orange, mandarin. *Lit:* 'red orange'. Rutaceae *Citrus reticulata* Blanco. In Naha', fruit matures in August and again in mid to late November. Introduced. *Loc:* **kolil nah** 'house gardens'; *Use:* **hanal, uk'ul** 'food, beverage'; *Part:* **wich** 'fruit'; *Prep:* **che'che' wa ayäts'ik uk'ab** 'raw or you squeeze out its juice'; *Gen:* **chinah** 'orange'. *Thes:* **che'**. *[Source:* CHX; CKD; KM; KP; NI*]* *[*\sd2 food*]*

chäk halol heliocarpus (sunfruit). *Lit:* 'red halol'. Malvaceae (Tiliaceae) *Heliocarpus appendiculatus* Turcz. This is a small or medium-sized tree 6–22 m (20–75′) in height. Leaves are broadly ovate with toothed margins, palmate venation (with five, conspicuous veins), and two, leafy auricles (small lobes) at the base. Small, greenish-yellow flowers in panicles are followed by 3 mm (0.11″), bristly red, slightly compressed pods. The bark is light grey-green and studded with numerous lenticels. The inner bark is fibrous and the wood is light-weight. (Allen, P. H. 1977:221, Pennington and Sarukhán 2005: 350). *Loc:* **paakche' kol** 'old fallow milpa'; *Use:* **k'an** 'cord'; *Part:* **uyoot'el** 'inner bark'; *Sim:* **säk halol** 'white halol (*Heliocarpus donnell-smithii*)'; *Sim:* **hach halol** 'authentic halol'; *Gen:* **halol**. *Thes:* **che'** (Fig. 5.24a, b). *[Source:* AM; BM; CKD*]* *[*\sd2 cordage*]*

chäk hut'kih morning glory. *Lit:* 'red hut'kih'. Convolvulaceae *Ipomoea* sp. This is a kind of morning glory

Fig. 5.24 (**a**) Chäk halol *Heliocarpus appendiculatus*. Photo Credit: Chan K'in Daniel (**b**) Chäk halol *Heliocarpus appendiculatus*. Photo Credit: Chan K'in Daniel

with reddish-purple, twining stems textured with small, dull thorns and pandurate (fiddle-shaped) leaves with rear-facing basal lobes. The vines ramble through lawns, over bushes, and up trees. Although BM said it was a distinct kind of *hut'kih*, AM said that it is the same plant that he calls *hach hut'kih* (*Ipomoea alba*). When BM pointed out that the leaves of *chäk hut'kih* were lobed and those of *hach hut'kih* were not, AM said that the leaves change their shape over time. As there were no blossoms on the observed specimen, it is impossible to tell whether or not it was *Ipomoea alba*. *Use:* **ti' ukinsik k'ik'** 'to kill latex (a vulcanizing agent)'; *Part:* **uyits uyol i mumuun uyak'il** 'the resin from the growing tips and tender, young vines'; *Prep:* **Upets'ik uyak'il i uyäts'ik uyits ich ha', pachil upulik k'ik' yetel i upeeksik** 'One crushes the vines and squeezes the resin into water, then puts in the latex from *Castilla elastica* and moves it around'; *Sim:* **hach hut'kih** 'authentic hut'kih (*Ipomoea alba*)'; *Sim:* **nukuch hut'kih** 'large hut'kih'; *Gen:* **hut'kih**. *Thes:* **ak'**. *[Note:* Botanical identification provided by John Mochrie MacDougal (pers. comm.).*] [Source:* AM; BM*] [*\sd2 other*]*

chäk 'ib bean, lima. *Lit:* 'red lima bean'. Fabaceae *Phaseolus lunatus* L. cv. 'Sieva'. A red variety of the sieva bean that is larger than *ek' 'ib* 'black sieva bean'. *Use:* **hanal** 'food'; *Part:* **nek'** 'seeds'; *Prep:* **chäkbil** 'boiled'; *Sim:* **ek' 'ib** 'black lima bean'; *Gen:* **'ib** 'lima bean'. *Thes:* **bu'ul**. *[Note:* b'ox ib', säk ib', chäk ib' [Itz.]. *Phaseolus lunatus* (Atran et al. 2004).*] [Source:* BM*] [*\sd2 food*]*

chäk ik chile pepper, red. Solanaceae *Capsicum annuum* L. *Use:* **hanal** 'food'; *Part:* **wich** 'fruit'; *Prep:* **che'che'** 'raw'; *Gen:* **ik**. *[*\sd2 food*]*

chäk 'is potato, sweet. *Lit:* 'red sweet potato'. Convolvulaceae *Ipomoea batatas* (L.) Lam. **Chäk upach usool**. This is a variety of sweet potato with red skin. *Use:* **hanal** 'food'; *Part:* **wi'** 'tuber'; *Prep:* **chäkbil, pokbil** 'boiled, roasted'; *Sim:* **säk 'is** 'white sweet potato'; *Sim:* **k'än 'is** 'yellow sweet potato'; *Gen:* **'is**. *Variant:* **chäk pach 'is**. *[Note:* chäk is. *Dioscorea* spp. yam (Nations and Nigh 1980).*] [Source:* AM; BM*] [*\sd2 food*]*

chäk k'o'och (äh) bean, castor (red). Euphorbiaceae *Ricinus communis* L. This is an evergreen, herbaceous or semi-woody tree with large, glossy reddish-green palmate leaves with deeply incised lobes and long, reddish leaf stems. It bears small greenish-yellow flowers in fat, 20.3–45.7 cm (8–18″) tall spikes near the tops of the stems. Flowers are followed by soft, oblong, reddish-brown fruit and bristly seed pods approximately 2.54 cm (1″) long. They split into three one-seeded parts. The smooth, glossy, black seeds are deadly poisonous. Although no uses were reported to me, Nations and Nigh (1980) claim that the tree is used medicinally. Levy et al. (2006) indicate various uses for this tree by the southern Lacandones, but the authors provide no details. *Sim:* **ya'ax k'o'och** 'green k'o'och'; *Gen:* **k'o'och (äh)**. *See:* **k'o'och (äh)**$_1$. *Thes:* **che'** (Fig. 5.25).

chäk mäkäl elephant ears (malanga, Caribbean cabbage, purple stem taro). Araceae *Xanthosoma mafaffa* Schott. *Use:* **hanal** 'food'; *Part:* **wi'** 'tuber'; *Prep:* **chäkbil** 'boiled'; *Part:* **le', ne muun** 'young, unfurled leaves'; *Prep:* **chäkbil** 'boiled'; *Part:* **wi'** 'tubers'; *Prep:* **chäkbil** 'boiled'; *Sim:* **säk mäkäl** 'white macal (*Xanthosoma mafaffa*)'; *Gen:* **hach mäkäl** 'authentic macal (*Xanthosoma mafaffa*)'. *See:* **mäkäl**. *Thes:* **xiiw** (Fig. 5.26). *[Note:* Taxon identification provided by Eduardo Gomes Goncalves, Depto. de Botânica, ICB, UFMG, Brazil (pers. comm.)*] [Source:* AM; BM; CKD*] [*\sd2 food*]*

chäk me'exk'in (äh) caliandra. *Lit:* 'red sun whiskers'. Fabaceae *Calliandra calothyrsus* Meisn. BM said that this is what his father, Chan K'in Viejo, called this verdant shrub

Fig. 5.25 Chäk k'o'och *Ricinus communis*. Photo Credit: Suzanne Cook

Fig. 5.26 Chäk mäkäl *Xanthosoma mafaffa*. Photo Credit: Suzanne Cook

with fine, fern-like leaves and red, wispy, pompom flowers. It produces flat seed pods about 10 cm (4″) long by 2.54 (1″) wide. The pods are dehiscent, splitting along the length of the seed joining the front and back halves of the pod. *Calliandra* is a genus that comprises 132 species primarily from Central and South America. They are usually shrubs or small trees, although some may become large trees. The genus is distinguished by the apical dehiscence of its pods and by flowers that have numerous, long filaments joined at the base. The flowers emerge in either axillary heads or terminal racemes. It is a common roadside weed occurring throughout humid areas in most parts of Mexico (Fig. 5.27). *[Note:* Other Lacandones whom I talked to were unable to provide a name for this species. According to Levy et al. (2006), *Calliandra calothyrsus* is an agricultural "indicator plant" for the southern Lacandones.*] [Source:* AM; BM; CKViejo*]*

Fig. 5.27 Chäk me'exk'in *Calliandra calothyrsus*. Photo Credit: Suzanne Cook

chäk nikte' frangipani (pagoda tree, plumeria, temple tree, West Indian jasmine). *Lit:* 'red blossom tree'. Apocynaceae *Plumeria rubra* f. *rubra*. This is a medium-sized, evergreen tree bearing fragrant, rose-coloured blossoms. It grows wild but is also cultivated as an ornamental. According to Tozzer (1907: 93), *Plumeria rubra* was the birth father of *nohochakyum (Hachäkyum)*, the Lacandones' supreme deity, but Bruce and my northern Lacandon consultants say that the birth flower of the principal deities was *baknikte' (Polianthus tuberosa)*. According to AM, the blossoms of the *chäk nikte'* are the perfume of the gods. *Sim:* **säk nikte'** 'white nikte' (*Plumeria rubra*)'; *Gen:* **nikte' (ah)**. *Thes:* **che'** (Fig. 5.28a, b). *[Note:* Botanical name from Pennington and Sarukhán (2005: 450). chak-nikte. *Plumeria rubra* (Nations 1992, in Atran et al. 2004); chak nicté [Yuc.]. *Plumeria rubra* (Roys 1931); chak-nik-te' [Yuc.]. frangipani. *Plumeria rubra* (Bricker et al. 1998); aj/ix chäk nikte' [Itz.]. *Plumeria rubra* (Atran et al. 2004).*] [Source:* AM; BM*] [*\sd2 other*]*

chäk 'ooh reed, common (giant cane). Poaceae *Gynerium sagittatum* (Aubl.) P. Beauv. **Chäk 'ooh uch'ihil ti' uxuul uk'um. Ne yi'h uyol, chäk uyol. Uyi'ih yan habanil. Uch'ihil tuhunal.** '*Chäk 'ooh* 'red 'ooh' grows at the edge of rivers. The stalk is hard and red. The flower spike is hollow. It grows on its own.' (BM) *Gynerium sagittatum* is a Neotropical reed that grows on river banks. Although it has numerous synonyms, recent findings show that the species is highly variable in the western parts of South America. Some botanists regard it as a species complex, whereas others (Clayton and Renvoize 1986) regard it as a single species. Kalliola and Renvoize (1994: 305) adopt the second perspective and have discovered two new varieties. *Loc:* **yaha' uk'um, Monte Líbano** 'the great river in Monte Líbano'; *Sim:* **hach 'ooh** 'authentic 'ooh (*Gynerium sagittatum*)'; *Gen:* **'ooh**. *[Source:* BM*]*

Fig. 5.28 (**a**) Chäk nikte' *Plumeria rubra f. rubra*. Photo Credit: Suzanne Cook (**b**) Chäk nikte' *Plumeria rubra* f. *rubra* blossom. Photo Credit: Suzanne Cook

chäk 'oon avocado, American (alligator pear). *Lit:* 'red avocado'. Lauraceae *Persea americana* Mill., syn. *Persea gratissima* Gaertn. *Use:* **hanal** 'food'; *Prep:* **mäk'bil che'che' yetel mulix** 'eaten raw with lime'; *Sim:* **ya'ax 'oon** 'green avocado'; *Gen:* **'oon** '*Persea americana*'. *Thes:* **che'**. *[*\sd2 food*]*

chäk 'oop bullock's heart (custard apple, Jamaican apple, netted custard apple, sugar apple). *Lit:* 'red custard apple'. Annonaceae *Annona reticulata* L. This is a low tree with an open, irregular crown and slender, smooth leaves. In some varieties, the leaves are long and narrow, 10–20 cm (3.9–7.8″) long and 2–7 cm (8–2.7″) wide, whereas in others, they are wrinkled and up to 10 cm (3.9″) wide (Mahdeem 1994). The fruit is heart-shaped, lopsided, or spherical, 8–15 cm (3–5.9″) in diameter, with thin, tough skin that may be faintly, moderately, or distinctly reticulated and yellow or brownish with a pink, reddish or brownish-red blush (Orwa et al. 2009: 1). The white flesh has a custard-like, somewhat granular texture. *Loc:* **kolil nah, kol** 'house gardens and milpas'; *Use:* **hanal** 'food'; *Part:* **wich** 'fruit'; *Prep:* **che'che'** 'raw'; *Sim:* **säk 'oop** 'white custard apple'; *Gen:* **hach 'oop** 'authentic custard apple'. *Variant:* **chäk uwich**. *Thes:* **che'**. *[Note:* chäk 'oop [Itz.]. *Annona reticulata* (Atran et al. 2004); chak óop [Yuc.]. *Annona purpurea* (in, Atran et al. 2004).*]* *[Source:* AM*]* *[*\sd2 food*]*

chäk 'oop che' *Clarisia biflora* (common name unknown). *Lit:* 'red custard apple tree'. Moraceae *Clarisia biflora* Ruiz & Pav. This is a tree that, when slashed, exudes copious, white latex. According to BM, it bears fruit similar to *hach 'oox* 'Mayan breadnut (*Brosimum alicastrum*)'; it is, approximately (2.54 cm) 1″ in diameter and red. Because the fruit exudes a white latex it is distasteful. *Use:* **ti' ubanik k'uch** 'to paint cord (for bow strings and arrow binding)'; *Part:* **uyits usool** 'resin from the bark'; *Prep:* **Uxa'ak'tik uyits yetel pom i kibil kab** 'One mixes the resin with copal and beeswax'; *Use:* **che'il k'ak' ti' uk'ak'tik ulekil k'uh** 'firewood for firing the godpots'; *Part:* **usool** 'bark'; *Use:* **ti' ubanik unook'** 'for dying fabric'; *Part:* **wich** 'fruit'; *Prep:* **Ha'li' ach'amik uwich ti' nook'** 'You just press the [cut] fruit on the cloth [like a stamp]'; *Sim:* **hach 'oop che'** 'authentic custard apple tree (*Annona reticulata, A. cherimoya*)?'; *Gen:* **'oop che'**. *Thes:* **che'**. *[Note:* chak' opche. *Clarisa biflora* (Nigh 2008).*]* *[Source:* BM; K'ayum Segundo; SK*]* *[*\sd2 fuel \sd2 craft*]*

chäk pach *Ternstroemia tepezapote* (common name unknown). *Lit:* 'red back'. Theaceae *Ternstroemia tepezapote* Schltdl. & Cham. This is a small, evergreen tree that grows to 20 m (65′) tall with a densely, branching habit. Leaves are simple, elliptic, and taper to a point at the apex. The texture is smooth and leathery with a prominent midrib, keeled on the underside, and inconspicuous, secondary venation. Leaves are arranged in whorls at the ends of branches. Flowers are white or pink and tipped with yellow. Blossoms are borne individually on long pedicels (flower stems) at leaf axils. Fruit is a round capsule, 1.8 cm (0.7″) in diameter, that tapers to a point. The fruit has a thick skin covered with small, pinkish-red blisters. Seeds are numerous, oblong, slightly flattened, and red with a powdery surface. In Naha', the fruit matures in October. The fruit is used to make a reddish-orange, impermanent dye (AM). According to Durán (1999), the northern Lacandones also use the resin to relieve mosquito bites. *Loc:* **petha'** 'lagoon'; *Use:* **ti' ubanik unook'** 'used to paint cloth'; *Part:* **wich** 'fruit'; *Prep:* **mäna'** 'none'; *Use:* **chibal kan** 'snakebite'; *Part:* **sool** 'bark'; *Prep:* **Achäkik ich ha', apulik yok'ol chibal** 'You boil it in water and put it on the bite'. *Thes:* **che'** (Fig. 5.29a, b). *[Note:* chak pach. *Ternstroemia tepezapote* (Durán 1999).*]* *[Source:* AM; BM; KP; SK*]* *[*\sd2 craft \sd2 medicinal*]*

a b

Fig. 5.29 (**a**) Chäk pach *Ternstroemia tepezapote*. Photo Credit: Suzanne Cook (**b**) Chäk pach *Ternstroemia tepezapote*. Photo Credit: Suzanne Cook

Fig. 5.30 (**a**) Chäk ts'in *Manihot esculenta*. Photo Credit: Suzanne Cook (**b**) Chäk ts'in *Manihot esculenta* roots. Photo Credit: Suzanne Cook

chäk pach 'is potato, sweet. *Lit:* 'red-skinned sweet potato'. Convolvulaceae *Ipomoea batatas* (L.) Lam. *Use:* **hanal** 'food'; *Part:* **wi'** 'tuber'; *Prep:* **pokbil, chäkbil** 'roasted, boiled'. *Variant:* **chäk 'is**. *[Source:* BM*]* *[*\sd2 food*]*

chäk pichik' guava. Myrtaceae *Psidium guajava* L. A variety of guava with deep pink pulp. Cultivated. *Use:* **hanal** 'food'; *Part:* **wich** 'fruit'; *Prep:* **che'che'** 'raw'; *Sim:* **säk pichik'** 'white pichik' (*Psidium guajava*)'; *Gen:* **pichik'**. *Thes:* **che'**. *[Source:* AM; BM*]* *[*\sd2 food*]*

chäk sukal sugarcane. *Lit:* 'red sugarcane'. Poaceae *Saccharum officinarum* L. *Use:* **hanal** 'food'; *Part:* **che'il** 'stalk'; *Prep:* **che'che'** 'raw'; *Sim:* **hach sukal** 'authentic sugarcane'; *Sim:* **ts'ibal sukal** 'striped sugarcane'; *Sim:* **mehen sukal** 'small sugarcane'; *Gen:* **sukal**. *[*\sd2 food*]*

chäk taw Jamaican nettletree. Ulmaceae *Trema micrantha* (L.) Blume. This is a shrub or small tree, 2–5.5(–10) m, with dark brown bark that is smooth when young but warty when mature. The flowers display a greenish-white calyx. They are followed by tiny, bright red-orange to yellow fruit, 1.5–3.5 mm (0.06–0.14″) in diameter. It is a pioneer tree that grows in hammocks, burned areas, along roadsides, canyons, near streams, in pastures, forest edges, and agricultural land. (Vázquez-Yanes et al. 1999). Although my Lacandon consultants reported only using the bark for its fibre, Nations and Nigh (1980) report that the Lacandones also use the timber for building materials. *Use:* **bilich** 'mattress'; *Part:* **uyoot'el** 'inner bark'; *Use:* **k'an** 'cord'; *Part:* **uyoot'el** 'inner bark'; *Sim:* **hach taw** 'authentic taw (*Belotia mexicana*)'; *Gen:* **taw**. *Thes:* **che'**. *[Note:* chak tao. balsa. *Trema micanthra* (Nations and Nigh 1980); chait pach. *Trema micanthra* (Durán 1999); u-pujan-il k'aax [Itz.]. *Trema micanthra* (Atran et al. 2004); sak pixoy [Yuc.]. *Trema micanthra* (in, Atran et al. 2004).*]* *[Source:* BM; SK*]* *[*\sd2 cordage \sd2 furniture*]*

chäk tsuk'in ak' smilax. Smilacaceae *Smilax* sp. This is a kind of herbaceous, thorny vine. *Use:* **ti' uyoomtik käkow** 'for frothing cocoa'; *Sim:* **ya'ax tsuk'in ak'** 'green tsuk'in ak''; *Gen:* **tsuk'in ak' (ah)**. *Thes:* **ak'**. *[Source:* BM*]* *[*\sd2 other*]*

chäk ts'in manioc (cassava). *Lit:* 'red manioc'. Euphorbiaceae *Manihot esculenta* Crantz. This is a variety of manioc that is similar to *hach ts'in* (= *säk ts'in*), except for its red petioles and tubers. *Use:* **hanal** 'food'; *Part:* **wi'** 'tuber'; *Prep:* **pokbil, chäkbil yetel ch'uhuk** 'roasted or boiled with sugar'; *Sim:* **hach ts'in** 'authentic manioc'; *Sim:* **yahche' ts'in** 'great tree manioc'; *Gen:* **ts'in** 'manioc' (Fig. 5.30a, b). *[*\sd2 food*]*

chäk ts'ulha' (äh) lily, giant spider. *Lit:* 'red water dove'. Amaryllidaceae *Crinum amabile* Donn ex Kew Gawler, syn. *Crinum asiaticum* L. This is a large, upright plant, 1 m (3′) tall with large, intensely fragrant flowers. The petals are long, thin, and slightly outward-curled, and striped in magenta and pink. Leaves are basal, long, and strap-shaped. It grows on dry land. Cultivated. *Use:* **tsoy uwich** 'ornamental'; *Sim:* **nukuch ts'ulha' (äh)**; *Sim:* **ts'ulha'il petha' (äh)**; *Gen:* **ts'ulha' (äh)**. *Variant:* **ts'ulil** (Fig. 5.31). *[Source:* AM; BM; BM's wife; CKD; CNK; KP*]* *[*\sd2 other*]*

chäk uwich (äh) apple, custard. *Lit:* 'red fruit'. Annonaceae *Annona reticulata* L. *Variant:* **chäk 'oop**. *Thes:* **che'**. *[Source:* KM*]* *[*\sd2 food*]*

chäk wawal [chäk wawal, chäk wawar, chäk wowo, chäk bobo] *Lit:* 'many red fans'. Marantaceae *Calathea* sp. *Use:* **yah upix** 'knee pain'; *Part:* **le'** 'leaves'; *Prep:* **Apulik**

Fig. 5.31 Chäk ts'ulha' *Crinum amabile*. Photo Credit: Chan K'in Daniel

ule' yok'ol upix 'You put the leaves on the knee'; *Sim:* **säk wawal** 'white wawal (*Calathea lutea*)'; *Sim:* **its'in säk wawal** 'little brother of white wawal (*Calathea crotalifera*)'; *Gen:* **wawal**. *See:* **muxan**. *Thes:* **xiiw**. *[Note:* Although the consultant failed to mention how the leaves are prepared, conceivably they are seared or boiled.*] [Source:* AM*] [*\sd2 medicinal*]*

chäk ya' chicle (chicle-tree, beef apple, níspero, sapotillo, sapodilla plum). *Lit:* 'red chicle'. Sapotaceae *Manilkara achras* (Mill.) Fosberg, syn. *Manilkara zapota* (L.) van Royen. *Use:* **puustah** 'wood shavings (for starting a fire)'; *Use:* **hälälte'** 'weaving sticks'; *Use:* **but'u'o'och** 'shuttle'; *Use:* **uyo'och tu' kubut'ik** 'beater'; *Use:* **xuunche'** 'warp end rods'; *Use:* **che'il tu' kuwätik** 'warping board'; *Use:* **säkal** 'loom'; *Use:* **chulul** 'bow'; *Use:* **k'ek' che'** 'barbed foreshaft'; *Use:* **k'ek' che', mulinuh** 'cocoa beater'; *Use:* **okman** 'pillars'; *Sim:* **säk ya'** 'caimitillo (*Chrysophyllum mexicanum*)'; *Sim:* **hach ya'** 'sapodilla (*Manilkara achras*)'; *Gen:* **ya'**. *See:* **hach ya'**. *Thes:* **che'**. *[Note:* chakia. *Dipholis salicifolia* (Durán 1999); chäk ya. red zapote. *Manilkara achras* (Nations and Nigh 1980); chäk ya'. *Manilkara achras* (Rätsch 1985a, b); chac-ya [Yuc.]. red zapote (Roys 1931); chäk ya', säk ya' [Itz.]. *Manilkara achras* (Atran et al. 2004).*] [Source:* AM; BM; KM; SK*] [*\sd2 fuel\sd housing\sd2 tools*]*

chäk'ax [chäk'ax, chähk'ax] Brasil blanco. *Lit:* 'red forest (?)'. Rubiaceae *Simira salvadorensis* (Standl.) Steyerm. *Loc:* **petha'** 'lagoon'; *Loc:* **Mensäbäk**; *Loc:* **Sos** 'Lacandón'; *Use:* **ti' ubanik uchäk hu'un, chäk** 'for painting ceremonial headbands red'; *Part:* **usool** 'bark'; *Use:* **ti' ut'äbik k'ak'** 'for starting fire'; *Part:* **che'il** 'wood'. *Morph:* **chäk-k'aax (?)**. *Thes:* **che'**. *[Note:* According to AM, the dye produced from this species is inferior to that produced from *puuna'* 'mahogany' (*Swietenia macrophylla*). chäk'ax. logwood (Bruce 1975); chak'ax. *Simira salvadorensis* (Diemont 2006); cháak(−k'aax) [Yuc.]. wild starch. *Maranta arundinacea* (Bricker et al. 1998). Common name from Tropicos, http://tropicos.org/Name/27906951?projectid=11. Retrieved May 2013.*] [Source:* AM; BM; KM; KP*] [*\sd2 craft \sd2 fuel*]*

chechem (äh)$_1$ [chechem, chechen] poisonwood tree. Anacardiaceae *Metopium brownei* (Jacq.) Urb. This tree is shunned because of its intensely poisonous properties (Standley and Steyermark 1949: 163). The bark oozes a milky white sap that contains urushiol, contact with which causes severe dermatitis and blindness, if gotten into the eyes. The Lacandones tell the story of a man in the village that, after losing his wife, became so despondent that he committed suicide by dousing himself with the sap. Another story Nations (2006) recounts is about a Lacandon man who murdered a rival by lacing his *posol* with the sap. To counteract the effects of the sap Lacandones immediately flush the area with the resin from the *pahsa' ak'* (*Costus pulverulentus*). Nations (2006: 101) reports that they also use the bark of the *Bursera simaruba*, to treat the burns caused by the sap. In the past, Lacandones used the resinous bark as fish poison. The smoke is caustic as well, although the hard wood is a choice fuel. Baer and Merrifield (1971: 153) report that poisonwood is one of the species used to burn the shells of river snails for lime, as the wood produces a hot flame and leaves very little ash. *Use:* **ti' ukinsik käy** 'to kill fish'; *Part:* **sool** 'bark'; *Prep:* **Umuxik usool, pachil uwahtik seis litros yok'ol ha'** 'One crushes the bark and then spreads six litres of it on the water'; *Use:* **che'il k'ak'** 'firewood'. *Thes:* **che'** (Fig. 5.32a, b). *[Note:* chechen. *Metopium brownei* (Nations and Nigh 1980); chéechem [Yuc.]. *Metopium brownei* (Bricker et al. 1998); chechem [Yuc.]. *Rhus tadicans* L., *Metopium brownei* (Roys 1931); chechem, b'ox chechem [Itz.]. *Metopium brownei*, säk chechem. *Sebastiania longicusis* (Atran et al. 2004); chechem., chechen [Itz.]. poisonwood tree, chechen tree. *Sebastiania* spp., *Metopium* sp., (Hofling and Tesucún 1997).*] [Source:* AM; BM; CKO; KyG; SK*] [*\sd2 other \sd2 fuel*]*

che'il bu'ul (ah) [che'i bu(')ul, che'i bu(')ur] beans, bush. *Lit:* 'stalk of beans'. Fabaceae *Phaseolus vulgaris* L. There is a black and a red variety. They have pole bean counterparts. *Use:* **hanal** 'food'; *Part:* **nek'** 'seeds'; *Prep:* **chäkbil wa tsähbil yetel seboya wa ich nahwah wa mukbil wah** 'boiled, fried with onions, put in ceremonial tamales, or baked in a large tamale'; *Spec:* **ek' che'il bu'ul** 'black bush bean'; *Spec:* **chäk che'il bu'ul** 'red bush bean'; *Ant:* **ak'il bu'ul (ah)** 'climbing beans (pole beans)'. *Thes:* **bu'ul**. *[Source:* BM*] [*\sd2 food*]*

che' chäk'an [che' chäk'an, che' chäk'äm] palo blanco 'white stick'. *Lit:* 'savanna tree'. Asteraceae *Vernonanthura patens* (Kunth) H. Rob., syn. *Vernonia patens* Kunth. **Che' ti' paakche' kol. Ak'anik ti' uhulu'che' uho'ol.** 'It's a tree

a b

Fig. 5.32 (**a**) Chechem *Metopium brownei*. Photo Credit: Suzanne Cook (**b**) Chechem *Metopium brownei*. Photo Credit: Suzanne Cook

of fallow milpas. You use it for the horizontal poles that insert through the ridgepole.' (BM) According to AM, the tree takes a long time to grow, thus making the soil around it hard as clay. The tree AM describes could be the *chich lu'um che'* 'hard earth tree' recorded by Rätsch (1985a: 150). I was shown two different species: one had opposite, decussant (arranged in pairs at right angles), lanceolate leaves, and stipules, and the other had alternate leaves, a straight, slender trunk with light beige, smooth bark, and a pagoda-like branching pattern. Although flowers were absent, the latter resembled *Vernonanthura patens*. *Loc:* **paakche' kol** 'fallow milpa'; *Use:* **säkche'il** 'rafters'; *Use:* **hiil** 'battens'; *Use:* **uhulu'che' uho'ol** 'horizontal poles that insert through the ridgepole to hold down the thatch roof'; *Part:* **uk'äbche', mehen che'** 'branches, saplings'. *See:* **chäk'an**. *Thes:* **che'**. *[Note:* chakan [Yuc.]. savannah (Roys 1931); chäk'an [Itz.]. sabana. savannah (Hofling and Tesucún 1997).*] [Source:* AM; BM; KA; K'ayum Segundo; SK*] [*\sd2 housing*]*

chiah (ah) dumbcane. Araceae *Dieffenbachia* sp. **Ne ki' utop'. Chan mihin utop'. Yan usool, säk, wolis.** 'The flowers are very tasty. The flowers are small. It has a round, white skin (spathe).' (AM) *Dieffenbachia* is a genus of approximately 30 species of tropical, evergreen, terrestrial plants from the Neotropics. They are usually found growing in river valleys and on steep slopes. They grow 1.5 m (5′) or taller. They are noted for their white-spattered leaves. All parts of the plant are considered poisonous if ingested, although AM says he sprinkles the pollen in his beverages to add fragrance. Cultivated. *Loc:* **kolil nah** 'house garden'; *Spec:* **hach chiah (ah)** 'authentic chiah (*Philodendron inaequilaterum*)' (Fig. 5.33). *[Note:* Botanical identification provided by Tom Croat, pers. comm.*] [Source:* AM*]*

chi'kam [chi'kam, chi'kan] jicama (yam bean, Mexican potato). Fabaceae *Pachyrhizus erosus* L. Urb. Jicama is a perennial vine native to Mexico, and Central and South America. It is cultivated for its large, crisp tubers. The rest of the plant is poisonous, particularly the fruit, a bean whose seeds contain rotenone (Duke 2008: 495): it is used as a fish poison by many indigenous groups, although the Lacandones are not reported to have done so. *Loc:* **kol** 'milpa'; *Use:* **hanal** 'food'; *Part:* **wi'** 'tuber'; *Prep:* **che'che'** 'raw'. *See:* **suli'** ($_1$). *Variant:* **suli'**. *From:* xicamatl (Nahautl). *[Note:* surí, chikam. *Pachyrhizus erosus* (Nations and Nigh 1980); ch'ikam. Pachyrhizus erosus (Rätsch 1985a); chikam. *Pachyrhizus erosus* (Baer and Merrifield 1971); chi'kam [Itz.]. *Pachyrhizus erosus* (Atran et al. 2004); chi'kam, k'uup [Yuc.]. *Pachyrhizus erosus* (in, Atran et al. 2004); chi̍'ikam [Yuc.]. (Bricker et al. 1998); chi'kam [Mop.]. *Pachyrhizus erosus* (in, Atran et al. 2004).*] [Source:* BM; SK*] [*\sd2 food*]*

chimon (äh) hicatee fig [Bel. Eng.] (white fig [Jam. Eng.]). Moraceae *Ficus maxima* Mill., syns. *F. glaucescens* (Liebm.) Miq., *Ficus radula* Humb. & Bonpl. ex Willd., *F. subscabrida* Warb. **Yan uyits. Ne ts'ak uyits. Yan uwich. Wolis uwich. Cuando täk'än, k'än usool uwich. Säk unoy. Yan unek', ne yaab unek'. Chan mihin unek'. Ek' unek'. Ha'li' ahbäk' i ch'iich' kuhantik.** 'It has resin that is very bitter. It has round

Fig. 5.33 Chiah *Dieffenbachia* sp. Photo Credit: Suzanne Cook

fruit. When ripe, the skin turns yellow. The pulp is white. It has numerous small, black seeds. Only the animals and birds eat it.' (BM) *Ficus maxima* is a tree that grows between 5 and 30 m (16–100′) tall. The leaves vary in shape from long and narrow to oval, 6–24 cm (2–9″) long and 2.5–12 cm (1–5″) wide. Leaf texture is smooth above and smooth or downy beneath. Fruits are solitary, spherical, and 1–2.5 cm (0.4–1″) in diameter. Fruits provide an important food source for birds and animals. Although my consultants reported no medicinal uses, Kashanipour and McGee (2004) note that some Lacandones apply the moistened leaves to snake bites. *Use:* **hu'un** 'bark cloth'; *Part:* **uyoot'el sool** 'inner bark'. *See:* **kopo' (ah)**. *Thes:* **che'**. *[Note:* chimon hun. *Ficus glaucescens* (Nations and Nigh 1980); chimo, jun, ak hun che. *Ficus petenensis* (Durán 1999); chi:mòn [S. Lac.]. amate (Canger 1970); chimun [Itz.]. higo, higuero. *Ficus radula* (Hofling and Tesucún 1997); chäk chimun, säk chimun [Itz.]. *Ficus radula* = *Ficus maxima* (Atran et al. 2004); chimun [Mop.]. *Ficus radula* (in, Atran et al. 2004).*] [Source:* BM*] [*\sd2 clothing*]*

chinah (ah) [china] orange, sweet. Rutaceae *Citrus sinensis* (L.) Osbeck. This is a cultivated citrus tree bearing large, relatively thinned-skinned oranges. Fruit ripens between October and January. It was introduced by the Spanish in the 16th century. *Loc:* **kolil nah, kol** 'house gardens, milpas'; *Use:* **hanal** 'food'; *Part:* **wich** 'fruit'; *Prep:* **che'che' wa yäts'bil ti' uk'ul** 'eaten raw or squeezed for a drink'; *Spec:* **chäk chinah** 'red chinah'. *Thes:* **che'**. *[Note:* chinah. *Citrus sinensis* (Nations and Nigh 1980); naraanjaj [Itz.]. *Citrus sinensis* (Atran et al. 2004); ch'uuk chiina [Mop.]. *Citrus sinensis* (in, Atran et al. 2004); ch'ujuk pak'áal, chiina [Yuc.]. *Citrus sinensis* (in, Atran et al. 2004).*] [*\sd2 food*]*

chintok' ironwood, black (leadwood). *Lit:* 'bend over flint'. Rhamnaceae *Krugiodendron ferreum* (Vahl) Urb. *Krugiodendron ferreum* is a small, drought tolerant tree, 5–10 m (16′-33′) tall, with small, greenish flowers and purplish-red drupes 5–7 mm (0.2–0.27″) long. As the name suggests, the timber is extremely dense. SK says this tree is type of *guayacán* (Sp.). *Loc:* **Mensäbäk**; *Use:* **chulul** 'bows'; *Part:* **che'il** 'timber'. *Morph:* **chin=tok'**. *See:* **hach chulul**. *Thes:* **che'**. *[Note:* chintok'. *Krugiodendron ferreum* (Rätsch 1985a); chim-tok [Yuc.]. Lit. flint-capsule. Probably *Krugiodendron ferreum*, (Vahl) Urban (Roys 1931); chintok' [Yuc.]. *Krugiodendron ferreum* (in, Atran et al. 2004); b'ox chintok' [Itz.]. *Krugiodendron ferreum* (Atran et al. 2004); chintok' [Itz.]. axbreaker tree. *Krugiodendron* sp. (Hofling and Tesucún 1997); chintok' [Mop.]. **Krugiodendron ferreum** (in, Atran et al. 2004).*] [Source:* BM; SK*] [*\sd2 hunting_and_fishing*]*

chi' (äh) nance. Malpighiaceae *Byrsonima* spp. My Lacandon consultants described two kinds of *chi': k'än chi'* 'yellow chi' and *ya'ax chi'* 'green chi'. *K'än chi'* is *Byrsonima crassifolia*, a small tree with edible, cherry-like yellow drupes. *Ya'ax chi'* is similar except that its fruits remain green at maturity. It is unclear whether these are different species or not. According to Pennington and Sarukhán (1968) and Sánchez-Vélez (1988), *Byrsonima crassifolia* produces fruits that vary in color and taste (cited in Bye and Linares 1990: 158). The nance fruit is significant in a number of Mesoamerican myths and tales. In Maya mythology, it is associated with the Underworld deity, *Vucub-Caquix*, in the Popol Vuh, (the "K'iche' Maya Book of Counsel" (Bruce 1975: 144). The nance also features in the northern Lacandon myth of *uyählehil bah* 'The Mole Trapper', a mortal who caught the attention of the beautiful daughter of *Kisin* (Lord of the Underworld) by throwing the fruits of the *chi'* at her before following her down into the Underworld. *Use:* **hanal** 'food'; *Part:* **wich** 'fruit'; *Prep:* **che'che'** 'raw'; *Spec:* **k'än chi'** 'yellow chi' (*Byrsonima crassifolia*)'; *Spec:* **ya'ax chi'** 'green chi' (*Byrsonima* sp.)'. *Thes:* **che'**. *[Note:* chi' (äh). *Byrsonima continifolia* (Bruce 1975); shinich. *Byrsonima crassifolia* (Nations and Nigh 1980); chi' [Itz.]. *Byrsonima crassifolia* (Hofling and Tesucún 1997); k'än chi'il kaj, chäk chi'il [Itz.]. *Byrsonima crassifolia* (Atran et al. 2004); chii [Yuc.] *Malpighia glabra* L. Nancén. [Barbados Cherry] (Roys 1931); chi' [Yuc.]. *Byrsonima crassifolia* (in, Atran et al. 2004); k'än chi', säk chi' [Mop.]. *Byrsonima crassifolia* (in, Atran et al. 2004).*] [Source:* BM; CKO*] [*\sd2 food*]*

chok$_1$ dog-killer (negro killer, Juan caliente). Connaraceae *Rourea glabra* Kunth. This is a vine, but the northern Lacandones consider it both a tree and a weed. *Rourea* is a pantropical genus of about 100 species 48 of which occur in the Neotropics. The fruits, seeds, and leaves of some species, including *R. glabra*, are used in Latin America to poison wild dogs and coyotes (Kalegari 2009; Schultes n.d.: 142), and mice and rats (Lockhart 1989). Lacandon hunters are careful to clean a curassow, a game bird that feds on the berries, and dispose its entrails, to prevent dogs from ingesting the poisonous innards (Baer and Merrifield 1971: 168). To purge the poison from a dog who has eaten *chok*, the Lacandones feed it *yo'och ahpek' (bubuhchi') Canavalia* spp. *Loc:* **Mensäbäk**. *Thes:* **che'**. *[Note:* chok. *Ardisia paschalis* (Nations 2006). Common names from Smithsonian Tropical Research Insltitute (http://biogeodb.stri.si.edu/herbarium/species/16479/?fam=Connaraceae). Retrieved February 2012.*] [Source:* AM; Baer and Merrifield (1971)*]*

choochel ak' (äh) [choochel ak', chooch ak', chäche'ak'] bean, twining snout. *Lit:* 'intestine vine'. Fabaceae *Rhynchosia erythrinoides* Schltdl. & Cham. **Ahchäche'ak' uch'ihil ich paakche' kol. Apäk'ik. Uch'ihil yetel ahmo'ak'.** 'Choochel ak' grows in the fallow milpas. You plant it. It grows with mo'ak'.' (KP) The species that KP identified as *choochel ak'* had angular stems with reddish blisters. Older vines were thick, compressed, woody, and draped down from tree branches. The leaves were pinnately trifoliate, leaflets were rhomboid and acuminate, and palmately veined. The inflorescence was a raceme of dark, coral-pink blossoms with beaked keels. Fruit was a compressed, round to sickle-shaped, velvety pod containing tiny, slightly compressed, bi-coloured dull orange-red and dark brown seeds. *Rhynchosia* is a large genus, consisting of approximately 200 species of perennial, twining vines that grow in brushwood, forest clearings, and jungle borders in tropical climates (Allen and Allen 1981; Grear 1978). Twelve species occur in Chiapas (Breedlove 1986), five of which are recorded from the northern Lacandon forest: *R. discolor, R. erythrinoides, R. longeracemosa, R. minima,* and *R. pyramidalis* (Comisión Nacional de Áreas Naturales Protegidas 2006). *Loc:* **paakche' kol** 'fallow milpa, house gardens, secondary forest'; *Use:* **uuh** 'necklaces'; *Part:* **nek'** 'seeds'; *Sim:* **mo'ak' (ah)** 'macaw vine (*Rhynchosia pyramidalis*)'. *Morph:* **chooch-el ak'**. *Thes:* **ak'** (Fig. 5.34). *[Note:* Botanical identification provided by Alfonso Delgado-Salinas, Department of Biology, Instituto de Biología, UNAM. chochi ak. *Rhynchosia erythrinoides* (Durán 1999).*] [Source:* AM, BM, CKO, CNK, KP*] [*\sd2 craft*]*

Fig. 5.34 Choochel ak' *Rhynchosia erythrinoides*. Photo Credit: Suzanne Cook

chuh calabash, bottle gourd. Cucurbitaceae *Lagenaria siceraria* (Molina) Standl. **Yan uyak'il, p'iis k'um, chen ma' ahantik unoy. Yan uhel chuh, mas mihin. Kis chuh uk'aba'. Lo'k'in, hach winik. La'eh chunil winik, uchik winik. Ahchuh utäk'man tuho'ol ti' mak uwich. Kubin 'aktun ich k'aax. Los amarran sus cabezas de sus ninos con palos. Kah lahkinsah'.** 'It has vines, like *k'um*, but you don't eat the flesh. There is another, smaller kind. Its name is 'fart gourd'. The *lo'k'in* were "true people". They were the first people, ancestors. They stuck the gourds over their heads to cover their faces. They would come from caves in the forest. They used to tie the heads of their children to boards (so that they would fit into the gourds). The villagers killed them all.' (AM, BM) This is a vine that produces gourds with smooth, light green, water-proof skin. Native to Africa, *Lagenaria siceraria* reached the Americas 8000–9000 years ago, the fruits presumably having floated across the Atlantic. After its arrival, the wild form was independently domesticated several times. The fruits take a variety of forms, from round and squat, to long, to an hour-glass shape. They also vary in size. The young fruit is edible; but the mature fruit, although inedible, is valued for its hard, water-resistant rind. In the Lacandon forest, seeds are planted when the *tsäk'ats* (*Licania platypus*) blooms (Baer and Merrifield 1971: 146). *Use:* **hama'** 'large drinking bowls'; *Use:* **so'ot** 'ceremonial rattle'; *Use:* **pak** 'bee hive'; *Spec:* **kis chuh (äh)** 'fart gourd'. *See:* **ha' chuh**. *Thes:* **ak'**. *[Note:* Bruce (1975: 188) surmises that the *lo'k'in* may have been a band of Lacandones or Acantunes [Ch'ol Lacandones] who were killed by federal troops under Venustiano Carranza

Fig. 5.35 Chukuch le' uxiiw *Heliconia librata* inflorescence. Photo Credit: Suzanne Cook

Fig. 5.36 Chukum che' *Croton draco*. Photo Credit: Suzanne Cook

(1914–1918) and subsequently mythologized as the survivors of the "Age of Wooden Men". luch; lek; chú. *Lagenaria siceraria* (Nations and Nigh 1980); chuj, säk lek, chäk lek [Itz.]. *Lagenaria siceraria* (Atran et al. 2004); chúuh [Yuc.]. canteen, bottle gourd. *Lagenaria siceraria* (Molina) Standley (Bricker et al. 1998).*]* *[Source:* AM; BM; CKO*]* *[*\sd2 containers \sd2 folklore_and_mythology*]*

chukuch le' xiiw heliconia. *Lit:* 'long leaf xiiw'. Heliconiaceae *Heliconia* sp., *Heliconia librata* Griggs. A large plant with enormous, paddle-shaped leaves. Grows in open, secondary forest along cleared roadsides north of Naha'. According to KyP, the inflorescences are red. *Use:* **ti' utap'ik ba'al** 'to cover and wrap up things'. *Thes:* **xiiw** (Fig. 5.35). *[Note:* chu kux xibi. *Heliconia librata* (Durán 1999).*]* *[Source:* BM; KyP*]* *[*\sd2 food*]*

chukum che' bloody croton. Euphorbiaceae *Croton draco* Schltdl.& Cham., syn. *C. lechleri* Muell.Arg. **Ne nuk ule'. K'än ule'. Yan uyits. Chäk uyits. Uk'anik ti' uts'ak k'ik' uchi', ts'uk uchi'. Ne ts'uts'.** 'The leaves are very large and yellow. (The tree) has red sap. One uses it to cure bleeding gums and mouth rot. It makes your mouth pucker.' (AM) *Croton draco* is a tree or bush between 2 (6.5′) and 18 m (15′) tall. Leaves are broadly oval or ovoid-deltoid from 8 cm (3″) to 28 cm (11″) long and from 5 cm (2″) to 18 cm (7″) wide. Leaf margins are entire or serrated. The leaf base is heart shaped. The petioles are generally long and glandular. Inflorescence is a raceme from 8 (3″) to 50 cm (19.5″) long. (Gonzales 2000) "Sangre de Draco", as it is called in Latin America, is widely used in traditional and urban medicine to relieve coughs, flu, diarrhea, and stomach ulcers; it is also used to heal wounds, and as an oral germicide (Murillo et al. 2001: 249). *Use:* **ts'uk uchi'** 'mouth rot'; *Part:* **uyits usool** 'resin from the bark'; *Prep:* **Ha'li' apaats'tik ti'** 'You just massage it in'. *Thes:* **che'** (Fig. 5.36). *[Note:* chukun. chimon tree. *Croton draco* (Nations and Nigh 1980); chukum [Itz.]. *Pithecolobium albicaule* (Atran et al. 2004); chukum [Itz]. chukum tree. chukum tree (Hofling and Tesucún 1997); chucum [Yuc.]. *Pithecolobium albicans* (Kunth.) Benth. A tree whose fruit yields black dye (Roys 1931: 239).*]* *[Source:* AM*]* *[*\sd2 medicinal*]*

chunup (äh) clusia (pitch apple). Clusiaceae *Clusia* spp. Clusia comprises about 300 woody species of shrubs, vines, and small to medium-sized trees (Vargas-Soto et al. 2009: 33). Leaves are thick, succulent, smooth, and paddle-shaped, approximately 5–20 cm (2–7.8″) long and 5–10 cm (2–3.9″) wide. Flowers are composed of 4–9 petals that range in colour from white to greenish-white, to yellow, to pink. The fruit is a leathery, greenish-brown, valvate capsule containing numerous red, fleshy seeds. According to BM, the flowers attract droves of birds, including: *ch'uhum* (zebra-backed woodpeckers); *tunsel* (lineated woodpecker); *mehen ch'iich'* (small birds); *k'ayum* (unidentified); and, *pupuch'* (long-billed thrasher). It grows at the edges of roads and in open forest. Some species, e.g., *C. flava*, are cultivated in gardens for their ornamental appeal. There was some confusion over the identification and naming of the plant in this photo (Fig. 5.37). KP referred to it as *ts'oy*, a name that was also applied to both *C. lundelli* and *C. salvinii* by other Lacandon consultants (Durán 1999). But AM strongly disagreed, saying that the species in the photo was *chunup*. According to him, *ts'oy* is a vine and *chunup* is a tree. In Mexico, *chunup* is one of many common names for *Clusia* species (Vargas-Soto and Andrade 2006: 38). *Variant:* **ts'oy (äh)**. *Thes:* **che'** (Fig. 5.37). *[Note:* chunup [Lac.]. *Clusia flava, C. salvinii* (Nations 1992, in Atran et al. 2004); chunuup [Itz.]. "tree companion", matapalo, strangler fig (Hofling and Tesucún 1997); chunuup [Itz.]. *Clusia flava, C. salvinii* (Atran et al. 2004); chunup [Yuc.]. *Clusia flava* (Bricker et al. 1998).*]* *[Source:* AM; KP*]*

Fig. 5.37 Chunup *Clusia lundellii*. Photo Credit: Chan K'in Daniel

Ch'–ch'

ch'alol$_1$ [ch'alol, ch'aroh, ch'aron] oak. Fagaceae *Quercus* spp. **Nukuch ch'alol, hach ch'alol la' uchik winik lehih kuhäxik uyaläk'te ch'alol.** 'The *nukuch ch'alol* 'large acorn' and the *hach ch'alol* 'true acorn' were used by our forefathers as a spinning top' (AM). A number of species of oaks are found in northern Chiapas: *Quercus sapotifolia* occurs in the Lacandon area (Golicher, pers. comm.), although it usually grows well above Naha', at altitudes between 1300 m (4265') and 2400 m (7874'); *Quercus corrugata* and *Q. skinneri* have been recorded from Naha' (Durán 1999). My botanical determinations are based on the nuts, the data from Durán's (1999) inventory, and on my Lacandon consultants' descriptions. An examination of the leaves, bark, and other plant organs is required before the botanical identities can be confirmed. *Use:* **alak'te'** 'spinning top'; *Part:* **nek'** 'nut'; *Spec:* **nukuch ch'alol** 'large oak (*Quercus corrugata*)'; *Spec:* **mehen ch'alol** 'small oak (*Quercus* sp.)'. *See:* **pixan k'ambul (äh)**. *Thes:* **che'**. *[Note:* charo, karote, avellano. *Quercus skinneri* (Durán 1999); ixim charo. *Quercus corrugata* (Durán 1999).*] [Source:* AM; BM*] [*\sd2 other*]*

ch'alol pixan k'ambul *Lit:* 'curassow's heart oak'. Fagaceae *Quercus skinneri* Benth. *See:* **pixan k'ambul**. *Variant:* **pixan k'ambul**. *[Source:* AM*]*

ch'ayok'$_1$ [ch'ayok', ch'ayuk'] nightshade. Solanaceae *Jaltomata, Lycianthe, Solanum.* This name refers to a number of species of vines, sub-shrubs, and shrubs with tomato-like blossoms and small, round, purple-black or red berries. One kind of *ch'ayok'* corresponds to *Solanum americanum*, a small, herbaceous plant with leaves that are medium blue-green on top and purple underneath. The margins are toothed. It bears small clusters of sweet, purple-black berries the size of blackcurrants. Another kind of *ch'ayok'* lacks the purple tint and the toothed margins. Still another kind is a small, slender tree with dark green, ovate leaves with entire (smooth) margins. All of these species are considered *lo'obil* 'weeds'. *Use:* **hanal** 'food'; *Part:* **le'** 'leaves'; *Prep:* **chäkbil, mäk'bil yetel mulix wa tsähbil yetel uye'el kax** 'boiled and eaten with lime juice or fried with eggs'. *[Note:* chayok. *Cnidoscolus aconitifolius* (Nations and Nigh 1980); ch'auk' [S. Lac.]. black nightshade. *Solanum americanum* (Diemont 2006); cY'a?-Uk' [S. Lac.]. hierba buena (Canger 1970); ch'a'yuk [Itz.]. *Solanum americanum* (Atran et al. 2004).*] [Source:* AM; KP*] [*\sd2 food*]*

ch'ayok'$_2$ *Jaltomata repandidentata* (common name unknown). Solanaceae *Jaltomata repandidentata* (Dunal) Hunz., syn. *Saracha repandidentata* (Dunal) Bitter. This is a weak stemmed herb or clambering, straggly shrub that grows to 1.5 m (4.9′) tall. Blossoms are pale green or white with green spots. Fruit are small, round, purple-black, edible berries. *Loc:* **kol** 'milpa'; *Use:* **hanal** 'food'; *Part:* **le'** 'leaves'; *Prep:* **chäkbil i mäk'bil yetel mulix wa tsähbil yetel uye'el kax** 'boiled and eaten with lime juice or fried with eggs'. *[Note:* Taxonomic identification provided by Armando Soto (pers. comm.).*] [Source:* KM; KP; CNK*] [*\sd2 food*]*

ch'ayok'$_3$ lycianthes, varying. Solanaceae *Lycianthes heteroclita* (Sendtn.) Bitter. This is a slender, soft-wooded shrub or small tree. The flower is characteristic of other *Solanum* species, with its tubular corolla and five appressed stamens that form a cone protruding from the center. The blossoms are usually purple. It is unknown whether or not the leaves are eaten. *Loc:* **kolil nah** 'house garden (over the septic field in KM and KP's backyard)'. *Variant:* **k'ulsek** (Fig. 5.38a). *[Note:* Taxonomic identification provided by Armando Soto (pers. comm.). English name from Grandtner (1997–2010).*]*

ch'ayok'$_4$ nightshade, rovirosa. Solanaceae *Solanum rovirosanum* Donn. Sm. This is a slightly branched shrub or treelet, growing up to 7 m (22.9′) tall. It bears white flowers with yellow anthers and nearly round, yellow-green berries. (Haber 2002–2006) No uses were reported. *Loc:* **kolil nah** 'house garden' (Fig. 5.38c). *[Note:* Taxonomic identification provided by Armando Soto (pers. comm.).*]*

ch'ayok'$_5$ nightshade, American black (glossy nightshade, small flowered nightshade, common nightshade, white nightshade). Solanaceae *Solanum amercianum* Mill., syn. *Solanum nigrum* L. var. *americanum* (Mill.) O. E. Schulz. This is an erect or spreading, short-lived perennial growing to 1.3 m (4.3′) tall. The stems and leaves are smooth and usually green or purple. Stems are often angled or narrowly winged. Blossoms are white or flushed with purple and have yellow-green centres. The fruit is an edible, round, purple-black berry. *Loc:* **kolil nah, kol, paakche' kol** 'house garden, milpa, fallow milpa'; *Use:* **hanal** 'food'; *Part:* **le'** 'leaves'; *Prep:* **chäkbil, a mäk'bil yetel mulix wa tsähbil yetel uye'el**

Fig. 5.38 (**a**) Ch'ayok' *Lycianthes heteroclite*. Photo Credit: Suzanne Cook (**b**) Ch'ayok' *Solanum americanum*. Photo Credit: Suzanne Cook (**c**) Ch'ayok' *Solanum rovirosanum*. Photo Credit: Suzanne Cook

Fig. 5.39 (**a**) Ch'äm *Bromelia pinguin*. Photo Credit: Suzanne Cook (**b**) Ch'äm *Bromelia pinguin*. Photo Credit: Suzanne Cook

kax 'boiled, eaten with lime or fried with eggs' (Fig. 5.38b). *[Note:* Taxonomic identification provided by Armando Soto (pers. comm.). ch'auk'. *Solanum americanum* (Diemont 2006); ix ch'a'yuk [Itz.]. *Solanum americanum* (Atran et al. 2004); chayuk [Mop.]. *Solanum americanum* (in, Atran et al. 2004).*] [Source:* KP*]* *[*\sd2 food*]*

ch'äm (äh) pinguin (wild pine). Bromeliaceae *Bromelia pinguin* L. This is a species of wild fruit related to the pineapple. It is cultivated for its edible fruit. While no medicinal uses were reported by my Lacandon consultants, the Itzaj drink the fruit juice for its diuretic properties and to expel intestinal worms (Atran et al. 2004). *Use:* **hanal** 'food'; *Part:* **wich** 'fruit'; *Prep:* **che'che', chäkbil** 'raw or cooked' (Fig. 5.39a, b). *[Note:* ch'am. Bromelia pinguin (Nations and Nigh 1980); ch'um. ch'äm [Itz.]. piñuela. wild pineapple. *Bromelia pinguin*, *Bromelia karatas* (Hofling and Tesucún 1997); ch'äm, säk ch'äm [Itz.]. *Bromelia pinguin* (Atran et al. 2004); chäk ch'äm [Itz.]. *Bromelia karatas* (Atran et al. 2004); ch'am [Yuc.]. Certain piñuelas of this land (Roys 1931: 233). Common English names from GRIN*] [Source:* KP; CKO*]* *[*\sd2 food*]*

ch'ibil wits podocarp, Jalisco. Podocarpaceae *Podocarpus matudae* Lund. *Variant:* **bey ule' tähte' (ah)**. *[Note:* chibix xi wits, bedetate. *Podocarpus matudai* (Durán 1999).*]* *[Source:* BM*]*

ch'ibix bo'oy (ah) palm, parlor (good-luck palm, neanthe bella, collinia, xate hembra). Arecaceae *Chamaedorea elegans* Mart. **Ch'ibix bo'oy, lati' ma' nuktal. Lati' chan mihin. Yan uchan che'il ule', ma' ne chukuch ule'. Ne yaab uchan mehen le' ti' uk'äb. P'isil bo'oy ule', chen ha'li' mas mihin, mas mihin, chan chukuch. Yan uwich, ya'ax uwich. Chan wolis uwich. Chan mihin uwich. Ne täk'än, p'isil (hach bo'oy) ek'. Ne yi'h uwich ne täk'än. Lati' ch'ibix ti' unuktal, mas ne nuk, mäna'. Yan ya'ax uche'il. Lati' ch'ibix.** *'Ch'ibix bo'oy* doesn't grow large. It's tiny. It has small leaf stalks, the leaves aren't very long. There are a lot small leaves

Fig. 5.40 Ch'ibix bo'oy (a) *Chamaedorea elegans*. Photo Credit: Suzanne Cook (b) Ch'ibix uk'um *Chamaedorea cataractarum. Photo Credit: Chan K'in Daniel*

on the branches. The leaves are like those of *bo'oy*, only (narrower) and shorter. It has tiny, round, green fruit. When they are ripe, they are black, like those of *hach bo'oy*. The fruits are very hard when ripe.' (BM) *Chamaedorea elegans* is a slender, erect, solitary-stemmed palm that grows to about 2 m (6.5′) tall. The stem has very closely spaced rings. Eleven to 21 pairs of leaflets fan out from the leaf shaft. Inflorescences are erect with red stems, followed by glossy black, fruits, 4–7 mm (0.16–0.27″) in diameter (Hodel 1992: 36–37). Leaves are exported to florist industry in the United States. Until recently, they were collected from the wild, but now the plants are cultivated in a nursery in the village. *Loc:* **k'aax** 'forest'; *Use:* **ti' konol** 'for sale'; *Part:* **le'** 'leaves'; *Spec:* **ch'ibix uk'um** 'cat palm (*Chamaedorea cataractarum*)'; *Spec:* **ch'ibix wits** (*Chamaedorea woodsoniana*); *Gen:* **bo'oy**. *Variant:* **ch'ibix bo'oy** (Fig. 5.40a). *[Note:* Botanical identification provided by Don Hodel (pers. comm.). English common names from Hodel (1992). chirixboy. *Chamaedorea elegans* (Nigh 2008); cambray. *Chamaedorea elegans* (Durán 1999); ix yaat [Itz.]. *Chamaedorea elegans* (Atran et al. 2004); mejen xiaat [Mop.]. *Chamaedorea elegans* (in, Atran et al. 2004); xyaat [Yuc.]. *Chamaedorea elegans* (in, Atran et al. 2004).*] [Source:* BM; K'ayum Segundo; KM; KyG; SK*] [*\sd2 industry*]*

ch'ibix uk'um palm, cat (cascade palm). Arecaceae *Chamaedorea cataractarum* Mart. **Uch'ibix uk'um p'isil ten lati' yan. Uch'ibil uk'um yan uwi', yan umots, yan uk'äb ule'. Yaab, yaab ule'. Lati' ahch'ibil. Uch'ibil uk'um chukuch uchan mehen ule'. Ma' ne kooch ule'. Tan ule' ma' kooch, chen ha'li' chukuch. Ne ya'ax ule'. Uch'ibix uk'um täk man hump'el metro chumuk ubakel. Uyaal, ne ki' uyaal. Inhantik uyaal yetel uye'el kax, tsähbil. Inchäkik, pachil p'älik. P'älik, quita usool. P'älik usool. Pachil xatik yetel cuchillo. Pachil utsähik. Pachil inpulik yetel uye'el kax. Hahe' inhantik. Ne p'enk'äch ki'. Ch'ibix uk'um. Asi es. Bex.** 'The river ch'ibix is the same as the one I have (in my garden). The river ch'ibix has a tuber. It has roots and a lot of leaf branches. There are a lot of leaves. The leaves of the river *ch'ibix* are long, narrow, and very green, and almost 1.5 m (14.9′) long. The immature inflorescences are really tasty. I eat the buds with fried eggs. I boil them, peel them, chop them up with a knife, fry them, and then put them in eggs. It's really delicious. That's how I will eat it. That's river *ch'ibix*.' (BM) *Chamaedorea cataractarum* is an upright, closely suckering species of palm that forms dense clumps on river banks and inundated areas (Hodel 1992: 242). *Use:* **hanal** 'food'; *Part:* **uyaal** 'inflorescence buds (male)'; *Prep:* **chäkbil yetel ch'ooch' yetel mulix, chäkbil i huytsähbil yetel uye'el kax** 'boiled and eaten with salt and lime, boiled and then scrammbled with eggs'; *Sim:* **ch'ibix bo'oy** 'parlor palm (*Chamaedorea elegans*)'; *Sim:* **ch'ibix wits** 'woodsonia palm (*Chamaedorea woodsoniana*)'; *Gen:* **ch'ibix (ah)**. *Variant:* **näkil hay ch'i'ib; ch'ibil uk'um** (Fig. 5.40b). *[Note:* Botanical identification provided by Don Hodel (pers. comm.). Spanish name from Hodel (1992: 242)*] [Source:* AM; BM*] [*\sd2 food*]*

ch'ibix wits [ch'ibix wits] palm, woodsonia. Arecaceae *Chamaedorea woodsoniana* L. H. Bailey. *Chamaedorea woodsonia* is the largest of the Chamaedorea palms, growing up to 12 m (40′) tall on a stout trunk of up to10 cm (4″) in diameter. Its solitary stem is erect or leaning. Leaves are erect and spreading, with up to 36 pairs of long, stiff, dull-green leaflets with prominant midribs. Male and female organs occur on separate plants. Female plants bear clusters of golden-orange fruit (Hodel 1992: 220). *Chamaedorea woodsonia* may be confused with *C. carchensis* and *C. costaricana*, but *Chamaedorea carchensis* has fewer pinnae than *C. woodsonia* and has black indument (hairs) on the petioles, and *Chamaedorea costaricana* has a clumping habit and bears fruit that matures from green to black without passing through an orange-yellow stage. All three are found in rainforest or montane vegetation in Chiapas and Central America. *Chamaedorea woodsonia* occurs between 800 m (2624.6') and 2000 m (6561.6') in elevation, *C. carchensis* between 900 m (2953') and 1600 m (5249'), and *C. costaricana*, between 50 m (164') and 2300 m (7546'). Of these three species, only *C. woodsonia* is reported from Chiapas (Breedlove 1986). *Loc:* **k'aax, wits** 'forest, hills'; *Use:* **hanal** 'food'; *Part:* **uyol** 'heart'; *Prep:* **ts'ilik usool i chäkik** 'peel it and boil it'; *Sim:* **ch'ibix bo'oy** 'parlor palm (*Chamaedorea elegans*)'; *Sim:* **ch'ibix uk'um** 'cat palm (*Chamaedorea cataractarum*)'; *Gen:* **ch'ibix (ah)**. *Variant:* **ch'ibil wits; ch'ilil wits**. *[Note:* Taxonomic identification provided by Don Hodel (pers. comm.). Common names from Palms Online. http://www.palmsonline.co.nz/Palms/chamaedorea-woodsoniana. Retrieved March 2013.*]* *[Source:* BM*]* */*\sd2 food*]*

ch'i'ib (äh) [ch'i'ib, ch'i'ip', ch'i'b, ch'i'p'] palm, pacaya. Arecaceae *Chamaedorea tepejilote* Liebm. **Chukuch ule'. Ne nuk uche'il, batak 7.6 cm (3″). Ya'ax uche'il. Uyaal ne ki' uhantal. Ule' ne chukuch, batak 1 m (3.3′). Ule' ma' uxantal. Yetel k'in, ne seeb ubabal. Lahih ubäho' ahpakyon. P'iis pakyon, chen mas mihin uche'il ubakel, batak 2.5 m (8.2′). Uyaal mas mihin. Uwich, ek', p'isil pakyon. Ne ts'u'. Pakyon as ne nuk uche'il, batak 5 m (16.4′) ubakel. Ne nuk uyaal, batak unukil 7.6 cm (3″). Ne yaab uyaal (AM). Uchuch uwich, chäk. Cuando uni' uyaal uwaakal, pachil yan uchan uwich. Cuando ne muun, säk. Ahantik.** 'The leaves are large. The trunk is very large, about 7.6 cm (3″) in diameter. The trunk is green. The flower buds are good to eat. The leaves are about 0.9 m (3′) long. The leaves don't last long. After they have been cut, they quickly roll up in the sun. It's the brother of *pakyon*. It's like *pakyon*, but its trunk is shorter, about 2.5 m (8.2′). Its flower buds are smaller. Its fruit is black, like that of *pakyon*, and soft. *Pakyon* has a somewhat larger trunk, about 5 m (16.4′). Its flower buds are very large, about 7.6 cm (3″) in diameter. There are a lot of buds. The stems of the fruit are red. When the tips of the buds split open there are small fruits. When they are very young, they are white. You eat them.' In the extant Lacandon ethnobotanical and agroforesty literature, *ch'i'ib* is invariably identified as *Chamaedorea alternans*. However, all the examples of *ch'i'ib* that the Lacandon consultants showed me corresponded to the multi-stemmed form of *C. tepejilote*. Although similar, the two species are differentiated by a few key features: *C. alternans* is mostly, or always, solitary, whereas *C. tepejilote* has both solitary and multistemmed forms; the inflorescences of *C. alternans* can occur in multiples at a node, like *C. tepejilote*, but they may also occur alternating up the stem or individually (Hodel 1992: 238); *C. alternans* has larger seeds and fruits, and delayed fruit maturation, relative to *C. tepejilote*. *Chamaedorea* is a dioecious plant, meaning that female and male reproductive organs occur on separate plants. Even though the Lacandones know the true botanical sex of each plant, they refer to the male plant (which produces the edible *uyaal* 'offspring') as *ch'uplal* 'female'. According to AM, the men give *ch'i'ib* to their wives, because they believe it is an aphrodisiac. The fruit is gathered from the wild or from plants that have been transplanted into house gardens. *Use:* **hanal** 'food'; *Part:* **uyaal** 'buds of male inflorescences'; *Prep:* **pokbil ich k'ak', chäkbil, wa chäkbil pachil tsähbil yetel uye'el kax** 'roasted, boiled, and eaten as is, or cooked and then fried with eggs'; *Sim:* **pakyon** 'pacaya palm (*Chamaedorea tepejilote*)'. *[Note:* Taxonomic identification provided by Don Hodel (pers. comm.). ch'i'b. a species of small palm tree (Bruce 1975); chiip. *Chamaedoea alternans* (Nigh 2008); ch'ib' [Itz.]. pacaya. pacaya palm (Hofling and Tesucún 1997); ch'ib' [Itz.]. *Chamaedorea tepejilote* (Atran et al. 2004); säk ch'ib' [Mop.]. *Chamaedorea tepejilote* (in, Atran et al. 2004); ch'ib [Yuc.]. *Chamaedorea tepejilote* (in Atran et al. 2004).*]* *[Source:* AM; BM; KP*]* */*\sd2 food*]*

ch'ilix bo'oy palm, parlor. Arecaceae *Chamaedorea elegans* Mart. *Variant:* **ch'ibix bo'oy**. *[Note:* chirixboy. *Chamaedorea elegans* (Nigh 2008).*]*

ch'ink'ak'il *Passiflora cookii* (common name unknown). *Lit:* 'fling coals/boils'. Passifloraceae *Passiflora cookii* Killip., syn. *P. hahnii* (E. Fourn.) Mast. This is a kind of vine that climbs high up into the trees. The fruit is similar to *pooch'in* (*Passiflora serratifolia*) and *ch'um ak'* (*Passiflora* aff. *ambigua*). It matures in March, according to BM. Although the plant was unavailable to inspect, it is probably *P. cookii*. *Passiflora cookii* is often confused with the closely related species *P. guatemalensis*, as both produce beautiful, white blossoms. Two features that set each species apart from one another are the number and character of the stipules and the colour of the flower centre, which is deep red in *P. cookii* and pale green in *P. guatemalensis* (Irvine 2005). To complicate matters, the synonym for *P. cookii* is *P. hahnii*, a species that demonstrates great variation. A taxonomic key has yet to be produced on these three species before conclusive identification can be made (MacDougal, pers. comm.). *Use:* **hanal** 'food'; *Part:* **wich** 'fruit'; *Prep:* **che'che'** 'raw'; *Use:* **ti' uk'äxik hiil yetel säkche'** 'to tie battens to rafters'; *Part:* **ak'** 'vines'; *Sim:* **pooch'in** '*Passiflora serratifolia*'; *Sim:* **ch'um ak' (äh)** '*Passiflora* aff. *ambigua*'. *Morph:* **ch'in-k'ak'-il**. *Thes:* **ak'**. *[Note:* chin ka kir, chi kan ki. *Passiflora cookii* (Durán 1999).*]* *[Source:* AM; BM*]* */*\sd2 food*]*

ch'it bo'oy chamaedorea (type of). Arecaceae *Chamaedorea* sp. **Hump'eli' uche'il. Ubakel, 2 metros, ket yetel pakyon. To'och ule', ukooch batak 7 cm (3″) uwaan batak 20 cm**

Fig. 5.41 Ch'it bo'oy *Chamaedorea* sp. Photo Credit: Suzanne Cook

(8″). Ne pim ule'. Ma' ne ya'ax. 'There is one stalk. It is about 2 m (6.5′) tall, the same size as *pakyon* (*Chamaedorea tepejilote*). The leaves are stiff, about 7 cm (3″) wide and 20 cm (8″) long, thick, and not very green.' (BM) According to BM, *ch'it* refers to the wavy character of the leaflets. AM says that *ch'it bo'oy* is another name for *ch'it k'än bo'oy* (which is not to be confused with the species called *k'än bo'oy* 'yellow bo'oy'). The species in the photo (Fig. 5.41) may be *C. pinnatifrons*, since the fruit displays a transition in colour from orange-red to black. *Use:* **ti' konol** 'to sell'; *Part:* **le'** 'leaves'; *Sim:* **hach bo'oy** 'authentic bo'oy'; *Sim:* **ch'ibix bo'oy**; *Sim:* **säk bo'oy** 'white bo'oy'; *Sim:* **k'än bo'oy** 'yellow bo'oy'; *Sim:* **chan mehen bo'oy** 'very small bo'oy'; *Sim:* **ton bo'oy** '*Chamaedorea* sp.'; *Gen:* **bo'oy**. *Variant:* **ch'it k'än bo'oy** (Fig. 5.41). *[Note:* ch'it [Yuc.]. *Thrinax argentea*, Lodd. (Millsp. I, 355). T. Wendlandiana, Becc. (Gaumer.) A palm with large fan-shaped leaves growing on the northern seacoast (Roys 1931: 237); ch'iit [Yuc.]. bamboo (Bricker et al. 1998).*] [Source:* AM; BM; BMjr; KyG*] [*\sd2 industry*]*

ch'it k'än bo'oy chamaedorea (type of). Arecaceae *Chamaedorea* sp. *Variant:* **ch'it bo'oy**. *[Source:* AM*]*

ch'obenche' bitterwood. *Lit:* 'red-earth tree (?)'. Meliaceae *Trichilia* sp. **Es che'. Ne chich uche'il. Nuk uche'il, batak 60.9 cm (24″). Utop', säk i chan mihin, wolis, much'umuch' utop'. Uwich, chan chukuch uwich, batak 1.27–2.54 cm**

Fig. 5.42 Ch'obenche' *Trichilia* sp. Photo Credit: Suzanne Cook

(0.5–1″). Unoy chäk, chich. Abril kutahal. Uyo'och ch'iich'. Uch'ihil yok'ol wits, ha'li'. Uk'anik ti' uyokman. 'It is a tree. Its wood is very hard. Its trunk is about 60.9 cm (24″) in diameter. Its flowers are small, round, and white. They grow in clusters. The fruit is small and a little bit long, about 1.27–2.5 cm (0.5–1″). The flesh is red and hard. It ripens in April. It's food for the birds. It only grows in the hills. The timber is used for house posts.' (BM) According to AM, the flowers grow in the leaf axils all the way up the stems. The tree blooms in January and sets fruit in April. The skin of the fruit is green. *Ch'obenche* is reported in Standley (1930: 311) as *Trichilia arborea* C. DC. and *T. terminalis* Millsp. This author says that the bark is used as an emetic, while the juice from the leaves is rubbed on the limbs to relieve convulsions. My consultant, AM, says that he uses the leaves to relieve bone and leg pain. Roys (1931: 238) says that the people in the Yucatan also exploited the tree for its medicinal properties. Breedlove (1986) records ten species of *Trichilia* from Chiapas, seven of which are reported from Naha': *T. havanensis; T. japurensis; T. minutiflora; T. montana; T. moschata; T. pallida;* and, *T. quadrijuga* (Comisión Nacional de Áreas Naturales Protegidas 2006). Durán (1999: 146) reports that the northern Lacandones eat the fruits of *T. japurensis* and *T. quadrijuga*. Levy et al. (2006: 90) record four species and their uses in the southern Lacandon territory: (1) *T. breviflora* provides materials for construction; (2) *T. havanensis* is used for fuel and soap; (3) *T. moschata* and, (4) *T. pallida* are used for fuel. *Loc:* **wits** 'hills'; *Use:* **k'anche'** 'chairs'; *Part:* **che'il** 'timber'; *Use:* **okman** 'pillars'; *Part:* **uk'äbche', mehen che'** 'branches, saplings'; *Use:* **che'il k'ak'** 'firewood'; *Use:* **ts'ak ti' uyah uyok i ubaakel** 'medicine for leg and bone pain'; *Part:* **le'** 'leaves'; *Prep:* **?** 'unknown'. *Thes:* **che'** (Fig. 5.42). *[Note:* ch'oben-che [Yuc.]. *Trichilia arborea* (Roys 1931);

ch'ob'enche' [Itz.]. *Trichilia havanensis* (Atran et al. 2004). The root, *ch'oben*, refers to red earth or red-ochre, according to Atran et al. (2004).*] [Source:* AM; BM*]* *[*\sd2 medicinal \sd2 fuel \sd2 housing*]*

ch'om ik chile pepper (type of). *Lit:* 'vulture chile'. Solanaceae *Capsicum* sp. *Use:* **hanal** 'food'; *Part:* **wich** 'fruit'; *Prep:* **che'che'** 'raw'; *Gen:* **ik**. *[Note:* Word not recognized by CKO or BM.*] [Source:* Boremanse (1981: 204)*]* *[*\sd2 food*]*

ch'om mäkäl elephant ears, capote. *Lit:* 'vulture mäkäl'. Araceae *Xanthosoma robustum* Schott., syn. *Xanthosoma roseum* Schott. This is an enormous variety of elephant ears that grows in the lagoons, on stream banks, and in boggy soil in clearings. It is most plentiful below 900 m (2553') in elevation (Standley and Steyermark 1958: 360–361). Although similar to *Colocasia esculenta* and *Alocasia macrorrhizos*, *Xanthosoma robustum* is distinguished by leaf lobes that deflex outward, the downward pointing leaf tip, and the whitish, waxy bloom on the underside of the blades and petioles (Bishop Museum 2013). According to my consultants, people do not eat the leaves, because they contain an irritating resin; furthermore, they say that the roots do not form edible tubers. *Use:* **booch'** 'umbrella'; *Part:* **le'** 'leaves'; *Sim:* **hach mäkäl** 'authentic macal (*Xanthosoma mafaffa*)'; *Gen:* **mäkäl**. *Variant:* **mäkäl ch'om, hach ch'om**. *Thes:* **xiiw** (Fig. 5.43). *[Note:* ch'om mäkal. elephant ear plant, skunk cabbage (Bruce 1975).*] [Source:* BM; KAG; SK*]* *[*\sd2 other*]*

ch'ulte' shaving brush tree. Bombacaeae *Pseudobombax ellipticum* (Kunth) Dugand. Native to southern Mexico, El Salvador, Guatemala, and Honduras, *Pseudobombax ellipticum* is a large tree that can attain a height of 18 m (60′) and a trunk girth of 1.3 m (4′). It is a deciduous tree with succulent stems and leaves that are palmately compound with five leaflets. It produces attractive, white or pink blossoms with numerous, erect stamens that extend above the curled petals, producing a wispy look. The fruit is an elongated, dehiscent capsule containing numerous seeds in a fluffy aril. According to Durán (1999), his northern Lacandon consultants said they use the wood to fashion kitchen utensils. *Use:* **che'il k'ak'** 'firewood'; *Use:* **k'an** 'cord'; *Part:* **uyoot'el usool** 'inner bark'. *[Note:* chute. *Pseudobombax ellipticum* (Durán 1999); sak-kuy che' [Yuc.]. *Pseudobombax ellipticum* (in, Atran et al. 2004); ix ch'ulte' [Itz.]. *Pseudobombax ellipticum* (Atran et al. 2004).*]* *[*\sd2 cordage \sd2 fuel*]*

Fig. 5.43 Ch'om mäkäl (hach ch'om) *Xanthosoma robustum*. Photo Credit: Suzanne Cook

ch'um ak' (äh) [ch'um ak', ch'un ak'] passion fruit vine (purple Jamaican honeysuckle). Passifloraceae *Passiflora* aff. *ambigua* Hemsl. This is a cultivated species that bears large, very sweet, yellow fruit in August and September. Inflorescences are large, approximately 6 cm (3″) long from the base to the tip of the flower; petals are bright purple with white on the exterior of the base and a red splotch on the interior of the base. Fruit is approximately 12 cm (7″) in diameter with a bumpy, yellow peel, white pulp, and a milky-white, gelatinous aril surrounding numerous grey seeds. The fruit looks similar to *P. ambigua*, only twice as large. Cultivated. *Loc:* **kolil nah** 'the Kohs' house garden'; *Use:* **hanal** 'food'; *Part:* **wich** 'fruit'; *Prep:* **che'che'** 'raw'. *Thes:* **che'** (Fig. 5.44a, b). *[Note:* chun ak'. *Passiflora edulis* (Nations and Nigh 1980)*] [Source:* BM; CKD; KP*]* *[*\sd2 food*]*

E–e

ek' logwood (bloodwood tree, inkwood). *Lit:* 'black'. Fabaceae *Haematoxylon campechianum* L. *Haematoxylon* grows in dense stands in around lagoons, lakes, and swamps. It is a small, thorny, leguminous tree, with bipinnate leaves, and small and paddle-shaped leaflets. Numerous fragrant blossoms are borne in terminal and axillary racemes. Sepals are rose coloured and petals are yellow. Fruits are flattened pods. The hard, dense, red heartwood yields dyes of bright red and shades of blue. There is no evidence, however, that the Lacandones exploited the tree for this purpose. *Loc:* **petha'** 'lagoon'; *Loc:* **Mensäbäk**; *Loc:* **Yahapetha'** 'Guineo'; *Use:* **che'il k'ak'** 'firewood'; *Use:* **okman** 'pillars'; *Part:* **uk'äbche' i mehen che'** 'branches and saplings'; *Prep:* **ha'li' ats'ilik usool** 'strip off the bark, that's all'; *Use:* **chuxte'** 'arrow foreshafts'. *Thes:* **che'**. *[Note:* ek'. a thorny, gnarled tree with extremely hard, jet-black wood (Bruce 1975); ek'. logwood. *Haematoxylon campechianum* (Nations 2006); ek'. logwood. *Haematoxylum campechiana* (Nations and Nigh 1980); ek [Yuc.]. palo de tinta. *Haematoxylon campechianum* (Roys 1931).*] [Source:* BM; Bruce (1976: 74)*]* *[*\sd2 fuel*]*

ek'el bu'ul bean. *Lit:* 'coloured/spotted bean'. Fabaceae *Phaseolus vulgaris* L. *Use:* **hanal** 'food'; *Part:* **nek'** 'seeds';

Fig. 5.44 (**a**) Ch'un ak' *Passiflora* aff. *ambigua* interior. Photo Credit: Chan K'in Daniel (**b**) Juliana and Omar with ch'un ak' fruit. Photo Credit: Chan K'in Daniel

Prep: **chäkbil, mäxbil, tsähbil** 'boiled, mashed and fried'; *Sim:* **box bu'ul** 'black bean'; *Sim:* **ek' bu'ul** 'black bean'; *Sim:* **chäk bu'ul** 'red bean'; *Sim:* **hach bu'ul** 'authentic bean'; *Gen:* **bu'ul**. *[Source:* AM*]* *[*\sd2 food*]*

ek' bahche' [ek' bahche', ek' bäche', ek' bache, ek' boche'] abnormal guatteria (black cork). Annonaceae *Guatteria anomala* R. E. Fries. This is a large, deciduous tree to 60 m (197′) tall and up to 4–5 m (13–16′) in diameter. Bark is finely fissured, sometimes flakey, and fibrous. Leaves are obovate to oblong-obovate, 8 × 3 cm- 20 × 7 cm (3 × 1″–7.8 × 2.7″), and arranged in a spiral. In April, small, fragrant, yellow flowers are borne between the internodes. These are followed by clusters of dark red or purple, ellipsoid berries, from June to September (Pennington and Sarukhán 2005: 168, Standley and Steyermark 1946: 286–287). A number of species of birds are drawn to the fruit, including *koox* (creasted guan), *kacho'* (mealy blue-crowned parrot), and *pän* (keel-billed toucan). Some sources (Diemont 2006; Levy et al. 2006) report that the Lacandones also eat the fruits, although my consultants failed to mention this. According to BM, the wood splits easily, making it a choice timber to work with. The tree is also good for the *milpa*, because it produces abundant leaf mould that nourishes and conditions the soil (AM). *Loc:* **k'aax** 'forest'; *Use:* **hi'che'** 'fence, pole wall'; *Use:* **hiil** 'battens'; *Use:* **pohche'** 'tables'; *Part:* **che'il** 'timber'; *Sim:* **säk bahche'** 'white bahche' (*Guarea glabra*)'; *Gen:* **bahche'**. *Thes:* **che'** (Fig. 5.45). *[Note:* Taxonomic identification provided by Roy Erkens (pers. comm.). ek' bache'. *Guatteria anomala* (Nations 2006; Nations and Nigh 1980); ek' bache. *Guatteria anomala* (Diemont 2006: 37); ek bache. *Guatteria anomala* (Durán 1999).*]* *[Source:* AM; BM; CKO; K'ayum Segundo; KP; SK*]* *[*\sd2 housing \sd2 agriculture \sd2 fuel \sd2 furniture*]*

ek' bu'ul bean. *Lit:* 'black bean'. Fabaceae *Phaseolus vulgaris* L. This is the common, medium-sized black bean. It is a bit smaller than the red and white varieties. There are bush and pole varieties. *Use:* **hanal** 'food'; *Part:* **nek'** 'seed'; *Prep:* **chäkbil, mäxbil, tsähbil** 'boiled, mashed and fried';

Fig. 5.45 Ek' bahche' *Guatteria anomala*. Photo Credit: Jean Iron

Sim: **box bu'ul** 'black bean'; *Sim:* **ek'el bu'ul** 'coloured/spotted bean'; *Sim:* **chäk bu'ul** 'red bean'; *Sim:* **hach bu'ul** 'authentic bean'; *Gen:* **bu'ul**. *See:* **ek'el bu'ul**. *[Source:* BM*]* *[*\sd2 food*]*

ek' che'il bu'ul beans, bush. *Lit:* 'black bush beans'. Fabaceae *Phaseolus vulgaris* L. *Use:* **hanal** 'food'; *Part:* **wich** 'fruit'; *Prep:* **chäkbil, mäxbil, tsähbil yetel säk seboya yetel axux** 'boiled, mashed and fried with onions and garlic'; *Sim:* **chäk che'il bu'ul** 'red bush beans'; *Gen:* **che'il bu'ul**. *[*\sd2 food*]*

ek' 'ib bean, lima. *Lit:* 'black lima bean'. Fabaceae *Phaseolus lunatus* L. cv. 'Sieva'. This is a variety that is smaller than chäk 'ib 'red lima bean'. *Use:* **hanal** 'food'; *Part:* **nek'** 'seeds'; *Prep:* **chäkbil** 'boiled'; *Sim:* **chäk 'ib** 'red lima bean'; *Gen:* **'ib** 'lima bean'. *Thes:* **bu'ul**. *[Note:*

b'ox ib', säk ib', chäk ib'. [Itz.]. *Phaseolus lunatus* (Atran et al. 2004).*] [Source:* AM; BM*] [*\sd2 food*]*

ek' muluch'₁ [ek' muluch', ek' muruch', ek' muuch'] fungi (types of). *Lit:* 'black mound/pile'. *Oudemansiella* aff. *steffendii, Pluteus harrisii*. Lacandones describe these kinds of fungi as small, black mounds that grow on the ground. AM, KM, and KP said they were fatally poisonous. BM, however, said that there were two kinds—one that was poisonous and the other that was not. Neither species were available to inspect, but Ruan-Soto et al. (2009) recorded two species of "ek' much'" from the southern Lacandon territory—*Oudemansiella* aff. *steffendii* and *Pluteus harrisii*. Apparently, both are edible. Thus, the identity of the inedible species to which BM, KM, and KP refer is uncertain. *Use:* **hanal** 'food'; *Prep:* **chäkbil** 'cooked'; *Sim:* **säk muluch' (ah)** 'white muluch' (*Oudemansiella canarii*)'; *Gen:* **muluch'**. *[Note:* ek' much' [S. Lac.]. *Oudemansiella* aff. *steffendii, Pluteus harrisii* (Ruan-Soto et al. 2009).*] [Source:* AM; BM; KM*] [*\sd2 food*]*

ek' muluch'₂ *Oudemansiella* aff. *steffendii* (common name unknown). Physalacriaceae *Oudemansiella* aff. *steffendii* (Rick) Sing. This is a kind of edible, gilled mushroom. It grows in clumps on rotting leaves and wood. *Use:* **hanal** 'food'; *Prep:* **chäkbil, k'olbil** 'cooked, soup thickened with corn dough'; *Sim:* **säk muluch'** 'white muluch' (*Oudemansiella canarii*)'; *Gen:* **muluch'**. *Variant:* **muluch'**. *[Source:* Ruan-Soto et al. (2009)*] [*\sd2 food*]*

ek' muluch'₃ *Pluteus harrisii* (common name unknown). Pluteaceae *Pluteus harrisii* Murr. *Use:* **hanal** 'food'; *Prep:* **chäkbil, k'olbil** 'cooked, soup thickened with corn dough, and often includes meat'; *Sim:* **säk muluch'** 'white muluch' (*Oudemansiella canarii*)'; *Gen:* **muluch' (ah)**. *Variant:* **muluch'**. *[Source:* Ruan-Soto et al. (2009)*] [*\sd2 food*]*

ek' näl maize. *Lit:* 'black maize'. Poaceae *Zea mays* L. This is the least preferred variety of maize, because it is drier and harder than the yellow and white varieties. *Use:* **hanal** 'food'; *Sim:* **säk näl** 'white maize'; *Sim:* **k'änk'än näl** 'yellow maize'; *Gen:* **näl**. *[Source:* AM*] [*\sd2 food*]*

ek' 'oonte' nectandra, globose. *Lit:* 'black avocado tree'. Lauraceae *Nectandra globosa* (Aubl.) Mez. This is a kind of wild avocado tree with very hard wood. It bears small, oval, black fruits similar to those of *ek' bahche'*. Fruits mature in October, attracting birds and animals, such as: *k'ambul* (curassow); *koox* (creasted guan); *kacho'* (mealy blue-crowned parrot); *hach k'ek'en* (white-lipped peccary); and, *kitam* (collared peccary). According to AM, this tree does not grow around Naha'. *Use:* **che'il k'ak'** 'firewoood'; *Sim:* **säk 'oonte'** 'white 'oonte' (*Licaria alata*)'; *Sim:* **mehen 'oonte'** 'small 'oonte' (*Nectandra* sp.)'; *Sim:* **nukuch 'oonte'** 'large 'oonte' (*Nectandra* sp.)'; *Sim:* **babah 'oonte'** '*Ocotea cernua*'; *Gen:* **'oonte'**. *Thes:* **che'**. *[Note:* econte. *Nectandra globosa* (Nigh 2008); ek' onté. *Nectandra* sp. (Nations and Nigh 1980); ek onte. *Compsoneura sprucei* (Durán 1999); aj tz'unu'unte' [Itz.]. *Nectandra globosa* (Atran et al. 2004). English names from Grandtner (1997–2010).*] [Source:* AM; BM; KA*] [*\sd2 fuel*]*

G–g

gorila bean, Saint Mary's (crucifixion bean, elephant creeper, woolly morning glory). *Lit:* 'gorilla'. Convolvulaceae *Merremia discoidesperma* Donn. Sm. The name might only refer to the seeds of this tree. *Use:* **uuh** 'neckaces'; *Part:* **nek'** 'seeds'. *Variant:* **ho'ol ma'ax**. *From:* gorila (Spanish). *[Source:* CXK; NI*] [*\sd2 craft*]*

H–h

ha'as marmalade plum, mammee sapote. Sapotaceae *Pouteria mammosa* (L.) Cronq., syns. *Manilkara zapota* (L.) P. Royen., *Pouteria sapota* (Jacq.) H. E. Moore & Stearn, *Calocarpum mammosum* Pierre. This is a large tree 7–20 m (22.9–65.6′) tall that bears masses of large, 8×20 cm (3×7.9″), oval fruits attached directly to the branches and trunk of the tree. The exterior rind of the fruit is yellowish-brown, rough, and thin. The flesh is dark orange, sweet, soft and slightly gritty, and encases one to four, glossy brown, spindle-shaped seeds. It usually grows at low elevations from southern Mexico to northern Nicaragua; but because it is widely cultivated, it may naturalize at elevations of up to 600 m (2000′) and occasionally higher (Morton 1987: 398–402). Conceivably, the naturalized forms correspond to *hach ha'as* 'authentic ha'as', which the Lacandones regard as wild. Cultivated. *Loc:* **kolil nah, kol** 'house garden, milpa'; *Use:* **hanal** 'food'; *Part:* **wich** 'fruit'; *Prep:* **che'che'** 'raw'; *Spec:* **hach ha'as** 'authentic ha'as'. *Thes:* **che'** (Fig. 5.46a, b). *[Note:* jaas. Pouteria sapota (Nigh 2008); ha'as. mamey. *Calocarpum mammosum* (Nations and Nigh 1980); ha'as [S. Lac.]. zapote (Canger 1970); haaz [Yuc.]. *Musa sapientum* L. (Millsp. I, 358). "Name orginally applied to the mamey which is now called chacal-haaz" (Roys 1931: 244); ix chäkälja'as, ix nukuch chäkälja'as [Itz.]. *Pouteria mammosa* (Atran et al. 2004); chäkäl-ja'as [Mop.]. *Pouteria mammosa* (in, Atran et al. 2004). Botanical synonyms from J. Morton (1987).*] [*\sd2 food*]*

haban uts'u' (ah) [haban uts'u', habän uts'u'] daisy tree, giant. *Lit:* 'hollow pith'. Asteraceae *Podachaenium eminens* (Lag.) Sch. Bip. **Uch'ihil paakche' kol. Uche'il batak 23 cm (9″). Ne ka'anal ubakel. Ule' nuk, batak utan ule'**

Fig. 5.46 (**a**) Ha'as *Pouteria mammosa*. Photo Credit: Suzanne Cook (**b**) Ha'as *Pouteria mammosa*. Photo Credit: Suzanne Cook

30 cm (12″), uchukuchil batak 30 cm (12″). Yan utop', k'än. Ne ki' ubok utop'. Much'umuch' utop'. Wolis, chen ma' ne nuk, batak 1.5 cm (1/2″). Yan ukab utop'. Uche'il ne tsoy ti' umentik uk'anche' näl. 'It grows in the fallow *milpas*. Its trunk is about 23 cm (9″) in diameter, and it is very tall. Its leaves are very large, about 30 cm (12″) across and 30 cm (12″) long. It has clusters of yellow, very fragrant flowers. The blossoms are round and not very large, about 1.5 cm (1/2″). The flowers have nectar. The stalks are good for making corncribs.' (BM) According to BM, the flowers bloom in March. *Use:* **hiil** 'battens'; *Use:* **hi'che' ti' k'anche' näl** 'pole wall for the corncrib'; *Part:* **che'il** 'stalks'. *Morph:* **hab-an u-ts'u'**. *Variant:* **ki' bok (äh)**$_3$. *Thes:* **che'** (Fig. 5.47). *[Note:* Botanical identification provided by Billie L. Turner (pers. comm.).*] [Source:* AM; BM; SK*] [*\sd2 housing*]*

habón wing-leaf soapberry. Sapindaceae *Sapindus saponaria* L. *See:* **subul**$_1$. *Variant:* **baba', habonsiyo, subul**. *From:* jabón (Spanish). *Thes:* **che'**. *[Source:* Bruce (1975: 153)*]*

habonsiyo wing-leaf soapberry. Sapindaceae *Sapindus saponaria* L. **Ma' uch'ihil way. Unek' ti' kah.** 'It doesn't grow here. The seeds are from the villages.' According to the Lacandon women I consulted, they gather the round, glossy, black seeds about 1/2 hour's drive north of Naha', near the town of Lacandón. *Loc:* **San Luis**; *Use:* **uuh** 'necklaces'; *Part:* **nek'** 'seeds'. *See:* **subul**$_1$. *Variant:* **baba', habón, subul**. *From:* jaboncillo (Spanish). *Thes:* **che'** (Fig. 5.48). *[Source:* KP; NI*] [*\sd2 craft*]*

Fig. 5.47 Haban uts'u' *Podachaenium eminens*. Photo Credit: Suzanne Cook

hach 'akte' palm, choco (choco palm, spiny palm). *Lit:* 'authentic 'akte''. Arecaceae *Astrocaryum mexicanum* Liebm. ex Mart., syn. *Bactris cohune* S. Watson. **Yan uche'il. Yan uk'i'ixel uche'il. Ule', yan uk'i'ixel. Ule' p'iis ahpahok, chukuch uwaan ule'. Uch'ihil ich k'aax, tseltselwits i äka'an. Uyaal uhantik. Ne ki' pokbil ich k'ak' wa tsähbil yetel uye'el kax. Uwich yan uk'i'ixel. Chich uwich. Uyaal p'iis top'. Wa muun uhantik. Wa ma' uhantik, pachil uyi'htal uyaal, uwaak'al utop' y uwaak'al uwich. Ch'i'ib bekil, casi ubäho'. Hay dos classes de hach 'akte': chäk 'akte' yan ule' uyool as chäk, uch'ihil pach wits; ya'ax 'akte' yan unuk uyool, ule' ma' ne ya'ax chen ts'ets'ele' chäk. Uch'ihil yaha'nah.** 'It has a trunk with long, spines. Its leaves have spines. The leaves long like those of *pahok*. It grows in the forest, on the hillsides, and in the low-lying flats. One eats the immature inflorescences. They're very tasty roasted in the fire or fried with eggs. The fruits have spines. They're hard. The immature inflorescences are the flowers. When they are tender one can eat them. If one doesn't eat them soon enough, the inflorescences become hard and split open, revealing the seeds. *Ch'i'ib* (*Chamaedorea tepejilote*) is similar; it's almost its brother. There are two kinds of *hach 'akte'*: red *'akte'* and green *'akte'*. Red *'akte'* has a reddish leaf sheath. It grows on the other side of the hills. Green *'akte'* has a large heart. Its leaves are dull green and a tiny bit red. It grows in Yaha'nah.' (BM) *Astrocaryum* is a genus of about 47 species of clumping or solitary palms with vicious spines on all parts, including the fruits of some species; the spines of *Astrocaryum mexicanum* are strongly flattened (Henderson et al. 1995). Their range extends from Mexico to eastern Brazil and the West Indies. Nations and Nigh (1980) report that the Lacandones eat the fruit. *Loc:* **pach wits, tseltselwits i äka'an, yaha'nah** 'on the other side of the hills, on the hillsides and in the lower, flat lands in Yaha'nah (La Cueva)'; *Use:* **hanal** 'food'; *Part:* **uyaal** 'male inflorescence'; *Part:* **uyool** 'heart'; *Spec:* **chäk 'akte'** 'red akte''; *Spec:* **ya'ax 'akte'** 'green akte''; *Gen:* **'akte'**. *See:* **ch'i'ib**. *Variant:* **'akte'** (Fig. 5.49a, b). *[Note:* chapay, akte. *Astrocaryum mexicanum* (Durán 1999); ak te'. *Astrocaryum mexicanum* (Nations and Nigh 1980).*] [Source:* AM; BM*] [*\sd2 food*]*

Fig. 5.48 Habonsiyo/habón *Sapindus saponaria*. Photo Credit: Suzanne Cook

hach 'ak (äh) lemongrass. Poaceae *Cymbopogon citratus* (DC.) Stapf. **Ähhach 'ak p'iis ahsu'uk. Yan uwi', wolis ukultal. Mäna' uche'il. Yan ule', chukuch, ma' kooch ule'. Yan ubok' ule', p'iis mulix. Umentik uchäkik ule' yetel ha' ti' uyuk'ik. Ne tsoy ti' se'em.** *'Hach 'ak* is the same as grass. It has a tuber that sits in a circle. It doesn't have a main stalk. It has long, narrow leaves. The leaves smell like lemon. It's good for a cold'. (BM). *Use:* **ts'ak ti' se'em** 'cold medicine'; *Part:* **le'** 'leaves'; *Prep:* **Achäkik seis ule', pachil awuk'ik** 'You boil six leaves in water and then drink it'. *Variant:* **'ak (ah)**. *Thes:* **su'uk**. *[Source:* AM; BM; KP*] [*\sd2 medicinal*]*

hach balche' lancepod. *Lit:* 'authentic balche''. Fabaceae *Lonchocarpus* sp. **Säk uche'il. Ule' ma' ya'ax. Ne ya'ax utop'.** 'Its trunk is white. Its leaves are not very green. Its flowers are very blue.' (BM) The Lacandones' principal deity, *Hachäkyum*, created *balche'* as a means for his mortals to venerate him. He gave them the tree along with instructions on how to ferment the bark and details on ritual protocol. It is said that the gods like to drink, and that they like to see their mortals happy and drunk as well; but *balche'* is only

a

b

Fig. 5.49 (**a**) Hach 'akte' *Astrocaryum mexicanum* thorns. Photo Credit: Suzanne Cook (**b**) Hach 'akte' *Astrocaryum mexicanum*. Photo Credit: Suzanne Cook

to be consumed in ritual contexts. In the extant Lacandon literature, *hach balche'* is associated with *Lonchocarpus longistylus*. Standley and Steyermark (1946) remark that *L. longistylus* occurs in the Yucatan, where it is known by the Maya name *balche'* and from which the Mayas made the fermented beverage by the same name. Tozzer (1907) witnessed the Lacandones in the Guatemalan Petén making *balche'*, giving Standley and Steyermark reason to believe that it too was *L. longistylus*, or one or more closely related species (1946: 279). Subsequent researchers in the Lacandon forest indicate that *hach balche'* corresponds to *L. longistylus* (Kashanipour and McGee 2004; Nations and Nigh 1980; Nigh 2008; Soustelle 1966). Rätsch (1994a) contends that it corresponds *to L. violaceus*, a species used in the Yucatan. *Lonchocarpus violaceus* is (or was) synonymous with *L. longistylus* Pitt. (Folan et al. 1979: 697, Lima et al. 1977: 195, cited in Hanelt et al. 2001: 652), but neither *L. longistylus* nor *L. violaceus* is recorded from the Lacandon forest (Breedlove 1986; Comisión Nacional de Áreas Naturales Protegidas 2006; Durán 1999; Levy et al. 2006). According to other researchers in the area, *hach balche'* also corresponds to *L. cruentus* (syn. *L. sericeus*) (Nations and Nigh 1980) and *L. castilloi* (Nations 1992, in Atran et al. 2004). *Loc:* **k'aax** 'forest'; *Loc:* **kol** 'milpa'; *Loc:* **paakche' kol** 'fallow milpa'; *Use:* **balche'** 'fermented beverage'; *Part:* **uyoot'el** 'inner bark'; *Prep:* **Ubäxik uche'il pachil uhakts'oltik upach pachil uts'ilik uyoot'el** 'One pounds a stick of wood, then rips off the outer bark, and then strips out the inner bark'; *Sim:* **ya'ax balche'** 'green balche' (*Lonchocarpus* spp.)'; *Gen:* **balche'** 'balche' (*Lonchocarpus* spp.)'. *Thes:* **che'** (Fig. 5.50a, b, c, d). *[Note:* hach balche'. *Lonchocarpus longistylus, Lonchocarpus cruentus* (Nations and Nigh 1980); machich che. *Lonchocarpus rugosus* (Durán 1999); män~chiich [Itz.]. center~hard = cabbage bark [Maya wine tree]. *Lonchocarpus castilloi* (Atran et al. 2004; Lois 1998); manchich [Mop.]. *Lonchocarpus castilloi* (in, Atran et al. 2004).*] [Source:* BM; AM; SK, and others*] [*\sd2 food*]*

hach bamax *Pseudolmedia oxyphyllaria* (common name unknown). *Lit:* 'authentic bamax'. Moraceae *Pseudolmedia oxyphyllaria* Donn. Sm. **Es che'. Yan utop', chan nuk, batak 2.45 cm (1″), i wolis. Säk utop'. Uwich ti' hach bamax as chukuch, batak 1.27 cm (1/2″). Ne yuul uwich. Ne chäk. Hump'eli' unek'. Ch'ihil ich k'aax, yok'ol wits.** 'It's a tree. It has flowers that are a little bit large, about 2.45 cm (1″), and round. The flowers are white. The fruit is somewhat long, about 1.27 cm (1/2″). The fruit is really smooth. There is one seed. It grows in the forest, in the hills.' (BM) The leaf venation of *hach bamax* is pinnate with more than 18 pairs of subhorizontal secondary veins, giving the surface a fine, smooth texture with a glossy finish. The leaf margins are slightly wavy. These features differentiate *hach bamax* from *tso'ots bamax*. According to SK, the leaves of *tso'ots bamax* are larger, and the fruit is somewhat smaller and furry. In addition to the edible fruit, the tree provides material for construction materials. Durán's consultants reported that the wood is also used to make tool handles (Durán 1999: 146). *Loc:* **wits** 'hills'; *Use:* **säkche'** 'rafters'; *Part:* **k'äbche'** 'branches'; *Use:* **hanal** 'food'; *Part:* **wich** 'fruit'; *Prep:* **che'che'** 'raw'; *Sim:* **tso'ots bamax** 'hairy bamax (*Pseudolmedia oxyphyllaria*)'; *Gen:* **bamax (äh)**. *Thes:* **che'**. *[Note:* Botanical identification provided by Servando Carvajal (pers. comm.). has bamax. *Pseudolmedia oxyphyllaria* (Durán 1999); bombash (Baer and Baer 1952); (ix) mä'na'ax [Itz.]. *Pseudolmedia oxyphyllaria* (Atran et al. 2004); mänax [Mop.]. *Pseudolmedia oxyphyllaria* (in, Atran et al. 2004).*] [Source:* BM; SK*] [*\sd2 food \sd2 housing*]*

hach boox banana, type of. *Lit:* 'real banana'. Musaceae *Musa acuminata* ×*Musa balbisiana* Coll. This is a variety that bears fruit about 15 cm (6″) long, stout, and slightly tapering at each end. The flesh is yellowish, firm, and somewhat tart. This is likely the 'Manzana' (Apple) variety, which has a sub-acid, apple flavour (California Rare Fruit Growers 1996, 1997). *Use:* **hanal** 'food'; *Part:* **wich** 'fruit'; *Prep:* **che'che' wa tsähbil wa huybil yetel ma'ats'** 'eaten raw, or fried, or stirred with posol'; *Sim:* **chäk boox**; *Gen:* **boox** (Fig. 5.51a, b, c). *[Note:* ix gineeyoj mansaanaj [Itz.]. "apple" banana (Atran et al. 2004).*] [Source:* BM; CKD; KM; KP*] [*\sd2 food*]*

hach bo'oy$_1$ xate macho (jade). *Lit:* 'authentic bo'oy'. Arecaceae *Chamaedorea oblongata* Mart. Many features that AM and BM use to distinguish *hach bo'oy* actually correspond to those of *C. oblongata*. The two plants that AM identified as *hach bo'oy* displayed fruits that were turning from green to black without passing through the orange and red stages. This is one distinguishing feature of *C. oblongata*. The other feature is the whitish undersides of the leaflets. It has a solitary habit and grows to 3 m (9.8′) in height. Chamaedorea *neurochlamys* is similar, except for its green leaf sheath and its unripe green fruit (Garwood and Bridgewater 2005). *Loc:* **k'aax** 'forest'; *Use:* **ti' utap'ik ba'al** 'to cover things'; *Use:* **pop** 'mats'; *Use:* **waal** 'fan'; *Use:* **luuch ti' balche'** 'spoon for the balche''; *Use:* **ti' konol** 'to sell'; *Part:* **le'** 'leaves'; *Sim:* **ch'ibix** (*Chamaedorea* spp.); *Sim:* **säk bo'oy** 'white bo'oy'; *Sim:* **k'än bo'oy** 'yellow bo'oy'; *Sim:* **chan mehen bo'oy** 'very small bo'oy'; *Sim:* **ch'it bo'oy** (*Chamaedorea* sp.); *Sim:* **ton bo'oy** (*Chamaedorea* sp.); *Gen:* **bo'oy**. *Variant:* **hade**. *[Source:* AM*] [*\sd2 industry \sd2 religion*]*

hach bo'oy$_2$ *Chamaedorea pinnatifrons* (common name unknown). *Lit:* 'authentic bo'oy'. Arecaceae *Chamaedorea pinnatifrons* (Jacq.) Oerst. **Es bo'oy, mas ne nuk. Che'il, ne chich. Ne ma' nuk uche'il. Ya'ax uche'il. Yan ule', yan**

Fig. 5.50 (**a**) Hach balche' *Lonchocarpus* sp. Photo Credit: Suzanne Cook (**b**) Hach balche' seeds. Photo Credit: Suzanne Cook (**c**) Hach balche' seed pod. Photo Credit: Suzanne Cook (**d**) Hach balche' trunk. Photo Credit: Suzanne Cook

uch'ibil uche' ule'. Yan uchan mehen che' ule'. Uk'äb, ma' ne tah uk'äb ule', i chukuch uwaan. Yaab uk'äb, yaab ule', su'– way hump'el, way hump'el, way hump'el, tatahan, tatahan yaab ule'! Chukuch uwaan ule' bo'oy. P'iis ch'i'ib. Yan uwich. Uwich, wa nen ya'ax, muun. Wa nen täk'än, ek'. Ele, ne täk'än umäk'ik ch'iich'. Ha'li' pän umäk'ik. Tuwolol ch'iich' ma' hach umäk'ik. Ne yaab unek', ne yaab. Creo que uxäkik, tint'an yan cien, o

Fig. 5.51 (**a**) Hach boox *Musa acuminata* ×*Musa balbisiana*. Photo Credit: Suzanne Cook (**b**) Hach boox *Musa acuminata* ×*Musa balbisiana*. Photo Credit: Suzanne Cook (**c**) Hach boox *Musa acuminata* ×*Musa balbisiana*. Photo Credit: Suzanne Cook

mas de cien, uwich. 'This is the largest of all the kinds of *bo'oy*. It has a very hard stalk. It isn't a very large stalk. It's green. There are a lot of leaves, the leaf stalks have veins. The leaf stems are small. The leaf stems aren't very straight, but they are long. There are many branches, thick with leaves – here is one, here is one, and here is one, one after another! The leaves of *bo'oy* are long, similar to *ch'i'ib* (*Chamaedorea tepejilote*). The fruits are green when immature. When they are mature, they are black. When they are very ripe, the toucan eats them. All the other birds don't really eat them much. There are a lot of fruits. I think if one were to count them, there would be as many as 100, or more.' (BM) In another interview, BM adds that the leaflets are very green on both surfaces, as opposed to those of *säk bo'oy*, which are whitish underneath. According to AM, this is why they also call it *ya'ax bo'oy* 'green bo'oy', or *ya'ax hach bo'oy* 'green authentic bo'oy'. AM and BM said that the fruits go through a yellow stage before turning black. Given this, the species must be *C. pinnatifrons*, since the fruits of this species change colour from green to yellow, orange, or even red, when soft ripe, and then age to brown or black (Hodel 1992: 106). However, AM and BM also said that *hach bo'oy* fruits turn from green to black without passing through the orange and red stages, and that the leaflets are thicker than those of *säk bo'oy*, features that are characteristic of *C. oblongata*. Given the Lacandones' intimate and lengthy relationship to their environment, it is surprising that they would mistake a species that plays such a central role in their traditional culture. Discrepancies in identification are likely due to the great variably *C. pinnatrifrons* displays. It is the most widespread and variable species of *Chamaedorea,* accounting for the innumerable synonyms and erroneous references to other *Chamaedorea* (particularly *C. oblongata* and *C. neurochlamys*) in herbaria and the botanical literature (Hodel 1992: 104, 106). Moreover, leaf thickness, venation, and colour vary within the species. According to Hodel (pers. comm.),

Fig. 5.52 (**a**) Hach bo'oy *Chamaedorea pinnatifrons*. Photo Credit: Suzanne Cook (**b**) Hach bo'oy *Chamaedorea pinnatifrons*. Photo Credit: Suzanne Cook

C. pinnatifrons and *C. oblongata* are almost impossible to tell apart without staminate flowers. This could account for misidentification and naming discrepancies, since Lacandones do not distinguish species of *bo'oy* on the basis of their flower structures. *Use:* **ti' u tap'ik ba'al** 'to cover things'; *Use:* **pop** 'mat'; *Use:* **waal** 'fan'; *Use:* **luuch ti' balche'** 'spoon for the balche'; *Use:* **ti' konol** 'to sell'; *Part:* **le'** 'leaves'; *Sim:* **ch'ibix (ah)** (*Chamaedorea elegans*); *Sim:* **säk bo'oy** 'white bo'oy (*Chamaedorea* spp.)'; *Sim:* **k'än bo'oy** 'yellow bo'oy (*Chamaedorea neurochlamys*)'; *Sim:* **chan mehen bo'oy** 'very small bo'oy (*Chamaedorea* sp.)'; *Sim:* **ch'it bo'oy** (*Chamaedorea* sp.); *Sim:* **ton bo'oy** (*Chamaedorea* sp.); *Gen:* **bo'oy**. *Variant:* **ya'ax hach bo'oy** (Fig. 5.52a, b). *[Source:* AM; BM; CKD; K'ayum Segundo; KM*]* *[*\sd2 industry \sd2 religion*]*

hach box bean. *Lit:* 'authentic bean'. Fabaceae *Phaseolus vulgaris* L. This is a vining plant that bears pods with large, black, edible beans. *Use:* **hanal** 'food'; *Part:* **nek'** 'seeds'; *Prep:* **chäkbil, mäxik i tsähbil, mukbil wah** 'boiled, mashed and fried, baked in a large tamale'; *Sim:* **chäk box** 'red box'; *Gen:* **box**. *Variant:* **nukuch box**. *Thes:* **bu'ul**. *[Source:* AM*]* *[*\sd2 food*]*

hach bu'ul bean. *Lit:* 'authentic bean'. Fabaceae *Phaseolus vulgaris* L. Beans are black and larger than *ek' bu'ul*, about 6.3 cm (0.25″) long, and slightly kidney-shaped. The name may be a synonym for *hach box*. *Use:* **hanal** 'food'; *Part:* **nek'** 'seeds'; *Prep:* **chäkbil, tsähbil** 'boiled, fried'; *Sim:* **box bu'ul** 'black bean'; *Sim:* **ek' bu'ul** 'black bean'; *Sim:* **ek'el bu'ul** 'coloured/spotted bean'; *Sim:* **chäk bu'ul** 'red bean'; *Gen:* **bu'ul**. *See:* **hach box**. *[*\sd2 food*]*

hach chiah (ah) abraza palo (English common name unknown). Araceae *Philodendron inaequilaterum* Liebm., syns. *P. coerulescens* Engl., *P. belizense* Standl., *P. guatemalense* Engl., *P. silvaticum* Engl., *P. zulianum* (G. S. Bunting) Croat & N. Lamb. **Mäna' uyek'el, ule' wolol ya'ax. Yan utop', säk, wolis. Yan usool utop'. Ne ki' ubok, p'iis buk'luch'. Uch'ihil ich k'aax. Yan uhel, ek'el ule'. Päk'bil.** 'The leaves are all green, without any patterns. The flower is white and round, and it has a skin (spathe). It smells very good, like vanilla. It grows in the forest. There is another kind with patterned leaves. That one is cultivated.' (AM) *Philodendron inaequilaterum* is a hemiepiphytic, slender vine that climbs appressed to its host tree. The stem has multiple branches that are long, thin 1.5 cm (0.5″) diameter, and spreading and almost leafless. Leaves are broadly ovate or elliptic-ovate, frequently oblique, abruptly acuminate at apex, and rounded or truncate at base. Inflorescences are born on short, stout peduncles between 2 and 4 cm (0.7–1.5″) long. The spathe is usually white or green and 12–20 cm (5–8″) long. The spadix is up to 9 cm (3.5″) long and 2.5 cm (1″) wide. Fruits are orange, angulate, up to 4 mm (0.15″) long and 2 mm wide (0.07″). Six to twenty oblong seeds are immersed in a clear, sweet, watery tissue. No uses were mentioned. *Loc:* **k'aax** 'forest'; *Sim:* **chiah (ah)** 'dumbcane (*Dieffenbachia* sp.)'. *Variant:* **chiah k'aax**. *Thes:* **ak'** (Fig. 5.53a, b). *[Source:* AM*]*

hach chulul [hach chulul, hachuruh, haxchuruh] guayacan trumpet tree (ipe, Brazilian walnut). *Lit:* 'real bow'. Bignoniaceae *Tabebuia guayacan* (Seem.) Hemsl., syn. *Handroanthus guayacan* (Seem.) S. O. Grose. The Lacandones distinguish two varieties of *hach chulul*. These

Fig. 5.53 (**a**) Hach chiah (chiah k'aax) *Philodendron inaequilaterum* leaves. Photo Credit: Suzanne Cook (**b**) Hach chiah *Philodendron inaequilaterum*. Photo Credit: Suzanne Cook

are *ek' chulul* 'black chulul' and *chäk chulul* 'red chulul'. BM says that *hach chulul* has more force and is therefore preferred over other woods, such as *chintok'* (*Krugiodendron ferreum*). However, he says it is not so good for arrows, because the wood is too heavy; when used for fishing, arrows made from the timber sink. KyP adds that *hach chulul* deposits a poisonous powder on the arrows, which aids in killing prey. The botanical identity of *hach chulul* is uncertain. In the extant Lacandon literature (e.g., Nations and Nigh 1980; Nations 1989; March 1998; Tozzer 1907), *hach chulul* corresponds to *Guaiacum sanctum* 'holywood lignum-vitae'. Based on the Janka hardness test, *Guaiacum* species are the hardest trade wood and sink in water. In addition, *Guaiacum* has been reported to cause skin irritation (http://www.wood-database.com/lumber-identification/hardwoods/lignum-vitae/). However, according to López-Toledo et al. (2008), in Mexico *Guaiacum sanctum* is generally confined to the coast and the Isthmus of Tehuantepec in Oaxaca, the Central Depression in Chiapas, and parts of Campeche, Quintana Roo, and the Yucatan (López-Toledo et al. 2008: 2, 6). Moreover, the species is not listed in the current botanical inventory from the northern Lacandon forest (Comisión Nacional de Áreas Naturales Protegidas 2006). Therefore, if the Lacandones used the wood from this species, they either traded for it or had to go far away to collect it. More likely, *hach chulul* corresponds to *Tabebuia guayacan* (Hernandez-Solano 1999:xix, Miranda 1953), which is found in the Lacandon community of Mensäbäk. Its timber is also among the most durable, with a high resistance to rot and insect attack; certain species of *Tabebuia* contain powdery yellow deposits within the wood, which cause skin, eye, and respiratory irritation. In addition, the heartwood varies from reddish brown to olive brown to blackish brown. *Loc:* **Mensäbäk**; *Use:* **chulul** 'bow'; *Spec:* **ek' hach chulul** 'black authentic chulul'; *Spec:* **chäk hach chulul** 'red authentic chulul'; *Gen:* **chulul**. *See:* **chintok'**. *Variant:* **chulul**. *Thes:* **che'**. *[Note:* hach chulul. *Guaiacum sanctum* (Nations and Nigh 1980; Nations 1989; March 1998). Taxonomic synonym from The Plant List. http://www.theplantlist.org/tpl/record/kew-347064 Retrieved April 2014.*] [Source:* AM; BM; CKO; KyP*] [*\sd2 hunting_and_fishing*]*

hach ch'alol oak, corrugated. *Lit:* 'authentic oak'. Fagaceae *Quercus corrugata* Hook., syns. *Q. cyclobalanoides* Trel., *Q. pilgeriana* Seem., *Q. reevesii* Trel. This is a large tree that may exceed 50 m (164′) in height, with a trunk diameter of up to 2.5 m (8′). Its acorn is nearly round or broadly ellipsoid, 3–5 cm (1.18–1.96″) in diameter, with a layer of short, matted, woolly hairs at the apex. The cupule is textured with wooly scales, enclosing 1/3 of the length of the nut. Leaves are deciduous, thick, leathery, and vary in shape from lanceolate to oblanceolate (broader above the middle and tapering toward the base), elliptical, or obovate (rounded at apex and tapering toward the base). *Quercus corrugata* occurs in Mexico (Chiapas, Guerrero, Hidalgo, Oaxaca, Puebla, Veracruz), Belize, Costa Rica, Guatemala, and El Salvador, between 850 and 2300 m (2789–7546′) in elevation. (Hélardot n.d.) According to Durán (1999), the northern Lacandones use the timber for house walls. *Loc:* **pachwits near Acte'**; *Use:* **polbilich** 'spinning top'; *Part:* **wich** 'fruit'; *Sim:* **mehen ch'alol** 'small ch'alol (*Quercus* sp.)'; *Sim:* **pixan k'ambul (äh)** 'curassow's heart (*Quercus skinneri*)'; *Gen:* **ch'alol** 'oak (*Quercus* spp.)'. *Variant:* **nukuch ch'alol**.

Thes: **che'**. *[Note:* English name from Grandtner 2005. ixim charo. *Quercus corrugata* (Durán 1999).*] [Source:* AM; BM*] [*\sd2 other*]*

hach ch'om elephant ears, capote. *Lit:* 'authentic ch'om'. Araceae *Xanthosoma robustum* Schott. This is a kind of *mäkäl* that lacks an edible tuber. It has enormous leaves and stems that contain an irritating, white resin. It ranges from Mexico to Panama. *Use:* **booch'** 'umbrella'; *Part:* **le'** 'leaves'; *Gen:* **mäkäl**. *Variant:* **ch'om mäkäl**. *Thes:* **xiiw**. *[Note:* ch'om is also the word for vulture*] [Source:* AM; BM*] [*\sd2 other*]*

hach ha'as sapote (type of). Sapotaceae *Pouteria* sp. **Nuk che'. Ma' amäk'ik. Uyo'och bäk'.** 'This is a large tree. You don't eat (the fruit). It is animal food'. (AM) According to AM and BM, the fruits are similar to *ha'as* (*Pouteria mammosa*), only much smaller. It could be *a wild form of P. mammosa*. *Gen:* **ha'as** 'marmalade plum (*Pouteria mammosa*)'. *Thes:* **che'**. *[Source:* AM*]*

hach halol majagua (sunfruit). Malvaceae *Heliocarpus donnell-smithii* Rose. This is a large tree whose fibre is used to make cord for bows, hanging baskets, and net bags. The processed fibre is called *hol*. *Use:* **k'an** 'cord'; *Sim:* **chäk halol** 'red halol'; *Gen:* **halol**. *Variant:* **säk halol**. *Thes:* **che'** (Fig. 5.54). *[Source:* BM*] [*\sd2 cordage*]*

hach häläl reed, common. *Lit:* 'real arrow'. Poaceae *Phragmites australis* (Cav.) Trin. ex Steud. **P'iis su'uk. Ma' ne nuk uche'il. Casi ket yetel 'ooh. Es lo'obil. Uch'ihil tuhunal, ma' päk'bil. Uch'ihil tu' ne lu'um, tu' ne hup, p'iis pantano. Uch'ihil ma' ne naach ti' naha', xok'ol uxuul petha'—puro pantanos, luk'. Uche'il ne 'aal. Cuando ne yi'h, as säk k'änk'än. Uche'il haban uts'u'. Yan umo'ok, p'iis cana, p'iis semet'.** 'It's the same as grass. It doesn't have a large stalk. It's almost the same as *'ooh*. It's a weed. It grows on its own; it's not cultivated. It grows in very soft soil, like marshes. It grows not far from Naha' on the banks of the lagoon, in swamps and mud. The stalk is very heavy. When it's mature, it's somewhat whitish yellow. The stalk is hollow. It has knots, the same as cane and *semet'* (*Rhipidocladum bartlettii*).' (BM) The species BM describes is likely *Phragmites australis*, a tall grass between 1.5 and 3 m (4.9–9.8′) tall. The hollow, rigid, woody stalks are nearly 2.54 cm (1″) in diameter. Leaves are narrow, 10–35 mm (0.4–1.4″) wide and 20–50 cm (7.9–20″) long. Flowers are produced in dense, dark purple panicles, 20–50 cm (7.87–19.68″) long. The feathery seed-heads turn light brown when mature. According to BM, *hach häläl* lacks the flower spike characteristic of *'ooh* (*Gynerium sagittatum*). In this respect, he says it resembles *wakäx su'uk* (*Pennisetum purpureum*). The stalks of *Phragmites australis* were once used for arrow shafts, but were supplanted by *'ooh* (AM). *Loc:* **Yaha'nah** 'La Cueva'; *Use:* **häläl, yolil** 'arrows, arrow shaft'; *Part:* **uche'il** 'stalk'; *Sim:* **häläl** '*Gynerium sagittatum*'; *Gen:* **häläl**. *See:* **'ooh**. *Variant:* **ho'te' che'**. *Thes:* **su'uk**. *[Note:* Prior to the publication of *Gynerium* (P. Beauvois 1812), most species of reed-like grass referred to *Arundo* L. Since then, research on the morphology of *Gynerium sagittatum* has uncovered a number of features, such as the presence or absence of pubescence on the sheath and differences in the colour and texture of leaf blade, that distinguish it from *Arundo*. (Kalliola and Renvoize 1994: 311)*] [Source:* AM; BM*] [*\sd2 hunting_and_fishing*]*

Fig. 5.54 Hach halol *Heliocarpus donnell-smithii*. Photo Credit: Suzanne Cook

hach hänan bayal (basket palm, liana palm). Arecaceae *Desmoncus orthacanthus* Mart. **Es ak'. Ne yan uk'i'ixel. Uwich, chak tahal. Ma' ahantik. Unek' ti' uuh. Ahantik uyool. P'iis ch'i'ib.** 'It is a vine. It has a lot of spines. The fruit turns red when it matures, but you don't eat it. The seeds are used for necklaces. You eat the infloresences. They're like *ch'i'ib* (*Chamaedorea tepejilote*).' (KP) This is a climbing palm with clustered stems of up to 12 m (39′) long. Numerous (10–50) leaves are arranged in two rows. Leaf sheathes are covered with short, black spines. Leaflets are eliptical to linear with black spines on the lower surface. Long, 6 cm (2.4″), black spines cover the petioles (leaf stems). Terminal leaflets

Fig. 5.55 (**a**) Hach hänan *Desmoncus orthocanthus*. Photo Credit: Suzanne Cook (**b**) Hach hänan *Desmoncus orthocanthus*. Photo Credit: Suzanne Cook (**c**) Hach hänan *Desmoncus orthocanthus* leaves and stem. Photo Credit: Suzanne Cook

are short, stiff, backward-pointing hooks that help the plant cling to surrounding vegetation. It bears clusters of red, orange or yellow-orange ellipsoid to round fruits of up to 2 cm (0.8″) in diameter. It is found growing in disturbed areas, secondary forests, and river margins at elevations below 1000 m (3280.8′) from the Atlantic slope of southern Mexico to Central America, South America, and Trinidad and Tobago. (Henderson et al. 1995: 199) The Lacandones have observed the local Tzeltales making baskets from the vines (BM, KA), something which the Lacandones do not do (BM, K'ayum Segundo). *Loc:* **kol, k'aax** 'milpa, forest'; *Use:* **chäk hu'un ti' pak** 'ceremonial headband for the balche' urn'; *Part:* **le'** 'leaves'; *Use:* **me'et** 'trivet'; *Part:* **che'il** 'stems'; *Use:* **uuh** 'necklaces'; *Part:* **nek'** 'seeds'; *Prep:* **Asusik usool ich ha'** 'You scrape off the outer skin in water'; *Use:* **hanal** 'food'; *Part:* **uyol** 'inflorescence buds'; *Prep:* **pokbil, chäkbil, atsähik yetel uye'el kax** 'roasted or boiled, you can fry it with eggs'; *Sim:* **ya'ax hänan** 'green hänan (*Chamaedorea elatior*)'; *Gen:* **hänan**. *Variant:* **hänan**. *Thes:* **ak'** (Fig. 5.55a, b, c). *[Note:* The southern Lacandones are reported to use *Desmoncus* sp. for crafts, food, and fodder (Levy et al. 2006).*] [Source:* AM; BM; CNK; KM; KP; K'ayum Segundo*] [*\sd2 craft \sd2 religion*]*

hach hoben Mexican pepperleaf (eared piper, rootbeer plant). *Lit:* 'authentic hoben'. Piperaceae *Piper auritum* Kunth, syn. *Piper sanctum* (Miq.) Schltdl. **Yan umokok. Uche'il mokmok. Nuk ule'. Wolis ule'. Ne muun uyool, p'iis umuunil ule'. Ne ki' yetel uk'ab aht'unu'. Utop' chukuch uchan top'. Ha'li' ch'iich' uhantik.** 'It has knots. The stalk is knotty. The leaves are large and round. The growing tip is very tender; it's like a new leaf. It's tasty cooked in snail broth. The flower (stalk) is long and the blossoms are small. Only birds eat (the seeds).' (BM) *Piper auritum* is a member of the pepper family, a genus of evergreen shrubs, small trees, and lianas in the Neotropics. The genus is distinguished by their white, cylindrical, erect inflorescences and winged petioles. *Piper auritum* is a 2.5–8 m (8.2–26′) tall tree, with a slender, 10 cm (3.9″), soft-wood, trunk. Leaves are large, ovate, 20–40 cm (7.8–15.7″) long and 12–27 cm (4.7–10.6″) wide, acuminate at the apex and cordate at base, forming unequal basal lobes, pinnately veined, and covered with a soft, short, white fuzz. When crushed, the leaves and stems smell strongly of root beer, hence the English common name "root beer plant". The leaves, seeds, and roots are used throughout Latin America as a diaphoretic, diuretic, emollient, stimulant, mild aesthetic, and anti-inflammatory (Duke 2003: 251). *Loc:* **paakche' kol** 'fallow milpas'; *Use:* **hanal** 'food'; *Part:* **mumuun ule', uyool ule'** 'young leaves and leaf shoots'; *Prep:* **che'che'** 'raw'; *Prep:* **chäkik yetel uk'ab aht'unu'** 'boiled in snail broth'; *Sim:* **säk hoben** 'white Piper (*Piper umbellatum*)'; *Gen:* **hoben**. *See:* **hach lo'obil**. *Variant:* **hoben**. *Thes:* **che'** (Fig. 5.56a, b). *[Note:* jover. *Piper auritum* (Nigh 2008); 'ob'eel. momo (Itz.). Santa Maria. *Piper auritum* (Atran et al. 2004; Hofling and Tesucún 1997); ob'eel [Mop.]. Piper auritum (in, Atran et al. 2004); ix makulan [Yuc.]. *Piper auritum* (in, Atran et al. 2004).*] [Source:* AM; BM; KM; KP; K'ayum segundo; SK*] [*\sd2 food \sd2 other*]*

hach hut'kih moonflower (moon vine). Convolvulaceae *Ipomoea alba* L. **Hach hut'kih, ele tukinsik ahk'ik'eh. Ne seeb ukinsik. Ahhach hut'kih, hump'eli' unek'. Ne yan, wolol yan uwich utop'. Ne yah utop'el. Wa ne uyak'il ne**

Fig. 5.56 (**a**) Hach hoben *Piper auritum*. Photo Credit: Suzanne Cook (**b**) Hach hoben *Piper auritum*. Photo Credit: Suzanne Cook

Fig. 5.57 (**a**) Hach hut'kih *Ipomoea alba* flower. Photo Credit: Suzanne Cook (**b**) Hach hut'kih *Ipomoea alba* stem. Photo Credit: Suzanne Cook (**c**) Hut'kih *Ipomoea alba* latex. Photo Credit: Suzanne Cook

yah utop'. Säk, wolol säk. Ne säk utop'. *'Hach hut'kih* is used to kill (vulcanize) the latex. It kills it very rapidly. *Hach hut'kih* has one seed. There are a lot of flowers and all the flowers have seeds. Its flowers are very large. If there are a lot of vines, the flowers are very large. They're completely white. (AM) AM says that as a child he would blow through the base of the flower, inflate it, and then release the air over the cooking fire, to fan the flames. He says the flower gets it name by the sound it makes when one blows through it—*HUT', HUT'*. *Ipomoea alba* is a perennial, herbaceous liana, 5–30 m (16–98′) tall with twining stems. Leaves may be entire or three-lobed, 5–15 cm (1.9–5.9″) long, with a 5–20 cm (1.9–7.9″) long stem. The flowers are fragrant, white or pink, and large, 8–14 cm (3–5.5″) in diameter. Like the Lacandones, the Mayan ancestors used the resin from *Ipomoea alba* to convert the latex from the *Castilla elastica* tree to rubber. *Use:* **ti' ukinsik k'ik'** 'to vulcanize rubber'; *Part:* **uyits uche'il** 'resin from the stems'; *Prep:* **Uyäts'tik uyits pachil puk'ik ich ha' yetel uyits ahk'ik'** 'One squeezes the resin and then dissolves it in water with the latex of *Castilla elastica*'; *Sim:* **chäk hut'kih** 'red hut'kih (*Ipomoea* sp.)'; *Sim:* **nukuch hut'kih** 'large hut'kih'; *Gen:* **hut'kih**. *See:* **ts'ak k'ik'**. *Variant:* **hut'kih**. *Thes:* **ak'** (Fig. 5.57a, b, c). *[Source:* AM (2010)*]* *[*\sd2 other*]*

hach hu'un ficus (type of). *Lit:* 'authentic bark cloth'. Moraceae *Ficus* sp. This is a large, free-standing ficus. *Use:* **hu'un, chäk hu'un, hach nook'** 'bark cloth, bark cloth headbands, authentic clothing'; *Part:* **uyoot'el sool** 'its inner bark'; *Sim:* **ak' hu'un** (*Poulsenia armata*); *Sim:* **bits'kal hu'un** 'strangle hu'un (*Ficus* sp.)'; *Gen:* **hu'un**. *See:* **chäk hu'un**. *Thes:* **che'** (Fig. 5.58a, b). *[Note: hach hu'un.* amate (Bruce 1968); hach hun. *Ficus* sp. (Nations and Nigh 1980).*]* *[Source:* AM*]* *[*\sd2 clothing*]*

hach ik chile, Tabasco. *Lit:* 'authentic chile'. Solanaceae *Capsicum frutescens* L. *Loc:* **kol, kolil nah** 'milpa, house garden'; *Use:* **hanal** 'food'; *Part:* **wich** 'fruit'; *Prep:* **che'che'** 'raw'; *Gen:* **ik**. *[Source:* KP*]* *[*\sd2 food*]*

hach 'ilon [hach 'ilon, hach 'iron] Swiss cheese plant. *Lit:* 'authentic 'ilon'. Araceae *Monstera* sp. **Yan uyak'il, yuul, chen no pegan al árboles.** 'It has smooth vines, but they

Fig. 5.58 (**a**) Hach hu'un *Ficus* sp. Photo Credit: Suzanne Cook (**b**) Hach hu'un inner bark. Photo Credit: Suzanne Cook

don't stick to trees.' (BM). *Use:* **xaak** 'baskets'; *Part:* **uyak'il** 'its vines'; *Prep:* **Uts'oolik usool uyak'il, ha'li'** 'One just strips the outer bark from the vines'; *Sim:* **säk 'ilon** (*Monstera* sp.); *Sim:* **kamis 'ilon** (*Monstera deliciosa*); *Gen:* **'ilon**. *Thes:* **ak'** (Fig. 5.59). *[Source:* AM; BM*]* *[*\sd2 cordage \sd2 containers*]*

hach k'ik'ni' balum coralberry (marlberry). *Lit:* 'authentic bleeding-nosed jaguar'. Myrsinaceae *Ardisia compressa* Kunth. This is a tree, 2–3 m (6.5–9.8′) tall with a trunk 15.24 cm (6″) in diameter. It bears panicles of small, pinkish-white flowers that are followed by red, 5 cm (2″) long, sweet fruit. The plant gets its name from a story in which the jaguar, after eating the fruit of this tree, bled profusely from his nose until he died. Because of this, the Lacandones believe that if you eat too many of the fruits your nose will bleed. The fruit is also savored by birds. *Use:* **hanal** 'food'; *Part:* **wich** 'fruit'; *Prep:* **che'che'** 'raw'; *Sim:* **mehen k'ik'ni' balum** 'small k'ik'ni' balum (*Eugenia* sp., *Ardisia compressa*)'; *Sim:* **nukuch k'ik'ni' balum** 'large k'ik'ni' balum'; *Gen:* **k'ik'ni' balum (ah)**. *See:* **mehen k'ik'ni' balum**$_2$. *Thes:* **che'**. *[Note:* Taxonomic identification provided by Francisco Morales (pers. comm.). kikin barun che. *Ardisia compressa* (Durán 1999).*]* *[*\sd2 food*]*

hach k'uuts tobacco. Solanaceae *Nicotiana tabacum* L. This is a large, cultivated plant grown for tobacco. Bundles of leaves are dried in the sun before rolling them into cigars. In the past, tobacco was smoked by men, women, and even children. Although it was offered to the gods, it did not have the

Fig. 5.59 Hach 'ilon *Monstera* sp. Photo Credit: Suzanne Cook

sacred importance that it had in other North American indigenous cultures. Today, very few Lacandones smoke home-grown tobacco, but they still cultivate it to sell to their Tzeltal neighbours and other passers-by. *Use:* **ti' ts'uuts'ul** 'for smoking'; *Part:* **le'** 'leaves'; *Use:* **ts'ak ti' yah upix** 'medicine for painful knees'; *Part:* **le'** 'leaves'; *Prep:* **Upokik yok'ol k'ak', utap'ik yok'ol pix** 'One roasts (10 leaves) in the fire and then covers the knee with them'; *Use:* **ti' ahts'ikin nok'ol** 'for the ts'ikin larvae'; *Part:* **uyits k'uuts** 'tobacco juice'; *Prep:* **Ukucha'achtik uk'uuts, wa ule' uk'uuts, pachil utubik uyits yok'ol ti', pachil kuhook'ol ahnok'ol kuhokik** 'One chews the cigar or some tobacco leaves then spits the juice on the affected area. Then when the worm emerges one pulls it out.'; *Sim:* **si'si' k'uuts** '*Baccharis trinervis*'; *Gen:* **k'uuts**. *[Note:* The *ts'ikin nok'ol* is the larvae of a type fly that lays its eggs under people's skin. k'uuts. tobacco (Bruce 1975); k'utz. *Nicotiana tabacum* (Nations and Nigh 1980); k'uutz [Mop.]. *Nicotiana tabacum* (in, Atran et al. 2004); k'uutz [Itz.]. *Nicotiana tabacum* (Atran et al. 2004); k'úutz [Yuc.]. *Nicotiana tabacum* (in, Atran et al. 2004).*] [Source:* AM; BM*]* *[*\sd2 medicinal \sd2 other*]*

hach luuch calabash. *Lit:* 'authentic calabash'. Bignoniaceae *Crescentia cujete* L. Calabash is an evergreen tropical tree, 7.3–12 m (25–40′) high, with spreading branches, a broad crown, and glossy, smooth, lanceolate leaves. Fragrant, funnel-shaped flowers grow on the woody branches and trunk. Flowers are followed by round, smooth, shiny-green gourds of up to 30–50 cm (12–20″) in diameter. The soft pulp is inedible, although it has medicinal applications in some cultures; the gourds are mainly grown for their use as containers. A similar species, *Crescentia alata*, differs in fruit size, flower colour, and leaf structure (http://ntbg.org/plants/). *Use:* **luuch** 'bowl; *Use:* **hama'** 'large drinking cup'; *Use:* **k'äb lu'uch 'hand cup'**; *Use:* **hach chaachib** 'colander'; *Sim:* **sina'an luuch (äh)** 'stretched out calabash (*Crescentia* sp., *C. alata*)'; *Sim:* **säk luuch** 'white calabash'; *Gen:* **luuch**. *Thes:* **che'** (Fig. 5.60). *[Note:* The process of preparing the cups entails roasting or boiling the gourds, to loosen and extract the pulp, and then setting them out in the sun to dry for a number of days. hach luch. *Crescentia alata* (Nations and Nigh 1980).*] [Source:* AM; BM*]* *[*\sd2 containers*]*

Fig. 5.60 Hach luuch *Crescentia cujete*. Photo Credit: Suzanne Cook

hach mäkäl elephant ears, malanga (Caribbean cabbage). *Lit:* 'authentic mäkäl'. Araceae *Xanthosoma mafaffa* Schott. This is an edible species of elephant ears. There are two kinds: *säk mäkäl* 'white macal' and *chäk mäkäl* 'red macal'. *Säk mäkäl* has greenish-white leaf stems, whereas the *chäk mäkäl* has reddish-purple stems. Numerous tubers encircle the main tuber. They have black skin and white flesh. *Loc:* **kol, kolil nah** 'milpas and house gardens'; *Use:* **hanal** 'food'; *Part:* **le', ne muun** 'young, unfurled leaves'; *Prep:* **chäkbil** 'boiled'; *Part:* **uyool** 'shoots'; *Prep:* **mukbil wah** 'buried tamale (a large, baked tamale)'; *Part:* **wi'** 'tubers'; *Prep:* **pokbil yetel k'ek'en, chäkbil yetel t'unu'** 'roasted with peccary, boiled with snails'; *Prep:* **k'oolbil** 'a soup thickened with corn dough'; *Spec:* **chäk mäkäl** 'red macal (*Xanthosoma mafaffa*)'; *Spec:* **säk mäkäl** 'white macal (*Xanthosoma mafaffa*)'; *Sim:* **ch'om mäkäl** 'vulture macal (*Xanthosoma robustum*)'; *Gen:* **mäkäl**. *See:* **suli'**$_2$. *Thes:* **xiiw**. *[Source:* AM; BM; CKD; KP*]* *[*\sd2 food*]*

hach mäkuuläm piper, matico (bamboo piper). Piperaceae *Piper aduncum* L., syn. *Piper angustifolium* Ruiz & Rav. Native to southern Mexico, the Caribbean, and most of tropical South America, *Piper aduncum* is a large, evergreen shrub or small tree, 6–8 m (20–23′) tall, with alternate, lance-shaped leaves 12–20 cm (5–8″) long. White inflorescences are recurved, like tails. According to BM, it is similar to *mehen mäkuuläm* 'small mäkuuläm', except that it is taller and has longer leaves. A number of indigenous groups in the Americas make a tea from the leaves and roots of *Piper aduncum*, to treat wide range of medical conditions, including diarrhea, dysentery, nausea, ulcers, and genito-urinary infections. The Lacandones use it to relieve gastro-intensinal problems, as well. *Piper aduncum* was introduced into western medicine in 1839 as a styptic and astringent (Taylor 2006). According to AM, the tender stalks of some species can be eaten raw, but most people do not eat them. *Use:* **chibal hämnen i hubul hämnen** 'stomach ache and diarrhea'; *Part:* **le'** 'leaves'; *Prep:* **Apets'ik ule' ich sisil ha' pachil awuk'ik, 'oxwäts'** 'You crush the leaves in cold water, then drink it, (do this) three times'; *Use:* **hanal** 'food'; *Part:* **muun uche'il ule'** 'tender leaf stalks'; *Prep:* **che'che'** 'raw'; *Sim:* **mehen mäkuuläm** 'small mäkuuläm'; *Gen:* **mäkuuläm**. *Variant:* **mäkuuläm**. *Thes:* **che'** (Fig. 5.61a, b, c). *[Source:* AM; BM*]* *[*\sd2 food \sd2 medicinal*]*

hach muxan *Heliconia librata* (common name unknown). Heliconiaceae *Heliconia librata* Griggs. *Use:* **ti' utap'ik**

Fig. 5.61 (**a**) Hach mäkuuläm *Piper aduncum*. Photo Credit: Suzanne Cook (**b**) Hach mäkuuläm *Piper aduncum*. Photo Credit: Suzanne Cook (**c**) Hach mäkuuläm *Piper aduncum* leaves. Photo Credit: Suzanne Cook

ba'al, mukbil wah 'for covering things and for the baked tamale'; *Part:* **le'** 'leaves'; *Sim:* **ya'ax ch'ib muxan xiiw** 'blue/green-spined muxan xiiw (*Calathea macrosepala*)'; *Gen:* **muxan**. *Variant:* **hach xiiw, säk muxan**. *Thes:* **xiiw**. *[\sd2 food]*

hach 'oochin *Philodendron donnell-smithii* (common name unknown). Araceae *Philodendron donnell-smithii* Engl. *Philodendron tripartitum* ranges from Mexico to Ecuador from near sea level to 1300 m (4000′) in elevation. In Mexico, it ranges from Veracruz to Tabasco and Chiapas, the Caribbean, Costa Rica, and from Panama down through Colombia, Ecuador, Bolivia and parts of Brazil. It is a vigorous, hemiepiphytic climber. It begins life as either a seed deposited on a tree branch by a bird that eventually sends roots down into the soil, or as a seed that takes root in the soil and subsequently climbs a host tree. As it grows it hugs closely to its host. The inflorescence is normally a solitary spathe but sometimes two or three occur at an axil. The spathes are typically 13–21 cm (5–8″) long, usually medium- to yellow-green or, rarely, creamy white with dark purple dots on the tube. Flowering normally occurs throughout the dry season and the first half of the rainy season (Lucas n.d.). No uses were recorded. *Sim:* **nukuch 'oochin** 'large 'oochin (*Philodendron sagittifolium, Syngonium chiapense, Syngonium podophyllum*)'; *Sim:* **mehen 'oochin** 'small 'oochin (*Philodendron hederaceum*)'; *Gen:* **'oochin**. *Variant:* **'oochin**. *Thes:* **ak'** (Fig. 5.62a, b). *[Note:* Identification of species provided by Tom Croat (pers. comm.).*] [Source:* AM*]*

hach 'ooh reed, common (giant cane). Poaceae *Gynerium sagittatum* (Aubl.) P. Beauv. **Hach 'ooh ne yi'h. Yan uts'ibal ek' i ya'ax. Uch'ihil tuhunal ti' yahaha'. Tupäk'ah uk'uk', tucha'ah tupäk'ah.** *'Hach 'ooh* is very hard. It is striped black and green. It grows in Yahaha'. (The people) took the shoots and transplanted them in Naha'. *Loc:* **tu chi' yahaha'** 'at the edge/mouth of Yahaha' (Monte Líbano River)'; *Use:* **häläl** 'arrows'; *Part:* **che'il top'** 'flower stalks'; *Use:* **hiil** 'battens'; *Part:* **uche'il** 'stalks'; *Sim:* **chäk 'ooh** 'red 'ooh (*Gynerium sagittatum*)'; *Gen:* **'ooh**. *Variant:* **säk 'ooh**. *[Source:* BM*] [\sd2 hunting_and_fishing]*

a b

Fig. 5.62 (**a**) Hach 'oochin *Philodendron donnell-smithii*. Photo Credit: Suzanne Cook (**b**) Hach 'oochin *Philodendron donnell-smithii* leaf. Photo Credit: Suzanne Cook

hach 'oop apple, custard. *Lit:* 'authentic custard apple'. Annonaceae *Annona* spp. This name refers to cultivated species of custard apple. *Loc:* **kolil nah, kol** 'house gardens and milpas'; *Use:* **hanal** 'food'; *Part:* **wich** 'fruit'; *Prep:* **che'che** 'raw'; *Spec:* **chäk 'oop** 'red custard apple (*Annona reticulata*)'; *Spec:* **säk 'oop** 'white custard apple (*Annona cherimoya*)'; *Gen:* **'oop**. *Variant:* **hach 'oop che'**. *Thes:* **che'**. *[\ sd2 food]*

hach 'oop che' custard apple tree. *Lit:* 'authentic custard apple tree'. Annonaceae *Annona* spp. *Sim:* **chäk 'oop che'** 'red custard apple tree (*Clarisia biflora*)'; *Gen:* **'oop che'**. *Variant:* **hach 'oop**.

hach 'oox breadnut, Mayan. *Lit:* 'authentic breadnut'. Moraceae *Brosimum alicastrum* Sw. ssp. *alicastrum* C.C. **Usool uwich k'än. Yuul usool uwich. Wolis, batak 25.4 mm (1″). Tahal ti' marzo.** 'The skin of the fruit is yellow. The fruit is smooth, round, and about 2.54 cm (1″) in diameter. It ripens in March.' (BM) The common name is Maya nut or *ramón*, in Spanish-speaking regions, and it generally refers to *Brosimum alicastrum*, a large, evergreen tree widely distributed throughout Mexico. It is a major component of the evergreen and semi-evergreen tropical forests of southern Mexico at elevations of 50–800 m (164–2624.6'). Because of its elevation preferences, *Brosimum alicastrum* should not be expected to grow in Naha', which lies between 825 m (2706.7') and 1200 m (3937'). Miguel Garcia, an educated northern Lacandon who grew up in Naha' and now works with conservationists in the area, said that *Brosimum alicastrum* does not grow in Naha', but that it can be found in Mensäbäk, at around 600 m (1968.5') elevation. BM confirmed this fact; however, the species is occasionally found at elevations up to 1000 m (3281') (Peters and Pardo-Tejeda 2001: 166). AM insisted that the tree in his yard was *hach 'oox*, which he also called *k'än 'oox* 'yellow 'oox'. BM disagreed, saying that the tree was actually *ya'ax 'oox* 'green 'oox', the only kind of *'oox* he claimed grows in Naha'. In the Lacandon botanical literature, *ya'ax 'oox* corresponds to both *Trophis racemosa* and *Brosimum alicastrum* (Nations and Nigh 1980; Nations 2006; Baer and Merrifield 1971). *Loc:* **Mensäbäk**; *Use:* **hanal** 'food'; *Part:* **wich** 'fruit'; *Prep:* **che'che'** 'raw'; *Part:* **unoy unek'** 'nutmeat'; *Prep:* **chäkbil i huch'bil yetel näl** 'boiled and ground with corn'; *Sim:* **chäk 'oox** 'red breadnut'; *Sim:* **k'än 'oox** 'yellow breadnut (*Brosimum alicastrum*)'; *Sim:* **ya'ax 'oox** 'green breadnut (*Brosimum alicastrum*)'; *Gen:* **'oox**. *Variant:* **k'än 'oox**. *Thes:* **che'**; **che'** (Fig. 5.63). *[Note:* Botanical identification provided by Servando Carvajal (pers. comm.). hach osh. *Brosimum alicastrum* (Nations and Nigh 1980: 21); hach osh, ya'ash osh.

Fig. 5.63 Hach 'oox *Brosimum* sp. leaves. Photo Credit: Suzanne Cook

Fig. 5.64 Hach pätan *Musa acuminata xMusa balbisiana*. Photo Credit: Suzanne Cook

Brosimum alicastrum (Nations 2006: 97); jach ox. *Brosimum alicastrum* (Baer and Merrifield 1971); óox [Yuc.]. *Brosimum alicastrum* (Bricker et al. 1998: 19); chäk 'oox, säk 'oox, k'än 'oox [Itz.]. *Brosimum alicastrum* (Atran et al. 2004); oox [Mop.]. *Brosimum alicastrum* (in, Atran et al. 2004); óox [Yuc.]. *Brosimum alicastrum* (in, Atran et al. 2004).*]* *[*\sd2 food*]*

hach pätan plantain (type of). *Lit:* 'authentic plantain'. Musaceae *Musa acuminata* ×*Musa balbisiana* Coll. *Use:* **hanal** 'food'; *Part:* **wich** 'fruit'; *Prep:* **chäkbil, wa ne täk'än, che'che'** 'cooked, if very ripe, raw'; *Gen:* **pätan**. *Variant:* **säk pätan** (Fig. 5.64). *[Source:* AM; BM; CKD*]* *[*\sd2 food*]*

hach sukal sugarcane. *Lit:* 'authentic sugarcane'. Poaceae *Saccharum officinarum* L. **Hach sukal es lo mas suave. Ahhach sukal uch'ihil mas que los otros. Hach sukal uch'ihil hasta nah punta.** 'Authentic sugarcane is the softest of the sugarcanes. It grows higher than the other kinds, as high as the peak of the house. ' (BM) This is the most popular variety used to make *balche'*. *Use:* **hanal** 'food'; *Part:* **che'il** 'stalk'; *Prep:* **che'che'** 'raw'; *Sim:* **ts'ibal sukal** 'striped sugarcane'; *Sim:* **mehen sukal** 'small sugarcane'; *Sim:* **chäk sukal** 'red sugarcane'; *Gen:* **sukal** (Fig. 5.65a, b). *[Source:* BM*]* *[*\sd2 food*]*

Fig. 5.65 (**a**) Hach sukal *Saccharum officinarum*. Photo Credit: Suzanne Cook (**b**) Hach sukal *Saccharum officinarum*. Photo Credit: Suzanne Cook

Fig. 5.66 (a) Hach suli' *Dioscorea alata*. Photo Credit: Suzanne Cook (b) Hach suli' *Dioscorea alata*. Photo Credit: Suzanne Cook

hach suli' yam, winged (greater yam, water yam, purple yam). Dioscoreaceae *Dioscorea alata* L., syns. *D. atropurpurea* Roxb., *D. purpurea* Roxb., *D. sativa* Del. **Nuk i wolis uwi', batak 45.7 cm (18″) ukooch. Usool p'iis ek' chäkchäk. Unoy, säk. Uch'ihil ich paakche' kol.** 'The tuber is large and round, about 45.7 cm (18″) in diameter. The skin is dark, reddish brown. The flesh is white. It grows in fallow *milpas*.' (BM) According to SK, the tubers can grow as large as soccer balls, with three or four to a plant. The skin is black and the flesh is white. *Dioscorea alata* can be distinguished by the compressed square stems, which look like "wings", and its large, 20 cm (8″) long, heart-shaped leaves with rounded, almost overlapping basal lobes. It has vigorous, twining vines that produce large, underground tubers that can weigh upwards of 10 kilos (22 lbs) each. Native to S. E. Asia, *Dioscorea alata* was introduced to the Americas by Portuguese and Spanish traders in the 16th century. *Use:* **hanal** 'food'; *Prep:* **chäkbil, pokbil** 'boiled, roasted'; *Sim:* **bilim suli'** '*Dioscorea bulbifera*'; *Gen:* **suli' (äh)** (Fig. 5.66a, b). *[Note:* ix pä'yak', chäk, säk, wolis [Itz.]. *Dioscorea bulbifera, D. alata* (Atran et al. 2004); b'awlyak [Yuc.]. *Dioscorea bulbifera, D. alata* (in, Atran et al. 2004); ix puyak, päyak [Mop.]. *Dioscorea bulbifera, D. alata* (in, Atran et al. 2004).*] [Source:* BM; SK*] [*\sd2 food*]*

Fig. 5.67 Hach su'uk *Digitaria setigera*. Photo Credit: Suzanne Cook

hach su'uk crabgrass (East Indian crabgrass, bristly crabgrass). *Lit:* 'authentic grass'. Poaceae *Digitaria setigera* Roth ex Roem. & Schult., syns. *Panicum sanguinale* L., *Syntherisma sanguinales* (L.) Dulac., *Paspalum sanguinale* (L.) Lam. **Mäna' ukanil.** 'It doesn't have runners'. Introduced into the area as cattle fodder (Hanan-Alipi & Mondragón-Pichardo 2009d), *Digitaria setigera* grows in fallow *milpas*, house gardens and other disturbed sites. Plants are approximately 80 cm (2.5′) tall, mostly upright, forming clumps with extensive but shallow roots. According to BM, this grass is easy to pull out, so it does not present much of a problem in the *milpa*. *Sim:* **wakäx su'uk** 'cow grass (*Pennisetum purpureum*)'; *Sim:* **'ak su'uk** 'grass grass (*Bothriochloa laguroides*)'; *Sim:* **winik su'uk** 'man grass (*Andropogon glomeratus*)'; *Gen:* **su'uk (ah)** (Fig. 5.67). *[Source:* AM; BM*]*

Fig. 5.68 Hach taw *Belotia mexicana*. Photo Credit: Suzanne Cook

hach taw cork. *Lit:* 'authentic cork'. Tiliaceae *Belotia mexicana,* syn. *Trichospermum mexicanum* (DC.) Baill. A tree of secondary forest, *Belotia mexicana* helps regenerate the soil in the fallow fields with its abundant leaves (AM, Cheng et al. 2011). *Use:* **bilich** 'mattress'; *Part:* **uyoot'el** 'inner bark'; *Use:* **chemil balche'** 'canoe for fermenting balche''; *Use:* **chem** 'canoe'; *Part:* **uche'il** 'trunk'; *Use:* **hi'che'** 'fence'; *Use:* **hi'che' ti' uk'anche' näl** 'walls for the corncrib'; *Part:* **uche'il** 'timber'; *Sim:* **chäk taw** 'red taw (*Trema micrantha*)'; *Gen:* **taw**. *Variant:* **taw**. *Thes:* **che'** (Fig. 5.68). *[Note:* tao. *Belotia mexicana* (Cheng et al. 2011); mahaua, tab che. *Belotia mexicana* (Durán 1999).*] [Source:* BM*] [*\sd2 furniture \sd2 housing*]*

hach tuch Thevetia, broadleaf (cascabel, dog's tongue). Apocynaceae *Thevetia ahouai* (L.) A. DC. *Spec:* **ya'ax tuch** 'green tuch'; *Spec:* **tuchil k'aax** 'forest tuch'; *Gen:* **tuch**. *Variant:* **tuch**. *Thes:* **che'**. *[Note:* tuch. *Ficus* sp. wild dogbane (Kashanipour and McGee 2004; Nations and Nigh 1980); tuch. *Thevetia ahouai* (Durán 1999).*] [Source:* BM*] [*\sd2 food*]*

hach ts'in manioc (cassava). *Lit:* 'authentic manioc'. Euphorbiaceae *Manihot esculenta* Crantz. **Uwi', batak 45.7 cm (18″) uwaan i 10.16 cm (4″) ukooch. Säk usool i säk unoy. Säk uche'il. Uka'anal ubakel batak 2 m (6.5′).** 'The tuber is about 45.7 cm (18″) long and 10.16 cm (4″) wide. The skin and the flesh are white. The stalks (and branches) are white. It's about 2 m (6.5′) tall.' (BM). *Use:* **hanal** 'food'; *Prep:* **pokbil wa chäkbil yetel ch'uhuk** 'Baked or boiled with sugar'; *Sim:* **chäk ts'in** 'red manioc'; *Sim:* **yahche' ts'in** 'great tree manioc'; *Gen:* **ts'in**. *Variant:* **säk ts'in** (Fig. 5.69a, b, c). *[Source:* AM; BM; KM; KP*] [*\sd2 food*]*

hach ts'us grape, Caribbean (water vine). Vitaceae *Vitis tiliifolia* Humb.& Bonpl., syn. *Vitis caribaea* DC. This is a climbing vine with thick, angulate to terete stems, which become woody with age. The larger vines contain quantities of potable water. Young branches are densely hairy becoming smooth as they mature. Leaves have a papery texture, are ovate to circular, usually with three shallow lobes. Leaf margins are finely to coarsely toothed. Lateral veins are

a

b

c

Fig. 5.69 (**a**) Hach ts'in *Manihot esculenta* leaves. Photo Credit: Suzanne Cook (**b**) Hach ts'in *Manihot esculenta*. Photo Credit: Suzanne Cook (**c**) Hach ts'in *Manihot esculenta* roots. Photo Credit: Suzanne Cook

conspicuous beneath. The top surface of the blades is hairy when young but relatively smooth, when mature. Pale or greenish-yellow blossoms are followed by large clusters of small, 4–6 mm (0.16–0.23″), purplish-black grapes. It occurs throughout Mesoamerica. (Elias 1968: 82–84) My consultants identified another variety they call *mehen ts'us* 'small ts'us'. It could be an immature form of *Vitis tiliifolia*, since the species observed had leaves and fruit similar to *hach ts'us*. *Use:* **hanal** 'food'; *Part:* **wich** 'fruit'; *Prep:* **che'che'** 'raw'; *Sim:* **mehen ts'us** 'small ts'us (*Vitis tiliifolia*)'; *Gen:* **ts'us (ah)**. *See:* **yuwin**. *Variant:* **ts'us**. *Thes:* **ak'**. *[*\sd2 food*]*

hach xiiw heliconia (type of). *Lit:* 'authentic xiiw'. Heliconiaceae *Heliconia* sp., *Heliconia librata* Griggs. *Use:* **ti' tap'ik ba'al, yo'och** 'food wrappers'; *Part:* **le'** 'leaves'; *Gen:* **xiiw**. *See:* **muxan**. *Variant:* **chukuch le' xiiw** (Fig. 5.70a, b). *[Source:* BM*]* *[*\sd2 food*]*

hach ya' chicle (chicle-tree, beef apple, níspero, sapotillo, sapodilla plum). Sapotaceae *Manilkara achras* (Mill.) Fosberg, syn. *Manilkara zapota* (L.) van Royen. **Ne chäk ubakel uche'il. Ne chich uche'il. Ne ts'u'uy. Ne nukuch che'. Ubakel batak 40 m (131′). Yan uwich. Chan säk utop'. Yan ubok. Nuk uwich, p'iis yek'el kafe. Wolis uwich, batak 7.6 cm (3″). Ne ch'uhuk. Yan 'oxp'el wa latunup' unek'. Unoy as k'änk'än ek'. Hutul ti' mayo, tahal.** 'The wood is very red, very hard, and very tough. It's a very large tree, about 40 m (131′) tall. It has fruit. The flowers are small and white. They have a fragrance. The fruit is large and coffee-coloured. The fruit is round, about 7.6 cm (3″) in diameter. It's very sweet. There are three or four seeds. The pulp is dark yellow. When ripe, in May, the fruit falls to the ground.' (BM) According to BM, the fruit is similar to *ha'as* (*Pouteria sapota*). The species BM describes is *Manilkara achras*, a tall evergreen tree that can reach 30.5 m (100′). The elliptic to obovate leaves are 5–8 cm (2″–5″) long and 2.5–5 cm (1″-2″) wide, medium green, and glossy. It bears inconspicuous, white bell-like flowers with a six-lobed corolla. Fruits are 5–10 cm (2″-4″) in diameter and vary in shape from globose to ovoid. The skin is thin, grey, or rusty brown and the pulp is yellowish brown, tender, and sweet. The pulp encases hard, black seeds. Fruiting occurs twice a year, although flowering may continue year-round. Trees were tapped for the white gummy latex used to make chewing gum. The wood is extensively used in construction and for a numerous artefacts. It was the timber of choice used in Mayan temples, particulary the lintels; the posts of the traditional Lacandon god house are only made of this wood. Although not mentioned by my consultants, Rätsch (1985a: 154) reports that the Lacandones use the bark to treat stomach and digestive problems. *Use:* **akam** 'foundation posts'; *Use:* **okman** 'pillars'; *Use:* **bahbil** 'horizontal planks for house walls'; *Use:* **che'il tu' kuwätik** 'warping board'; *Use:* **but'u'o'och** 'shuttle'; *Use:* **chulul** 'bows'; *Use:* **chuste'** 'arrow foreshaft'; *Use:* **k'ek' che'** 'cocoa beater'; *Use:* **hälälte'** 'loom, weaving sticks'; *Use:* **ho'olan che'il** 'ridge-pole'; *Part:* **che'il** 'timber'; *Use:* **che'il k'ak'** 'firewood'; *Use:* **kunyah ti' ch'upul uwich** 'curing prayer to help reduce

a b

Fig. 5.70 (**a**) Hach xiiw (chukuch le uxiiw) *Heliconia* sp. Photo Credit: Suzanne Cook (**b**) Hach xiiw *Heliconia* sp. Photo Credit: Suzanne Cook

the swelling of the face'; *Part:* **Ha'li' at'änik uk'aba'** 'You only mention its name'; *Use:* **hanal** 'food'; *Part:* **wich** 'fruit'; *Prep:* **che'che'** 'raw'; *Sim:* **ha'as** 'marmalade plum (*Pouteria mammosa*)'; *Sim:* **säk ya'** 'caimitillo (*Chrysophyllum mexicanum*)'; *Sim:* **o'ot' ya'** 'canistel (*Pouteria campechiana*)'; *Gen:* **ya'**. *Variant:* **ya', chäk ya' (?)**. *Thes:* **che'**. *[Note:* ya'. chico zapote (Durán 1999); hach ya'. *Manilkara zapota* (Nations and Nigh 1980); ya', säk ya', chäk ya' (jach ya') [Itz.]. *Manilkara achras* (Atran et al. 2004); chäk ya'aj, säk ya'aj [Mop.]. *Manilkara achras* (in Atran et al. 2004)*] [Source:* BM*] [*\sd2 housing \sd2 medicinal \sd2 craft \sd2 hunting_and_fishing \sd2 food*]*

hade xate macho (jade). Arecaceae *Chamaedorea oblongata* Mart. *See:* **hach bo'oy**$_1$. *Variant:* **säk bo'oy**. *From:* jade (Spanish). *[*\sd2 industry*]*

halol [hoor, hor, haror] heliocarpus (sunfruit). Malvaceae (Tiliaceae) *Heliocarpus* spp. These are small trees noted for their soft, light weight wood and strong, durable fibre. They are found in secondary forests. Three species have been recorded from the region: *Heliocarpus donnell-smithii* Rose, *Heliocarpus appendiculatus* Turcz, and *Heliocarpus americanus* L. (Lay 1949: 509). The Lacandones distinguish two kinds: *säk halol* 'white halol', named after its white bark, and *chäk halol* 'red halol', named after its red bark (AM, BM). *Halol* is a soil builder and conditioner, due to the abundant leaves it sheds. It is also an "indicator" species; in the old days, the Lacandones would begin to clear the *milpas* when the *halol* was in bloom. *Activity:* **halalch'äktik** 'cut down brush with machete'; *Use:* **hi'che'** 'fence, vertical house poles'; *Use:* **k'anche' näl** 'corncrib'; *Part:* **che'il** 'timber'; *Part:* **k'äbche'** 'branches'; *Use:* **häxan** 'rolled or twisted cord'; *Use:* **k'an** 'cord'; *Use:* **suum** 'rope'; *Use:* **ba'ay** 'woven bags'; *Use:* **ch'uyu'** 'hanging basket'; *Use:* **hach k'an** 'authentic hammock'; *Part:* **uyoot'el** 'inner bark'; *Prep:* **Abälik uyoot'el yok'ol awok** 'You roll the inner bark down your leg'; *Use:* **hapak** 'quiver'; *Part:* **hach uyoot'el** 'outer bark'; *Spec:* **chäk halol** 'red halol (*Heliocarpus appendiculatus*)'; *Spec:* **hach halol** 'authentic halol (*Heliocarpus donnell-smithii*)'. *Variant:* **halon**. *Thes:* **che'**. *[Note:* halol. a tree of the ceiba family (Spanish palo de corcho, "cork tree") (Bruce 1975); ha:ro:r [S. Lac.]. majagua (Canger 1970); halo. *Heliocarpus donnell-smithii* (Nations and Nigh 1980); halo. *Heliocarpus donnell-smithii* (Marion 1991, in March 1998); halo che. *Heliocarpus donnell-smithii* (Durán 1999); jaror. *Heliocarpus appendiculatus* (Durán 1999); jolol [Itz.]. *Hampea tomentosa* (Atran et al. 2004).*] [Source:* AM; BM; CKO; KA; SK*] [*\sd2 housing \sd2 cordage \sd2 agriculture*]*

halo'och k'i'ix [halo'(o)ch k'i'ix, haro'(o)ch k'i'ix, haräxk'ix] blackberry; brambles. Rosaceae *Rubus* sp. This is a kind of bramble that resembles *Rubus nebulosus*, with its thick, ridged, arching canes armed with short, recurved thorns. Leaves alternate on long, thorny petioles, terminating in five, palmately arranged, elliptical leaflets that taper to a point. Leaf blades are bright green, palmately veined, with toothed margins. According to BM, it produces edible berries. He recalls one visitor who made it into jam. Lacandones whom I interviewed did not mention any use for the plant, not even as a major source of food. *Thes:* **ak'** (Fig. 5.71). *[Note:* Jarochkiix. *Mimosa ervendbergii* (Nigh 2008).*] [Source:* BM; CKO; K'ayum Segundo*] [*\sd2 food*]*

ha' chuh calabash, bottle gourd. *Lit:* 'water gourd'. Cucurbitaceae *Lagenaria siceraria* (Molina) Standl. **Ne tsoy ti' uch'a'ik ha'. Uchik uk'anik ti' balche'.** 'Very good for carrying water in. A long time ago, *balche'* was made in it.'.

Fig. 5.71 Halo'och k'i'ix *Rubus* sp. Photo Credit: Suzanne Cook

Use: **ti' chuyik ha'** 'for carrying water'. *See main entry:* **chuh**. *[Source:* AM*]* *[*\sd2 containers*]*

hälälı [häläl, härär] reed. *Lit:* 'arrow'. Poaceae. *Use:* **häläl** 'arrows'; *Part:* **che'il utop'** 'flower stalk'; *Spec:* **hach häläl** 'common reed (*Phragmites australis*)'; *Spec:* **häläl** 'giant cane (*Gynerium sagittatum*)'. *See:* **hach häläl**. *[*\sd2 hunting_and_fishing*]*

häläl₂ giant cane. Poaceae *Gynerium sagittatum* (Aubl.) P. Beauv. *Use:* **häläl** 'arrows'; *Part:* **che'il utop'** 'flower stalk'; *Sim:* **hach häläl** 'authentic arrow (*Phragmites australis*)'; *Gen:* **häläl**. *Variant:* **'ooh**. *[*\sd2 hunting_and_fishing*]*

hänan [hänan, hänah] bayal (basket palm, liana palm). Arecaceae *Desmoncus orthacanthus* Mart., *Chamaedorea elatior*. Lacandones identify two kinds of *hänan*: *hach hänan* 'authentic hänan' and *ya'ax hänan* 'green hänan'. *Hach hänan* is *Desmoncus orthocanthus*, a vine with large, hooked spines on long stems. *Ya'ax hänan* is *Chamaedorea elatior*, the only climbing species in the genus, with smooth stems and clusters of small, round fruit. *Use:* **me'et** 'trivet'; *Part:* **ak'il** 'vines'; *Use:* **hanal** 'food'; *Part:* **uyool** 'heart'; *Prep:* **chäkbil, p'iis ch'i'ib** 'boiled, like *ch'i'ib*'; *Spec:* **ya'ax hänan** 'green hänan (*Chamaedorea elatior*)'; *Spec:* **hach hänan** 'authenic hänan (*Desmoncus orthacanthus*)'. *Thes:* **ak'**. *[Note:* henen. Bayal. *Desmoncus orthocanthos* [sic.] (Nations 2006); *Desmoncus* sp. (Levy et al. 2006).*]* *[Source:* AM; BM; CHX; CNK; K'ayum Segundo; KM; KP; NI*]* *[*\sd2 craft \sd2 cordage \sd2 food*]*

häxal k'ak'₁ *Hamelia calycosa* (common name unknown). *Lit:* 'firedrill'. Rubiaceae *Hamelia calycosa* Donn. Sm., syns. *Hamelia boyacana* Standl., *Hamelia chiapensis* Brandegee, *Hamelia klugii* Standl. *Use:* **che'il häxbil k'ak'** 'firedrill'. *See:* **ak' tsup**. *Variant:* **häxk'ak'**. *Thes:* **che'**. *[Note:* jaxa kak. *Hamelia calycosa* (Durán 1999). Botanical synonyms from The Plant List. http://www.theplantlist.org/tpl/record/kew-95996. Retrieved January 2013.*]* *[Source:* Durán (1999)*]* *[*\sd2 tools*]*

hoben [hoben, hobeh] piper. Piperaceae *Piper* spp. **Uka'anal ubakel 5 metros. Ne ki' umuun ule'.** 'It is 5 m (16.5′) tall. The tender leaves are delicious.' (BM) *Hach hoben* 'authentic hoben' is an herbaceous shrub with cordate to oval, velvety leaves. It grows mainly in fallow *milpas* and other disturbed areas. The Lacandones whom I consulted regard it as *lo'obil* 'a weed'. *Loc:* **paakche' kol** 'fallow milpa'; *Use:* **hanal** 'food'; *Part:* **uyol ule', ule'** 'growing tips, leaves'; *Prep:* **chäkbil yetel t'unu'** 'boiled with snails'; *Spec:* **hach hoben** 'authentic hoben (*Piper auritum*)'; *Spec:* **säk hoben** 'white hoben (*Piper umbellatum*)'. *See:* **mäkuuläm**. *Variant:* **hach hoben**. *Thes:* **che'**. *[Note:* jover. *Piper auritum* H.B.K. (Nigh 2008).*]* *[Source:* BM; KA; KP; K'ayum Segundo; SK*]* *[*\sd2 food*]*

holop' [holop', horop'] philodendron. Araceae *Philodendron* spp. **Uyanchunil p'iis orquídea, p'iis uk'uk'che'. Pachil, ki'ki' ts'aabil uyaak'il u k'äb bin uhasta ich lu'um. Pachil, yan umots.Yan utop', chen ma' ahantik. Hay dos clases: yuyul holop' i k'i'ilix holop'. Uk'anik yuyul holop' ti' hitik kun.** 'It's a vine. In the beginning it is like an orchid, like a sprout in the branches of a tree. Then, very slowly it sends vines down to the ground where they take root. It has flowers, but you don't eat them. There are two classes: smooth holop' and bumpy holop'. One uses the vines of the smooth holop' for tying the leaves of the fan palm (*Cryosophila stauracantha*).' (BM) *Philodendron smithii* is usually a hemiepiphytic plant. The stem clings to the tree as it climbs. It occurs in a number of forest formations in Chiapas, Tabasco, and Quintana Roo (Mexico), Guatemala, Honduras, and Nicaragua on the Atlantic slope between 40 and 1630 m (131–5348′) in elevation (http://www.aroid.org). *Use:* **k'an ti' uhitik kun** 'rope to tie thatch'; *Use:* **xaak** 'baskets'; *Part:* **uyak'il** 'its vines'; *Spec:* **yuyul holop'** 'smooth holop' (*Philodendron smithii, P. tripartitum*)'; *Spec:* **k'i'ilix holop'** 'bumpy holop' (*Philodendron radiatum*)'. *Thes:* **ak'** (Fig. 5.72). *[Note:* jolob. *Philodendron*

Fig. 5.72 Holop' *Philodendron smithii*. Photo Credit: Suzanne Cook

smithii (Durán 1999). Durán (1999) reports that the northern Lacandones use *Philodendrom smithii* as a poison to kill rats and head lice.*]* *[Source:* BM*]* *[*\sd2 cordage \sd2 containers*]*

ho'ol ma'ax [ho'ol ma'ax, ho' ma'ax] bean, Saint Mary's (crucifixion bean, elephant creeper, woolly morning glory). *Lit:* 'monkey head'. Convolvulaceae *Merremia discoidesperma* Donn. Sm. This is a high climbing woody liana of Chiapas (Mexico), Guatemala, Costa Rica, Hispaniola, and Cuba. The vine produces small seeds that are thick, hard, black, and lustrous. Their most distinguishing feature is the grooved cross on one side of the seed. *Use:* **uuh** 'necklaces'; *Part:* **nek'** 'seeds'. *Variant:* **gorila**. *[Source:* NI*]* *[*\sd2 craft*]*

huhup [huhup, huhu'] mombin (wild hog plum). Anacardiaceae *Spondias* sp., *Spondias radlkoferi* Donn. Sm. This is an enormous, deciduous tree with odd pinnate leaves and loose clusters of elongated, green, hard, slightly sweet fruits. Children sometimes eat the shoots. The trunk is fissured and grey, and the wood is soft. Because of this, the timber is not useful. In the extant Lacandon botanical literature, *huhup* corresponds to *Spondias mombin*. This tree grows 20 m (65′) tall, has a somewhat buttressed trunk between 60 and 75 cm (2–2.6′) in diameter, and thick, fissured bark. Yellow, plum-like fruit hang in loose clusters at the ends of stems. The skins are thin and tough, and the meagre pulp is fibrous but juicy (Morton 1987: 245). According to Croat (1974, 1978), *Spondias radlkoferi* has long been confused with *Spondias mombin*. Features that differentiate the two species include: the period of maturation of the fruit; fruit texture; pubescence type; the shape of the endocarp; and, bark texture. Whereas *Spondias mombin* bears dark yellow-orange fruit during the rainy season, from April to November, *Spondias radlkoferi* begins to ripen between September and November, although its fruit remains green and hard at maturity (Croat 1974: 135, Croat 1978: 542). Conceivably, *huhup* corresponds to both *Spondias radlkoferi* and a variety of *Spondias mombin*. Further investigation is required. *Use:* **hanal** 'food'; *Part:* **k'uuk'**; *Prep:* **che' che'** 'raw'. *See:* **'abäl; k'inim**. *Variant:* **ya'ax 'abäl**. *Thes:* **che'** (Fig. 5.73a, b, c). *[Note:* jujup. *Spondias mombin* (Diemont 2006; Nigh 2008); ju'ju', k'inim [Itz.]. *Spondias mombin* (Atran et al. 2004); juujub' [Yuc.]. *Spondias mombin* (in Atran et al. 2004).*]* *[Source:* BM*]* *[*\sd2 food*]*

hunbuhk'äbwinik pätan [hunbuhk'äbwinik pätan, humbuhk'äbwinik, humbuhk'ä'winik, humuk'ä'winik] plantain (type of). *Lit:* 'one hundred plantains'. Musaceae *Musa acuminata* ×*Musa balbisiana* Coll. This is a variety that produces long, yellow plantains in large bunches. *Use:* **hanal** 'food'; *Part:* **wich** 'fruit'; *Prep:* **chäkbil, chen wa ne täk'än che'che'** 'cooked, but if very ripe, raw'; *Sim:* **ka'buhk'äbwinik pätan** 'two hundred plantains'; *Gen:* **pätan** (Fig. 5.74a, b). *[Source:* BM*]* *[*\sd2 food*]*

hut'kih morning glory. Convolvulaceae *Ipomoea* spp., *Ipomoea alba* L. **Hut'kih es ak'. Mihin ak' wolis uyak'il. Ek' uyak'il, batak 6.35 mm (1/4″). Ubin ne chukuch uyak'il, bin yok'ol lu'um o yok'ol uk'äb lo'obil. Uyak'il yan uyits, säk uyits. Säk utop', nuk utop', batak 10.16 cm (4″). Yan uwich, batak 5.08 cm (2″) uwaan. Wolis uwich. Hut'kih es para kinsik k'ik'. Ukinsik k'ik' como hacer muñecos para los dioses. Es único hut'kih. Hay dos. El otro se llama hut'kih ne nuk. Es mas grand y tiene ne nuk uyak'il. Pero no es mero hut'kih. Las hojas son muy chiquitas, mas mihin ule', ma' nuk ule', chan mihin. Hay un otro se llama, uts'ak k'ik'. Asi es para kinsik k'ik'. Hut'kih es igual uts'ak k'ik'. Uts'ak k'ik' parece casi como hut'kih. Los dos se usa para kinsik k'ik'. Häts'ik yetel che', yetel ti' umäxal. Uyahbäxal, häts'ik. Kuhäts'ik kuyäts'ik uya'lil kuyäts'ik uts'u'. Man uts'u' ubaakeleh. Bix ahhut'kih. Uyak'il ya'lil ne yäken. Tawilah puk'ik. Ne täk'ik. Ne tah toh witsbil ma'ats'. Ele, kakinsikeh ya muuch'ikeh seeb, seeb kukimil. P'iis ahuntul ahhut'kih. Lehih ne tsoy kamuuch'ik ti'—uts'ak k'ik'. Ahäts'ik lah uts'u' baakel. Te' way hook'ol uyäk. Chen yak'il usool mäna' yits, mäna' ma' yäk, ma' yäk uyits. Ha'li' uts'u'. Te' kapuleh ts'ak kinsikeh.** '*Hut'kih* is a vine. It's vines are small and round. The vines are black, about 6.35 mm (1/4″) [in diameter]. The vines are long, they travel on the ground or on the branches of surrounding brush. The vines have a resin, a white resin. The flowers are white and large, about 10.16 cm (4″). It has fruit (seeds) about 5.08 cm (2″) long. *Hut'kih* is used to "kill" latex. It "kills" latex for making the figurines for the gods. It is the only *hut'kih*. There are two; the other *hut'kih* is very large. It's much larger [than the real *hut'kih*] and has very large vines. But it isn't the real *hut'kih* (*Ipomoea alba*). It has very small leaves, much smaller [than the real *hut'kih*]. There is still another one called *uts'ak k'ik'* (*Merremia tuberosa?*). This is also used to "kill" latex.

Fig. 5.73 (**a**) Huhup *Spondias* sp. Photo Credit: Suzanne Cook (**b**) Huhup *Spondias* sp. (**c**) Huhu' (huhup) bark. Photo Credit: Suzanne Cook

Hut'kih is the same as *uts'ak k'ik'*. *Uts'ak k'ik'* looks almost the same as *hut'kih*. Both are used to "kill" the latex. You pound [the vines] with a stick to mash them. First one smashes (them), pounds them. When one pounds them one squeezes out the juice, one squeezes the pith. [The juice] passes through the pith of the stem. It's the same with *hut'kih*. The juice from the vines is really strong. You saw it dissolve [the latex]. It really congeals it. It's very good for the *witsbil ma'ats'* ceremony. And so, when you "kill" it, you quickly gather [the latex] together, and then it "dies" quickly. It's the

Fig. 5.74 (**a**) Hunbuhk'ä'winik pätan *Musa acuminata* ×*Musa balbisiana*. Photo Credit: Chan K'in Daniel (**b**) Hunbuhk'ä'winik pätan *Musa acuminata* ×*Musa balbisiana* fruit. Photo Credit: Chan K'in Daniel

same for the other kind of *hut'kih*. *Uts'ak k'ik'* is really good for congealing [the latex]. You pound its stem all over. This here is where the poison comes out [AM points to the tender stem.] But the bark of the vine doesn't have any resin, there isn't any poison, the resin isn't strong. It's only in the pith. This is medicine you put in to "kill" [the latex]. *Use:* **ti' ukinsik k'ik'** 'to vulcanize rubber'; *Part:* **uyits uts'u'** 'resin from the pith of the vines'; *Prep:* **Ahäts'ik ayäts'ik uya'lil ti' uts'u', pachil amuuch'ik seeb yetel che'il** 'You mash it and squeeze out the resin from the pith and then you quickly bring it together with a stick'; *Sim:* **ts'ak k'ik'** 'morning glory (*Merremia tuberosa?*)'; *Spec:* **chäk hut'kih** 'red hut'kih (*Ipomoea* sp.)'; *Spec:* **hach hut'kih** 'authentic hut'kih (*Ipomoea alba*)'; *Spec:* **nukuch hut'kih** 'large hut'kih'. *See:* **ts'ak k'ik'**. *Variant:* **hach hut'kih**. *Thes:* **ak'**. *[Note: Witsbil* is a kind of Lacandon ceremony in which *ma'ats'*, a type of corn gruel, and *tulis k'ik'*, rubber figurines, are offered to the gods.*] [Source:* AM; BM*] [*\sd2 other*]*

hu'un ficus. Moraceae *Ficus* sp. *Use:* **nook'** 'clothes'; *Spec:* **ak' hu'un** 'vine hu'un (*Poulsenia armata*)'; *Spec:* **bits'kal hu'un** 'strangle hu'un (*Ficus* sp.)'; *Spec:* **hach hu'un** 'authentic hu'un (*Ficus* sp.)'. *[*\sd2 clothing*]*

I–i

'ib ['ib, 'ip'] bean, lima. Fabaceae *Phaseolus lunatus* L. cv. 'Sieva'. This is a small-seeded type of *Phaseolus lunatus*. It bears short, wide pods containing three, slightly flattened, oval beans. *Loc:* **kolil nah, paakche' kol** 'house garden, fallow milpas'; *Use:* **hanal** 'food'; *Prep:* **chäkbil** 'boiled'; *Spec:* **chäk 'ib** 'red lima'; *Spec:* **ek' 'ib** 'black lima'. *Thes:* **bu'ul**. *[Note:* ip'. *Phaseolus leucanthus* (Nations and Nigh 1980); b'ox 'ib', säk 'ib', chäk 'ib' [Itz.]. *Phaseolus lunatus* (Atran et al. 2004); 'ìib' [Yuc.]. *Phaseolus lunatus* (Bricker et al. 1998); ib' [Mop.]. *Phaseolus lunatus* (in, Atran et al. 2004).*] [Source:* AM; BM*] [*\sd2 food*]*

'ibil k'aax ['ibil k'aax, 'ip'il k'aax] bean, "Little Marble". *Lit:* 'lima bean of the forest'. Fabaceae *Oxyrhynchus trinervius* (J. D. Smith), syns. *Dioclea trinervia* Donn. Sm., *Monoplegma sphaerospermum* Piper, *Monoplegma trinervium* (Donn. Sm.) Piper. **Es ak'. Chukuch uwich. Tso'otsol uwich. 'Oxp'el unek'. Utop' p'iis bu'ul. Ya'ax utop'. Tahal ti' kolil.** 'It's a vine with long, fuzzy fruit. There are three seeds. The flowers are like those of beans. They are blue. The (beans) begin to mature in the "felling season". (KP, CXK). *Loc:* **k'aax** 'forest'; *Use:* **uuh** 'necklaces'; *Part:* **nek'** 'seeds'; *Gen:* **'ib**. *Thes:* **ak'** (Fig. 5.75). *[Source:* CNK; CXK; KP*] [*\sd2 craft*]*

iich'ak tolok mata rama (God bush). *Lit:* 'lizard claw'. Loranthaceae *Struthanthus orbicularis* (Kunth) Eichler. **Ak', ne buluk. Ma' tsoy.** 'It's a very invasive vine. It's very bad.' (KP) This is a vigorous parasitic, green vine with smooth stems and opposite pairs of oval-elliptic, smooth, somewhat thick leaves up to 5 cm (2″) long and 4 cm (1.6″) wide. Petioles are u-shaped, 1 cm (0.4″) long. Fragrant, greenish or greenish-yellow flowers are borne in sets of 3 on slender, spike-like inflorescences. Fruits are ellipsoid, 1 cm (0.4″) long, orange, rusty-red or purple, and contain latex. This is a widespread and ecologically variable species (Croat 1978:

369). *Variant:* **nukuch tak'lan che'**. *Thes:* **ak'** (Fig. 5.76a, b, c). *[Note:* Botanical identification provided by Daniel Nickrent (pers. comm.).*] [Source:* KP*]*

ik pepper, chile. Solanaceae *Capsicum* spp. The Lacandones distinguish a number of kinds of chile peppers. Most are the pungent, *Capsicum annuum* varieties. Chile is among the most important [food] species, with the greatest diversity in both the wild and under cultivation (Landa et al. n.d.: 23): Nations and Nigh (1980: 10) say that the Lacandones cultivate six varieties of *Capsicum annuum* and one variety of *Capsicum frutescens*, Baer and Baer (1952) mention the Diente de Perro (*Capsicum annuum*), Breedlove (1986) lists seven species from Chiapas, and my consultants named nine kinds. Chiles are usually eaten raw as a condiment with beans and other main dishes. They are also used to treat oral diseases, especially toothaches, "mouth rot" and, possibly, stomatitis (which may be the same as mouth rot). *Ahkunyah ik*, the curing chant of the chile, is usually performed at the same time as the chile is chewed. Three of the hottest chilies—*säk yolte' ik* 'white walking-stick pepper', *säk chawah ik* 'white dangling pepper', and *ya'ax chawah ik* 'green dangling pepper'—are part of a ritual called *k'ämkih*, which is performed during the changing of the god pots ceremony. These are offered to lower the tempers of the gods, who are angry that their incensories have become encrusted with soot (AM and CK admit that the god pots are not renewed as frequently as they should be). *Use:* **hanal** 'food'; *Part:* **wich** 'fruit'; *Prep:* **che'che'** 'raw'; *Use:* **chibal koh, ts'uk chi', säkpähk'ak'** 'toothache, mouth rot, stomatitis (?)'; *Part:* **wich** 'fruit'; *Prep:* **che'che'** 'raw'; *Use:* **sis(s)ik uyol ti' ahk'uh** 'cool the tempers of the gods'; *Sit:* **k'ämkih, maasik k'uh** 'kind of ritual during the changing of the god pots'; *Part:* **wich** 'fruit'; *Prep:* **che'che'** 'raw'; *Spec:* **nok'ol ik** 'worm chile (*Capsicum frutescens*)'; *Spec:* **chäk ik** 'red chile (*Capsicum annuum*)'; *Spec:* **k'än ik (äh)** 'yellow chile (*Capsicum annuum*)'; *Spec:* **ya'ax ik** 'green chile (*Capsicum annuum*)'; *Spec:* **säk yolte' ik (äh)** 'white walking-stick chile (*Capsicum* sp.)'; *Spec:* **ta' ch'o' ik** 'rat excrement chile (*Capsicum annuum* var. *glabriusculum*)'; *Spec:* **hach ik** 'authentic chile (*Capsicum frutescens*)'; *Spec:* **chawah ik (ah)** 'dangling chile (*Capsicum frutescens*)'; *Spec:* **ch'om ik** 'vulture chile (*Capsicum* sp.)'. *[Note:* Vernacular names and botanical synomyms from USDA Plants Database. http://plants.usda.gov/java/profile?symbol=CAANG. Retrieved March 2012. ik, iik [Mop., Yuc., Itz.] *Capsicum annuum* (in, Atran et al. 2004).*] [Source:* AM; CK*] [*\sd2 food \sd2 medicinal \sd2 religion*]*

Fig. 5.75 'Ibil k'aax *Oxyrhynchus trinervius* seed. Photo Credit: Suzanne Cook

'ilon ['iron] Swiss cheese plant. Araceae *Monstera* spp. AM and BM describe this as a kind of evergreen vine that sticks to trees as it climbs. The leaves are large and *kalkal* 'holey' or *chälchäl* 'transparent'. These descriptions refer to the *Monstera*, a genus that comprises 41 species found only in Mexico, Central and South America. The name of the genus comes from the Latin word meaning "monstrous" or "abnormal". The vines climb to heights of 20 m (65′) by means of

Fig. 5.76 (**a**) Iich'ak tolok (nukuch tak'lan che') *Struthanthus orbicularis* vine. Photo Credit: Suzanne Cook (**b**) Iich'ak tolok (nukuch tak'lan che') *Struthanthus orbicularis* vine. Photo Credit: Suzanne Cook (**c**) Iich'ak tolok (nukuch tak'lan che') *Struthanthus orbicularis* fruit. Photo Credit: Suzanne Cook

Fig. 5.77 'Is *Ipomoea batatas*. Photo Credit: Suzanne Cook

aerial roots. These roots descend to the ground where they take hold for additional support. The inflorescence is a 5–45 cm long spadix that bears a cluster of white berries, which are edible in some species. Breedlove (1986) lists six species from Chiapas: *Monstera acacoyaguensis; Monstera acuminata; Monstera deliciosa; Monstera punctulata; Monstera siltepecana;* and, *Monstera tuberculata. Monstera deliciosa* is known for its delicious fruit, which tastes like a combination of banana and pineapple. However, the Lacandones believe that none of the kinds of *'ilon* are edible. This could be because the inflorescences are not readily available, especially on plants found below 1000 m (3281') in elevation, where *Monstera deliciosa* climbs high up in to the forest canopy (Michael Grayum, pers. comm.). It could also be that the Lacandones mistake the species. According to aroid specialist, Tom Croat, "Monstera can sometimes be so variable as to appear like several species" (Croat, cited in Lucas, n.d.). *Use:* **xaakil chukkäy** 'fish trap'; *Use:* **xaak** 'baskets'; *Use:* **me'et** 'trivets'; *Part:* **ak'** 'vines'; *Spec:* **hach 'ilon** 'authentic 'ilon (*Monstera* sp.)'; *Spec:* **kamis 'ilon** 'centipede 'ilon (*Monstera deliciosa*)'; *Spec:* **säk 'ilon** 'white 'ilon (*Monstera* sp.)'. *Thes:* **ak'**. *[Note:* ilo. basket vine. *Monstera* sp. (Nations and Nigh 1980); camis irom, jach iram. *Monstera deliciosa* (Durán 1999).*] [Source:* AM; BM; KM; K'ayum Segundo*] [*\sd2 containers \sd2 cordage*]*

'is potato, sweet. Convolvulaceae *Ipomoea batatas* (L.) Lam. Sweet potatoes produce long, tapering roots with smooth skin. The skin colour ranges from red to purple, brown, or beige. The flesh may be beige, yellow, orange, or purple. Lacandones distinguish three kinds: *chäk* 'red', *k'än* 'yellow', and *säk* 'white'. They are propagated by stem cuttings in a separate plot in the *milpa*. *Loc:* **kol** 'milpa'; *Use:* **hanal** 'food'; *Part:* **wi'** 'tubers'; *Prep:* **chäkbil, pokbil** 'boiled, roasted'; *Spec:* **säk 'is** 'white sweet potato'; *Spec:* **chäk 'is** 'red sweet potato'; *Spec:* **k'än 'is** 'yellow sweet potato' (Fig. 5.77). *[Note:* is. sweet potato, yam (Bruce 1975); is. *Ipomoea* (Nations and Nigh 1980); 'iis [Yuc.]. sweet potato. *Ipomoea batatas* (Bricker et al. 1998); 'is [Itz.]. calabacita. soft-skinned squash (Hofling 1997; kamuut [Itz.]. *Ipomoea batatas* (Atran et al. 2004).*] [Source:* BM*] [*\sd2 food*]*

'itsänte' glorybower (bagflower, bleeding-heart). Lamiaceae *Clerodendrum ligustrinum* (Jacq.) R. Br. This is a deciduous shrub with opposite or whorled leaves and flowers with tubular corollas formed by five white lobes. *C. ligustrinum* was observed in KM and KP's garden, although the species is reported to grow at altitudes much lower than Naha'. As these plants were only observed in house gardens, it is likely they were purchased cultivars. The leaves are used to season fish, as they are in the Petén (Standley and Williams 1970: 193). In addition, they are boiled and applied to snakebites by the Mayas in the Yucatan (Roys 1931: 248). *Use:* **ts'ak ti' uyo'och käy i t'unu'** 'food for fish and snails'; *Part:* **le'** 'leaves'; *Prep:* **chäkbil yetel käy** 'cooked with fish'; *Use:* **tsoy uwich** 'ornamental'; *Part:* **top'** 'blossoms'. *Thes:* **che'**. *[Note:* itzimté [Yuc.]. *Clerodendrum ligustrinum* (Roys 1931).*] [Source:* BM*] [*\sd2 food*]*

Fig. 5.78 (**a**) Its'in säk wawal *Calathea crotalifera* flowers. Photo Credit: Chan K'in Daniel (**b**) Its'in säk wawal *Calathea crotalifera* leaf. Photo Credit: Suzanne Cook

its'in säk wawal yellow rattle shaker (rattlesnake calathea). *Lit:* 'younger brother of *säk wawal*'. Marantaceae *Calathea crotalifera* S. Watson, syn. *Calathea insignis* Petersen. *Loc:* **k'aax** 'forest'; *Sim:* **säk wawal** 'white wawal (*Calathea lutea*)'; *Sim:* **chäk wawal** 'red wawal (*Calathea* sp.)'; *Gen:* **wawal**. *Thes:* **xiiw** (Fig. 5.78a, b). *[Note:* This is not the common name for this plant. At the time of collection, neither AM nor BM could recall its name.*] [Source:* AM*]*

iximche' [iximche', ixinche'] *Pleuranthodendron lindenii* (common name unknown). Flacourtiaceae *Pleuranthodendron lindenii* (Turcz) Sleumer. This is a small tree that grows to 6.09 m (20′) tall. It produces masses of fragrant, white flowers. Yellow fruits follow in November, drawing parrots and toucans. It grows in mountainous terrain. According to my consultants, the wood is too soft to be useful. *Thes:* **che'**. *[Note:* ixim che. *Pleuranthodendron lindenii* (Durán 1999); ixim che'. *Pleuranthodendron lindenii* (Diemont 2006); iximche [Yuc.]. *Casearia nitida*. café cimarrón. Indian boxwood, *Citharexylum Schottii.* (Roys 1931); ixi'imche' [Itz.]. *Casearia nitida* (Atran et al. 2004).*] [Source:* BM*]*

K–k

ka'buhk'äbwinik pätan plantain (type of). *Lit:* 'two hundred plantains'. Musaceae *Musa acuminata* ×*Musa balbisiana* Coll. A variety of plantain similar to *hunbuhk'ä'bwinik pätan* 'one hundred plantains', but with two bunches of bananas growing on a stalk instead of one. *Use:* **hanal** 'food'; *Part:* **wich** 'fruit'; *Prep:* **chäkbil, chen wa ne täk'än, che'che'** 'cooked, but if very ripe, raw'; *Sim:* **hunbuhk'äbwinik pätan** 'one hundred plantains'; *Gen:* **pätan**. *[Source:* BM*] [*\sd2 food*]*

kahanche' [kahanche', kähänche'] apple, monkey. Clusiaceae *Clusia flava* Jacq. **Es che'. Uche'il 45.7 cm (18″). Uka'anil 20 metros (65.6′) . Ne ki' ubok utop'. Uchik, uhulik yetel puts' ubuhkintik uche'il ukaal. P'iis ubokbal jabonil.** 'It's a tree. The trunk is 45.7 cm (18″) in diameter and 20 m (65.6′) high. The flowers are very fragrant. In the past (the women) wore the flowers around their neck. They smell like soap.' (BM) The specimen observed was a small to medium-sized evergreen shrub with large, spathulate, slightly succulent leaves with closely parallel secondary veins. Small, fragrant, waxy, butter-yellow flowers with numerous, cream-coloured filaments were borne in clusters at the tips of thick, flexible stems. It differs from *Clusia lundelli*, which bears single, slightly larger flowers on each stem. *Loc:* **k'aax, kolil nah** 'forest, house garden'; *Use:* **uuh** 'necklaces'; *Part:* **top'** 'flower'. *See:* **chunup (äh); ts'oy (äh)**. *Thes:* **che'** (Fig. 5.79). *[Note:* kajan che. *Clusia rosea* (Durán 1999).*] [Source:* AM; BM; KM; KP*] [*\sd2 craft*]*

kakache' ocotea, drooping (timber sweet, cayenne rosewood). Lauraceae *Ocotea cernua* (Nees) Mez. This is a large, evergreen tree that bears clusters of small, 15 cm (0.5″) drupes cradled in a cupule, called *ulaak'* 'its spouse' (KP). It is a large, branching tree with wide, elliptic, dark green, glossy leaves with entire margins and widely spaced secondary veins. There are several trunks. The bark is dark brown. The cupule is tan coloured tinged red in areas. The margin of the cupule is rolled inward. The drupe is oblong, green with surface striations and scant flesh. The seed is almost as large as the endocarp. *Use:* **okman** 'house pillars'; *Part:* **che'il** 'timber'. *See:* **mehen 'oonte'**.

Fig. 5.79 Kahanche' *Clusia flava*. Photo Credit: Suzanne Cook

Variant: **babah 'oonte'**. *Thes:* **che'**. *[Note:* kaki che. *Nectandra coriacea* (Durán 1999). English names from English from Botany On-line (http://irapl.altervista.org/botany/main.php?taxon=Ocotea+cernua) and Grandtner (1997–2010).*]* *[Source:* AM; BM; KP*]* *[*\sd2 housing*]*

kakawat peanut. Fabaceae *Arachis hypogaea* L. The peanut is a legume that likely orginated in Brazil and Peru (National Peanut Board, USA). The Lacandones cultivate them in small quantities. *Loc:* **kol** 'milpa'; *Use:* **hanal** 'food'; *Part:* **nek'** 'seeds'; *Prep:* **k'elbil, pokbil** 'toasted, roasted'; *Use:* **chibal hämnen** 'stomach ache'; *Part:* **le'** 'leaves'; *Prep:* **te'** 'tea'. *See:* **sikilte' lu'um**. *From:* cacahuate (Spanish). *[Note:* Remedy provided by Kashanipour and McGee (2004: 61). kakawat. *Arachis hypogaea* (Nations and Nigh 1980); kakawate [Yuc.]. *Arachis hypogaea* (in, Atran et al. 2004); kakawaatej [Itz.]. *Arachis hypogaea* (Atran et al. 2004); manäya [Mop.]. *Arachis hypogaea* (in, Atran et al. 2004).*]* *[Source:* KM; KP*]* *[*\sd2 food\sd2 medicinal*]*

kamis 'ilon (ah) Swiss cheese plant. *Lit:* 'centipede 'ilon'. Araceae *Monstera deliciosa* Liebm. This kind of *'ilon* has huge leaves with holes in their blades. The larger blades appear shredded. The word *kamis* 'centipede' refers to the numerous aerial roots along the stems (BM). The infloresence is a spadix 5–45 cm (2–17.7″) long, which produces a cluster of white sweet, berries. No uses were reported, although the fruit is edible. *Loc:* **k'aax** 'forest'; *Sim:* **säk 'ilon** '*Monstera* sp.'; *Sim:* **hach 'ilon** '*Monstera* sp.'; *Gen:* **'ilon**. *Thes:* **ak'**. *[Note:* camis iron, jach iram, chup. *Monstera deliciosa* (Durán 1999).*]* *[Source:* AM; BM*]*

kat candle tree. Bignoniaceae *Parmentiera edulis* DC. This is a medium sized tree of primary forests, growing up to 10 m tall. Large, pinkish-white flowers grow directly on the branches, followed by slender and long, greenish-yellow fruits with a waxy skin and fibrous, sweet pulp. It is native to Mexico and Guatemala. *Use:* **hanal** 'food'; *Part:* **wich** 'fruit'; *Prep:* **che'che'** 'raw'. *See:* **'anis₁**. *Variant:* **'anis**. *[Note:* kat, änis. *Parmentiera edulis* (Nations and Nigh 1980); (aj) kat [Itz.]. *Parmentiera edulis* (Atran et al. 2004); cat, ixkat ku'uk [Yuc.]. *Parmentiera edulis* (in, Atran et al. 2004).*]* *[Source:* Nations and Nigh (1980).*]* *[*\sd2 food*]*

Fig. 5.80 Kawe' *Coffea* L. Photo Credit: Suzanne Cook

kawe' coffee. Rubiaceae *Coffea* L. Lacandones rarely drink coffee, and did not traditionally cultivate it. However, recently some people have begun to grow it for commercial purposes. *From:* café (Spanish) (Fig. 5.80). *[Source:* BM*]* *[*\sd2 industry*]*

käkow [käkow, käkaw, kakaw, käka, kakah] cacao. Sterculiaceae *Theobroma cacao* L. *Theobroma cacao* is a small understory tree with leathery-skinned, pendulous, bullet-shaped pods. Each pod contains numerous almond-shaped seeds that are encased in a sweet, mucilaginous aril. In the past, the tree was cultivated in house gardens where they could be close at hand for harvesting. Cacao is the name for the unprocessed seeds and the tree, whereas cocoa is the name of the product made from the cacao seeds. Cacao had immense symbolic value throughout Mesoamerica. First and foremost, it was an important religious food: to the Mayas, cacao was sacred and therefore was only consumed in ritual contexts; to the Aztecs, it was a precious commodity, because it had to be transported from the humid Maya lowlands. As such, it was a luxury item consumed only by aristocrats (Coe and Coe 2007: 81). It was also used as currency in the Mayan economy during the Classic period. The historical value of cacao in Mesoamerican religion and commerce is reflected in the word itself. One theory holds that the word *kakaw* originated in the Mixe-Zoquean family and spread throughout the Olmec heartland and into the southeastern Mesoamerican languages, between 200 B.C. and 400 A.D. (Kaufman and Justeson 2007: 193). Another theory is that the word derives from the Nahuatl (Aztec) word *cacahuatl* (/ kakawatl/) (Dakina and Wichmann 2000: 58, Roys 1931: 222). There are a number of varieties of cacao; the most esteemed is *Theobroma cacao* var. *cacao*. *Theobroma cacao* was cultivated at one time, but escaped into the wild and hybridized with other species creating three main forms: (1) Criollo 'native' (*T. cacao* var. *cacao*) is a wild form of *Theobroma cacao* that ranges from southern Mexico to South America, and north and east of the Andes. The fruits are oblong to ovoid, tapering to a point, with five or ten longitudinal ridges. Within this group, three varieties are recognized: (i) "pentagonum", with five ridges and considered the best quality; (ii) "leiocarpum", with five shallower furrows and a more obtuse tip; and, (iii) "lacandonense", a rare form from the Lacandon region that has an acute tip and ten furrows. (2) Forestero (*T. cacao* var. *sphaerocarpum*) is a high yielding and more vigorous variety than the criollos but with inferior quality. It was introduced to Mesoamerica from the Amazon basin. Fruits are ellipsoid to round, lacking a pointed tip, and smooth. (3) Trinitario is a hybrid of criollo and forestero that originated in Trinidad. Additionally, two relatives of *Theobroma cacao–T. bicolor* (*pataxte'* < pataxtli [Nahuatl]) and *T. angustifolium*–are used as cocoa substitutes or adulterants (Rieger 2006: 119). Among the Mayas, however, *Theobroma bicolor* was, and still is, as esteemed as *Theobroma cacao*. The Lacandones distinguish two, possibly three, kinds of cacao—*käkow* (or *hach käkow*), *balumte'*, and *balumchan*. *Hach käkow* 'authentic cacao' is *Theobroma cacao* var. *cacao* and likely the "lacandonense" form of criollo. *Balumte'*, lit. 'jaguar tree', is likely *Theobroma bicolor*. *Balumchan* may be a third kind or a synonym of *balumte'*. *Balumte'* and *balumchan* are said to grow deep in the forest, whereas *hach käkow* is cultivated. *Use:* **uk'ul, ubo'otik ti' k'uh** 'beverage, one offers it to the gods'; *Part:* **nek'** 'seeds'; *Prep:* **Ak'elik yok'ol xämäch. Kuts'o'ok ak'elik, awälä(k') tik. Pachil, ahuch'ik tu ka', awalak. Pachil apuk'ik yok'ol ma'ats'. Pachil ahäxik. Päytan apulik uyits ahtsuk'in ak', pachil ahäxik. Ulik'il uyoom at'änik ut'anil. At'änik uneehel ahch'eel ti' umentik ulik'il uyoom.** 'You toast it on the comal. After you toast it you turn it over. Then, you grind it on a metate, you turn (it). Then you dissolve it in the corn gruel. Then you beat it. First you put in the resin from the *tsuk'in ak'* and then you beat it. As the foam rises up, you recite the secret. You mention the tail of the black penelopina, to help make the foam rise'; *Sim:* **balumte'** 'jaguar tree (*Theobroma* sp.)'; *Sim:* **balumchan** '*Theobroma* sp.' *See:* **tsuk'in ak' (ah)**. *Thes:* **che'**. *[Note:* käkow. chocolate (Bruce 1975); käkäw, kakaw. *Theobroma cacao* (Nations and Nigh 1980); balum té. *Theobroma* sp. (Nations and Nigh 1980); käkäw [S. Lac]. cacao (Canger 1970); xau. *Theobroma cacao* L. (Roys 1931); kakaw (k'än, säk, chäk, b'ox) [Itz.]. *Theobroma cacao* (Atran et al. 2004); kakaw (ajchäk kakaw, ajk'än kakaw, ajsäk kakaw) [Itz.]. cacao. *Theobroma cacao* (Hofling and Tesucún 1997); cacau, xau, haa [Yuc.]. *Theobroma cacao* (Roys 1931).*] [Source:* AM; KM; KP*] [*\sd2 food*]*

kib lu'um [kib lu'um, kip' lu'um] fungus, red cup. *Lit:* 'earth candle'. Sarcoscyphaceae *Cookeina tricholoma* (Mont.) Kuntze. The name refers to a kind of inedible mushroom classified by my Lacandon consultants as *kuxum che'*, lit. 'tree mould'. These small fungi consist of a deep, cup-shaped, brightly coloured cap that seems to glow on the dark, forest floor. Children use them as whistles. The southern Lacandones whom Ruan-Soto et al. (2009) interviewed also acknowledged the existence of poisonous mushrooms, which they called "kibrum". They identified three kinds: "kibrum" *Agaricus* sp., *Pluteus albostipitatus*; "ubojop kibrum aire utob che" (a companion of *kib lu'um* that looks like a flower) *Polyporus* sp., and "el compañero del kibrum" (the companion of *kib lu'um*) *Marasmiellus* sp. They say that young men eat these to get intoxicated (Ruan-Soto et al. 2009: 68). *Use:* **xuxu'** 'whistle'. *Variant:* **kuxum che'**. *[*\sd2 other*]*

kih sisal (henequen). Asparagaceae *Agave sisalana* Perrine. One of two types of agave cultivated for its fibre, *Agave sisalana* is a 1.5–2 m (4.9–6.56′) tall cactus with long, rigid, sword-like, succulent leaves that grow in a dense rosette. Young leaves are armed with minute teeth along their margins. The fibre is extracted by rasping the leaves and then brushing out the fibre. Sisal originated in Chiapas (Gentry 1982: 628–631). *Use:* **häxan** 'rolled cord'; *Use:* **k'an** 'cord'; *Use:* **ch'im** 'purse'; *Use:* **leh** 'trap'; *Use:* **lutsil** 'fishing line' (Fig. 5.81). *[Note:* kih. sisal plant, sisal fiber (Spanish henequén) (Bruce 1975); kih. henequen. *Agave sisilana* (Nations and Nigh 1980); *Agave sisalana* (March 1998); kih [S. Lac.]. pita, helenque (Canger 1970); maw [S. Lac.]. a variety of agave (Baer and Merrifield 1971); kij [Itz.]. henequén. agave. maguey. *Agave* spp. (Hofling 1997); säk kij [Itz.]. *Agave fourcroydes*; ya'ax kij. *Agave angustifolia* (Atran et al. 2004); Ci [Yuc.]. *Agave* sp. Maguey (Roys 1931).*] [Source:* BM*] [*\sd2 cordage \sd2 hunting_and_fishing*]*

kimil sukun spiny pod (milkweed). *Lit:* 'older brother is dead'. Asclepiadaceae *Gonolobus* sp. This is a kind of thin, green, climbing vine that produces large, 10 cm×3.8

Fig. 5.81 Kih *Agave sisalana*. Photo Credit: Chan K'in Daniel

Fig. 5.82 Kimil sukun Asclepiadaceae. Photo Credit: Suzanne Cook

(4×1.5″), cone-shaped, dehiscent pods covered with stiff, green prickles. Leaves are elliptical, rounded at the base, and acuminate at the apex. No uses were reported. *Loc:* **petha'** 'lagoon'. *Thes:* **ak'** (Fig. 5.82). *[Note:* Botanical identification provided by Carlos Beutelspacher Baights (pers. comm.).*] [Source:* BM 2007*]*

kis chuh (äh) bottle gourd. *Lit:* 'fart gourd'. Cucurbitaceae *Lagenaria siceraria* (Molina) Standl. This is similar to *chuh*, only longer. According to AM, it is not good as a container, because of its long shape it falls over when it is set down. *Use:* **uchuhil ha'** 'water bottle'; *Use:* **kab, pak** 'bee (hive), hive'; *Part:* **wich** 'fruit'; *Spec:* **kis chuh mehen** 'small fart gourd'; *Gen:* **chuh**. *See:* **ha' chuh**. *[Note:* luch, lek, chú. *Lagenaria siceraria* (Nations and Nigh 1980), säk lek, chäk lek [Itz.]. *Lagenaria siceraria* (Atran et al. 2004).*] [Source:* AM; BM*] [*\sd2 containers*]*

kis chuh mehen calabash, bottle gourd. *Lit:* 'little fart gourd'. Cucurbitaceae *Lagenaria siceraria* (Molina) Standl. *Use:* **ubäxal ti' mehen paalal** 'child's toy'; *Gen:* **chuh**. *See main entry:* **kis chuh (äh)**. *[*\sd2 other*]*

ki' bok (äh)$_1$ daisy family. *Lit:* 'the fragrant one'. Asteraceae. The name refers to a number of species in the Aster family. Most kinds of *ki' bok* are found growing in fallow *milpas* and disturbed areas. All are considered *lo'obil* 'weeds'. They include: a vine that grows in the lagoons and bears five small, bright yellow petals (probably *Coreopsis mutica*); *Podachaenium eminens*, a tall, herbaceous shrub with clusters of small, white blossoms at the apex of branching stems and large, velvety, dark green leaves; *Clibadium arboreum*, a woody-stemmed, branching shrub with corymbs of small, white blossoms with black anthers; *Verbesina giganteoides*, a tall shrub or small tree with a smooth, purplish-red stalk, large, deeply lobed leaves on long leaf stems, and corymbs of small, white flowers (which, according to AM, smell like garlic); and, *Calea urticifolia*, a shrub 1–3 m (3–9.8′) tall with opposite, rough, sandpaper-like leaves edged with small teeth, and yellow-orange blossoms. *[Note:* kibok. (S. Lac.) *Podachaenium eminens* (Diemont 2006: 83).*] [Source:* AM; BM; CNK; KM; KP*]*

ki' bok (äh)$_2$ *Clibadium arboreum* (common name unknown). Asteraceae *Clibadium arboreum* Donn. Sm. **Uyo'och ähchäklanpix. Es lo'obil.** 'It's the food of the *chäklanpix* (tiger heliconian). It's a weed.' (KM) This tall, 1–3 m (3–9′) perennial shrub is covered with matted hairs. Leaves are opposite, simple, elliptical to oblong, 6–18 cm (2–4″) long and 1.5–6 cm (0.5–2″) wide, acuminate (pointed) at apex and obtuse (rounded) at the base. Blades are fuzzy on both sides. Leaf margins are serrated. Inflorescences are borne in corymbs at the ends of the branches. Black anthers extend beyond the white, tubular corollas. Flowers emit a slight lemon and honey fragrance. It occurs in fallow *milpas*, roadsides, and other disturbed areas. *Clibadium* species are reportedly used as fish poison and medicine by indigenous peoples throughout Latin America. As a fish poison, it is commonly referred to as "barbasco" (Arriagada 1995: 328). My Lacandon consultants did not report any uses for the plant. *Loc:* **paakche' kol** 'fallow milpa'. *Variant:* **chäklanpix (ah)**. *[Note:* Botanical identification provided by Dr. J. E. Arriagada (pers. comm.).*] [Source:* AM; KM*]*

ki' bok (äh)$_3$ daisy tree, giant. Asteraceae *Podachaenium eminens* (Lag.) Sch. Bip. *Use:* **hiil** 'battens'; *Use:* **hi'che' ti' k'anche' näl** 'pole wall for the corncrib'; *Part:* **che'il** 'stalks'. *Variant:* **haban uts'u' (ah)**. *Thes:* **che'**. *[Note:* kibok. *Podachaenium eminens* (Nigh 2008).*] [Source:* BM; KM: KP; CNK*]*

koke' ak' (ah) sarsaparilla (greenbriar). Smilacaceae *Smilax lanceolata* L. This is a kind of green vine with sharply angled

Fig. 5.83 (a) Koke' ak' *Smilax luculenta*. Photo Credit: Suzanne Cook (b) Koke' ak' *Smilax luculenta*. Photo Credit: Suzanne Cook

stems and thick thorns. Leaves are narrow, lanceolate, smooth, and acuminate at the apex and cuneate at the base (base width smaller than the rest of the leaf before the apex). Blades are unarmed, green on both sides, and lack conspicuous secondary veins. It is found growing at the edges of clearings and fallow *milpas*. Flowers and fruits were absent when the photograph (Fig. 5.83a, b) was taken, making a positive identification difficult. *Koke' ak'* is mentioned in the 'Song for Frothing Cacao', although it is not used in the preparation of the chocolate beverage (AM). *Loc:* **paakche' kol** 'fallow milpas'; *Sim:* **'op' ak'** '*Smilax* sp.'; *Sim:* **tsuk'in ak' (ah)** '*Smilax domingensis*'. *Thes:* **ak'** (Fig. 5.83a, b). *[Note:* Identification provided by Oscar Farrera Sarmiento. coceeh. zarzaparilla. *Smilax mexicana* Griseb.(Landa [1937]); coc-ak [Yuc.]. Lit. asthma-vine (Roys 1931); ixkokej-ak' [Yuc.]. *Smilax mollis, S. lundellii* (in, Atran et al. 2004).*] [Source:* AM; JK*] [*\sd2 folklore_and_mythology*]*

Fig. 5.84 Komo' (te' 'usin) *Renealmia* sp. Photo Credit: Suzanne Cook

komo' (äh) ginger, ostrich plume (pine cone ginger). Zingiberaceae *Renealmia* sp. Flowers were not present when the photo (Fig. 5.84) was taken, but the fruit of this species is that of either *Renealmia alpinia* or *R. thyrsoidea*. Both species grow in the area. *Renealmia alpinia* is a large herb, up to 4 m (6.5′) tall, that grows in colonies from underground stems. Leaves are alternate and double-ranked, 0.3–1.5 m (11.8–59″) long, 6–23 cm (2.4–9″) wide, elliptical, acuminate, with numerous secondary veins. The inflorescences grow from the base of plant. Red and yellow, tubular flowers are borne in small clusters along the stalks, followed by red-orange, 3-valved fruits containing numerous, white seeds in a sticky, orange aril. It prefers wet areas in disturbed sites, second growth, abandoned fields, and forest edges from sea level to 1300 m (4265′). (Gargiullo et al. 2008: 330) The inflorescences are more erect and the pedicels are much longer than those of the species in the photo, which leads P. M. J. Maas to believe that the species is *Renealmia thyrsoidea*; although this species is unknown north of Costa Rica (Maas, pers. comm.). Further examination of the plant is required. The northern Lacandones only use the leaves for food preparation. Levy et al. (2006) report that another species, *R. mexicana*, is found in the southern Lacandon territory where it is used for medicinal purposes. *Use:* **mukbil wah** 'large, baked tamale'; *Use:* **ti' päk'ächtik wah** 'to pat out tortillas'; *Part:* **le'** 'leaf'. *See:* **te' 'usin**. *Thes:* **xiiw** (Fig. 5.84). *[Note:* Botanical identification provided by Axel Dalberg Poulsen, Oslo Botanical Garden

Fig. 5.85 (**a**) Kopo' *Ficus* sp. Photo Credit: Suzanne Cook (**b**) Kopo' wits *Ficus* sp. Photo Credit: Suzanne Cook

Natural History Museum University of Oslo, and Paul Maas.*] [Source:* AM; KP*]* [\sd2 food*]*

kopo' (ah)$_1$ fig. Moraceae *Ficus* spp. **Yan ka'p'eli': mehen kopo' i nukuch kopo'. Yan uhel uk'aba' kopo' wits. Mehen kopo' päk'bil i nukuch kopo' uch'ihil tuhunal ich k'aax. Uwich ahmehen kopo', chäk uwich. Uwich ahnukuch kopo', k'än. Kopo' wits uch'ihil yok'ol wits. Una'akal uyak'il umots.** 'There are two kinds: small *kopo'* and large *kopo'*. There is another one, its name is *'kopo'* of the hills'. The small *kopo'* is cultivated and the large *kopo'* grows on its own in the forest. The fruit of the small *kopo'* is red. The fruit of the large *kopo'* is yellow. *Kopo' wits* grows in the hills. The vines of the roots climb.' (BM) Despite these two, broad divisions, *kopo'* designates a number of species of strangler figs belonging to Ficus subg. Urostigma sect. Americanae (described in Berg 2007). *Kopo'* is mainly exploited for its bark fibre, which is pounded into cloth. It was one of the species that the ancient Maya used for the Mayan codices, the accordion-style books with hieroglyphs, the paper having been first primed with a coating of lime calcium before being printed (Nations 2006: 95). In addition to bark cloth, the southern Lacandones use the foliage of *Ficus cotinifolia* for forage and the wood of *Ficus obtusifolia* (syn. *F. involuta*) for construction (Levy et al. 2006). One species that grows around the lagoon releases the fruit into the water whereupon fish rise to the surface and feed in a frenzy. The Lacandones take advantage of this event and drop in their fishing lines (Bruce 1974: 308). The Lacandon song, 'Song of the Fish', alludes to this event: fishermen sing it to ensure an abundant catch (AM). The leaves of some species are used in cooking. *Use:* **hu'un ti' xikul** 'bark cloth for traditional tunic'; *Part:* **uyoot'el sool** 'inner bark'; *Use:* **ti' mukbil uyo'och** 'baking pouch'; *Part:* **le'** 'leaves'; *Sim:* **chimon** *'Ficus maxima*'; *Spec:* **mehen kopo'** 'small kopo' (*Ficus pertusa*)'; *Spec:* **nukuch kopo'** 'large kopo''; *Spec:* **kopo' wits** 'kopo' of the hills (*Ficus* sp.)'. *Thes:* **che'** (Fig. 5.85a). *[Note:* kopoh. *Ficus cookii, F. involuta* (Nations 2006); kopo'. *Ficus* spp. (Nations 2006); kopo, una tzoy. *Clusia salvinii* (Durán 1999); ak'jun (Nations 1992, in Atran et al. 2004); kóopo' [Yuc.]. *Ficus cotinifolia* (Bricker et al. 1998); copó [Yuc.]. *Ficus cotinifolia"* a sort of fig tree, 45 feet high" (Roys 1931: 226); kopo', aj säk kopo', aj chäk kopo' [Itz.]. *Ficus aurea, F. cotinifolia, F. involuta, F. obtusifolia* (Atran et al. 2004); kopo' [Mop.]. *Ficus aurea, F. cotinifolia, F. involuta, F. obtusifolia* (in, Atran et al. 2004); ixkopo', ju'un [Yuc.]. *F. continfolia* (in, Atran et al. 2004).*] [Source:* AM; BM; CKO; KM*]* [\sd2 clothing \ sd2 food*]*

kopo' wits fig, rock (strangler fig). Moraceae *Ficus* sp. The *kopo' wits* is a large ficus with an extensive, twining root system, characteristic of strangler figs. It bears small, round, berry-like fruits. As the name suggests, it grows in the hills. No uses were reported. *Loc:* **yok'ol wits** 'in the hills'; *Sim:* **mehen kopo'** 'small kopo' (*Ficus pertusa*)'; *Sim:* **nukuch kopo'** 'large kopo''; *Gen:* **kopo' (ah)**. *Thes:* **che'** (Fig. 5.85b). *[Source:* BM*]*

korason de mono sea heart. Fabaceae *Entada gigas* (L.) *Fawc. & Rendle*. A species of flowering liana, *Entada gigas* is prized for its large, woody, heart-shaped seeds, which measure 12 cm (4.7″) wide and 2 m (6.6′) long. Ten to fifteen seeds are produced in a single pod. The seeds contain a hollow cavity, making them bouyant. They are commonly referred to as drift seeds, because ocean currents carry them

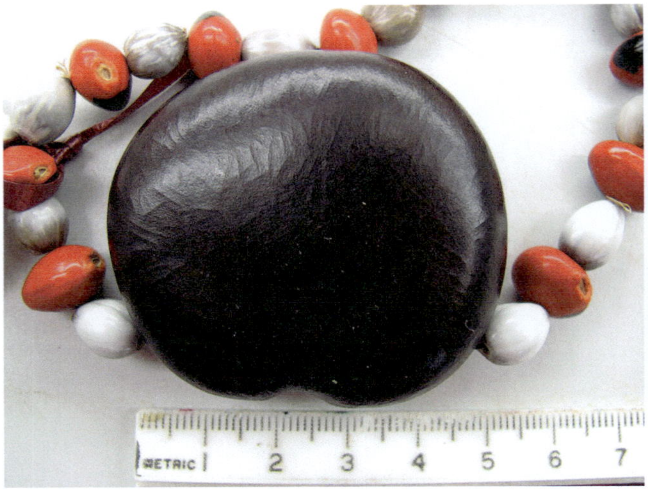

Fig. 5.86 Korason de mono *Entada gigas*. Photo Credit: Suzanne Cook

Fig. 5.87 Kubuh *Pachira aquatica* leaves. Photo Credit: Suzanne Cook

across the oceans and deposit them on distant shores. *Entada gigas* is native to Central America, the Caribbean, northern South America, and Africa. It also grows in some parts of the Tropical Rainforest and the Lower Montane Rain Forest, along the banks of fresh water rivers and mangrove swamps (Beutelspacher-Baigts 2009: 50). According to my Lacandon consultants, CNK and NI, the species does not grow around Naha'. They purchase the seeds from neighbouring Tzeltal villagers. *Loc:* **kah** 'villages'; *Use:* **uuh** 'necklaces'; *Part:* **nek'** 'seeds'. *From:* corazón de mono (Spanish) (Fig. 5.86). *[Note:* Identification of seed provided by Beutelspacher Baigts (pers. comm.).*]* *[*\sd2 craft*]*

kubuh (ah) [kubuh] chestnut, Guiana. Malvaceae *Pachira aquatica* Aubl. This is a large tree that can grow up to 18 m (59′) in the wild. It has shiny green, palmate leaves and lanceolate leaflets. Its showy, white flowers are strap-like, cream or greenish petals, which open and curl back to reveal numerous, white or yellow, whispy, erect stamens. Flowers are followed by large 20–30 cm (8–12″) long, fuzzy, brown, ovoid pods that contain white, soft flesh surrounding numerous seeds. Fruit matures in November in Naha'. (BM) Both the fruit and the large, 4–5 cm (1.5–2″) long, angular seeds are edible. *Use:* **hanal** 'food'; *Part:* **wich** 'fruit'; *Part:* **nek'** 'seeds'; *Prep:* **che'che'** 'raw'. *Thes:* **che'** (Fig. 5.87). *[Note:* kubuh. *Pachira aquatica* (Nations and Nigh 1980); wakut' [Mop.]. *Pachira aquatica* (in Atran et al. 2004); (ix) wakut [Itz.]. *Pachira aquatica* (Atran et al. 2004); ixk'uy che' [Yuc.]. *Pachira aquatica* (in, Atran et al. 2004).*]* *[Source:* BM*]* *[*\sd2 food*]*

kulantro [kulantro, kurantro] coriander, long (Mexican coriander, fitweed, spiritweed, duck-tongue herb, sawtooth or saw-leaf herb, sawtooth coriander). Apiaceae *Eryngium foetidum* L., syn. *Eryngium antihystericum* Rottbøll. This is a tropical, perennial herb with slender, long, tapered leaves

Fig. 5.88 Kulantro *Eryngium foetidum*. Photo Credit: Suzanne Cook

with spiny margins, and greenish-white flowers with leafy bracts. It is native to Mexico and South America. The Lacandones cultivate it for its aromatic leaves. *Loc:* **kolil nah, kol** 'house gardens, milpas'; *Use:* **ts'ak ti' o'och** 'spice for food'; *Part:* **le'** 'leaves'; *Prep:* **chäkbil yetel bäk'** 'cooked with meat'. *See:* **perehir** (Fig. 5.88). *[Source:* BM; KM*]* *[*\sd2 food*]*

kulix [kulix, kurix] cabbage. Brassicacea *Brassica oleracea* var. *alba* L. An introduced species. Cultivated. *Use:* **hanal** 'food'; *Part:* **le'** 'leaves'; *Prep:* **chäkbil** 'boiled'. *[Source:* KM; KP; CNK*]* *[*\sd2 food*]*

kunkunche' icacina. Icacinaceae *Oecopetalum mexicanum* Greenm. & C. H. Thomps. This is a tree or bush 2–25 m (6.5–82′) tall, with elliptic-lanceolate leaves, 15–25 cm (5.9–9.8″) long and 7–10 cm (2.7–3.9″) wide. White, 8 mm (0.3″) long flowers are borne in racemes in leaf axils or at the end of branches. The masses of flowers in February attract droves of honey bees. Flowers are followed by oblong fruit,

Fig. 5.89 (**a**) Kun *Cryosophila stauracantha* plants. Photo Credit: Suzanne Cook (**b**) Kun *Cryosophila stauracantha* fruit. Photo Credit: Suzanne Cook

2–3 cm (0.79–1.2″) long and 1–2 cm (0.4–0.79″) wide, smooth and green when young, but wrinkled and brown at maturity. A hard, brown endocarp encases an oily, white, bitter, protein-rich seed. (Gutiérrez 1994) Despite its bitterness, the nutmeat is a popular snack throughout Latin America. It is equally savored by parrots. *Loc:* **xok'ol petha'** 'near the lagoon'; *Part:* **noy** 'nutmeat'; *Prep:* **Achäkik pachil apulik uch'ooch'** 'You boil it and then you add salt'. *Variant:* **kunkunte'**. *Thes:* **che'** (Fig. 5.90). *[Note:* This is not the same tree as kunkunche' kab, according to BM.*] [Source:* BM; CKD; KP*] [*\sd2 food*]*

kunkunche' kab pepperbush. Clethraceae *Clethra* sp. This is a very tall tree that bears clusters of small, white, fragrant blossoms that attract droves of honey bees, in August. Durán (1999) reports that his Lacandon consultants collect honey from the hives and use the wood in house construction, for corncribs, and as fuel. *Use:* **k'anche' näl** 'corncrib'; *Use:* **umentik nah** 'to make houses'; *Part:* **che'** 'wood'. *Thes:* **che'** (Fig. 5.91). *[Note:* yox kun che apto kap, kukun che ak tum kap. *Clethra suaveolens* (Durán 1999). Common name from http://www.fs.fed.us/ne/morgantown/4557/gmoth/gtr/table3.html. Retrieved March 2012.*] [Source:* BM*] [*\sd2 housing*]*

kun (ah) [kun, kum] palm (fan palm, give-and-take palm, rootspine palm). Arecaceae *Cryosophila stauracantha* (Heynh.) R. J. Evans, syn. *C. argentea* Bartlett. *Cryosophila* is a genus of nine, medium-sized fan palms that are distinguished by their downward pointing stem-spines, or modified roots. Most species grow in lowland humid or wet forests from western Mexico to north-western Colombia 0–1700 m (0-5577′). In the Lacandon literature, *kun* corresponds to *C. nana* (syn. *Acanthorrhiza mocini*) and *C. stauracantha*. Two main features that distinguish the two species are differences in height and the colour of the leaves. *Cryosophila stauracantha* can reach a height of 9 m (30′) with a diameter of 10–12.7 cm (4–5″). The leaves are whitish underneath. *Cryosophila nana* is a shorter tree whose leaves are green on both sides (Henderson et al. 1995: 41). Furthermore, their environmental requirements are different: *Cryosophila nana* is a tree of dry, deciduous forests and pine-oak woodlands on the Pacific coast of Mexico and Guatemala, between 200 and 1000 m (656–3281′) in elevation. *Cryosophila stauracantha* is a lowland rain forest species, growing from sea level to 600 m (1968.5′) (Henderson et al. 1995: 41). According to the distributional maps in Henderson et al., *C. nana* is on the west coast of Chiapas, whereas *C. stauracantha* grows in north-eastern Chiapas, Campeche, Belize, and northern Guatemala. Based on this information, *C. stauracantha* is likely the species growing in Naha', although the community is above the elevation range for the species. This is the only species recorded from the area (Comisión Nacional de Áreas Naturales Protegidas 2006). In addition to thatch and food, *kun* provides fuel: pieces of the leaf stems are burned to coals and then placed under hammocks at night for warmth. *Loc:* **ts'u' k'aax** 'dense forest'; *Use:* **hanal** 'food'; *Part:* **uyool** 'heart'; *Prep:* **pokbil** 'roasted'; *Use:* **ch'ooch'** 'salt'; *Part:* **uchuch ule'** 'leaf stem'; *Prep:* **Ak'ak'tik uchuch ule', pachil amalik uta'anil i abut'ik ich kum wa luuch** 'You roast the stems in the fire and then you collect the ashes and put them in a kettle or bowl (for later use)'; *Use:* **pix nah** 'house cover'; *Part:* **le'** 'leaves'; *Use:* **chuuk k'ak' ti' yalam k'an ich 'aak'a'** 'coals for beneath the hammock at night' (Fig. 5.89a, b). *[Note:* Although thatch made of *kun* usually needs to be replaced every two years, wood smoke from cooking fires makes it

Fig. 5.90 Kunkunche' *Oecopetalum mexicanum*. Photo Credit: Chan K'in Daniel

Fig. 5.91 Kunkunche' kab *Clethra* sp. Photo Credit: Suzanne Cook

last a bit longer (BM). Species names from Henderson et al. (1995). kun. *Acanthorrhiza mocini* (Bruce 1975); kun. spined salt palm. *Acanthorrhiza mocini* (Nations and Nigh 1980); kun. *Crysophila nana* (Marion 1991, in March 1998); kum. *Crysophila argentea* (Durán 1999); kùn [S. Lac.]. mata de escoba (Canger 1970); aj kuum, aj miis [Itz.]. *Cryosophila stauracantha* (Areca.) (Atran et al. 2004); miis [Mop.]. *Cryosophila stauracantha* (in, Atran et al. 2004); kuum, mı́is [Yuc.]. *Cryosophila stauracantha* (in, Atran et al. 2004).*]* *[Source:* AM; BM*]* */*\sd2 housing \sd2 food \sd2 fuel*]*

kuti' magnolia, heartflower tree. Magnoliaceae *Talauma mexicana* (DC.) G. Don. This is a large tree with large, ovate leaves and large, oval fruit with a tan to dark brown skin. It bears small, white flowers in late January. Fruits begin to

Fig. 5.92 Kuti' *Talauma mexicana* fruit. Photo Credit: Chan K'in Daniel

form in March and ripen in October or November. The fruits split open, scattering numerous red seeds that draw flocks of birds, such as: *ahch'iluk'* (*Odontophorus guttatus*); *ahwan* (*Dentrortyx* sp.); *ähts'ul* (*Leptotila jamaicensis*); *ähpän* (*Ramphastos sulfuratus*); and, numerous varieties of small birds (BM). The bark consists of two layers, which are easily split into thinner sheets. When they are dried they curve up slightly. *Use:* **che'il k'ak'** 'firewood'; *Use:* **nah, k'anche' näl** 'houses, and corncrib *Part:* **uche'il** 'timber'; *Use:* **hopil k'ak'** 'ash pan'; *Part:* **sool** 'bark'; *Use:* **hubul hämnen** 'diarrhea'; *Part:* **sool** 'bark'; *Prep:* **te'** 'tea'. *Thes:* **che'** (Fig. 5.92). *[Note:* kuti. *Talauma mexicana* (Nations and Nigh 1980); kutik. *Magnolia* sp. (Durán 1999). Medicinal remedy provided by Durán (1999).*]* *[Source:* BM; K'ayum Segundo*]* */*\sd2 other \sd2 fuel \sd2 housing \sd2 medicinal*]*

kuti'il wits [kutiwits] magnolia, heartflower tree. *Lit:* 'kuti' of the hills'. Magnoliaceae *Talauma mexicana* (DC.) G. Don. *See main entry:* **kuti'**. *Thes:* **che'**. *[Note:* cu'ti huitz. *Talauma mexicana* (Programa de Conservación y Manejo Área de

Fig. 5.93 Ku'uchnook' Bidens odorata. Photo Credit: Suzanne Cook

Protección de Flora y Fauna Nahá 2006); kutik. *Magnolia* sp. (Durán 1999).*] [Source:* BM; KP*]*

ku'uchnook' [ku'uchnook', kuchnook', kuxnook'] hairy beggarticks (Spanish needle). *Lit:* 'sticks on clothes'. Asteraceae *Bidens odorata* Cav. This is a short, branching herb bearing cheerful, little white daisies on stiff stems and numerous burrs, which stick to pant-legs and skirt hems. Considered *lo'obil* 'a weed', it grows in open, disturbed areas, such as fallow *milpas*, road-sides, and house gardens. *Use:* **ts'ak ti' chibal** 'medicine for insect bites'. *Morph:* **ku'uch=nook'**. *See:* **hach lo'obil** (Fig. 5.93). *[Note:* Medicinal remedy provided by Kashanipour and McGee (2004). kuch nok. (Kashanipour and McGee 2004); kuxnok'. *Bidens ordorata* (Nigh 2008).*] [Source:* K'ayum Segundo; SK*]*

kuut' (ah) *Calathea macrosepala* (common name unknown). Marantaceae *Calathea macrosepala* K. Schum. This is a low-growing, herbaceous plant with long, broad leaves; blades are bright green on the surface and lighter green underneath, with two maroon stripes that run parallel to the midrib. Inflorescence is like a cylinder, composed of bright green bracts and pale yellow, tubular blossoms. No uses were reported. *Gen:* **muxan** 'Zingiberales (in part)'. *Variant:* **ya'ax ch'ib muxan xiiw**. *Thes:* **xiiw** (Fig. 5.94). *[Note:* kut' [Itz.]. *Calathea allouia* (Atran et al. 2004).*]* *[Source:* AM*]*

Fig. 5.94 Kuut' *Calathea macrosepala*. Photo Credit: Suzanne Cook

kuxum che'$_2$ artist's bracket (artist's conch, flacher lackporling). Ganodermataceae *Ganoderma applanatum* (Pers.) Pat. This is a tough, inedible fungus that grows on dead wood and living sapwood of older trees. The bracket-shaped, fruiting bodies grow in horizontal rows, resembling plates. Like other species of bracket fungi, these plates are believed to be the tortillas of *Kisin* (the Devil). (Fig. 5.95a)

kuxum che'$_3$ coral fungi. Gomphaceae *Ramaria* Fr. ex Bonord. *[Source:* AM*]*

kuxum che'$_4$ fungus, red cup. Sarcoscyphaceae (Pezizales) *Cookeina tricholoma* (Mont.) Kuntze. This is a cup fungus that grows on twigs and rotten tree limbs. It looks like a sea anemone glowing on the forest floor. Children use them as whistles. *Cookeina tricholoma* and *C. sulcipes* are also used as toys by southern Lacandon children (Ruan-Soto et al. 2009: 68). *Variant:* **kib lu'um** (Fig. 5.95b). *[*\sd2 other*]*

K'–k'

k'ak'che'$_1$ *Aegiphila monstrosa* (common name unknown). *Lit:* 'fire tree'. Verbenaceae *Aegiphila monstrosa* Moldenke. *Aegiphila monstrosa* is a small tree that grows in wet thickets in sparse or second growth forest. Mature stems are quadrangular, woody, and rangy. Leaves are petiolate, simple, elliptic-acuminate with entire margins. The blades are smooth, bright green, and glossy. Clusters of fragrant, white, cauliflorous blossoms grow along the branchlets. Four long anthers extend beyond the cruciform, white corolla. It is an

a

b

Fig. 5.95 (**a**) *Ganoderma applanatum* (**b**) Kuxumche' (kib lu'um) *Cookeina tricholoma*. Photo Credit: Suzanne Cook

endangered species (Nelson 1998). No uses were reported. *Loc:* **uxuul paakche' kol ti' Koh** 'the edge of KM's fallow milpa'. *Thes:* **che'** (Fig. 5.96a, b, c). *[Note:* AM said this tree was not called *k'ak'che'*, but he could not remember the name for it.*] [Source:* KP*]*

k'ak'che'$_2$ bauhinia, rio rubeleruz. *Lit:* 'fire tree'. Fabaceae *Bauhinia rubeleruziana* J. D. Smith, syns. *B. emarginella* Standl., B. *palenquensis* Lund., *Casparea rubeleruziana* (Donn. Sm.) Britton & Rose. One of many non-endemic *Bauhinia* species in Mexico, *Bauhinia rubeleruziana* is distinguished by its 20 or more brick-red blossoms and yellowish-brown hairs on the lower surface of young leaves (Wunderlin 1983: 87). Flowers are followed by large, green, slightly curved, flattened pods. My consultants only reported using the timber for building material, although Durán (1999) mentions that his consultants use it for house walls, chairs, and other household items. *Use:* **akam** 'foundation posts'; *Part:* **uche'il** 'the trunk'. *Variant:* **säkche'**. *Thes:* **che'** (Fig. 5.96d, e). *[Note:* kan che. *Bauhinia rubeleruziana* (Durán 1999). English name from Grandtner (1997–2010). Botanical synonyms from The Plant List.http://www.theplantlist.org/*] [Source:* AM*] [*\sd2 housing*]*

k'alok' che' (äh) [k'alok' che', k'arok' che'] billia. *Lit:* 'wild boar tree'. Sapindaceae *Billia colombiana* Planch. & Lindl. This forest tree bears large, elliptical fruit with a hard, thin shell covered in a brownish-yellow scurf. A thin layer of pulp encases a large, reddish-purple, dehiscent seed, approximately 7.6 cm (3″) in diameter. The fruit and seeds are eaten by squirrels, peccaries, and horses (AM, BM). *K'alok'* is also the name for the male wild boar. SK said that the Tzeltales make toy wheels from the two halves of the seeds. *Billia colombiana* is found in the mountainous mesophilous forest (a.k.a. Pine-Oak-Liquidambar Forest [Breedlove 1981]) in the northern Lacandon communities of Naha' and Mensäbäk (Hernández-Nava 2003). *Billia hippocastanum* is found in the understory of Montane Rain Forest, between 900 and 2200 m elevation (Breedlove 1981: 10). The timber is used in house construction. *Use:* **bahbil** 'plank house walls'; *Part:* **che'il** 'timber'. *Thes:* **che'** (Fig. 5.97a, b). *[Note:* Taxonomic identification provided by Reinaldo Aguilar (pers. comm.). karopché. *Billia colombiana* (Hernández 2003: 7).*] [Source:* AM; BM; SK*] [*\sd2 housing*]*

k'ante' (ah) [k'ante', k'änte'] coral tree (coral bean tree). *Lit:* 'yellow tree'. Fabaceae *Erythrina berteroana* Urb. This is a small to medium-sized tree, 10 m (32.8′) tall, armed with thorns. Leaves are alternate, trifoliate, and 10–35 cm (3.9–13.8″) long. Leaflets are ovate or deltoid, 5–12.5 cm (1.96–4.9″) long, 4–12.5 cm (1.6–4.9″) wide, entire, and shortly acute or acuminate at the apex. Scarlet, plume-like flowers are borne in terminal racemes. Each blossom is approximately 5–10 cm (1.9–3.9″) long. Fruits are slender, dehiscent pods that contain numerous small, hard, red seeds, which are released in October (in Naha'). The seeds resemble those of *'op' ch'uplal* (*Ormosia isthmensis*), and because of this, they are also referred to as *'op' ch'uplal*. Seeds are used in jewellery. The roots are scraped and soaked to extract a yellow dye. Additionally, the northern Lacandones use them as living fences (Durán 1999). Although the flower buds, young leaves, and tender twigs are eaten like stringbeans in Guatemala and other parts of Central America, my Lacandon consultants did not report eating any part of the plant. *Use:* **ti' banik nook'** 'dye for cloth'; *Part:* **mots** 'roots'; *Prep:* **Aho'ochtik usool umots pachil asulik uxot' umots ich ha'** 'You grate the bark off the roots and then you soak pieces of the root in water'; *Use:* **uuh** 'necklaces'; *Part:* **nek'** 'seeds'. *See:* **xoyok (ah)**; **k'ante' ak'**. *Thes:* **che'** (Fig. 5.98a, b, c). *[Note:* c'ante'. *Erythrina americana* (Baer and Merrifield 1971: 156); kante. *Erythrina berteroana* (Durán 1999); aj/ix k'änte' [Itz.]. *Gliricidia sepium* (Atran et al. 2004).*] [Source:* AM; KM*] [*\sd2 craft*]*

Fig. 5.96 (**a**) K'ak'che' *Aegiphila monstrosa* flowers. Photo Credit: Suzanne Cook (**b**) K'ak'che' *Aegiphila monstrosa* leaves. Photo Credit: Suzanne Cook (**c**) K'ak'che' *Aegiphila monstrosa* growing tip. Photo Credit: Suzanne Cook (**d**) K'ak'che' *Bauhinia rubeleriziana* leaves. Photo Credit: Suzanne Cook (**e**) K'ak'che' *Bauhinia rubeleriziana* seed pods. Photo Credit: Suzanne Cook

Fig. 5.97 (**a**) K'alok' che' *Billia colombiana* seed. Photo Credit: Chan K'in Daniel (**b**) K'alok' che' *Billia colombiana* seed. Photo Credit: Chan K'in Daniel

Fig. 5.98 (**a**) K'ante' *Erythrina berteroana* seeds. Photo Credit: Suzanne Cook (**b**) K'ante' *Erythrina berteroana* flowers. Photo Credit: Suzanne Cook (**c**) K'ante' *Erythrina berteroana* plant. Photo Credit: Suzanne Cook

k'ante' ak' [k'ante' ak', k'änte' ak'] coral tree. Fabaceae *Erythrina* sp. This is described in the botanical literature as a shrub with spines and long 10 cm×2.54 cm (4×1″), red fruit. BM says that it is a tree; yet, it may be a scandent form of *Erythrina*. Only two *Erythrina* species have been recorded from Naha'—*E. berterona* and *E. folkersii* (Comisión Nacional de Áreas Naturales Protegidas 2006; Durán 1999). Kashanipour and McGee (2004) report that *Erythrina coralloides* is used by some Lacandones, who chew the bark and ingest the sap to relieve stomach aches. None of my Lacandon consultants mentioned any uses for it; according to AM, if the sap gets into your eyes it can cause blindness. Research shows, however, a number of *Erythrina* species have medicinal applications throughout the world: ehanolic extract from the stem bark of *Erythrina caffra* has antibacterial properties (Olajuyigbe and Afolayan 2012); the flowers of some species are brewed as a tea, which is regularly used as a relaxant in Mesoamerica; in the West, the alkaloid b-Erythroidine was used as a muscle relaxant in surgery and in the treatment of schizophrenia. This alkaloid may be the poison that fishermen in South America exploit to stun fish (Kass n.d.). As far as can be determined, the Lacandones did not use *Erythrina* as a fish poison. *Use:* **uuh** 'necklaces'; *Part:* **nek'** 'seeds'; *Use:* **hubul hämnen** 'diarrhea'; *Part:* **uyits usool** 'sap from the bark'; *Prep:* **mäna'** 'none'. *See:* **k'ante' (ah)**. *Thes:* **ak'**. *[Note:* Medicinal use provided by Kashanipour and McGee (2004). kanti ak'. *Erythrina coralloides* (Kashanipour and McGee 2004). Remedy provided by Kashanipour and McGee. Lacandon translations provided by the author.*] [Source:* BM*]* *[*\sd2 craft \sd2 medicinal*]*

k'ayoch' (ah) [k'ayoch', k'äyoch'] mushroom, pink oyster. Pleurotaceae *Pleurotus djamor* (Fr.) Boedijn. *K'ayoch'* is much larger than *muluch'* (*Oudemansiella* spp.). It is frequently eaten. Two species have been reported from the Lacandon for-

Fig. 5.99 (**a**) K'äb ta'k'in *Oreopanax obtusifolius* plant. Photo Credit: Suzanne Cook (**b**) K'äb ta'k'in *Oreopanax obtusifolius* inflorescence. Photo Credit: Suzanne Cook

est: *Pleurotus djamor* is found in the southern Lacandon community of Lacanha' Chan Sayab (Ruan-Soto et al. 2009), and *P. ostreatus* is reported from the northern Lacandon area (Comisión Nacional de Áreas Naturales Protegidas 2006). *Use:* **hanal** 'food'; *Prep:* **k'olbil** 'added to soup thickened with corn dough'; *Sim:* **muluch'** '(*Oudemansiella* spp., *Pluteus harrisii*)'; *Sim:* **xikin** 'ear (*Schizophyllum commune*), chicken-breast fungus (*Favolus tenuiculus*)'. *[Note:* Taxonomic identification from Ruan-Soto et al. (2009). Compare k'óol [Yuc.]. thick sauce or gravy [made of cornstarch] (Bricker et al. 1998: 156).*] [Source:* AM; KM; KP*] [*\sd2 food*]*

k'äb ta'k'in$_1$ purslane. *Lit:* 'hand money'. Portulacaceae *Portulaca oleracea* L. This is a spreading, prostrate plant with small, thick, roundish leaves. It is regarded as *lo'obil* 'a weed'. *Use:* **hanal** 'food'; *Part:* **le', che'il** 'leaves, stems'; *Prep:* **chäkbil** 'boiled'. *[Note:* ix b'erd'olaagaj [Itz.]. *Portulaca oleracea* (Atran et al. 2004); ajxukul, kab'al chunup [Yuc.]. *Portulaca oleracea* (in, Atran et al. 2004).*] [Source:* AM; BM; KM*] [*\sd2 food*]*

k'äb ta'k'in$_2$ [k'äb ta'k'in, k'ä' ta'k'in] obtuse leaf oreopanax. *Lit:* 'hand money'. Araliaceae *Oreopanax obtusifolius* L. O. Williams, syn. *Oreopanax guatemalensis* (Lem. ex Bosse) Decne. & Planch. These may be large, epiphytic shrubs or trees growing to 10 m (32.8′) tall, with a slender trunk up to 10 cm (3.9″) in diameter. Leaves are simple, alternate, broadly obovate to ovate-suborbicular ovate, rounded at the apex, and rounded, obtuse, to subcordate at the base. The texture ranges from papery to rather leathery. Blades are 8–20 (−35) cm (3–7.8″) long and 3–18 (−25) cm (1–7″) wide, with a dark green, glossy lamina. Leaves are palmately nerved. Small, white flowers, borne in panicles up to 25 cm (9.8″) long are followed by tiny clusters of small, white, round berries. (Standley and Williams 1966: 17) The fruit attracts a variety of birds, including *pän* (keeled-billed toucan), *t'ut'* (small parrots, white-crowned parrot), and *kacho'* (mealy blue-crowned parrot). No uses were reported. *Variant:* **k'ä' ta'k'in che'**. *Thes:* **che'** (Fig. 5.99a, b). *[Note:* kaa ta kin che. *Oreopanax obtusifolius* (Durán 1999). English name from Grandtner (2005).*] [Source:* AM*]*

k'änk'än näl maize. *Lit:* 'very yellow maize'. Poaceae *Zea mays* L. The second most valued maize, yellow maize is used to make tamales and tostadas. It is also used for tortillas and corn gruels when white maize is unavailable. *Use:* **hanal** 'food'; *Part:* **wich** 'fruit'; *Use:* **tap'ik wah** 'tamales'; *Sim:* **säk näl** 'white corn'; *Sim:* **ek' näl** 'black corn'; *Gen:* **näl**. *[Source:* AM*] [*\sd2 food*]*

k'än'ol (äh) maiden plum. Anacardiaceae. This is a tall, spindly tree with glossy green leaves and bark pimpled with lenticels (raised pores). Characteristic of the family is the presence of resinous channels in the bark; the resin is clear but turns black when exposed to air (Martínez-Millán et al. 2005). When touched, the resin from these pores causes severe itching. According to BM, a solution of lime juice and salt applied to the affected area will bring some relief. *Spec:* **mehen k'än'ol** 'small k'än'ol (*Astronium graveolens*)'; *Spec:* **nukuch k'än'ol** 'big k'än'ol (*Mosquitoxylum jamaicense*)'. *Variant:* **mehen k'än'ol, ahk'än**. *Thes:* **che'**. *[Source:* BM*]*

k'änxa'an bullywood (commercial name for the timber). Combretaceae *Terminalia amazonia* (J. F. Gmel.)., syn. *T. obovata* (Ruiz & Pav.) Steud. A dominant species in the evergreen rainforest, *Terminalia amazonia* can reach a height of 42.7 m (140′) with a trunk diameter of 1.2–1.4 m (4–5′). The long, bare, symmetrical trunk rises 18.3–21.3 (60–70′) above large buttresses. The heartwood varies from yellowish olive to golden brown, and is sometimes striped reddish brown. The timber is durable to both white-rot and brown-rot fungus, and dry-wood termites. It occurs in areas receiving high levels of precipitation in southern Mexico through to Central America, northern South America, Brazil, Peru, and Trinidad and Tobago at altitudes between 20 and 1200 m (66 and 3937′) (Chudnoff 1984). It is commonly found along the rivers and creeks (Rodríguez-Sánchez 2003–2012). *Use:* **hi'che'** 'exterior wall boards'; *Part:* **che'il** 'timber'; *Use:* **che'il k'ak'** 'firewood'. *See:* **pukte'**. *Variant:* **pukte'**. *[Note:* canhan, pucte. *Terminalia amazonia* (Durán 1999); k'änxa'an [Mop.]. *Terminalia amazonia* (in, Atran et al. 2004); k'an xa'an [Yuc.]. (in, Atran et al. 2004); aj k'änxa'an [Itz.]. *Terminalia amazonia* (Atran et al. 2004).*]* *[*\sd2 housing \sd2 fuel*]*

k'än 'abäl mombin, yellow (Spanish plum, hog plum, scarlet plum). Anacardiaceae *Spondias purpurea* forma *lutea* F. & R. **Es che'. Umäk'ik uwich. K'än uwich. Cuando täk'än, ch'uhuk i päh chichin. Ne haay usool uwich. K'än unoy. Chukuch unek', wolis. Hump'eli' unek'. Päk'bil. Septiembre, Octobre utahal.** 'This is a tree. One eats the fruit. The fruit is yellow. When it's ripe, the fruit is sweet and a little bit sour. The skin of the fruit is very thin. The pulp is yellow. The seed is long and round. There is one seed. It is cultivated. The fruit ripens in September-October.' (BM) In the extant botanical inventories from the Lacandon forest, *k'än 'abäl* corresponds to *Spondias mombin* (Diemont 2006; Levy et al. 2006; Nations and Nigh 1980; Nigh 2008). However, the tree that my Lacandon consultants referred to as *k'än 'abäl* resembles the yellow-fruited form of *S. purpurea*, with its plum-like fruits that are borne singly or in groups of two or three on the branches. *S. purpurea* forma *lutea* is easily mistaken for *Spondias mombin* "the true yellow mombin" (Morton, J. 1987). *S. mombin* is distinguished from *S. purpurea* by its erect, stately form, growing to 20 m (65′) tall, and its somewhat buttressed, 60–75 cm (2–2.5′) diameter trunk with thick, fissured bark, and blunt-pointed spines or knobs on young trees (Morton, J. 1987). Its leaflets are larger than those of *S. purpurea* forma *lutea* and the fruits are less fleshy and hang in numerous, branched clusters at the ends of stems. Cultivated. *Loc:* **haläknikil ti' uyatoch Koh** 'KM's yard'; *Use:* **hanal** 'food'; *Part:* **wich** 'fruit'; *Prep:* **che'che'** 'eaten raw'; *Sim:* **ya'ax 'abäl** 'green mombin'; *Sim:* **chäk 'abäl** 'red hogplum (*Spondias purpurea*)'; *Gen:* **'abäl**. *See:* **k'inim**; **huhup**. *Thes:* **che'** (Fig. 5.100a, b, c). *[Note:* aj chäk joob'oj ('abäl), aj k'än joob'oj ('abäl) [Itz.]. *Spondias purpurea* (Atran et al. 2004).*]* *[Source:* BM*]* *[*\sd2 food*]*

k'än bo'oy [k'än bo'oy, k'äm bo'oy] false jade (monkey-tail). *Lit:* 'yellow bo'oy'. Arecaceae *Chamaedorea neurochlamys* Burret. *Chamaedorea neurochlamys* is a solitary, erect palm 1.5–4.5 m (4.9–14.7′) tall with 3–5 pinnate leaves, which are glossy above and dull below, with 6–8 pairs of sickle-shaped leaflets and a distinctly white leaf sheath at the apex. The fruits are bi-lobed or sickle-shaped, yellowish orange or orange, when unripe or soft ripe, aging to dark brown when fully ripe (Hodel 1992: 92). The yellow-orange fruits are the distinctive feature that earns the plant its Lacandon name. According to AM, the leaves are of no practical value, because they quickly dry out. *Sim:* **hach bo'oy** 'authentic bo'oy'; *Sim:* **ch'ibix** '*Chamaedorea* spp.'; *Sim:* **säk bo'oy** 'white bo'oy'; *Sim:* **chan mehen bo'oy** 'very small bo'oy'; *Sim:* **ch'it bo'oy**; *Sim:* **ton bo'oy**; *Gen:* **bo'oy**. *Variant:* **ch'it k'än bo'oy** (Fig. 5.101a, b). *[Note:* Botanical identification provided by Don Hodel (pers. comm.). Common English names from Garwood and Bridgewater (2005). kan boy. *C. oblongata* (Durán 1999).*]* *[Source:* AM; BM; K'ayum Segundo*]*

k'än chi' nance. *Lit:* 'yellow nance'. Malpighiaceae *Byrsonima crassifolia* (L.) H. B. K., syn. *Byrsonima cubensis* Juss., *Byrsonima cotinifolia* Kunth. *Byrsonima crassifolia*

Fig. 5.100 (**a**) K'än 'abäl *Spondias purpurea forma lutea* ripe fruit. Photo Credit: Suzanne Cook (**b**) K'än 'abäl *Spondias purpurea* f. *lutea* leaves. Photo Credit: Suzanne Cook (**c**) K'än 'abäl *Spondias purpurea* f. *lutea* tree. Photo Credit: Suzanne Cook

Fig. 5.101 (**a**) K'än bo'oy *Chamaedorea neurochlamys* leaves . Photo Credit: Suzanne Cook (**b**) K'än bo'oy *Chamaedorea neurochlamys* fruit. Photo Credit: Suzanne Cook

is a large shrub or tree reaching up to 10 m (33′), or even 20′ m (66′). It bears yellow drupes that look like cherries with a similar texture and acidity, although the flavour is somewhat insipid and smells like cheese. A native of the Neotropics, it grows wild in open pine forests and grassy savannas from sea-level to 1800 m (60009) on rocky or sandy, alkaline soil (Morton, J. 1987:207–209). The fruit ripens in September, in Naha'. As soon as they fall to the ground, the fruits are scooped up by children and quickly devoured. Cultivated. *Loc:* **kolil nah, kol, paakche' kol** 'house garden, milpa, fallow milpa'; *Use:* **hanal** 'food'; *Part:* **wich** 'fruit'; *Prep:* **che'che** 'raw'; *Sim:* **ya'ax chi'** 'green chi''; *Gen:* **chi' (äh)**. *Thes:* **che'** (Fig. 5.102a, b, c, d). *[Note:* chi'. nantze, nanche. *Byrsonima cotinifolia* (Bruce 1975); shinich. *Byrsonima crassifolia* (Nations and Nigh 1980). ajk'än chi' [Itz.]. yellow nance tree. *Byrsonima crassifolia* (Hofling and Tesucún 1997); chi' [Yuc.]. *Byrsonima crassifolia* (Bricker et al. 1998: 69).*] [Source:* BM; KM; KP; OM; SK*] [*\sd2 food*]*

k'än chululche' [k'än chululche', k'än churuche'] *Wimmeria bartlettii* (common name unknown). *Lit:* 'yellow heart wood'. Celastraceae *Wimmeria bartlettii* Lund. **K'än uts'u'**. 'The pith is yellow.' (BM) This is a species with yellowish-white, pliable wood. According to AM, this tree is not a kind of *chulul* ('bow (tree)'). *Use:* **okman** 'house posts'; *Use:* **che'il k'ak'** 'firewood'. *Morph:* **k'än chul-ul=che'**. *Thes:* **che'**. *[Note:* kan churun che. *Wimmeria bartlettii* (Durán 1999).*] [Source:* AM; BM*] [*\sd2 fuel \sd2 housing*]*

k'än ik (äh) chile pepper, yellow. *Lit:* 'yellow chile pepper'. Solanaceae *Capsicum annuum* L. *Use:* **ts'ak** 'spice, condiment'; *Part:* **wich** 'fruit'; *Prep:* **che'che'** 'raw'; *Gen:* **ik**. *[Source:* Bruce (1975: 177)*] [*\sd2 food*]*

k'än 'is potato, sweet. *Lit:* 'yellow sweet potato'. Convolvulaceae *Ipomoea batatas* (L.) Lam. **K'än uts'u'.** 'It's yellow inside' (BM) A variety of sweet potato with yellow flesh. *Use:* **hanal** 'food'; *Part:* **wi'** 'tuber'; *Prep:* **chäkbil, pokbil** 'boiled or roasted'; *Sim:* **chäk 'is** 'red sweet potato'; *Sim:* **säk 'is** 'white sweet potato'; *Gen:* **'is**. *[Source:* BM*] [*\sd2 food*]*

k'än luuch calabash, Mexican (winged calabash). *Lit:* 'yellow gourd'. Bignoniaceae *Crescentia alata* Kunth. *Use:* **luuch** 'cup, bowl'; *Part:* **upach** 'shell'; *Gen:* **luuch**. *See:* **sina'an luuch (äh)**. *[Source:* AM*] [*\sd2 containers*]*

k'än 'oox breadnut, Mayan. *Lit:* 'yellow breadnut'. Moraceae *Brosimum alicastrum* Swartz ssp. *alicastrum* C.C. This is a kind of breadnut tree that bears 3.8 cm (1.5″) round, sweet, yellow fruits in March. According to BM, the fruit is much smaller than that of the *hach 'oox* and the nutmeats are not as tasty. AM says that *k'än 'oox* and *hach 'oox* are the same species. *Loc:* **ho' yahaha'** 'headwaters of the Santo Domingo River'; *Use:* **hanal** 'food'; *Part:* **wich** 'fruit'; *Prep:* **che'che'** 'raw'; *Part:* **nek'** 'seed'; *Prep:* **chäkbil yetel wah, säkan** 'cooked with corn gruel, or made into dough'; *Sim:* **hach 'oox** 'authentic breadnut (*Brosimum alicastrum*)'; *Sim:* **chäk 'oox** 'red breadnut'; *Sim:* **ya'ax 'oox**

Fig. 5.102 (**a**) K'än chi' *Byrsonima crassifolia* ripe fruit. Photo Credit: Suzanne Cook (**b**) K'än chi' *Byrsonima crassifolia* unripe fruit. Photo Credit: Chan K'in Daniel (**c**) K'än chi' *Byrsonima crassifolia* tree. Photo Credit: Suzanne Cook (**d**) K'än chi' *Byrsonima crassifolia* leaves. Photo Credit: Suzanne Cook

'green breadnut (*Brosimum alicastrum*)'; *Gen:* **'oox**. *See:* **hach 'oox**. *Thes:* **che'**. *[Source:* AM; BM*]* *[*\sd2 food*]*

k'än pätan plantain (type of). *Lit:* 'yellow plantain'. Musaceae *Musa acuminata* ×*Musa balbisiana* Coll. *Use:* **hanal** 'food'; *Part:* **wich** 'fruit'; *Prep:* **chäkbil** 'eaten cooked'; *Sim:* **hach pätan** 'authentic plantain'; *Sim:* **päh pätan (äh)** 'tart plantain'; *Sim:* **pätan ma' ka'anal ubakel** 'short (dwarf) plantain'; *Sim:* **hunbuhk'äbwinik pätan** 'one hundred plantains'; *Sim:* **ka'buhk'äbwinik pätan** 'two hundred plantains'; *Gen:* **pätan**. *[Source:* SK*]* *[*\sd2 food*]*

k'äxeex (äh) wormseed (epazote, Mexican tea). Amaranthaceae *Chenopodium ambrosioides* L., syn. *Teloxys ambrosioides* L. This is an herbaceous plant approximately 1.2 m (3.9′) tall, with light green, lanceolate leaves up to 12 cm (4.7″) long with serrated margins. It bears small, green flowers produced in branched panicles at the apices of stems. Native to the Neotropics, it is a weed found in disturbed areas of cultivated fields and waste places. Lacandones cultivate it in house gardens for use as a food for chicken soup and stews, and as a de-wormer for children (AM, BM, Kashanipour and McGee 2004). According to AM, this plant is not traditional Lacandon medicine, but was introduced a long time ago by a man from Reál. *Loc:* **kolil nah** 'house gardens'; *Use:* **ts'ak ti' o'och** 'spice'; *Part:* **le'** 'leaves'; *Use:* **ts'ak ti' nok'ol** 'medicine for worms'; *Part:* **mots** 'roots'; *Prep:* **Achäkik umots ich ha',**

pachil awuk'ik 'You boil the roots in water and then drink it'. *[Note:* k'oshesh. *Chenopodium ambrosioides* (Nations and Nigh 1980); aj 'apasootej [Itz.]. *Chenopodium berlandieri, C. ambrosioides* (Atran et al. 2004); lucum xiw [Yuc.]. *Chenopodium ambrosioides* (Roys 1931).*] [Source:* AM; BM; KP*] [*\sd2 food \sd2 medicinal*]*

k'eben [k'eben, k'ebeh, k'eweh] fish-tail palm (Ernest August's palm). Arecaceae *Chamaedorea ernesti-augusti* H. Wendl. *Chamaedorea ernesti-augusti* is a solitary-stemmed plant that grows up to 2 m (6.5′) tall. Leaves are bifid (notched at the apex), 60.9 cm (2′) long and 30 cm (1′) wide, corrogated, and stiff. Inflorescences are produced on erect, 60 cm (24″) tall flower stalks. Flowers are yellow-orange, and sunken in shallow depressions. They are followed by nearly round to ellipsoid berries, which are green to bluish green when unripe and black when mature (Hodel 1992: 42). *Use:* **ti' konol** 'to sell'; *Part:* **le'** 'leaves'; *Use:* **ti' utap'ik pom** 'to wrap up pine resin'; *Part:* **le'** 'leaves'; *Sim:* **bo'oy**; *Gen:* **bo'oy** (Fig. 5.103). *[Note:* k'ewen. *Chamaedorea ernesti-augusti* (Nigh 2008); kebe. *Chamaedorea metallica* (Durán 1999).*] [Source:* AM; BM; KA*] [*\sd2 other \sd2 industry*]*

k'i'ilix holop' [k'i'ilix holop', k'irix horop', k'ixer hoop', k'ik'ix horop'] philodendron, split leaf. *Lit:* 'spiny holop''. Araceae *Philodendron radiatum* Schott. This is a vine with large, broad, deeply cut leaves and strong, flexible vines. Juvenile stems are textured with numerous, blunt thorns called *tuch'ul*. New leaves are simple and cordate (heart-shaped), flushed pink with green mottling. AM says the plant looks like *nukuch 'oochin*. The vines are used for lashing, although they are not preferred because of the bumps, which exude an irritating sap. According to Tom Croat (pers. comm.), the sap contains tannins that can be quite caustic. It also contains a high concentration of calcium oxalate crystals, which can cause severe reactions if touched. One remedy to relieve the pain and itchiness is the application of human breast milk (personal observation). Another is the application of the resin from *pahsa' ak'* (*Costus pulverulentus*) (AM). The species ranges from Chiapas, Mexico through Central America and into most of the north-western portion of South America, Peru, and parts of Brazil. *Use:* **ti' uk'äxik hiil, säkche', kun** 'for tying together house poles, rafters, and lashing thatch to roofs'; *Sim:* **yuyul holop'** 'smooth holop' (*Philodendron smithii, P. tripartitum*)'; *Gen:* **holop'**. *Thes:* **ak'** (Fig. 5.104a, b, c, d, e). *[Note:* a kix jub. *Philodendron radiatum* (Durán 1999). According to my consultants, *nukuch 'oochin* refers to *Philodendron sagittifolium, Syngonium chiapense,* and *Syngonium podophyllum] [Source:* AM; BM; KA; K'ayum Segundo*] [*\sd2 cordage*]*

Fig. 5.103 K'eben *Chamaedorea ernesti-augusti*. Photo Credit: Suzanne Cook

k'i'ilix p'ix pear, vegetable (pear squash). *Lit:* 'spiny p'ix'. Cucurbitaceae *Sechium edule* (Jacq.) Sw. *Use:* **hanal** 'food'; *Part:* **wich, le', nek'** 'fruit, leaves, seed'; *Prep:* **chäkbil** 'boiled'; *Sim:* **säk p'ix** 'white vegetable pear'; *Sim:* **ya'ax p'ix** 'green vegetable pear'; *Gen:* **p'ix**. *Thes:* **ak'**. *[Source:* KyP*] [*\sd2 food*]*

k'i'ix 'uk'unche' turkey berry (devil's fig, prickly nightshade). *Lit:* 'thorny 'uk'unche''. Solanaceae *Solanum torvum* Sw. *Thes:* **che'** (Fig. 5.105a, b, c). *[Note:* Used medicinally by the southern Lacandones (Levy et al. 2006).*] [Source:* AM; BM*]*

k'ik'$_1$ Panama rubber tree. *Lit:* 'blood'. Moraceae *Castilla elastica* Sessé, syns. *Castilla gummifera* Pittier, *Castilla lactiflua* O. F. Cook, *Castilla panamensis* O. F. Cook, *Ficus gummifera* Bertol. This is a large, forest tree 20–50 m (65.6–164′) tall with a pyramidal or open crown. The trunk is straight with smooth to slightly fissured, grey-brown bark and abundant lenticels that run horizontally down the trunk. Copious, white latex is exuded in a bark slash. Leaves are simple, thick, 20–45 cm (7.9–17.7″) long and10–20 cm (3.9–7.9″ wide, oblong, with wavy margins and hairs on the underside of the blades. Male flowers are typically borne in clusters of four; female flowers are borne in solitary clusters. Petal and sepals are bright orange or red. Fruit is a conical, four-angled drupe, 4–5 cm (1.6–1.96″) in diameter, cartilaginous, juicy, and insipid to slightly sour (Vàzquez-Yanes et al. 1999). *Use:* **k'ik'** 'rubber'; *Prep:* **ahäts'ik**

Fig. 5.104 (**a**) K'i'ilix holop' *Philodendron radiatum* leaves. Photo Credit: Suzanne Cook (**b**) K'i'ilix holop' *Philodendron radiatum* aerial roots. Photo Credit: Suzanne Cook (**c**) K'i'ilix holop' *Philodendron radiatum* inflorescence. Photo Credit: Suzanne Cook (**d**) K'i'ilix holop' *Philodendron radiatum tuch'ul*. Photo Credit: Suzanne Cook (**e**) K'i'ilix holop' *Philodendron radiatum* plant. Photo Credit: Suzanne Cook

Fig. 5.105 (**a**) K'i'ix 'uk'unche' *Solanum torvum* leaves and unripe fruit. Photo Credit: Suzanne Cook (**b**) K'i'ix 'uk'unche' *Solanum torvum* plant. Photo Credit: Suzanne Cook (**c**) K'i'ix 'uk'unche' *Solanum torvum* flowers. Photo Credit: Suzanne Cook

uyak'il ahhut'kih ayäts'ik uyits pachil apuk'ik yetel k'ik' yetel ha' 'You pound the vines of the hut'kih, squeeze out the resin, and then dissolve it with the latex and water'. *See:* **tulis k'ik' (ah)**. *Variant:* **'ule (< hule [Sp.])**. *Thes:* **che'** (Fig. 5.106a, b, c, d). *[Note:* k'ik'. *Castilla elastica* (Nations 2006); kik-che [Yuc.]. *Castilla elastica*, Cerv. Hule. (Roys 1931); kikche, k'i'ik' che' [Yuc.]. *Castilla elastica* (in, Atran et al. 2004); uule [Mop.]. *Castilla elastica* (in, Atran et al. 2004); 'uulej [Itz.]. *Castilla elastica* (Atran et al. 2004).*] [Source:* AM: BM; KyP*] [*\sd2 other*]*

k'ik'che' [k'ik'che', k'iche'] virola. *Lit:* 'blood tree'. Myristicaceae *Virola guatemalensis* (Hemsl.) Warb. When the trunk is cut, the bark slash exudes a red resin. Branches emerge in a whorl around the trunk, giving the tree a pagoda-like structure. Leaves are arranged alternately in two opposite vertical rows. They are petiolate, oblong, and simple with entire or slightly undulate margins and pinnate venation, attenuate at the apex and cuneate at the base. Fruit is an oval, two-valved, dehiscent drupe about 5.08 cm (2″) long, with greenish-yellow skin. The black and purple streaked seed is encased in a red aril. Fruits mature toward the end of October, attracting droves of birds including: the collarded araçari toucan; crested guan; curassow; emerald toucanet; mealy blue-crowned parrot; partridge; and, tinamou (BM). Timber is used for construction. Durán (1999) reports that the wood is also used to make *"chuistes"* (*chuste'* 'foreshafts') for arrows. *Loc:* **k'aax** 'forest'; *Use:* **hi'che'** 'vertical house poles, fence'; *Part:* **che'il** 'timber'. *Thes:* **che'** (Fig. 5.107a, b, c). *[Note:* kiche, gik che.*Virola guatemalensis* (Durán 1999).*] [Source:* AM; BM; Bor Solorzano*] [*\sd2 housing*]*

k'ik'ni' balum (ah)$_1$ [k'ik'ni' balum, k'ik'ni' barum] limoncillo. *Lit:* 'bleeding-nosed jaguar'. Celastraceae *Rhacoma eucymosa* (Loes. & Pittier) Standl., syn. *Crossopetalum parviflorum* (Hemsl.) Lund. *Spec:* **nukuch k'ik'ni' balum** 'large k'ik'ni' balum'; *Spec:* **mehen k'ik'ni' balum** 'small k'ik'ni' balum (*Eugenia* sp., *Ardisia compressa*)'; *Spec:* **hach k'ik'ni' balum** 'authentic k'ik'ni' balum (*Ardisia compressa*)'. *Morph:* **k'ik'-ni' balum**. *Thes:* **che'**. *[Note:* kini barun, kikin balun. *Rhacoma eucymosa* (Comisión Nacional de Áreas Naturales Protegidas 2006; Durán 1999); k'ik'ni-och [Yuc.]. nose-bleed' plant (Roys 1931).*] [Source:* AM*] [*\sd2 folklore_and_mythology*]*

k'ik'ni' balum (ah)$_2$ coralberry (marlberry). Myrsinaceae *Ardisia compressa* Kunth (Fig. 5.108). *[Source:* KP*]*

k'inim mombin (yellow mombin, Spanish plum, gully plum). Anacardiaceae *Spondias mombin* L., syn. *S. lutea*. This is a deciduous tree that grows up to 20 m (60′) tall with a trunk girth of up to 1.5 m (5′). Bark is thick, corky, and deeply fissured. Leaves are simple, imparapinnate (with a terminal leaflet). Leaflets are oblong or oblong-lanceolate, broadly acuminate, and smooth. The mature yellow-orange fruit has a leathery skin and a thin layer of edible pulp. It is native to the Neotropics and is rarely cultivated. Throughout the Spanish-speaking communities in the Caribbean and in Mexico it is called *jobo*, from the Carib language spoken in South America. According to Nations and Nigh (1980: 22), the Lacandones eat the fruit and roots. *See:* **huhup**; **k'än 'abäl**. *[Note:* This word is uncommon in the northern Lacandon community, but it has been recorded by Nations and Nigh (1980) in the speech of their Lacandon consultants, some of which were southern Lacandones. According to Stross (2008: 20), *k'inim* is the word for hog plum (tree) in several Mayan languages, and he surmises that it is based on the proto-Cholan root *k'in* 'sun, day'. The -im suffix may be the archaic morpheme meaning 'sacred' (Bruce 1968). Atran et al. (2004), however, analyze *k'in=im* as 'sun=breast' (2004:88). k'inim. *Spondias mombin* (Nations and Nigh 1980); x-kinin-hobon [Yuc.]. hog plum. *Spondias lutea*

Fig. 5.106 (**a**) K'ik' *Castilla elastica* sapling. Photo Credit: Suzanne Cook (**b**) K'ik' *Castilla elastica* latex. Photo Credit: Suzanne Cook (**c**) K'ik' latex from bark slash. Photo Credit: Suzanne Cook (**d**) K'ik' fruit. Photo Credit: Suzanne Cook

Fig. 5.107 (**a**) K'ik'che' *Virola guatemalensis* fruit. Photo Credit: Suzanne Cook (**b**) K'ik'che' *Virola guatemalensis* sapling. Photo Credit: Suzanne Cook (**c**) K'ik'che' *Virola guatemalensis* leaves. Photo Credit: Suzanne Cook

Fig. 5.108 K'ik'ni' balum *Ardisia compressa*. Photo Credit: Suzanne Cook

(Roys 1931); k'inim [Itz.]. *Spondias mombin* (Atran et al. 2004); k'inim. [Itz.]. wild hogplum (type). *Spondias* sp. (Hofling and Tesucún 1997).*] [Source:* Nations and Nigh (1980: 22)*]*

k'oben (äh)$_1$ water lily. Nymphaeaceae *Nymphaea odorata* Aiton. *[Source:* BM*]*

k'o'och (äh)$_1$ bean, castor. Euphorbiaceae *Ricinus communis* L. *Ricinus communis* is a fast growing bush or small tree that can attain a height of 2–3 m (6.6–9.8′) in a single year. It is the most poisonous in the world (Guinness Book of World Records), possessing the toxin, ricin, in the raw seeds; four to eight seeds are considered lethal to humans. There are many horticultural varieties of *Ricinus communis*: a green and a red form grow in the Yucatan. The green form is called *blanco* and the red form, *rojo, colorado, negro*, or *morado* (Standley and Steyermark 1949: 157). The Lacandones also differentiate a red form, which they call *chäk k'o'och*, and a green form, *ya'ax k'o'och*. Nations and Nigh (1980: 17) report that *chäk k'o'och* is used medicinally, but every Lacandon whom I interviewed said that they cultivate the trees solely for shade and their ornamental value. *Use:* **bo'oy** 'shade'; *Spec:* **ya'ax k'o'och** 'green k'o'och'; *Spec:* **chäk k'o'och (äh)** 'red k'o'och'. *Thes:* **che'**. *[Note:* chäk k'och. castor bean. *Ricinus communis* (Nations and Nigh 1980); chäk k'o'och, säk k'o'och [Itz.]. *Ricinus communis* (Atran et al. 2004); X-koch [Yuc.]. *Ricinus communis* L. Higuerilla (Roys 1931).*] [Source:* BM; KP*]* */*\sd2 other*]*

k'o'och (äh)$_2$ [k'o'och, k'och] pumpwood (trumpet tree). Urticaceae (Moraceae) *Cecropia* spp. **P'iis ahk'o'ocheex pero ma' la'. K'o'och uk'aba', ähk'o'och. Uyo'och ch'iich'**. 'The [word] is the same as the word for throat, but it doesn't mean that. The tree's name is *ähk'o'och* and it provides food for birds [especially *pän* (keel-billed toucan)].' (AM) Native to the Neotropics, *Cecropia* is recognizable by its large, deeply divided leaves, which consist of 7–11 lobes. The slender, straight trunks are hollow and segmented, while the internodes are filled with a soft pith. These are fast growing trees that rapidly colonize cleared forests. Two species are found in the area—*C. obtusifolia* and *C. peltata*. Perhaps, the only morphological feature that differentiates them is the inflorescence, in addition to environmental preferences: *C. obtusifolia* is often found in wet habitats, whereas *C. peltata* is common in open, drier areas. According to BM, there is a species of fierce ant called *sinikil k'o'och*, literally 'ant of the *Cecropia*', that lives inside the branches. He says its bite is very painful. This must be the fierce Azteca ant, which feeds upon the glycogen produced by the tree while protecting the tree from other predators. *Cecropia* species are valued for their medicinal properties throughout the Neotropics: The highly esteemed *C. peltata* is used to fight obesity and to treat asthma, liver ailments, and diabetes; *C. obtusifolia* is used as a heart tonic and diuretic (Martínez 1969; Morton 1981); in Naha', the Lacandones make an infusion of the growing tips of *C. obtusifolia*, to treat urinary infections in children and nerve disorders in adults, and an infusion of the leaves of *C. peltata* is used to treat rheumatism, diabetes, and bedwetting (Durán 1999); a tea made with the leaves of *C. peltata* is drunk to relieve chills, fatigue, and general listlessness (Kashanipour and McGee 2004: 55), whereas a poultice made from the young leaves and tender shoots is applied to skin ulcers caused by leishmaniasis (chiclero disease) (AM). In the southern Lacandon community, the stems of *C. obtusifolia* are used as tobacco and fuel (Diemont 2006: 83). The Lacandones use the timber of *C. peltata* for construction (Nations and Nigh (1980: 16). *Use:* **chiclero** 'leishmaniasis (chiclero's disease)'; *Part:* **muun ule', uyool** 'tender leaves and the growing tips'; *Prep:* **Uyächik ule', uyol i usulik ich ha', pachil uts'ik ti'** 'One crushes the leaves and the shoots, steeps them in water, and then places them on the (affected area)'; *Use:* **yah ubaakel** 'sore bones'; *Part:* **uyool** 'growing tips'; *Prep:* **te'** 'tea'; *Spec:* **nukuch k'o'och** 'large pumpwood'; *Spec:* **mehen k'o'och** 'small pumpwood'. *Thes:* **che'** (Fig. 5.109). *[Note:* k'och, k'o'och. *Cecropia obtusifolia* (Nigh 2008); co'och. *Cecropia peltata, Cecropia obtusifolia* (Comisión Nacional de Áreas Naturales Protegidas 2006); k'och. shield-leaf k'och. *Cecropia peltata* (Nations and Nigh 1980); k'o'och. *Cecropia obtusifolia* (Diemont 2006); koch. *Cecropia peltata, C. obtusifolia* (Durán 1999); X-k'och-lé [Yuc.]. *Cecropia obtusa* [Millsp. I, 358; Standl.; Gaumer.] (Roys 1931); k'o'ochche' [Itz.]. *Cecropia obtusifolia, C. peltata*. (Atran et al. 2004); ix-k'o'ochche' [Itz.]. castor bean~tree, trumpet tree. *Cecropia peltata* (Lois 1998).*] [Source:* AM; BM; K'ayum Segundo*]* */*\sd2 medicinal*]*

k'o'och ba'ats' (äh)$_1$ louteridium, Donnell Smith. Acanthaceae *Louteridium donnell-smithii* S. Wats. This is a stout, erect shrub or small tree, 2–4 m (6.5–13') tall (up to 9 m [29.5']) that grows in well shaded and wet areas in mixed

Fig. 5.109 K'o'och *Cecropia* spp. Photo Credit: Suzanne Cook

forest or dense thickets from sea level to 1,600 m (5,249′) (Standley and Steyermark 1946: 407). Its brittle branches support light green, densely villous, ovate or broadly ovate leaves, 10–36 cm (3.93–14.17″) long and 6–28 cm (2.36–11″) wide, with undulate-dentate (wavy-toothed) margins. The characteristic leaf shape and pubescence distinguish it from the other *Louteridium* species in Chiapas (Erin Tripp, pers. comm.). *K'o'och ba'ats'* typically designates orchids with throats (e.g., *Stanhopea, Maxillania, Lycaste*). The shape of the tubular flowers of *Louteridium donnell-smithii* are similar and thus may account for the Lacandon name. Flowers are born in rather large, terminal racemes or open, narrow panicles. Petals are usually dull purplish or brownish red flushed with green. No uses were reported. *Loc:* **k'aax** 'forest'. *See:* **k'o'och (äh)$_2$**. *Thes:* **che'** (Fig. 5.110a, b, c). *[Note:* Botanical identification provided by Erin Tripp, Rancho Santa Ana Botanic Garden, CA.*] [Source:* AM; BM; K'ayum Segundo*]*

k'o'och ba'ats' (äh)$_2$ orchid. *Lit:* 'throat of the howler monkey'. Orchidaceae. A general term for varieties of orchids with a "throat", e.g., *Stanhopea, Maxillania, Lycaste*. Large and small varieties are referred to as *nukuch k'o'och ba'ats'* and *mehen k'o'och ba'ats'*, respectively. Some examples recorded include: *Lycaste bradeorum* Schltr.; *Maxillaria aciantha* Rchb.f.; *M. cucullata* Lindl.; *M. histrionica* (Rchb.f.) L.O. Williams; *M. pulchra* (Schltr.) L.O. Williams; and, *M. variabilis* Bateman ex Lindl. Other species recorded in the Naha' include: *Bletilla striata* (Thumb.) Reichenb.; *Cattleya* Lindl.; *Encyclia cochleata* (L.) Lemee; *Encyclia neurosa* (Ames) Dressler & Pollard; *Encyclia radiata* (Lindl.) Dressler; *Epidendrum atroscriptum* Hágsater; *Epidendrum polyanthum* Lindl.; *Epidendrum ramosum* Jacq.; *Epidendrum santaclarense* Ames; *Isochilus carnosiflorus* Lindl.; *Jacquiniella equitantifolia* (Ames) Dressler; *Lockhartia tzeltal* Soto Arenas; *Maxillaria anceps* Ames & C. Schweinf.; *Nidema boothii* (Lindl.) Schltr.; *Sobralia decora* Batem.; and, *Stanhopea tigrina* Bateman ex Lindl. (Durán 1999). No use was reported to me for these orchids, but Levy et al. (2006) note that the southern Lacandones use *Isochilus carnosiflorus* in ceremonies. *Variant:* **top' che'** (Fig. 5.110d). *[Note:* chak koch bats. *Maxillaria cucullata* Lindl. (Durán 1999); koch bats. *Lycaste* sp., koch. *Lycaste bradeorum* Schltr. (Durán 1999).*] [Source:* AM; BM; K'ayum Segundo; NI*]*

k'uche' (äh) [k'uche', k'unche'] cedar, Spanish (Mexican cedar, Barbados cedar). Meliaceae *Cedrela odorata* L., syn. *Cedrela mexicana* M. Roem. This is a large, evergreen or dry-season, deciduous tree that is indigenous to the Neotropics. It occurs in both moist and seasonally dry forests, roadsides, pastures, and disturbed areas with well-drained soils. The tree ranges in height from 10 to 30 m (32.08–94.4′) and has a trunk diameter as large as 1.82 m (6′). Its straight, cylindrical, often buttressed trunk is clear of branches up to 12.19–18. 28 m (40′-60′) above the forest floor. Bark is grey-brown bark with deep fissures. Leaves are pinnately compound, grouped towards the end of the branches, 15–50 cm (5.9–19.68″) long. Five to 14 pairs of leaflets, 8–20 cm (3.14–7.87″) long, 2.5–8 cm (0.98–3.14″) wide, ranging in shape from ovate to lanceolate. Inflorescences are hanging panicles of small greenish-white flowers. Fruits are brown, woody capsules containing numerous, winged seeds. (http://www.hear.org/pier/species/cedrela_odorata.htm) My consultants listed a few uses for the timber, i.e., spoons, walls, and *balche'* canoe. Previous researchers reported more uses: timber is used for lumber and carvings (Diemont 2006: 83), crafts, household items, tool shafts, doors, and tables (Durán 1999: 146); leaves are used in a poultice to treat insect bites and irritated skin (Kashanipour and McGee 2004: 62); and, honey and wax are collected from wild bee hives in the tree (Durán 1999: 146). *Use:* **huyub** 'spoon (large), ladle'; *Use:* **hi'che'** 'fence, vertical house poles'; *Use:* **chemil balche'** 'ceremonial canoe in which *balche'* is fermented.'. *Thes:* **che'** (Fig. 5.111). *[Note:* k'uche'. *Cedrela odorata*. (Diemont 2006); k'uh che'. *Cedrela odorata* (Kashanipour and McGee 2004); kuche. *Cedrela*

Fig. 5.110 (**a**) K'o'och ba'ats' *Louteridium donnell-smithii* plant. Photo Credit: Suzanne Cook (**b**) K'o'och ba'ats' *Louteridium donnell-smithii* leaves. Photo Credit: Suzanne Cook (**c**) K'o'och ba'ats' *Louteridium donnell-smithii* flower. Photo Credit: Naomi Fraga (**d**) K'o'och ba'ats' *Stanhopea tigrina*. Photo Credit: Suzanne Cook

odorata (Nigh 2008); kun che. *Cedrela odorata* (Durán 1999); k'u-che' [Itz.]. *Cedrela odorata* (Hofling and Tesucún 1997: 399); säk k'uche', chäk k'uche' [Itz.]. *Cedrela mexicana, C. odorata* (Atran et al. 2004); k'uche' [Mop.]. *Cedrela mexicana* (in Atran et al. 2004); k'u'che', ku'unche', kulche' [Yuc.]. *Cedrela mexicana* (in Atran et al. 2004).*] [Source:* BM*] [*\sd2 housing \sd2 religion \sd2 tools*]*

k'ulsek lycianthes, varying. *Lit:* 'sacred bat'. Solanaceae *Lycianthes heteroclita* (Sendtn.) Bitter. This is an herbaceous, twining shrub with light green, smooth, lanceolate leaves entire margins and palmate venation. Leaves alternate along the rounded, slightly rough stem. The vine bears 1 cm (0.6″) diameter purple fruit. According to KP, bats feed on this plant. *Variant:* **uyo'och ähsek**; **ch'ayok'**$_3$. *Thes:* **ak'**. *[Note:* Levy et al. (2006) report that the fruit is eaten by the southern Lacandon group.*] [Source:* AM; KP*]*

k'ul ak' barbasco. *Lit:* 'god vine'. Dioscoreaceae *Dioscorea bartlettii* C. V. Morton. This is a wild yam with a large, exposed, deeply fissured, woody, basal burl and thick, heart-shaped leaves. Long stems are covered with straight, sharp thorns. Before they used clay, the Lacandones' ancestors used the burl for god pots. According to Levy et al. (2006), the southern Lacandones eat the roots and use the burl for crafts. Nations (2006: 96) reports that the Lacandones once used the roots for poisoning fish. *Use:* **lekil k'uh** 'god pot'; *Part:* **bola** 'burl'. *See:* **'op' ak'**. *Thes:* **ak'** (Fig. 5.112a, b). *[Note:* kul ak'. *Magnolia mexicana* (Kashanipour and McGee 2004: 62); op ak' [Lac.]. *Dioscorea bartlettii* (in Atran et al. 2004).*] [Source:* AM; BM*] [*\sd2 containers*]*

k'um [k'um, k'uum] calabasa (squash, pumpkin). Cucurbitaceae *Cucurbita moschata* (Duch. ex Lam.) Duch. ex Poir ., *Cucurbita pepo* L. This is a large, edible squash with a hard rind, cultivated in *milpas*. The fruit varies greatly in size, shape, and texture. It may be smooth or ribbed, light green to uniform dark green or with light spots, or beige; the flesh is yellow or greenish yellow and soft with little fibre (Saade and Montes-Hernández 1994). It is cut from the stem when ripe and left for many months to cure. During this time, the pulp develops a blackish-green slime. Young fruits are called *ya'ax k'um* 'green squash'. *Loc:* **kolil nah, kol** 'house garden, milpa'; *Use:* **hanal** 'food'; *Part:* **wich** 'fruit'; *Prep:* **chäkbil yetel ch'uhuk** 'boiled with sugar'; *Part:* **nek'** 'seeds'; *Prep:* **k'elbil** 'toasted'; *Spec:* **ya'ax k'um** 'green squash'. *Variant:* **box** (Fig. 5.113a, b). *[Note:* k'um. pumpkin, squash, calabash (Bruce 1975: 184); k'um. squash. *Curcurbita moschata* (Nations and Nigh 1980: 10); k'úum [Yuc.]. *Curcurbita moschata* (Bricker et al. 1998: 159); k'uum [Itz.]. *Curcurbita moschata* (Atran et al. 2004).*] [*\sd2 food*]*

Fig. 5.111 K'uche' *Cedrela odorata*. Photo Credit: Suzanne Cook

k'um sek (ah) passion flower (bat-leaved type). *Lit:* 'bat's calabasa'. Passifloraceae *Passiflora biflora* Lam. Considered *lo'obil* 'a weed', this thin vine is seen rambling along fences and climbing through the brush in unkempt house gardens. Leaves are shaped like a bat's wings, and have yellow dots on the lamina. These yellow dots are laminar nectaries, which function as a defense against *Heliconine* butterflies, by mimicing the yellow eggs of butterflies that develop into a type of cannibalistic caterpillar (Y.R. Kuethe, pers. comm.). Miniature, inedible purple fruits, approximately 2.54 cm (1″) in diameter, follow tiny, whitish-yellow flowers. According to KM and KP, the flowers provide food for bats. *Thes:* **ak'** (Fig. 5.114a, b, c, d). *[Note:* kumi cho. *Passiflora cobanensis* (Durán 1999).*] [Source:* KM; KP*]*

k'unche' monkeytail (shadow palm). Arecaceae *Geonoma oxycarpa* Mart., syn. *G. interrupta* var. *interrupta* (Ruiz & Pav.) Mart. *See:* **k'unche' pahok**.

k'unche' pahok [k'unche' pahok, k'uche' pahok] monkeytail (shadow palm). Arecaceae *Geonoma oxycarpa* Mart., syn. *G. interrupta* var. *interrupta* (Ruiz & Pav.) Mart. *G. oxycarpa* grows 4.7–9 m (15–30′) tall. It has enormous, elongated leaves, and sickle-shaped, partially and irregularly split leaflets. Stems are usually solitary. Two botanical varieties are recognized: *Geonoma interrupta* var. *euspatha* and *Geonoma interrupta* var. *interrupta*. The differences are subtle (Henderson et al. 1995: 227). Only the latter appears to occur in the Lacandon forest. According to BM, it prefers soft soil. It grows wild, but is also cultivated for its leaves and immature inflorescences. It bears clusters of round, black fruit, which is savored by *wan* (partridge), *pän* (keel-billed toucan), and *nok'ol* (tinamou). *Loc:* **Jardin**; *Use:* **upix**

Fig. 5.112 (**a**) K'ul ak' *Dioscorea bartlettii* burl. Photo Credit: Suzanne Cook (**b**) K'ul ak' *Dioscorea bartlettii* leaves. Photo Credit: Suzanne Cook

Fig. 5.113 (**a**) K'um *Cucurbita moschata*. Photo Credit: Suzanne Cook (**b**) K'um *squash blossoms*. Photo Credit: Suzanne Cook

Fig. 5.114 (**a**) K'um sek *Passiflora biflora* fruit. Photo Credit: Suzanne Cook (**b**) K'um sek *Passiflora biflora* fruit interior. Photo Credit: Suzanne Cook (**c**) K'um sek *Passiflora biflora* leaves and immature fruit on vine. Photo Credit: Suzanne Cook (**d**) K'um sek *Passiflora biflora* leaf. Photo Credit: Suzanne Cook

chemil balche', upix nah 'the cover for the balche' canoe, roof thatch'; *Part:* **le'** 'leaves'; *Use:* **hanal** 'food'; *Part:* **uyaal** 'immature inflorescences'; *Prep:* **pokbil** 'roasted'; *Sim:* **hach pahok** 'authentic pahok'; *Sim:* **chäk pahok** 'red pahok'; *Gen:* **pahok (ah)**. *Variant:* **k'unche'** (Fig. 5.115). *[Note:* Common names from Grandtner (2005). kunchepajok. *Geonoma oxycarpa* (Nigh 2008).*] [Source:* AM; BM*] [*\sd2 housing \sd2 other \sd2 food*]*

k'uuts tobacco. Solanaceae *Nicotiana tabacum* L. Leaves are dried and rolled into cigars. Cigars were smoked not only for enjoyment but also to keep the mosquitoes away. According to BM, the plant helps control the weeds in the *milpa*. Lacandones rarely smoke tobacco anymore, but bundles are stored and sold to passers-by. *Use:* **ti' uts'uuts'** 'to smoke'; *Spec:* **hach k'uuts** 'authentic tobacco (*Nicotiana tabacum*)'; *Spec:* **si'si' k'uuts** '*Baccharis trinervis*'. *Variant:* **hach k'uuts**. *[Note:* kutz [Yuc.]. *Nicotiana tabacum*, L. (Roys 1931: 259); k'uutz [Itz.]. *Nicotiana tabacum* (Atran et al. 2004).*] [Source:* BM*] [*\sd2 other*]*

k'uxub [k'uxub, k'uxu', k'uxuk] annatto (lipstick tree). Bixaceae *Bixa orellana* L. This is a medium sized bush with crimson-pink or white flowers and spiky seed pods that contain numerous small, hard, brown seeds encased in a red, gelatinous aril. It is native to the Neotropics. The Spanish common name, *achiote*, derives from the Nahuatl word *achiotl*. Large quantities of the seed pods are harvested, the seeds removed, and the aril rendered down to a paste. The paste is then patted into small cakes and dried in the sun. When needed, chunks are broken off from the cake and reconstituted. Today, *k'uxub* is used mainly as a food additive in soups, stews, and some ceremonial offerings. Before, when the Lacandones wove their own clothes, they used *k'uxub* as a dye for fibre. Due to its brilliant red colour, *k'uxub* represents blood. The Lacandones believe that *Hachäkyum*, the principal deity, made the red paint that was used in the Mayan murals with a mixture of human blood and *k'uxub* (Bruce 1975: 185). In their own ceremonies, the Lacandones painted themselves with the dye in the same manner that the gods had painted themselves with the blood of sacrificed victims. Believing that the gods relished the sight and smell of blood, the Lacandones sacrificed their own blood in ritual blood letting ceremonies, up until the middle of the 20th century. When the traditional religion began to disintegrate, blood-letting ceased and human blood was replaced by *k'uxub* (McGee 2002: 142). In colonial documents the Mayas of the 16th century are said to have used the seeds to both flavour and dye the cocoa beverage (Hellmuth 2009). *K'uxub* also has medicinal uses in some Mayan communities. The Itzaj, for example, apply it to chicken pox and drink a tea made from the fruits to relieve a sore throat (Atran et al. 2004: 95). Among the Lacandones of Naha', the plant is invoked during a ritual to cure swelling of the face (AM). *Use:* **häxal k'ak', uche'il häxbil k'ak'** 'firedrill, firedrill sticks'; *Use:* **ts'ak ti' o'och** 'spice'; *Use:* **ti' banik nook'** 'fabric paint'; *Part:* **wich** 'fruit'; *Prep:* **Achäkik uwich ich kum uhalik unek', uch'uktik wa sisil, upätik umehen muuch'** 'You boil the fruits in a kettle, remove the seeds, wait until they cool off, then form the paste into small mounds'; *Use:* **chupul** 'inflammation'; *Part:* **Ha'li' kunyah** 'Only its name is mentioned in a curing ritual'; *Sym:* **k'ik'** 'blood'. *Thes:* **che'** (Fig. 5.116a, b, c). *[Note:* k'uxu'. *Bixa orellana* (Bruce 1975); k'ushub. *Bixa orellana* (Nations and Nigh 1980); k'uxu' [Itz.]. annatto, achiote tree. *Bixa orellana* (Hofling and Tesucún 1997); b'itz'k'uxu', kiwi', tzo'otzk'uxu' [Itz.]. *Bixa orellana* (Atran et al. 2004); ah-k'uxu' [Itz.]. 'anota'. *Bixa orellana* (Lois 1998); k'uxu' [Yuc.]. *Bixa orellana* (Bricker et al. 1998); k'uxub', kiwi' [Yuc.]. *Bixa orellana* (in, Atran et al. 2004).*] [Source:* AM; BM; KM; KP*] [*\sd2 food \sd2 craft \sd2 other*]*

Fig. 5.115 K'unche' pahok *Geonoma oxycarpa*. Photo Credit: Suzanne Cook

k'u' kan spikemoss. *Lit:* 'snake's nest'. Selaginellaceae *Selaginella silvestris* Aspl. This is a prostrate plant with bright green, scale-like leaves on branching stems. It grows in the moist understory. The name refers to the bed for the snake, because snakes like to curl up on this plant (AM, BM). This is an example of *kuxum lu'um* 'earth mould'. *See:* **kuxum lu'um**. *Variant:* **k'u' ahwech** (Fig. 5.117). *[Source:* AM; BM; CKD*]*

L–l

läkte' prickly-ash (Hercules' club). *Lit:* 'destroy tree'. Rutaceae *Zanthoxylum* sp. This is a tall 30 m (98′) tree with a straight trunk 80 cm (32″) in diameter and covered with large, conical spines (Pennington and Sarukhán 2005: 280). According to SK, the tree produces a strong resin which the Lacandones use to prevent cockroaches from entering homes. A number of studies in African and Indian countries report on *Zanthoxylum* chemical compounds and essential oils, which, among other things, are used to discourage or destroy grain insects (Negi et al. 2011). Several *Zanthoxylum* species have been recorded in the Lacandon forest (Comisión

Fig. 5.116 (**a**) K'uxub *Bixa orellana* flowers. Photo Credit: Chan K'in Daniel (**b**) K'uxub *Bixa orellana* flowers. Photo Credit: Suzanne Cook (**c**) K'uxub *Bixa orellana* plant. Photo Credit: Suzanne Cook

Fig. 5.117 K'u' kan *Selaginella silvestris*. Photo Credit: Suzanne Cook

Nacional de Áreas Naturales Protegidas 2006). Among them, *Z. kellermanii* P. Wilson is used for construction materials and *Z. mayanum* Standl., for construction materials and crafts (Levy et al. 2006). Given its insecticidal properties the timber is likely used in house construction in Naha' as well, although none of my Lacandon consultants reported on this. Durán (1999) recorded *Z. procerum* Donn. Sm. in Naha' but he does not indicate any uses for the tree. *Use:* **yäk ba'al ti' uk'insik tu'bok** 'poison to kill cockroaches'; *Part:* **uyits** 'its resin'. *Thes:* **che'** (Fig. 5.118). *[Note:* lacte. *Zanthoxylum procerum* (Durán 1999); *Zanthoxylum belizence, Z. kellermani, Z. microcarpum,* and *Z. procerum* (Comisión Nacional de Áreas Naturales Protegidas 2006); *Zanthoxylum kellermanii* P. Wilson, and *Zanthoxylum mayanum* Standl. (Levy et al. 2006).*] [Source:* BM; SK*] [*\sd2 other*]*

lek$_2$ calabash. Cucurbitaceae *Lagenaria siceraria* (Molina) Standl. This is a large, rounded, squat, sometimes ribbed variety of inedible squash cultivated in house gardens and milpas. The flowers are white and bloom at night. The skin is light green and smooth. The interior comprises gauzy, white flesh and numerous, large, brownish-black seeds. *Loc:* **paakche' kol, kol** 'fallow milpa, milpa'; *Use:* **lekil wah** 'tortilla plate'. *See:* **chuh** (Fig. 5.119a, b). *[Note:* luch; lek; chú. *Lagenaria siceraria* (Nations and Nigh 1980); lek [Itz.]. *Lagenaria siceraria* (Atran et al. 2004); lek, leek [Yuc.]. *Lagenaria siceraria* (in, Atran et al. 2004).*] [Source:* AM; KM; KP*] [*\sd2 containers*]*

le'mo' cactus, climbing (orchid cactus, leaf cactus). *Lit:* 'creased leaf (?)'. Cactaceae *Epiphyllum* sp. **Es ak'. Yan uyak'il uk'äb. Yan uwich, ne chäk uwich, t'elt'el uwich. As chukuch uwich, batak 7.6 cm (3″) uwaan. Säk unoy. Ne yaab unek', chan mihin unek', ek'. Amäk'ik.** 'This is a vine. It has branching stems. It has very red fruit with ridges. The fruit is about 7.6 cm (3″) long. The pulp is white and contains numerous, small, black seeds. You eat it.' (KP) *Epiphyllum* is a genus of epiphytic cactae from the Neotropics. They are characterized by their non-prickly, flat, usually lobed, stems and ornate, highly aromatic flowers. The species described by KP is likely Dutchman's pipe (*E. oxypetalum*), a sprawling, branching species that can climb to 6.1 m (20′) high. Its dark green stems begin as terete and then flatten on successive secondary stems, producing a slightly-raised ridge on the upper surfaces. The oval, tapering stems are 0.9 m (3′) long and 10–12 cm (4–5″) wide. Stem margins are scalloped and wavy. It bears large, white, upward-curving flowers that bloom at night, attracting moths with its heady perfume. Flowers are followed by long, 12 × 8 cm (4.7–3.14″) angular, red fruits. *Loc:* **k'aax, kolil nah** 'forest, house garden'; *Use:* **hanal** 'food'; *Part:* **wich** 'fruit'; *Prep:* **che'che'** 'raw'. *Morph:* **le'-mo'**. *Thes:* **ak'** (Fig. 5.120a, b). *[Note:* limo. unidentified epiphyte Nations and Nigh (1980: 11). Although my Lacandon consultants did not provide the translations for each morpheme, the word is likely a compound le' 'leaf' = mo' 'wrinkled, pleated leaf'. Compare mo'ol [Itz.]. pleat, fold, wrinkle (Hofling and Tesucún 1997: 452).*] [Source:* AM; BM; KP*] [*\sd2 food*]*

Fig. 5.118 Läkte' *Zanthoxylum* sp. Photo Credit: Suzanne Cook

a b

Fig. 5.119 (**a**) Lek *Lagenaria siceraria*. Photo Credit: Suzanne Cook (**b**) Lek *Lagenaria siceraria* seeds. Photo Credit: Suzanne Cook

a b

Fig. 5.120 (**a**) Le'mo' *Epiphyllum* sp. Photo Credit: Suzanne Cook (**b**) Le'mo' *Epiphyllum* sp. Photo Credit: Suzanne Cook

lo'k'in parathesis (type of). *Lit:* 'bad day/time'. Myrsinaceae *Parathesis* sp. This is a rangy shrub or treelet with alternate, petiolate leaves. Blades are lanceolate, entire, slightly wavy, acuminate, and smooth. Leaf tissue folds inward at the base creating a winged petiole. Secondary veins are pinnate, angled 45 degrees, and fairly inconspicuous. Leaves are born at branch tips. Stems are woody. Clusters of red-ripe berries are born in terminal panicles. It grows in the shady understory, near lagoons and the margins of secondary forest. KM called this tree *xinich*, but AM disagreed, saying that *lo'k'in* is similar to *xiilich* (= xinich?), although not as big. *Parathesis lanceolata* and *Parathesis donnell smithii* are reported to occur in the northern Lacandon territory (Comisión Nacional de Áreas Naturales Protegidas 2006), whereas *Parathesis chiapensis* occurs in the southern territory where its fruit is eaten by the southern Lacandones (Levy et al. 2006). Although no uses were reported, Kashanipour and McGee (2004: 66) say that a tea made from the bark of *"lol k'in"* is taken to alleviate diarrhea. *Use:* **hubul hämnen** 'diarrhea'; *Part:* **sool** 'bark'; *Prep:* **te'** 'tea'; *Sim:* **xinich (äh)**. *Thes:* **che'** (Fig. 5.121a, b). *[Note:* Taxonomic identification provided by Rolando Perez (pers. comm.). lol k'in (Kashanipour and McGee 2004). Medicinal remedy is reported by these authors. It is unclear whether or not the name of the plant is related to *lo'k'in*, the cannibals in Lacandon myths*] [Source:* AM*] [*\sd2 medicinal*]*

lo'loh (äh) fungus, jelly. *Lit:* 'bad flower (?)'. Auriculariaceae *Auricularia cornea* (Ehrenb) Ehrenb ex Endl., *A. delicata* Henn. This is a kind of gelatinous fungi that grow in clusters

Fig. 5.121 (**a**) Lo'k'in *Parathesis* sp. plant. Photo Credit: Suzanne Cook (**b**) Lo'k'in *Parathesis* sp. leaves and fruit. Photo Credit: Suzanne Cook

on dead wood. *Use:* **hanal** 'food'; *Prep:* **chäkbil yetel bu'ul** 'boiled with beans'. *[Note:* Ruan-Soto et al. (2009) list a number of *Auricularia* species in the southern Lacandon community of Lacanja Chan Sayab: lo'ro, *A. cornea* (Ehrenb) Ehrenb ex Endl.; sak're lo'ro, *A. fuscosuccinea* (Mont.) Farl. (forma *albina*); choch e wakax lo'ro, yuyo lo'ro, *A. delicata* (Fr.) Henn., *A. fuscosuccinea* (Mont.) Farl. Compare lòol [Yuc.]. flower, squash blossom (Bricker et al. 1998).*]* *[Source:* KM*]* *[*\sd2 food*]*

luuch₁ [luuch, luch] calabash. Bignoniaceae *Crescentia* spp. These are tropical trees that grow to 10 m (35′) tall. They have spreading branches and a broad crown. Glossy, bright green leaves grow on thick branches. The tubular, fragrant flowers grow directly on the branches and trunk, followed by smooth, round or oval gourds with thin, hard rinds and soft pulp. Fruit shape and size vary considerably, even within a species. The Lacandones distinguish two kinds—*hach luuch* 'authentic calabash' and *sina'an luuch* 'stretched out calabash'. Although inedible, the fruits are cultivated for their rinds. *Loc:* **kolil nah, kol, paakche' kol** 'house garden, milpa, fallow milpa'; *Use:* **luuch, hach chaachib, so'ot** 'cups, bowls, authentic colander, rattle'; *Prep:* **Asulik yok'ol uk'um ti' hunts'it na' ti' ts'ukik unoy, pachil alahoytik unoy, pachil atihsik yok'ol k'in** 'You soak the gourds in a stream for at least a month, to loosen the pulp from the rind, then you scoop out the pulp, and then you dry it in the sun'; *Spec:* **hach luuch** 'authentic calabash (*Crescentia cujete*)'; *Spec:* **sina'an luuch (äh)** 'stretched out calabash (*Crescentia* sp., *C. alata*)'; *Spec:* **säk luuch** 'white calabash'. *Variant:* **luch che'**. *Thes:* **che'**. *[Note:* hach luch. *Crescentia alata*; sinah luch. *Crescentia cujete* (Nations and Nigh 1980); luch [Itz.]. water gourd. *Crescentia* spp., *C. cujete* (Hofling and Tesucún 1997); luch [Itz.]. *Crescentia cujete*; ix was = ix mehen luch, ix tutz'uk luch. *Crescentia alata* (Atran et al. 2004); lùuch [Yuc.]. *Crescentia cujete*; ajwas *Crescentia alata* (in, Atran et al. 2004); luch [Yuc.]. *Crescentia cujete* (Roys 1931).*]* *[Source:* AM;BM*]* *[*\sd2 containers*]*

luuch che' calabash tree. Bignoniaceae *Crescentia* spp. *Variant:* **luuch**.

luwin bullhoof. Ulmaceae *Ampelocera hottlei* (Standl.) Standl. *See:* **yuwin**. *Thes:* **che'**. *[Note:* sitz muk. luin. *Ampelocera hottlei*. Fruit is eaten (Nations and Nigh 1980); luin. *Ampelocera hottlei* (March 1998); rubin. *Ampelocera hottlei* (Nigh 2008); lwin [Itz.]. *Ampelocera hottlei* (Hofling and Tesucún 1997).*]* *[Source:* Davis (1978)*]*

M–m

maasam (äh) [maasam, maasäm, maasa, maasan] banana 'Lady Finger' (date banana, fig banana, sugar banana). Musaceae *Musa acuminata* Coll. (*AA grp*) cv. 'Sucrier', syn. *musa xpardisiaca* L. cv. 'Lady Finger'. This variety produces stubby, fat, yellow bananas with a sweet, fruity flavour. According to KM, the fruit must be very ripe before eating or else it will cause constipation. *Use:* **hanal** 'food'; *Part:* **wich** 'fruit'; *Prep:* **che'che'** 'raw'; *Gen:* **boox**. *Variant:* **maasam boox**. *[Note:* masaam boox (äh). lady finger bananas (Bruce 1975: 192).*]* *[Source:* BM; CNK; KM*]* *[*\sd2 food*]*

maasan kitan scouringrush horsetail. Equisetaceae *Equisetum hyemale* var. *affine* (Engelm.) A. A. Eaton. This is a spreading, reed-like perennial with cylindrical, evergreen

Fig. 5.122 Maasan kitan *Esquisetum hyemale*. Photo Credit: Suzanne Cook

stems that are jointed, hollow, straight, with rough, longitudinal ridges. The tiny leaves encircle the stem, forming a narrow blackish-green band around each joint. It is found in moist forests, forest edges, stream banks, and swamps. *Loc:* **petha'** 'lagoon'; *Use:* **chul ti' mehen paalal** 'childrens' flute'; *Part:* **uche'il** 'stalk'; *Use:* **ts'ak ti' uyah ubaakel** 'medicine for bone pain'; *Part:* **che'il** 'stalks'; *Prep:* **?** 'unknown'. *Variant:* **ha' maasan kitan** (Fig. 5.122). *[Source:* BM; KyP; SK*]* *[*\sd2 other \sd2 medicinal*]*

ma'ax ak' (äh)$_1$ slipplejack (?). Sapindaceae *Serjania* sp. This is a woody liana with tendrils and compound leaves. Flowers are borne in racemes. Fruit have winged margins. A few species—*S. atrolineata* Wright, *S. triquetra* Radlk., *S. caracasana* (Jacq.) Willd., and *S. macrocarpa*—are used as fish poison (Atran et al. 2004; Durán 1999; Levy et al. 2006). The southern Lacandones use *S. atrolineata* as medicine (Diemont 2006: 82). The flowering of the vine indicates the time to sow late maize in the northern Lacandon territory (Nations and Nigh 1980: 12). *Loc:* **k'aax** 'forest'; *Activity:* **päybil näl** 'replant maize'. *See:* **ma'ax ak'$_1$**. *[Note:* max ak, a kura ak. *Serjania atrolineata* (Durán 1999); marxak. *Serjania atrolineata* (Diemont 2006: 82); aj 'oxlet'ak' [Itz.]. *Serjania triquetra* (Atran et al. 2004.).*]* *[Source:* Durán (1999)*]* *[*\sd2 other*]*

ma'ax ak' (äh)$_2$ soapberry. *Lit:* 'spider monkey vine'. Sapindaceae *Paullinia* sp. This is a liana approximately 5 cm (2″) thick, with twisting and fused vines. It bears clusters of small, white flowers, in December (BM). The flowers are born in racemes. Orange-red capsules split open, revealing black or greenish seeds covered at the base with a white aril. *Paulinia pinnata* L. and *P. costata* Schl. are reported from Naha' (Comisión Nacional de Áreas Naturales Protegidas 2006). *Paulinia* and *Serjania* are closely-related genera: both are robust, woody climbing vines with doubly compound leaves. *Serjania* species in the area include: *S. atrolineata* Sauv. & Wr.; *S. caracasana* Jacq.; *S. gonoiocarpa* Radlk.; *S. macrocarpa* Standley & Steyerm.; *S. mexicana* (L.) Willd.; and, *S. paucidentata* DC. (Comisión Nacional de Áreas Naturales Protegidas 2006). The tough stems of both genera are used as lashing material and fish poison. Durán (1999) reports that the northern Lacandones use *Serjania atrolineata* and *S. caracasana* for poison, whereas Nations and Nigh (1980: 12) report that *Paulinia pinnata* and *P. costata* are used. Kashanipour and McGee (2004: 61) note that a decoction made from the stalk and leaves of *Paulinia pinnata* is drunk to reduce fever. The blossoms of ma'ax ak' signal the time for *päybil näl* 'torna milpa' (Sp.). *Use:* **ts'ak ti' chäkwilil** 'medicine for fever'; *Part:* **che'il, le'** 'stems, leaves'; *Prep:* **te'** 'tea'. *Variant:* **ak' ma'ax**. *Thes:* **ak'**. *[Note:* mäsh ak'. barbasco. *Paullinia costata*. (Nations and Nigh 1980); ma'ax ak'. *Paullinia pinnata* (Kashanipour and McGee 2004); *Paullinia costata* (Levy et al. 2006); max ak. *Serjania atrolineata* (Durán 1999). Medicinal remedy provided by Kashanipour and McGee (2004: 61).*]* *[Source:* AM; BM*]* *[*\sd2 agriculture \sd2 medicinal*]*

mahaas (ah) [mahaas, mähas] funeral tree (funeral swizzlestick tree). Malvaceae (Bombacaceae) *Quararibea funebris* (La Llave) Vischer. *Loc:* **k'aax** 'forest'; *Use:* **pakat** 'bird bolt'; *Part:* **che'il** 'wood'; *Use:* **ch'am** 'grill rack'; *Part:* **uxet' usool** 'sections of bark'; *Use:* **hi'che' ahpek'** 'dog enclosure'; *Part:* **xa'ay** 'forked branches'; *Use:* **ch'uyu' ti' ba'al wah bäk' wah nook'** 'hanger for things, like meat and clothes'; *Part:* **xa'ay** 'forked branches'; *Use:* **k'ek' che', mulinuh** 'cocoa beater'; *Part:* **che'il** 'wood'; *Use:* **tukil** 'spindle whorl'; *Part:* **che'il** 'wood'; *Sim:* **ton ku'uk** 'ear flower (*Cymbopetalum* sp.)'. *Thes:* **che'** (Fig. 5.123). *[Note:* The fragrant flowers were used to flavour cocoa by the Aztecs and are still the principal ingredient in *tejate*, a traditional Mayan chocolate-maize beverage. mahas. *Quararibea funebris* (Nations and Nigh 1980); majs (Baer and Merrifield 1971); aj-maja'as [Itz.]. beater. *Quararibea guatemalteca* (Lois 1998: 260).*]* *[Source:* AM; BM; SK*]* *[*\sd2 hunting_and_fishing \sd2 food \sd2 other \sd2 craft*]*

Fig. 5.123 Mahaas *Quararibea funebris*. Photo Credit: Suzanne Cook

mankoh mango. Anacardiaceae *Mangifera indica* L. A species introduced from South Asia, the mango tree bears large, deep orange, sweet, syrupy fruit with a stringy texture. The juicy pulp surrounds a single, large, slightly flattened seed. The skins are smooth, thin, leathery, either green or yellow and often tinged with red. Size and shape vary from nearly round, to tear drop, and range between 10 and 15 cm (4 and 6″) long. Those cultivated in Naha' fit into the palm of an adult's hand. Cultivated. *Use:* **hanal** 'food'; *Part:* **wich** 'fruit'; *Prep:* **che'che'** 'raw'. *From:* manga (Portuguese). *Thes:* **che'**. *[Note:* ix maangoj [Itz.]. *Mangifera indica* (Atran et al. 2004); maango [Mop.]. *Mangifera indica* (in, Atran et al. 2004); ixmango [Yuc.]. *Mangifera indica* (in, Atran et al. 2004).*]* *[*\sd2 food*]*

mäkäl [maka, mäkär] elephant ears. Araceae *Xanthosoma* spp. *Mäkäl* is a general term for edible and inedible species of elephant ears. The two edible kinds are *säk mäkäl* 'white xanthosoma' and *chäk mäkäl* 'red xanthosoma'. They are distinguished on the basis of the colour of their stalks and corms. Collectively, they are called *hach mäkäl* 'authentic xanthosoma' (*Xanthosoma mafaffa*). A third kind, deemed inedible by the Lacandones, is *ch'om mäkäl* 'vulture mäkäl' (*Xanthosoma robustum*). *Spec:* **ch'om mäkäl** 'vulture macal (*Xanthosoma robustum*)'; *Spec:* **hach mäkäl** 'authentic macal (*Xanthosoma mafaffa*)'. *Thes:* **xiiw**. *[Note:* makal. ñame. Chinese yam. *Dioscorea alata* (Soustelle 1966); makäl. malanga. *Xanthosama* sp. (Nations and Nigh 1980); makal, màakal [Yuc.]. elephant ears. *Xanthosoma yucatanense* (Bricker et al. 1998); macal [Yuc.]. *Dioscorea alata* L. ñame. (Standl.) (Roys 1931); mäkäl [Itz.]. *Xanthosoma yucatanense* (Atran et al. 2004); mäkäl [Mop.]. *Xanthosoma yucatanense* (in, Atran et al. 2004).*]* *[Source:* AM; BM*]* *[*\sd2 food*]*

mäkuuläm [mäkuuläm, mäkuuräm] pepper. Piperaceae *Piper* spp. *Mäkuuläm* refers to large shrubs, treelets, or vines that bear slender, white inflorescences that look like curved or erect tails. Leaves vary across species, from heart- to lance-shaped, some are velvety in texture, whereas others are fuzzy or crinkled. Piper grows along roadsides, in forest clearings, yards, and fallow *milpas*. At least 16 species have been recorded in the northern Lacandon territory: *Piper geniculatum; P. pseudoasperifolium; P. yucatananensis; P. aduncum; P. aequale; P. aeroginosibaccum; P. amalago; P. auritum; P. decurrens; P. hispidum; P. sancti-felicis; P. umbellatum; P. uspantanense; P. variabile; P. yzabalanum;* and, *P. psilorachis* (Comisión Nacional de Áreas Naturales Protegidas 2006). The northern Lacandones distinguish at least two species that they call *mäkuuläm* and two more that they call *hoben*. Some *mäkuuläm* species are used to treat snakebite and gastro-intestinal problems (Durán 1999; Kashanipour and McGee 2004). *Spec:* **mehen mäkuuläm** 'small mäkuuläm'; *Spec:* **hach mäkuuläm** 'authentic mäkuuläm'; *Sim:* **hoben**. *See:* **hach lo'obil**; **hoben**. *Thes:* **che'**. *[Note:* m'k'uram, makurum. *Piper aduncum* L. (Nigh 2008); mäk u lin. *Piper* spp. (Nations 2006); makurum. Piper aduncum (Diemont 2006); makurum. *Piper hispidum* (Durán 1999); ix makulan [Yuc.]. *Piper auritum* (in, Atran et al. 2004); mak'olan xiw [Yuc.]. *Piper amalgo, P. gaumeri, P. psilorhachis* (in Atran et al. 2004); ob'eel [Mop.]. *Piper auritum* (in Atran et al. 2004). According to Durán (1999), the species used to treat snakebite is *P. hispidum.]* *[Source:* AM; AT; BM; KM; SK*]* *[*\sd2 housing \sd2 medicinal \sd2 food \sd2 other*]*

mäya' (äh) plantain (type of). Musaceae *Musa acuminata* ×*Musa balbisiana* Coll. This variety of plantain produces only ten or so very large fruits that measure about 45.7 cm (18″) long (BM). In Spanish, it is called *plátano macho* 'male plantain'. According to Bruce (1975: 90), it is the largest type of plantain that the Lacandones cultivate. *Use:* **hanal** 'food'; *Part:* **wich** 'fruit'; *Prep:* **tsähbil** 'fried'; *Gen:* **pätan**. *[Source:* BM; CNK; KM*]* *[*\sd2 food*]*

mehen bä'ch'ich' seagrape, Honduras. Polygonaceae *Coccoloba hondurensis* Lund., syns. *Coccoloba tenuis* Lund., *Coccoloba verapazensis* Lund. This is a medium sized tree growing along river banks in shady places. White flowers are followed by clusters of round, fleshy fruits. It is native to southern Mexico, Belize, Guatemala, and Honduras (Grandtner 2005: 229). *Loc:* **k'aax** 'forest'; *Use:* **hanal** 'food'; *Part:* **wich** 'fruit'; *Prep:* **che'che'** 'raw'; *Use:* **nah** 'houses'; *Part:* **che'il** 'timber'; *Use:* **che'il** 'firewood'; *Sim:* **nukuch bä'ch'ich' (ah)** 'large bä'ch'ich' (*Coccoloba barbadensis*)'; *Gen:* **bä'ch'ich'** '*Coccoloba* spp.'. *Thes:* **che'**. *[Note:* English name from Grandtner (2005: 229). Botanical synonyms from The Plant List. 2010. Version 1. http://www.theplantlist.org/. Retrieved June 2013.*]* *[Source:* mejen bochich, bochich. *Coccoloba* aff. *hondurensis* (Durán 1999); bor ich. *Coccoloba hondurensis* (Comisión Nacional de Áreas Naturales Protegidas 2006).*]* *[*\sd2 food \sd2 housing*]*

mehen bits' inga (type of). Fabaceae *Inga* sp. **Mehen bits' uch'ihil ich k'aax wa paakche' kol. Crecen solito. Uwich, kupäk'ik ähch'iich'.** '*Mehen bits'* grows in the forest or in the fallow *milpas*. It is not cultivated. Birds disperse the seeds.' (BM). *Loc:* **k'aax, paakche' kol** 'forest, fallow milpa'; *Use:* **hanal** 'food'; *Part:* **wich** 'fruit'; *Prep:* **che'che** 'raw'; *Sim:* **suumil bits'** 'rope bits' (*Inga oerstediana*)'; *Sim:* **t'elen bits'** 'ridged bits' (*Inga sapindoides*)'; *Sim:* **bits'il k'aax** 'forest bits' (*Inga belizensis*)'; *Sim:* **nukuch bits'** 'large bits' (*Inga punctata*)'; *Sim:* **bits' petha'** 'lagoon bits' (*Inga spuria*)'; *Gen:* **bits'**. *Thes:* **che'**. *[Note:* mejen tete bits. *Inga punctata* (Durán 1999). English name from ZipZoo.com, http://zipcodezoo.com/Plants/I/Inga_densiflora/#common*]* *[Source:* BM*]* *[*\sd2 food*]*

mehen bu'ulche' cojoba. *Lit:* 'small bean tree'. Fabaceae *Cojoba* sp., *C. arborea* L. (Britton& Rose), *C. graciliflora* (S. F. Blake) Britton & Rose, syn. *Cojoba donnell-smithii* Britton & Rose. **Mehen bu'ulche' uch'ihil 15 metros (49′). Unek', ek'. Säsäk utop'.** '*Mehen bu'ulche'* is a small tree that grows to 15 m (49′) tall. The seeds are black. The flowers are somewhat white.' (BM) *Cojoba* trees are commonly found in open sites and transition zones. Two species are reported from the northern Lacandon territory: *C. arborea* and *C. graciliflora* (Comisión Nacional de Áreas Naturales Protegidas 2006). Either one could be *mehen bu'ulche'*. Further research is required. No uses were reported. *Spec:* **nukuch bu'ulche'** 'large bean tree'; *Gen:* **bu'ulche'**. *Thes:* **che'**. *[Source:* BM*]*

mehen chäk k'o'och bean, castor. Euphorbiaceae *Ricinus communis* L. This species was unavailable to inspect. Conceivably, the name refers to a sapling. Further investigation is required. *Sim:* **nukuch chäk k'o'och** 'large castor bean'; *Gen:* **chäk k'o'och (äh)**. *Thes:* **che'**. *[Source:* AM*]*

mehen ch'alol oak (type of). Fagaceae *Quercus* sp., *Q. oleoides* (?). *Loc:* **ya'ax petha'** 'green lagoon'; *Use:* **uuh** 'necklaces'; *Part:* **wich** 'fruit'; *Sim:* **nukuch ch'alol** 'authentic ch'alol (*Quercus corrugata*)'; *Sim:* **pixan k'ambul (äh)** 'curassow's heart (*Quercus skinneri*)'; *Gen:* **ch'alol**. *Thes:* **che'** (Fig. 5.124). *[Source:* AM; KyM*]* *[*\sd2 craft*]*

mehen ch'eel koox [mehen ch'eel koox, mehen ch'e koox] fatpork. *Lit:* 'small female crested guan'. Rubiaceae *Chione chiapasensis* Standl., syn. *C. venosa* var. *venosa*. A small tree. Fruit is eaten by the *koox* (*Penelope purpurascens*). *See:* **päh-pähche'**. *Thes:* **che'** (Fig. 5.125). *[Note: Chione chiapasensis* (Breedlove 1986; Durán 1999).*]* *[Source:* AM*]*

mehen kopo' fig, Sonoran strangler (perforated fig, plum-leaved fig tree, strangler fig). Moraceae *Ficus pertusa* L.f., syn. *Ficus padifolia* Kunth. This is a small variety of *kopo'* with bright green, ovate leaves about 5.08 cm (2″) long by 2.54 cm (1″) wide, acuminate at apex. Leaf margins curl up slightly, giving the appearance of shallow spoons. It bears small, 0.63 cm (1/4″), round fruit in clusters near leaf axils.

Fig. 5.124 Mehen ch'alol. Photo Credit: Suzanne Cook

The skin is speckled red and yellowish white when immature, turning red when ripe. In Naha', the fruits begin to ripen in late September. Found growing in house gardens and yards. Although the fruits are inedible, the tree is nevertheless cultivated, likely to provide shade. No uses were reported. *Use:* **bo'oy** 'shade'; *Sim:* **nukuch kopo'** 'large kopo' (undetermined)'; *Gen:* **kopo' (ah)**. *Thes:* **che'** (Fig. 5.126a, b). *[Note:* Botanical identification was provided by Alvaro Campos (pers. comm.). Common English name from http://arboretum.arizona.edu/taxa/Ficus_pertusa.html and http://zipcodezoo.com/Plants/F/Ficus_pertusa/. mehen kopo. *Ficus pertusa* (Durán 1999).*]* *[Source:* AM; AM's wife; KM; KP*]* *[*\sd2 other*]*

mehen koyoh coyo persea (wild pear). Lauraceae. *Use:* **hanal** 'food'; *Part:* **wich** 'fruit'; *Prep:* **che'che'** 'raw'; *Sim:* **nukuch koyoh** 'large koyoh'; *Gen:* **koyoh**. *Variant:* **hach koyoh**. *[*\sd2 food*]*

mehen k'än'ol glassy wood [Bel. Eng.] (tigerwood). Anacardiaceae *Astronium graveolens* Jacq. *Astronium graveolens* is a medium or large tree, 8–35 m (26–114′) tall. Its trunk is straight, approximately 35–100 cm (13.7–39″) in diameter, with small, narrow buttresses. The crown is dense.

Fig. 5.125 Mehen ch'eel koox *Chione chiapasensis*. Photo Credit: Suzanne Cook

Fig. 5.126 (**a**) Mehen kopo' *Ficus pertusa* fruit. Photo Credit: Suzanne Cook (**b**) Mehen kopo' *Ficus pertusa* cultivated tree. Photo Credit: Suzanne Cook

Young twigs are grey, green, or greyish brown with numerous, dark, lenticels. Flowers occur in axillary or terminal panicles up to 20 cm (7.8″) long. Female flowers are yellow-green, which are followed by clusters of small, 10–15 mm (0.4–0.6″), oval fruits that ripen to black, in October. The pulp is yellow and resinous. Parrots are fond of them. The northern Lacandones also call this tree *mala mujer* 'evil woman', because it exudes an irritating resin through its *tuchil* 'lenticels'.This resinous material likely contains urushiols (alkenyl phenols) (Marín and Flores n.d.). BM says that one can relieve the irritation caused by the resin by applying a solution of lime juice and salt to the affected areas. None of my Lacandon consultants reported any uses for the tree, although Levy et al. (2006) say that the southern Lacandones use the timber for crafts. *Loc:* **k'aax** 'forest'; *Sim:* **nukuch k'änol** 'large k'änol (*Mosquitoxylum jamaicense*)'; *Gen:* **k'än'ol (äh)**. *Variant:* **k'än'ol**. *Thes:* **che'** (Fig. 5.127). *[Source:* BM; KyP; SK*]*

mehen k'än luuch calabash tree (type). Bignoniaceae *Crescentia* sp. *Gen:* **k'än luuch**. *See:* **sina'an luuch (äh)**. *Thes:* **che'**. *[Note:* It could be *C. alata*, but without having seen the tree, the term could very well refer to a small gourds of *C. cujete.] [Source:* AM*] [*\sd2 containers*]*

Fig. 5.127 Mehen k'än'ol *Astronium graveolens*. Photo Credit: Suzanne Cook

mehen k'ik'ni' balum$_1$ eugenia (type). *Lit:* 'small bleeding-nosed jaguar'. Myrtaceae *Eugenia* sp. The observed species was a woody shrub with glossy, medium green, broad leaves. Neither fruit nor flowers were present, making a positive identification difficult. The southern Lacandones eat the fruits of *Eugenia tikalana* Lund. (Levy et al. 2006), a species which Durán's northern Lacandon consultants call *"xin che"*. They too report eating the fruits. *Use:* **hanal** 'food'; *Part:* **wich** 'fruit'; *Prep:* **che'che'** 'fresh'; *Sim:* **nukuch k'ik'ni' balum** 'large k'ik'ni' balum'; *Sim:* **hach k'ik'ni' balum** 'authentic k'ik'ni' balum (*Ardisia compressa*)'; *Gen:* **k'ik'ni' balum (ah)**. *Thes:* **che'**. *[Source:* AM; BM*]* *[*\sd2 food*]*

mehen k'ik'ni' balum$_2$ coralberry (marlberry). *Lit:* 'small bleeding-nosed jaguar'. Myrsinaceae *Ardisia compressa* Kunth. *Use:* **hanal** 'food'; *Part:* **wich** 'fruit'; *Prep:* **che'che'** 'raw'; *Sim:* **nukuch k'ik'ni' balum** 'large k'ik'ni' balum'; *Gen:* **k'ik'ni' balum (ah)**. *Variant:* **hach k'ik'ni' balum**. *Thes:* **che'**. *[Source:* AM; BM; KP*]* *[*\sd2 food*]*

mehen k'o'och pumpwood (trumpet tree). *Lit:* 'small k'o'och'. Urticaceae *Cecropia* sp. *Loc:* **k'aax** 'forest'; *Sim:* **nukuch k'o'och** 'large k'o'och'; *Gen:* **k'o'och (äh)** 'pumpwood (*Cecropia* spp.)'. *See main entry:* **k'o'och$_2$**. *[Source:* AM*]*

mehen mäkuuläm pepper, Jamaican (cordoncillo). Piperaceae *Piper aduncum* L., *P. hispidum* Sw. The two species BM and AM referred to as mehen mäkuuläm correspond to both *Piper aduncum* and *P. hispidum*. According to BM, *mehen mäkuuläm* is similar to *hach mäkuuläm*, only with shorter leaves. Although they did not mention any uses for the tree, Durán's consultants said that they prepared an infusion from the leaves to treat snakebite and gastritis. *Use:* **chibal kan** 'snakebite'; *Part:* **le'** 'leaves'; *Prep:* **te'** 'tea'; *Use:* **chibal hämnen** 'stomach ache'; *Part:* **le'** 'leaves'; *Prep:* **te'** 'tea'; *Sim:* **hach mäkuuläm** 'authentic mäkuuläm (*Piper aduncum*)'; *Gen:* **mäkuuläm**. *Variant:* **ts'ak kan ak'**. *Thes:* **che'** (Fig. 5.128a, b, c). *[Note:* English common name from the Natural Resource Conservation Service (NRCS) United States Department of Agriculture (USDA). http://plants.usda.gov/java/profile?symbol=PIHI2. makurum. *Piper hispidum* (Durán 1999); m'k'uramik ak. *Piper hispidum* (Nigh 2008).*]* *[Source:* AM; AT; BM*]* *[*\sd2 medicinal*]*

mehen 'oochin$_1$ philodendron, heart leaf (velvet philodendron). Araceae *Philodendron hederaceum* (Jacq.) Schott., syns. *P. scandens, P. oxycardium, P. micans, P. Brazil. Philodendron hederaceum* is a hemiepiphytic vine, which begins life on a tree branch or as a seed dropped on the ground. Its vines often grow high up in the canopy and hang down towards the jungle floor. As it climbs, the stem and blades cling to the host tree. Leaves are heart-shaped and, when young, velvety. As the plant matures, mature leaf blades change their appearance and often measure 11–50 cm (4–19.5″) long and 8–49 cm (3.25–13.33″) wide. It bears a single inflorescence per axil. The spathe is 9–24 cm (3.5–9.4″) long, dark green or yellowish green and tinged purple on the outside and red on the inside. (Lucas n.d.) My Lacandon consultants also referred to an immature form of *Syngonium podophyllum* as *mehen 'oochin*, suggesting that the word may simply refer to a juvenile form of *'oochin*, rather than a botanical species. *Loc:* **k'aax** 'forest'; *Sim:* **hach 'oochin** 'authentic 'oochin (*Philodendron donnell-smithii*)'; *Sim:* **nukuch 'oochin** 'large 'oochin (*Philodendron sagittifolium, Syngonium chiapense, Syngonium podophyllum*)'; *Gen:* **'oochin** (Fig. 5.129a). *[Source:* AM; BM*]*

mehen 'oochin$_2$ arrowhead vine (goosefoot plant). Araceae *Syngonium podophyllum* Schott. According to AM and BM, this is a terrestrial kind of *'oochin* that never climbs. *Sim:* **hach 'oochin** 'authentic 'oochin (*Philodendron donnell-smithii*)'; *Sim:* **nukuch 'oochin** 'large 'oochin'; *Gen:* **'oochin** (Fig. 5.129b). *[Note:* Botanical identification provided by Tom Croat (pers. comm.).*]* *[Source:* AM; BM*]*

mehen 'oonte' nectandra (type of). Lauraceae *Nectandra* sp. This is a tree that bears clusters of small (2.5 cm) oval, dark purple drupes nestled in reddish cupules. The fruit attracts animals and flocks of birds. According to my Lacandon consultants, it is similar to *babah 'oonte'*, except for its larger leaves. *Sim:* **säk 'oonte'** 'white 'oonte' (*Licaria alata*)'; *Sim:* **ek' 'oonte'** 'black 'oonte' (*Nectandra globosa*)'; *Sim:* **nukuch 'oonte'** 'large 'oonte' (*Nectandra* sp.)'; *Sim:* **babah 'oonte'** '*Ocotea cernua*'; *Gen:* **'oonte'**. *Thes:* **che'**. *[Note:* mehen on te. *Nectandra raticulata* (Nations 2006); mejen onte. *Nectandra coriacea, Nectandra membrenacea, Ocotea cernua* (Durán 1999).*]* *[Source:* AM; BM*]*

Fig. 5.128 (**a**) Mehen mäkuuläm plant. Photo Credit: Suzanne Cook (**b**) Mehen mäkuuläm leaves and fruit. Photo Credit: Suzanne Cook (**c**) Mehen mäkuuläm leaves. Photo Credit: Suzanne Cook

Fig. 5.129 (**a**) Mehen 'oochin *Philodendron hederaceum*. Photo Credit: Suzanne Cook (**b**) Mehen 'oochin *Syngonium podophyllum*. Photo Credit: Suzanne Cook

Fig. 5.130 Mehen sukal *Saccharum officinarum*. Photo Credit: Suzanne Cook

Fig. 5.131 Mehen tak'lan che' *Phoradendron crassifolium*. Photo Credit: Suzanne Cook

Fig. 5.132 Mehen tuch *Thevetia ahouai*. Photo Credit: Suzanne Cook

mehen p'akan ground cherry. Solanaceae *Physalis* sp. *Use:* **hanal** 'food'; *Part:* **wich** 'fruit'; *Prep:* **che'che'** 'raw'; *Gen:* **p'akan (ah)**. *[Source:* AM*]* *[*\sd2 food*]*

mehen sukal sugarcane. *Lit:* 'small sugarcane'. Poaceae *Saccharum officinarum* L. *Use:* **hanal** 'food'; *Part:* **che'il** 'stalk'; *Prep:* **che'che'** 'raw'; *Sim:* **chäk sukal** 'red sugarcane'; *Sim:* **hach sukal** 'authentic sugarcane'; *Sim:* **ts'ibal sukal** 'striped sugarcane'; *Gen:* **sukal** (Fig. 5.130). *[Source:* AM; BM*]* *[*\sd2 food*]*

mehen tak'lan che' mistletoe. Viscaceae *Phoradendron crassifolium* (Pohl ex DC) Eichler. This is a kind of mistletoe that produces yellow fruits. It ranges from southern Chiapas to Bolivia, southern Brazil, and the Lesser Antilles (Job Kuijt, pers. comm., Kuijt 2003: 168–170). The Lacandones note that it kills trees, especially citrus species. *Loc:* **k'aax** 'forest'; *Sim:* **nukuch tak'lan che'** 'small tak'lah che' (*Struthanthus orbicularis*)'; *Gen:* **tak'lan che'** 'mistletoe' (Fig. 5.131). *[Source:* AM; BM*]*

mehen tuch thevetia, broadleaf (cascabel, dog's tongue). Apocynaceae *Thevetia ahouai* (L.) A. DC. *Sim:* **nukuch tuch** 'large tuch'; *Gen:* **tuch**. *See:* **tuch** (Fig. 5.132). *[Source:* BM*]*

mehen ts'us grape, Caribbean (water vine). Vitaceae *Vitis tiliifolia* Humb. & Bonpl., syn. *Vitis caribaea* DC.(?) The Lacandon distinguish a larger variety they call *hach ts'us* 'authentic ts'us'. As far as can be determined, it is the mature form of *mehen ts'us*, since the species observed had similar leaves and fruit. AM mentioned that he crushes the fruit in water and applies it to his hair to prevent it from turning grey. *Use:* **hanal** 'food'; *Part:* **wich** 'fruit'; *Use:* **habon ti' tso'otsel ho'ol** 'shampoo'; *Prep:* **mäxik uwich yok'ol ha'** 'crush the fruit in water'; *Sim:* **hach ts'us** 'authentic ts'us

Fig. 5.133 Mehen ts'us *Vitis tiliifolia*. Photo Credit: Suzanne Cook

(*Vitis tiliifolia*)'; *Gen:* **ts'us (ah)**. *See:* **hach ts'us**. *Thes:* **ak'** (Fig. 5.133). *[Source:* AM*]* *[*\sd2 other \sd2 food*]*

menta mint. Lamiaceae *Mentha citrata* Ehrh. *Use:* **ts'ak ti' kax** 'spice for chicken'. *See:* **ts'ak kax**. *From:* menta (Spanish). *[*\sd2 food*]*

miis palm, Panama hat. *Lit:* 'broom'. Cyclanthaceae *Carludovica labela* R. E. Schult., syn. *Asplundia labela* (R. E. Schult.) Harling. **Casi lo mismo que k'eben, chen ne chukuch ule'. Ne yaab uch'ibil ule'. Ma' ak'anik.** 'It's almost the same as *k'eben* (*Chamaedorea ernesti-augustii*), except that its leaves are longer and it has numerous veins. We don't use it.' (BM) *Carludovica labela* is a species of low-growing plants with a mass of short stems and corrugated leaves shaped like large fans growing from the roots. It may be mistaken for a palm, but at least four features set it apart: (1) veins always run to the end of the leaf, as opposed to terminating at the margins; (2) stems are always round in cross section, instead of flattened on top; (3) leaves have more than one main vein, instead of a single midrib; (4) petioles do not thicken where they meet the leaf, as opposed to palmate-leaved palms, which have a thickened ridge on the apex of the petiole (Shumway et al. 2009). In addition, the inflorescence is "conspicuously staminiod, with several, usually deciduous, spathes subtending the simple spadix and fleshy fruits with numerous, small seeds" (Croat 1978: 178) (Fig. 5.134). *[Note:* kum, miis [Itz.]. *Crysophila stauracantha* (Atran 1999; Atran et al. 2004); metché ca. *Carludovica labela* (Comisión Nacional de Áreas Naturales Protegidas 2006); nekambor. *Carludovica labela* (Durán 1999). Synonyms from The Plant List (2010). Version 1. http://www.theplantlist.org/. Retrieved February 2012.*]* *[Source:* BM*]*

mo'ak' (ah) vine, virility (snout bean). *Lit:* 'macaw vine'. Fabaceae *Rhynchosia pyramidalis* (Lam.) Urb., syns. *Dolicholus pyramidalis* (Lam.) Britton & Wilson, *Dolichos pyramidalis* Lam. **Uch'ihil ich paakche' kol, yetel choochel**

Fig. 5.134 Miis *Carludovica labela*. Photo Credit: Suzanne Cook

ak'. Apäk'ik. 'It grows in the fallow *milpa* with choochel ak'. You plant it.' (KP) Mo'ak is a twining vine with spikes of yellow, pea-like blossoms. These are followed by pods containing glossy, bi-coloured red and black seeds. Stems are 5 m (16.4′) or longer, which grow from a woody base. Found growing in fallow *milpas*, house gardens, and disturbed areas. Other *Rhynchosia* species in the area include: *R. erythrinoides* Schltdl. & Cham.; *R. longeracemosa* (Mart. et Gal.) Rose; *R. minima* (L.) DC.; and, *R. discolor* Mart. & Galeotti (Comisión Nacional de Áreas Naturales Protegidas 2006). *Loc:* **kolil nah** 'house garden'; *Use:* **uuh** 'necklaces'; *Part:* **nek'** 'seeds'. *Thes:* **ak'** (Fig. 5.135a, b). *[Note:* Botanical identification provided by Alfonso Delgado-Salinas, (pers. comm.). mo ak. *Rhynchosia pyramidalis* (Durán 1999).*]* *[Source:* BM; KM; KP*]* *[*\sd2 craft*]*

mulix [murix] lime, Mexican (key lime). Rutaceae *Citrus aurantifolia* (Christm.) Swingle, syns. *C. acida* Roxb., *C. lima* Lunan, *C. medica* var. *ácida* Brandis, *Limonia aurantifolia* Christm. This is a small, evergreen tree, 2–4 m (6.5–13′) high, armed with numerous, very sharp, axillary spines. It bears clusters of green, aromatic fruit that turns yellow when very ripe. Although called limón 'lemon', in Spanish, the fruit looks and tastes like a lime. Ubiquitous in Lacandon *milpas* and house gardens, the Mexican lime is native to the Indo-Malayan region. It was brought to Mexico and the Caribbean by the Spanish in the 16th century (Morton, J. 1987). Some Lacandones also cultivate the sweet lime (*Citrus limetta*), the Lacandon name of which is unknown. According to one Lacandon consultant, there is a *hach mulix* 'authentic mulix', which may refer to *Citrus aurantifolia*. More research is required. *Loc:* **kol, kolil nah** 'milpa, house garden'; *Use:* **uk'ul** 'beverage'; *Part:* **wich** 'fruit'; *Prep:* **uyäts'ik uk'ab i uhuytik yetel ha' yetel sukal** 'one squeezes the juice and stirs it with water and sugar'; *Use:* **ts'ak ti' se'em, säk se'em** 'medicine for the common cold, catarrh'; *Part:* **k'ab** 'juice';

Fig. 5.135 (a) Mo'ak' *Rhynchosia pyramidalis* seeds. Photo Credit: Chan K'in Daniel (b) Mo'ak' *Rhynchosia pyramidalis* vine. Photo Credit: Suzanne Cook

Prep: **chäkbil** 'boiled'; *Use:* **k'ik' ha'** 'sweating'; *Use:* **chäkwilil** 'fever'; *Part:* **uyits ule'** 'resin from the leaves'; *Prep:* **Apulik yok'ol awich, tsem, ho'ol** 'You put it on your face, chest, and head'. *Thes:* **che'**. *[Note:* mulix. lemon (Bruce 1975); mälish. *Citrus aurantifolia* (Nations and Nigh 1980); mulix. *Citrus aurantifolia* (Kashanipour and McGee 2004); ix päj limoon [Itz.]. *Citrus aurantifolia* (Atran et al. 2004). Synonyms from Morton (1987: 168).*] [Source:* AM*] [*\sd2 medicinal \sd2 beverage*]*

muluch' (ah)$_1$ [muluch', muruch', muuch'] fungi, bonnets. *Lit:* 'mound, pile'. Physalacriaceae *Oudemansiella* spp. **Muluch', yan säk muluch' yok'ol che'. Ahantik. Muluch'il lu'umeh es ek' muluch'. Ma' ahantik.** 'There is a white muluch', which grows in or on tree trunks. You eat it. The kind of *muluch'* that grows on the soil is called black *muluch'*. You don't eat it.' (KP) *Use:* **hanal** 'food'; *Prep:* **chäkbil yetel mulix yok'ol xiiw** 'cooked with lime in a broad leaf'; *Spec:* **ek' muluch'** 'black muluch' (*Oudemansiella* aff. *steffendii, Pluteus harrisii*)'; *Spec:* **säk muluch' (ah)** 'white muluch' (*Oudemansiella canarii*)'; *Sim:* **k'ayoch'** 'pink oyster mushoom (*Pleurotus djamor*)'; *Sim:* **xikin che'** 'wood ears (*Schizophyllum commune*)'; *Sim:* **lo'loh** '*Auricularia cornea*'. *[Source:* AM; BM; KM; KP*] [*\sd2 food*]*

muluch' (ah)$_2$ *Pluteus harrisii* (common name unknown). Pluteaceae *Pluteus harrisii* Murr. *Use:* **hanal** 'food'. *Variant:* **ek' muluch'**. *[*\sd2 food*]*

mumun che'$_1$ [mumun che', mumu che'] *Saurauia yasicae* (common name unknown). *Lit:* 'very tender tree'. Actinidiaceae *Saurauia yasicae* Loes., syn. *Saurauia belizensis* Lund. This is a small tree or large shrub with long, sagging branches, and alternate, petiolate leaves that cluster at the ends of branchlets. The leaves are fuzzy underneath, with numerous lateral veins diverging parallel to the midvein. Leaf margins are serrated. Inflorescences are panicles of terminal cymes (similar to lilacs). Blossoms have five overlapping sepals and white petals. The fruit is a soft, white or pale green berry with minute, brown seeds. No uses were reported. *Thes:* **che'** (Fig. 5.136b). *[Note:* Taxonomic identification provided by Rolando Pérez (pers. comm.).*] [Source:* KP; K'ayum Segundo*]*

mumun che'$_2$ giant groundsel. *Lit:* 'tender tree'. Asteraceae *Telanthophora grandifolia* (Less.) H. Rob. & Brettell, syn. *Senecio grandifolius* Less. This is a large shrub with reddish-purple stems covered with pale, blunt prickles. Leaves are lacerate (jagged, deeply cut edges) and glossy. It bears apical inflorescences with yellow florets and rays. No uses were reported. *Loc:* **petha'** 'lagoon' (Fig. 5.136a). *[Note:* Taxonomic identification provided by Billie L. Turner, University of Texas at Austin (pers. comm.). mumuche. Tetrochidium rotundatum (Nigh 2008: 239); baron che. *Senecio grandifolius* (Durán 1999).*] [Source:* AM*]*

mumun che'$_3$ [mumunche', mumuche'] gonzalagunia, thyrsoid. *Lit:* 'tender tree'. Rubiaceae *Gonzalagunia thyrsoidea* (Donn. Sm.) B. L. Rob., syns. *Gonzalagunia tacanensis* Lund., *Duggena thyrsoidea* (Donn. Sm.) Standl. *See:* **muxan che'$_2$**. *Thes:* **che'** (Fig. 5.136c). *[Note:* Botanical identification provided by Charlotte Taylor (pers. comm.). mumun che. *Psychotria uliganosa* (Durán 1999); amaman che. *Psychotria pubescens, P. horizontalis* (Durán 1999).*] [Source:* AM; BM; K'ayum Segundo*]*

mumun sikil squash seeds; summer squash. *Lit:* 'tender seed'. Cucurbitaceae *Cucurbita* sp., *C. pepo* L. ssp. *pepo*. *Use:* **hanal** 'food'; *Part:* **wich** 'fruit'; *Prep:* **chäkbil** 'boiled'; *Use:* **hanal** 'food'; *Part:* **nek'** 'seeds'; *Prep:* **k'elbil** 'toasted'. *See:* **sikil**. *[Note:* mumun sikil. tender green edible gourd (Bruce 1975); sikil [Yuc.]. pie-pan squash seeds. *Cucurbita moschata* (Bricker et al. 1998).*] [Source:* BM; KyP; KM*] [*\sd2 food*]*

Fig. 5.136 (**a**) Mumun che' *Telanthophora grandifolia*. Photo Credit: Suzanne Cook (**b**) Mumun che' *Saurauia yasicae*. Photo Credit: Suzanne Cook (**c**) Mumun che' (muxan che') *Gonzalagunia thyrsoidea*. Photo Credit: Suzanne Cook

muxan [muxan, muxän] zingiberales. Zingiberales (in part). *Muxan* refers to members in the Zingiberaceae, Marantaceae, Heliconiaceae families, and, according to Baer and Merrifield (1971), Cannaceae. The name optionally includes *xiiw*, e.g., *säk muxan xiiw, hach muxan xiiw*. *Use:* **ti' utep'ik ba'al, o'och** 'for covering things and food'; *Spec:* **säk muxan (ah)** 'white muxan (*Heliconia librata*)'; *Spec:* **hach muxan** 'authentic muxan (*Heliconia librata*)'; *Spec:* **ya'ax ch'ib muxan xiiw** 'blue/green spined muxan xiiw (*Calathea macrosepala*)'. *Variant:* **muxan xiiw**. *Thes:* **xiiw**. *[Note:* muxan [Itz.]. muxan palm. *Calathea* spp. (Hofling 1997); muxam. platanillo leaves. *Canna indica* (Baer and Merrifield 1971); säk muxan, ya'ax muxan [Itz.]. *Calathea lutea* (Atran et al. 2004); le'che' [Mop.]. *Calathea lutea* (in, Atran et al. 2004).*]* *[*\sd2 food*]*

muxan che'$_1$ dogwood (christmasbush). Euphorbiaceae *Alchornea latifolia* Sw. This is a tree or perennial shrub 8–25 m (30 m) tall, with a straight trunk up to 60 cm (2′) in diameter. The exterior bark is smooth or finely fissured and reddish brown with whitish splashes. Branches are spreading, sometimes pendulous. Leaves are 7–18 cm (2.7–7″) long and 9–28 cm (0.6–11″) wide, elliptic to oval, arranged in a spiral at the ends of the small branches. Three main nerves extend from the leaf base. Leaf margins are crenate (notched). Blades are dark green above and pale green underneath. Small, yellowish- or greenish-white flowers are borne in axillary shoots. These are followed by two-valved, brown capsules. In Chiapas, *Alchornea latifolia* occurs in secondary forest, pine and oak forests, lower montane rain forest, and tropical perennial and seasonal forests. It is abundant on the slopes of ravines and gorges between 400 m (1312') and 800 m (2624.6') in elevation. (Vázquez-Yanes et al. 1999). *Loc:* **k'aax** 'forest'. *Variant:* **mumun che'**. *Thes:* **che'** (Fig. 5.137). *[Note:* English common name from Grandtner (1997–2010). kusu che, musan che, ukun che. *Alchornea latifolia* (Durán 1999).*]* *[Source:* AM; KP*]*

Fig. 5.137 Muxan che' *Alchornea latifolia*. Photo Credit: Suzanne Cook

muxan che'$_2$ gonzalagunia, thyrsoid. *Lit:* 'calathea tree'. Rubiaceae *Gonzalagunia thyrsoidea* (Donn. Sm.) B. L. Rob., syn. *G. tacanensis* Lund. This is a shrub or small tree 3–6 m (9.8–19.7′) tall with stout branches covered with stiff hairs. Leaves are elliptic-oblong, acuminate at both ends, 20–23 cm (7.8–9″) long, lustrous green above and sparsely covered with fine, soft hairs beneath. Small groups of flowers are produced at the apical growth of the floral axes and further down the branches. Petals form a white and pink tube 14 mm (0.5″) long. The fruit is a red, four-celled, flattened berry about 8 mm (0.3″) in diameter. (Standley and Williams 1971: 86) When ripe in June and July, the fruits attract droves of birds (AM). The wood is extremely soft and burns quickly (BM). It is found growing in moist or wet, mixed mountain forest or thickets between 800 and 1500 m (2624–4921′) in elevation. No uses were reported. *Loc:* **k'aax** 'forest'. *Variant:* **mumun che'**. *Thes:* **che'** (Fig. 5.136c). *[Note:* sac oor che. *Gonzalagunia tacanensis* (Durán 1999). Botanical identification provided by David Lorence and Charlotte Taylor (pers. comm.). English name from Grandtner (1997–2010).*] [Source:* AM; BM*]*

N–n

naba' balsam of Peru. Fabaceae *Myroxylon balsamum* (L.) Harms var. *pereirae* (Royle) Harms. *Myroxylon balsamum* is a large 40 m (131′) tall tree with fragrant bark, racemes of white flowers with yellow stamens, and fruit pods containing a single seed. Its distributional range includes the Lacandon forest below 600 m (1968′) in elevation (Pennington and Sarukhán 2005). AM says that the Lacandones used to pulverize the bark and then roll it into cigars, to give the tobacco a pleasant taste and aroma. Nations and Nigh (1980: 23) and Nations (2006) also report this practice among the Lacandones. The plant is invoked in a *kunyah* (curing chant, healing prayer) that is recited to cure *chupul wich* 'swelling of the face' (AM). *Loc:* **Mensäbäk**; *Use:* **Uts'uts'ik yetel hach k'uuts** 'One smokes it with tobacco'. *Variant:* **nab che'**. *Thes:* **che'**. *[Note:* Other species mentioned in this *kunyah* are *k'uxub* 'annatto', *ya'axche'* 'ceiba', *axana* (type of bird), and sensi' '(green) jay'. naba. balsam. *Myroxylon balsamum* (Nations and Nigh 1980); nawa (Nations 1992, in Atran et al. 2004); (aj)nab'a' [Itz.]. balsam. *Myroxylon balsamum* (Atran et al. 2004; Lois 1998); ix nab'a [Mop.] *Myroxylon balsamum* (in, Atran et al. 2004); nab'a [Yuc.]. *Myroxylon balsamum* (in, Atran et al. 2004). Spanish name from March (1998).*] [Source:* AM; BM*] [*\sd2 other*]*

näl [näl, när] maize. Poaceae *Zea mays* L. In the mythology of many Mesoamerican cultures, the original people were made of maize, and was the first food that the gods created for them. The four colours of maize—red, black, white, yellow—represent the four cardinal directions in Mayan cosmology. The Lacandones regard maize as a living being; it has a heart and a spirit, they say. Thus, if a kernel should fall one must pick it up and keep it safe. One must take good care of it, as if it were a child. If it is thrown away it cries, thinking that it is unloved; but, if it is given to the hens, it is happy (KyP). In many Mayan cultures, maize is regarded as a protector against evil spirits, ghosts, and other malevolent forces. The Lacandones place ears in houses assumed to be haunted; parents, should they have to leave their infant in the house alone, place an ear of maize to watch over it. Similarly, hosts of ceremonies place two ears on the north and south ends of the *xikal* 'copal board' during rituals, to prevent the incense offerings from being carried away by spirits (Tozzer 1907: 128). Maize surpasses all other foods in importance, including beans. According to Lacandon elders, M and CK, maize is their sustenance, whereas all the other foods are simply snacks. Maize is eaten in a variety of ways, but predominantly in the form of gruels and tortillas. While several types of gruels are made with fresh maize, most corn-based foods are made with corn that has been boiled in quicklime (calcium oxide), which softens the grain and releases the available niacin, riboflavin, and proteins. This process is called nixtamalization, a word derived from Nahautl *nextli* 'ashes' tamalli 'tamal, corn dough'. Quicklime is obtained by burning the shells of river snails or from limestone found in dried beds of calcium rich rivers. Three traditional varieties of maize are grown in Naha', named according to the colour of the kernels: *ek'* 'black', *k'änk'än* 'yellow', and *säk* 'white'. There is also *chäk näl* 'red maize', which is considered a kind of white maize (AM, KP). Typically, the different coloured varieties belong to the same race (Perales et al. 2005: 950). According to Ortega, the races of maize grown by the Lacandones are likely *tepecintle, tuxpeño, nal tel, and olotón* (cited in Durán 1999: 98). Most of the data on Lacandon maize production comes from research in the southern Lacandon territory. Some of the varities cultivated there are related to *tepecintle*, while *tuxpeño* is common throughout the lowlands of Mexico, from Veracruz to Quintana Roo and Chiapas (Perales, pers. comm.). This is one of the most productive and successful Mexican races of white maize cultivated in the lowland tropics (Srinivasan et al.). Maize types related to *nal tel*, the original maize grown in Guatemala, are seldom found in their pure form today, but the influence of *nal tel* upon other races is evident throughout Middle America at a wide range of altitudes (Wellhausen et al. 1957). *Nal tel* is white or yellow-seeded. It is the preferred maize for fall and winter planting (Dr. Rafel Ortega, in Durán 1999: 98). Other types of maize in the area may be related to *Zapalole Chico* or *Z. grande*, but these are rare (Perales, pers. comm.). According to one Lacandon consultant, he and some of his neighbours also grow type of maize from Guatemala (race undetermined) that reaches 4 m (13′) high. They find that the kernels are softer than those of the Lacandon maize, making them easier to remove. *Use:* **hanal** 'food'; *Prep:* **säkan** 'corn dough'; *Prep:* **pokbil** 'roasted'; *Spec:* **k'änk'än näl** 'very yellow corn'; *Spec:* **ek' näl** 'black corn'; *Spec:* **säk näl** 'white corn'; *Spec:* **chäk näl** 'red maize'; *Gen:* **näl**. *[Note:* ixi'im [Itz.]. Zea mays (Atran et al. 2004); ixi'im [Mop.]. Zea mays (in Atran et al. 2004); ixi'im, sak tux [Yuc.]. Zea mays (in, Atran et al. 2004). According to BM, maize plants develop a disease referred to as *chauiste* (or *chahuistle*), a Nahuatl word that means to be assaulted by a disease. It can be anything with the property of an infectious agent or even a malignant spirit. The word is applied to both animals and plants. When applied to plants it usually refers to a fungal infection; but it could also be applied to other diseases in the plant (Dr. Filiberto Malagon, Laboratorio de Malariología Facultad de Medicina Universidad Nacional Autónoma de México, pers. comm.). According to BM, symptoms include premature yellowing of the leaves and stunted ears. The process begins at the top of the plant and quickly moves down, eventually killing the entire plant. The causes are unknown, but BM believes it is due to poor soil fertility. He takes this as a sign to rotate his *milpas.] [Source:* AM; BM; KM; KyP*]* *[*\sd2 food*]*

neeh k'ambul (ah) *Anthurium schlechtendalii* (bird's nest type). *Lit:* 'curassow tail'. Araceae *Anthurium schlechtendalii* Kunth. **Ya'ax utop', chäk unek'. Uyo'och ch'iich', ahxet'et' y ahk'ayum. Tipo ubak'che'. Una'akal yok'ol läche'. Ne ka'anal una'akal yok'ol che'. Ne chukuch ule'. Yan utop'. Cuando ne yi'h unek' ne ek'.** 'The flower is purple and the seeds are red. They provide food for birds, such as the red-throated ant-tanager (*Habia fuscicauda*) and k'ayum (undetermined).' (AM) 'It's a kind of epiphyte. It climbs up in very high up into the trees. The leaves are very long. It has flowers. When the seeds are very ripe and hard they are very black.' (BM) *Anthurium schlechtendalii* is type of bird's nest *Anthurium*. Bird's nest *Anthuriums* are distinguished by their rosettes of erect-spreading, lanceolate leaves and short petioles (leaf stems). They are usually epiphytic, but sometimes epipetric (growing on rocks). *Anthurium schlechtendalii* has erect leaves with wavy margins and prominent secondary veins, a spike-like spadix that is subtended by a maroon-brown spathe, and bright red fruit. (International Aroid Society). *Loc:* **k'aax** 'forest'. *[Note:* Botanical identification provided by Tom Croat (pers. comm.). ne kambor. *Carludovica labela* (Durán 1999); u-nej k'ämb'ul [Itz.]. *Anthurium schlechtendalii* (Atran et al. 2004).*] [Source:* AM; BM*]*

ne nuk k'ik'ni' balum Ardisia (type of). *Lit:* 'very large bleeding nose of the jaguar'. Myrsinaceae *Ardisia* sp. **Ket p'iis uk'ik'el balum uni' balum. Uchehik uwich, ne chäk. Chen uk'aba' uk'ik'el ubalum uni'. Ne chäk uwich, batak 2.54 cm (1″) uwaan. Chukuch unek'. Usool ne hay; unoy säk y ch'uhuk. Säk utop'. As mihin utop'. Huhump'el huhump'el utop' y uwich. Ma' umäk'ik yaab wa pachil uhook'ol uk'i'k'el ani'. Ch'ihil as wits.** 'It's the same as the k'ik'ni' balum. The fruit forms very red bunches. They're called 'the blood of the jaguar's nose'. The fruits are very red, about 2.54 cm (1″) long. The seeds are long. The peel is very thin. The nutmeat is white and sweet. The flowers are white and somewhat small. Each flower produces a fruit. You don't eat many of them, because if you do, your nose will bleed later. It grows in the hills.' (BM). *Loc:* **wits** 'hills'; *Use:* **hanal** 'food'; *Part:* **wich** 'fruit'; *Prep:* **che'che'** 'raw'; *Sim:* **mehen k'ik'ni' balum** 'small k'ik'ni''; *Gen:* **k'ik'ni' balum (ah)**. *Variant:* **nukuch k'ik'ni' balum, hach k'ik'ni' balum**. *[Source:* AM; BM*]* *[*\sd2 food*]*

nikte' (ah)$_2$ frangipani (pagoda tree, plumeria, temple tree, West Indian jasmine). *Lit:* 'flower-tree'. Apocynaceae *Plumeria rubra* L. *Spec:* **chäk nikte'** 'red flower tree (*Plumeria rubra* forma *rubra*)'; *Spec:* **säk nikte'** 'white flower tree (*Plumeria rubra*)'. *Morph:* **nik-te'**. *See:* **hach nikte'**. *Thes:* **che'**. *[Note:* nicté [Yuc.]. *Plumeria* sp. Lit. flower-tree. frangipani. The generic name for Plumeria, see Chac-nicte, Zac-nicte, Zabac-nicte (Roys 1931: 269); sak-nik-te' [Yuc.]. white frangipani (Bricker et al. 1998); nikte'

Fig. 5.138 Nikte' kisin *Bouvardia longiflora*. Photo Credit: Chan K'in Daniel

Fig. 5.139 Ni' ah'och *Anthurium* sp. Photo Credit: Suzanne Cook

[Itz.] *Plumeria rubra* (Atran et al. 2004).*] [Source:* BM; KA; SK*]*

nikte' kisin bouvardia, scented. Rubiaceae *Bouvardia longiflora* (Cav.) Kunth. **P'iis lo'obil. Yan uchan mehen che'il. Chan mihin ule', ne yaab uchan mehen k'äb. Utop', chan mihin. Säk utop'. Yan ubok, ne chichin. Cuando ne tantal uwaak'al utop', yan ubok yok'ol k'in, uhook'ol ubok' utop'.** 'It's like a weed. It has a small trunk. The leaves are small, with many small branches. The flowers are small and white. They have a slight fragrance. When the flowers open up in the sun, the perfume comes out'. (BM) According to AM, there is no name for this plant; it is referred to as Kisin's flower, because it is his perfume. Cultivated. *Loc:* **kolil nah** 'house gardens' (Fig. 5.138). *[Source:* AM; BM*] [*\sd2 folklore_and_mythology \sd2 other*]*

ni' ah'och *Anthurium huixtlense* (common name unknown). *Lit:* 'opossum's nose'. Araceae *Anthurium huixtlense* Matuda. This is a typical member of the Calomystrium section of Anthuriums, with its cordate leaves, prominent basal lobes, pink spadix, and light green spathe. No uses were reported. *Loc:* **k'aax** 'forest' (Fig. 5.139). *[Note:* Botanical identification provided by Tom Croat (pers. comm.).*] [Source:* AM*]*

ni' ts'ul deer eye (ox eye). *Lit:* 'foreigner's nose'. Fabaceae *Mucuna argyrophylla* Standl. **Uyits ule' ubanik uk'anil luts. Uyek'el p'iis kafe. Uchik umentik uk'anil luts yetel kih. Pachil, ubanik uk'anil yetel uyits ni' ts'ul.** 'Resin from the leaves is used to paint fishing lines. It is coffee-coloured. A long time ago, they made the fishing lines from agave fibre and then they painted the lines with the resin of "the foreigner's nose".' (BM) *Mucuna argyrophylla* is a woody, climbing vine. Like other *Mucuna* species, it bears waxy, pea-like blossoms, which dangle down in clusters from long pedicils. Flowers are greenish, yellowish, or slightly purple. The upper petal of the blossom and the blade are 2 cm (0.8″) wide and the keel is 4.5 cm (1.8″). Flowers are followed by large, flat pods, 15–25 cm (5.9–9.8) long and 4.5 cm (1.7″) wide, flat, and covered with a rusty-brown fur. Each pod contains two or three large, compressed, shiny black seeds. The hilum lacks the light colouration of other types of *Mucuna* seeds (Rojas-Chávez 2009). According to Levy et al. (2006), the southern Lacandones eat the seeds of *M. argyrophylla* and use the plant (parts not mentioned) in making dye. *Loc:* **k'aax, ubel** 'forest (secondary), road sides'; *Use:* **uuh** 'necklaces'; *Part:* **nek'** 'seeds'; *Use:* **Abanik kih, k'uch yetel** 'You paint agave or (cotton) thread with it'; *Part:* **uyits** 'resin'; *Prep:* **Ahuytik yetel säbäk yetel kibil kab** 'You stir it in with soot and wax'. *See:* **wich ahyuk (ah)**. *Thes:* **ak'** (Fig. 5.140). *[Source:* AM; BM; KP*] [*\sd2 craft*]*

nok'ol ik pepper, chile. *Lit:* 'worm chile'. Solanaceae *Capsicum frutescens* L. *Use:* **ts'ak ti' chi'** 'oral medicine'; *Part:* **wich** 'fruit'; *Prep:* **che'che'** 'raw'; *Use:* **hanal** 'food'; *Part:* **wich** 'fruit'; *Prep:* **che'che'** 'raw'; *Gen:* **ik**. *[Source:* Boremanse (1981: 204)*] [*\sd2 food*]*

nukuch bä'ch'ich' (ah) seagrape, Barbados. Polygonaceae *Coccoloba barbadensis* Jacq., syns. *C. barbadensis* var. *mexicana* Meisn., *C. shiedeana* Lindau, *C. mayana* Lund. *Loc:* **k'aax** 'forest'; *Use:* **hanal** 'food'; *Part:* **wich** 'fruit'; *Prep:* **che'che'** 'raw'; *Sim:* **mehen bä'ch'ich'** 'small bä'ch'ich' (*Coccoloba hondurensis*)'; *Gen:* **bä'ch'ich'**. *Thes:* **che'**. *[Note:* nukux bochich. *Coccoloba barbadensis* (Durán 1999); bolchich. *Coccoloba barbadensis* (Pennington and Sarukhán 2005). English name from Grandtner (2005: 227)*] [Source:* AM*] [*\sd2 food*]*

Fig. 5.140 Ni' ts'ul *Mucuna argyrophylla*. Photo Credit: Suzanne Cook

nukuch bits' Inga punctata (common name unknown). Fabaceae *Inga punctata* Willd., syn. *Inga leptoloba* Schltdl. This fruit tree has a small, cylindrical trunk about 20 cm (7.9″) in diameter and covered with dark, grey-brown bark. Profuse, thin limbs give rise to a thick, irregular crown. Raised, rust-colored, lenticels cover the surfaces of the small branches and twigs. Leaves are lanceolate, acuminate at the apex with a developed drip-tip about 20 × 12 cm (7.9 × 4.7″). They are pinnately compound and alternate, and occur in pairs of three or four. Large stipules, up to 8 mm (0.3″) long, are present at the base of the peduncles (flower stem). The flowers occur in dense, axillary spikes, resembling white pompoms. Fruits are flattened, smooth or slightly striated, green pods 4–20 × 1–2.8 cm (1.6–7.9 × 0.4–1.1″). Large, glossy brown seeds are surrounded by a gauzy, white aril, which is somewhat dry and very sweet. The fruit is similar to that of *I. spectabilis* in size and shape but with more seeds (>10). According to BM, this is the largest of all the *bits'*. *Loc:* **kol, kolil nah** 'milpa, house garden'; *Use:* **hanal** 'food'; *Part:* **wich** 'fruit'; *Prep:* **che'che'** 'raw'; *Sim:* **mehen bits'** 'small bits' (*Inga* sp.)'; *Sim:* **t'elen bits'** 'ridged bits' (*Inga sapindoides*)'; *Sim:* **suumil bits'** 'rope bits' (*Inga oerstediana*)'; *Sim:* **bits'il k'aax** 'forest bits' (*Inga belizensis*)'; *Sim:* **bits' petha'** 'lagoon bits' (*Inga spuria*)'; *Gen:* **bits'**. *Thes:* **che'** (Fig. 5.141a, b, c). *[Note:* mejen tete bits. *Inga punctata* (Durán 1999).*] [Source:* BM*] [*\sd2 food*]*

Fig. 5.141 (**a**) Nukuch bits' *Inga punctata* pod. Photo Credit: Suzanne Cook (**b**) Nukuch bits' *Inga punctata* sweet aril. Photo Credit: Suzanne Cook (**c**) Nukuch bits' *Inga punctata* seeds. Photo Credit: Suzanne Cook

nukuch box bean. *Lit:* 'large bean'. Fabaceae *Phaseolus vulgaris* L. *Variant:* **hach box**. *Thes:* **bu'ul**. *[Source:* AM*]* *[*\sd2 food*]*

nukuch bu'ulche' *Lit:* 'large bean tree'. Fabaceae *Cojoba* sp.(?). **Nukuch bu'ul che' uche'il 20–30 metros (65.6–98.4′). Ne mihin ule', batak 6.35 mm (0.25″). Utop', säk. Uwich, p'iis bu'ul. Ne tah uwich. Yan 6–7 unek'. Chäk usool unek'.** '*Nukuch bu'ul che'* is a tree 20–30 m (65.6–98.4′) in height. The leaves are very small, about 6.35 mm (0.25″) long. The flower is white. The fruit is like a bean. The fruit is very straight. There are 6–7 seeds (in a pod). The skin of the seeds is red.' (BM) The species may belong to *Cojoba*, although the pods of this genus are not straight. Two *Cojoba* species recorded from the northern territory are *C. arborea* and *C. graciliflora* (Comisión Nacional de Áreas Naturales Protegidas 2006). Both produce coiled or twisted pods. Further research is required. *Loc:* **xok'ol ya'ax petha'** 'near the green lagoon'; *Spec:* **mehen bu'ulche'** 'small bean tree'; *Gen:* **bu'ulche'**. *Thes:* **che'**. *[Source:* BM*]*

nukuch chäk k'o'och bean castor. *Lit:* 'large k'o'och'. Euphorbiaceae *Ricinus communis* L. *Sim:* **mehen chäk k'o'och** 'small castor bean'; *Gen:* **chäk k'o'och** 'castor bean'. *Thes:* **che'**. *[Source:* AM*]*

nukuch ch'alol oak, corrugated. *Lit:* 'large oak'. Fagaceae *Quercus corrugata* Hook. The name refers to the large sized acorn of this species. According to BM, the fruits have *ts'ibal* 'stripes'. *Loc:* **yaha'nah (La Cueva)**; *Use:* **polbilich** 'spinning top'; *Part:* **nek'** 'seed'; *Sim:* **mehen ch'alol** 'small oak (*Quercus* sp.)'; *Sim:* **pixan k'ambul (äh)** 'curassow's heart (*Quercus skinneri*)'; *Gen:* **ch'alol**. *Variant:* **hach ch'alol**. *Thes:* **che'**. *[Note:* English name from Grandtner 2005.*]* *[Source:* AM*]* *[*\sd2 other*]*

nukuch hut'kih *Lit:* 'large hut'kih'. Convolvulaceae *Ipomoea* sp.(?) *Use:* **ti' kinsik k'ik'** 'to vulcanize rubber'; *Part:* **uyits uts'u' uyak'il** 'resin from the pith of the vines'; *Sim:* **hach hut'kih** 'authentic hut'kih (*Ipomoea alba*)'; *Sim:* **chäk hut'kih** 'red hut'kih (*Ipomoea* sp.)'; *Gen:* **hut'kih**. *Thes:* **ak'**. *[Source:* AM*]* *[*\sd2 other*]*

nukuch ik coffee, wild (typical Panama balsamo). Rubiaceae *Psychotria panamensis* var. *panamensis* Standl. *Variant:* **yo'och ik nukuch**. *[Source:* AM*]* *[*\sd2 medicinal*]*

nukuch koyoh coyo persea (wild pear). *Persea schiedeana* Nees. A cultivated relative of *Persea americana* Mill., the fruit of this tree is large and long, with green, smooth skin and beige, stringy pulp that is dry and bland. *Use:* **hanal** 'food'; *Part:* **wich** 'fruit'; *Prep:* **che'che'** 'raw'; *Sim:* **hach koyoh** 'authentic koyoh'; *Gen:* **koyoh**. *Thes:* **che'** (Fig. 5.142). *[Source:* KM*]* *[*\sd2 food*]*

Fig. 5.142 Nukuch koyoh *Persea schiedeana*. Photo Credit: Suzanne Cook

Fig. 5.143 Nukuch k'än'ol *Mosquitoxylum jamaicense*. Photo Credit: Suzanne Cook

nukuch k'än'ol mahogany, bastard (wild mahogany, mosquito wood). Anacardiaceae *Mosquitoxylum jamaicense* Krug & Urb. This is a tree or large shrub. Leaflets number 11–17, vary in shape from obovate to oblong, rounded or obtuse at the apex, leathery, and almost smooth. Flowers are borne in large, axillary panicles. In October, they are followed by bright red berries, 8 mm (0.3″) long. The fruit attracts squirrels and all kinds of birds. The wood is pale reddish brown tinged with yellow. It is hard, heavy, and moderately durable (Standley and Record 1936: 226). The species lacks the caustic resin present in *mehen k'än'ol* (*Astronium graveolens*). No uses were reported, but, according to Durán's Lacandon consultants, the wood is used in construction, and honey and wax are harvested from bee hives found in the trees. *Sim:* **mehen k'än'ol** 'small k'än'ol (*Astronium graveolens*)'; *Gen:* **k'än'ol (äh)**. *Thes:* **che'** (Fig. 5.143). *[Note:* Common name from Standley and Record (1936) and the Smithsonian Tropical Research Institute, http://biogeodb.stri.si.edu/herbarium/species/13765/?page=11. Retrieved January 2012. nukux kan or. *Mosquitoxylum jamaicense* (Durán 1999).*] [Source:* BM*]*

nukuch k'o'och pumpwood (trumpet tree). *Lit:* 'large k'o'och'. Urticaceae *Cecropia* sp. *Sim:* **mehen k'o'och** 'small k'o'och'; *Gen:* **k'o'och (äh)** 'pumpwood (*Cecropia* spp.)'. *See main entry:* **k'o'och$_2$**. *Thes:* **che'**. *[Source:* AM*]*

nukuch le' si'si' k'uuts American burnweed (fireweed, pilewort). *Lit:* 'big leaf dry tobacco'. Asteraceae *Erechtites hieracifolia* (L.) Raf. ex DC. **Uch'ihil ich kol. Ya'ax ule', chukuch ule', batak 15.25 cm (6″). Säk utop', much'much' utop'. Päytan utop' pachil yan unek' utal ik' uyustik ubin. Ma' tsoy. Ulahtaxik uwich ich kol, uhook'ol ne yaab. Ule' p'iis k'uuts.** 'It grows in the *milpas*. The leaves are green and long, about 15.25 cm (6″). The flowers are white and clustered. First there are flowers and then there are seeds, which the wind blows all over the place. It's really bad. It spreads the seeds throughout the *milpas*. Its leaves are the same as tobacco.' (BM) From BM's description, the plant is likely *Erechtites hieracifolia*, a tall (2 m [6.5′]), annual grass with alternate, lanceolate leaves that are 10 (to 30) cm (3.9 [to12″]) long and up to 8 cm (3″) wide. The leaf margins are toothed. Stems are erect, solid, and slighly hairy. Inflorescences are panicles of 20–25 yellowish-green to white flowers, borne on the tips of the stems and at leaf axils. Each round flower head is 3–10 mm (0.1–0.4″) in diameter. Fruit is a nutlet contained in a feathery pappus. It is a pioneer plant that can become invasive. *Loc:* **paakche' kol** 'fallow milpa'; *Gen:* **si'si' k'uuts** '*Baccharis trinervis*'. *See:* **k'uuts**. *[Source:* BM; K'ayum Segundo*]*

nukuch 'oochin$_1$ *Philodendron sagittifolium* (common name unknown). Araceae *Philodendron sagittifolium* Liebm. According to my Lacandon consultants, this aroid produces sweet, red inflorescences that contain an irritating resin. AM says that, while one may eat the inflorescences borne lower down the stem, those higher up cause indigestion. Birds eat the fruit as well. *Philodendron sagittatum* grows throughout most of Central America in tropical forests and premontane wet rain forests, from near sea level to 1800 m (5500′). Normally a hemiepiphytic climbing vine, its distinctive features include its numerous, erect inflorescences born at each axil, 4–5 cm long (1.5–2″) long peduncle (the stem that supports the spathe), and the somewhat flattened, green spathe (a leaf-like bract that encloses the spadix [flower cluster]) with an interior that is often tinged red. However, because it is a morphologically variable and an ecologically versatile species in Central America, it can be difficult to identify. (Lusas, n.d.) *Philodendron* species, with their trifoliolate leaf-blades, are often mistaken for *Syngonium* (Tom Croat, pers. comm.). *'Oochin* generally refers to both *Syngonium* and *Philodendron* species, in Lacandon. Like the Lacandones, the Tzotzil Maya lump species from each genus into a single category called *noch'och'*. In the Tzotzil language, *noch'och'* not only designates *Philodendron* and *Syngonium* species, but also species of *Monstera* (Breedlove and Laughlin 1993: 118). It appears that the word derives from *noch'*, which means 'clinging (of a vine to a tree or cliff)' (Laughlin 1975: 415). In Lacandon, *'oochin* could be a metaphorical extension of 'opossum', with reference to the clinging behaviour of an opossum's offspring; but, AM conjectures that the name refers to the fact that opossums eat the fruits. *Use:* **hanal** 'food'; *Part:* **wich** 'fruit'; *Prep:* **che'che'** 'raw'; *Sim:* **hach 'oochin** 'authentic 'oochin (*Philodendron donnell-smithii*)'; *Sim:* **mehen 'oochin** 'small 'oochin (*Philodendron hederaceum, Syngonium podophyllum*)'; *Gen:* **'oochin**. *Thes:* **ak'** (Fig. 5.144a, b). *[Note:* AM also identified a related species which he calls *ya'ax nukuch 'oochin*. It is unclear whether or not this is a variety of *nukuch 'oochin*, since no other *nukuch 'oochin* varieties were identified. Taxonomic identification provided by A.C. Aceby and Tom Croat (pers. comm.).*] [Source:* AM*] [*\sd2 food*]*

nukuch 'oochin$_2$ *Syngonium chiapense* (common name unknown). Araceae *Syngonium chiapense* Matuda. *Sim:* **mehen 'oochin** 'small 'oochin'; *Sim:* **hach 'oochin** 'authentic 'oochin (*Philodendron donnell-smithii*)'; *Gen:* **'oochin** (Fig. 5.144c). *[Note:* Taxonomic identification provided by Tom Croat (pers. comm.).*] [Source:* AM; BM*]*

nukuch 'oochin$_3$ arrowhead vine (goosefoot plant). Araceae *Syngonium podophyllum* Schott. This is a climbing epiphytic vine with milky sap. Leaves are borne on petioles 10–60 cm (4–23″) long. Young leaves are heart-shaped, oblong, ovate, triangular, or hastate (arrowhead shaped). Mature leaves are deeply lobed, with 5–11 lobes. Inflorescences are fleshy spathes, often clustered. Green flowers are borne on a round column contained in the spathe, with male flowers on the upper column and female flowers lower down. The spathes turns bright red at maturity. Fruit is a berry. (Langeland et al. 1979) No uses were reported. *Sim:* **hach 'oochin** 'authentic 'oochin (*Philodendron donnell-smithii*)'; *Sim:* **mehen 'oochin** 'small 'oochin'; *Gen:* **'oochin** (Fig. 5.144d). *[Note:* Identification of species in the photo provided by Tom Croat (pers. comm.). ochi. *Syngonium podophyllum* (Durán 1999).*]*

nukuch 'oonte' nectandra (type of). *Lit:* 'large avocado tree'. Lauraceae *Nectandra* sp. *Loc:* **k'aax** 'forest'; *Sim:* **säk**

Fig. 5.144 (**a**) Nukuch 'oochin *Philodendron sagittifolium*. Photo Credit: Suzanne Cook (**b**) Nukuch 'oochin *Philodendron sagittifolium*. Photo Credit: Suzanne Cook (**c**) Nukuch 'oochin *Syngonium chiapense*. Photo Credit: Suzanne Cook (**d**) Nukuch 'oochin *Syngonium podophyllum*. Photo Credit: Suzanne Cook

'oonte' 'white 'oonte' (*Licaria alata*)'; *Sim:* **ek' 'oonte'** 'black 'oonte' (*Nectandra globosa*)'; *Sim:* **mehen 'oonte'** 'small 'oonte' (*Nectandra* sp.)'; *Sim:* **babah 'oonte'** '*Ocotea cernua*'; *Gen:* **'oonte'**. *Thes:* **che'**.

nukuch säk bahche' American muskwood. *Lit:* 'large white bahche''. Meliaceae *Guarea* sp., *G. grandifolia* (?). The species that my consultants called nukuch *säk bahche'* was a young tree with leaflets longer than those of *Guarea glabra*, pinnate with 5–7 elliptic-oblong leaflets per rachis (leaf stem). The trunk was straight with reddish-brown, relatively smooth bark, except for patches where the leaves had detached. Neither flowers nor fruit were present, making identification uncertain, but it could be *Guarea grandifolia*, which is reported from the area (Comisión Nacional de Áreas Naturales Protegidas 2006). No uses were mentioned.

Sim: **mehen säk bahche'** 'small white bahche''; *Gen:* **säk bahche'**. *Thes:* **che'**. *[Note:* nukuch sac bah che, nukuch sawache, puna, caoba, saquache. *Guarea glabra* (Durán 1999).*]* *[Source:* BM*]*

nukuch tak'lan che' mata rama (God bush). Loranthaceae *Struthanthus orbicularis* (Kunth) Eichler. *Loc:* **kolil nah** 'house garden'; *Sim:* **mehen tak'lan che'** 'small tak'lah che' (*Phoradendron crassifolium*)'; *Gen:* **tak'lan che'**. *Variant:* **iich'ak tolok**. *[Note:* English names from Croat (1978).*]* *[Source:* AM*]*

O–o

'oochin ['o(o)chin, 'o(o)ch] arum. Araceae. **Ak'. Yan uorquídeas. Nuk ule'. Ukinsik che', por ejemplo, china wa mankoh. Ne chich umuk. Ma' tsoy ti' che'. Chuwak uheek'el ule'. Uwich p'iis zanahoria. Chukuch i chäk-chäk. Yuul usool. Cuando täk'än ne chäk usool uwich. Ne ch'uhuk. Yan uyits, säk uyits. Uyol ne sak'. Uyo'och ah'ooch.** 'It's a vine. It has orchids. The leaves are very large. It kills trees, such as orange or mango. It's very strong. The lobes of the leaf blades are longish. The fruit looks like a carrot. It's long with smooth skin. When it's ripe, the skin is very red and sweet. It has a white resin. The resin in the growing tip makes your skin itch. It's food for opossums.' (BM) *'Oochin* refers to a number of *Syngonium* species and a few *Philodendron* species. According to AM, there are two kinds—*hach 'oochin* 'authentic 'oochin' and and *nukuch 'oochin* 'large 'oochin'. The description by BM may correspond to *Syngonium donnell-smithii. Syngonium donnell-smithii* and (*Monstera deliciosa*) produces sweet, edible fruit. *Spec:* **nukuch 'oochin** 'large 'oochin (*Syngonium podophyllum, S. chiapense, Philodendron sagittifolium*)'; *Spec:* **hach 'oochin** 'authentic 'oochin (*Philodendron donnell-smithii*)'; *Spec:* **mehen 'oochin** 'small 'oochin (*Philodendron hederaceum, Syngonium podophyllum*)'. *Thes:* **ak'**. *[Note:* ochi. *Syngonium podophyllum, Philodendron standleyi* (Durán 1999); ixoochil [Yuc.]. *Syngonium podophyllum* (in, Atran et al. 2004); ix pegap-eegaj [Itz.]. *Syngonium podophyllum* (Atran et al. 2004).*]* *[Source:* AM; BM; K'ayum Segundo; KM; KP*]*

'ooh ['ooh, 'oh] cane, giant (common reed). Poaceae *Gynerium sagittatum* (Aubl.) P. Beauv. **Hay dos clases: chäk 'ooh i hach 'ooh. Chäk 'ooh uch'ihil ti' uxuul uk'um. Ne yi'h uyool, chäk uyool. Uyiih yan habanil. Hach 'ooh ne yi'h. Yan uts'ibal ek' i ya'ax. Uch'ihil tuhunal ti' yahaha'. Uchik tupäk'ah uk'uuk', tucha'ah tupäk'ah. Uche'il ne tsoy ti' hiil ti' nah. Uyiih ti' häläl.** 'There are two kinds: *chäk 'ooh* 'red 'ooh' and *hach 'ooh* 'authentic 'ooh'. *Chäk 'ooh* grows on the river banks. The growing tip is hard and red. The flower spike is hollow. *Hach 'ooh* is very hard. It has black and green stripes. It has a white flower stalk. It grows on its own on the banks of the Monte Líbano River. A long time ago, the people took some shoots and planted them in Naha'. The stems are very good for house battens. The flower stalks are used for arrows.' (BM) According to BM, *'ooh* has replaced *hach häläl* 'authentic arrow' (*Phragmites australis*), because it is lighter and the arrows go farther. Nevertheless, according to AM, *hach häläl* made far superior arrows. *Gynerium sagittatum* is a rhizomatous reed-grass that forms dense stands along river banks and swampy areas. It has stout culms and long leaf blades that fan out at the top of the stems. de Kroon and Kalliola (1995) describe two varieties of *Gynerium sagittatum* that co-occur in the same areas. *Hach 'ooh* (a.k.a. *säk 'ooh* 'white 'ooh') and *chäk 'ooh* may correspond to these different varieties. When *Gynerium sagittatum* flowers in August, it is cut down and prepared for arrows. *Loc:* **Naha'**; *Loc:* **Yahaha'** 'Monte Libano River'; *Loc:* **Sa'am** 'Monte Libano'; *Use:* **häläl, yolil** 'arrows, arrow shaft'; *Part:* **uche'il utop'** 'flower stalk'; *Use:* **hiil** 'battens'; *Part:* **uche'il** 'stalks'; *Sim:* **hach häläl** 'common reed (*Phragmites australis*)'; *Spec:* **hach 'ooh** 'authentic 'ooh (*Gynerium sagittatum*)'; *Spec:* **chäk 'ooh** 'red 'ooh (*Gynerium sagittatum*)'. *Variant:* **häläl$_2$** (Fig. 5.145a, b, c). *[Note:* Botanical identification provided by Gilberto Cortes and Risto Kalliola (pers. comm.). oh. *Phragmites australis* (Nations and Nigh 1980; Nations 2006); oh'. reed (McGee 2002); oh [S. Lac.]. caña brava (Canger 1970).*]* *[Source:* AM; BM; KyP; SK*]* *[*\sd2 hunting_and_fishing \sd2 housing*]*

'oon avocado, American (alligator pear). Lauraceae *Persea americana* Mill., syn. *Persea gratissima* Gaertn. The Lacandones cultivate a red and a green variety of avocado. The fruit is round with relatively smooth skin and a substantial amount of creamy pulp. Three races of avocado evolved in different climatic environments in the Chiapas-Guatemala-Honduras region: the Mexican race evolved in the highlands of south-central Mexico; the Guatemalan race, in the valleys of the Guatemalan highlands; and, the West Indian race either originated in the Pacific lowlands or evolved from the Mexican race. The Mexican race is the hardiest and the source of most California avocados (Morton 1987). *Use:* **hanal** 'food'; *Part:* **wich** 'fruit'; *Prep:* **che'che' yetel mulix** 'raw with lime'; *Spec:* **chäk 'oon** 'red avocado'; *Spec:* **ya'ax 'oon** 'green avocado'. *See:* **'oonte'**; **koyoh**. *[Note:* oon. alligator pear (Bruce 1975); on. avocado. *Persea americana* (Nations and Nigh 1980); mehen on. avocado. *Persea gratissima* (Nations and Nigh 1980); 'om, aj/ix ya'ax 'om [Itz.]. *Persea americana* (Atran et al. 2004); 'òom [Yuc.] var. 'òon. avocado. *Persea americana* (Bricker et al. 1998); on [Mop.].

Fig. 5.145 (**a**) 'Ooh *Gynerium sagittatum* plants. Photo Credit: Suzanne Cook (**b**) 'Ooh *Gynerium sagittatum* leaves. Photo Credit: Suzanne Cook (**c**) 'Ooh Gynerium sagittatum flower stalks. Photo Credit: Suzanne Cook

Persea americana (in, Atran et al. 2004).*] [Source:* BM*] [*\ sd2 food*]*

'oonte' avocado tree. Lauraceae. The name designates a number of *Nectandra* species, *Ocotea cernua*, and *Licaria alata*. All are large trees that produce small, single-seeded, globose or oblong drupes in cupules (shallow cups). They look like minature avocado fruits, which explains the Lacandon name. Inflorescences are loose panicles of small flowers with white or greenish tepals. Eleven *Nectandra* species have been recorded from Naha': *N. ambigens; N. cissiflora; N. coriacea; N. cuspidate; N. globosa; N. loeseneri; N. membranacea; N. nitida; N. reticulate; N. salicifolia;* and, *N. sinuata*. Only one *Ocotea* species, *O. cernua*, has been recorded from the area. It is distinguished by its widely spaced secondary leaf veins and wavy leaf margins. The fruits are relished by a variety of birds. *Licaria alata* produces similar fruit to the other species, and is edible. *Spec:* **säk 'oonte'** 'white 'oonte' (*Licaria alata*)'; *Spec:* **ek' 'oonte'** 'black 'oonte' (*Nectandra globosa*)'; *Spec:* **mehen 'oonte'** 'small 'oonte' (*Nectandra* sp.)'; *Spec:* **nukuch 'oonte'** 'large 'oonte' (*Nectandra* sp.)'; *Spec:* **babah 'oonte'** (*Ocotea cernua*). *See:* **'oon**. *Variant:* **'oonche'**. *[Note:* on te *Nectandra* sp. (Nations 2006).*] [Source:* BM; SK*]*

'oop apple, custard. Annonaceae *Annona* spp. This is a general term for a class of round fruits with a leathery, warty peel and soft, sweet, pineapple-flavoured flesh surrounding large, hard, glossy, dark seeds. The Lacandones differentiate two cultivated species and three wild species. The cultivated species are *chäk 'oop* 'red custard apple' (*Annona reticulata*) and *säk 'oop* 'white custard apple' (*Annona cherimoya*). Together, they are referred to as *hach 'oop* 'authentic custard apple'. *Spec:* **hach 'oop** 'authentic custard apple'; *Spec:* **'oop ma'ax** 'spider monkey custard apple'; *Spec:* **'oopil k'aax** 'custard apple of the forest, wild custard apple'; *Sim:* **'oop che'** 'custard apple tree'. *Thes:* **che'**. *[Note:* oop. custard apple (Bruce 1975); op. *Annona cherimoya* (Nations and Nigh 1980); op. *Annona reticulata* (Nations 1992, in Atran et al. 2004); 'óop [Yuc.]. custard apple. *Annona squa-*

mosa (Bricker et al. 1998); 'oop, chäk 'oopil kaj, k'än 'oopil kaj, säk 'oopil kaj [Itz.]. *Annona reticulata* (Atran et al. 2004).*] [Source:* AM; BM; CKO*] [*\sd2 food \sd2 folklore_and_mythology*]*

'oopil k'aax ['oopil k'aax, 'opil k'aax] *Annona scleroderma* (common name unknown). *Lit:* 'wild annona'. Annonaceae *Annona scleroderma* Lam. According to my Lacandon consultant, SK, *'oopil k'aax* is similar to the cultivated custard apple (*hach 'oop* 'authentic 'oop'), only larger—approximately 20 cm (8″) in diameter. He describes the peel as greyish green, and the pulp as whitish brown and sweet. The taste is similar to *'oop*, but it has a stronger odor and a slightly grainy texture. The species SK describes may correspond to *Annona scleroderma*: Nations and Nigh note that *"k'ashi op"* corresponds to *Annona scleroderma* (1980: 21), which is the only wild *annona* species recorded in Naha' (Comisión Nacional de Áreas Naturales Protegidas 2006). However, the fruit of this species is much smaller than what SK describes. *Annona primigenia* is a kind of wild custard apple that the Itzaj call *'oopil k'aax* (Atran et al. 2004). Its fruit is heart shaped and blood-red. The Yucatec name for this species is *chujun óop* (Atran et al. 2004: 89). In Mopan, a language closely related to Itzaj, Yucatec, and Lacandon, this species is called *poxte'* (Atran et al. 2004: 89). *Poxte'* (or *poshe-te*) is a common Yucatec name for *Annona scleroderma*. More research is required to confirm the identity and range of the species that *'oopil k'aax* designates. *Loc:* **ts'u' k'aax** 'dense forest'; *Use:* **hanal** 'food'; *Part:* **wich** 'fruit'; *Prep:* **che'che'** 'raw'; *Sim:* **hach 'oop** 'authentic custard apple'; *Sim:* **'oop ma'ax** 'spider monkey custard apple'; *Gen:* **'oop** (Fig. 5.146). *[Note:* k'ashi op. wild annona. *Annona scleroderma* (Nations and Nigh 1980); opikax, anona. *Desmopsis stenopetala* (Durán 1999); 'oopil k'aax, ix ch'ejun k'aax [Itz.]. *Annona primigenia* (Atran et al. 2004).*] [Source:* AM; CKO; KM; SK*] [*\sd2 food*]*

Fig. 5.146 'Oopil k'aax *Annona scleroderma*. Photo Credit: Chan K'in Daniel

'oop che' custard apple tree. *Lit:* 'annona tree'. Annonaceae *Annona* spp. *Loc:* **kolil nah, kol** 'house garden, milpa'; *Spec:* **hach 'oop che'** 'authentic custard apple tree (*Annona* spp.)'; *Spec:* **chäk 'oop che'** 'red custard apple tree (*Clarisia biflora*)'; *Sim:* **'oop** 'custard apple (Annona spp.)'; *Gen:* **'oop che'** 'custard apple (*Annona* spp.)'. *Variant:* **'oop.** *[Note:* 'oopche' [Itz.]. *Cordia collococca* [Muñeco] (Atran et al. 2004).*] [Source:* AM; SK*]*

o'ot' ya' canistel (yellow sapote, egg fruit). Sapotaceae *Pouteria campechiana* (H.B.K.) Baehni, syns. *Lucuma campechiana* H.B.K., *Lucuma nervosa* A. DC. Native to southern Mexico, Yucatan, Belize, Guatemala, and El Salvador, *Pouteria campechiana* is an evergreen tree, 6–12 m (20–40′) tall, and in some cases, 30.48 m (100′). Leaves are slender, glossy, and acuminate at the base. Branches contain a gummy latex. Fruits vary in size and shape from round to oval, with deep, yellow-orange flesh with the consistency of a hard-boiled egg yolk. BM says the fruit is like *ha'as* (*Pouteria mammosa*). *Use:* **hanal** 'food'; *Part:* **wich** 'fruit'; *Prep:* **che'che'** 'raw'; *Sim:* **säk ya'** 'caimitillo (*Chrysophyllum mexicanum*)'; *Sim:* **hach ya'** 'sapodilla (*Manilkara achras*)'; *Sim:* **chäk ya'** 'red ya' (*Manilkara achras*)'; *Gen:* **ya'**. *Thes:* **che'**. *[Note:* ot ya. yellow zapote. *Pouteria campechiana* (Nations and Nigh 1980); hotlia wits. *Pouteria campechiana* (Durán 1999); k'änixte' [Itz.]. *Pouteria campechiana* (Atran et al. 2004); k'aniste' [Yuc.]. *Pouteria campechiana* (in, Atran et al. 2004).*] [Source:* AM; BM*] [*\sd2 food*]*

'oox ['oox, 'ox] breadnut, Mayan. Moraceae *Brosimum* spp. **Hay tres clases: hach 'oox, chäk 'oox, i ya'ax 'oox. Hach 'oox yan usool uwich k'än. Yuul uwich, wolis, batak 2.5 cm (1″). Chäk 'oox, mas mihin ke hach 'oox. As chäk-chäk. Yuul usool. Utahal ti' marso.** 'There are three kinds of *'oox*: authentic 'oox, red *'oox*, and green *'oox*. Authentic *'oox* has fruit with yellow skin. The fruit is smooth, round, and about 2.5 cm (1″) long. The fruit of red *'oox* is much smaller than that of authentic *'oox*. The skin is somewhat red and smooth. It ripens in March.' (BM) According to both AM and BM, the fruit of *ya'ax 'oox* is similar to that of *hach 'oox*, only it is larger and remains green at maturity. In Spanish, *'oox* is called *ramón*. It refers to a number of species of *Brosimum* and related members of Moraceae that produce large, edible fruits and starchy seeds. The protein-rich seeds were milled into flour for tortillas when stocks of maize ran low. For this reason, another Mayan name for *ramón* is *iximche'* 'maize tree'. The abundance of breadnut trees growing among the Mayan ruins suggests that they were cultivated by the ancient Maya, who likely stored the seeds as a hedge against starvation (Cook 1935: 616). Even today, there is some stigma to eating *ramón*. According to AM, *mäna' kolil, umalik 'oox* 'people too lazy to make *milpa* have to collect *ramón*.' The tree also provides latex, with which the Lacandones used to plug cavities in their teeth. According to other sources, the Lacandones use the wood for construction (Diemont 2006: 84). Because the timber produces a hot flame and leaves very little ash, the wood is preferred for burning the shells of river snails, which are then slaked and used to nixtamalize maize kernels (Baer and Merrifield 1971: 153). The botanical identity of each variety of *'oox* is a challenge to sort out. According to Rätsch (1985a) and Nations (2006), *hach 'oox* and *ya'ax 'oox* correspond to *Brosimum alicastrum*, whereas Nations and Nigh (1980) contend that *ya'ax 'oox* corresponds to *Trophis racemosa*. In the forest industry, the term *ramón* refers to *Trophis* spp., not *Brosimum alicastrum* (Chudnoff 1984, cited in Simpson 1994: 1, fn. 1). Burger (1962: 31) notes that *"cha cox"* corresponds to *Trophis racemosa* ssp. *ramon* (Schlecht. & Cham.) W. Burger, whereas Atran et al. (2004) say it corresponds to *Brosimum alicastrum. Brosimum alicastrum* shares many similarities with *Trophis racemosa*, especially in fruit, which may account for the confusion. However, *Brosimum alicastrum* can be distinguished by its leaves, which are neither deeply lobed nor serrated compared to those of *T. racemosa*, and by its secondary veins and collecting vein, which are more even and symmetrical (Enquist and Sullivan 2001: 65). Numerous species of *Brosimum* have been recorded from the northern Lacandon territory, including *B. alicastrum, B. costaricanum, B. lactescens*, and *B. panamense* (Comisión Nacional de Áreas Naturales Protegidas 2006). Of these, *B. alicastrum* naturally occurs in Mensäbäk and other locations that are lower in elevation than Naha' (Miguel Garcia, pers. comm.). However, *B. alicastrum* is cultivated in Naha' without difficulty. *Use:* **hanal** 'food'; *Part:* **wich** 'fruit'; *Prep:* **che'che'** 'raw'; *Use:* **hanal** 'food'; *Part:* **nek'** 'seeds'; *Prep:* **Achäkik uluksik usool yetel k'ulta'an, ta'anil k'ak', pachil uhuch'ik yetel wah** 'You boil them with quicklime (or) wood ash, to remove the skins, and then grind the seeds with tortilla dough'; *Use:* **ts'ak ti' koh, ti' unok'ol ukoh** 'medicine for teeth, for caries'; *Part:* **uyits usool** 'latex from the bark'; *Prep:* **Axaak'tik uyits yetel ch'ooch' yetel tämän, pachil apulik ich ukoh** 'You mix the latex with salt and cotton, and then you put it in the tooth'; *Spec:* **hach 'oox** 'authentic 'oox (*Brosimum alicastrum*)'; *Spec:* **chäk 'oox** 'red 'oox'; *Spec:* **k'än 'oox** 'yellow 'oox (*Brosimum alicastrum*)'; *Spec:* **ya'ax 'oox** 'green 'oox (*Brosimum alicastrum*)'. *[Note:* hach ox, ya'ax ox. *Brosimum alicastrum* (Rätsch 1985a); ox. *Brosimum alicastrum* (Diemont 2006); ox. *Brosimum alicastrum* (Nigh 2008); ya'ash osh, hach osh. *Brosimum alicastrum* (Nations 2006); ya'ash osh. *Trophis racemosa* (Nations and Nigh 1980); chäk 'oox, säk 'oox, k'än 'oox [Itz.]. *Brosimum alicastrum* (Atran et al. 2004); aj ya'ax 'oox [Itz.]. *Trophis racemosa* (Atran et al. 2004); ya'ax ox, kupusi' [Mop.]. *Trophis racemosa* (in, Atran et al. 2004); ya'ax ox, chak ox [Yuc.]. *Trophis racemosa* (in, Atran et al. 2004).*] [Source:* BM*] [*\sd2 food*]*

P–p

pach' pineapple. Bromeliaceae *Ananas comosus* (L.) Merr. An introduced species. *Loc:* **kol** 'milpa'; *Use:* **hanal** 'food'; *Part:* **wich** 'fruit'; *Prep:* **che'che'** 'raw'. *[Note:* p'ach[sic]. *Ananas sativus* (Nations 1992, in Atran et al. 2004); pach' [S. Lac.]. piña (Canger 1970); piinyaj [Itz.]. *Ananas sativus* (Atran et al. 2004).*] [Source:* BM; BMjr*] [*\sd2 food*]*

pahok (ah) [pahok, paho] palm, monkeytail (shadow palm). Arecaceae *Geonoma oxycarpa* Mart., syn. *G. interrupta* var. *interrupta* (Ruiz & Pav.) Mart.*; Synechanthus fibrosus* H. Wendl.(?) **Yan uche'il. Ne chich uche'il. Uk'anik ti' upäk'ik ti' uxul. Ma' uk'i'ixel ule'. Ule' chukuch ule'. Ule' ti' umentik nah. Ne ki' uyol. K'unche pahok as ne nuk. Ya'ax utop'. Ne nuk uyool. Uch'ihil tu' ne 'äkan. Hach pahok, as ts'ets'el, as mihin. Ya'ax utop'. Uche'il batak 7.6 cm (3″). Ule' batak uwaan u le' 76 cm (2.5′). Ule' ne tsoy ti' nah. Hach pahok uch'ihil tu' ne 'äkan. Chäk pahok, chäk utop'. Uwich, chäk. Chäk usool uyol. Uhantik uyol ha'li' chan mihin. Uch'ihil ich wits.** *'Pahok* has a stalk. The stalk is very hard. One uses it as a planting stick. The leaves do not have any spines. The leaves are long. The leaves are used for houses. The heart is delicious. *K'unche' pahok* is a little bigger. Its flowers are green. Its heart is very large. It grows in the low lying flats. *Hach pahok* is a little bit smaller [than *k'unche' pahok*], it's somewhat small. Its flowers are green. It has a stalk about 7.6 cm (3″) in diameter. Its leaves (leaflets) are about 76 cm (2.5′) long. The leaves are very good for houses. *Hach pahok* grows in very low, level ground. *Chäk pahok* has a red flower. Its fruit is red. The skin of the shoots is red. One can eat the shoots only when they are very small. It grows in the hills.' (BM) *Pahok* refers to three kinds of low, understory palms whose leaves are used for thatch. The Lacandon names for each—*hach pahok, k'unche' pahok*, and *chäk pahok*—suggest that these palms are distinct. *K'unche' pahok* refers to *Geonoma oxycarpa.* According to Henderson et al. (1995), it is the only *Geonoma* species in Chiapas. The taxonomic identities of the other two are more difficult to determine. In the extant Lacandon literature, *pahok, hach pahok*, and *chäk pahok* are simply called *pahok* and glossed as *guatapil* (*Chamaedorea* sp.). However, approximately 25 *Chamaedorea* species occur in the Lacandon forest, 6 of which the Lacandones refer to as *bo'oy* (*ch'ibix bo'oy, ch'it bo'oy, hach bo'oy, k'än bo'oy, mehen bo'oy*, and *säk bo'oy*), and 2 others are *k'eben* and *ya'ax hänan*. Moreover, *Geonoma* is distinguished from *Chamaedorea* by the presence of pits in the flower stems, a feature which *Chamaedorea* species lack. In addition, as high levels of polymorphism (multiple structures) are evident within the *Geonoma oxycarpa* (Stauffer et al. 2002: 220), the Lacandon varieties of *pahok* may refer to these various forms. As for *chäk pahok* 'red pahok', it might be *Synechanthus fibrosus*, a species that is often mistaken for *Chamaedorea* (Henderson et al. 1995: 79) and which resembles *Geonoma oxycarpa*, save for its red fruit. This species is omitted in the botanical inventories from the Lacandon forest, but according to the distributional map in Henderson et al. (1995) and its environmental preferences, it should be found in the area. More investigation is required to determine the botanical range of *pahok*. According to previous researchers in the area, the Lacandones chew a peeled stem for a number of days to reduce a fever (Kashanipour and McGee 2004: 60). The Lacandones are also reported to have made barbed foreshafts from the stems for fishing spears (Nations 2006). The leaves are ideal thatching material, since, according to BM, they provide a lot of coverage and thus are quicker to assemble than the leaves from other species, such as *kun* (*Cryosophila stauracantha*). *Pahok* is one of the species invoked in the 'Song of the Flint', which a Lacandon hunter recites when breaking in a new antler punch to chip flint. *Loc:* **k'aax** 'forest'; *Use:* **pix nah** 'thatch'; *Use:* **pixil chem** 'canoe cover'; *Part:* **le'** 'leaves'; *Use:* **hanal** 'food'; *Part:* **uyol** 'heart'; *Prep:* **pokbil** 'roasted'; *Use:* **xuul** 'planting stick'; *Part:* **uche'il** 'stalk'; *Spec:* **chäk pahok** 'red pahok'; *Spec:* **k'unche' pahok** '*Geonoma oxycarpa*'; *Spec:* **hach pahok** 'authentic pahok'. *Variant:* **hach pahok**. *[Note:* pahoh/pahok. guatapil (Bruce 1975); pahok. guatipil [sic]. *Chamaedorea* spp. (Kashanipour and McGee 2004); paho. Guatapil palm. *Chamaedorea* sp. (Nations 2006); paho. *Chamaedorea* sp. guatapil palm (Nations and Nigh 1980).*]* *[Source:* AM; BM; Bruce (1976: 72; 1975: 205)*]* *[*\sd2 housing \sd2 food \sd2 tools*]*

pahsa' ak' (äh) [pahsa' ak', pähsäk'in, pasa' ak', päse' ak', päsa' ak', päsak'] ginger, spiral. Costaceae *Costus pulverulentus* C. Presl. **Lati' uts'ak ich ähchechem.** 'That is medicine (to treat eyes) affected by the *chechem* resin.' (AM) *Pahsa' ak'* is a short bush, approximately 0.9 m (3′) tall. Long, shiny, succulent leaves approximately 30 cm (12″) long and 10 cm (4″) wide fan out in a spiral. The spongy, reddish-brown stems are banded with green rings. Large, spiky, crimson bracts that resemble pine cones are born at the ends of the stems. From between the bracts emerge large, white, crepe-like labella. The family Costaceae traces its origins to Africa and contains four genera. The genus *Costus* includes 100 species, 51 of which grow in the Neotropics. Seven species are reported from Chiapas: *Costus bakeri* Schumann; *C. guanaiensis* Rusby var. *macrostrobilus* (K. Schum.) Maas; *C. pulverulentus* C. Presl.; *C. ruber* Griseb.; *C. sanguineus* J. D. Smith; *C. spicatus* (Jacq.) Sw. Shapiro, and *C. villosissimus* Jacq. *Loc:* **k'aax** 'forest'; *Use:* **ts'ak ti' chechem** 'medicine for chechem (sap) (that has fallen into the eyes)'; *Part:* **uyits che'il** 'sap from the stems'; *Prep:* **Ha'li' ak'upik uche'il i ach'ahik uyits yok'ol uwich** 'You just cut the stem and drip the resin into the eyes'; *Use:* **ts'ak ti' sak'il ti' 'oochin** 'medicine for itchiness from contact with the 'oochin plant'; *Part:* **uyits** 'resin'; *Prep:* **Mäna'** 'None'; *Use:* **uk'ul** 'beverage'; *Part:* **uyits** 'resin'; *Prep:* **Mäna'** 'None'. *Thes:* **ak'** (Fig. 5.147). *[Note:* päsak. *Costus speciosus* (Kashanipour and McGee 2004; March 1998); nukuch pasa. *Costus spicatus* (Nations and Nigh 1980); pasai. *Costus pulverulentus* (Durán 1999); nukuch pasa' [Itz.]. caña cristo. *Costus pictus* (Atran 1999). Species identification provided by Paul Maas (pers. comm.). *[Source:* AM; BM; K'ayum Segundo; KyP; SK*]* *[*\sd2 medicinal \sd2 food*]*

Fig. 5.147 Pahsa' ak' *Costus pulverulentus*. Photo Credit: Suzanne Cook

pakay stinktoe (sausage tree, cuapinol). Fabaceae *Hymenaea courbaril* L. *Use:* **tukil** 'spindle whorl'; *Part:* **ubox unek'** 'seed shell'; *Use:* **hanal** 'food'; *Part:* **wich** 'fruit'. *[Note:* English common names from Stross (1997). pakai. *Hymenaea courbaril* (Nations and Nigh 1980).*]* *[Source:* Nations and Nigh (1980); SK*]* *[*\sd2 craft \sd2 food*]*

pakyon (äh) [pakyon, pakyo'] palm, pacaya. Arecaceae *Chamaedorea tepejilote* Liebm. **P'isil ahch'i'ib chen as nuk uche'il i uyaal. Cinco metros ubakel. Uchuch, chäk. Ahantik uyaal. Cuando uyaal uwaakal, ne yan uchan uwich. Cuando ne muun, säk. Uche'il uch'ihil tuhunal, hump'eli' uche'il.** 'It's the same as *ch'i'ib*, except for its larger stalk and buds. It grows about 5 meters tall. The peduncle is red. You eat the buds. When the buds burst open there are numerous little fruits. When they are young they are white. It grows by itself. It has one stem.' (BM). According to AM, *pakyon* and *ch'i'ib* are brothers. The fruit of *pakyon* sets in October, earlier than *ch'i'ib*. *Chamaedorea alternans* is similar to *C. tepejilote*, but differs in that it has

Fig. 5.148 (**a**) Pakyon *Chamaedorea tepejilote* plant. Photo Credit: Suzanne Cook (**b**) Pakyon *Chamaedorea tepejilote* ripe berries. Photo Credit: Suzanne Cook (**c**) Pakyon *Chamaedorea tepejilote* male inflorescence. Photo Credit: Suzanne Cook (**d**) Pakyon *Chamaedorea tepejilote* male inflorescence. Photo Credit: Suzanne Cook

multiple inflorescences at a single node and is mostly or always solitary, whereas *C. tepejilote* has both solitary and suckering forms (Don Hodel, pers. comm.). Although a wild forest species, *pakyon* is also cultivated. *Use:* **hanal** 'food'; *Part:* **uyaal** '(male) inflorescence buds'; *Prep:* **pokbil, chäkbil, tsähbil yetel ye'el kax** 'roasted, boiled, fried with eggs' (Fig. 5.148a, b, c, d). *[Source:* AM; BM; KM; KP*]* *[*\sd2 food*]*

papas potato. Solanaceae *Solanum tuberosum* L. *Use:* **hanal** 'food'. *From:* papas (Spanish). *[Note:* papa [Yuc.]. *Solanum tuberosum* (in, Atran et al. 2004); paapaj [Itz.]. *Solanum tuberosum* (Atran et al. 2004).*]* *[*\sd2 food*]*

pap uwi' ginger. *Lit:* 'spicy tuber'. Zingiberaceae *Zingiber officinale* Roscoe. *Use:* **ts'ak** 'medicine'; *Part:* **wi'** 'rhizome'; *Prep:* **Achäkik ich ha' pachil awuk'ik** 'You boil it in water and then you drink it'. *Variant:* **sensi'**. *[Note:* pap uwi'. *Zingiber officinale* (Nations and Nigh 1980); 'enjiib'lej [Itz.]. *Zingiber officinale* (Atran et al. 2004); enji'le [Mop.]. *Zingiber officinale* (in, Atran et al. 2004); enjib'le [Yuc.]. *Zingiber officinale* (in, Atran et al. 2004).*]* *[Source:* KM*]* *[*\sd2 medicinal*]*

patoh ak' (ah) [patoh ak', patah ak'] nickernut. *Lit:* 'duck vine'. Fabaceae *Caesalpinia major* (Medik.) Dandy & Exell, syn. *Bonduc majus* Medik., *Caesalpinia globulorum* Bakh.f. & P. Royen. This is a kind of sprawling shrub with stout branches armed with sharp, recurved thorns. Bipinnate leaves are also prickly. Racemes of small, yellow flowers produced in leaf axils are followed by clusters of spiny, two-seeded pods. Seeds are large, 3 by 2.5 cm (1.25×1″), flattened, glossy, and brown. It is native to Madagascar and throughout tropical Asia, the Pacific, and the Neotropics. Seeds are collected for jewellery and tied to the wrists of babies, to protect them from the "evil eye". *Loc:* **k'aax** 'in the forest'; *Use:* **uuh** 'necklaces'; *Use:* **ti' 'aak'ä' ich** 'for the 'evil eye'. *Thes:* **ak'** (Fig. 5.149). *[Note:* jai patan ak. *Senna racemosa* (Durán 1999); xpàatoh-'àak' [Yuc.]. duck vine. *Centrosema schottii* (Bricker et al. 1998: 210).*]* *[Source:* CNK; KP*]* *[*\sd2 craft \ sd2 folklore_and_mythology*]*

Fig. 5.149 Patoh ak' *Caesalpinia major*. Photo Credit: Suzanne Cook

pähk'ol [pähk'ol, pohk'ol] *Spathiphyllum phryniifolium* (common name unknown). Araceae *Spathiphyllum phryniifolium* Schott. This is a large verdant, clumping plant found growing in the lagoon and soggy soils in deep shade. It has wide, glossy, medium-green leaves and bears a yellow or cream-coloured spadix. The spathe is green, lanceolate, acuminate at apex, pinnately veined. Margins are undulate (wavy). *Use:* **hanal** 'food'; *Part:* **top'** 'flower'; *Prep:* **mukbil wah, wa tsähbil yetel uye'el kax, wa pokbil, wa mak'bil che'che'** 'baked inside a tamale of corn dough, or fried with eggs, or roasted, or eaten raw'; *Use:* **päk'äch** 'surface to pat out tortillas'; *Part:* **le'** 'leaf'. *Thes:* **xiiw** (Fig. 5.150a, b). *[Note:* The spadices are also eaten in Quezaltenango and elsewhere in Guatemala (Standley and Steyermark 1952: 352–353); poko. *Spathiphyllum phryniifolium* (Durán 1999).*] [Source:* BM; SK*] [*\sd2 food*]*

pähpähche' fat pork. *Lit:* 'sour sour tree'. Rubiaceae *Chione chiapasensis* Standl., syn. *C. venosa* var. *venosa*. This is a small tree with a slender trunk with light brownish-grey bark splotched with white patches. Branches are thin and crooked, displaying an overall rangy habit with an uneven crown. Leaves are alternate, lanceolate, acuminate, acute at apex. Leaf margins are entire or dentate, pinnately nerved, with the midrib raised underneath. The lamina is relatively smooth and coriaceous. Clusters of fuzzy, oval to round fruit are borne at the ends of branches. No uses were indicated. *Loc:* **petha' Naha'** 'Lake Naha''. *See:* **mehen ch'eel koox**. *Thes:* **che'** (Fig. 5.151a, b). *[Note:* popojche. *Cordia stellinifera* [sic] (Nigh 2008); pupu che. *Ardisia nigrescens*; pupute. *Ardisia compressa* (Durán 1999); papa che. *Photinia microcarpa* (Durán 1999). English common names from Grandtner (2005).*] [Source:* BM; KP*]*

Fig. 5.150 (**a**) Pähk'ol *Spathiphyllum phryniifolium* plants. Photo Credit: Suzanne Cook (**b**) Pähk'ol *Spathiphyllum phryniifolium* inflorescence. Photo Credit: Suzanne Cook

Fig. 5.151 (**a**) Pähpähche' *Chione chiapasensis* tree. Photo Credit: Suzanne Cook (**b**) Pähpähche' *Chione chiapasensis* fruit and leaves. Photo Credit: Suzanne Cook

Fig. 5.152 (**a**) Pähpäh xibil *Ardisia* sp. fruit and leaves. Photo Credit: Chan K'in Daniel (**b**) Pähpäh xibil *Ardisia* sp. fruit and leaves. Photo Credit: Suzanne Cook

pähpäh xibil [pähpähxibih, pähpähchibi, pähpähxiwbih, pähpähxirih] *Ardisia* sp. (common name unknown). *Lit:* 'very sour leaf'. Myrsinaceae *Ardisia* sp. This is a forest tree that bears clusters of sour fruit resembling grapes. According to BM, the fruit matures to reddish purple in September. AM says that eating the fruits prevents hair from turning grey. *Use:* **hanal** 'food'; *Part:* **wich** 'fruit'; *Prep:* **che'che'** 'raw'; *Use:* **xuul** 'planting stick'; *Part:* **k'äbche'** 'branches'. *See:* **ts'us (ah)**. *Variant:* **pähpähxilil**. *Thes:* **che'** (Fig. 5.152a, b). *[Note:* popo xibi. *Ardisia paschalis* (Durán 1999); popo shibi. *Ardisia paschalis* (Comisión Nacional de Áreas Naturales Protegidas 2006); (ix) xilil [Itz.]. *Ardisia paschalis* (Atran et al. 2004); xilil [Mop.]. *Ardisia paschalis* (in, Atran et al. 2004); ton che' [Yuc.]. *Ardisia paschalis* (in, Atran et al. 2004).*] [Source:* AM; BM; KP; SK*] [*\sd2 food \sd2 tools \sd2 other*]*

pähpox [pähpox, pohpox, popoox] soursop. *Lit:* 'sour annona'. Annonaceae *Annona muricata* L. This is a low-branching and bushy evergreen tree about 7.5–9 m (25–30′) that bears large, oval, heart-shaped, or slightly elongated, prickly fruit ranging from 10–30 cm (4–12″) long, up to 15 cm (6″) wide, and weighing between 4.5 and 6.8 kg (10–15 lbs). It is similar to the custard apple, with its white, fibrous, tangy-sweet segments, each encasing a large, black seed. Flowers and fruit develop from the trunk and larger branches. In Naha', the fruit ripens in August. Cultivated. *Use:* **hanal** 'food'; *Part:* **wich** 'fruit'; *Prep:* **che'che'** 'raw'. *Thes:* **che'** (Fig. 5.153a, b, c, d). *[Note: Pox* has only been registered in this compound. päh pox. wild guava (Bruce 1975); p'op'osh. *Annona muricata* (Nations and Nigh 1980); p'opox [Yuc.]. *Annona muricata* (in, Atran et al. 2004); gwanaab'anaj [Itz.]. *Annona muricata* (Atran et al. 2004); pòosh [Yuc.]. *Annona squamosa* (Bricker et al. 1998); pox [Yuc.]. *Annona cherimola* Mill. (Standl. 1920–26, p. 283; Seler, 1902–08, III, 568); zuli-pox [Yuc.]. a certain large and spiny fruit, a species of annona (Motul.) (Roys 1931: 279, 317); poxte' [Mop.]. *Annona primigenia* (in, Atran et al. 2004).*] [Source:* KM; KP*] [*\sd2 food*]*

päh ak' (äh) *Disciphania calocarpa* (common name unknown). *Lit:* 'sour vine'. Menispermaceae *Disciphania calocarpa* Standl. This is a liana that bears hanging bunches of oval, glossy-red fruits the size of grapes. The fruits are very sour. It grows in Mexico, Guatemala, Costa Rica, Nicaragua, Belize, and Honduras. *Loc:* **k'aax** 'forest'; *Use:* **hanal** 'food'; *Part:* **wich** 'fruit'; *Prep:* **che'che' yetel ch'ooch'** 'raw with salt'. *Thes:* **ak'** (Fig. 5.154a, b). *[Note:* pa ak. *Disciphania calocarpa* (Durán 1999)*] [Source:* AM; BM; CXK; KP*] [*\sd2 food*]*

päh pätan (äh) banana, Orinoco (tall). *Lit:* 'tart banana'. Musaceae *Musa acuminata* ×*Musa balbisiana* Coll. This variety of banana is angular, stubby, and fat, with a thick skin that ripens to a golden yellow. The plant is approximately 2.26 m (14′) tall. *Use:* **hanal** 'food'; *Part:* **wich** 'fruit'; *Prep:* **mäk'bil chäkbil chen wa ne täk'än, mäk'bil che'che'** 'eaten cooked, but if very ripe, eaten raw'; *Gen:* **pätan**. *Variant:* **päh pätan boox** (Fig. 5.155a, b). *[Source:* BM; CNK; KP; SK*] [*\sd2 food*]*

pätan [pätan, pätän] plantain. Musaceae *Musa acuminata* ×*Musa balbisiana* Coll. *Pätan* refers to the firm "cooking" bananas, or plantains, which are generally boiled, roasted, or fried. Only extremely ripe fruits of some varieties may be eaten raw. When cooked, *pätan* retains its firmness, unlike the softer *boox* "dessert bananas". Despite these distinctions, there still seems to be considerable confusion among the Lacandones over whether *pätan* or *boox* is the general cover term for plantains and bananas. Some of my consultants say that *pätan* is the general term and subsumes *boox* and *pätan*, whereas other consultants say the reverse is true. After maize, plantains and bananas are the most important food staple in the Lacandon diet. In the past, they served as famine food when maize supplies were low. Plantains and bananas

Fig. 5.153 (**a**) Pähpox *Annona muricata* tree. Photo Credit: Suzanne Cook (**b**) Pähpox *Annona muricata* immature fruit. Photo Credit: Suzanne Cook (**c**) Pähpox *Annona muricata* inflorescence. Photo Credit: Chan K'in Daniel (**d**) Pähpox *Annona muricata* leaves. Photo Credit: Suzanne Cook

Fig. 5.154 (**a**) Päh ak' *Disciphania calocarpa* vine with fruit. Photo Credit: Suzanne Cook (**b**) Päh ak' *Disciphania calocarpa* fruit

are still added to corn dough and *posol*. *Use:* **hanal** 'food'; *Part:* **wich** 'fruit'; *Prep:* **pokbil, tsähbil** 'roast, fried'; *Prep:* **säkan** 'dough'; *Use:* **ti' utap'ik ba'al, mukbil wah i bu'ulil wah, ti' upäk'ächtik wah, ti' utsähik uye'el kax** 'for wrapping things, like buried tamales and bean tamales, for patting out tortillas, and for frying eggs'; *Sim:* **boox** 'banana'; *Sim:* **mäya' (äh)** 'plantain type'; *Spec:* **hach pätan** 'authentic plantain'; *Spec:* **k'än pätan** 'yellow plantain'; *Spec:* **päh pätan (äh)** 'tart plantain'; *Spec:* **pätan ma' ka'anal ubakel** 'short (dwarf) plantain'; *Spec:* **hunbuhk'äbwinik pätan**

Fig. 5.155 (**a**) Päh pätan *Musa acuminata* ×*Musa balbisiana* fruit. Photo Credit: Suzanne Cook (**b**) Päh pätan *Musa acuminata* ×*Musa balbisiana* plant with fruit. Photo Credit: Suzanne Cook

'one hundred plantains'; *Spec:* **ka'buhk'äbwinik pätan** 'two hundred plantains'. *Thes:* **xiiw**. *[Note:* pätan. banana, plantain, cooking banana (Bruce 1975); patan. plantain (Nations and Nigh 1980); pätam, pa:tam [S. Lac]. plátano (Canger 1970); ja'as [Itz.]. plátano. *Musa acuminata* × *M. balbisiana* (Atran et al. 2004); já'as [Yuc.]. *Musa acuminata* × *M. balbisiana* (in, Atran et al. 2004); ja'as [Mop.]. *Musa acuminata* × *M. balbisiana* (in, Atran et al. 2004).*]* *[Source:* BM; KM; CNK*]* */*\sd2 food*]*

pätan ma' ka'anal ubakel plantain (type of). *Lit:* 'not a tall plantain'. Musaceae *Musa acuminata* ×*Musa balbisiana* Coll. *Use:* **hanal** 'food'; *Part:* **wich** 'fruit'; *Prep:* **chäkbil** 'cooked'; *Gen:* **pätan**. *[Source:* SK*]* */*\sd2 food*]*

peesache' (äh) [pe(e)sache', pesa] allspice tree (Jamaican pepper). Myrtaceae *Pimenta dioica* (L.) Merr. Allspice is native to the West Indies, southern Mexico and Central America. It is a small 12–30 m (40–100′) tall evergreen tree with smooth, whitish bark and fragrant, spicy leaves and berries, which smell like a combination of clove, nutmeg, cinnamon, and black pepper. Spanish explorers confused it with black pepper, calling it *pimiento*, which is another word the Lacandones use for this tree. Fruits mature in June, July and August, drawing pimenteros (pimiento harvesters) throughout the southern lowland forest in Chiapas, Guatemala and Belize (Nations 2006: 94). Less inclined to climbing the forest trees, Lacandones pluck the berries and leaves whenever needed from the trees they cultivate in their house gardens. Lacandones primarily use the berries and leaves for flavouring soups and meats, especially armadillo. The Mayas living in the Yucatan during the 16th century used it to season toasted corn gruel (Landa 1978: 34, cited in Nations 2006: 95). In addition to its use in the cuisine, the berries were used by the ancient Maya to embalm their leaders (Rosengarten 1973: 101, cited in Nations 2006: 95). The Itzaj make a tea from the seeds to relieve stomach aches, colic, dysentery, and headaches (Atran et al. 2004: 118). *Use:* **uts'ak ah wech'** 'armadillo spice'; *Part:* **le', wich** 'leaves and fruit'; *Use:* **ts'ak ti' hubul hämnen, chibal hämnen** 'medicine for diarrhea, stomach ache'; *Part:* **le'** 'leaves'; *Prep:* **te'** 'tea'; *Use:* **ts'ak ti' chibal koh** 'medicine for toothache'; *Part:* **le'** 'leaves'; *Prep:* **Ha'li' achaach'tik ule'** 'You just chew the leaves'. *Variant:* **pimyenta**. *Thes:* **che'** (Fig. 5.156). *[Note:* pesa. clavo silvestre. *Eugenia caryophyllata* (Nations and Nigh 1980); pesa. wild clove. *Eugenia caryophyllata* (Kashanipour and McGee 2004); pesa che, pimienta. *Pimenta dioica* (Durán 1999); nab'a'ku'uk [Itz.]. allspice seeds, lit. 'squirrel's incense' (Hofling and Tesucún 1997; Atran et al. 2004); nab'aku'uk [Mop.]. *Pimenta dioica* (in Atran et al. 2004); nab'akuk, nukuch pool [Yuc.]. *Pimenta dioica* (in, Atran et al. 2004). Remedy for colic reported by Durán (1999). Nations and Nigh (1980) and Kashanipour and McGee (2004) must be referring to *Pimienta dioica*, since *Eugenia caryophyllata* originated in the Spice Islands and is now grown in Zanibar, Indonesia, India, Pakistan, and other countries in the Old World tropics. If it does exist in Mexico, it was introduced relatively late. Furthermore, the species is not listed in any of the botanical inventories from Chiapas.*]* *[Source:* BM; KM; KP; KyP*]* */*\sd2 food*]*

Fig. 5.156 Peesache' *Pimenta dioica*. Photo Credit: Suzanne Cook

a

b

Fig. 5.157 (**a**) Pets'k'in *Schizolobium parahybum* seed. Photo Credit: Suzanne Cook (**b**) Pets'k'in *Schizolobium parahybum* trunk. Photo Credit: Chan K'in Daniel

perehir cilantro (coriander, Chinese parsley). Apiaceae *Coriandrum sativum* L., *Eryngium foetidum* L. Native to Europe, Africa and Asia, *Coriandrum sativum* is an annual, leafy plant similar to parsley in habit and form, growing to 50 cm (20″) tall with broad, deeply lobed leaves on top of long, somewhat wiry stems. The seeds are known as coriander, whereas the leaves are known as cilantro. An introduced species, Lacandones apply the Spanish word *perihil* 'parsley' to it, as it resembles parsley. *Coriandrum sativum* is similar in taste to the native plant *Eryngium foetidum*. Both are cultivated for the fresh leaves. *Loc:* **kolil nah** 'house garden'; *Use:* **ts'ak ti' o'och** 'food for food'; *Part:* **le'** 'leaves'; *Prep:* **chäkbil, che'che'** 'cooked, raw'. *See:* **kulantro**. *From:* perejil (Spanish). *[Note:* hon k'ish. *Coriandrum sativum* (Nations and Nigh 2008); aj/ix kulaantroj [Itz.]. *Coriandrum sativum* (Atran et al. 2004).*] [Source:* KM; KP*]* *[*\sd2 food*]*

pets'k'in₁ Brazilian fern tree (Brazilian fire tree, Mexican fern tree). Fabaceae *Schizolobium parahybum* (Vell.) Blake. **Ne nuk u che'il. Uch'ihil ich k'aax. Mäna' way. Ne naach. K'än utop'. Mehen. Wolis uwaakal.** 'It has a very large trunk. It grows in the forest. There aren't any here (in Naha'). They are very far away. The flowers are yellow, small, and round when they open up.' (AM) *Schizolobium parahybum* is a stately tree with a straight, cylindrical bole, high buttresses, and a wide, spreading, open crown. Bright green, fern-like, pinnate leaves fan out from the branches. Golden yellow flowers are produced in axillary racemes or terminal panicles. Seeds are brown, smooth, semi-glossy, flattened, oval disks, approximately 2 cm (0.79″) long and 1.5 cm (0.6″) wide. In January, the tree explodes into masses of golden flowers, heralding in the new year. Although *Schizolobium parahybum* is recorded from the Lacandon forest (Levy et al. 2006; Comisión Nacional de Áreas Naturales Protegidas 2006), it is not native to the northern Lacandon territory, according to AM. *Loc:* **k'aax** 'forest'; *Loc:* **yaha'nah, mensäbäk**; *Use:* **uuh** 'necklaces'; *Part:* **nek'** 'seeds'. *Thes:* **che'** (Fig. 5.157a, b). *[Note:* petskin. *Schizolobium parahybum* (Nigh 2008).*] [Source:* AM; BM; KM; KP*]* *[*\sd2 craft*]*

pets'k'in$_2$ monkeysoap (guanacaste). Fabaceae *Enterolobium cyclocarpum* (Jack) Griseb., syns. *Albizia longipes* Britton & Killip, *Feuilleea cyclocarpa* (Jacq.) Kuntze, *Inga cyclocarpa* (Jacq.) Willd., *Mimosa ciclocarpum* (Jacq.) Griseb., *Mimosa cyclocarpa* Jacq., *Mimosa parota* Sessé & Moc., *Pithecellobium cyclocarpum* (Jacq.) Mart., *Prosopis dubia* Kunth. This is another stately tree standing 20–30 m (65.6–98.4′) or more in height with a straight and sometimes buttressed trunk up to 3 m (9.8) in diameter and a wide, spreading crown. Exterior bark is smooth or slightly cracked, light grey to brownish grey with numerous longitudinal lenticels. Leaves are bipinnate with 15–30 pairs of bright green leaflets. Tubular, light green flowers are borne in axils. Fruit is a circular, indehiscent pod, 7–15 cm (2.7– 5.9″) in diameter, that contains numerous, dark, shiny, sweet seeds. (Vázquez-Yanes et al. 1999) No uses were reported by my consultants, but, according to Nations and Nigh (1980), the timber is used in construction. *Use:* **uuh** 'necklaces'; *Part:* **nek'** 'seeds'. *[Note:* petz'k'in. guanacastle. *Enterolobium cyclocarpum* (Nations and Nigh 1980); pich [Itz.]. *Enterolobium cyclocarpum* (Atran et al. 2004).*]* *[Source:* CNK; KM; KP*]* *[*\sd2 craft*]*

pichik' [pichik', pichi'] guava. Myrtaceae *Psidium guajava* L. This is a small tree that produces round fruits with soft flesh, yellowish-green skins, and numerous hard, white seeds. Two varieties are cultivated in Naha', one with deep pink flesh, called *chäk pichik'* and the other with creamy white pulp, called *säk pichik'*. The trees bear fruit from late summer to early fall. *Use:* **hanal** 'food'; *Part:* **wich** 'fruit'; *Prep:* **che'che'** 'raw'; *Spec:* **säk pichik'** 'white guava'; *Spec:* **chäk pichik'** 'red guava'. *Thes:* **che'** (Fig. 5.158). *[Note:* pichik, pul. guava. *Psidium guajava* (Nations and Nigh 1980); pichi' [Itz.]. guava. *Psidium guajava* (Hofling and Tesucún 1997); k'än/säk pichi', chäk pichi' [Itz.]. *Psidium guajava* (Atran et al. 2004); pichib', pichi' [Yuc.]. guava. *Psidium guajava* (Bricker et al. 1998).*]* *[Source:* AM; BM*]* *[*\sd2 food*]*

Fig. 5.158 Pichik' (chäk, säk) *Psidium guajava* red and white forms. Photo Credit: Chan K'in Daniel

pimyenta allspice (Jamaican pepper). Myrtaceae *Pimenta dioica* (L.) Merr. *Loc:* **k'aax** 'forest'; *Use:* **ts'ak** 'spice and medicine'; *Part:* **nek', le'** 'seeds, leaves'. *See:* **peesache' (äh)**. *From:* pimienta (Spanish). *Thes:* **che'**. *[Source:* BM; CNK*]* *[*\sd2 food*]*

pixan k'ambul (äh) [pixan k'ambul, pixan k'ambu] oak, chicarro (Mexican red oak). *Lit:* 'curassow's heart'. Fagaceae *Quercus skinneri* Benth. *Quercus skinneri* is found in wet montane forests, between 700 m (2296') and 2200 m (7218') in elevation. It grows rapidly, reaching 15–35 m (49.2–114.8′) in height. Leaves are obovate or broadly lanceolate, thin, sometimes semi-leathery, with a bristle-tipped apex. The acorn is ovoid, 2–5 cm (0.78–1.96″) long, and light brown. The cupule encloses less than 1/4 of the nut, with appressed, pubescent scales, and unrolled at the rim. (Hélardot n.d.) The Lacandones describe the acorn as large with thinner walls than the acorns of the *hach ch'alol* (*Quercus corrugata*). Although no uses were reported, the Lacandon consultants that Durán (1999) interviewed say they use the nut as a spinning top. My consultant, AM, disagreed. According to BM and KyM, the nuts are soft. *Sim:* **hach ch'alol** 'authentic ch'alol (*Quercus corrugata*)'; *Sim:* **mehen ch'alol** 'small ch'alol (*Quercus* sp.)'; *Gen:* **ch'alol**. *Variant:* **ch'alol pixan k'ambul**. *Thes:* **che'** (Fig. 5.159). *[Note:* karote, charo. *Quercus skinneri* (Durán 1999). Common names from Mudry, B. *Quercus skinneri*. Quercus skinneri. *Woods of the world.* woodsoftheworld.org/displayspecies.php?&species_name=Quercus%20skinneri.*]* *[Source:* AM; BM; KyM*]*

ponche' [ponche', pohche'] flame tree. Bombacaceae *Bernoullia flammea* Oliv. This is a fast-growing tree that bears bright red blossoms, in May, and 30 cm (12″) long brown, inedible pods. It is one of the preferred trees in the *acahual* and the forest, because it grows quickly and adds

Fig. 5.159 Pixan k'ambul *Quercus skinneri*. Photo Credit: Suzanne Cook

Fig. 5.160 Pooch'in *Passiflora serratifolia*. Photo Credit: Suzanne Cook

abundant leaf litter to the soil. *Indicator:* **ya'axk'in** 'flowers signal the beginning of the dry season'; *Use:* **hi'che'** 'fence, vertical house poles'; *Use:* **ho'olan che'il** 'ridgepole'; *Use:* **ch'aak** 'beds'; *Use:* **k'anche'** 'chairs, benches, and the traditional low bench used for patting out tortillas'. *Thes:* **che'**. *[Note:* pon che. *Bernoullia flammea* (Durán 1999).*] [Source:* AM; BM*] [*\sd2 housing \sd2 agriculture \sd2 furniture*]*

pooch'in passion fruit, wild (common name unknown). Passifloraceae *Passiflora serratifolia* L., syns. *Granadilla serratifolia* (L.) Medik., *Passiflora denticulata* Sessé & Moc. **Ne ya'ax utop'. Nuk, batak 5.08–7.8 cm (2–3″). Tuhunal ch'ihil. Octobre tähäl uwich.** 'The flowers are very blue, approximately 5.08–7.8 cm (2–3″) in diameter. It grows on its own. The fruits begin to ripen in October.' (AM, BM) *Passiflora serratifolia* belongs to the subgenus Passiflorae, section Serratifoliae. Leaves are medium green, alternate, long, elliptical, acuminate at the apex, rounded or subcordate at the base. Its reflexed, reddish-purple flowers, are intensely fragrant. It bears large, oval to round fruit that ripen to a deep yellow. The vines ramble through fallow *milpas* and house gardens. It is indigenous to the area (Yero Kuethe pers. comm.). *Loc:* **kolil nah** 'house garden'; *Use:* **hanal** 'food'; *Part:* **wich** 'fruit'; *Prep:* **che'che'** 'raw'; *Sim:* **ch'um ak' (äh)** '*Passiflora* aff. *ambigua*'; *Sim:* **ch'ink'ak'il** '*Passiflora cookii*'. *Thes:* **ak'** (Fig. 5.160). *[Note:* x-poch-kak [Yuc.]. *Passiflora ciliata*, Ait. Pasionaria. (Gaumer; Millsp. I, 311), lit. 'fire-cluster'. Described as a vine, 6 feet, white flowers with crimson pistils and tinged with purple, common on waste lands near Izamal (Roys 1931: 277). Botanical identification provided by Yero Rheyner Kuethe and John MacDougal (pers. comm.).*] [Source:* AM; BM; KP*] [*\sd2 food*]*

puhan Jamaica cherry. Muntingiaceae *Muntingia calabura* L. This is a fast growing, slender tree, 7.5–12 m (25–40′) tall, with spreading, nearly horizontal branches. The 5–12.5 cm (2–5″) long, dark-green, leaves are evergreen, alternate, lanceolate or oblong, long-pointed at the apex and oblique at the base, minutely hairy on the upper surface, with grey or brown hair on the underneath, and irregularly toothed margins. Flowers are borne singly, in pairs, or in threes in the leaf axils. The white, 1.25–2 cm (0.5–0.75″) blossoms resemble those of strawberries, with numerous, prominent yellow stamens. Abundant, round, 1.25 cm (0.5″) red fruits are soft, juicy, and sweet. The yellow seeds are minute and numerous. (Morton 1987: 65–69) In Naha', the fruit is produced in March and ripens in May (BM). The tree occurs in dry to moist thickets and secondary forests along slopes and valleys. It is a pioneer species, but it is also cultivated. Kashanipour and McGee (2004: 54) report that the Lacandones eat the leaves to relieve stomach aches; yet, BM says no part of the plant is used as medicine. In Lacandon dream interpretation, it is associated with illness: *puhan, ba'ik chi', u k'in mehen k'ak'il* 'puhun, like chi' [*Byrsonima* spp.], foretells small pox, chicken pox, and measles' (Chan K'in Viejo, in Bruce 1975: 143–144). *Use:* **hanal** 'food'; *Part:* **wich** 'fruit'; *Prep:* **che'che'** 'raw'; *Use:* **chibal hämnen** 'stomach ache'; *Part:* **le'** 'leaves'; *Prep:* **Ha'li' achaach'tik** 'You just chew the leaves'. *Thes:* **che'** (Fig. 5.161a, b). *[Note:* puhan. tropic species of huckleberry (Bruce 1975); puham. cherry. *Prunus serotina* (Nations and Nigh 1980); puham. *Muntingia calabura* (Kashanipour and McGee 2004); pujam. *Muntingia calabura* (Nations 1992, in Atran et al. 2004); *pujan* [Yuc.]. *Muntingia calabura* (in, Atran et al. 2004); pujan [Itz.] *Muntingia calaura* (Atran et al. 2004); kapulin [Mop.]. *Muntingia calabura* (in, Atran et al. 2004). Remedy from Kashanipour and McGee (2004).*] [Source:* BM*] [*\sd2 food \sd2 folklore_and_mythology*]*

pukte' bullywood (commercial name for the timber). Combretaceae *Terminalia amazonia* (J. F. Gmel.) Exell. *Terminalia amazonia* is the most widely distributed Neotropical species in the genus, ranging from the Gulf of Mexico to the Guyanas in South America, and the Antilles. It is a dominant species in the evergreen rainforest and wet forest, attaining heights of above 50 m (164′), and in some forests, as high as 70 m (230′). The trunk can grow as large as 150 cm (5′) in diameter, supported by buttresses. The heartwood is yellowish red. The timber is heavy, durable, and resistant to fungal disease and termites. (Rodríguez-Sánchez 2003–2012.). *Use:* **puche'** 'bridges'; *Use:* **hi'che'** 'fence, pole wall'; *Use:* **pohche'** 'tables'; *Spec:* **säk pukte'** 'white bullet tree'. *See:* **k'änxa'an**. *Variant:* **k'änxa'an**. *Thes:* **che'** (Fig. 5.162a, b). *[Note:* pucte, canhan.*Terminalia amazonia* (Durán 1999); pok te'. *Bucida buceras* (Nations and Nigh 1980); pucté [Yuc.]. *Bucida buceras* L. (Roys 1931); b'ox pukte', säk/k'än pukte' [Itz.]. *Bucida buceras* (Atran et al. 2004); puk-te'. puxté, pucté [Itz.]. bullet tree (Hofling and

Fig. 5.161 (**a**) Puhan *Muntingia calabura* tree. Photo Credit: Chan K'in Daniel (**b**) Puhan *Muntingia calabura* flowers and leaves. Photo Credit: Suzanne Cook

Fig. 5.162 (**a**) Pukte' *Terminalia amazonia* tree. Photo Credit: Suzanne Cook (**b**) Pukte' *Terminalia amazonia* seeds. Photo Credit: Suzanne Cook

Tesucún 1997). English names from http://www.fincaleola.com/roble%20coral.htm.*]* *[Source:* BM; K'ayum Segundo*]* *[*\sd2 housing \sd2 furniture*]*

put papaya. Caricaceae *Carica papaya* L. This is a large herb, reaching 6–9 m (19.6–29.5′) in height, with a hollow stem 30–40 cm (12-16″) in diameter. It produces melon-like fruit, 15–50 cm (6-20″) long and 10–20 cm (4-8″) wide, and up to 9 kg (20 lbs). The orange-red flesh is succulent, aromatic, and sweet. Numerous small, black, oval seeds are encased in a transparent, gelatinous aril. The fruit ripens in October. Baer and Baer (1952) report two varieties, *säk put* 'white papaya' and *chäk put* 'red papaya', although no details are provided. According to my consultants, there is only one kind. Cultivated. *Loc:* **kol, paakche' kol, kolil nah** 'milpa, fallow milpa, house garden'; *Use:* **hanal** 'food'; *Part:* **wich** 'fruit'; *Prep:* **che'che', chäkbil** 'raw or cooked'; *Spec:* **säk put** 'white papaya'; *Spec:* **chäk put** 'red papaya' (Fig. 5.163a, b, c). *[Note:* put. papaya, pawpaw (Bruce 1975); put [Itz.]. papaya. *Carica papaya* (Hofling and Tesucún 1997); k'än put, säk put [Itz.]. *Carica papaya* (Atran et al. 2004); pùut [Yuc.]. papaya. *Carica papaya* (Bricker et al. 1998).*]* *[Source:* CKD; KP*]* *[*\sd2 food*]*

pu'uhte' ardisia. Myrsinaceae *Ardisia* sp. **Chan mihin uwich, ya'ax uwich. Uchihil ich k'aax, yok'ol petha'. Uche'il ne nuk. Chan mihin utop'. Chan mihin uwich. Amäk'ik. Abril uwaakal. Ne ya'ax uwich. Mehen ch'iich' kuhantik.** 'The fruit is small and blue. It grows in the forest, around the lagoon. The trunk is very large. Its flowers are tiny.

a
b
c

Fig. 5.163 (**a**) Put *Carica papaya* tree. Photo Credit: Chan K'in Daniel (**b**) Put *Carica papaya* flowers. Photo Credit: Suzanne Cook (**c**) Put *Carica papaya* fruit. Photo Credit: Suzanne Cook

Its fruits are tiny. You can eat them. They appear in April. They are very blue. The little birds eat them.' (AM, BM). *Use:* **hanal** 'food'; *Part:* **wich** 'fruit'; *Prep:* **che'che'** 'raw'. *Variant:* **pupuhte'**. *Thes:* **che'**. *[Note:* pupute. *Ardisia compressa*; pupu che. *Ardisia nigrescens* (Durán 1999).*] [Source:* AM; BM; KM*] [*\sd2 food*]*

puuna' (äh) [puuna', puna'] mahogany. Meliaceae *Swietenia macrophylla* King. **Kuluk'ul utop', apäk'ik näl, pachil tuwolol apak'al.** 'When the flowers fall off, it is time to plant maize; then after that, all your other crops' (AM). In addition to signalling agricultural actitives, mahogany plays an indispensable role in Lacandon material culture. The timber is used extensively in house construction and for artefacts. In the old days, the bark was mashed and mixed with water to create a solution for tanning animal hides (Baer and Merrifield 1971: 150). It was also burned to yield lye, which was used to nixtamalize maize (Tozzer 1907: 51). In addition, the juice of the boiled fruit is drunk to relieve toothaches and fever (Kashanipour and McGee 2004: 62). The southern Lacandones are reported to have used the bark as a fish poison (Baer and Merrifield 1971: 247). *Indicator:* **ulubul utop'** 'the falling blossoms'; *Activity:* **päk'ik näl** 'plant maize'; *Use:* **chem** 'dugout canoe'; *Part:* **chunche'** 'trunk'; *Use:* **ba'ab** 'paddle'; *Part:* **che'il** 'timber'; *Use:* **p'ätälche'** 'board for supporting god pots'; *Use:* **pakche'il kih** 'board for removing agave fibre'; *Use:* **pohche'** 'tables'; *Use:* **k'anche'** 'chairs, benches'; *Use:* **ch'aak** 'beds'; *Use:* **tsuy** 'nock (of arrow)'; *Use:* **hi'che'** 'fence, wall of poles'; *Use:* **hool** 'door'; *Use:* **xikal** 'incense board'; *Part:* **che'il** 'timber'; *Use:* **ti' ubänik nook', hu 'un, chäk** 'for dyeing clothes and bark cloth red'; *Part:* **usool** 'bark'; *Use:* **chibal koh** 'toothache'; *Part:* **wich** 'fruit'; *Prep:* **Achäkik uwich, awuk'il, apaats'tik uwich ich achi'** 'You boil the fruit and drink it, and rub the fruit in your mouth'; *Use:* **chäkwilil** 'fever'; *Part:* **uk'a' uwich** 'fruit juice'; *Prep:* **chäkbil** 'boiled'. *See:* **chäk'ax**. *Thes:* **che'** (Fig. 5.164a, b, c). *[Note:* puuna'. mahogany (Bruce 1975); puna. *Swietenia macrophylla* (Kashanipour and McGee 2004); puna. *Swietenia macrophylla* (Nigh 2008); yax jub che. *Swietenia macrophylla* (Durán 1999); punab [Yuc.]. *Swietenia macrophylla*. caoba (Standl.) (Roys 1931); punab' [Yuc.]. *Swietenia macrophylla* (in, Atran et al. 2004); chäkälte' [Itz.]. *Swietenia macrophylla* (Atran et al. 2004); chäkälte' [Mop.]. *Swietenia macrophylla* (in, Atran et al. 2004). *Puuna'* remedy from Kashanipour and McGee (2004).*] [Source:* AM; BM; SK*] [*\sd2 housing \sd2 furniture \sd2 transportation \sd2 craft \sd2 agriculture \sd2 other*]*

P'–p'

p'ak tomato. Solanaceae *Lycopersicon esculentum, L. esculentum* var. *cerasiforme* (Dunal) D. M. Spooner, G. J. Anderson & R. K. Jansen. This is the common name for tomato, but the species most widely cultivated is a red, cherry tomato. It is quite tart and acidic. The fruits are always cooked before eating. *Use:* **hanal** 'food'; *Part:* **wich** 'fruit';

Fig. 5.164 (**a**) Puuna' *Swietenia macrophylla* tree. Photo Credit: Suzanne Cook (**b**) Puuna' *Swietenia macrophylla* seeds. Photo Credit: Suzanne Cook (**c**) Puuna' *Swietenia macrophylla* sapling. Photo Credit: Suzanne Cook

Prep: **chäkbil, tsähbil** 'boiled, fried'. *[Note:* p'àak [Yuc.]. *Lycopersicon esculentum* (Bricker et al. 1998); p'ak [Itz.]. *Lycopersicon esculentum* (Atran et al. 2004); p'ak [Mop.]. *Lycopersicon esculentum* (in, Atran et al. 2004).*]* *[*\sd2 food*]*

p'akan (ah) tomato, Mexican husk (Mayan husk tomato, ground cherry). Solanaceae *Physalis pruinosa* L., syn. *P. maxima* Mill. This is a woody stemmed, prostrate annual, with trifoliate, ovate leaves, pointed at the apex and wedge-shaped at the base. Flowers are small and yellow with dark spots on the throat. Fruits are less than 2 cm (0.79″) yellow berries encased in a papery husk. Plants are seen rambling through the brush at the margins of yards, *milpas*, and other previously cultivated sites. *Use:* **hanal** 'food'; *Part:* **wich** 'fruit'; *Prep:* **che'che', chäkbil** 'eaten raw or boiled'; *Spec:* **mehen p'akan** 'small p'akan' (Fig. 5.165). *[Source:* AM; KM; KP*]* *[*\sd2 food*]*

Fig. 5.165 P'akan *Physalis pruinosa*. Photo Credit: Suzanne Cook

p'ix (ah) pear, vegetable (pear squash). Cucurbitaceae *Sechium edule* (Jacq.) Swartz. This is a robust, green vine that bears fruit the size and shaped of pears. There is a prickly and a smooth-skinned variety. When cooked, they are starchy and bland with a consistency similar to boiled potatoes.

Cultivated in *milpas* and house gardens, and found rambling through fallow *milpas*. The widest variation of *S. edule* under cultivation is found between southern Mexico and Guatemala, while the geographical distribution of the wild relatives extends throughout Mesoamerica. Nations and Nigh (1980) report that the Lacandon cultivate six varieties, whereas Baer and Merrifield (1971: 202) report four. My consultants reported three: *säk p'ix* 'white vegetable pear', *ya'ax p'ix* 'green vegetable pear', and *k'i'ilix p'ix* 'spiny vegetable pear'. *Loc:* **kol, kolil nah, paakche' kol** 'milpas, house garden, fallow milpa'; *Use:* **hanal** 'food'; *Part:* **wich, nek'** 'fruit, seed'; *Prep:* **chäkbil** 'boiled'; *Part:* **le'** 'leaves'; *Prep:* **chäkbil** 'boiled'; *Spec:* **ya'ax p'ix** 'green vegetable pear'; *Spec:* **k'i'ilix p'ix** 'spiny vegetable pear'; *Spec:* **säk p'ix** 'white vegetable pear' (Fig. 5.166a, b, c). *[Note:* p'ish. *Sechium edule* (Nations and Nigh 1980); p'ix [S. Lac.]. chayote (Canger 1970); k'i'ix pach k'úum [Yuc.]. *Sechium edule* (in, Atran et al. 2004); ix wiskil [Itz.]. *Sechium edule* (Atran et al. 2004).*] [Source:* KM; KyP*] [*\sd2 food*]*

Fig. 5.166 (**a**) P'ix *Sechium edule* fruit. Photo Credit: Suzanne Cook (**b**) P'ix *Sechium edule* leaves. Photo Credit: Suzanne Cook (**c**) P'ix *Sechium edule* vines. Photo Credit: Suzanne Cook

S–s

sanyah [sanyah, sonyah] watermelon. Cucurbitaceae *Citrullus lanatus* (Thunb.) Matsum. & Nakai. Originally from central and south Africa (Whitaker and Davis 1962, Robinson and Decker-Walters), the fruit was introduced into the Americas by European colonists and African slaves. Cultivated. *Loc:* **kol** 'milpa'; *Use:* **hanal** 'food'; *Part:* **wich** 'fruit'; *Prep:* **che'che'** 'raw'. *From:* sandía (Spanish). *[Note:* sànyàh [S. Lac.]. sandía (Canger 1970); sand'iiyaj [Itz.]. *Citrullus* spp. (Hofling and Tesucún 1997); ix k'uumja', sand'iiyaj [Itz.]. *Citrullus vulgaris* (Atran et al. 2004).*] [*\sd2 food*]*

sanyah ahch'o' (ah) cucurbit (Guadeloupe cucumber, meloncito, speckled gourd). *Lit:* 'rat's watermelon'. Cucurbitaceae *Melothria pendula* L. This creeping cucumber is a perennial vine with thin, smooth stems and palmately-lobed, alternately-arranged leaves, 7 cm (2.75″) long, with toothed margins. It bears small, yellow flowers with five petals, approximately 0.8 cm (0.3″) in diameter. Fruits resemble miniature watermelons, 1 cm (0.4″), striped in two shades of green. Found rambling through thickets at the edges of swamps and brush. It is an important food source for bats. No uses were reported. *Use:* **uyo'och ahsek** 'bats' food'. *Thes:* **ak'** (Fig. 5.167a, b, c). *[Note:* kumi cho. *Passiflora cobanensis* (Durán 1999).*] [Source:* KM; KP*]*

san lorenso San Lorenzo (air plant). Crassulaceae *Kalanchoe pinnata* (Lam.) Pers., syn. *Bryophyllum pinnatum* (Lam.) Oken. This is a small shrub, 91–152 cm (3–5′) tall, with hollow stems, large, succulent, light green leaves that are scalloped along the margins and edged in red, and bell-like pendulous flowers. Introduced into Naha' by the Tzeltales, *Kalanchoe pinnata* is now cultivated in some Lacandon house gardens. The leaves contain a potent antihistamine and anti-allergic properties, which would explain its use for treating upper respiratory conditions and coughs

Fig. 5.167 (**a**) Sanyah ahch'o' *Melothria pendula* vine with leaves and immature fruit. Photo Credit: Suzanne Cook (**b**) Sanyah ahch'o' *Melothria pendula* flower and fruit. Photo Credit: Suzanne Cook (**c**) Sanyah ahch'o' *Melothria pendula* flower, tendril and stem. Photo Credit: Suzanne Cook

Fig. 5.168 San Lorenso (ts'ak se'em) *Kalanchoe pinnata*. Photo Credit: Suzanne Cook

(Taylor 2005). *Loc:* **ti' kah, kolil nah** 'the villages, house garden'; *Use:* **ts'ak ti' se'em** 'medicine for a common cold'; *Part:* **le'** 'leaves'; *Prep:* **Apokik ule', pachil abut'ik ich chakal ha', pachil awuk'ik** 'You sear the leaves over an open fire, then put them in hot water, and then you drink it'. *From:* San Lorenzo (Spanish) (Fig. 5.168). *[Note:* siyempreb'iib'aj [Itz.]. *Kalanchoe pinnata* (Atran et al. 2004).*] [Source:* CKD's wife; KP*] [*\sd2 medicinal*]*

sasakche' [sasakche', sasäkche'] angelica tree (potatowood, sweetwood, white gumbo limbo). *Lit:* 'whitish tree'. Araliaceae *Dendropanax arboreus* (L.) Decne. & Planch. *Dendropanax arboreus* is an evergreen, epiphytic tree, 14–25 m (45.9–82′) tall. It has an open, irregular, dense crown. Leaves are alternate and simple, varying in shape from ovate to elliptic, oblong or ovate-oblong. They are green, lustrous, and smooth. Umbels of flowers are produced on racemes at the ends of stems. The calyx is cupular, whitish yellow, and the petals are yellow. Fruit is a small berry, 4–8 mm (0.15–0.3″) long, ribbed, flattened at the apex, shiny, and purple when ripe. Each fruit contains numerous seeds. The species ranges from the Gulf of Mexico to Chiapas and the Yucatan Peninsula from sea level to 1500 m (4921'). (Vázquez-Yanes et al. 1999) The small, black fruit is relished by numerous kinds of birds. The *kacho'* (mealy blue-crowned parrot), *pän* (keel-billed toucan), and *ts'ul* (dove) are particularly fond of it. No uses were reported: according to BM, the white heartwood is very soft and breaks down quickly, making it unsuitable for construction. *Thes:* **che'** (Fig. 5.169a, b). *[Note:* saja jak che, sac sac che, jeke che. *Dendropanax arboreus* (Durán 1999); sak chakaj [Yuc.]. *Dendropanax arboreus* (in, Atran et al. 2004); tzup [Itz.]. *Dendropanax arboreus* (Atran et al. 2004). Taxonomic identification provided by Alvaro Campos (pers. comm.). English names from http://www.ars-grin.gov/misc/mmpnd/Dendropanax.html.*] [Source:* AM; BM; KP*]*

sa'yok' (ah) [sa'yok', sayok'] milady, my lady. Apocynaceae *Aspidosperma megalocarpon* Muell. Arg., syn. *Aspidosperma macrocarpum* Mart. *Aspidosperma megalocarpum* is a timber tree native to Brazil. It is also reported from Mexico, Central America, the high forests in the lower Amazon, and the Guianas. The timber is prized for its strong, hard wood, straight grain, and pest-repellant properties. My Lacandon

Fig. 5.169 (**a**) Sasakche' *Dendropanax arboreus* fruit. Photo Credit: Chan K'in Daniel (**b**) Sasakche' *Dendropanax arboreus* inflorescence and fruit. Photo Credit: Suzanne Cook

consultants only reported using it for construction, but other investigators note that the sap and the pulverized bark are used as an insecticide and cockroach repellant (Durán 1999; Nations and Nigh 1980; Nations 2006). Kashanipour and McGee (2004) report that the Lacandones make a decoction from the bark to relieve diarrhea (2004: 60). *Use:* **säkche'** 'rafters'; *Use:* **taan** 'girt (front)'; *Use:* **akam tanil** 'front pillars'; *Use:* **waan** 'girt (side)'; *Part:* **uk'äbche', ne tah** 'branches, the very straight ones'; *Use:* **hiil** 'battens'; *Part:* **k'äbche'** 'saplings, branches'; *Use:* **ts'ak ti' hubul hämnen** 'medicine for diarrhea'; *Part:* **sool** 'bark'; *Prep:* **te'** 'tea'. *Thes:* **che'**. *[Note:* sayuk. *Aspidosperma* spp. (Kashanipour and McGee 2004); tzayok. bayo. *Aspidosperma megalocarpon* (Nations 2006; Nations and Nigh 1980); atzoyoc'che. *Aspidosperma macrocarpon* (Comisión Nacional de Áreas Naturales Protegidas 2006); A sa yok che. *Aspidosperma megalocarpon* (Durán 1999); sa'yuk' [Itz.]. *Aspidosperma stegomeris, A. cruenta* (Atran et al. 2004); sa'yuk [Itz.]. malerio tree. *Aspidosperma* spp. (Hofling and Tesucún 1997). English common name from Nations (2006).*] [Source:* BM*] [*\sd2 housing \sd2 medicinal \sd2 other*]*

säha' (ah) *Doliocarpus dentatus* (common name unknown). Dilleniaceae *Doliocarpus dentatus* (Aubl.) Standl. This is a kind of forest vine with small, ovate, fuzzy leaves with serrated margins. *Loc:* **k'aax** 'forest'; *Use:* **ti' uhi'ik uchulul** 'for sanding bows'; *Part:* **le'** 'leaf'. *Thes:* **ak'** (Fig. 5.170). *[Note:* a sa ha. *Doliocarpus dentatus* (Durán 1999).*] [Source:* AM*] [*\sd2 hunting_and_fishing*]*

säkche'$_1$ orchid tree. *Lit:* 'white tree'. Fabaceae *Bauhinia rubeleruziana* Donn. Sm. *Use:* **akam** 'foundation posts'. *See:* **k'ak'che'$_2$**. *Thes:* **che'**. *[Note:* sak che. *Eupatorium nubigenum* (Diemont 2006); uyok wakax [Itz.]. 'cow's foot'. *Bauhinia divaricata* (Atran et al. 2004); yok wakax, maay wakax, tz'ulub' took' [Yuc.]. *Bauhinia divaricata* (in, Atran et al. 2004).*] [Source:* AM*] [*\sd2 housing*]*

Fig. 5.170 Säha' (deleted text) *Doliocarpus dentatus*. Photo Credit: Suzanne Cook

säknikil uuh Job's tears. *Lit:* 'white blossom necklace'. Poaceae *Coix lacryma-jobi* L. *Use:* **uuh** 'necklace'; *Part:* **nek'** 'seeds'. *Variant:* **säkpahen (äh)**. *[Source:* KM*] [*\sd2 craft*]*

säkpahen (äh) [säkpähen, säkpaheh, sakbahe] Job's tears. *Lit:* 'the pale white (one)'. Poaceae *Coix lacryma-jobi* L. This is a tall, green, shrub with long, lanceolate leaves that

Fig. 5.171 (**a**) Säkpahen *Coix lacryma-jobi* plants. Photo Credit: Suzanne Cook (**b**) Säkche' *Coix lacryma-jobi* flowers. Photo Credit: Suzanne Cook

Fig. 5.172 (**a**) Säkpähk'ak' che' *Psychotria poeppigiana*. Photo Credit: Suzanne Cook (**b**) Säkpähk'ak' che' *Psychotria poeppigiana*. Photo Credit: Suzanne Cook

resemble corn leaves. It is cultivated for its glossy, grey seeds, which begin to ripen in October (in Naha'). *Loc:* **kolil nah, kol** 'house gardens and milpas'; *Use:* **uuh** 'necklace'; *Part:* **nek'** 'seeds'. *Morph:* **säk-pal-e'en**. *Variant:* **säknikil uuh** (Fig. 5.171a, b). *[Note:* sukpaen (Tozzer 1907); sac pajen che'. *Miconia dodecandra, Croton pyramidalis*; sac pan che. *Croton guatemalensis* (Durán 1999).*] [Source:* CNK; KM; KP; NI*] [*\sd2 craft*]*

säkpähk'ak' che' sore-mouth bush (hot lips). Rubiaceae *Psychotria poeppigiana* Müll. Arg., syn. *Cephaelis tomentosa* (Aubl.) Vahl. **Ne chäk ule'. Yan uyits. Naksik uyool achi' uyits. Chan mihin ule', chukuch, batak 5.8 × 10 cm (2″ × 4″). Ubäxik ubäh yok'ol pukte', ek' bahche'.** 'The leaves are very red. They have a resin that irritates your mouth. The leaves are tiny and long, about 5.8 × 10 cm (2″ × 4″). It presses itself onto bullywood (*Terminalia amazonia*) and the black cork tree (*Guatteria anomala*).' (AM) *Psychotria poeppigiana* is an evergreen shrub, 3–4.6 m (10–15′) tall, that bears two red bracts that look like large, painted lips. The inconspicuous, yellow flowers and sepals form a narrow tube. Flowers are followed by bright blue berries. According to AM, eating the berries of this plant causes the mouth disease called *säkpähk'ak'*, which exhibits white pustules in the mouth and on the lips. To cure this, he merely invokes the plant's name in a curing ritual. The plant does, however, possess medicinal properties that are effective in treating many conditions, including thrush (Coe 2008). *Psychotria poeppigiana* is found in Chiapas to Oaxaca, Tabasco, Veracruz, and down through South America as far as northern Argentina. Pollinators are mainly hummingbirds, particularly the small hermit Phaethornithinae subfamily of Trochilidae (Rodríguez-Flores and Stiles 2005). *Loc:* **k'aax** 'forest'; *Use:* **kunyah ti' säkpähk'ak'** 'a curing ritual for säkpähk'ak''. *Thes:* **che'** (Fig. 5.172a, b). *[Note:* Taxon identification provided by John R. Paul, Department of Biology Colorado State University.*] [Source:* AM; BM; Bruce (1975)*] [*\sd2 medical conditions*]*

säkyuuhche' (äh) *Guatemala blepharidium* (common name unknown). Rubiaceae *Blepharidium mexicanum* Standl., syns. *Blepharidium guatemalense* Standl., *Tocoyena* *tabascensis* Standl. **P'iis ya'. Säk usool. Ya'ax ichil. Chukuch utop', as k'än uyek'el. Mix ubok.** 'It's like chicle (*Manilkara achras*). Its bark is white outside and blue/green inside. The flowers are long with faint yellow markings. They don't have any fragrance.' (AM). *Use:* **ho'olan che'il** 'ridgepole'; *Use:* **che'il k'ak'** 'firewood'. *Thes:* **che'**. *[Note:* Word unrecognized by BM. Sak yuste. *Blepharidium mexicanum* (Nigh 2008); sagisché'. *Blepharidium mexicanum* (Pennington and Sarukhán 2005: 486); säk yush che. *Blepharidium mexicana* [sic] (Nations and Nigh 1980). Synonyms from Zipcodezoo, http://zipcodezoo.com/Plants/B/Blepharidium_guatemalense/*] [Source:* AM*] [*\sd2 fuel \sd2 housing*]*

säk ak' *Arrabidaea verrucosa* (common name unknown). *Lit:* 'white vine'. Bignoniaceae *Arrabidaea verrucosa* (Standl.) A. Gentry. *Loc:* **k'aax** 'forest'; *Use:* **ti' uk'äxik utaan yetel uyokman i ti' uk'äxik usäkche' yetel ukuch säkche'il** 'to tie the girt to the main house posts and to tie the rafters to the load-bearing beams'; *Part:* **uyak'il** 'the vines'; *Sim:* **ya'ax ak' (ah)** 'blue/green vine (*Arrabidaea* sp.)'; *Sim:* **nikte' ak'** 'blossom-tree vine'. *Thes:* **ak'**. *[Note:* Baer and Merrifield describe säk ak' as "a thorny vine used as combs" (1971: 149). sak ak. *Arrabidaea verrucosa* (Durán 1999).*] [Source:* AM; BM*] [*\sd2 cordage*]*

säk bahche' [säk bahche', säk bache'] alligatorwood (muskwood). Meliaceae *Guarea glabra* Vahl., syns. *Guarea excelsa* Kunth, *Guarea bijuga* C. DC. This is an evergreen tree 25–30 m (82′) in height and 50 cm-1 m (19.7″) in girth, with a straight trunk and small buttresses. The crown is dense and rounded. Young branches are pubescent becoming glabrous (smooth), greyish brown to pale grey, or white, with age. The bark is soft and vertically fissured or scaly. The inner bark is cream coloured or pink and aromatic with an apple scent. Leaves are pinnately compound, up to 60 cm (2′) long, alternate with opposing leaflets, dark green, smooth on the upper surface and light green underneath. Fragrant, cream-coloured, pink, or greenish-white flowers are borne in erect, axillary panicles. Fruit is a smooth, round, brown or red valved capsule, 1.5–2.0 cm (0.6–0.79″) in diameter. (Pennington and Sarukhán 2005: 296) According to BM, the fruit contains numerous red seeds, which draw flocks of birds, especially *pän* 'keeled-billed toucan' (*Ramphastos sulfuratus*). The northern Lacandones use the timber for firewood, whereas the southern Lacandones also use it for carving and construction (Diemont 2006; Levy et al. 2006). My Lacandon consultants say there is a small and a large variety of *säk bahche'*. Having not seen the species, I am uncertain as to whether or not they are distinct species or simply large and small forms of the same species. According to Durán (1999), *nukuch bahche'* corresponds to *Guarea glabra. Loc:* **k'aax, xok'ol uk'un, äka'an wits** 'forest, near rivers, and the flat, low-lying areas in the hills'; *Loc:* **Sa'am** 'Monte Líbano'; *Use:* **che'il k'ak'** 'firewood'; *Spec:* **nukuch säk bahche'** 'large white bahche' (*Guarea* sp.)'; *Spec:* **mehen säk bahche'** 'small white bahche''; *Sim:* **ek' bahche'** 'black bahche' (*Guatteria anomala*)'; *Gen:* **bahche'**. *Thes:* **che'** (Fig. 5.173). *[Note:* sa'bajche'. *Guarea glabra* (Diemont 2006); nukuch sawua che, nukuch sacba che, saguache. *Guarea glabra* (Durán 1999); sed'riiyoj, säk k'uche' [Itz.]. *Guarea excelsa* (Atran et al. 2004).*] [Source:* AM; BM*] [*\sd2 fuel*]*

Fig. 5.173 Säk bahche' *Guarea glabra* leaves. Photo Credit: Suzanne Cook

säk bobo (ah) [säk bobo, säk wowo] calathea 'Cigar'. Marantaceae *Calathea lutea* (Aubl.) Schultes. *See:* **säk wawal**. *[Source:* AM; NX*]*

säk bo'oy$_1$ [säk bo'oy, säk booy] chamaedorea. Arecaceae *Chamaedorea* spp. Lacandones differentiate *säk bo'oy* from all other *bo'oy* by the texture and colour of its leaflets. According to BM, the leaf stems are longer, wider, and not as thick or as straight as those of *hach bo'oy* 'authentic bo'oy', and they have numerous veins running the length of the blades. More importantly, the blades are glossy-green on top and whitish underneath. This feature distinguishes *C. oblongata*, in addition to its thicker leaflets and ovoid-ellipsoid fruit, which turn shiny-black at maturity (Hodel 1992: 164). The leaves of *C. oblongata* are harvested and sold through local intermediaries to US and European floral industries. The species my Lacandon consultants identified as *säk bo'oy* also corresponded to *C. pinnatifrons* and *C. arenbergiana. C. pinnatifrons* is a highly variable species and resembles *C. oblongata* in many respects; the only reliable way to tell them apart is by their flowers (Don Hodel pers. comm.). Since the Lacandones do not distinguish *Chamaedorea* species on this basis, they are liable to misidentify the species. *Use:* **huyub ti' balche'** 'spoon for balche' (ceremonial mead)'; *Use:* **pop** 'mat'; *Use:* **ti' konol** 'to sell'; *Sim:* **hach bo'oy** 'authentic bo'oy'; *Sim:* **ch'ibix** (*Chamaedorea* spp.); *Sim:* **k'än bo'oy** 'yellow bo'oy'; *Sim:* **chan mehen bo'oy**

Fig. 5.174 (a) Säk bo'oy *Chamaedorea arenbergiana* plant. Photo Credit: Suzanne Cook (b) Säk bo'oy *Chamaedorea arenbergiana* leaves. Photo Credit: Suzanne Cook

'very small bo'oy'; *Sim:* **ch'it bo'oy** '*Chamaedorea* sp.'; *Sim:* **ton bo'oy** (*Chamaedorea* sp.); *Gen:* **bo'oy**. *Variant:* **hade, macho hade**. *[Source:* AM; BM; KM*]* *[*\sd2 religion \ sd2 industry*]*

säk bo'oy$_2$ *Chamaedorea arenbergiana* (common name unknown). Arecaceae *Chamaedorea arenbergiana* H. A. Wendl. *Chamaedorea arenbergiana* is one of the taller species, growing to 4 m (13′). It is a solitary-stemmed palm that grows in wet forests. Mature plants have very wide, pinnate leaflets that are somewhat leathery and prominantly ridged. Its leaves are much larger than those of *C. oblongata* and *C. neurchlamys*. A distinguishing feature of this species is its densely packed fruit, which resembles a black corncob. It is not clear why my consultants referred to this as *säk bo'oy*, since *säk bo'oy* typically refers to *Chamaedorea* species with a whitish cast on the underside of the leaflets. Moreover, Durán's Lacandon consultants called *Chamaedorea arenbergiana "yax boy"* ['green bo'oy'] (Durán 1999). According to AM, *ya'ax bo'oy* is another name for the green form of *hach bo'oy*. *Gen:* **bo'oy**. *See:* **hach bo'oy$_2$** (Fig. 5.174a, b). *[Note:* Botanical identification provided by Don Hodel (pers. comm.).*]* *[Source:* AM; BM; KM*]* *[*\sd2 industry*]*

säk bo'oy$_3$ xate macho (jade). *Lit:* 'white bo'oy'. Arecaceae *Chamaedorea oblongata* Mart. In the Lacandon literature, *säk bo'oy* has been identified as *Chamaedorea oblongata*. The white undersides of its leaves is the defining feature, hence the name. Vegetatively, however, it is similar to *Chamaedorea neurochlamys*. The two features that help to differentiate *C. oblongata* from other Chamaedorea species are its thicker leaflets and ovoid-ellipsoid fruit, which turn shiny-black at maturity (Hodel 1995: 164). *Gen:* **bo'oy**. *See:* **säk bo'oy$_1$**. *Variant:* **hade**. *From:* jade (Spanish). *[Note:* sac bori. *Chamaedorea oblongata* (Durán 1999); sacboy. *Chamaedorea oblongata* (Nigh 2008); u-xib'-al ix yaat [Itz.]. *Chamaedorea oblongata* (Sp. jade) (Atran et al. 2004); mach xiaat [Mop.]. *Chamaedorea oblongata* (in, Atran et al. 2004).*]* *[*\sd2 industry*]*

säk bo'oy$_4$ *Chamaeorea pinnatifrons* (common name unknown). *Lit:* 'white bo'oy'. Arecaceae *Chamaedorea pinnatifrons* (Jacq.) Oerst. Taxonomic identification was made on the basis of its fruit, which was turning from red orange to black. This is a characteristic of the species. *Gen:* **bo'oy**. *See:* **hach bo'oy**. *[*\sd2 industry*]*

säk chawah ik [säk chawah] chile pepper (type of). Solanaceae *Capsicum annuum* L. A very hot, white variety of *chawah*. *Use:* **ts'ak ti' o'och** 'condiment'; *Use:* **ts'ak ti' chi'** 'oral medicine'; *Part:* **wich** 'fruit'; *Prep:* **che'che'** 'raw'; *Sim:* **ya'ax chawah ik** 'green chawah'; *Gen:* **chawah ik (ah)**. *[Source:* AM*]* *[*\sd2 food \sd2 religion \sd2 medicinal*]*

säk chulul (äh) [säk chulul, säk churuh] macawood (granadillo). *Lit:* 'white bow'. Fabaceae *Platymiscium dimorphandrum* (J. D. Smith) Donn. Sm., syns. *Platymiscium pinnatum* (Jacq.) Dugand, *Amerimnon pinnatum* Jacq., *Lonchocarpus amerimnum* DC., *Platymiscium polystachyum* Benth. **Ne ts'u'uy.** 'It's very tough.' (BM) The bark of the tree observed was relatively smooth and grey with white splotches, similar to the bark of *P. dimorphandrum*. Yet, according to Pennington and Sarukhán (2005), *Platymiscium dimorphandrum* does not occur in the northern Lacandon forest. Gutiérrez-Carvajal and Dorantes-López (2003–2004) report that it is found between 42 and 700 m. (138′-2296.5′) above sea level, which is lower than Naha'. Nevertheless, the species is listed in the Comisión Nacional de Áreas Naturales Protegidas (2006) botanical inventory for Naha'. Jiménez (1999) describes two varieties

Fig. 5.175 (**a**) Säk chulul *Platymiscium dimorphandrum* trunk. Photo Credit: Suzanne Cook (**b**) Säk chulul *Platymiscium dimorphandrum* seeds. Photo Credit: Chan K'in Daniel

of *Platymiscium pinnatum* (syn. *Platymiscium dimorphandrum*): (1) *P. pinnatum* var. *pinnatum* is found in rainforests and semi-deciduous forests on the slopes of Talamanca mountain range in Costa Rica up to 1500 m (4921′) in elevation; (2) *P. pinnatum* var. *polystachyum* grows in rainforests up to 600 m (1968.5′) in elevation. Due to these environmental and elevational factors, *säk chulul* may correspond to *P. pinnatum* var. *pinnatum*. Nations and Nigh (1980) report that *säk chulul* corresponds to *Platymiscium yucatanum*. Although very similar to *P. dimorphandrum, P. yucatanum* is only reported from Veracruz to Tabasco and in northern Oaxaca (Pennington and Sarukhán 2005: 266). *Use:* **chulul ti' mehen paalal** 'bows for children'; *Sim:* **hach chulul** 'authentic chulul'; *Gen:* **chulul**. *Thes:* **che'** (Fig. 5.175a, b). *[Note:* säk chulul. *Platymiscium yucatanum* (Nations and Nigh 1980); kuklenche, sak chulujche *Platymiscium dimorphandrum* (Nations 1992, in Atran et al. 2004); sakchuru. *Platymiscium dimorphandrum* (Nigh 2008); sac churu che. *Fraxinus uhdei* (Durán 1999); zac chulul, chulul [Yuc.]. *Apoplanesia paniculata*, Presl. (Roys 1931); sub'inche' [Itz.]. *Platymiscium dimorphandrum* (Atran et al. 2004).*] [Source:* BM; K'ayum Segundo*] [*\sd2 hunting_and_fishing*]*

säk halol [säk halol, säk haror, säk haro] majagua, white. Malvaceae (Tiliaceae) *Heliocarpus donnell-smithii* Rose. A forest pioneer, *Heliocarpus donnell-smithii* is a small tree 15 m (49′) in height. Leaves are simple, ovate, acuminate at the apex and obtuse-truncated-cordate at the base. Leaf margins are serrated. They lack the two leafy auricles present in *Heliocarpus appendiculatus*. The leaves are arranged in a spiral. Greenish-yellow flowers are borne in axillary and terminal panicles, followed by 4 mm (0.15″) bristly, brownish, slightly compressed pods. The bristles of young fruits are rose-coloured. Bark is smooth, greenish grey that becomes fissured with age. (Pennington and Sarukhán 2005: 352) *Use:* **k'an ti' k'an, ba'ay, ch'uyu'** 'cord for hammocks, bags, and hanging baskets'; *Part:* **uyoot'el** 'inner bark'; *Sim:* **chäk halol** 'red halol (*Heliocarpus appendiculatus*)'; *Gen:* **halol**. *Variant:* **hach halol**. *Thes:* **che'** (Fig. 5.176a, b, c). *[Note:* s'akjaror. jarum. *Heliocarpus appediculatus* (Nigh 2008); harum. *Heliocarpus appediculatus* (Diemont 2006).*] [Source:* AM; BM; KM; KP; K'ayum Segundo*] [*\sd2 cordage*]*

säk hoben piper, monkey hand. Piperaceae *Piper umbellatum* L., syns. *Lepianthes peltata* (L.) Raf. This is a perennial, scrambling shrub or small tree that may reach 2 m (6.5′) in height. It grows from a woody rootstock with numerous branches near the base. Young branches are covered with soft hairs. Leaves are almost circular, 30 cm (11.8″) in diameter; the leaf base is deeply cordate, and the apex is acuminate to rounded. Blades are palmately veined, dark green above and greyish with glandular black spots underneath. The inflorescence is a cluster of erect flower spikes that resemble fingers of a clasped hand. *Piper umbellatum* is native to Mexico, South America, and the Antillies. It is usually found growing at the edges of clearings and in disturbed areas. Although none of the Lacandones whom I consulted mentioned any medicinal uses for the plant, Kashanipour and McGee (2004) report that some Lacandones use the leaves topically to reduce inflammation. In other regions of Mexico the leaves are used to flavour assorted dishes, whereas the juice from the leaves is applied to the skin to repel insects, including bedbugs. It is also used as an anti-inflammatory and as a remedy for dyspepsia, constipation, migraine headaches, colic, rheumatism, hypertension, and intestinal worms. (Vibrans 2009). *Use:* **hanal** 'food'; *Part:* **le', yool** 'the leaves and growing tips'; *Prep:* **chäkbil yetel äht'unu'** 'boiled with snails'; *Use:* **ti' utäp'ik k'ät** 'to cover (wet) clay'; *Part:* **le'** 'leaves'; *Use:* **ch'upul** 'swelling'; *Part:* **le'** 'leaves'; *Prep:* **Apokik ule' ich k'ak', pachil apulik yok'ol uch'upul** 'You roast the leaves over the fire and then you apply them to the swelling'; *Sim:*

Fig. 5.176 (**a**) Säk halol *Heliocarpus donnell-smithii* tree. Photo Credit: Suzanne Cook (**b**) Säk halol *Heliocarpus donnell-smithii* leaves. Photo Credit: Suzanne Cook (**c**) Säk halol *Heliocarpus donnell-smithii* trunk. Photo Credit: Suzanne Cook

Fig. 5.177 (**a**) Säk mäkäl *Xanthosoma mafaffa*. Photo Credit: Chan K'in Daniel (**b**) Säk mäkäl *Xanthosoma mafaffa*. Photo Credit: Suzanne Cook

hach hoben 'authentic Piper (Piper auritum)'; *Gen:* **hoben**. *See:* **mäkuuläm**. *Thes:* **che'**. *[Note:* säk hoben. *Piper umbellatum* (Kashanipour and McGee 2004). Remedy from Kashanipour and McGee (2004: 63). Lacandon translation provided by author. English common name from Vibrans (2009).*] [Source:* AM; BM; K'ayum Segundo*] [*\sd2 food \ sd2 other \sd2 medicinal*]*

säk 'ilon [säk 'ilon, säk 'iron] Swiss cheese plant. *Lit:* 'white/false 'ilon'. Araceae *Monstera* sp. BM says this species is not found in Naha'. It is unclear what species BM was referring to, since the species that both KM and K'ayum Segundo identified as *säk 'ilon* was found at the edge of the forest just south of the village. *Use:* **xaak** 'baskets'; *Part:* **ak'** 'vine'; *Sim:* **hach 'ilon** '*Monstera* sp.'; *Sim:* **kamis 'ilon** '*Monstera deliciosa*'; *Gen:* **'ilon**. *Thes:* **ak'**. *[Source:* KM; K'ayum Segundo*] [*\sd2 cordage*]*

säk 'is potato, sweet. *Lit:* 'white sweet potato'. Convolvulaceae *Ipomoea batatas* (L.) Lam. **Säk usool**. A white-skinned variety. Cultivated. *Use:* **hanal** 'food'; *Part:* **wi'** 'tuber'; *Prep:* **chäkbil, pokbil** 'boiled, roasted'; *Sim: chäk 'is 'red sweet potato'; Gen:* **'is**. *[Source:* AM; BM; CNK; KM; KP; SK*] [*\sd2 food*]*

säk mäkäl elephant ears (malanga, Caribbean cabbage). *Lit:* 'white macal'. Araceae *Xanthosoma mafaffa* Schott. **Säk uchuch ule'.** 'The leaf stem is white.' One of two varieties of *hach mäkäl* that is cultivated for its large, white tubers and leaf stems. The other is *chäk mäkäl* 'red macal'. *Use:* **hanal** 'food'; *Part:* **wi'** 'tuber'; *Prep:* **chäkbil** 'boiled'; *Part:* **le', ne muun** 'young, unfurled leaves'; *Prep:* **chäkbil** 'boiled'; *Part:* **uyol** 'the shoots'; *Prep:* **mukbil wah** 'baked in a leaf (baked tamale)'; *Sim:* **chäk mäkäl** 'red macal (*Xanthosoma mafaffa*)'; *Gen:* **hach mäkäl**. *Thes:* **xiiw** (Fig. 5.177a, b). *[Source:* AM; BM; CKD*] [*\sd2 food*]*

Fig. 5.178 (**a**) Säk muxän *Heliconia librata* inflorescence. Photo Credit: Suzanne Cook (**b**) Säk muxän (chukuch ule' xiiw) *Heliconia* sp. leaves. Photo Credit: Suzanne Cook

säk muluch' (ah) fungi, white bonnets. *Lit:* 'white mound/ pile'. Physalacriaceae *Oudemansiella canarii* (Jungh.) Höhn. This is a white capped, gilled mushroom that grows in clumps on rotting leaves and wood. *Use:* **hanal** 'food'; *Prep:* **chäkbil yok'ol xiiw** 'cooked in a leaf wrapper'; *Sim:* **ek' muluch'** 'black muluch' (*Oudemansiella* aff. *steffendii, Pluteus harrisii*)'; *Gen:* **muluch' (ah)**. *[Note:* säk much', el compañero del much' [S. Lac.]. *Oudemansiella canarii* (Jungh.) Höhn (Ruan-Soto et al. 2009). Taxonomic identification based on Ruan-Soto et al (2009).*] [Source:* AM; KM; KP*] [*\sd2 food*]*

säk muxan (ah) *Heliconia librata* (common name unknown). Heliconiaceae *Heliconia librata* Griggs. *Helicona librata* is an evergreen herb that grows in the forest understory or at the forest periphery. It grows from rhizomes, rapidly colonizing tree-fall gaps, open stream beds, and disturbed areas. It grows to heights that range from 45 cm to 1.5–4.6 m (1.5–15′). The relatively inconspicuous flowers are borne in extravagant, orange-yellow bracts. Leaves are alternate, simple, paddle-shaped blades averaging 0.9 m (36″) long, but may be longer. *Use:* **ti' täp'ik ba'al, yo'och** 'for wrapping up things and food'; *Sim:* **ya'ax ch'ib muxan xiiw** '*Calathea macrosepala*'; *Gen:* **muxan**. *Variant:* **chukuch le' xiiw, hach xiiw**. *Thes:* **xiiw** (Fig. 5.178a, b). *[Note:* Scientific names from Schoenhals (1988). s'kre. *Heliconia librata* (Nigh 2008); säk muxan, ya'ax muxan [Itz.] *Calathea lutea* (Atran et al. 2004).*] [Source:* AM*] [*\sd2 food*]*

säk näl maize. *Lit:* 'white maize'. Poaceae *Zea mays* L. This is the preferred variety of maize for making tortillas and corn gruels, because the soft kernels produce a finer dough than other varieties of maize. *Use:* **wah** 'tortillas'; *Sim:* **ek' näl** 'black maize'; *Sim:* **k'änk'än näl** 'yellow maize'; *Gen:* **näl**. *[Source:* AM; KP*] [*\sd2 food*]*

säk nikte' frangipani (pagoda tree, plumeria, temple tree, West Indian jasmine). *Lit:* 'white blossom tree'. Apocynaceae *Plumeria rubra* L., syn. *Plumeria acuminata* W. T. Aiton. *Plumeria rubra* is a partially deciduous tree with a single trunk and multiple branches of similar length that support an open, spreading crown. The tree is between 5 and 8 m (16.4–24′) tall, but may reach up to 25 m (82′). Inflorescences grow in dense panicles. The fragrant, white or pink blossoms are composed of five-petals arranged in a pinwheel. Fruits are smooth, yellowish green or orange-green pods, 25–30 cm (9.8–12″) long and 3 cm (1.2″) in diameter. Seeds are numerous, dark brown, and winged. (NTBG n.d.) Cultivated. *Loc:* **kolil nah** 'house gardens'; *Use:* **tsoy uwich** 'ornamental'; *Sim:* **chäk nikte'** 'red nikte' (*Plumeria rubra* f. *rubra*)'; *Gen:* **nikte' (ah)**. *Thes:* **che'**. *[Note:* säk nikte'. *Plumeria alba* (Tozzer 1907); säk nikte' [Itz.]. *Plumeria rubra* var. *pudica* (Atran et al. 2004); sak nikte' [Yuc.]. *Plumeria rubra* var. *pudica* (in, Atran et al. 2004).*] [*\sd2 ornamental*]*

säk 'ooh reed, common (giant cane). Poaceae *Gynerium sagittatum* (Aubl.) P. Beauv. *Loc:* **tu chi' yahaha'** 'at the edge/ mouth of Yahaha' (Monte Líbano River)'; *Use:* **häläl** 'arrows'; *Part:* **che'il top'** 'flower stalks'; *Use:* **hiil** 'battens'; *Part:* **uche'il** 'stalks'; *Sim:* **chäk 'ooh**; *Gen:* **'ooh**. *Variant:* **hach 'ooh**. *[Source:* BM*] [*\sd2 house \sd2 hunting_and_fishing*]*

säk 'oonte' [säk 'oonte', säk 'onte'] *Licaria alata* (common name unknown). Lauraceae *Licaria alata* Miranda, syn. *Licaria excelsa* Kosterm. This is a large tree bearing oval, purple fruit, approximately 6.35 cm (2.5″) long, that are nestled in red cupules. The single, large seed is almost the same size as the fruit. The fruit matures in November. Both the pulp and the seeds are eaten. According to my consultants, the seeds are oily and are especially tasty when toasted. Fruits and seeds are also enjoyed by the wildlife. *Loc:* **ya'ax petha'** 'green lagoon'; *Use:* **hanal** 'food'; *Part:* **nek'** 'seeds'; *Prep:* **k'elbil** 'toasted'; *Sim:* **ek' 'oonte'** 'black 'oonte' (*Nectandra globosa*)'; *Sim:* **mehen 'oonte'** 'small 'oonte' (*Nectandra* sp.)'; *Sim:* **nukuch 'oonte'** 'large 'oonte' (*Nectandra* sp.)'; *Sim:* **babah 'oonte'** '*Ocotea cernua*'; *Gen:* **'oonte'**. *Variant:* **nukuch oonte'**. *Thes:* **che'**. *[Note:* säk oon-te'. a species of

Fig. 5.179 (**a**) Säk 'oop *Annona cherimoya* immature fruit. Photo Credit: Suzanne Cook (**b**) Säk 'oop *Annona cherimoya* fruit interior. Photo Credit: Chan K'in Daniel (**c**) Säk 'oop *Annona cherimoya* cultivated tree. Photo Credit: Chan K'in Daniel

wild avocado (Bruce 1976); sac onte, ek onte. *Compsoneura sprucei*; sac onte. *Licaria alata* (Durán 1999).*]* *[Source:* AM; BM; KA*]* *[*\sd2 food*]*

säk 'oop cherimoya (custard apple). *Lit:* 'white custard apple'. Annonaceae *Annona cherimola* Mill. This is similar to the red custard apple, with its spherical, conical, or heart-shaped fruit, 10–20 cm (4–8″) long and up to 10 cm (4″) wide. The skin may be thick or thin and is relatively smooth with slight thumb print depressions or covered with protuberances (Morton 1987: 65–69). The flesh is white, soft, aromatic, and tangy-sweet. It surrounds numerous, hard, brown or black, glossy seeds. In Naha', fruit matures from August through October. Cultivated. *Loc:* **kolil nah, kol** 'house gardens and milpas'; *Use:* **hanal** 'food'; *Part:* **wich** 'fruit'; *Prep:* **che'che'** 'raw'; *Sim:* **chäk 'oop** 'red custard apple'; *Gen:* **hach 'oop**. *Thes:* **che'** (Fig. 5.179a, b, c). *[Note:* oop. custand apple (Bruce 1975); op. custard apple. *Annona cherimoya* (Nations and Nigh 1980); 'óop [Yuc.]. *Annona squamosa* (Bricker et al. 1998); 'oop[Itz.]. *Annona* spp. (Hofling and Tesucún 1997).*]* *[Source:* AM*]* *[*\sd2 food*]*

säk pätan plantain (white); yellow skin and white flesh. *Lit:* 'white plantain'. Musaceae *Musa acuminata* ×*Musa balbisiana* Coll. *Part:* **wich** 'fruit'; *Prep:* **chäkbil, wa ne täk'än, che'che'** 'cooked, if very ripe, raw'; *Gen:* **pätan**. *Variant:* **hach pätan**. *[Note:* According to the Lacandones, you may eat it raw, but it will give you worms.*]* *[Source:* BM; BMwife; CNK; KM; KP*]* *[*\sd2 food*]*

säk pichik' guava. *Lit:* 'white guava'. Myrtaceae *Psidium guajava* L. A variety of guava with creamy, white flesh. Cultivated. *Loc:* **kolil nah, kol** 'house garden, milpa'; *Use:* **hanal** 'food'; *Part:* **wich** 'fruit'; *Prep:* **che'che'** 'raw'; *Sim:* **chäk pichik'** 'red guava'; *Gen:* **pichik'**. *[Source:* AM; BM*]* *[*\sd2 food*]*

säk p'ix pear, vegetable (pear squash). Cucurbitaceae *Sechium edule* (Jacq.) Sw. *Use:* **hanal** 'food'; *Part:* **wich, le', nek'** 'fruit, leaves, seed'; *Prep:* **chäkbil** 'boiled'; *Sim:* **ya'ax p'ix** 'green vegetable pear'; *Sim:* **k'i'ilix p'ix** 'spiny vegetable pear'; *Gen:* **p'ix**. *[Source:* KyP*]* *[*\sd2 food*]*

säk seboya (äh) onion, white. *Lit:* 'white onion'. Amaryllidaceae *Allium cepa* L. *Use:* **hanal** 'food'; *Part:* **wi'** 'bulb'; *Prep:* **chäkbil** 'cooked'. *Variant:* **seboya**. *From:* cebolla (Sp.). *[Source:* BMjr's mother*]* *[*\sd2 food*]*

säk su'uk grass, napier (merker grass, cane grass, elephant grass). *Lit:* 'white grass'. Poaceae *Pennisetum purpureum* Schumach. *Gen:* **su'uk (ah)**. *Variant:* **wakäx su'uk**. *[Source:* BM*]*

säk ts'in manioc. *Lit:* 'white manioc'. Euphorbiaceae *Manihot esculenta* Crantz. *Use:* **hanal** 'food'; *Part:* **wi'** 'tuber'; *Prep:* **pokbil** 'roasted'; *Sim:* **chäk ts'in** 'red manioc'; *Sim:* **yahche' ts'in** 'great tree manioc'; *Gen:* **ts'in**. *See:* **hach ts'in**. *Variant:* **hach ts'in**. *[Source:* BM*]* *[*\sd2 food*]*

säk wawal [säk wawar, säk wawah, säk wowo, säk bobo] calathea 'Cigar'. *Lit:* 'many white fans'. Marantaceae *Calathea lutea* (Aubl.) Schultes. *Calathea lutea* is a 2–3 m. (6.5–9.8′) tall herb. It is distinguished by its smooth, broadly elliptic, 50–100 cm (19.6–39″) long, greenish-blue, paddle-shaped leaves held aloft on 1.2–2 m (3.9–6.5′) long petioles with a whitish bloom that covers the undersides of the leaves. Inflorescences emerge from a sheath below the leaves. Spikes are oblong, more or less flattened, 10–30 cm (4–12″) long and 3–5 cm (1.2–2″) wide. Yellowish bracts turn reddish bronze at maturity age. Petals are 4–4.5 cm (1.6–1.8″) long; stamens are pale yellow or whitish. Fruit capsules contain orange seeds nestled in brilliant orange arils. An aggressive species, it colonizes tree-fall areas, creek beds, and shorelines. (Croat 1978: 267) *Loc:* **k'aax, paakche' kol, petha'** 'forest, fallow milpa, lagoon'; *Use:* **ti' utäp'ik ba'al** 'to cover things'; *Part:* **le'** 'leaves'; *Sim:* **chäk wawal** 'red wawal'; *Sim:* **its'in säk wawal** 'younger brother of white wawal (*Calathea crotalifera*)'; *Gen:* **wawal**. *Variant:* **säk**

Fig. 5.180 (**a**) Säk wawal *Calathea lutea*. Photo Credit: Chan K'in Daniel (**b**) Säk wawal *Calathea lutea*. Photo Credit: Suzanne Cook

bobo (ah). *Thes:* **xiiw** (Fig. 5.180a, b). *[Note:* AM said everyone erroneously pronounces this word as [säk wowo] or [säk bobo]. According to him, the correct pronunciation is [säk wawal]. He said he did not know the meaning of *wawal*, but according to BM, *wawal*, refers to the way the leaves curve inward. Conceivably, it derives from the root *wal*, 'fan'. säk wowoh. canna lily. *Canna edulus* (Nations and Nigh 1980); säk muxan, ya'ax muxan [Itz.]. *Calathea lutea* (Atran et al. 2004).*] [Source:* AM; CKO*] [*\sd2 food*]*

säk ya' apple, wild star. *Lit:* 'white/false chicle'. Sapotaceae *Chrysophyllum mexicanum* Brandegee ex Standl., syns. *Micropholis sericea* L. O. Williams, *Cynodendron mexicanum* (Brandegee) Baehni. In addition to the list of uses provided below, Durán (1999: 149) reports that the shoots are made into a tea to treat diarrhea, whereas Kashanipour and McGee (2004: 64) say that the leaves are chewed to relieve a headache. *Use:* **hanal** 'food'; *Part:* **wich** 'fruit'; *Prep:* **che'che'** 'raw'; *Use:* **chuste'** 'arrow foreshaft'; *Use:* **chulul** 'bow (for children)'; *Use:* **che'il k'ak'** 'firewood'; *Use:* **akam** 'foundation posts'; *Use:* **ho'olan che'il** 'ridgepole'; *Sim:* **chäk ya'** 'red ya''; *Sim:* **hach ya'** 'sapodilla (*Manilkara achras*)'; *Sim:* **o'ot' ya'** 'canistel (*Pouteria campechiana*)'; *Gen:* **ya'**. *Thes:* **che'** (Fig. 5.181a, b). *[Note:* yo och uk.*Chrysophyllum mexicanum* (Durán 1999); chi ke'. *Chrysophyllum mexicanum* (Kashanipour and McGee 2004); uni'keejil k'aax, ix siki(l)ya' [Itz.]. *Chrysophyllum mexicanum* (Atran et al. 2004). English common name from Multilingual Multiscript Plant Name Database. http://www.plantnames.unimelb.edu.au/Sorting/Chrysophyllum.html. Synonyms from http://www.theplantlist.org/tpl/record/kew-53480.*] [Source:* SK; Tozzer (1907)*] [*\sd2 hunting_and_fishing \sd2 food \sd2 fuel*]*

säk yolte' ik (äh) [säk yolte' ik, säk yote' ik] pepper, chile. *Lit:* 'white walking-stick pepper'. Solanaceae *Capsicum* sp. This is a kind of long, white, chile pepper and one of three hot chiles used to make *k'ämkih*, an offering given to cool off the tempers of angry gods during the changing of the god pots ceremony. It is also one of several kinds of chile pepper used to treat oral infections. *Use:* **ts'ak ti' o'och** 'condiment'; *Part:* **wich** 'fruit'; *Prep:* **che'che'** 'raw'; *Use:* **ts'ak ti' k'uho' ti' usiskuntik uyoolo'** 'spice for the gods to cool their tempers'; *Part:* **wich** 'fruit'; *Prep:* **che'che'** 'raw'; *Use:* **ts'ak uchi', ts'ak ti' säkpähk'ak'** 'medicine for oral conditions, including *säkpähk'ak*''; *Part:* **wich** 'fruit'; *Prep:* **mäna'** 'none'; *Gen:* **ik**. *See:* **k'ämkih**. *[Note:* äh säk yolte' ik. type of long pepper appropriately called "walking stick peppers" (Bruce 1976: 148).*] [Source:* AM*] [*\sd2 religion \sd2 medicinal*]*

seboya onion. Amaryllidaceae *Allium cepa* L. *Use:* **hanal** 'food'; *Part:* **wi'** 'bulb'; *Prep:* **chäkbil** 'boiled'. *See:* **ts'ak k'ek'en**. *Variant:* **säk seboya**. *From:* cebolla (Spanish). *[Note:* seb'ooyaj [Itz.]. *Allium cepa* (Atran et al. 2004).*] [*\sd2 food*]*

semet' *Rhipidocladum bartlettii* (common name unknown). Poaceae *Rhipidocladum bartlettii* (McClure) McClure. A genus of Neotropical woody bamboo, *Rhipidocladum bartlettii* grows in small, dense clumps or tufts of up to 20 culms, 200–1000 cm (6.6– 32.8′) long and 2–8 mm (0.08–0.3″) in diameter. Numerous branchlets arise in a fan-like array. Leaves are deciduous, smooth, and linear, 6–15 cm (2.4–5.9″) long by 2–11 mm (0.08–0.43″) wide. (Clayton et al. 2006 onwards) It is found growing at the edges of cleared forest and roadsides backing onto second-growth

Fig. 5.181 (**a**) Säk ya' *Chrysophyllum mexicanum* leaves. Photo Credit: Suzanne Cook (**b**) Säk ya' *Chrysophyllum mexicanum* young trees. Photo Credit: Suzanne Cook

Fig. 5.182 (**a**) Semet' *Rhipidocladum bartlettii*. Photo Credit: Suzanne Cook (**b**) Semet' *Rhipidocladum bartlettii*. Photo Credit: Suzanne Cook

forest. *Use:* **lama', lomo' käy** 'harpoon, fish spear'; *Part:* **che'il** 'stalk'; *Sim:* **bambu** '*Bambusa vulgaris* 'Vittata'; *Sim:* **ämäy** '*Guadua amplexifolia* (?)' (Fig. 5.182a, b). *[Note:* Botanical identification provided by Gib Cooper, Tradewinds Bamboo Nursery (pers. comm.). semet. *Merostrachys* sp. (Nations and Nigh 1980; Nations 2006: 80); senet' [Itz.]. *Arthrostylidium* sp. (Atran et al. 2004).*] [Source:* BM*] [*\sd2 hunting_and_fishing*]*

sensi' (ah) ginger. Zingiberaceae *Zingiber officinale* Roscoe. An introduced species. Cultivated. *Loc:* **kol, kolil nah** 'milpa, house garden'; *Use:* **ts'ak** 'spice'; *Part:* **wi'** 'rhizome'; *Prep:* **che'che'** 'raw'; *Use:* **ts'ak ti' hubul hämnen** 'medicine for diarrhea'; *Part:* **wi'** 'rhizome'; *Prep:* **uk'ul chäkbil** 'boiled beverage'; *Use:* **chibal koh** 'toothache'; *Part:* **wi'** 'rhizome'; *Prep:* **Che'che', ha'li' ahupik ich akoh, chi'** 'You just insert a raw piece in your tooth or mouth'; *Use:* **ts'ak ti' säk se'em** 'medicine for catarrh'; *Part:* **wi'** 'rhizome'; *Prep:* **te'** 'tea'. *Variant:* **pap uwi'**. *[Note:* Term not recognized by BM.*] [Source:* AM; CNK; SK*] [*\sd2 medicinal \sd2 food*]*

siib [siib, siip', sip'] grass, saw (fen-sedge, twig-sedge). Cyperaceae *Cladium jamaicense* Crantz, syn. *Mariscus jamaicensis* (Crantz) Britt. This is a black and white, stiff, cane-like, aquatic grass. No uses were reported. *Loc:* **petha'** 'lagoon'. *Variant:* **siibal**. *Thes:* **su'uk** (Fig. 5.183). *[Note:* siib' [Itz.]. *Panicum bartlettii* (Atran et al. 2004); siib' [Itz.]. water grass. (Hofling 1997)*] [Source:* SK*]*

Fig. 5.183 Siib *Cladium jamaicense*. Photo Credit: Suzanne Cook

siip che' (ah) bunchosia (peanut butter tree). *Lit:* 'sin tree'. Malpighiaceae *Bunchosia* sp. According to AM, this is a forest tree that bears fruit in *nah ya'axk'in* 'main season'. The fruit is approximately 2.5 cm (1″) round and has three seeds. Its pulp is red and sweet. The species AM describes may be a species of *Bunchosia*. This is a genus of shrubs or small trees with short-petioles, entire leaves, yellow flowers borne in racemes or small panicles in the leaf axils, and red or orange, 2– 3-lobed drupes. The name Bunchosia comes from bunchos, the Arabic word for coffee, and refers to the fruit. Fruits of many species are edible, some attaining the size of small plums (Anderson et al. 2006). *Bunchosia* is also called the peanut butter tree, due to its similar texture and flavour. The three species recorded from Naha' are: *B. guatemalensis, Bunchosia* sp., and *B. lanceolata* (Comisión Nacional de Áreas Naturales Protegidas 2006). The fruit of *B. lanceolata* is eaten in the southern Lacandon territory (Levy et al. 2006). *Loc:* **Sa'am** 'Monte Líbano'; *Use:* **hanal** 'food'; *Part:* **wich** 'fruit'; *Prep:* **che'che'** 'raw'. *Thes:* **che'**. *[Note:* sipche, siki kax. *Deherainia smaragdina* (Durán 1999); sìipche' [Yuc.]. 'sin tree'. *Bunchosia swartziana* (Bricker et al. 1998); zipché [Yuc.]. *Bunchosia grandulosa* (Roys 1931); sipche' [Itz.]. *Bunchosia swartziana* (Atran et al. 2004).*] [Source:* AM*] [\sd2 food]*

sikil [sikir] cushaw pumpkin (pipian). *Lit:* 'seeds'. Cucurbitaceae *Cucurbita argyrosperma* K. Koch, syn. *C. mixta* Pang. This is a vining squash with prickly stems and globose-ovoid fruit that is striped grey, green, and white. Although eaten, the flesh is of relatively low quality compared to that of *C. moschata* and *C. pepo*. Its large seeds and more than 7000 years of domestication suggest that this species was raised mainly for its protein-rich seeds (Saade and Montes-Hernández 1994: 63). *Use:* **hanal** 'food'; *Part:* **ha'li' unek'** 'seeds only'; *Prep:* **k'elbil, wa huch'bil pachil huybil yetel ma'ats'** 'toasted, or ground and then stirred with *posol*'. *[Note:* sikil. pipián [*Cucurbita argyrosperma*] (Baer and Merrifield 1971); sikir. *Cucurbita mixta* (Nations 1992, in Atran et al. 2004); sikil [Itz.]. *Cucurbita mixta* (Atran et al. 2004); sikil-k'úum [Yuc.]. *Cucurbita mixta* (in, Atran et al. 2004); sicil [Yuc.] squash seed (Roys 1931); sikil [Yuc.]. pie-pan squash seeds (Bricker et al. 1998).*] [Source:* KM; KP*] [\sd2 food]*

sikilte' che' nut, physic. *Lit:* 'seed tree'. Euphorbiaceae *Jatropha curcas* L. **Che', batak 3 wah 4 metros (9.8–13′) ubakal i batak 12.7 cm (5″) ukooch. Uts'u' ule'. Ya'ax, utop', as ya'ax utop', kuchkuchkih. Ne yan uwich. Chukuch uwich. Ek' unek'. Ahantik unek'. Yan tsats, tsatsil ti' hach winik.** 'This is a tree, about 3 or 4 m. (9.8–13′) tall and about 12.7 cm (5″) around. It is dense with leaves. The flowers are somewhat green and very tough. There are a lot of fruits. The fruits are long. They're black. You eat the seeds. They have oil, which the *hach winik* 'True People' used for cooking.' (AM) The physic nut is an erect, evergreen shrub, 4 m (13′) high with spreading branches, thick, sappy stems, and grey bark. Leaves are smooth, dark green, and rounded with 3–5 lobes and five prominent, palmate veins. Yellow-green flowers form in clusters on stalks in the upper axils. Fruit is a pod, which changes in colour from green to dark brown at maturity and then splits into three valves. Each valve contains 1–3 black, oblong seeds, approximately 2 cm (0.78″) long and 1 cm (0.4″) wide. Native to the Neotropics, *Jatropha curcas* is cultivated in (sub)tropical regions around the world. Despite the high levels of toxalbumin curcin present in all parts of the plant, the oily nuts are consumed. *Loc:* **kol** 'milpa'; *Use:* **hanal** 'food'; *Part:* **nek'** 'seeds'; *Prep:* **k'elbil i pachil huch'bil yetel ka'** 'toasted on the griddle, and then ground with toasted corn dough'; *Use:* **tsats ti' chäkik uyo'och** 'cooking oil'; *Use:* **pats'bil ts'al** 'salve'. *Morph:* **sikil-te' che'**. *Thes:* **che'** (Fig. 5.184a, b). *[Note:* sikir té. *Jatropha curcas* (Nations and Nigh 1980); siklil-te' [Yuc.]. *Jatropha curcas*. A small tree whose fruit is roasted and eaten like peanuts (Bricker et al. 1998); pinyonsiiyoj [Itz.]. *Jatropha curcas* (Atran et al. 2004).*] [Source:* AM*] [\sd2 food]*

sikilte' lu'um peanut. *Lit:* 'seed tree of the earth'. Fabaceae *Arachis hypogaea* L. *Loc:* **kol** 'milpa'; *Use:* **hanal** 'food'; *Part:* **nek'** 'seed'; *Prep:* **k'elbil** 'toasted'. *Morph:* **sikil-te' lu'um**. *[\sd2 food]*

a

b

Fig. 5.184 (**a**) Sikilte' che' *Jatropha curcas* fruit and seeds. Photo Credit: Santiago Fernandez http://www.arbolesymedioambiente.es/jatrofa.html (**b**) Sikilte' che' *Jatropha curcas* leaves. Photo Credit: Jim Conrad

sik' safflower (false saffron). Asteraceae *Carthamus tinctorius* L. This is a medium-sized plant bearing fragrant, yellow flowers and small, round fruits. It is an erect, branching, annual that grows up to 180 cm (70″) tall. The simple, sessile leaves are oblong to ovate-lanceolate, thistle-like with spiny toothed margins. Brilliant yellow, orange, or red flowers are borne among a whorl of spiny bracts at the tips of branches. The entire inflorescence is shaped like an urn, 2.5–4 cm (1–1.5″) in diameter. Probable origin is the Middle East. *Loc:* **Yahapetha'** 'Guineo'; *Use:* **bok** 'fragrance'; *Use:* **ti' banik nook'** 'for dying cloth'; *Part:* **top'** 'flower'. *[Source:* Baer and Merrifield (1971); BM*]* *[*\sd2 craft \sd2 ornamental*]*

sina'an luuch (äh) [sina'an luuch, sina' luuch] calabash. *Lit:* 'stretched out calabash'. Bignoniaceae *Crescentia* sp., *C. alata*. According to AM, this is an old variety of *luuch* that is not grown much anymore. He described it as a small tree, approximately 8 m (26′) tall, bearing fruits similar to those of *hach luuch* (*C. cujete*), only smaller and longer. Previous researchers in the area identified it as *C. alata*, with its characteristic small, cannon-ball like fruit and trifoliate leaves, which appear in the shape of a cross formed by winged petioles and three leaflets. The tree AM referred to as sina'an luuch lacked this leaf structure, and hence was likely *C. cujete*. Arango et al. (2004) report that fruit forms of *C. cujete* vary considerably in shape and size. *Sina'an luuch* could be an example of such variation. A number of cultures around the world cultivate *C. alata* for its sweet flesh and licorice-flavoured, protein-rich seeds (Gilman and Watson 1993: 1), but the flesh of *C. cujete* is poisonous. The fruit is primarily used for making bowls, cups, jugs, water containers, ornaments, and musical instruments. *Use:* **suulil lek ti' uyoom käkow** 'cup for chocolate froth'; *Sim:* **hach luuch** 'authentic calabash (*Crescentia cujete*)'; *Sim:* **säk luuch** 'white calabash'; *Gen:* **luuch**. *Variant:* **k'än luuch**. *[Note:* sinah luch. *Crescentia cujete* (Nations and Nigh 1980); sina' ruch [S. Lac.]. *Crescentia alata* (Baer and Merrifield 1971).*]* *[Source:* AM*]* *[*\sd2 containers*]*

si'si' k'uuts three-nerved baccharis. *Lit:* 'very dry tobacco'. Asteraceae *Baccharis trinervis* (Lam.) Pers., syn. *Eupatorium trinervatum* Sessé & Moc. **Montecito ich kol. Ule' bex ahk'uuts. Es lo'obil.** 'This is a small bush that grows in the *milpa*. The leaves are like those of the tobacco plant. It's a weed.' (BM) *Baccharis trinervis* is a scandent or subscandent bush up to 4 m (13′) tall. Leaves are alternate, palmately-nerved, 1–10 cm (0.4–4″) long and 0.2–3.5 cm (0.08-1.4″) wide, and elliptical; leaf margins are entire. Inflorescences are panicles of small, white or pinkish-white flowers, 2.5 × 3 mm (0.09-0.1″) long, borne in the leaf axils. Fruit is a nutlet contained in a feathery pappus. *Loc:* **kol** 'milpa'; *Sim:* **nukuch le' si'si' k'uuts** 'big leaf dry tobacco (*Erechtites hieracifolia*)'; *Sim:* **hach k'uuts** 'authentic tobacco (*Nicotiana tabacum*)'; *Gen:* **k'uuts**. *See:* **hach lo'obil**. *[Note:* sisicus. *Baccharis trinervis, Erechtites hieracifolia* (Nigh 2008); sisk'utz. *Baccharis trinervis*. assa-peixe fino (Diemont 2006); sisi lobi. *Eupatorium collinum* (Durán 1999). English name from Grandtner (1997–2010).*]* *[Source:* BM*]*

sit Florida tibisee (smallcane). Poaceae *Lasiacis* sp. A. Hitchc. This is a tall, perennial grass native to the Neotropics. It has wide, strap-shaped leaves and black, oily fruit, which provides a valuable food source for birds.

It grows in disturbed areas. Four species are reported from Naha': *Lasiacis nigra; L. grisebachii; L. standleyi;* and, *Lasiacis* sp. (Comisión Nacional de Áreas Naturales Protegidas. 2006). No uses were reported. *[Note:* sit. *Olyra latifolia* (Durán 1999).

subinte' (ah) acacia, bull horn. Fabaceae *Acacia collinsii* Saff., syn. *Vachellia collinsii* (Safford) Seigler & Ebinge. **Se parece ket yetel ah'akunte', chen mas chäk uk'i'ixel. Yan usinik, p'iis 'akunte'. 'Akunte' ubäho'.** 'It looks the same as *'akunte'* (*Acacia mayana*), except that its thorns are redder. Like *'akunte'*, it has ants inside.' (BM) *Acacia collinsii* has the most extensive geographical distribution of all New World "ant-acacias". Like *'akunte'* (*Acacia mayana*), *A. collinsii* exhibits a symbiotic relationship with several species of ant, including *Pseudomyrmex spinicola* and *P. ferruginea*. These fierce, biting ants live in the hollow thorns, deriving nourishment from the tree while protecting the plant from herbivours. It is a small tree, 10 m (30′) in height, with dark grey-brown, shallowly furrowed bark. Twigs are reddish to dark brown and smooth. Leaves are alternate, 40–195 mm (1.6–7.8″) long with 11–29 pairs of leaflets. Stipular spines are light reddish brown to dark brown, terete (more or less cylindrical), straight, stout, inflated, and smooth. The inflorescence is a cylindrical spike of yellow, sessile (stemless) flowers. Fruits are dark brown to black legumes. It grows in dry to moderately wet pastures, fields, and open, shrubby vegetation from sea level to 1000 m (3281′). (Miller 2008) Although my consultants reported no uses for the tree, Baer and Baer (1952) note that the wood is used to make the bird bolt. According to Levy et al. (2006), the timber is used for fuel in the southern Lacandon territory. *Sim:* **'akunte'** '*Acacia mayana*'. *Variant:* **subin (ah)**. *Thes:* **che'**. *[Note:* subin "dragon" (Bruce 1976); sub'in, (aj) sinikche', aj säk sub'in [Itz.]. *Acacia collinsii* (Atran et al. 2004); aj chäk subin [Itz.]. *A. cornigera* (Atran et al. 2004); sùub'in-che' [Yuc.]. "four species of Acacia called by this name" (Bricker et al. 1998: 250); sub'in, sinikte' [Mop.]. *Acacia collinisii* (in, Atran et al. 2004).*] [Source:* BM*]*

subin (ah) acacia, bull horn. *Lit:* 'subin tree'. Fabaceae *Acacia collinsii* Saff. *See:* **subinte' (ah)**.

subul₁ soapberry. Sapindaceae *Sapindus saponaria* L. This is a tree of up to 15 m (49′) tall with a girth of up to 40 cm (15.7″), a straight trunk, and smooth, dark greyish- to dark yellow bark. The grooved branches are horizontal or ascending. Leaves are pinnate with 3–5 pairs of oval or lanceolate leaflets and arranged in a spiral. Flowers are borne in terminal panicles of up 20 cm (7.9″) long. The fragrant blossoms are yellowish green with five petals surrounded by five, overlapping sepals. Bloom time is from August to March. Fruits are berries, either solitary or in groups of 2–3, bright green, and fleshy with 1–3 black seeds. Fruits mature from January to September. The trees are abundant in secondary forests from sea level to 600 or 700 m (1968–2296′). (Pennington and Sarukhán 2005: 340) Nations and Nigh (1980: 23) report that the Lacandones use the timber for house beams. *Loc:* **Lacandón** 'Lacandón'; *Use:* **uuh** 'necklaces'; *Part:* **nek'** 'seeds'. *Variant:* **baba', habonsiyo, habon**. *Thes:* **che'** (Fig. 5.185a, b, c). *[Note:* subul. amole. *Sapindus saponaria* (Nations and Nigh 1980); sijun [Lac.]. *Sapindus saponaria* (Nations 1992, in Atran et al. 2004); zubul [Yuc]. Pennington and Sarukhán (2005); sub'úul, sib'úul [Yuc.]. *Sapindus saponaria* (Bricker et al. 1998); tz'ibuul, sib'ul, sijum [Yuc.]. *Sapindus saponaria* (in, Atran et al. 2004); su'bu'ul [Itz.]. *Sapindus saponaria* (Atran et al. 2004).*] [Source:* K'ayum Segundo; NI*] [*\sd2 craft*]*

subul₂ jungleplum (bustic). Sapotaceae *Dipholis* sp. The photo of the species that K'ayum Segundo called *subul* looks like a *Dipholis* species, with its characteristic straight trunk, uneven crown, grey bark with uneven patches of white, and

Fig. 5.185 (**a**) *Sapindus saponaria* fruit and seed. Photo Credit: Jim Conrad (**b**) *Sapindus saponaria* blossom. Photo Credit: Suzanne Cook (**c**) *Sapindus saponaria* suds. Photo Credit: Suzanne Cook

Fig. 5.186 Subul *Dipholis* sp. Photo Credit: Suzanne Cook

elliptic, petiolate, spirally arranged leaves at the ends of branches. Fruit and flowers were not present when the photo was taken, making a positive identification difficult. Both *D. stevensonii* and *D. salicifolia* occur in Naha' (Comisión Nacional de Áreas Naturales Protegidas 2006; Durán 1999). Although my consultants did not report any uses, *D. stevensonii* is used for tool handles, rifle butts, tourist bows, and fuel by the northern Lacandones (Durán 1999: 149). *D. salicifolia* is used for fuel, crafts, and construction by the southern Lacandones (Levy et al. 2006). Another species, *Dipholis minutiflora,* is used for house construction, according to Levy et al. (2006) and Nigh (2008). *Loc:* **lago del Ocotalito** 'Lake Ocotalito'. *Variant:* **guaité**. *Thes:* **che'** (Fig. 5.186). *[Note:* subur. *Dipholis minutiflora* (Nigh 2008); wuaite, chakia, huech che. *Dipholis salicifolia* (Durán 1999); tz'itzya', sak ya' [Yuc.]. *Dipholis salicifolia* (in, Atran et al. 2004); tz'itz' ya' [Itz.]. *Dipholis salicifolia* (Atran et al. 2004).*] [Source:* K'ayum Segundo*]*

sukal [sukal, sukar] sugarcane. Poaceae *Saccharum officinarium* L. **Sukal p'iis caña. Hay cuatro clases de sukal: hach sukal es lo mas suave; chäk sukal es mas chich; ts'ibal sukal es blanco i negro, tiene rayas; mehen sukal es k'än, chan mihin sukalel, uche'il no crece, es chan mihin. Ahhach sukal uch'ihil mas. Chäk sukal, ooo, ne nuk. Ts'ibal sukal es un poquito menos. Hach sukal hasta nah punta. Chäk sukal, hasta mas de la punta ho'olnah. Chäk sukal i ts'ibal sukal son casi dos metros. Ts'ibal sukal es dos y media mas kabal. No crecen mucho, pero si yan nuk. Yan uya'lil i ch'uhuk. Ma' ubäho' ti' 'ooh tumen sukal uchich uche'il yan uya'li ch'uhuk.** '*Sukal* is the same as sugarcane. There are four kinds: 'authentic sugarcane' is the softest; 'red sugarcane' is the hardest; 'striped sugarcane' is striped white and black; and, 'small sugarcane' is yellow, it's a tiny sugarcane, its stalk doesn't grow (tall). The 'authentic sugarcane' grows taller. 'Red sugarcane', ooo, it's very large. The 'striped sugarcane' is a little smaller (than this). 'Authentic sugarcane' is as high as a rooftop, while 'red sugarcane' grows higher than the roof. The 'striped sugarcane' is 2.5 m (8′) high, but it's large enough. They all have sweet juice. Sugarcane is not related to *'ooh* (*Gynerium sagittatum*), because the stalk of the sugarcane is hard and has sweet juice.' (BM) Sugarcane is a perennial, clumping grass, with stout, jointed, fibrous stalks 2–6 m (6–19′) tall. It likely traces its origin to Papua New Guinea, one of the major centres in the world for sugarcane varieties. It was introduced to the New World by Christopher Columbus. Sugarcane was important to the Lacandones when the traditional religion was still practiced, as it was used to ferment *balche'*, but when refined white sugar was introduced, sugarcane became a secondary crop. Before the introduction of sugarcane, wild honey was used. *Use:* **ch'uhuk** 'sweetener'; *Part:* **che'il** 'stalk'; *Prep:* **Ha'li' amäk'ik unoy uche'il, wa ayäts'ik uya'lil** 'You just chew the pulp or squeeze out the juice'; *Spec:* **hach sukal** 'authentic sugarcane'; *Spec:* **chäk sukal** 'red sugarcane'; *Spec:* **ts'ibal sukal** 'striped sugarcane'; *Spec:* **mehen sukal** 'small sugarcane'. *From:* azúcar (Spanish). *[Note:* sukal. *Saccharum officinarum* (Nations and Nigh 1980); kaanyaj [Itz.]. *Saccharum officinarum* (Atran et al. 2004); ch'uuk [Mop.]. *Saccharum officinarum* (in, Atran et al. 2004); kaanya, newech [Yuc.]. *Saccharum officinarum* (in, Atran et al. 2004).*] [Source:* BM*] [*\sd2 food \sd2 other*]*

suli'$_2$ elephant ears, malanga (Caribean cabbage). Araceae *Xanthosoma mafaffa* Schott. *Use:* **hanal** 'food'; *Part:* **wi'** 'tuber'; *Prep:* **chäkbil** 'boiled'. *Variant:* **hach mäkäl**. *[Note:* suri, makäl (Nations 1992, in Atran et al. 2004); mäkäl [Itz.]. *Xanthosoma yucatanense* (Atran et al. 2004).*] [*\sd2 food*]*

suli' (äh)$_3$ yam (barbasco). Dioscoreaceae *Dioscorea* spp. A third species, *Dioscorea trifida* "cush-cush yam", was observed in a recently developed demonstration garden in Naha', but it is unclear whether this species is normally cultivated, as my Lacandon consultant did not recognize it. According to Nations (2006), indigenous groups in the area stunned fish by casting tubers of the barbasco vine into the water. It is uncertain whether Lacandones also did so, as none of my Lacandon consultants mentioned the practice. *Use:* **hanal** 'food'; *Part:* **wi'** 'tuber'; *Prep:* **chäkbil, pokbil** 'boiled, roasted'; *Spec:* **hach suli'** 'winged yam (*Dioscorea alata*)'; *Spec:* **bilim suli'** 'air potato (*Dioscorea bulbifera*)'. *[Note:* ix pä'yak' [Itz.]. *Dioscorea bulbifera, D. alata* (Atran et al. 2004).*] [Source:* BM; KM; KP; SK*] [*\sd2 food*]*

suli' (äh)$_4$ potato, white. Solanaceae *Solanum tuberosum* L. This is an introduced species of white potato that is purchased from the market or travelling vendors. Some of my consultants said they tried to grow potatoes, but failed. *Use:* **hanal** 'food'; *Part:* **wi'** 'tubers'; *Prep:* **chäkbil, pokbil, tsähbil** 'boiled, fried'; *Sim:* **hach suli'** 'authentic suli' (*Dioscorea alata*)'; *Gen:* **suli'**. *Variant:* **papas**. *[Note:* paapaj [Itz.]. *Solanum tuberosum* (Atran et al. 2004); papa [Yuc.]. *Solanum tuberosum* (in, Atran et al. 2004).*] [Source:* KM; KP; SK*] [*\sd2 food*]*

suumil bits' coffee-shade tree. Fabaceae *Inga oerstediana* Benth. ex Seem. Boege, syn. *Inga minutula* Mart. This is a deciduous tree 6–18 m (9–60′) tall. Smaller branches are densely covered in matted hairs. Leaves with 3–4 pairs of leaflets, oblong to obovate, obtuse acuminate at apex, rachises (leaflet stems) are wide or winged, petiole wide, winged, or cylindrical. Inflorescences occur in spikes. Whispy, white corollas are 9–15 mm long. Fruits are cylindrical, ribbed, fuzzy pods, 8–24 cm (3–9.5″) long by 1–2.5 cm (0.4–1″) wide. (Zamora 1999) The gauzy, white aril is edible. *Loc:* **paakche' kol** 'fallow milpa'; *Use:* **hanal** 'food'; *Part:* **wich** 'fruit'; *Prep:* **che'che'** 'raw'; *Sim:* **mehen bits'** 'small bits' (Inga sp.)'; *Sim:* **t'elen bits'** 'ridged bits' (*Inga sapindoides*)'; *Sim:* **bits'il k'aax** 'forest bits' (*Inga belizensis*)'; *Sim:* **nukuch bits'** 'large bits' (*Inga punctata*)'; *Sim:* **bits' petha'** 'lagoon bits' (*Inga spuria*)'; *Gen:* **bits'**. *Thes:* **che'** (Fig. 5.187a, b). *[Note:* English common name from http://www.plantnames.unimelb.edu.au/Sorting/Inga.html. Retrieved November 2011. Taxonomic identification provided by Nelson Zamora (pers. comm.).*] [Source:* BM*] [*\sd2 food*]*

T–t

taw taw. Ulmaceae *Trema micrantha,* Tiliaceae *Belotia mexicana*. Lacandones distinguish two kinds of taw: *chäk taw* 'red taw' and *hach taw* 'authentic taw'. Both are pioneer species found in secondary forests. *Use:* **bilich** 'mattresses'; *Part:* **hach uyoot'el** 'inner bark'; *Use:* **chemil balche'** '*balche*' canoe'; *Part:* **uche'il** 'trunk'; *Spec:* **chäk taw** 'red taw (*Trema micrantha*)'; *Spec:* **hach taw** 'authentic taw (*Belotia mexicana*)'. *Thes:* **che'** (Fig. 5.188a, b, c). *[Note:* taw. *Belotia mexicana* (Diemont 2006); tao. *Belotia mexicana* (Cheng et al. 2011); mahaua, tabche. *Belotia mexicana* (Durán

Fig. 5.187 (**a**) Suumil bits' *Inga oerstediana*. Photo Credit: Suzanne Cook (**b**) Suumil bits' *Inga oerstediana*. Photo Credit: Suzanne Cook

1999); tao. *Trema micanthra* (Nations 1989).*] [Source:* AM; BM; K'ayum Segundo; SK*] [*\sd2 religion \sd2 furniture \sd2 agriculture*]*

ta' ch'o' ik pepper, bird (chiltepe, chile tepin, bird's eye). *Lit:* 'rat excrement chile'. Solanaceae *Capsicum annuum* var. *glabriusculum* (Dunal) Heiser & Pickersgill. This is type of

Fig. 5.188 (**a**) Taw leaves. Photo Credit: Chan K'in Daniel (**b**) Taw inner bark. Photo Credit: Chan K'in Daniel (**c**) Chäk taw *Trema micrantha* tree. Photo Credit: Chan K'in Daniel

Fig. 5.189 (**a**) Tähte' *Pinus* sp. bark slash. Photo Credit: Suzanne Cook (**b**) Tähte' *Pinus* sp. wood chips. Photo Credit: Suzanne Cook

very small, very hot, red chile pepper. *Use:* **hanal** 'food'; *Part:* **wich**; *Prep:* **che'che'** 'raw'; *Gen:* **ik**. *[Source:* BM; K'ayum Ma'ax*]* *[*\sd2 food*]*

tähte' [tähte', tahte'] pitch pine. *Pinaecae Pinus* spp. There are several species of pine that are known collectively as *ocote* (Spanish). All are high in turpentine and rosins. Two species reported in the Lacandon territory are *Pinus pseudostrobus*, an extremely variable species (Mirov 1954) with a wide geographical and environmental range (Earle 2013a), and *Pinus oocarpa*, which occurs in southern Mexico above 100 m (328') in elevation. Both species are found in the hills surrounding Lake Naha' (Domínguéz-Válzquez and Islebe 2008: 331). *Pinus tenuifolia* is another species that occurs in Lake Ocotalito, and to the west and northeast of Lake Naha' (Comisión Nacional de Áreas Naturales Protegidas 2006:). Still other species found in the same area are *Pinus strobus* (Durán 1999: 30) and *Pinus maximinoi* (Comisión Nacional de Áreas Naturales Protegidas 2006). The former is likely *Pinus strobus* var. *chiapensis* Mart. (See Standley and Steyermark 1958: 55). According to AM, the reason why the northern Lacandones settled around Lake Naha' was due to the abundance of *tähte'* growing there: the pine resin is a vital component in Lacandon traditional religion, serving as one of the main offerings to the gods. Harvesting the resin entails scarring the side of the base of a tree to induce the flow of resin. A leaf is placed under the slash and is left there for a few days, to allow sufficient time for the resin to accumulate. The resin is then collected and brought back to the god house where it is mashed in a small mahogany canoe, called *chemil pom*. The resin is either formed into balls, called *päk'anil pom*, for offering immediately or stored in corn husks until needed. Shards of the wood are sliced off the side of the trunk and used as fire brands and kindling. *Loc:* **petha'** 'lagoon'; *Loc:* **Sa'am** 'Monte Líbano'; *Use:* **kib** 'torches'; *Part:* **che'il** 'wood'; *Use:* **pom** 'incense'; *Part:* **uyits** 'resin'. *See:* **pom**. *Thes:* **che'** (Fig. 5.189a, b). *[Note:* tähte'. pine (Bruce 1975);

täte. pitch pine. *Pinus pseudostrobus* (Nations and Nigh 1980; March 1998); jujub' [Itz.]. 'okootej [Itz.]. *Pinus caribaea* var. *hondurensis* (Atran et al. 2004); huhub [Yuc.]. *Pinus caribaea* (Roys 1931).*] [Source:* AM; BM; KM*]* [\sd2 fuel \sd2 religion*]*

tämän cotton. Malvaceae *Gossypium hirsutum* L. Cotton was widely cultivated when the Lacandones wove their own fabric. But when commercial fabric became available, cotton became a secondary crop. A few plants are still grown for handspun thread used for curing strings. *Gossypium hirsutum* is native to Mexico and Central America. In the USA, it is referred to as American Upland cotton (Cotton Incorporated). *Use:* **k'uch** 'thread'. *[Note:* tämän. *Gossypium* L. (Nations and Nigh 1980); yuryur tuxer, tusir tämän, ax tux [S. Lac.] (Baer and Merrifield 1971); tämän [Itz.]. *Gossypium hirsutum* (Atran et al. 2004); tämän [Mop.]. *Gossypium hirsutum* (in, Atran et al. 2004); ajtaman [Yuc.]. *Gossypium hirsutum* (in, Atran et al. 2004).*]* [\sd2 cordage*]*

tämänil (äh) hibiscus. Malvaceae *Hibiscus rosa-sinensis* L. This is an introduced species, ubiquitous in hedge rows and Lacandon house gardens. The Lacandon name refers to the flower, because it resembles that of *tämän* 'cotton plant'. *[Note:* tz'utz'la' [Itz.]. *Hibiscus rosa-sinensis* (Atran et al. 2004); tulipan [Yuc.]. *Hibiscus rosa-sinensis* (in, Atran et al. 2004).*] [Source:* AM*]*

tämän che' Turkcap (Turk's turban, wax mallow, ladies teardrop, Scotchman's purse). *Lit:* 'cotton tree'. Malvaceae *Malvaviscus arboreus* Cav. This is a species of flowering tree in the hibiscus family. It bears small, red, crepe-like blossoms similar to hibiscus flowers. *See:* **tämän**; **tämänil (äh)**. *Thes:* **che'** (Fig. 5.190). *[Note:* ix tulipanil k'aax [Itz.]. *Malvaviscus arboreus* (Atran et al. 2004); tämän che' [Itz.]. *Torrubia petenensis (T. linearibracteata)* (Atran et al. 2004); ixb'isil, taman ch'iich' [Yuc.]. *Malvaviscus arboreus* (in, Atran et al. 2004).*] [Source:* KAG*]*

tekoox (äh) pokeweed, tropical (red inkplant). Phytolaccaceae *Phytolacca purpurascens* A. Braun & Bouché, syns. *Phytolacca icosandra* L., *Phytolacca octandra* L. *Phytolacca* are species of flowering plants native to the Neotropics. *P. purpurascens* is an annual or short-lived perennial, up to 2 m (6.5′) tall with branching, hollow stems and an angular stalk. Leaves are elliptical or oval-elliptical, 7–20 cm (2.7–7.8″) long and 2.5–9.5 cm (1–3.7″) wide. White or pink flowers are borne in racemes 8–15 cm (3.1–5.9″) long, followed by clusters of small, black berries 6–8 mm (0.27–0.3″) in diameter. The fruit is slightly lobed and contains 12–16 seeds. The species is similar to *Phytolacca rivinoides*, whose flowers are borne on relatively long pedicels 7–12 mm (0.27–0.5″). (Mondragón-Pichardo & Vibrans 2009) AM says that the plant is useful in the *milpa*, because the abundance of leaves it sheds nourishes the soil. In the past, the berries were used as laundry detergent. *Use:* **jabon ti' nook'** 'laundry soap'; *Part:* **wich** 'fruit'; *Prep:* **Ahixtik anook' yetel uwich yetel chichin ha'** 'You scrub your clothes with the fruit and a little water'. *Thes:* **che'**. *[Note:* x-tel-kox [Yuc.]. Phytolacca purpurascens (Martínez 1979). Synonyms from The Plant List. 2010. Version 1. Common names from Mondragón-Pichardo & Vibrans (2009).*] [Source:* AM; BM; KM; KyP*]* [\sd2 agriculture \sd2 other*]*

Fig. 5.190 Tämän che' *Malvaviscus arboreus* flower. Photo Credit: Chan K'in Daniel

te' 'usin ginger, red. *Lit:* 'king vulture tree'. Zingiberaceae *Alpinia purpurata* K. Schum. This is an herbaceous plant of the ginger family. It is between 0.9 4.57 m (3-15′) tall and 0.6–1.2 m (2–4′) wide. The leafy, cane-like stems arise from rhizomes, spreading laterally in thick clusters and producing aerial shoots. The deep green leaves are alternate and sessile (lacking a petiole) with long sheaths that wrap around the stem. Leaf blades are oblong, 30.5–81 cm (12–32″) long and 10–22.8 cm (4–9″) wide and have a pointed apex. Inflorescence stems can grow to 20.5 cm (12″) long. Fruits are nearly round capsules approximately 10–15.24 cm (4–6″) long and 1.9 cm (0.75″) in diameter. Seeds are small, long, black, and oily (Kobayashi et al. 2007: 1). The species was introduced from the Pacific (Axel Poulsen, pers. comm.). Most Lacandones cultivate it for its ornamental appeal and its leaves. *Use:* **ti' päk'äch wah** 'for patting out tortillas'; *Use:* **nahwah** '(ceremonial) tamales'; *Part:* **le'** 'leaves'; *Use:* **tsoy uwich** 'ornamental'. *See:* **komo' (äh)**. *Thes:* **xiiw** (Fig. 5.191). *[Note:* te usir. *Maranta divaricata* (Durán 1999).*] [Source:* AM; BM; KP*]* [\sd2 food \sd2 other*]*

ton bo'oy chamaedorea (type of). *Lit:* 'male bo'oy'. Arecaceae *Chamaedorea* sp. This is a short kind of *bo'oy* with

Fig. 5.191 Te' usin *Alpinia purpurata* inflorescence. Photo Credit: Suzanne Cook

small leaves. No uses were reported. *Loc:* **k'aax** 'forest'; *Sim:* **hach bo'oy** 'authentic bo'oy'; *Sim:* **ch'ibix** (*Chamaedorea spp.*); *Sim:* **säk bo'oy** 'white bo'oy'; *Sim:* **k'än bo'oy** 'yellow bo'oy'; *Sim:* **chan mehen bo'oy** 'very small bo'oy'; *Sim:* **ch'it bo'oy** (*Chamaedorea* sp.); *Gen:* **bo'oy**. *[Source:* AM*]*

ton ku'uk [ton ku'uk, ton kuuk, ton kuk] ear flower. *Lit:* 'male squirrel/squirrel's testicles'. Annonaceae *Cymbopetalum* sp. **P'isil maahas. Uche'il säk. Ma' chich. Uch'ihil ich k'aax. Ubakel ka'anal como 20 metros. Umentik pakat. Yan uwich. Ya'ax uwich. Chukuch, batak 5 cm (2″). Ch'iich' uhantik uwich.** 'It's the same as *mahaas* (*Quararibea funebris*). Its trunk is white and its wood is soft. It grows in the forest. It's about 20 m (65′) tall. It's used to make the bird bolt. It has green fruit, about 5 cm (2″) long. The birds eat the fruit.' (BM) *Cymbopetalum* are trees or shrubs with leaves arranged alternately along the branches. Flowers are large with an inner whorl of thick, fleshy, pouch-shaped petals. Solitary blossoms are borne terminally or in axils, followed by dry, multi-seeded fruits. It grows in the perennial tropical forests of Campeche, Chiapas, Oaxaca, Tabasco, and Veracruz. (Bye and Linares 1990:161) *Cymbopetalum penduliflorum* and *C. baillonii* have been recorded from the northern Lacandon forest (Durán 1999; Pennington and Sarukhán 2005). Although my consultants reported no medicinal or food uses for the species, Kashanipour and McGee (2004) note that Lacandones drink a tea made from the leaves to reduce fevers. The wood is used in house construction, for oars, handles, and planting sticks, and the flowers are eaten with various kinds of meat dishes (Durán 1999). *Loc:* **k'aax** 'forest'; *Use:* **pakat** 'bird bolt'; *Part:* **che'il** 'wood'; *Sim:* **mahaas (ah)** 'funeral tree (*Quararibea funebris*)'. *Thes:* **che'**. *[Note:* k'uk (Baer and Baer 1952); ton kuk, ton k'uk (Kashanipour and McGee 2004); ton kuk. *Cymbopetalum penduliflorum* (Durán 1999); tan cu'uc. *Cymbopetalum baillonii* (Pennington and Sarukhán 2005).*] [Source:* BM*] [*\sd2 hunting_and_fishing*]*

ton tsimin huevos de caballo (English name unknown). *Lit:* 'male horse, horse's testicles'. Apocynaceae *Tabernaemontana amygdalifolia* Jacq. This is a small understory tree 5–8 m (16.4–26.25′) tall with smooth branches. White flowers, 5 × 3 cm (2 × 1″), are borne in cymes from leaf axils. Each blossom is composed of a narrow tube that flares out into five, spatulate petals spirally arranged around the anther. Flowers are followed by large, 10 × 7 mm (3 × 1″), asymmetrically curved pairs of hanging fruits, which split open exposing bright orange arils encasing numerous, small seeds. Leaves are opposite, simple, elliptical, entire, and acuminate. According to BM, when the fruits split open in July, birds flock to feed on the seeds. The southern Lacandones are reported to use the wood for fuel, and the latex (Diemont 2006; Levy et al. 2006). *Loc:* **k'aax** 'forest'; *Use:* **che'il k'ak'** 'firewood'. *Thes:* **che'**. *[Note:* ton tzimin. *Stemmadenia donnell-smithii* (Nations 1992, in Atran et al. 2004); ton simin. *Tabernaemontana amygdalifolia* (Diemont 2006; Levy et al. 2006); ton simin. *Psychotria flava* (Durán 1999); ix tontsimin [Itz.]. *Stemmadenia donnell-smithii* (Atran et al. 2004).*] [Source:* BM*] [*\sd2 fuel \sd2 other*]*

tuch thevetia, broadleaf (cascabel, dog's tongue). *Lit:* 'navel'. Apocynaceae *Thevetia ahouai* (L.) A. DC. This is an evergreen shrub or tree with glossy, dark green, ovate leaves and bright red, slightly lobed fruit with a milky sap. According to my Lacandon consultants, the wood is extremely hard and is therefore useful for construction purposes. Four species have been recorded from Chiapas (Breedlove 1986): *Thevetia ahouai* (L.); *T. ovata* (Cav.) A. DC.; *T. peruviana* (Pers.) Schum.; and, *T. plumeriaefolia* Benth. Only *T. ahouai* is recorded from the Lacandon forest (Comisión Nacional de Áreas Naturales Protegidas 2006; Durán 1999; Levy et al. 2006). According to my consultants, the species corresponds to both *mehen tuch* 'small tuch' and *hach tuch* 'authentic tuch'. They identify another kind, *ya'ax tuch* 'green tuch', which was unavailable for inspection. *Loc:* **k'aax** 'forest'; *Use:* **okman** 'house posts'; *Part:* **che'il** 'wood'; *Use:* **k'anche' näl** 'corncrib'; *Part:* **che'il** 'wood'; *Use:* **ts'ak ti' chibal koh** 'medicine for toothache'; *Use:*

ts'ak ti' yah uchi' 'medicine for sore mouth'; *Part:* **uyits, ule'** 'resin and leaves'; *Prep:* **Asuulik ich chäkal ha', pachil achalik achi' yetel** 'You steep it in hot water and then you rinse your mouth with it'; *Use:* **ts'ak ti' yahil** 'medicine for wounds'; *Part:* **uyits, ule'** 'resin and leaves'; *Prep:* **Amuxik i pulik yok'ol yahil** 'You crush it and put it on the wound'; *Use:* **hanal** 'food'; *Part:* **wich** 'fruit'; *Prep:* **che'che'** 'raw'; *Spec:* **hach tuch** 'authentic tuch'; *Spec:* **mehen tuch** 'small tuch'; *Spec:* **ya'ax tuch** 'green tuch'; *Spec:* **tuchil k'aax** 'forest tuch'. *Thes:* **che'**. *[Note:* tuch. *Thevetia ahouai* (Durán 1999).*] [Source:* BM*] [*\sd2 housing \sd2 food \sd2 medicinal*]*

tuchil k'aax thevetia, broadleaf (cascabel, dog's tongue). *Lit:* 'forest tuch'. Apocynaceae *Thevetia ahouai* (L.) A. DC. *See:* **tuch**. *Variant:* **hach tuch**. *[Source:* BM*]*

tuk' palm, coyol (grugru palm, macaw palm, macauba palm). Arecaceae *Acrocomia mexicana* Karw. ex Mart., syn. *Acrocomia aculeata* (Jacq.) Lodd. ex Mart. This is a large, tall palm with a spiny trunk. It bears clusters of yellowish-green fruits, 2.5–5 cm (1–2″) in diameter, with a smooth, brittle skin surrounding a softish, white pulp. It grows in disturbed areas, savannahs, and open woodlands in areas of seasonal rainfall at low altitudes (Henderson et al. 1995: 166). The spines contribute to a microhabitat, collecting fallen leaves from the surrounding trees; funnel-web spiders anchor their webs among the spines, in an effort to trap insects; in turn, the insects and spiders attract birds and larger creatures, whose droppings add phosphorus and other nutrients to the leaf litter. *Loc:* **k'aax** 'forest'; *Use:* **hanal** 'food'; *Part:* **wich** 'fruit'; *Use:* **tukil** 'spindle whorl'; *Part:* **usool uwich** 'fruit shell'; *Sim:* **tuts** '*Attalea butyracea* (?)'. *[Note:* tuk'. *Acrocomia mexicana* (Bruce 1975); tuk' [Itz.]. cocoyol palm. *Acrocomia mexicana* (Hofling and Tesucún 1997); ya'ax tuk', k'än tuk', sustuk' [Itz.]. *Acrocomia mexicana* (Atran et al. 2004); tuk' [Yuc.]. palm tree (Bricker et al. 1998).*] [Source:* BM; SK*] [*\sd2 craft \sd2 food*]*

tuts palm, corozo. Arecaceae *Attalea butyracea* (Mutis *ex.* L.*f.*) Wess Boer.(?) According to BMjr, tuts is similar to *tuk'* but without the spines. *Loc:* **Yaxchilan**; *Use:* **hanal** 'food'; *Part:* **uyool** 'heart'; *Prep:* **pokbil** 'roasted'; *Use:* **Akäxik unek' ti' uk'äb ti' mehen paalal'** 'You tie the seeds to babies' wrists (to amuse them)'; *Use:* **uuh** 'necklaces'; *Part:* **nek'** 'seeds'; *Use:* **miis** 'broom'; *Sim:* **tuk'** '*Acrocomia mexicana*'. *[Note:* tuch. Scheelea preussi (Nations 1992, in Atran et al. 2004); tutz . *Scheelea liebmannii* (Nations and Nigh 1980); tuts. *Attalea cohune* (Nations 2006); k'än tutz [Itz.]. *Scheelea liebmannii* or *Elaeis malanocca* (Bruce 1975); tutz, aj/ix säk tutz = jach tutz [Itz.]. *Orbignya cohune* (Atran et al. 2004); aj/ix k'än tutz [Itz.]. *Scheelea lundellii* (Atran et al. 2004); mop, ajtutz' [Yuc.]. *Orbignya cohune* (in, Atran et al. 2004); säk tutz [Mop.]. *Orbignya cohune* (in, Atran et al. 2004). Attalea is a large and complex genus, often split into four genera; the only way to distinguish species is by the male inflorescence. (Henderson et al. 1995: 154)*] [Source:* BM; BMjr; NI; SK*] [*\sd2 food \sd2 craft \sd2 folklore_and_mythology \sd2 other*]*

Fig. 5.192 Tu'xikin *Ipomoea* sp. flowers. Photo Credit: Suzanne Cook

tu'xikin morning glory. *Lit:* 'rotten ear'. Convolvulaceae *Ipomoea* spp. The name refers to a number of species of morning glory with purple, purple-red, or pink flowers, because they are believed to cause ear infections if one plays with them too much. The Yucatec Maya share the same belief: Bricker et al. (1998: 258) explain that children are warned not to play with the flowers lest their ears rot. The origin of this belief is unclear, but it has been suggested that it goes back to a time when the vine was used for psychedelic purposes by religious specialists; blue morning glories, particularly *Ipomea violacea*, are used for their hallucinogenic properties by a number of Mexican indigenous groups (PsychoTropia 2011). In northern Lacandon cosmology, the morning glory is associated with *Kisin* (the Devil), who wore the flower as a hat. *Spec:* **nukuch tu'xikin** 'large rotten ear'; *Spec:* **mehen tu'xikin** 'small rotten ear'. *Thes:* **ak'** (Fig. 5.192). *[Note:* tu' xikin (*Ipomoea violacea*), mehen tu' xikin (*Aristolochia foetida*), nukuch tu'xikin (*Aristolochia grandiflorax*) (Rätsch 1994a); tu'u-xiknil [Yuc.]. morning glory. *Ipomoea crinicalyx* (Bricker et al. 1998). The species in the photo (Fig. 5.192) was identified as *Ipomoea tuxtlensis* by Dan Austin, Andrew MacDonald, and Eleazar Carranza (pers.comm.).*] [Source:* AM; BM; KM; KP*] [*\sd2 folklore_and_mythology*]*

T'–t'

t'elen bits' mountain bribri [Bel. Eng.]. *Lit:* 'ridged bits''. Fabaceae *Inga sapindoides* Willd., syn. *Inga pavoniana* G. Don. This is a small, 6–16 m (19.7–53.5′) tree. Branches are pubescent and angular. Leaflets are opposite, usually 3–4 pairs per stem, lanceolate, 9–28 × 5–13 cm (3.5–11 × 1.96–5″) with obtuse to acute apices, glossy blades, and winged rachises (leaflet stems). Petioles may also be winged. Pompom-like flowers are borne on long flower spikes. Fruit is a smooth, yellowish-green, ridged pod 11–30 × 2–3 × 1.5 2.5 cm (4.3–12 × 0.8–1.2 × 0.6 × 1″). The pod is filled with a sweet, white, gauzey aril encasing numerous, large seeds. It grows along road sides, forest edges, riverbanks, and in secondary vegetation. Its range is from southern Mexico to Peru and northern Venezuela. *Loc:* **el campamento** 'Frans and Trudi Blom's old base camp'; *Use:* **hanal** 'food'; *Part:* **wich**; *Prep:* **che'che'** 'raw'; *Use:* **che'il k'ak'** 'firewood'; *Sim:* **suumil bits'** 'rope bits' (*Inga oerstediana*)'; *Sim:* **mehen bits'** 'small bits' (*Inga* sp.)'; *Sim:* **bits'il k'aax** 'forest bits' (*Inga belizensis*)'; *Sim:* **nukuch bits'** 'large bits' (*Inga punctata*)'; *Sim:* **bits' petha'** 'lagoon bits' (*Inga spuria*)'; *Gen:* **bits'**. *Variant:* **t'elt'el bits'**. *Thes:* **che'** (Fig. 5.193). *[Source:* BM*]* *[*\sd2 food*]*

Ts–ts

tsaah (äh) devil nettle (bull nettle). Euphorbiaceae *Cnidoscolus multilobus* (Pax.) I. M. Johnston, syn. *Jatropha multiloba* Pax. This is a tall shrub that bears a large flower head with a cluster of small white flowers, composed of five petal-like sepals united at the base borne on stiff, upright, branching stems. Leaves are up to 25 cm long and 30 cm (12″) wide, deeply divided into 5 lobes, rounded-cordate at the base, and coarsely toothed on the margins. At the apex of the petiole (leaf stalk) is a gland. Numerous stinging hairs cover the veins, petioles, and inflorescences. These hairs deliver an intense sting that can last for hours. According to SK, a charm is recited to relieve the sting of the plant. A pioneer plant, it colonizes jungle clearings from sea level to 2600 m (8530′). Flowers and fruits from February to September (D. R. Biblioteca Digital de la Medicina Tradicional Mexicana 2009; Hanan-Alipi & Mondragón-Pichardo 2009c). No uses were reported. *See:* **p'ap'ax**. *Variant:* **mala mujer**. *Thes:* **che'** (Fig. 5.194a, b, c, d, e). *[Source:* AT; BM; KA; KyP; SK*]* *[*\sd2 folklore_and_mythology*]*

Fig. 5.193 T'elen bits' *Inga pavoniana* fruit and leaves. Photo Credit: Suzanne Cook

tsaayentsaay (ah)$_1$ bean, rice (small red bean). *Lit:* 'spliced together here and there'. Fabaceae *Vigna umbellata* (Thunb.) Ohwi & H. Ohashi., syn. *Phaseolus calcaratus* Roxb. This is a vigourous, herbaceous climber with thin stems and small, butter-yellow, pea blossoms. The vines tend to become weedy and invasive. According to AM, the seed is red and very small, about the size of a grain of rice. Because of this, he calls the seeds (and the plant) *chäk bu'ul* 'red bean'. KP says they are good to eat, but stresses that they are not beans. *Vigna* and *Phaseolus* are closely-related genera that form a complicated taxonomic group called the Phaseolus-Vigna complex. *Vigna umbellata* is believed to have been domesticated in Southeast Asia (Ohashi et al. 1988). Pods are smooth, cylindrical, and curved, 6–13 cm × 0.3–0.6 cm, containing 6–8 tiny, oblong seeds in a variety of colours—yellow, green, dark red, brown, black, and speckled. *Loc:* **kolil nah** 'house garden'; *Use:* **chäkbil** 'boiled'; *Part:* **nek'** 'seeds'. *Morph:* **tsaay-en-tsaay**. *Variant:* **chäk bu'ul** (Fig. 5.195a). *[Note:* arrozbur. rice bean. *Phaseolus calcaratus* (Diemont 2006).*]* *[Source:* AM; KP*]* *[*\sd2 food*]*

tsaayentsaay (ah)$_2$ ticktrefoil, Dixie. Fabaceae *Desmodium* spp. *Desmodium* is a rampent, herbaceous plant with small, pea blossoms, sticky stems, and fuzzy strings of loments, each encasing a single seed. The loments break off easily on clothing and fur, the main vehicles by which the seeds are dispersed. Considered *lo'obil* 'a weed', it is found in unkempt house gardens, *milpas*, and other disturbed sites. No uses were reported. *Loc:* **kol, kolil nah, paakche' kol** 'milpa, house garden, fallow milpa' (Fig. 5.195b). *[Note:* tzap pu xibi, tza tza le ak. *Desmodium macrodesmum* (Durán 1999); tzayaltzay [Yuc.]. *Anoda triangularis* DC.; tzayaltzay. "A plant that clings to clothing." (Roys 1931: 290); tsáayentsáay [Yuc.] "spliced here and there (several objects)." (Bricker et al. 1998: 41). English common name and scientific synonyms from the USDA Plants database (http://plants.usda.gov/java/profile?symbol=DETO)*]* *[Source:* KM; KP; NI*]*

tsatsel pom [tsatsel pom, tsatse pom taasi pom] copal. *Lit:* 'pom's grease'. Burseraceae *Protium copal* (Schltdl. & Cham.) Engl. **Ubäho' ti' hach pom.Yan uyits.** 'It's the brother of

Fig. 5.194 (**a**) Tsaah *Cnidoscolus multilobus* leaves. Photo Credit: Suzanne Cook (**b**) Tsaah *Cnidoscolus multilobus* leaf, top side. Photo Credit: Suzanne Cook (**c**) Tsaah *Cnidoscolus multilobus* unopened flowers. Photo Credit: Suzanne Cook (**d**) Tsaah *Cnidoscolus multilobus* plant. Photo Credit: Suzanne Cook (**e**) Tsaah *Cnidoscolus multilobus* stinging hairs. Photo Credit: Suzanne Cook

Fig. 5.195 (a) Tsaayentsaay *Vigna umbellata*. Photo Credit: Suzanne Cook (b) Tsaayentsaay *Desmodium tortuosum*. Photo Credit: Suzanne Cook

hach pom 'authentic copal'. It has resin.' (AM) According to BM, the tree is very tall, perhaps as tall as *hach pom*. The leaves are long. When crushed, they emit a pungent but pleasant odour. Like *hach pom*, it has small, round fruit, about 12.7 mm (1/2″) in diameter, with red skin and white flesh. Whereas the species observed is *Protium copal*, Nations and Nigh (1980) and Baer and Merrifield (1971) note that "tasi pom" corresponds to *Bursera simaruba*. My consultants and those of Durán referred to *Bursera simaruba* as *chäklah*. *Use:* **pom** 'incense'; *Part:* **uyits** 'resin'; *Use:* **hanal** 'food'; *Part:* **wich** 'fruit'; *Prep:* **che'che'** 'raw'; *Sim:* **hach pom** 'authentic pom'; *Gen:* **pom**. *Variant:* **taasil pom**. *Thes:* **che'** (Fig. 5.196a, b). *[Note:* tatsi pom. *Protium copal* (Durán 1999); tasi pom. *Bursera simaruba* (Nations and Nigh 1980); tasi' pom. *Bursera simaruba* (Baer and Merrifield 1971).*] [Source:* AM; BM; SK*] [*\sd2 food \sd2 religion*]*

tsäk'ats [tsäk'ats, tsäk'äts] sansapote. Chrysobalanaceae *Licania platypus* Fritsch., syn. *Moquilea platypus* Hemsl. **Es che'. Uwich ne ch'uhuk. Ahantik.** 'It's a tree. The fruit is very sweet. You eat it.' (AM) This is a large, deciduous tree, 30–50 m (100–160′) tall. Abundant, fragrant flowers are borne in broad, terminal panicles 10–35 cm (4–14″) long. Blossoms are small, densely hairy with recurved petals and numerous protruding stamens. Flowers are followed by one to three obovoid fruit, 13–20 cm (5–8″) long and 10–14 cm (4–5.6″) wide. Fruit has a thin, dark-brown or reddish, warty rind. The flesh is orange-yellow, soft, fibrous, dry or juicy and somewhat pumpkin-scented. Each fruit has one, rounded or ovate-oblong, flattened seed, 6–10 cm (2.5–4″) long. (Morton 1987: 113–114) In the Lacandon forest, the tree is generally found between 300 and 600 m (984–1968.5′) in elevation. The tree blooms from July to September and the fruits ripen from August to December. (Pennington and Sarukhán 2005: 196) *Use:* **hanal** 'food'; *Part:* **wich** 'fruit'; *Prep:* **che'che'** 'raw'; *Use:* **ts'ak ti' hubul hämnen** 'medicine for diarrhea'; *Part:* **nek'** 'seed'; *Prep:* **Achäkik uwich ich ha' pachil awuk'ik. Ne k'ah.** 'You boil the seed in water and then you drink it. It's very bitter.'; *Use:* **ts'ak ti' chibal hämnen** 'medicine for stomach ache'; *Part:* **le'** 'leaves'; *Prep:* **te'** 'tea'. *Thes:* **che'** (Fig. 5.197a, b, c, d). *[Note:* Medicinal treatment for stomach ache provided by Kashanipour and McGee (2004). säk'atz. *Licania platypus* (Kashanipour and McGee 2004); säk'ätz. *Licania platypus* (Nations 2006); säkatz. mesonsapote. *Licania platypus* (Nations and Nigh 1980); tzakatz. *Licania platypus* (Nations 1992, in Atran et al. 2004); tzok'otz [Itz.]. *Licania platypus* (Atran et al. 2004).*] [Source:* AM; BM; K'ayum Segundo*] [*\sd2 medicinal \sd2 food*]*

tsäläm (ah) [tsäläm, tsäräm, tsärän] white leadtree (jumbay, white popinac). Fabaceae *Leucaena leucocephala* (Lam.) de Wit var. *glabrata* (Rose) S. Zárate, syns. *Acacia leucocephala* (Lam.) Link, *Leucaena glauca* Benth., *Mimosa glauca* sensu L., *Mimosa leucocephala* Lam., *Delonix regia* (Boj. ex Hook.) Raf. The name refers to two species of tree. One is *Leucaena leucocephala*, a large tree with bipinnate leaflets and small, yellow-cream, pompom-like flowers and rusty-brown, pendant pods containing numerous, small, glossy-brown, slightly flattened seeds marked with a faint, dark halo inside the seed margin. The other, *Delonix regia*, is a large, open-crown tree with fern-like leaves and scarlet blossoms; fruit is a large, woody pod containing elongated, brown and beige seeds; it is known as royal poinciana, flamboyant, and flame tree, and blooms from June to early July, in Naha'. Ornamental. *Loc:* **kolil nah, k'aax** 'house garden, forest'; *Use:* **uuh** 'necklaces'; *Part:* **wich** 'seed'; *Use:* **chem** 'canoe'; *Part:* **chunche'** 'trunk'. *Thes:* **che'** (Fig. 5.198a, b, c, d, e). *[Note:* saran. *Leucaena pulverulenta* (Durán 1999); tzalam [Yuc.]. *Lysiloma bahamensis* Benth., *L. latisiliqua* L. (Roys

a

b

Fig. 5.196 (**a**) Tsatsel pom *Protium copal* leaves. Photo Credit: Suzanne Cook (**b**) Tsatsel pom *Protium copal* fruit. Photo Credit: Suzanne Cook

1931); tzälam [Itz.]. *Lysiloma bahamensis* (Atran et al. 2004); b'ayniiyoj [Itz.]. *Leucaena leucocephala* (Atran et al. 2004); waxim [Yuc.] *Leucaena leucocephala* (in, Atran et al. 2004).*] [Source:* AM; BM; KM; KP; NI*]* [\sd2 transportation \sd2 craft*]*

tsits shrimp plant. Acanthaceae *Justicia* sp. **Chäk utop', as chukuch. Yan uyak'il utop'. Utop' estan como plumas**. 'It has red, somewhat long flowers. The flowers have vines. The flowers are like feathers.' (BM) The species BM describes is an evergreen shrub with opposite, lanceolate, glossy leaves that have strongly nerved, bullate (puckered) blades. Lacandones cultivate it for its showy, tubular flowers and the leaves, which yield a dark, purplish-blue dye. *Loc:* **kolil nah** 'house garden'; *Use:* **ti' banik nook'** 'dying fabric'; *Part:* **le'** 'leaves'; *Prep:* **upokik ule' ich k'ak', pachil usulik ich ha'** 'One roasts the leaves in the fire and then soaks them in water'. *Thes:* **ak'** (Fig. 5.199). *[Note:* tz'itz'. Trichospermum mexicanum (Nations 2006); tzitz [Yuc.]. "Probably *Jacobina spicigera*, (Schlecht) Bailey. (Standl.) … A shrub 3–5 feet high. Leaves yield a purple dye. … Tzitz-haa means to sprinkle holy water, the blue color […] used in connection with ceremonies" (Roys 1931: 290); tzitz, sik'in [Itz.]. *Jacquinia aurantiaca* (Atran et al. 2004); tzitz [Itz]. palo de añil, chaperno. bribri tree. *Inga spp., I. vera, I. leptoloba* (Hofling and Tesucún 1997).*] [Source:* BM; CNK; KM; KP*]* [\sd2 craft*]*

tso'ots$_1$ loquat (Japanese medlar, Japanese nispero). *Lit:* 'hair'. Rosaceae *Eriobotrya japonica* (Thunb.) Lindl. This is a medium-sized evergreen tree with large, long, dark green, lanceolate leaves with serrated margins. It bears compact clusters of fuzzy, yellow to orange, oval fruits about the size of hog plums. Yellow-orange, juicy pulp surrounds one, large, smooth seed. In Naha', the fruit begins to ripen in October. Cultivated in house gardens, the loquat is indigenous to southeastern China and southern Japan. The Japanese variety is the one grown in Naha', according to Morton's description (Morton 1987: 103–108). *Loc:* **kolil nah** 'house garden'; *Use:* **hanal** 'food'; *Part:* **wich** 'fruit'; *Prep:* **che'che'** 'raw'. *Thes:* **che'** (Fig. 5.200a, b, c, d, e). *[Source:* CKD*]* [\sd2 food*]*

tso'ots bamax cherry (Bel. Eng.). *Lit:* 'hairy bamax'. Moraceae *Pseudolmedia oxyphyllaria* Donn. Sm., syn. *Pseudolmedia glabrata* (Liebm.) C. C. Berg. **Tso'ots bamax yan uwich, wolis uwich, batak 1.27-2.45 cm (1/2–1″). Chan tso'otsel upach. Yan ka'p'eli': k'än i chäk. Abril utahal. Uch'ihil ich k'aax yok'ol 'äkan lu'um. Ba'ats', k'ambul, tuwolol ch'iich' umäk'ik uwich. Hale' i tsub umäk'ik unek'.** '*Tso'ots bamax* has round fruits, about 1.27–2.45 (1/2″-1″) in diameter. Their peels have small hairs. There is a red and a yellow kind. Both ripen in April. The

Fig. 5.197 (**a**) Tsäk'ats *Licania platypus* cultivated tree. Photo Credit: Suzanne Cook (**b**) Tsäk'ats *Licania platypus* fruit. Photo Credit: Chan K'in Daniel (**c**) Tsäk'ats *Licania platypus* leaves. Photo Credit: Suzanne Cook (**d**) Tsäk'ats *Licania platypus* small tree. Photo Credit: Suzanne Cook

tree grows in the forest, in the low-lying flat areas. Howler monkeys, curassows, and all the birds eat the fruit. The spotted cavy and the squirrel eat the seeds.'(BM). According to SK, the fruit is a little bit sour. When slashed, the trunk exudes copious amounts of yellow latex. SK says that the yellow latex distinguishes *bamax* from other kinds of trees with latex. *Tso'ots bamax* and *hach bamax* are both identified as *Pseudolmedia oxyphyllaria* in Durán (1999). The species in the photograph (Fig. 5.203a) here may be *Pseudolmedia oxyphyllaria*, according to Servando Carvajal (pers. comm.). *Use:* **hanal** 'food'; *Part:* **wich** 'fruit'; *Prep:* **che'che'** 'raw'; *Use:* **säkche'** 'rafters'; *Part:* **mehen che'** 'saplings'; *Spec:* **k'än tso'ots bamax** 'yellow bamax'; *Spec:* **chäk tso'ots bamax** 'red bamax'; *Sim:* **hach bamax** '*Pseudolmedia*

Fig. 5.198 (**a**) Tsäläm *Leucaena leucocephala* trees. Photo Credit: Suzanne Cook (**b**) Tsäläm *Leucaena leucocephala* leaves, seeds, and flowers. Photo Credit: Suzanne Cook (**c**) Tsäläm *Leucaena leucocephala* seed pods. Photo Credit: Suzanne Cook (**d**) Tsäläm delonix regia. Photo Credit: Suzanne Cook (**e**) Tsäläm delonix regia. Photo Credit: Suzanne Cook

oxyphyllaria'; *Gen:* **bamax (äh)**. *Thes:* **che'** (Fig. 5.201a, b). *[Note:* tzot bamax, has bamax. *Pseudolmedia oxyphyllaria* (Durán 1999). Common name from Grandtner (2005).*]* *[Source:* BM; SK*]* *[*\sd2 housing \sd2 food*]*

tsuk'in ak' (ah) [tsuk'in ak', tsuk'ir ak', tsuk'i 'ak'] sarsaparilla. Smilacaceae *Smilax domingensis* Willd. **Hay dos clases: chäk tsuk'in ak' i ya'ax tsuk'in ak'. Uk'anik ti' häxik käkow. Ne yaab uyoom. Ne chich umuk' ahya'ax tsuk'in. Yan uk'i'ixel. Uch'ihil ich paakche' kol wo ich k'aax.** 'There are two kinds: 'red *tsuk'in* vine' and 'green *tsuk'in* vine'. Both are used to froth cacao. They make a lot of foam. The green *tsuk'in* vine is very strong. It has thorns. It grows in fallow *milpas* and in the forest.' (BM) According to Durán (1999), the root of *"suki ak"* (*Gouania lupuloides*) is fermented with honey to make an alcoholic beverage. *Loc:* **k'aax, kol** 'forest, milpa'; *Use:* **ti' uhäxik käkow** 'for beating cacao (to make it foam)'; *Part:* **uyits uyol** 'resin from the growing tips'; *Prep:* **Ubäxik uyol, uyäts'ik uyits, uhäxik yetel käkow** 'One pounds the growing tips and squeezes out the resin, and then beats it with cocoa'; *Spec:* **chäk tsuk'in ak'** 'red tsuk'in vine (*Smilax* sp.)'; *Spec:* **ya'ax tsuk'in ak'** 'green tsuk'in ak' (*Smilax* sp.)'. *Thes:* **ak'** (Fig. 5.202a, b). *[Note:* shukur. *Smilax domingensis* (Nigh 2008); suki ak. *Gouania lupuloides* (Durán 1999); suqir (Baer and Merrifield 1971); äksukir. grass (Davis 1978).*]* *[Source:* AM; BM; JK*]* *[*\sd2 food*]*

Ts'–ts'

ts'ak ahch'up *Cissus biformifolia* (common name unknown). *Lit:* 'medicine for the (sting of) the *ch'up*'. Vitaceae *Cissus biformifolia* Standl. This is a weedy liana with tendrils and alternate, simple leaves with toothed margins. It bears clusters of small, red flowers along the stem. Found in secondary forest. The resin is used to counteract the venom of the *ch'up*, a type of furry, venomous caterpillar. *Loc:* **k'aax** 'forest'; *Use:* **ts'ak ahch'up** 'medicine for the *ch'up*'; *Part:* **yits** 'resin'; *Prep:* **Apaats'tik uyits yok'ol** 'You

Fig. 5.199 Tsits *Justicia* sp. leaves. Photo Credit: Suzanne Cook

rub the resin on (the area)'. *[Note:* stak chup. *Cissus biformifolia* (Durán 1999).*] [Source:* AM*] [*\sd2 medicinal*]*

ts'ak kan *Strychnos tabascana* (common name unknown). *Lit:* 'snakebite medicine'. Loganiaceae *Strychnos tabascana* Sprague & Sandw., syn. *Strychnos panamensis* var. *hirtiflora* Standl. *Strychnos tabascana* is an unarmed, woody vine, with hairy branchlets, tendrils, and petioles. Leaves are 1–3 mm. long, the blades ovate to elliptic-lanceolate, 5–11 cm (2–4") long, 2.5–5 cm (1–2") wide, acuminate, cuneate to obcordate (heart-shaped, with the attachment at the pointed end) at the base, often fuzzy beneath when young, turning glabrous with age. Inflorescences are produced in terminal cymes. Flowers are white or cream, 11–17 mm (.4-.6") long. Fruit is round, 4–8 cm (1.5-3") in diameter, with many seeds. (Standley and Williams 1969: 300–301) A number of *Strychnos* species contain poisonous, indole alkaloids, e.g., strychnine, in the roots, stems, and leaves. Seed extract of *Strychnos nux-vomica* contains caffeic acid and monomeric caffeic acid, an antidote for snake bites. (Gupta and Peshin 2012) It is unclear whether or not *Strychnos tabascana* is as effective. More research on Lacandon snakebite medicine is required. *Use:* **ts'ak uchibal kan** 'snakebite'; *Part:* **mots** 'root'; *Prep:* **uk'ul** 'beverage'; *Prep:* **Amuxik umots pachil apulik yok'ol uchibal** 'You mash the root and then you place it on the bite'. *See:* **ts'ak kan ak'**. *[Note:* uzan kan. *Strychnos tabascana* (Durán 1999).*] [Source:* Durán (1999)*] [*\sd2 medicine/curing/prevention*]*

ts'ak kan ak'₁ lemon pepper (joint wood, buttonwood). *Lit:* 'snake medicine vine'. Piperaceae *Piper amalago* Kunth. This is a bush or tree with bright, green, simple leaves, elliptical-ovate, acuminate at apex and at the base. Leaf margins are entire to undulate (wavy). Three inner veins are subparallel, arising from or near the base, while two additional outer main veins connect with the secondary veins. Main veins are suppressed and keeled on the underside of the blade. Secondary veins are reticulate (net-like). The lamina between the veins is bullate (puckered). Leaves slightly arch downward. Texture is smooth, glossy, and somewhat papery. Leaves are arranged opposite each other. The petiole (leaf stalk) is geniculate (bent like a knee). Tendrils are absent. *Piper amalgo* is also used as snakebite by the Itzaj (Atran et al. 2004: 120) and traditional healers in Veracruz (Hernández-Ramos et al. 2007). Durán (1999) reports that *Piper hispidum* is also used to treat snakebite in Naha'. *Use:* **chibal kan** 'snakebite medicine'; *Part:* **mots** 'roots'; *Prep:* **?** 'tea (?)'. *See:* **mäkuuläm**. *Variant:* **uyak'il ts'ak kan**. *Thes:* **ak'** (Fig. 5.203a). *[Note:* A third but unrelated species is also used for snakebite (Fig. 5.203b, c, d). ix pu'ukche', ix pu'uktzu' [Itz.]. *Piper amalgo* (Atran et al. 2004)*] [Source:* AT*] [*\sd2 medicinal*]*

ts'ak kax [ts'ahkax, ts'akax] mint. *Lit:* 'chicken spice'. Lamiaceae *Mentha citrata* Ehrh., syn. *M. aquatica* var. *citrata* (Ehrh.) Benth. An introduced species from Europe, its 90 cm (35″) tall stems are green or purple and variably hairy to almost hairless. Leaves are lanceolate, green or purplish, toothed, hairy to nearly hairless. Flowers are tubular, purple, pink, or lilac. *Parts:* **le'** 'leaves'. *[Note:* ts'ak kax. chicken medicine (spice) (Bruce 1975); tz'akash. *Mentha citrata* (Nations and Nigh 1980); tz'akax. *Mentha citrata* (Nations 1992, in Atran et al. 2004); ix yerb'ab'weenaj [Itz.]. *Mentha citrata* (Atran et al. 2004); ixak'il xiw [Yuc.]. *Mentha citrata* (in, Atran et al. 2004).*] [Source:* BM; SK*] [*\sd2 food*]*

ts'ak kitam [ts'ak kitam, ts'ah kitam] onion, wild (scallion). *Lit:* 'collared peccary spice'. Amaryllidaceae *Allium glandulosum* Link & Otto. This is a perennial wild onion with flattened, grass-like leaves and a small, white, round bulb. Flowers are purple. Cultivated in house gardens and *milpas*. According to BM, this plant is called *ts'ak kitam*, because it is cooked with peccary meat. *Use:* **hanal** 'food'; *Part:* **wi', uche'il** 'bulb, stalks'; *Prep:* **chäkbil yetel bu'ul, bäk', sopa** 'cooked with beans, meat, soup'. *Variant:* **ts'ak k'ek'en**. *[Source:* BM; BMjr; CKO*] [*\sd2 food*]*

ts'ak k'ek'en [ts'ak k'ek'en, ts'a k'ek'en] onion, wild (scallion). *Lit:* 'white-lipped peccary spice'. Amaryllidaceae *Allium glandulosum* Link & Otto. *Loc:* **kolil nah, kol** 'house

Fig. 5.200 (**a**) Tso'ots *Eriobotrya japonica* inflorescence. Photo Credit: Suzanne Cook (**b**) Tso'ots *Eriobotrya japonica*. Photo Credit: Chan K'in Daniel (**c**) Tso'ots *Eriobotrya japonica* ripe fruit on tree. Photo Credit: Chan K'in Daniel (**d**) Tso'ots *Eriobotrya japonica* fruit interior. Photo Credit: Chan K'in Daniel (**e**) Tso'ots *Eriobotrya japonica* leaves. Photo Credit: Chan K'in Daniel

Fig. 5.201 (**a**) Tso'ots bamax *Pseudolmedia oxyphyllaria* leaves. Photo Credit: Suzanne Cook (**b**) Tso'ots bamax *Pseudolmedia oxyphyllaria* sap. Photo Credit: Suzanne Cook

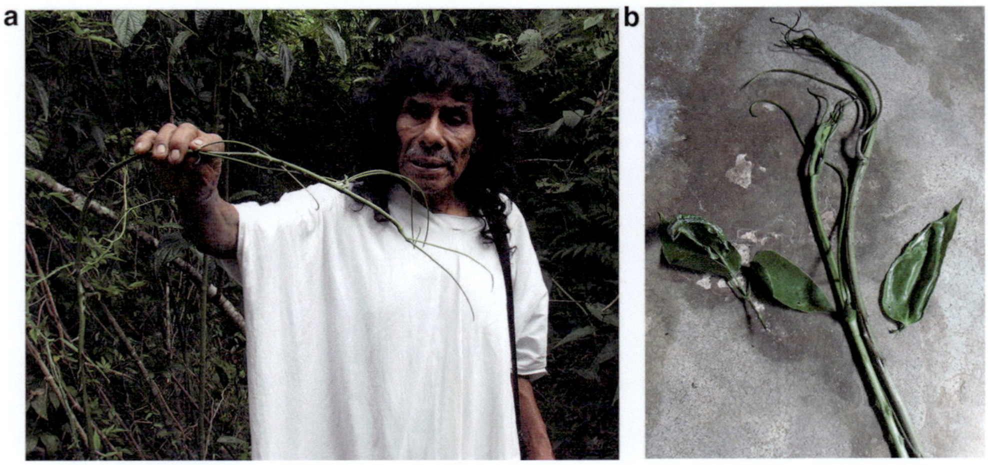

Fig. 5.202 (**a**) Tsuk'in ak' *Smilax* sp. vine with leaves. Photo Credit: Suzanne Cook (**b**) Tsuk'in ak' *Smilax* sp. stem and leaves. Photo Credit: Suzanne Cook

garden, milpa'; *Use:* **hanal** 'food'; *Prep:* **chäkbil** 'cooked'. *See:* **ts'ak kitam**. *Variant:* **ts'ak kitam**. *[Note:* tz'ak'ek'en. scallion. *Allium porrum* (Nations and Nigh 1980).*]* *[*\sd2 food*]*

ts'ak k'ik' Hawaiian wood rose (Spanish arborvine, Spanish woodbine, yellow morning glory) (?). *Lit:* 'rubber venom'. Convolvulaceae *Merremia tuberosa* (L.) Rendle., syns. *Ipomoea tuberosa* L., *Operculina tuberosa* (L.) Meisn. **Esta es para kinsik k'ik'. Es igual hut'kih. Uts'ak k'ik' parece casi como hut'kih. Los dos se usa para kinsik k'ik'. Ule', chan mihin, chan wolis ule'. Tuwolol ule' p'isil chan heek'el, ti' bin uheek'el. Ne yaab. 'Oxheek'el. Ch'ik yaab umuunil uyool. Ne yan uyits, p'iis hut'kih. Utop', k'än, p'iis hut'kih.** 'This is used for killing *k'ik'* (*Castilla elastica*). It's the same as *hut'kih* (*Ipomoea* sp., *I. alba*). *Uts'ak k'ik'* looks almost the same as *hut'kih*. The two are used for killing *k'ik'*. The leaves are small and round. All the leaves have small lobes; their lobes (stretch) out. There are a lot of them. The growing tips are small and tender. There are a lot of tender growing tips. They contain a lot of resin, the same as *hut'kih*. The yellow flowers are similar to those of *hut'kih*. It has three seeds.' (AM) Although this species was unavailable, from AM's description it sound like the species could be *Merremia tuberose*. Native to Mexico and Central America, *Merremia tuberosa* is a pantropical weed, with 10 m (33′) long stems and morning glory-like flowers, 6 cm (2.33″) across. Leaves are green, smooth, alternate, and palmately lobed with usually 7 lanceolate to elliptic lobes with pointed, tapering tips. The stems are slender, smooth, twining, and green near the tips. (Kinsey 2013) The resin from the vine is extracted and mixed with the latex of the *k'ik'* tree (*Castilla elastica*) to produce rubber. *Use:* **ti' kinsik k'ik'** 'to "kill" rubber'; *Loc:* **Sibal** 'Santo Domingo'. *See:* **hut'kih**. *Thes:* **ak'**. *[Source:* AM*]* *[*\sd2 other*]*

Fig. 5.203 (**a**) Ts'ak kan ak' snake medicine vine *Piper amalago* leaves. Photo Credit: Suzanne Cook (**b**) Ts'ak kan ak' snake medicine stems (undetermined species). Photo Credit: Suzanne Cook (**c**) Ts'ak kan ak' snake medicine leaves. Photo Credit: Suzanne Cook (**d**) Ts'ak kan ak' snake medicine vine. Photo Credit: Suzanne Cook

ts'ibal sukal sugarcane. *Lit:* 'striped sugarcane'. Poaceae *Saccharum officinarum* L. *Use:* **hanal** 'food'; *Part:* **che'il** 'stalk'; *Prep:* **che'che'** 'raw'; *Sim:* **chäk sukal** 'red sugarcane'; *Sim:* **hach sukal** 'authentic sugarcane'; *Sim:* **mehen sukal** 'small sugarcane'; *Gen:* **sukal**. *[Source:* BM*]* *[*\sd2 food*]*

ts'in manioc (cassava). Euphorbiaceae *Manihot esculenta* Crantz. **'Oxhek' ule'. Ya'ax ule'. Yan uwi', much'much' uwi'. Nuk uwi'.** 'The green leaves have three lobes. The plant produces bunches of large tubers.' (BM) In Naha', three kinds of manioc are cultivated: (1) *hach ts'in* 'authentic manioc' (also known as *säk ts'in* 'white manioc'); (2) *chäk ts'in* 'red manioc'; and, (3) *yahche' ts'in* 'great manioc'. KyP describes each as follows: "*Ya'axche' ts'in* is green and is the strongest of them all. The tuber is thick and long, about one meter. We eat it salted. We also make tortillas out of them. *Chäk ts'in* and *säk ts'in* are softer. They are delicious when cooked with sugar. None of the plants produce seeds, so we must plant the little shoots. We plant them when the moon is half-full, which is something that our people have done for thousands of years. They take about two years to produce tubers large enough to eat." *Manihot esculenta* is a tall shrub that ranges in height from 0.9 to 4.58 m (3–15′). It is cultivated worldwide for its starchy, swollen roots. It originated in South America, before being introduced into Mesoamerica. Although highly variable, key features of all *Manihot esculenta* include a shrubby, highly branched habit with woody, brittle stems and palmate leaves with five to nine lobes. The fruit is a dehiscent capsule with three locules (chambers within an ovary). There are two main groups: one has very light grey, silvery stems, whereas the other has yellow, orange, or brown stems. The roots of grey-stemmed varieties are smooth, whereas those of the coloured-stemmed varieties are rough. Because *Manihot* is pollinated by insects, it easily hybridizes. (Rogers 1964: 370–371) *Use:* **hanal** 'food'; *Part:* **wi'** 'tubers'; *Prep:* **pokbil, chäkbil yetel ch'uhuk** 'roasted, boiled with sugar'; *Prep:* **säkan** 'dough'; *Spec:* **hach ts'in** 'authentic manioc'; *Spec:* **yahche' ts'in** 'large tree manioc'; *Spec:* **chäk ts'in** 'red manioc'. *[Note:* ts'in. yucca, sweet casava (Bruce 1975); tz'in. manioc. *Manihot esculenta* (Nations and Nigh 1980). c'íin, c'íim [Yuc.]. *Manihot esculenta* (Bricker et al. 1998); tz'iim [Itz.]. *Manihot esculenta* (Atran et al. 2004).*] [Source:* AM; BM; JK; KyP*]* *[\sd2 food]*

ts'oy (äh)$_1$ clusia (pitch apple). Clusiaceae *Clusia* spp. **Es ak'. Ne nuk ule'. Ne pim ule'. To'och ule'. Crece de una bola, como orquídea. K'uuk'che'. Ule', bin uyak'il hasta ich lu'um. Yan umots. Uche'il uyak'il, ek'. Ne yi'h. Hay tres clases: mehen ts'oy, nukuch ts'oy, i ne nuk uts'oy. Yan uyits, k'än uyits.** 'It's a vine with very large leaves. The leaves thick and stiff. It grows from a ball, like an orchid. This is called *k'uuk'che'*. From the leaves vines emerge and go to the ground where they take root. The stems of the vines are black and very hard. There are three classes: 'small *ts'oy*', 'large *ts'oy*', and 'very large *ts'oy*'. The trunk and vines exude a yellow resin.' (BM) Clusias are trees or shrubs, often epiphytic; sometimes they strangle their hosts with their epiphytic roots. Many clusias begin life as a seedling high in the branches of a canopy tree or in a crevice on steep cliffs. Some become woody vines, with thick, sticky, resinous, yellow sap. All have smooth, thick, succulent, paddle-shaped leaves with numerous, parallel, lateral veins. Inflorescences are branched, with several flowers. Rounded sepals number 4–6; petals number 4–10 with numerous stamens. Fruit is a thick-walled, leathery or fleshy, dehiscent capsule. (Standley and Williams 1961: 39) *Clusia flava, C.guatemalensis, C. lundellii,* and *C. salvinii* have been recorded from the northern Lacandon forest (Durán 1999; Comisión Nacional de Áreas Naturales Protegidas 2006). Clusias are rarely exploited by the Lacandones, although the fragrant flowers of at least one species, *Clusia flava* Jacq., were incorporated into necklaces to add fragrance. *Spec:* **mehen ts'oy** 'small clusia'; *Spec:* **nukuch ts'oy** 'large clusia'; *Spec:* **ne nuk ts'oy** 'very large clusia'. *See:* **chunup (äh)**. *Thes:* **ak'**. *[Note:* tzoy. *Clusia lundelli, C. salvinii* (Durán 1999).*] [Source:* BM; SK*]*

ts'oy (äh)$_2$ hubu. Clusiaceae *Clusia lundellii* Standl. An evergreen liana with thick, medium-green, obovate leaves that grow in an opposite, decussate arrangement along thick stems. The yellow flower comprises four, succulent, waxy, overlapping sepals and 6 petals. Found growing at the edge of secondary forest beside the main road into Naha'. No uses were reported. *Loc:* **k'aax** 'forest'. *See:* **chunup (äh)**. *Thes:* **ak'** (Fig. 5.204a, b). *[Note:* tzoy. *Clusia lundellii* (Durán 1999). Common name from the Smithsonian Collections Search Centre. http://collections.si.edu/search/

Fig. 5.204 (**a**) Ts'oy *Clusia lundellii* plant. Photo Credit: Suzanne Cook (**b**) Ts'oy *Clusia lundellii* flowers. Photo Credit: Chan K'in Daniel

Fig. 5.205 (**a**) Ts'ubtok *Hampea stipitata* flowers. Photo Credit: Suzanne Cook (**b**) Ts'ubtok *Hampea stipitata* leaves, stem, and flowers. Photo Credit: Suzanne Cook (**c**) Ts'ubtok *Hampea stipitata* leaf, underside. Photo Credit: Suzanne Cook (**d**) Ts'ubtok fruit. Photo Credit: Suzanne Cook

record/nmnhbotany_11005845. Retrieved April 2014.*]* *[Source:* AM; BM; KP*]*

ts'ubtok [ts'ubtok, ts'u'tok] mountain moho [Bel. Eng.]. Malvaceae *Hampea stipitata* S. Watson, syn. *Hampea euryphylla* Standl. **Es che'. Crece solito ich k'aax wo baja k'aax. Ma' ne nuk. Ule' utop', säk, much'umuch' utop'. Adentro es como algodón, tiene yaab unek', ek'. Usool uk'anik ti' kuchik che', ti' k'äxik nah.** 'This is a tree. It grows on its own in the forest or in the lower forest. It's not very large. The petals are white and the flowers are clustered together. In the middle (of the fruit) there is something like cotton with numerous, black seeds. You use the bark for carrying wood and house lashing.' (SK). *Use:* **k'an ti' ba'ay, ti' häxik hiil** 'cord for bags and lashing battens'; *Part:* **usool** 'bark'. *Thes:* **che'** (Fig. 5.205a, b, c, d). *[Note:* ts'uk tok. *Hampea stipitata* (Diemont 2006); ts'ub to'. majagua. *Hibiscus tiliaceus* (Baer and Merrifield 1971); ts'ub-tok' [Yuc]. *Bauhinia divaricata* L. (Roys 1931); tz'ulub' took' [Yuc.]. *Bauhinia divaricata* (in, Atran et al. 2004). English name from Grandtner 2005.*]* *[Source:* BM; K'ayum Segundo; SK*]* *[*\sd2 cordage*]*

ts'ulha'il petha' (äh) [ts'ulha'il petha', ts'uli petha'] lily, cahaba (shoals spider lily). *Lit:* 'water dove of the lagoon'. Amaryllidaceae *Hymenocallis littoralis* (Jacq.) Salisb. This is a species of *Amaryllis* with lush, dark-green, upright foliage. The flower is held aloft, like a frilly white collar with streamers of long, thin petals. Extremely fragrant, the flowers attract nocturnal moths and a species of swallowtail butterfly. According to my Lacandon consultants, a kind of large grasshopper called *nuxib* 'old man' feeds on the foliage. *Loc:* **petha'** 'lagoon'; *Sim:* **nukuch ts'ulha' (äh)** 'large ts'ulha''; *Sim:* **chäk ts'ulha' (äh)** 'red ts'ulha''; *Gen:* **ts'ulha' (äh)**. *Morph:* **ts'ul-ha'-il**

Fig. 5.206 (**a**) Ts'ulha'il petha' *Hymenocallis littoralis* flower. Photo Credit: Suzanne Cook (**b**) Ts'ulha'il petha' *Hymenocallis littoralis* flower. Photo Credit: Suzanne Cook (**c**) Ts'ulha'il *Hymenocallis littoralis*. Photo Credit: Suzanne Cook

petha'. *See:* **chäk ts'ulha' (äh)**. *Variant:* **ts'ulil petha'** (Fig. 5.206a, b, c). *[Source:* AM; KP*]*

ts'ulha' (äh) [ts'ulha', ts'ula', ts'uli] amaryllis and lilies. *Lit:* 'water dove'. *Amaryllidoideae, Liliaceae. Spec:* **chäk ts'ulha' (äh)** 'red ts'ulha''; *Spec:* **ts'ulha'il petha' (äh)** 'ts'ulha' of the lagoon'; *Spec:* **nukuch ts'ulha' (äh)** 'large ts'ulha''. *[Note:* utsura'ir ja' (Baer and Merrifield 1971).*]* *[Source:* BM; CKP; KP*]*

ts'us (ah) grape, wild. Vitaceae *Vitis* spp. *Vitis bourgaeana* and *Vitis tiliifolia* have been recorded from the Lacandon forest (Durán 1999; Comisión Nacional de Áreas Naturales Protegidas 2006; Levy et al. 2006). These members of the grape family are native to Mesoamerica. Although none of my Lacandon consultants reported any medicinal value, Kashanipour and McGee (2004: 64) note that some Lacandones drink a tea made from the roots of *Vitus* spp. to relieve coughs. *Loc:* **paakche' kol** 'fallow milpa'; *Use:* **hanal** 'food'; *Part:* **wich** 'fruit'; *Prep:* **che'che yetel ch'ooch'** 'raw with salt'; *Use:* **ti' tso'otsel ho'ol** 'hair tonic'; *Prep:* **Umäxik uwich yok'ol ha' i pulik yok'ol ho'ol** 'One crushes the fruit in water and puts it on the head'; *Spec:* **hach ts'us** 'authentic ts'us (*Vitis tiliifolia*)'; *Spec:* **mehen ts'us** 'small ts'us' (Fig. 5.207a, b). *[Note:* thus. wild grape.*Vitis* spp. (Kashanipour and McGee 2004); tzuts. *Vitis bourgaeana* (Durán 1999).*]* *[Source:* AM; BM; KP*]* *[*\sd2 other \sd2 food*]*

U–u

'ukunte' (äh) royal fig (lateralflower sapium). Euphorbiaceae *Sapium lateriflorum* Hemsl. This is a fast-growing tree with long, thick, glossy, dark green leaves that end in a slight point. Branches radiate from a slender trunk,

Fig. 5.207 (**a**) Ts'us *Vitis* sp. unripe fruit. Photo Credit: Suzanne Cook (**b**) Ts'us *Vitis* sp. leaves and fruit. Photo Credit: Suzanne Cook

like spokes around a hub. Soft, tender stems produce copious amounts of a white latex. Research (Levy-Tacher 2004; Levy and Golicher 2004; Diemont et al. 2006) indicates that *Sapium lateriflorum* restores soil fertility in Lacandon fallow *milpas*, provides seeds for birds, and lumber for construction. According to these authors, the southern Lacandones do not remove the saplings, but allow them to develop as part of their fallow management system. *Loc:* **k'aax, paakche' kol, kolil nah** 'forest, fallow milpas, house gardens'; *Use:* **ts'ak ti' sarna** 'medicine for scabies'; *Part:* **uyits** 'latex'; *Prep:* **Ha'li' apulik uyits yok'ol awoot'el** 'You just apply the latex to your skin.' *Variant:* **ukante'**. *Thes:* **che'** (Fig. 5.208a, b, c, d). *[Note:* u'cunte. *Sapium lateriflorum* (Diemont 2006; Diemont et al. 2006).*] [Source:* AM; KA; KP*] [*\sd2 medicinal*]*

'uk'unche'₁ potato tree (mullein nightshade, tobacco tree, big eggplant). Solanaceae *Solanum erianthum* D. Don. A native of Mexico, *Solanum erianthum* is an evergreen shrub, 2–8 m (6.5–26′) tall, woody or herbaceous, with branches that are unarmed and lenticulate (raised pores). Leaves are ovate-elliptical and the margin is entire or slightly undulate (wavy), 10–25 cm (4–9.4″) long and 3–15 cm (1.2–6″) wide. Leaves and stems are covered with a dense fuzz. Inflorescences are branched and densely hairy. Small, 11–17 mm (0.4–0.7″) diameter, white, five-petaled blossoms with yellow anthers are borne in lateral cymes. The flowers emit a foul odor, probably to attract bats, which eat the 9–12 mm (0.3–0.5″) round, pubescent orange fruit. (Nee 1993: 64). It is found growing in disturbed sites. *Loc:* **paakche' kol, kolil nah** 'fallow milpa, house gardens'; *Use:* **ch'am** 'grill'; *Part:* **xaay** 'forked branches'. *Thes:* **che'** (Fig. 5.209b, c). *[Note:* Used medicinally by the southern Lacandones (Levy et al. 2006). These authors do not provide details.*] [Source:* KM; KP*]*

'uk'unche'₂ turkeyberry nightshade (devil's fig, prickly solanum). Solanaceae *Solanum torvum* Sw. Native to southern Mexico, Central American, and northern South American, *Solanum torvum* is a spreading or scrambling shrub to 3 m (10′) tall, with dark green, pubescent leaves, stems, and pedicels. Stems and leaves are armed with short prickles. Leaves are broadly oval-ovate, 10–15 cm (4–6″) long and 8–10 cm (3–4″) wide. Inflorescence is branched, bearing 50–100 flowers on 5–10 mm long pedicels. Blossoms are white, deeply incised, and 20–25 mm (0.8–1″) in diameter. Round, 10–15 mm (0.4–0.6″) berries ripen from green to yellow and dry to black. It is found at sea level to 2000 m (6561.6′). (Hanan-Alipi & Mondragón-Pichardo 2009f). No uses were reported. *Loc:* **paakche' kol** 'fallow milpa'. *Variant:* **k'i'ix 'uk'unche'**. *Thes:* **che'** (Fig. 5.209a). *[Source:* AM*]*

utop' ti' kisin night-blooming cestrum (night-blooming jasmine, night jessamine). *Lit:* 'the flower of Kisin'. Solanaceae *Cestrum nocturnum* L. This is a fragrant, night-blooming vine. According to Lacandon mythology, it is the birth flower of *Kisin* 'the Devil'. *See:* **'aak'älyoom**. *[Source:* AM; CK*] [*\sd2 folklore_and_mythology*]*

'uuch (äh) chocolate pudding fruit. Ebenaceae *Diospyros digyna* Jacq., syn. *Diospyros ebenaster* Retz. This is a black persimmon, native to Mexico. It is approximately the size of a medium tomato. At maturity, the pulp turns black, while the skin remains green. It matures in November. Cultivated. *Loc:* **kolil nah, kol** 'house garden, milpa'; *Use:* **hanal** 'food'; *Part:* **wich** 'fruit'; *Prep:* **che'che** 'raw'. *Thes:* **che'** (Fig. 5.210a, b). *[Note:* uuch. zapote negro (Bruce 1975); uch'. *Diospyros digyna* (Nations and Nigh 1980); x ta'-'uch, x ta'-lùuch [Yuc.]. *Diospyros digyna* (Bricker et al. 1998).*] [Source:* AM; BM; CKD*] [*\sd2 food*]*

Fig. 5.208 (**a**) 'Ukunte' *Sapium lateriflorum* tree. Photo Credit: Suzanne Cook (**b**) 'Ukunte' *Sapium lateriflorum* tree. Photo Credit: Suzanne Cook (**c**) 'Ukunte' *Sapium lateriflorum* growing tip. Photo Credit: Suzanne Cook (**d**) 'Ukunte' *Sapium lateriflorum* latex. Photo Credit: Suzanne Cook

W–w

waalka'an$_1$ fern tree. Cyatheaceae *Alsophila salvinii* Hook. *Alsophila salvinii*, usually found at altitudes between 900 m (2953') and 2600 m (8530'), occurs within the elevational range of Naha'. (Robbin Moran, pers. comm.). *Loc:* **k'aax** 'forest'.

waalka'an$_2$ fern, bracken. Dennstaedtiaceae *Pteridium pseudocaudatum,* syn. *P. aquilinum* var. *pseudocaudatum* (Clute) A. Heller. This is a climbing fern armed with thorns

Fig. 5.209 (**a**) 'Uk'unche' *Solanum torvum*. Photo Credit: Suzanne Cook (**b**) 'Uk'unche' *Solanum erianthum* flowers. Photo Credit: Suzanne Cook (**c**) 'Uk'unche' *Solanum erianthum* leaves. Photo Credit: Suzanne Cook

along the petioles; internodes are long. It sends up single fronds throughout cleared areas. The Lacandones consider it *lo'obil* 'a weed' and a great nuisance in the *milpa*. *Loc:* **paakche' kol** 'fallow milpa'. *[Note:* Taxonomic identification by Robbin Moran (pers. comm.)*]* *[Source:* KM*]*

waalka'an$_3$ fern, shield. Dryopteridaceae *Polystichum* sp. This is a tall, bushy, multi-stemmed species. Stout, creeping rootstocks form a crown and a ring of long, evergreen fronds. It is a terrestrial or rock-dwelling fern. Lacandones consider it *lo'obil* 'a weed'. *[Note:* Taxonomic identification by Robbin Moran (pers. comm.)*]* *[Source:* AM*]*

wakäx su'uk grass, napier (merker grass, cane grass, elephant grass). *Lit:* 'cow grass'. Poaceae Pennisetum *purpureum* Schumach. This is a tall, 2–6 m (6.5–19.6′), perennial, clumping grass with a rangy habit and razor-sharp, bright green, relatively narrow leaves, 20–40 mm (0.8–1.6″) wide. It produces dense spikes of numerous, fluffy spikelets on 25 cm (9.8″) long, yellow or purple stems. Native to tropical climates, this introduced species has naturalized in many parts of Mexico at altitudes between 300 and 1600 m (984 and 5249′) (McVaugh 1983; Villaseñor and Espinosa 1998). It grows on the sides of the roads and throughout pastures, fallow fields, and cultivated land. Although highly invasive, it continues to be cultivated by ranchers for fodder. The species may be easily confused with the common reed (*Phragmites australis*) and the giant reed (*Arundo donax*) (University of Queensland 2011). *Sim:* **winik su'uk** 'man grass (*Andropogon glomeratus*)'; *Sim:* **'ak su'uk** 'grass grass (*Bothriochloa laguroides*)'; *Sim:* **hach su'uk** 'authentic grass (*Digitaria setigera*)'; *Gen:* **su'uk (ah)**. *Variant:* **säk su'uk**. *Thes:* **su'uk** (Fig. 5.211a, b). *[Source:* BM; CKD*]* */*sd2 other*]*

wawal [wawa(l), wowo, bobo] calathea. Marantaceae *Calathea* spp. *Use:* **ti' makik ba'al, yo'och, ti päk'äch** 'for covering and wrapping food and other things, and for a non-stick surface on which to pat out tortillas'; *Part:* **le'** 'leaves'; *Spec:* **säk wawal** 'white wawal (*Calathea lutea*)'; *Spec:* **chäk wawal** 'red wawal (*Calathea* sp.)'; *Spec:* **its'in säk wawal** 'younger brother of white wawal (*Calathea crotalifera*)'; *Sim:* **muxan**. *Thes:* **xiiw**. */*sd2 food*]*

wäch' wild tamarind. Fabaceae *Dialium guianense* (Aubl.) Sandwith. Unrelated to the true tamarind (*Tamarindus indica*), this species is a variety of ironwood with edible, bittersweet fruit (Schoenhals 1988: 54–55). It is a tall tree found in the upper-most stratum of the forest, attaining a height of 35 m (115′), with a buttressed, cylindrical trunk clear of branches to 15.24 m (50′); the heartwood is brown or reddish brown, and the sapwood, whitish or yellowish (USDA Forest Service n.d.). AM reports that the timber is a choice firewood, because it burns all night. *Use:* **hanal, uk'ul** 'food, beverage'; *Part:* **wich** 'fruit'; *Prep:* **che'che'** 'raw'; *Use:* **che'il k'ak'** 'firewood'. *Thes:* **che'**. *[Note:* wäch'. wild tamarind. *Dialium guianense* (Nations and Nigh 1980); wech'. *Dialium guíñense*. (Diemont 2006); wuech, guapake. *Dialium guianense* (Durán 1999).*]* *[Source:* AM*]* */*sd2 fuel*]*

wich ahyuk (ah) velvet bean (horse-eye bean, deer eye). *Lit:* 'deer eye'. Fabaceae *Mucuna sloanei* Fawc. & Rendle. The name refers to the seeds of an invasive vine bearing umbrella-like clusters of large pea blossoms with curved petals. Seed pods contain round, slightly flattened, brown seeds with a black seam around the circumference, making them look miniature hamburgers. The different species are differentiated on the basis of the colour and shape of their "smiley"—a peculiar scar produced at one end of the hilum: *M. urens* has a thinner, up curved, reddish smiley and *M. sloanei* has a thick, blocky one (Sullivan and Williams 2008). The buoyant seeds drift on ocean currents to distant shores around the globe. As such, they are also called "sea-beans". *Mucuna sloanei* is found in tropical areas from Southern Mexico to Guatemala and El Salvador (Standley and Steyermark 1946). *Use:* **uuh** 'necklaces'; *Part:* **nek'** 'seed'.

Fig. 5.210 (**a**, **b**) 'Uuch *Diospyros digyna*. Photo Credit: Chan K'in Daniel

See: **ni' ts'ul**. *Variant:* **bubuhchi' (ah)**$_2$. *Thes:* **che'** (Fig. 5.212a, b). *[Note:* uyich keej [Itz.]. *Mucuna argyrophylla* (Atran et al. 2004); uyich kej [Yuc.]. *Mucuna argyrophylla* (in, Atran et al.).*] [Source:* NI*] [*\sd2 craft*]*

winik su'uk [winik suuk] grass, bushy bluestem (bushy beardgrass). *Lit:* 'man/person grass'. Poaceae *Andropogon glomeratus* (Walter) Britton, Sterns & Poggenb., syns. *Andropogon corymbosus* (Hack.) Nash, *Andropogon densus* Desv., *Andropogon glaucopsis* (Elliott) Steud. A native to the region, this tall grass topped with white, fluffy seed heads is common in swampy areas along the banks of rivers and lakes. Lacandones call it *winik su'uk*, because it looks like a man in a *xikul* (traditional Lacandon tunic) standing out in the field. *Spec:* **wakäx su'uk** 'cow grass'; *Spec:* **hach su'uk** 'authentic grass (*Digitaria setigera*)'; *Spec:* **'ak su'uk** 'grass grass (*Bothriochloa laguroides*)'; *Gen:* **su'uk (ah)**. *Variant:* **säk su'uk**. *Thes:* **su'uk** (Fig. 5.213a, b). *[Source:* AM; BM*]*

wolche' (äh) [wolche', worche', wo'che'] sapote, white. Rutaceae *Casimiroa* sp (?). This is a forest tree with a slender trunk, approximately 30 cm (12″) in diameter. Leaves are large and lanceolate. It bears 7.6–10 cm (3″–4″) round, edible fruit with a rough, hard, rusty-brown skin. Inside, a yellow, spongy pulp clings to a number of large seeds. The fruit begins to mature in late October. BM says the fruit is similar to sapodilla. It provides food for forest animals, including *ku'uk* 'squirrel', *aak'ma'ax* 'night monkey', and *hale'* 'spotted cavy'. The species was observed growing in secondary forest, close to the road leading into Naha'. It is possibly *Casimiroa sapota* (Douglas Stevens, pers. comm.). *Loc:* **k'aax** 'forest'; *Loc:* **Sa'am** 'Monte Líbano'; *Loc:* **Naha'**; *Use:* **hanal** 'food'; *Part:* **wich** 'fruit'; *Prep:* **che'che'** 'raw'; *Use:* **che'il k'ak'** 'firewood'. *See:* **ya'**. *Thes:* **che'** (Fig. 5.214a, b). *[Note:* wo'che'. "type of wild fruit of the zapodilla family with large seeds and scant, edible pulp" (Bruce 1975: 243).*] [Source:* AM; BM*] [*\sd2 fuel \sd2 food*]*

X–x

xa'an [xa'an, xaan, xaam] palm, fan (sabal, Mexican palmetto, Texas palmetto). Arecaceae *Sabal mexicana* Mart. My Lacandon consultant said that this tree does not have spines. Therefore, it must be *Sabal mexicana*, a large palm distinguished from other "fan palms" by its 90–120 cm (35–47″) long, unarmed petioles, downward arching midrib, and threads that dangle along the margins of the leaves. It produces long spikes 1.2–1.8 m (3.9–5.9′) of small, bisexual flowers, which are followed by black drupes, 12 mm (0.47″) in diameter. *Loc:* **k'aax** 'forest'; *Use:* **pix nah** 'thatch'; *Part:* **le'** 'leaves'; *Use:* **ch'ooch'** 'salt'; *Part:* **uchuch ule'** 'heart, stems'; *Prep:* **Atookik uchuch ule'** 'You burn the leaf stalks'; *Use:* **uuh** 'necklaces'; *Part:* **nek'** 'seed'; *Spec:* **hach xa'an** 'authentic xa'an'; *Spec:* **booxa'an** '*Acoelorraphe wrightii*'. *See:* **kun (ah)**. *Variant:* **hach xa'an** (Fig. 5.215). *[Note:* xa'an. a certain palm tree (guano) (Bruce 1975); sha'an. unspined salt palm. *Sabal mexicana* (Nations and Nigh 1980); xa'an, b'otan [Itz.]. *Sabal mauritiiforme* (Atran et al. 2004); aj b'on xa'an [Itz.]. *Sabal mexicana* (Atran et al. 2004); b'otan xá'an [Yuc.]. *Sabal mauritiiforme* (in, Atran et al. 2004); xa'an [Mop.]. *Sabal mau-*

a b

Fig. 5.211 (**a**) Wakäx su'uk *Pennisetum purpureum* leaves, stems. Photo Credit: Chan K'in Daniel (**b**) Wakäx su'uk *Pennisetum purpureum* flower stalks. Photo Credit: Suzanne Cook

a b

Fig. 5.212 (**a**) Wich ahyuk *Mucuna sloanei hilum*. Photo Credit: Suzanne Cook (**b**) Wich ahyuk *Mucuna sloanei* seed. Photo Credit: Suzanne Cook

a b

Fig. 5.213 (**a**) Winik su'uk *Andropogon glomeratus* seed head. Photo Credit: Suzanne Cook (**b**) Winik su'uk *Andropogon glomeratus* plant. Photo Credit: Suzanne Cook

Fig. 5.214 (**a**) Wolche' *Cordia dodecandra* fruit. Photo Credit: Suzanne Cook (**b**) Wolche' *Cordia dodecandra* leaves. Photo Credit: Suzanne Cook

Fig. 5.215 Xa'an *Sabal mexicana*. Photo Credit: Suzanne Cook

ritiiforme (in, Atran et al. 2004).*]* *[Source:* AM; BM*]* *[*\sd2 food \sd2 housing \sd2 craft*]*

xänä' mook pillpod sand-mat. Euphorbiaceae *Chamaesyce hirta* (L.) Millsp. This is a spreading, prostrate plant with small, dark green leaves with faintly serrated edges and tiny flowers clustered at leaf axials. Grows on previously cultivated land. Considered *lo'obil* 'a weed'. *Loc:* **kolil nah** 'house garden' (Fig. 5.216). *[Note:* English name from Herrera et al. (2010).*]* *[Source:* KM*]*

Fig. 5.216 Xänä' mook *Chamaesyce hirta*. Photo Credit: Suzanne Cook

xiilich (äh) coralberry (marlberry). Primulaceae (Myrsinaceae) *Parathesis* sp. *See:* **xinich (äh)**. *[Source:* AM*]*

xikin (ah)$_1$ fungus, split-gill. *Lit:* 'ear'. Schizophyllaceae *Schizophyllum commune* Fr. Primarily a saprobic wood-rotting fungus, *Schizophyllum commune* covers damaged areas of bark on the trunks, or dead or dying branches of diseased trees. Despite its unpleasant texture, this fungus is widely eaten in Mexico (Ruan-Soto et al. 2006). *Use:* **hanal** 'food'; *Prep:* **hoybil** 'stirred into a thin gruel of moistened pieces of tortillas'. *Variant:* **xikin che'**. *[*\sd2 food*]*

xikin (ah)$_2$ Polyporaceae *Polyporus tenuiculus* (Beauv.) Fr. *Use:* **hanal** 'food'. *[*\sd2 food*]*

xikin che' split gill. *Lit:* 'tree/wood ear'. Schizophyllaceae *Schizophyllum commune* Fr. **Ahxikin, che'. Ne ki'. Es ek'.** '*Xikin* grows on trees. It's very tasty. It's black.' (AM) *Schizophyllum commune* is the most widely distributed mushroom. It has a greyish-white, shell-shaped cap, often wavy and lobed, with pale red or grey gills. The species grows on trees, hence the name. *Use:* **hanal** 'food'; *Prep:* **hoybil** 'stirred into a thin gruel of moistened pieces of tortillas'; *Sim:* **muluch'**; *Sim:* **k'ayoch'**. *Variant:* **xikin** (Fig. 5.217). *[Note:* xikin che' much' [S. Lac.]. *Schizophyllum commune* (Ruan-Soto et al. 2009); xikin che' wakax [S. Lac.]. *Favolus tenuiculus* (Ruan-Soto et al. 2009); xicin-che [Yuc.]. Lit. tree-ear. a tree-fungus (Roys 1931).*] [Source:* AM; KM; KP*]* *[*\sd2 food*]*

xinich (äh) parathesis (type of). Primulaceae (Myrsinaceae) *Parathesis* sp. According to AM, *xinich* is like *lo'k'in* (*Parathesis* sp.), only larger. The Lacandones regard it as *lo'obil* 'a weed'. Although the plant has no culinary or medicinal value, Lacandones pluck off the ripe drupes in passing and eat them out of hand. *Loc:* **k'aax** 'forest'; *Use:* **hanal** 'food'; *Part:* **wich** 'fruit'; *Prep:* **che'che'** 'raw'; *Sim:* **lo'lk'in** '*Parathesis* sp.'. *Variant:* **xiilich (äh)**. *Thes:* **che'** (Fig. 5.218a, b). *[Note:* shinich. nanche. *Byrsonima crassifolia* (Nations and Nigh 1980); xili. *Ardisia karwinskyana* (Durán 1999); xilil [Itz.]. xilil tree. *Ardisia paschalis* (Hofling and Tesucún 1997). Botanical identification provided by Rolando Perez (pers. comm.).*] [Source:* KM; NI*]* *[*\sd2 food*]*

xoyok (ah) [xoyok, xoyoh] yellow wood. Rubiaceae *Morinda panamensis* Seem. Two trees were identified as xoyok: one was a slender tree and the other was a tall tree with a trunk diameter of approximately 60 cm (24″). The inner bark of both species is yellow. The tall tree is *Morinda panamensis*. It can grow as tall as 9 m (29′) with a trunk diameter of 15 cm (5.9″).

Fig. 5.217 Xikin *Schizophyllum commune*. Photo Credit: Suzanne Cook

Fig. 5.218 (**a**) Xinich (xilich) *Parathesis* sp. fruit. Photo Credit: Suzanne Cook (**b**) Xinich *Parathesis* sp. leaves. Photo Credit: Suzanne Cook

Although fruits and flowers were unavailable, the botanical literature describes the inflorescences as sessile, round clusters of white blossoms borne in the leaf axils, followed by numerous yellow fruits (Standley 1930). *Loc:* **k'aax** 'forest'; *Use:* **ti' ubanik nook' k'än** 'to dye clothes yellow'; *Part:* **mots** 'roots'; *Prep:* **Aho'ochtik umots, pachil asulik ich siis ha' yetel k'ulta'an** 'You scrape the roots then you soak [the scrapings] in cold water with lime calcium'. *See:* **k'ante' (ah)**. *Thes:* **che'** (Fig. 5.219a, b, c, d, e). *[Note:* According to AM, *xoyok* produces a yellow dye that is superior to that of *k'ante'* (*Erythrina berteroana*). shoyok. *Morinda yucatanensis.* red dye plant (Nations 2006); xo'yoc. red dye made from the (root), it pro-

Fig. 5.219 (**a**) Xoyok *Morinda panamensis* trunk. Photo Credit: Suzanne Cook (**b**) Xoyok *Morinda yucatanensis* leaves. Photo Credit: Suzanne Cook (**c**) Xoyok *Morinda panamensis* inner bark. Photo Credit: Suzanne Cook (**d**) Xoyok *Morinda yucatanensis* leaves. Photo Credit: Suzanne Cook (**e**) Xoyok young tree. Photo Credit: Suzanne Cook

duces a brighter red than mahogany bark (Baer and Merrifield 1971: 156); xo yok che. *Morinda panamensis, Psychotria chiapensis* (Durán 1999); x-hoyoc [Yuc.]. *Morinda yucatanensis,* Greenm. (Standl. 1920–26, p.1393; Loesener, 1923, p.339). Piñuela. A subscandent shrub whose root is used by the Indians to dye thread a scarlet color (Pio Pereze, 1866–77, in Roys 1931: 245); x hó'oyok [Yuc.]. *Morinda yucatanensis.* A vine used to treat warts. Fruit smashed in water to prepare red dye for house painting (Bricker et al. 1998: 112). English common name from the Smithsonian Collections Search Center (http://collections.si.edu).*] [Source:* AM; KM*]* /\sd2 craft*]*

Y–y

ya' chicle. Sapotaceae *Manilkara, Pouteria, Chrysophyllum. Loc:* **k'aax** 'forest'; *Use:* **akam** 'foundation posts'; *Use:* **bahbil** 'horizontal wall boards'; *Use:* **chuste'** 'foreshaft'; *Use:* **chulul** 'bows'; *Spec:* **hach ya'** 'sapodilla (*Manilkara achras*)'; *Spec:* **chäk ya'** 'red chicle (*Manilkara achras*)'; *Spec:* **o'ot' ya'** 'canistel (*Pouteria campechiana*)'; *Spec:* **säk ya'** 'caimitillo (*Chrysophyllum mexicanum*)'. *See:* **chäk ya'**. *Thes:* **che'**. *[Note:* ya' [Itz.]. chicle tree. sapodilla plum. *Manilkara zapota* (Hofling and Tesucún 1997); ya'. chicozapote, chicle (Nations 2006); hach ya' (*Manilkara zapota*), chäk ya' (*Manilkara achras*), ot' ya' (*Manilkara* sp.) (Rätsch 1985a, b; 1994b); ya, chico zapote. *Manilkara zapota* (Durán 1999); ya. *Manilkara achras* (Comisión Nacional de Áreas Naturales Protegidas 2006).*] [Source:* AM; BM; SK*]* /\sd2 hunting_and_fishing \sd2 food \sd2 housing*]*

ya'axche' [ya'axche', ya'ache', ya'che'] ceiba (cotton tree, floss silk). Bombacaceae *Ceiba pentandra* (L.) Gaertn. A grove of ceiba trees is an impressive sight, rising above the forest canopy looking like large umbrellas held aloft on thick, columnar trunks. Young trees are bright green and armed with large, conical spines. Fruits are large and ellipsoidal up to 20 cm (7.9″) long. Depending on the pronunciation of the word, [ya'axche', ya'che', yahche'], it can mean 'green tree', 'first tree', or 'great tree', respectively (Bruce 1975: 249). The ceiba represents the Mayan tree of life: In ancient Maya mythology, it stood at the center of the earth, connecting the terrestrial world to the cosmos. One belief held in the Yucatan is that the trunk pierces through the successive heavens to the ultimate, seventh heaven where El Gran Dios, the god of the Spaniards, lives (Tozzer 1907: 154). It is represented by a cross and serves as an important architectural motif in the Temple of the Cross Complex at the Mayan ruins of Palenque (Houston 1996). According to the Lacandones, the centre of the world is Yaxchilan, located on the banks of the Usumacinta River. It so happens that there is a great ceiba tree standing in the main courtyard of these ruins. In Lacandon mythology, Yaxchilan became the home of the Lacandones' principal deity, *Hachäkyum*, after he had left his home in Palenque. Although he subsequently left Yaxchilan for another area closer to the northern Lacandon communities, the northern Lacandones continued to make pilgrimages to the site to venerate him and to retrieve stones from his temple for their god pots. The ceiba is a tree of old, second growth and primary forest. Although my Lacandon consultants did not mention any uses for the tree, the southern Lacandones made clothes from it (Levy et al. 2006). Atran et al. (2004: 95) note that the Itzaj use the leaves, bark, and pieces of the trunk for a number of ailments. In ancient Mayan times, the soft, white fluff of the fruit capsules was used to stuff pillows and, more recently, parachutes. *Loc:* **k'aax** 'forest'. *Thes:* **che'** (Fig. 5.220a). *[Note:* ya'che'. ceiba (tree) (Bruce 1975); yaajche. *Ceiba pentandra* (Nigh 2008); yax che'. *Ceiba pentandra* (Diemont 2006); ya'a'-che [Lac.].*Ceiba pentandra* (in, Atran et al. 2004); ya'axche' [Itz.]. *Ceiba pentandra* (Atran et al. 2004).*]* *[Source:* AM*]*

ya'ax 'abäl mombin, green (hog plum). Anacardiaceae *Spondias* sp. This species may refer to *Spondias radlkoferi* or the green, unripe fruit of *Spondias mombin*, or both. *Sim:* **k'än 'abäl** 'yellow mombin (*Spondias purpurea* forma *lutea*)'; *Sim:* **chäk 'abäl** 'red mombin (*Spondias purpurea*)'; *Gen:* **'abäl** 'hog plum (*Spondias* spp.)'. *Variant:* **huhup**. *Thes:* **che'**. *[Note:* ajya'ax 'ab'äl [Itz.]. green (unripe) hog plum (Hofling and Tesucún 1997).*] [Source:* BM; NI*]*

ya'ax ak' (ah) arrabidaea (type of). *Lit:* 'green vine'. Bignoniaceae *Arrabidaea* sp. **Es ak'. Uch'ihil ich k'aax. Nikte' ak' ubäho'.** 'It's a vine. It grows in the forest. *Nikte' ak'* is its companion.' (BM). *Loc:* **k'aax** 'forest'; *Use:* **ti' käxik utaan yetel uyokman** 'to tie the girt to the house posts'; *Sim:* **nikte' ak'** 'blossom-tree vine'; *Sim:* **säk ak'** 'white vine (*Arrabidaea verrucosa*)'. *Thes:* **ak'**. *[Note:* yash ak. vine for lashing (Baer and Baer 1952).*] [Source:* BM; KM; KP*]* /\sd2 cordage*]*

ya'ax balche' lancepod. Fabaceae *Lonchocarpus* spp. **Ne ya'ax ule'. Uche'il ne ya'ax usool. Säk utop'. Ne yäk uyits usool.** 'The leaves are very green. The bark of the trunk is very green. The flowers are white. The resin in the bark is very strong.' (BM) Legend has it that *Hachäkyum* set out to create *hach balche'* 'authentic balche' from which his mortals would ferment a mead that they would then offer to him and the other gods. But he made a mistake on his first try, creating *ya'ax balche'* 'green balche'' instead. His assistants brewed up a batch of the ceremonial mead using the bark of this tree, drank it, and died. Although mildly toxic to humans and other mammals, Lancepod is more toxic to insects and aquatic life. It's roots are a source of rotenone, a broad-spectrum insecticide and pesticide. It is also poisonous to fish; one traditional use of lancepod was fish poison (Higbee 1947): pulverized roots or bark from certain species are

Fig. 5.220 (**a**) Ya'axche' *Ceiba pentandra*. Photo Credit: Suzanne Cook (**b**) Ya'ax balche' *Lonchocarpus* sp. Photo Credit: Suzanne Cook (**c**) Ya'ax balche' *Lonchocarpus rugosus* flowers. Photo Credit: Chan K'in Daniel (**d**) Ya'ax balche' *Lonchocarpus* sp. leaves, top side. Photo Credit: Suzanne Cook (**e**) Ya'ax balche' *Lonchocarpus* sp. leaves, underside. Photo Credit: Suzanne Cook (**f**) Ya'ax balche' *Lonchocarpus rugosus* tree. Photo Credit: Suzanne Cook (**g**) Ya'ax balche' bark. Photo Credit: Suzanne Cook (**h**) Ya'ax balche' leaves. Photo Credit: Suzanne Cook

scattered on the surface of the water, depriving the fish of oxygen thus forcing them to the surface where they are easily caught (Standley and Steyermark 1946). This method is still used throughout the Neotropics, especially in Peru and the Amazon. The Lacandones no longer do this, saying that the resin harms the environment (SK). As far as can be determined, *ya'ax balche'* corresponds to at least three Lonchocarpus species: *L. guatemalensis* Benth., *L. rugosus* Benth., and *L. verrucosus* M.Sousa. *Loc:* **k'aax, paakche' kol** 'forest, fallow milpas'; *Use:* **k'anche' näl** 'corncrib'; *Use:* **pa'te' kax** 'chicken coop'; *Sim:* **hach balche'** 'authentic balche' (*Lonchocarpus* sp.)'; *Gen:* **balche'**. *Thes:* **che'** (Fig. 5.220b, c, d, e, f, g, h). *[Note:* yaaxbache. *Lonchocarpus guatemalensis* (Nigh 2008); jobillo, yax pa che. *Lonchocarpus verrucosus* (Durán 1999); ya'ax b'ache. *Lonchocarpus guatemalensis* (Nations 1992, in Atran et al. 2004); aj/ix ya'axmojen [Itz.]. *Lonchocarpus guatemalensis* (Atran et al. 2004); ya'ax mujan [Mop.]. *Lonchocarpus guatemalensis* (in, Atran et al. 2004).*] [Source:* AM; BM; KA; SK*] [*\sd2 folklore_and_mythology*]*

ya'ax bo'oy chamaedorea (type of). *Lit:* 'green bo'oy'. Arecaceae *Chamaedorea* sp., *C. pinnatifrons* (Jacq.) Oerst.? *Ya'ax bo'oy* was unavailable to inspect, so its botanical identity remains uncertain. AM explained that it is similar to *hach bo'oy*, only greener. BM was unfamiliar with the term. *Loc:* **k'aax** 'forest'; *Spec:* **hach bo'oy** 'authentic bo'oy'; *Spec:* **ch'ibix** (*Chamaedorea* spp.); *Spec:* **säk bo'oy** 'white bo'oy'; *Spec:* **k'än bo'oy** 'yellow bo'oy'; *Spec:* **chan mehen bo'oy** 'very small bo'oy'; *Spec:* **ch'it bo'oy** (*Chamaedorea* sp.); *Spec:* **ton bo'oy** (*Chamaedorea* sp.); *Gen:* **bo'oy**. *See:* **hach bo'oy**$_2$. *Variant:* **ya'ax hach bo'oy**. *[Note:* yax boy. *Chamaedorea erenbergiana* [sic.] (Durán 1999).*] [Source:* AM*]*

ya'ax chawah ik chile (type of). Solanaceae *Capsicum annuum* L. *Use:* **hanal** 'food'; *Part:* **wich** 'fruit'; *Prep:* **che'che'** 'raw'; *Use:* **ts'ak ti' chi'** 'oral medicine'; *Part:* **wich** 'fruit'; *Prep:* **che'che'** 'raw'; *Sim:* **säk chawah ik** 'white chawah'; *Gen:* **chawah ik (ah)**. *[Source:* AM*] [*\sd2 food*]*

ya'ax chi' locustberries (serrets). Malpighiaceae *Byrsonima* sp. BM says that this tree is a relative of *k'än chi'* (*Byrsonima crassifolia*), but its fruit stays green when ripe. The taxonomic identity of *ya'ax chi'* is undetermined. On the one hand, it could be *Byrsonima bucidaefolia* Standl.; the Itzaj call this *aj/ix ya'axjom chi'*, which they cook before eating (Atran et al. 2004: 113). On the other hand, it could be a variety of *Byrsonima crassifolia*: according to Pennington and Sarukhán (1968) and Sánchez-Vélez (1988), the fruits of this species vary in colour and taste (cited in Bye and Linares 1990: 158). *Use:* **hanal** 'food'; *Part:* **wich** 'fruit'; *Prep:* **che'che'** 'raw'; *Sim:* **k'än chi'** 'yellow nance'; *Gen:* **chi' (äh)**. *Thes:* **che'**. *[Note:* sakpah [Yuc.]. *Byrsonima bucidaefolia* (in, Atran et al. 2004); aj/ix ya'axjom chi, säkpäj chi' [Itz.]. *Byrsonima bucidaefolia* (Atran et al. 2004).*] [Source:* BM*] [*\sd2 food*]*

ya'ax ch'ib muxan xiiw *Calathea macrosepala* (common name unknown). *Lit:* 'blue vein(ed) muxan xiiw'. Marantaceae *Calathea macrosepala* K. Schum. This is a low-growing, herbaceous plant with long, broad leaves with sigmoid margins and transverse veins. Blades are bright green on the surface, and lighter green underneath with two maroon stripes that run parallel to the midrib. Inflorescence is cylindrical and composed of bright green bracts and pale yellow, tubular blossoms. *Sim:* **hach xiiw** 'authentic xiiw'; *Sim:* **säk muxan (ah)** 'white muxan'; *Gen:* **muxan**. *Variant:* **kuut' (ah)**. *Thes:* **xiiw** (Fig. 5.221a, b). *[Source:* AM*]*

ya'ax hach bo'oy chamaedorea (type of). *Lit:* 'green authentic bo'oy'. Arecaceae *Chamaedorea* sp., *C. pinnatifrons* (Jacq.) Oerst. ? *Gen:* **hach bo'oy** 'authentic bo'oy'. *Variant:* **ya'ax bo'oy**. *[Note:* yax boy. *Chamaedorea erenbergiana* [sic.] (Durán 1999).*] [Source:* AM*]*

ya'ax hälal che' [ya'ax hälal che', ya'ax härа che'] petunia, Mexican (Mexican bluebell). *Lit:* 'blue glow tree'. Acanthaceae *Ruellia brittoniana* Leonard. This is a cultivated, low shrub, 0.9 m (3′) tall, with narrow, dark green, lanceolate leaves and vibrant, lavender-blue trumpet-shaped flowers borne at the tips of strong, semi-woody, vertical stems. *Use:* **tsoy uwich** 'ornamental'. *Thes:* **che'**. *[Source:* AM*] [*\sd2 other*]*

ya'ax hänan *Chamaedorea elatior* (common name unknown). Arecaceae *Chamaedorea elatior* Mart. **Es ak'. Ma' ne nuk. Mäna' uk'i'ixel. Uwich, ek'. Ma' uhantik. Uyak'il ti' ch'uyu'.** 'It's a vine. It's not very big. It doesn't have thorns. Its fruit is black. You don't eat them. The vines are used to make hanging baskets.' (BM). A variety of climbing palm with flexible stems, *Chamaedorea elatior* is the only climbing *Chamaedorea* species. Its slender, occasionally branching, cane-like stems can reach a length of up to 20 m (65′). It scrambles up trees aided by stiff, reflexed leaflets near the tip of the stems. Its scandent habit, bamboo-like leaves, and barbed tip are features shared by *hach hänan* (*Desmoncus orthocanthus*). Unlike *hach hänan*, however, the stems and petioles are smooth. Fruits are round 8–11 cm (<0.5″) and black when mature. It grows in moist or wet, dense forests from 100–1500 m (328–4921′) in elevation (Hodel 1992: 60–61). *Loc:* **k'aax** 'forest'; *Use:* **me'et** 'trivet'; *Part:* **che'il** 'stem'; *Use:* **chäk hu'un ti' pak** 'ceremonial headband for pak (the *balche'* urn)'; *Part:* **le'** 'leaves'; *Sim:* **hach hänan** 'basket tie-tie (*Desmoncus orthocanthus*)'; *Gen:* **hänan**. *Thes:* **ak'** (Fig. 5.222a, b, c). *[Source:* AM; BM*] [*\sd2 craft*]*

ya'ax ik chile (type of). *Lit:* 'green chile'. Solanaceae *Capsicum annuum* L. *Use:* **hanal** 'food'; *Part:* **wich** 'fruit'; *Prep:* **che'che** 'raw'; *Gen:* **ik**. *[Source:* AM*] [*\sd2 food*]*

Fig. 5.221 (**a**) Ya'ax ch'ib muxan xiiw *Calathea macrosepala*. Photo Credit: Chan K'in Daniel (**b**) Ya'ax ch'ib muxän xiiw *Calathea macrosepala*. Photo Credit: Suzanne Cook

Fig. 5.222 (**a**) Ya'ax hänan *Chamaedorea elatior*. Photo Credit: Suzanne Cook (**b**) Ya'ax hänan *Chamaedorea elatior*. Photo Credit: Suzanne Cook (**c**) Ya'ax hänan *Chamaedorea elatior*. Photo Credit: Suzanne Cook

Fig. 5.223 Ya'ax'oon. Photo Credit: Suzanne Cook

Fig. 5.224 Ya'ax 'oox *Brosimum alicastrum* ssp. *Alicastrum*. Photo Credit: Suzanne Cook

ya'ax k'o'och bean, castor (green). Euphorbiaceae *Ricinus communis* L. *Spec:* **chäk k'o'och (äh)** 'red k'o'och'; *Gen:* **k'o'och (äh)**. *See:* **k'o'och (äh)**$_1$.

ya'ax k'um calabasa (squash, pumpkin). *Lit:* 'green squash'. Cucurbitaceae *Cucurbita moschata* (Duch. ex Lam.) Duch. ex Poir . *Use:* **hanal** 'food'; *Part:* **wich** 'fruit'; *Prep:* **chäkbil** 'boiled'; *Gen:* **k'um**. *[*\sd2 food*]*

ya'ax 'oon avocado, American (alligator pear). *Lit:* 'green avocado'. Lauraceae *Persea americana* Mill., syn. *Persea gratissima* Gaertn. *Use:* **hanal** 'food'; *Part:* **wich** 'fruit'; *Prep:* **che'che' yetel mulix** 'raw with lime'; *Sim:* **chäk 'oon** 'red avocado'; *Gen:* **'oon** (Fig. 5.223). *[*\sd2 food*]*

ya'ax 'oox breadnut, Mayan. *Lit:* 'green breadnut'. Moraceae *Brosimum alicastrum* Sw. ssp. *alicastrum* C.C. **Xubal ya'ax uwich. Mas nuk que hach 'oox. Ne hup uwich. Säk uyits.** 'The fruits remain green. They are bigger than the fruits of *hach 'oox* and very soft. The latex is white.' (AM) In the Mayan botanical literature, *ya'ax 'oox* corresponds to at least three species: (1) *Trophis racemosa* (Atran et al. 2004, Hofling 1997, Nations and Nigh 1980); (2) *Trophis mexicana* (Baer and Merrifield 1971; Nations and Nigh 1980); and, (3) *Brosimum alicastrum* (Nations 2006). The leaves and fruit of the tree BM referred to as *ya'ax 'oox* correspond to those of *Brosimum alicastrum* ssp. *alicastrum* (Servando Carvajal, pers. comm.). According to my Lacandon consultants, the mature fruits of *ya'ax 'oox* remain green at maturity. By comparison, the fruit of *Trophis mexicana* is red and that of *T. racemosa* is black (Enquist and Sullivan 2001: 65). The distinction between *ya'ax 'oox* and all other kinds of *'oox* is also made by the Itzaj; according to them, *ya'ax 'oox* refers to *Trophis racemosa* and *T. mexicana* (Hofling and Tesucún 1997). *Loc:* **Naha'**; *Use:* **hanal** 'food'; *Part:* **wich** 'fruit'; *Prep:* **Achäkik uluk'sik usool yetel ta'anil k'ak', pachil ahuch'ik yetel wah** 'You boil them and remove the skins with the ashes from the hearth. Then you grind them with corn dough'; *Use:* **ts'ak ti' chibal koh** 'toothache'; *Part:* **uyits ti' sool** 'latex from the bark'; *Prep:* **Axaak'tik uyits yetel ch'ooch', pachil asulik tämän yetel chen abut'ik yok'ol akoh** 'You mix together the latex and salt, then you soak a piece of cotton with it, and then you stuff it in your tooth'; *Sim:* **hach 'oox** 'authentic breadnut (*Brosimum alicastrum*)'; *Sim:* **chäk 'oox** 'red breadnut'; *Sim:* **k'än 'oox** 'yellow breadnut (*Brosimum alicastrum*)'; *Gen:* **'oox**. *Thes:* **che'** (Fig. 5.224). *[Note:* ya'ash osh. *Trophis mexicana* (Nations and Nigh 1980); ya'ash osh, hach osh. *Brosimum alicastrum* (Nations 2006); ya'x ox. *Trophis mexicana* (Baer and Merrifield 1971); ya'ax 'oox [Itz.]. *Trophis racemosa* (Atran et al. 2004).*] [Source:* AM; CKD*] [*\sd2 food \sd2 medicinal*]*

ya'ax p'ix pear, vegetable (pear squash). Cucurbitaceae *Sechium edule* (Jacq.) Sw. *Loc:* **kolil nah, kol** 'house garden, milpa'; *Use:* **hanal** 'food'; *Part:* **wich, le', nek'** 'fruit, leaves, seed'; *Prep:* **chäkbil** 'boiled'; *Sim:* **säk p'ix** 'white vegetable pear'; *Sim:* **k'i'ilix p'ix** 'spiny vegetable pear'; *Gen:* **p'ix (ah)**. *[Source:* KyP*] [*\sd2 food*]*

ya'ax tsuk'in ak' smilax. *Lit:* 'green tsuk'in ak''. Smilacaceae *Smilax* sp. A kind of herbaceous, extremely thorny vine. The specimen observed was found at the edge of clearing and secondary forest. *Use:* **ti' mentik uyoom ti' käkow** 'used to froth cocoa'; *Part:* **uyits ti' uyool, mumuun uyak'il** 'the resin from the growing tips and young vines'; *Prep:* **Ayäch'ik, a muxik, pachil apulik yok'ol käkow** 'You crush and mash it, and then add it to cacao'; *Sim:* **chäk tsuk'in ak'** 'red tsuk'in vine (*Smilax* sp.)'; *Gen:* **tsuk'in ak' (ah)**. *Thes:* **ak'**. *[Source:* BM*] [*\sd2 food*]*

yahah ts'in manioc. Euphorbiaceae *Manihot esculenta* Crantz. *Use:* **hanal** 'food'; *Gen:* **ts'in**. *See:* **yahche' ts'in**. *Variant:* **yahche' ts'in**. *[Source:* AM*] [*\sd2 food*]*

Fig. 5.225 Yahche' ts'in *Manihot esculenta*. Photo Credit: Suzanne Cook

yahche' ts'in [yahche' ts'in, ya'axche' ts'in] manioc. *Lit:* 'great tree manioc/green manioc tree'. Euphorbiaceae *Manihot esculenta* Crantz. **Uche'il as chäkchäk. Uchuch ule', chäk. Ne chukuch uwi', batak 1 m (3.28′) uwaan. Usool, ek' chäkchäk. Unoy, ne säk.** 'The stalk is somewhat reddish (coffee colour). The leaf stem is red. The tuber is very long, about 1 m (3.28′) long. The skin is dark reddish brown. The flesh is very white.' (BM) AM says that the tubers are white inside and out, and that they are larger than those of hach ts'in, up to 1 m (3.29′) long and 30 cm (12″) in diameter. The large tubers are always boiled, because it would take too long to roast them. The specimen referred to as *yahche' ts'in*, shown in the photo, has green petioles, a greyish-tan stalk, and palmate leaves with seven lobes. AM says this variety is not cultivated much anymore. *Use:* **hanal** 'food'; *Prep:* **chäkbil, amäk'ik yetel ch'ooch'** 'boiled, you eat it with salt'; *Prep:* **yetel säkän ti' wah** 'with dough for tortillas'; *Sim:* **hach ts'in** 'authentic manioc'; *Sim:* **chäk ts'in** 'red manioc'; *Gen:* **ts'in**. *Variant:* **yahah ts'in** (Fig. 5.225). *[Note:* KyP pronounces the word as [ya'axche'] ts'in and says that the tubers are large and green.*] [Source:* AM; BM; KyP*] [*\sd2 food*]*

yo'och ahpek' beans, Jack (Chickasaw lima bean). Fabaceae *Canavalia* spp. *Canavalia* is a genus of tropical legume vine, comprising approximately 70–75 species. Characteristic of the genus, the pods have three ribs along their length. At least two or three species occur in Naha'. One is *C. ensiformis* (L.) DC., which bears light pink or mauve blossoms followed by 30 cm (1′) long, straight pods containing large, smooth, white seeds. The identity of the other is more tenuous: it bears showy, bright pink blossoms similar to those of *C. villosa* Benth. and *C. rosea* (Sw.) DC. The latter is commonly found on tropical beaches and adjacent areas (Vibrans 2011). According to AM, there is one species that produces slightly compressed, coffee-coloured seeds. This may be *C. villosa*. *Canavalia* is native to Mexico, Central America, Bolivia, and Ecuador, and may be one of the earliest cultivars in the area. The earliest domestication date is around 900 A.D., in Oaxaca, however, charred seeds believed to be *Canavalia* found in the Yucatan date back to 300 B.C. (Sauer and Kaplan 1969). The seeds of *Canavalia enformis* are eaten, but the genus, as whole, plays a more important role in local pharmacopeia, as well. For example, in Naha', the seeds of *C. enformis* are fed to dogs who have ingested *chok* berries, or that have eaten the innards of birds who have ingested these berries. *Loc:* **kolil nah** 'house garden'; *Use:* **ts'ak ti' chok** 'medicine for dogs that have ingested chok'; *Part:* **nek'** 'seeds'; *Prep:* **k'elbil** 'toasted'; *Gen:* **box bu'ul**. *Variant:* **bubuhchi' (ah)**; **bu'ul boox**. *Thes:* **bu'ul**. *[Note:* frijol 'ab'oonoj [Itz.]. *Canavalia ensiformis* (Atran et al. 2004).*] [Source:* AM; NukAM; KM; KP*] [*\sd2 medicinal*]*

yo'och ähbaach' coffee, wild (limoncillo). *Lit:* 'food of the chachalaca'. Siparunaceae *Siparuna thecaphora* (Poepp. & Endl.) A.DC., syns. *Siparuna nicaraguensis* Hemsl., *Siparuna andina* (Tul.) A. DC. This is an evergreen, aromatic shrub or treelet, 3–7 m (9.8–23′) tall, with a slim trunk. Leaves are opposite or borne in whorls of three. They vary in shape from oblong to lanceolate to elliptic to obovate, 8–25 cm (3–10″) long and 5–8 cm (2–3″) wide. Flowers are yellow, cream, or orange-red. The small, 1–1.5 cm (0.5–0.9″) round fruit are sparsely pubescent (downy) to smooth and red with pale brown or green spots. Fruits contain numerous, bright red seeds in a gelatinous aril. As the Lacandon name implies, the *chachalaca* (*Ortalis vetula*) feeds on the berries. *Use:* **ts'ak ti' aak'ä' ich** 'medicine for the evil eye'; *Part:* **le'** 'leaves'; *Prep:* **Ak'elik ule' ich k'ak', pachil apulik alkohol yok'ol ti', pachil apulik yok'ol uho'ol ti' mehen paalal** 'You sear the leaves over an open fire, then you douse them with alcohol, and then you place them on the head of the child'; *Use:* **che'il k'ak'** 'firewood'. *Variant:* **tu'tu' che'**. *Thes:* **che'** (Fig. 5.226a, b). *[Note:* ya che kap, leulin. *Siparuna andina* (Durán 1999).*] [Source:* AM; KP; SK*] [*\sd2 fuel*]*

yo'och ik psychotria. *Lit:* 'chile food'. Rubiacea *Psychotria* spp., syn. *Cephaelis* spp. *Psychotria* are understory shrubs or trees. Fifty-three species have been reported from Chiapas (Breedlove 1986). Twenty-three species are found in Naha' (Comisión Nacional de Áreas Naturales Protegidas 2006). According to AM, there are only two kinds: *nukuch yo'och* ik 'large yo'och ik' and *mehen yo'och ik* 'small yo'och ik'. The large kind is used to treat adult skin conditions, whereas the small kind is used to treat baby rashes. Durán notes that *P. panamensis* is used to treat babies that are "*muy molestos*", while *P. horizontalis* is used to help newborns "*crezcan bien*" (1999: 148). Two other species used are *P. chiapensis*, whose fruit is eaten and used as a dye plant (plant parts unspecified),

Fig. 5.226 (**a**) Yo'och ähbaach' *Siparuna thecaphora*. Photo Credit: Suzanne Cook (**b**) Yo'och ähbaach' *Siparuna thecaphora*. Photo Credit: Suzanne Cook

Fig. 5.227 (**a**) Yo'och ik mehen *Psychotria* sp. Photo Credit: Suzanne Cook (**b**) Yo'och ik mehen *Psychotria* sp. Photo Credit: Suzanne Cook

and *P. galeottiana*, which is used in house construction (Durán 1999: 149). *Use:* **ts'ak ti' sak'il** 'for itching (skin)'; *Part:* **le'** 'leaves'; *Prep:* **chäkbil** 'boiled'; *Spec:* **yo'och ik mehen** 'small yo'och ik (*Psychotria* sp.)'; *Spec:* **yo'och ik nukuch** 'large yo'och ik (*Psychotria panamensis* var. *panamensis*)'. *Thes:* **che'**. *[Source:* AM*]* *[*\sd2 medicinal*]*

yo'och ik mehen psychotria, type of. Rubiaceae *Psychotria* sp. *Yo'och ik mehen* is a small tree or large bush with woody stems and a scandent habit. Leaves are ovate, acuminate, thick, somewhat succulent, bright green, glossy, with inconspicuous, secondary venation. It produces small, round, bright red, fleshy berries. It is found growing in filtered light in the forest. The species could be *P. fruticetorum* (Charlotte Taylor pers. comm.). AM says that the leaves are steeped in water, and then the water used to soothe rashes on new-born babies (neonatal acne, erythema toxicum neonatorum (ETN)?). According to Durán (1999), *P. panamensis* is used to treat babies that are "*muy molestos*", while *P. horizontalis* is used to help newborns "*crezcan bien*" 'grow well'. *Loc:* **k'aax** 'forest'; *Use:* **ts'ak ti' uyoot'el mehen paalal** 'medicine for baby's skin'; *Part:* **le'** 'leaves'; *Prep:* **Achäkik ule' ich ha', pachil apulik ich uyichkil ti' mehen paalal** 'You boil the leaves in water, then you put it into the baby's bath'; *Sim:* **yo'och ik nukuch** 'large yo'och ik (*Psychotria panamensis* var. *panamensis*)'; *Gen:* **yo'och ik**. *Variant:* **mehen yo'och ik, mehen ik**. *Thes:* **che'** (Fig. 5.227a, b). *[Note:* yox ik mejen, amamanche', mehen wech. *Psychotria horizontalis* (Durán 1999).*]* *[Source:* AM*]* *[*\sd2 medicinal*]*

yo'och ik nukuch Panama balsamo. Rubiaceae *Psychotria panamensis* var. *panamensis* Standl., syns. *Psychotria catillicalyx* Dwyer, *Psychotria durilancifolia* Dwyer, *Psychotria molinae* Standl., *Psychotria nubiphila* Dwyer, *Psychotria yunckeri* Standl. **Mehen che'. Ma' ne nuk utan ule'. Ya'ax ule'. Chukuch uwaan ule'. Uch'ihil k'aax, tuwolol, tuyäkanil, ich wits. Chäk uwich, bexil uwich ti' mehen ik.** 'It's a small tree. Its leaves are not very wide. The leaves are green and long. It grows in the ravines and hills. Its fruit is red, similar to the fruit of *yo'och ik mehen* (*Psychotria* sp.).' (AM) The botanical identification of this species is tentative, because the plants observed lacked flowers and fruits. *Use:* **ts'ak ti' uyoot'el** 'medicine for skin (rashes)'; *Part:* **le'** 'leaves'; *Prep:* **Achäkik ule' ich ha', pachil awichkil yok'ol ha'** 'You boil the leaves in water and then you bathe in the water'; *Sim:* **yo'och ik mehen** 'small yo'och ik (*Psychotria* sp.)'; *Gen:* **yo'och ik**. *Variant:* **nukuch ik**. *Thes:* **che'** (Fig. 5.228a, b). *[Note:* Botanical identification provided by John R. Paul (pers. comm.). xo yok, xo yok che. *Psychotria panamensis* (Durán 1999). Synonyms from http://www.theplantlist.org. English name from Grandtner (1997–2010).*]* *[Source:* AM*]* *[*\sd2 medicinal*]*

Fig. 5.228 Yo'och ik nukuch *Psychotria panamensis* var. *panamensis*. Photo Credit: Suzanne Cook

yuwin Caribbean grape. Vitaceae *Vitis tiliifolia* Humb. & Bonpl. ex Schult, syn. *Vitis caribaea* DC. **Ak'. Ne nuk, säk uyak'il. Yan utop'. Kuwaak'al utop' yan ahweel.** 'This is a vine. It is very large with white vines. It has flowers. When the blossoms burst open there is a kind of small blackfly inside.' (AM) This is a Neotropical liana whose large vines contain potable water. No uses were reported, but Baer and Merrifield say that *"yuwir"* is used as fuel in pit roasting (1971: 159). *Loc:* **Villa Rosa**. *See:* **hach ts'us**. *Variant:* **yuhi**. *Thes:* **ak'**. *[Note:* yuhi. water vine. *Vitis tilliifolia* (Nations and Nigh 1980); yuwir. water vine (Baer and Merrifield 1971); sayab'ak' [Itz.].*Vitis caribaea* (Atran et al. 2004).*]* *[Source:* AM*]*

yuyul holop'$_1$ [yuyul holop', yuyur holop', yuyuhorop'] *Philodendron tripartitum* (common name unknown). *Lit:* 'bumpy holop''. Araceae *Philodendron tripartitum* (Jacq.) Schott. This is type of philodendron with smooth vines and three-lobed, glossy leaves. Vines are covered with raised, blunt lenticels. *Loc:* **k'aax, kolil nah** 'forest, house garden'; *Use:* **ti' uk'äxik kun** 'for tying palm thatch'; *Use:* **xaak** 'baskets'; *Part:* **ak'** 'vines'; *Sim:* **k'i'ilix holop'** 'spiny holop' (*Philodendron radiatum*)'; *Gen:* **holop'** (Fig. 5.229a). *[Note:* Botanical identification provided by Tom Croat (pers. comm.).*]* *[Source:* AM; BM*]* *[*\sd2 cordage*]*

yuyul holop'$_2$ *Philodendron smithii* (common name unknown). Araceae *Philodendron smithii* Engl. *Loc:* **k'aax, kolil nah** 'forest, house gardens'; *Use:* **ti' uk'äxik kun** 'for tying palm thatch'; *Use:* **xaak** 'baskets'; *Part:* **ak'** 'vines'; *Sim:* **k'i'ilix holop'** '*Philodendron radiatum*'; *Gen:* **holop'**. *Thes:* **ak'** (Fig. 5.229b). *[Note:* Botanical identification provided by Tom Croat (pers. comm.).*]* *[Source:* AM; BM*]* *[*\sd2 cordage \sd2 containers*]*

a

b

Fig. 5.229 (**a**) Yuyul holop' *Philodendron tripartitum*. Photo Credit: Suzanne Cook (**b**) Yuyul holop' *Philodendron smithii*. Photo Credit: Suzanne Cook

Botanical-Lacandon Index

Acanthaceae	
Justicia sp.	**tsits**
Louteridium donnell-smithii S. Wats.	**k'o'och ba'ats' (äh)**
Ruellia brittoniana Leonard	**ya'ax hälal che'**
Actinidiaceae	
Saurauia yasicae Loes., syn. *Saurauia belizensis* Lundell	**mumun che'**
Amaranthaceae	
Chenopodium ambrosioides L., syn. *Teloxys ambrosioides* L.	**k'äxeex**
Amaryllidaceae	
Allium cepa L.	**säk seboya (äh) seboya**
Allium glandulosum Link & Otto	**ts'ak kitam, ts'ak k'ek'en**
Allium sativa L.	**'axux**
Crinum amabile Donn ex Kew Gawler, syn. *Crinum asiaticum* L.	**chäk ts'ulha' (äh)**
Hymenocallis littoralis (Jacq.) Salisb.	**ts'ulha'il petha' (äh)**
Anacardiaceae	**k'än'ol (äh)**
Astronium graveolens Jacq.	**mehen k'än'ol**
Mangifera indica L.	**mankoh**
Metopium brownei (Jacq.) Urb.	**chechem (äh)**
Mosquitoxylum jamaicense Krug & Urb.	**nukuch k'än'ol**
Spondias mombin L., syn. *S. lutea* L.	**k'inim**
Spondias purpurea L.	**chäk 'abäl**
Spondias purpurea, S. purpurea forma *lutea* F. & R.	**k'än 'abäl**
Spondias sp.	**huhup**
Spondias sp.	**ya'ax 'abäl**
Spondias spp.	**'abäl**
Annonaceae	
Annona cherimola Mill.	**säk 'oop**
Annona muricata L.	**pähpox**
Annona reticulata L.	**chäk 'oop, chäk uwich (äh)**
Annona sp.	**mehen 'oop che'**
Annona sp.	**'oopil k'aax**
Annona spp.	**'oop**
Annona spp.	**'oop che'**
Annona reticulata, A. cherimoya	**hach 'oop**
Cymbopetalum sp.	**ton ku'uk**
Desmopsis stenopetala (Donn. Sm.) R.E. Fries	**'oopil k'aax**
Guatteria anomala R. E. Fries	**ek' bahche'**
Apiaceae	
Coriandrum sativum L.	**perehir**
Eryngium foetidum L., syn. *Eryngium antihystericum Rottbøll*	**kulantro (perehil)**
Apocynaceae	
Aspidosperma megalocarpon Muell. Arg., syn. *Aspidosperma macrocarpum* Mart.	**sayok (ah)**
Plumeria rubra f. *rubra*	**chäk nikte'**
Plumeria rubra L.	**nikte' (ah)**
	säk nikte'
Tabernaemontana amygdalifolia Jacq., *syn. Cerbera ahouai* Linn.	**ton tsimin**
Thevetia ahouai (L.) A. DC.	**tuchil k'aax**
	hach tuch
	tuch
	mehen tuch

(continued)

(continued)

Araceae	
Anthurium huixtlense Matuda	**ni' ah'och**
Anthurium schlechtendalii Kunth	**neeh k'ambul (ah)**
Dieffenbachia sp.	**chiah (ah)**
Monstera deliciosa Liebm.	**kamis 'ilon (ah)**
Monstera sp.	**hach 'ilon**
Monstera sp.	**säk 'ilon**
Monstera spp.	**'ilon**
Philodendron donnell-smithii Engl.	**hach 'oochin**
Philodendron hederaceum (Jacq.) Schott., syn. *P. scandens*	**mehen 'oochin**
Philodendron inaequilaterum Liebm., syn. *P. coerulescens* Engler	**hach chiah (ah)**
Philodendron radiatum Schott.	**k'i'ilix holop'**
Philodendron sagittifolium Liebm.	**nukuch 'oochin**
Philodendron smithii Engl.	**holop'**
	yuyul holop'
Philodendron tripartitum (Jacq.) Schott.	**yuyul holop'**
Spathiphyllum phryniifolium Schott.	**pähk'ol**
Syngonium cf. angustifolium Schott.	**'oochin**
Syngonium chiapense Matuda	**nukuch 'oochin**
Syngonium donnell-smithii Engl., syn. *Syngonium salvadorense* Schott.	**'oochin**
Syngonium podophyllum Schott.	**mehen 'oochin**
	nukuch 'oochin
Syngonium sp.	**'oochin**
Xanthosom robustum Schott.	**hach ch'om**
Xanthosoma mafaffa Schott.	**chäk mäkäl**
	hach mäkäl
	säk mäkäl
	suli'
Xanthosoma robustum Schott.	**ch'om mäkäl**
Xanthosoma spp.	**mäkäl**
Araliaceae	
Dendropanax arboreus (L.) Decne. & Planch.	**sasakche'**
Oreopanax obtusifolius L. O. Williams	**k'ä' ta'k'in**
Arecaceae	
Acoelorraphe wrightii (Griseb. & H.Wendl.) H. Wendl. ex Becc., syn. *Paurotis wrightii* (Griseb. & Wendl) Britton ex Britton & Schafer	**booxa'an**
Acrocomia mexicana Karw. ex Mart., syn. *Acrocomia aculeata* (Jacq.) Lodd. ex Mart.	**tuk'**
Astrocaryum mexicanum Liebm. ex Mart., syn. *Bactris cohune* S. Watson	**'akte'**
	hach 'akte', ya'ax 'akte'
Attalea butyracea (Mutis ex. L.f.) Wess Boer. (?)	**tuts**
Chamaedorea pinnatifrons (Jacq.) Oerst.	**säk bo'oy**
Chamaedorea arenbergiana H. A. Wendl.	**säk bo'oy**
Chamaedorea cataractarum Mart.	**ch'ibix uk'um**
Chamaedorea elatior Mart.	**ya'ax hänan**
Chamaedorea elegans Mart.	**ch'ibix bo'oy**
	ch'ilix bo'oy, ch'ibix (ah)
Chamaedorea ernesti-augusti H. Wendl.	**k'eben**
Chamaedorea neurochlamys Burret.	**k'än bo'oy**
Chamaedorea oblongata Mart.	**hade**
	hach bo'oy
	säk bo'oy
Chamaedorea pinnatifrons (Jacq.) Oerst.	**hach bo'oy**
Chamaedorea sp.	**chan mehen bo'oy**

(continued)

(continued)

Chamaedorea sp.	**ch'it bo'oy**
Chamaedorea sp.	**ch'it k'än bo'oy**
Chamaedorea sp.	**ton bo'oy**
Chamaedorea sp.	**ya'ax bo'oy**
Chamaedorea sp.	**ya'ax hach bo'oy**
Chamaedorea spp.	**bo'oy (äh)**
Chamaedorea tepejilote Liebm.	**ch'i'ib' (äh)**
	pakyon (äh)
Chamaedorea woodsoniana L. H. Bailey	**ch'ibix wits**
Cryosophila stauracantha (Heynh.) R.J. Evans, syn. *C. argentea* Bartlett.	**kun (ah)**
Desmoncus orthacanthus Mart.	**hach hänan**
	hänan
Geonoma interrupta var. *interrupta* (Ruiz & Pav.) Mart., syn. *G. oxycarpa Mart.*	**k'unche'**
	k'unche' pahok, pahok
Sabal mexicana Mart.	**xa'an**
Asclepiadaceae	
Gonolobus sp.	**kimil sukun**
Asparagaceae	
Agave sisalana Perrine	**kih**
Polianthes tuberosa L.	**bäknikte'**
Asteraceae	
Baccharis trinervis (Lam.) Pers., syn. *Eupatorium trinervatum* Sessé & Moc.	**si'si' k'uuts**
Bidens odorata Cav.	**ku'uchnook'**
Carthamus tinctorius L.	**sik'**
Clibadium arboreum Donn. Sm., syn. *C. donnell-smithii* J. M. Coult.	**chäklanpix (ah)**
Erechtites hieracifolia (L.) Raf. ex DC.	**nukuch le' si'si' k'uuts**
Podachaenium eminens (Lag.) Sch. Bip.	**ki' bok (äh)**
	haban uts'u' (ah)
Schistocarpha eupatorioides (Fenzl) Kuntze	**chäklanpix (ah)**
Telanthophora grandifolia (Less.) H. Rob. & Brettell, syn. *Senecio grandifolius* Less.	**mumun che'**
Vernonanthura patens (Kunth) H. Rob., *syn. Vernonia patens* Kunth.	**che' chäk'an**
Auriculariaceae	
Auricularia cornea (Ehrenb) Ehrenb ex Endl., *A. delicata* Henn.	**lo'loh (äh)**
Bignoniaceae	
Crescentia alata Kunth	**k'än luuch**
Crescentia cujete L.	**hach luuch**
Crescentia sp.	**mehen k'än luuch**
Crescentia sp.	**sina'an luuch (äh)**
Crescentia spp.	**luuch, luuch che'**
Cydista aequinoctialis (L.) Miers, syn. *Bignonia aequinoctialis* L.	**'axux ak'**
Parmentiera edulis DC., syn. *P. aculeata* (Kunth) Seem.	**kat**
	'anis
Tabebuia guayacan (Seem.) Hemsl.	**hach chulul**
Bixaceae Bixa orellana L.	**k'uxub**
Bombacaceae	
Bernoullia flammea Oliv.	**ponche'**
Ceiba pentandra (L.) Gaertn.	**ya'axche'**
Pseudobombax ellipticum (Kunth) Dugand	**ch'ulte'**
Boraginaceae	
Cordia alliodora (Ruiz & Pav.) Oken	**bahun che'**
Brassicacea	
oleracea var. *alba* L.	**kulix**

(continued)

(continued)

Bromeliaceae	
Ananas comosus (L.) Merr.	**pach'**
Bromelia pinguin L.	**ch'äm (äh)**
Burseraceae	
Bursera simaruba (L.) Sarg.	**chäklah**
Protium copal (Schltdl. & Cham.) Engl.	**tsatse pom**
Cactaceae	
Epiphyllum sp.	**le'mo'**
Cannabaceae (Ulmaceae)	
Trema micanthra (L.) Blume	**taw**
Cannaceae	
Canna indica L., syns. *C. edulis* Ker Gawl, *C. lutea* Larrañaga (nom. illeg.)	**chank'ala'**
Caricaceae	
Carica papaya L.	**put**
Celastraceae	
Rhacoma eucymosa (Loes. & Pittier) Standl.	**k'ik'ni' balum (ah)**
Wimmeria bartlettii Lund.	**k'än chululche'**
Chrysobalanaceae	
Licania platypus Fritsch., syn. *Moquilea platypus* Hemsl.	**tsäk'ats**
Clethraceae	
Clethra sp.	**kunkunche' kap'**
Clusiaceae	
Calophyllum brasiliense Camb. var. *rekoi* Standl.	**babah**
Clusia flava Jacq.	**kahanche'**
Clusia lundellii Standl.	**ts'oy (äh)**
Clusia spp.	**chunup (äh)** **ts'oy (äh)**
Combretaceae	
Terminalia amazonia (J. F. Gmel.) Exell., syn. *T. obovata* (Ruiz & Pav.) Steud.	**k'änxa'an, pukte'**
Connaraceae	
Rourea glabra Kunth	**chok**
Convolvulaceae	
Ipomoea alba L.	**hach hut'kih**
Ipomoea batatas (L.) Lam.	**chäk 'is** **chäk pach 'is** **'is** **k'än 'is** **säk 'is**
Ipomoea sp.	**chäk hut'kih**
Ipomoea sp.	**tu'xikin**
Ipomoea sp.	**nukuch tu'xikin**
Ipomoea spp., *Ipomoea alba* L.	**hut'kih**
Merremia discoidesperma Donn. Sm	**ho'ol ma'ax, gorila**
Merremia tuberosa (L.) Rendle., syn. *Ipomoea tuberosa* L.	**ts'ak k'ik'**
Costaceae	
Costus pulverulentus C. Presl	**pahsa' ak' (äh)**
Crassulaceae	
Kalanchoe pinnata (Lam.) Pers., syn., *Bryophyllum pinnatum* (Lam.) Oken	**san lorenso, ts'ak ti' se'em**
Cucurbitaceae	
Citrullus lanatus (Thunb.) Matsum. & Nakai.	**sanyah**
Cucurbita argyrosperma K. Koch, syn. *C. mixta*	**sikil**
Cucurbita moschata (Duch. ex Lam.) Duch. ex Poir.	**ya'ax k'um**
Cucurbita moschata (Duch. ex Lam.) Duch. ex Poir., *Cucurbita pepo* L.	**k'um**
Cucurbita sp.	**box**

(continued)

(continued)

Cucurbita sp., *C. pepo* L. ssp. *pepo*	**mumun sikil**
Lagenaria siceraria (Molina) Standl.	**chuh**
	ha' chuh
	kis chuh mehen
	kis chuh (äh)
	lek
Melothria pendula L.	**sanyah ahch'o' (ah)**
Sechium edule (Jacq.) Sw.	**k'i'ilix p'ix**
	säk p'ix
	ya'ax p'ix
	p'ix (ah)
Cyatheaceae	
Alsophila cf. *cuspidata, Alsophila salvinii* Hook.	**waalka'an**
Cyclanthaceae	
Carludovica labela R. E. Schult., syn. *Asplundia labela* (R. E. Schult.) Harling	**miis**
Cyperaceae	
Cladium jamaicense Crantz, syn. *Mariscus jamaicensis* (Crantz) Britt.	**ts'iibal, ts'iib**
Dennstaedtiaceae	
Pteridium pseudocaudatum, syn. *P. aquilinum* var. *pseudocaudatum* (Clute) A. Heller	**waalka'an**
Dilleniaceae	
Doliocarpus dentatus (Aubl.) Standl.	**säha' (ah)**
Dioscoreaceae	
Dioscorea alata L., syn. *D. atropurpurea* Roxb.	**hach suli'**
Dioscorea bartlettii C. V. Morton	**k'ul ak'**
Dioscorea bulbifera L.	**bilim suli'**
Dioscorea spp.	**suli' (äh)**
Dryopteridaceae	
Polystichum sp.	**waalka'an**
Ebenaceae	
Diospyros digyna Jacq., syn. *Diospyros ebenaster* Retz.	**'uuch (äh)**
Equisetaceae	
Equisetum hyemale var. *affine* (Engelm.) A. A. Eaton	**maasan kitan**
Euphorbiaceae	
Alchornea latifolia Sw.	**muxan che'**
Chamaesyce hirta (L.) Millsp.	**xänä' mook**
Cnidoscolus multilobus (Pax) I. M. Johnston, syn. *Multilobar Jatropha* Pax	**tsaah (äh)**
Croton draco Schltdl.& Cham., syn. *C. lechleri* Muell. Arg.	**chukum che'**
Euphorbia pulcherrima Willd. ex Klotzsch	**chäkche' (äh)**
Jatropha curcas L.	**sikilte' che'**
Manihot esculenta Crantz.	**chäk ts'in**
	hach ts'in
	säk ts'in
	ts'in
	yahah ts'in
	yahche' ts'in
Ricinus communis L.	**chäk k'o'och**
	mehen chäk k'o'och
	nukuch chäk k'o'och
Sapium lateriflorum Hemsl.	**'ukunte' (äh)**
Fabaceae	
Acacia collinsii Saff.	**subin (ah), subinte' (ah)**
Acacia mayana Lundell, syn. *Vachellia mayana* (Lundell) Seigler & Ebinger	**akunte'**
Arachis hypogaea L.	**kakawat**
	sikilte' lu'um

(continued)

(continued)

Bauhinia rubeleruziana Donn. Sm., syns. B. emarginella Standl.	**k'ak'che', säkche'**
Caesalpinia major (Medik.) Dandy & Exell, syn. *Bonduc majus* Medik.	**patoh ak' (ah)**
Calliandra calothyrsus Meisn.	**chäk me'exk'in (äh)**
Canavalia spp.	**bubuhchi' (ah)**
	yo'och ahpek'
Cojoba sp., *C. arborea* L. (Britton & Rose) or *C. graciliflora* (S. F. Blake) Britton & Rose, syn. *Cojoba donnell-smithii* Britton & Rose	**mehen bu'ulche'**
Cojoba spp.	**bu'ulche'**
Desmodium tortuosum (Sw.) DC., syn. *Desmodium annuum* A. Gray p.p.	**tsaayentsaay (ah)**
Dialium guianense (Aubl.) Sandwith	**wäch'**
Entada gigas (L.) Fawc. & Rendle	**korason de mono**
Enterolobium cyclocarpum (Jack.) Griseb, syn. *Albizia longipes* Britton & Killip.	**pets'k'in**
Erythrina berteroana Urb.	**k'ante' (ah)**
Erythrina sp.	**k'ante' ak'**
Haematoxylon campechianum L.	**ek'**
Hymenaea courbaril L.	**pakay**
Inga oerstediana Benth. ex Seem. Boege, syn. *Inga minutula* Mart.	**suumil bits'**
Inga punctata Willd., syn. *Inga leptoloba* Schltdl.	**nukuch bits'**
Inga sapindoides Willd., syn. *Inga pavoniana* G. Don	**t'elen bits'**
Inga belizensis Standl.	**bits'il k'aax**
Inga sp.	**mehen bits'**
Inga spp.	**bits'**
Inga spuria Humb. & Bonpl. ex Willd, syn. *Inga vera* ssp. *spuria* (Humb. & Bonpl. ex Willd.) J. León	**bits' petha'**
Lonchocarpus sp.	**hach balche'**
Lonchocarpus spp.	**balche'**
	ya'ax balche'
Mucuna argyrophylla Standl.	**ni' ts'ul**
Mucuna pruriens (L.) DC., syn. *Stizolobium pruriens* (L.) Medik.	**bubuhchi' (ah)**
Mucuna sloanei Fawc. & Rendle	**bubuhchi' (ah), wich ahyuk (ah)**
Myroxylon balsamum (L.) Harms var. *pereirae* (Royle) Harms	**naba'**
Ormosia isthmensis Standl.	**'äm ch'uplal**
Ormosia Jacks.	**'äm**
Ormosia schippii Pierce ex Standl. & Steyerm.	**'äm ton**
Oxyrhynchus trinervius (J.D. Smith) Rudd., syn. *Dioclea trinervia* Donn. Sm.	**'ip'il k'aax**
Pachyrhizus erosus (L.) Urb.	**chi'kam**
Phaseolus lunatus L. cv. 'Sieva'	**ek' ip'**
	chäk 'ip'
	'ip'
Phaseolus vulgaris L.	**ak'il bu'ul (ah)**
	bu'ul
	chäk box
	chäk che'il bu'ul
	che'il bu'ul (ah)
	ek'el bu'ul
	ek' bu'ul
	ek' che'il bu'ul
	hach box
	hach bu'ul
	nukuch box
	chäk bu'ul
Platymiscium dimorphandrum (J. D. Smith) Donn. Sm., syn. *Platymiscium pinnatum* (Jacq.) Dugand.	**säk chulul (äh)**
Rhynchosia erythrinoides Schltdl. & Cham.	**choochel ak' (äh)**

(continued)

(continued)

Rhynchosia pyramidalis (Lam.) Urban. (?), syn. *Dolicholus phaseoloides* (DC.) Kuntze.	**mo'ak' (ah)**
Schizolobium parahybum (Vell.) Blake	**pets'k'in**
Vigna umbellata (Thunb.) Ohwi & H. Ohashi., syn. *Phaseolus calcaratus* Roxb.	**chäk bu'ul, tsaayentsaay (ah)**
Vigna unguiculata (L.) Walp.	**box bu'ul**
Fagaceae	
Quercus corrugata Hook.	**nukuch ch'alol**
Quercus corrugata Hook., syn. *Q. cyclobalanoides* Trel.	**hach ch'alol**
Quercus skinneri Benth.	**pixan k'ambul (äh)**
Quercus sp. *Q. oleoides* Schltdl. & Cham. (?)	**mehen ch'alol**
Quercus spp.	**ch'alol**
Flacourtiaceae (Salicaceae)	
Pleuranthodendron lindenii (Turcz) Sleumer.	**iximche'**
Lacistema aggregatum (Bergius) Rusby (?)	**payok che' (ah)**
Ganodermataceae	
Ganoderma applanatum (Pers.) Pat.	**kuxum che'**
Gomphaceae	
Ramaria sp., spp.(?)	**kuxum che'**
Heliconiaceae	
Heliconia librata Griggs.	**chukuch le' xiiw, hach xiiw, säk muxan (ah)**
Icacinaceae	
Oecopetalum mexicanum Greenm. & C.H. Thomps.	**kunkunche'**
Lamiaceae	
Clerodendrum sp.	**itsänte'**
Mentha citrata Ehrh., syn. *M. aquatica* L.	**menta, ts'ak kax**
Lauraceae	
Licaria alata Miranda, syn. *Licaria excelsa* Kosterm.	**säk 'oonte'**
Nectandra globosa (Aubl.) Mez.	**ek' 'oonte'**
Nectandra sp.	**mehen 'oonte'**
Nectandra sp.	**nukuch 'oonte'**
Ocotea cernua (Nees) Mez.	**babah oonte', kakache'**
Persea americana Mill., syn. *Persea gratissima* Gaertn.	**chäk 'oon, ya'ax 'oon, 'oon**
Loranthaceae	
Struthanthus orbicularis (Kunth) Eichler.	**iich'ak tolok, nukuch tak'lan che'**
Magnoliaceae	
Talauma mexicana (DC.) G. Don.	**kuti', kuti'il wits**
Malpighiaceae	
Bunchosia sp.	**siip che' (ah)**
Byrsonima crassifolia (L.) H. B. K., syn. *Byrsonima cubensis* Juss.	**k'än chi'**
Byrsonima sp.	**ya'ax chi'**
Byrsonima spp.	**chi' (äh)**
Malvaceae	
Quararibea funebris (La Llave) Vischer	**mahaas (ah)**
Heliocarpus appendiculatus Turcz.	**chäk halol**
Heliocarpus donnell-smithii Rose	**säk halol**
Heliocarpus spp.	**halol**
Abelmoschus moschatus Medik, syn. *Hibiscus abelmoschus* L.	**'anis**
Gossypium hirsutum L.	**tämän**
Hampea stipitata S. Watson, syn. *Hampea euryphylla* Standl.	**ts'ubtok**
Heliocarpus donnell-smithii Rose	**hach halol**
Hibiscus rosa-sinensis L.	**tämänil (äh)**
Malvaviscus arboreus Cav.	**tämän che'**
Pachira aquatica Aubl.	**kubuh (ah)**

(continued)

(continued)

Marantaceae	
Calathea sp.	**chäk wawal**
Calathea crotalifera S. Watson, syn. *Calathea insignis* Petersen	**its'in säk wawal**
Calathea lutea (Aubl.) Schultes.	**säk bobo (ah), säk wawal**
Calathea macrosepala K. Schum.	**kuut' (ah), ya'ax ch'ib muxan xiiw**
Calathea spp.	**wawal**
Heliconia librata Griggs	**hach muxän, säk muxan (ah)**
Meliaceae	
Cedrela odorata L., syn. *Cedrela mexicana* M. Roem.	**k'uche' (äh)**
Guarea glabra Vahl, syn. *Guarea excelsa* Kunth, *Guarea bijuga* C.DC.	**säk bahche'**
Guarea sp.	**nukuch säk bahche'**
Swietenia macrophylla King	**puuna' (äh)**
Trichilia sp.	**ch'obenche'**
Menispermaceae	
Disciphania calocarpa Standl.	**päh äk' (äh)**
Moraceae	
Brosimum alicastrum Sw. ssp. *alicastrum* C.C.	**hach 'oox, k'än 'oox, ya'ax 'oox**
Brosimum spp.	**'oox**
Castilla elastica Sessé.	**k'ik'**
Clarisia biflora Ruiz & Pav.	**chäk 'oop che'**
Ficus maxima Mill., syn. *F. glaucescens* (Liebm.) Miq.	**chimon (äh)**
Ficus pertusa L.f., syn. *Ficus padifolia* Kunth	**mehen kopo'**
Ficus sp.	**bits'kal hu'un**
Ficus sp.	**hach hu'un**
Ficus sp.	**hu'un**
Ficus sp.	**kopo' wits**
Ficus spp.	**kopo' (ah)**
Poulsenia armata (Miq.) Standl.	**ak' hu'un**
Pseudolmedia oxyphyllaria Donn. Sm., syn. *Pseudolmedia glabrata* (Liebm.) C. C. Berg	**hach bamax, tso'ots bamax**
Pseudolmedia spp.	**bamax (äh)**
Muntingiaceae	
Muntingia calabura L.	**puhan**
Musaceae	
Musa acuminata ×*Musa balbisiana* Coll.	**hach pätan**
	hach boox
	säk pätan
	boox
	chäk boox
	hunbuhk'ä'winik pätan
	ka'buhk'ä'winik pätan
	k'än pätan
	mäya' (äh)
	päh pätan (äh)
	pätan
	pätan ma' ka'anal ubakel
Musa acuminata Coll. (AA grp) cv. 'Sucrier', syn. *musa x pardisiaca* L. cv. 'Lady Finger'	**maasam (äh)**
Myristicaceae	
Virola guatemalensis (Hemsl.) Warb.	**k'ik'che'**
Ardisia compressa Kunth	**hach k'ik'ni' balum, mehen k'ik'ni' balum, pähpäh xibil (?)**
Ardisia sp.	**ne nuk k'ik'ni' balum**
Ardisia sp.	**pu'uhte'**

(continued)

(continued)

Parathesis sp.	**xiilich (äh)**
Parathesis sp.	**lo'k'in**
Parathesis sp.	**xinich (äh)**
Myrtaceae	
Eugenia sp.	**mehen k'ik'ni' balum, pimyenta**
Pimenta dioica (L.) Merr.	**peesache' (äh)**
Psidium guajava L.	**chäk pichik', säk pichik', pichik'**
Nymphaeaceae	
Nymphaea odorata Aiton.	**k'oben (äh)**
Orchidaceae	
Orchidaceae	**k'o'och ba'ats' (äh)**
Vanilla planifolia Jacks. ex Andrews, syn. *Vanilla fragrans* Ames	**buk'luch' (äh)**
Passifloraceae	
Passiflora aff. *ambigua* Hemsl.	**ch'um ak' (äh)**
Passiflora biflora Lam.	**k'um sek (ah)**
Passiflora cookii Killip., syn. *P. hahnii* (E. Fourn.) Mast.	**ch'ink'ak'il**
Passiflora serratifolia L., syn. *Granadilla serratifolia* (L.) Medik.	**pooch'in**
Physalacriaceae	
Oudemansiella aff. *steffendii* (Rick) Sing.	**ek' muluch'**
Oudemansiella canarii (Jungh.) Höhn	**säk muluch' (ah)**
Phytolaccaceae	
Phytolacca purpurascens A. Braun & Bouché, syn. *Phytolacca icosandra* L.	**tekoox (äh)**
Pinaecae	
Pinus spp., *P. oocarpa* Schiede ex Schltdl., *P. pseudostrobus* Lindl., *P. tenuifolia* Benth	**tähte'**
Piperaceae	
Piper aduncum L., syn. *Piper angustifolium* Ruiz & Rav.	**hach mäkuuläm**
	mehen mäkuuläm
Piper amalago Kunth	**ts'ak kan ak'**
Piper auritum Kunth, syn. *Piper sanctum* (Miq.) Schltdl.	**hach hoben**
Piper hispidum Sw.	**mehen mäkuuläm**
Piper spp.	**hoben**
	mäkuuläm
Piper umbellatum L., syn. *Lepianthes peltata* (L.) Raf.	**säk hoben**
Pleurotaceae	
Pleurotus djamor (Fr.) Boedijn	**k'ayoch' (ah)**
Pluteaceae	
Pluteus harrisii Murr.	**ek' muluch'**
Poaceae	
Poaceae	**'ak**
Gynerium sagittatum (Aubl.) P. Beauv.	**häläl, 'ooh**
Andropogon glomeratus (Walter) Britton, Sterns & Poggenb., syn. *Andropogon corymbosus* (Hack.) Nash	**winik su'uk**
Bambusa vulgaris 'Vittata' Schrad. ex J.C. Wendl. 'Vittata'	**bambu**
Bambuseae	**'ämäy**
Bothriochloa laguroides (DC.) Herter	**'ak lo'obil (äh)**
	ak' su'uk
Coix lacryma-jobi L.	**säknikil uuh**
	säkpahen (äh)
Cymbopogon citratus (DC.) Stapf	**'ak (ah), hach 'ak (äh), ts'ak ti' se'em**
Digitaria setigera Roth ex Roem. and Schult., syns., *Panicum sanguinale* L., *Syntherisma sanguinales* (L.) Dulac., *Paspalum sanguinale* (L.) Lam.	**hach su'uk**

(continued)

(continued)

Gynerium sagittatum (Aubl.) P. Beauv.	**chäk 'ooh, hach 'ooh, säk 'ooh**
Panicoideae e.g., Andropogoneae Dumort. (in part), Paniceae Haller	**su'uk (ah)**
Pennisetum purpureum Schumach.	**säk su'uk**
	wakäx su'uk
Phragmites australis (Cav.) Trin. ex Steud.	**hach häläl**
Rhipidocladum bartlettii (McClure) McClure	**semet'**
Saccharum officinarium L.	**sukal**
	chäk sukal
	hach sukal
	mehen sukal
	ts'ibal sukal
Zea mays L.	**näl**
	ek' näl
	k'änk'än näl
	säk näl
Podocarpaceae	
Podocarpus matudae Lund., syn. *P. matudae* Lund. var. *reichei* (J. Buchholz and N. E. Gray) de Laub. and Silba	**bey ule' tähte' (ah)**
Podocarpus matudae Lund.	**ch'ibil wits**
Polygonaceae	
Coccoloba barbadensis Jacq., syn. *C. barbadensis* var. *mexicana* Meisn.	**nukuch bä'ch'ich' (ah)**
Coccoloba hondurensis Lund.	**mehen bä'ch'ich'**
Coccoloba spp.	**bä'ch'ich'**
Polypodiales (Order)	**waalka'an**
Portulacaceae	
Portulaca oleracea L.	**k'ä' ta'k'in**
Primulaceae (see Myrsinaceae)	
Rhamnaceae	
Krugiodendron ferreum (Vahl) Urb.	**chintok'**
Rosaceae	
Eriobotrya japonica (Thunb.) Lindl.	**tso'ots**
Rubus sp.	**halo'och k'i'ix**
Rubiaceae	
Psychotria spp., syn. *Cephaelis*	**yo'och ik**
Blepharidium mexicanum Standl., syn. *Blepharidium guatemalense* Standl., *Tocoyena tabascensis* Standl.	**säkyuuhche' (äh)**
Bouvardia longiflora (Cav.) Kunth	**nikte' kisin**
Chione chiapasensis Standl., syn. *C. venosa* var. *venosa*	**mehen ch'eel koox, pähpähche'**
Coffea L.	**kawe'**
Gonzalagunia thyrsoidea (Donn. Sm.) B. L. Rob., syn. *Gonzalagunia tacanensis* Lund.	**mumun che', muxan che'**
Hamelia calycosa Donn. Sm., syn. *Hamelia boyacana* Standl.	**häxal k'ak'**
Morinda panamensis Seem.	**xoyok (ah)**
Psychotria panamensis var. *panamensis* Standl. syn. *Psychotria catillicalyx* Dwyer	**nukuch ik, yo'och ik nukuch**
Psychotria poeppigiana Müll. Arg., syn. *Cephaelis tomentosa* (Aubl.) Vahl	**säkpähk'ak' che'**
Psychotria sp.	**yo'och ik mehen**
Rutaceae	
Casimiroa sp.	**wolche' (äh)**
Citrus aurantifolia (Christm.) Swingle	**mulix**
Citrus reticulata Blanco.	**chäk chinah**
Citrus sinensis (L.) Osbeck	**chinah (ah)**
Zanthoxylum sp.	**läkte'**
Santalales (Order)	**tak'lan che'**

(continued)

(continued)

Sapindaceae	
Billia colombiana Planch. & Lindl.	**k'alok' che' (äh)**
Paullinia sp.	**ma'ax ak' (äh)**
Sapindus saponaria L.	**baba', habón, habonsiyo, subul**
Serjania sp.	**ma'ax ak' (äh)**
Sapotaceae	
Chrysophyllum mexicanum Brandegee ex Standl., syn. *Micropholis sericea* L. O. Williams	**säk ya'**
Dipholis sp.	**subul**
Manilkara achras (Mill.) Fosberg, syn. *Manilkara zapota* (L.) van Royen.	**chäk ya', hach ya', ya'**
Manilkara, Pouteria, Chrysophyllum	**ya'**
Pouteria campechiana (H.B.K.) Baehni, syn. *Lucuma campechiana* H.B.K.	**o'ot' ya'**
Pouteria mammosa (L.) Cronq., syn. *Manilkara zapota* (L.) P. Royen., *Pouteria sapota* (Jacq.) H. E. Moore & Stearn, *Calocarpum mammosum* Pierre	**ha'as**
Pouteria sp.	**hach ha'as**
Sarcoscyphaceae (Pezizales)	
Cookeina tricholoma (Mont.) Kuntze	**kib lu'um, kuxum che'**
Schizophyllaceae	
Schizophyllum commune Fr.	**xikin (ah), xikin che'**
Selaginellaceae	
Selaginella silvestris Aspl.	**k'u' kan**
Siparunaceae	
Siparuna thecaphora (Poepp. & Endl.) A.DC., syn. *Siparuna nicaraguensis* Hemsl.	**yo'och ähbaach'**
Smilacaceae	
Smilax domingensis Willd.	**tsuk'in ak' (ah)**
Smilax lanceolata L.	**koke' ak' (ah)**
Smilax sp.	**chäk tsuk'in ak'**
Smilax sp.	**chäk ak'**
Smilax sp.	**ya'ax tsuk'in ak'**
Solanaceae	
Capsicum annuum L.	**k'än ik**
	ya'ax chawah ik
	chäk ik
	k'än ik (äh)
	säk chawah ik
	ya'ax ik
Capsicum annuum var. *glabriusculum* (Dunal) Heiser and Pickersgill	**ta' ch'o' ik**
Capsicum frutescens L.	**chawah ik (ah)**
	hach ik
	nok'ol ik
Capsicum sp.	**ch'om ik**
Capsicum sp.	**säk yolte' ik (äh)**
Capsicum spp.	**ik**
Cestrum nocturnum L.	**'aak'älyoom, utop' ti' kisin**
Jaltomata repandidentata (Dunal) Hunz., syn. *Saracha repandidentata* (Dunal) Bitter	**ch'ayok'**
Lycianthes heteroclita (Sendtn.) Bitter	**ch'ayok', k'ulsek**
Lycopersicon esculentum L., *L. esculentum* var. *cerasiforme* (Dunal) D. M. Spooner, G. J. Anderson & R. K. Jansen	**p'ak**
Nicotiana tabacum L.	**hach k'uuts, k'uuts**
Physalis pruinosa L., syn. *P. maxima* Mill.	**p'akan (ah)**
Physalis sp.	**beyanchan**
Physalis sp.	**mehen p'akan**
Solanum amercianum Mill., syn. *Solanum nigrum* L. var. *americanum* (Mill.) O. E. Schulz	**ch'ayok'**
Solanum erianthum D. Don *and S. torvum* Sw.	**'uk'unche'**
Solanum rovirosanum Donn. Sm.	**ch'ayok'**

(continued)

(continued)

Solanum torvum Sw.	**k'i'ix 'uk'unche'**
Solanum tuberosum L.	**papas, suli' (äh)**
Sterculiaceae *(Malvaceae)*	
Theobroma cacao L.	**käkow**
Theobroma bicolor Humb. & Bonpl., syns. *Theobroma ovatifolia* Moc & Sessé ex DC., *Cacao bicolor* (Humb. & Bonpl.) Poir, *Tribroma bicolor* (Humb. & Bonpl.) Cook, *Theobroma cordata* Rufz & Pavón	**balumte'**
Theobroma sp.	**balumchan**
Theaceae	
Ternstroemia tepezapote Schltdl. & Cham.	**chäk pach**
Tiliaceae	
Belotia mexicana K. Schum., syn. *Trichospermum mexicanum* (DC.) Baill.	**hach taw**
Ulmaceae	
Ampelocera hottlei (Standl.) Standl.	**luwin**
Trema micrantha (L.) Blume	**chäk taw**
Urticaceae (Moraceae)	
Cecropia Loefl., *C. obtusifolia* Bertol., *C. peltata* L.	**k'o'och (äh)**
Cecropia sp.	**mehen k'o'och**
Cecropia sp.	**nukuch k'o'och**
Ustilaginales	
Ustilago maydis (Persoon) Roussel	**soobil näl**
Verbenaceae	
Aegiphila monstrosa Moldenke	**k'ak'che'**
Lantana trifolia L., syn. *Lantana maxima* Hayek	**chäkchob**
Viscaceae	
Phoradendron crassifolium (Pohl ex DC) Eichler	**mehen tak'lan che'**
Vitaceae	
Cissus biformifolia Standl.	**ts'ak ahch'up**
Vitis spp.	**ts'us (ah)**
Vitis tilaefolia Humb. & Bonpl. ex Schult, syn. *Vitis caribaea* DC	**hach ts'us, mehen ts'us, yuhi, yuwin**
Zingiberaceae	
Alpinia purpurata K. Schum.	**te' 'usin**
Renealmia sp.	**komo' (äh)**
Zingiber officinale L.	**pap uwi', sensi' (ah)**
Zingiberales (Order) (in part)	**muxan**

English-Lacandon Index

A–a

abnormal guatteria (black cork) **ek' bahche'.**

abraza palo (English common name unknown) hach chiah (ah).

acacia, bull horn subinte' (ah); 'akunte'; subin (ah).

Aegiphila monstrosa (common name unknown) k'ak'che'$_{1}$.

air plant san lorenso.

alligator pear (avocado, American) chäk 'oon; 'oon; ya'ax 'oon.

alligatorwood säk bahche'.

allspice peesache' (äh); pimyenta.

amaryllis and lilies ts'ulha' (äh).

American burnweed nukuch le' si'si' k'uuts.

American muskwood nukuch säk bahche'.

angelica tree sasakche'.

annatto (lipstick tree) k'uxub.

annona (type of) ‘oopil k’aax.

Anthurium (birds nest type) neeh k’ambul (ah).

Anthurium (common name unknown) ni’ ah’och.

apple, custard chäk uwich (äh); hach ‘oop; ‘oop.

apple, monkey kahanche’.

apple, wild star säk ya’.

ardisia pähpäh xibil; pu’uhte’.

Arrabidaea (type of) ya’ax ak’ (ah).

Arrabidaea verrucosa säk ak’.

arrowhead vine mehen ‘oochin$_2$; nukuch ‘oochin$_3$.

Artist’s bracket kuxum che’$_2$.

arum ‘oochin.

avocado, American (alligator pear) chäk ‘oon; ‘oon; ya’ax ‘oon.

avocado tree ‘oonte’.

B–b

bagflower ‘itsänte’.

balsam of Peru naba’.

bamboo, golden bambu.

bamboo piper hach mäkuuläm.

banana boox.

banana (type of) hach boox.

banana, Cuban red chäk boox$_2$.

banana, Orinoco päh pätan (äh).

Barbados cedar k’uche’ (äh).

barbasco k’ul ak’; suli’ (äh)$_3$.

basket palm hach hänan; hänan.

bauhinia (rio rubeleruz) k’ak’che’$_2$.

bayal hach hänan; hänan.

bead tree ‘äm.

bean bu’ul; chäk box; chäk bu’ul; ek’el bu’ul; ek’ bu’ul; hach box; hach bu’ul; nukuch box.

bean, castor k’o’och (äh)$_1$; nukuch chäk k’o’och; mehen chäk k’o’och; ya’ax k’o’och

bean, Jack bubuhchi’ (ah)$_1$; yo’och ahpek’

bean, lima chäk ‘ib; ek’ ‘ib.

bean, “Little Marble” ‘ibil k’aax.

bean, rice tsaayentsaay (ah)$_1$.

bean, Saint Mary’s ho’ol ma’ax; gorila.

bean, shelling box$_1$; box bu’ul$_2$.

bean twining snout choochel ak’ (äh).

beans, bush che’il bu’ul (ah); ek’ che’il bu’ul.

beans, climbing ak’il bu’ul (ah).

beardgrass, silver ‘ak lo’obil (äh); ‘ak su’uk.

beef apple (sapodilla plum, sapatillo); hach ya’.

billia k’alok’ che’ (äh).

bitterwood ch’obenche’.

black cork (abnormal guatteria) ek’ bahche’.

black eyed pea box bu’ul$_1$.

blackberry halo’och k’i’ix.

bleeding-heart ‘itsänte’.

bloodwood tree ek’.

bloody croton chukum che’.

bottle gourd chuh; ha’ chuh; kis chuh mehen; kis chuh (äh).

bouvardia, scented nikte’ kisin.

brambles halo’och k’i’ix.

Brasil blanco chäk’ax.

Brazil beauty-leaf babah.

Brazilian fern tree pets’k’in$_1$.

Brazilian fire tree pets’k’in$_1$.

Brazilian walnut hach chulul.

breadnut, Mayan hach ‘oox; k’än ‘oox; ‘oox.

breadnut, Mayan ya’ax ‘oox.

bristly crabgrass hach su’uk.

bull nettle tsaah (äh).

bullhoof luwin.

bullock’s heart chäk ‘oop.

bullywood pukte’.

bullywood (commercial name for the timber) k’änxa’an.

bunchosia siip che’ (ah).

bush beans chäk che’il bu’ul.

bushy beardgrass winik su'uk.

bustic subul$_2$.

buttonwood ts'ak kan ak'$_1$.

C–c

cabbage kulix.

cacao käkow.

cacao mocambo balumte'.

cacao wild balumchan$_1$.

cactus, climbing le'mo'.

calabasa ya'ax k'um.

calabash box$_3$; chuh; hach luuch; ha' chuh; kis chuh mehen; k'um; lek$_2$; luuch$_1$; sina'an luuch (äh).

calabash, Mexican k'än luuch.

calabash tree luuch che'.

calabash tree (small yellow variety) mehen k'än luuch.

calathea säk bobo (ah); wawal.

Calathea, "Cigar" säk wawal.

Calathea macrosepala (common name unknown) kuut' (ah); ya'ax ch'ib muxan xiiw.

caliandra chäk me'exk'in (äh).

candle tree 'anis$_1$; kat.

cane giant 'ooh.

cane grass säk su'uk.

canistel o'ot' ya'.

Caribbean cabbage chäk mäkäl; säk mäkäl.

Caribbean grape yuwin.

cascabel hach tuch; tuch; tuchil k'aax.

cascade palm ch'ibix uk'um.

cassava ts'in.

catalina chäklanpix (ah)$_3$.

cayenne rosewood kakache'.

cedar, Spanish (Mexican cedar) k'uche' (äh).

ceiba ya'axche'.

chamaedorea säk bo'oy$_1$; ya'ax bo'oy.

chamaedorea bo'oy (äh)$_2$; chan mehen bo'oy.

chamaedorea (type of) ch'it bo'oy; ch'it k'än bo'oy; ton bo'oy; ya'ax hach bo'oy.

Chamaedorea arenbergiana (common name unknown) säk bo'oy$_2$.

chamaedorea (common name unknown) hach bo'oy$_2$; ya'ax hänan.

Chamaeorea pinnatifrons (common name unknown) säk bo'oy$_4$.

cherimoya säk 'oop.

cherry tso'ots bamax.

chestnut, Guiana kubuh (ah).

lima bean, Chickasaw bubuhchi' (ah)$_1$.

lima bean, Chickasaw yo'och ahpek'.

chicle (sapodilla plum) chäk ya'; hach ya'; ya'.

chile (type of) ya'ax chawah ik; ya'ax ik.

chile pepper (type of) ch'om ik; säk chawah ik.

chile pepper (red) chäk ik.

chile pepper (yellow) k'än ik (äh).

chile, Tabasco hach ik.

Chinese parsley perehir.

chocolate pudding fruit 'uuch (äh).

Christmas flower chäkche' (äh).

Christmas star chäkche' (äh).

christmasbush muxan che'$_1$.

cilantro perehir.

Cissus biformifolia (common name unknown) ts'ak ahch'up.

Clarisia biflora (common name unknown) chäk 'oop che'.

Clibadium arboreum (common name unknown) chäklanpix (ah)$_2$; ki' bok (äh)$_2$.

clusia (pitch apple) chunup (äh).

clusia (pitch apple) ts'oy (äh)$_1$.

coccoloba bä'ch'ich'.

coffee kawe'.

coffee, wild yo'och ähbaach'.

coffee wild (typical Panama balsamo) nukuch ik.

coffee-shade tree suumil bits'.

cojoba bu'ulche'; mehen bu'ulche'.

collinia ch'ibix bo'oy (ah).

colorado chäk $boox_2$.

common reed 'ooh.

copal tsatsel pom.

coral bean tree (coral tree) k'ante' (ah); k'ante' ak'

coral fungi kuxum che'_3.

coralberry hach k'ik'ni' balum; k'ik'ni' balum $(ah)_2$; mehen k'ik'ni' $balum_2$; xiilich (äh).

cordoncillo mehen mäkuuläm.

coriander perehir.

coriander, long kulantro.

cork hach taw.

cotton tämän.

cotton tree ya'axche'.

cow okra $'anis_1$.

cow pea box $bu'ul_1$.

cowhage bubuhchi' $(ah)_2$.

cowitch bubuhchi' $(ah)_2$.

coyo persea mehen koyoh.

crabgrass hach su'uk.

cuapinol pakay.

cucumber tree $'anis_1$.

cucurbit (Guadeloupe cucumber, meloncito, speckled gourd) sanyah ahch'o' (ah).

cushaw pumpkin sikil.

custard apple chäk 'oop; säk 'oop.

custard apple tree hach 'oop che'; 'oop che'.

D–d

daisy family chäklanpix $(ah)_1$; ki' bok $(äh)_1$.

daisy tree, giant haban uts'u' (ah); ki' bok $(äh)_3$.

date banana maasam (äh).

deer eye ni' ts'ul; wich ahyuk (ah).

devil nettle tsaah (äh).

devil's fig k'i'ix 'uk'unche'; $'uk'unche'_2$.

Disciphania calocarpa (common name unknown) päh ak' (äh).

dogwood muxan che'_1.

Doliocarpus dentatus (common name unknown) säha' (ah).

dumbcane chiah (ah).

E–e

ear flower ton ku'uk.

eared piper hach hoben.

East Indian crabgrass hach su'uk.

egg fruit o'ot' ya'.

elephant ears chäk mäkäl; mäkäl; säk mäkäl; $suli'_2$.

elephant ears (Caribbean cabbage) hach mäkäl.

elephant ears capote ch'om mäkäl; hach ch'om.

elephant grass säk su'uk.

epazote k'äxeex (äh).

Ernest August's palm k'eben.

eugenia (type of) mehen k'ik'ni' $balum_1$.

F–f

false ginger chank'ala'.

false jade k'än bo'oy.

false mamery babah.

false saffron sik'.

fat pork pähpähche'.

fatpork mehen ch'eel koox.

fen-sedge siib; siibal.

fern, bracken $waalka'an_2$.

fern, shield $waalka'an_3$.

fern tree $waalka'an_1$.

ficus hu'un.

ficus (type of) hach hu'un.

fig kopo' $(ah)_1$.

fig banana maasam (äh).

fig, rock kopo' wits.

fig, Sonoran strangler mehen kopo'.

fig, strangler bits'kal hu'un.

fireweed nukuch le' si'si' k'uuts.

fish-tail palm k'eben.

flame tree ponche'.

Florida tibisee sit.

floss silk ya'axche'.

frangipani chäk nikte'; nikte' (ah)$_2$; säk nikte'.

funeral swizzlestick tree mahaas (ah).

funeral tree mahaas (ah).

fungi (types of) ek' muluch'$_1$.

fungi, bonnets muluch' (ah)$_1$.

fungi, white bonnets säk muluch' (ah).

fungus, jelly lo'loh (äh).

fungus, red cup kib lu'um; kuxum che'$_4$.

fungus, split-gill xikin (ah)$_1$.

G–g

garlic 'axux.

garlic vine 'axux ak'.

giant cane chäk 'ooh; hach 'ooh; häläl$_2$.

giant groundsel mumun che'$_2$.

giant cane säk 'ooh.

ginger pap uwi'; sensi' (ah).

ginger, ostrich plume komo' (äh).

ginger, red te' 'usin.

ginger, spiral pahsa' ak' (äh).

give and take palm kun (ah).

glassy wood mehen k'än'ol.

glorybower 'itsänte'.

goat's milk tree bamax (äh).

God bush iich'ak tolok; nukuch tak'lan che'.

gonzalagunia, thyrsoid mumun che'$_3$; muxan che'$_2$.

good-luck palm ch'ibix bo'oy (ah).

goosefoot plant mehen 'oochin$_2$; nukuch 'oochin$_3$.

granadillo säk chulul (äh).

grape, Caribbean hach ts'us; mehen ts'us.

grape, wild ts'us (ah).

grass 'ak$_1$.

grass, bushy bluestem winik su'uk.

grass, napier säk su'uk.

grass, napier wakäx su'uk.

grass, saw siib; siibal.

greater yam hach suli'.

greenbriar koke' ak' (ah).

ground cherry mehen p'akan.

grugru palm tuk'.

guaba inga bits' petha'.

guanacaste pets'k'in$_2$.

Guatemala blepharidium (common name unknown) säkyuuhche' (äh).

guava chäk pichik'; pichik'.

guava säk pichik'.

guayacan trumpet tree hach chulul.

gully plum k'inim.

gumbo limbo chäklah.

H–h

hairy beggarticks ku'uchnook'.

Hamelia calycosa (common name unknown) häxal k'ak'$_1$.

Hawaiian wood rose ts'ak k'ik'.

heartflower tree kuti'; kuti'il wits.

heliconia (type of) chukuch le' xiiw; hach xiiw.

Heliconia librata (common name unknown) hach muxan; säk muxan (ah).

heliocarpus chäk halol; halol.

henequen kih.

Hercules' club läkte'.

hibiscus tämänil (äh).

hicatee fig [Bel.], white fig [Jam.Eng.]) chimon (äh).

hog plum 'abäl; ya'ax 'abäl.

hog plum, scarlet plum k'än 'abäl.

horse-eye beans wich ahyuk (ah).

horse-eye beans 'äm.

hot lips säkpähk'ak' che'.

hubu ts'oy (äh)$_2$.

huevos de caballo ton tsimin.

I–i

icacina kunkunche'.

Indian shot chank'ala'.

inga bits'.

inga (type of) mehen bits'.

inga, Belize bits'il k'aax.

Inga punctata (common name unknown) nukuch bits'.

inkwood ek'.

ipe hach chulul.

ironwood black chintok'.

J–j

jade hach bo'oy$_1$; säk bo'oy$_3$.

Jaltomata repandidentata (common name unknown) ch'ayok'$_2$.

Jamaican apple chäk 'oop.

Jamaican cherry puhan.

Jamaican nettletree chäk taw.

Jamaican pepper peesache' (äh).

Jamaican pepper pimyenta.

Jamaican red chäk boox$_2$.

Japanese medlar tso'ots$_1$.

Japanese nispero tso'ots$_1$.

jicama chi'kam.

Job's tears säknikil uuh; säkpahen (äh).

joint wood ts'ak kan ak'$_1$.

Juan caliente chok$_1$.

jumbay tsäläm (ah).

jungleplum subul$_2$.

K–k

key lime (Mexican lime) mulix.

L–l

ladies teardrop tämän che'.

'Lady Finger' banana maasam (äh).

lancepod balche'$_1$; hach balche'; ya'ax balche'.

lantana (threeleaf) chäkchob.

lateralflower sapium 'ukunte' (äh).

lavender popcorn chäkchob.

leadwood chintok'.

leaf cactus le'mo'.

lemon pepper ts'ak kan ak'$_1$.

lemongrass 'ak (ah)$_2$; hach 'ak (äh).

liana palm hach hänan; hänan.

Licaria alata (common name unknown) säk 'oonte'.

lily, cahaba ts'ulha'il petha' (äh).

lily, canna chank'ala'.

lily, giant spider chäk ts'ulha' (äh).

lima bean 'ib.

lime, Mexican (key lime) mulix.

limoncillo k'ik'ni' balum (ah)$_1$; yo'och ähbaach'.

lipstick tree (annatto) k'uxub.

locustberries (serrets) ya'ax chi'.

logwood ek'.

loquat tso'ots$_1$.

Louteridium donnell smithii k'o'och ba'ats' (äh)$_1$.

lycianthes, varying ch'ayok'$_3$; k'ulsek.

M–m

macauba palm tuk'.

macaw palm tuk'.

macawood säk chulul (äh).

magnolia kuti'; kuti'il wits.

mahogany puuna' (äh).

mahogany, bastard nukuch k'än'ol.

maiden plum k'än'ol (äh).

maize ek' näl; k'änk'än näl; näl; säk näl.

majagua hach halol.

majagua, white säk halol.

malanga chäk mäkäl; säk mäkäl.

mammee sapote ha'as.

manax tso'ots bamax.

mango mankoh.

manioc (cassava) chäk ts'in; hach ts'in; säk ts'in; ts'in; yahah ts'in; yahche' ts'in

marlberry hach k'ik'ni' balum; k'ik'ni' balum (ah)$_2$; mehen k'ik'ni' balum$_2$; xiilich (äh).

marmalade plum ha'as.

mastate ak' hu'un.

mata cachorro (mata negro) chok$_1$.

mata rama iich'ak tolok; nukuch tak'lan che'.

merker grass säk su'uk; wakäx su'uk.

Mexican bluebell ya'ax hälal che'.

Mexican fern tree pets'k'in$_1$.

Mexican pepperleaf hach hoben.

Mexican potato chi'kam.

Mexican red oak pixan k'ambul (äh).

Mexican tea k'äxeex (äh).

milady sa'yok' (ah).

milkweed kimil sukun.

mint menta; ts'ak kax.

mistletoe mehen tak'lan che'.

mombin 'abäl; huhup; k'inim.

mombin, green ya'ax 'abäl.

mombin, red chäk 'abäl.

mombin, yellow k'än 'abäl.

monkeysoap pets'k'in$_2$.

monkey-tail k'unche'; k'unche' pahok; pahok (ah).

monkey-tail k'än bo'oy.

moon vine hach hut'kih.

moonflower hach hut'kih.

morning glory chäk hut'kih; hut'kih; tu'xikin.

mosquito wood nukuch k'än'ol.

mountain bribri t'elen bits'.

mountain moho [Bel. Eng.] ts'ubtok.

mullein nightshade 'uk'unche'$_1$.

mushroom (pink oyster) k'ayoch' (ah).

mallow (musk okra, musky-seeded hibiscus, ornamental okra, tropical jewel hibiscus) 'anis$_2$.

muskwood säk bahche'.

my lady sa'yok' (ah).

N–n

nance chi' (äh); k'än chi'.

neanthe bella ch'ibix bo'oy (ah).

nectandra (type of) mehen 'oonte'.

nectandra (type of) nukuch 'oonte'.

nectandra, globose ek' 'oonte'.

netted custard apple chäk 'oop.

nickernut patoh ak' (ah).

night jessamine 'aak'älyoom; utop' ti' kisin.

night-blooming cestrum (night-blooming jasmine) 'aak'älyoom; utop' ti' kisin.

nightshade ch'ayok'$_1$.

nightshade, American black ch'ayok'$_5$.

nightshade, rovirosa ch'ayok'$_4$.

nut, physic sikilte' che'.

O–o

oak ch'alol$_1$.

oak (type of) mehen ch'alol.

oak, chicarro pixan k'ambul (äh).

oak, corrugated hach ch'alol; nukuch ch'alol.

obtuse leaf oreopanax k'äb ta'k'in.

ocote pine tähte'.

ocotea, drooping (timber sweet, cayenne rosewood) babah mehen 'oonte'; babah 'oonte'; kakache'.

onion seboya.

onion cordia bahun che'.

onion, white säk seboya (äh).

onion, wild (scallion) ts'ak kitam; ts'ak k'ek'en.

orange, mandarin chäk chinah.

orange, sweet chinah (ah).

orchid k'o'och ba'ats' (äh)$_2$.

orchid cactus le'mo'.

orchid tree säkche'$_1$.

ornamental okra 'anis$_2$.

Oudemansiella aff. steffendii (common name unknown) ek' muluch'$_2$.

ox eye ni' ts'ul.

P–p

pagoda tree nikte' (ah)$_2$.

painted bamboo bambu.

palm, cat ch'ibix uk'um.

palm, chocho (choco palm, spiny palm) 'akte'; hach 'akte'.

palm, corozo tuts.

palm, coyol tuk'.

palm, Everglades booxa'an.

palm, fan kun (ah); xa'an.

palm, pacaya ch'i'ib (äh); pakyon (äh).

palm, Panama hat miis.

palm, parlor ch'ibix bo'oy (ah); ch'ilix bo'oy.

palm, woodsonia ch'ibix wits.

palo blanco 'white stick' che' chäk'an.

Panama balsamo yo'och ik nukuch.

Panama rubber tree k'ik'$_1$.

papaya put.

parathesis (type of) lo'k'in; xinich (äh).

Passiflora cookii (common name unknown) ch'ink'ak'il.

passion flower, bat-leaved type k'um sek (ah).

passion flower vine hut'kih.

passion fruit vine (purple Jamaican honeysuckle) ch'um ak' (äh).

Passiflora serratifolia (common name unknown) pooch'in.

peanut kakawat; sikilte' lu'um.

peanut butter tree siip che' (ah).

pear, vegetable (pear squash) k'i'ilix p'ix; p'ix (ah); säk p'ix; ya'ax p'ix.

pepper, Jamaican mäkuuläm; mehen mäkuuläm.

pepper, chile ik.

pepper, chile chawah ik (ah); nok'ol ik; säk yolte' ik (äh).

pepper, bird ta' ch'o' ik.

pepperbush kunkunche' kap'.

perforated fig mehen kopo'.

petunia, Mexican ya'ax hälal che'.

philodendron holop'.

Philodendron smithii (common name unknown) yuyul holop'$_2$.

Philodendron donnell-smithii (common name unknown) hach 'oochin.

philodendron, heart leaf mehen 'oochin$_1$.

Philodendron sagittifolium (common name unknown) nukuch 'oochin$_1$.

philodendron, split leaf k'i'ilix holop'.

Philodendron tripartitum (common name unknown) yuyul holop'$_1$.

pilewort nukuch le' si'si' k'uuts.

pillpod sand-mat xänä' mook.

pimento pimyenta.

pine cone ginger komo' (äh).

pineapple pach'.

pinguin ch'äm (äh).

piper hoben.

piper, matico hach mäkuuläm.

piper, monkey hand säk hoben.

pipian sikil.

pitch pine tähte'.

plantain pätan; hach pätan; hunbuhk'äbwinik pätan; ka'buhk'äbwinik pätan; k'än pätan; mäya' (äh); pätan ma' ka'anal ubakel; säk pätan

Pleuranthodendron lindenii (common name unknown) iximche'.

plumeria säk nikte'.

plum-leaved fig tree mehen kopo'.

Pluteus harrisii (common name unknown) ek' muluch'$_3$; muluch' (ah)$_2$.

podocarp, Jalisco bey ule' tähte' (ah); ch'ibil wits.

poinsettia chäkche' (äh).

poisonwood tree chechem (äh)$_1$.

pokeweed, tropical tekoox (äh).

pole beans ak'il bu'ul (ah).

potato papas.

potato, air bilim suli'.

potato, sweet chäk 'is; chäk pach 'is; 'is; k'än 'is; säk 'is.

potato tree 'uk'unche'$_1$.

potato, white suli' (äh)$_4$.

prickly nightshade k'i'ix 'uk'unche'.

prickly solanum 'uk'unche'$_2$.

prickly-ash läkte'.

Pseudolmedia oxyphyllaria (common name unknown) hach bamax.

psychotria (type of) yo'och ik; yo'och ik mehen.

pumpwood k'o'och (äh)$_2$; mehen k'o'och; nukuch k'o'och.

purple stem taro chäk mäkäl.

purple yam hach suli'.

purslane k'äb ta'k'in$_1$.

R–r

rattlesnake calathea its'in säk wawal.

red dacca chäk boox$_2$.

red gumbo limbo chäklah.

red inkplant tekoox (äh).

reed häläl$_1$.

reed, common chäk 'ooh; hach häläl; hach 'ooh; säk 'ooh.

Rhipidocladum bartlettii (common name uknown) semet'.

rootbeer plant hach hoben.

rootspine palm kun (ah).

royal fig 'ukunte' (äh).

S–s

sabal xa'an.

safflower sik'.

Saint Mary's (crucifixion bean, elephant creeper, woolly morning glory) gorila.

San Lorenzo san lorenso.

sansapote tsäk'ats.

Santa Maria babah.

sapodilla plum (sapotillo) chäk ya'; hach ya'

sapote (type of) hach ha'as.

sapote, white wolche' (äh).

sapotillo (sapodilla plum) chäk ya'; hach ya'.

sarsaparilla koke' ak' (ah); tsuk'in ak' (ah).

Saurauia yasicae (common name unknown) mumun che'$_1$.

sausage tree pakay.

Schipp ormosia (John Crow bead [Bel.]) 'äm ch'uplal; 'äm ton.

Scotchman's purse tämän che'.

scouringrush horsetail maasan kitan.

sea heart korason de mono.

seagrape, Barbados nukuch bä'ch'ich' (ah).

seagrape, Honduras mehen bä'ch'ich'.

serrets (locustberries) ya'ax chi'.

shadow palm k'unche'; k'unche' pahok; pahok (ah).

shaving brush tree ch'ulte'.

shimbillo bits'.

shoals spider lily ts'ulha'il petha' (äh).

shrimp plant tsits.

sisal kih.

slipplejack (?) ma'ax ak' (äh).

small red bean tsaayentsaay (ah)$_1$.

smallcane sit.

smilax chäk tsuk'in ak'; ya'ax tsuk'in ak'.

snout bean mo'ak' (ah).

soapberry baba'; ma'ax ak' (äh)$_1$; subul$_1$.

sore-mouth bush säkpähk'ak' che'.

soursop pähpox.

Spanish elm bahun che'.

Spanish arborvine ts'ak k'ik'.

Spanish needle ku'uchnook'.

Spanish plum k'än 'abäl; k'inim.

Spanish woodbine ts'ak k'ik'.

Spathiphyllum phryniifolium (common name unknown) pähk'ol.

spikemoss k'u' kan.

spiny palm (choco palm, spiny palm) hach 'akte'.

spiny pod kimil sukun.

split gill xikin che'.

squash k'um; ya'ax k'um.

squash seeds mumun sikil.

stinktoe pakay.

strangler fig kopo' wits; mehen kopo'.

strawberry tree puhan.

striped bamboo bambu.

Strychnos tabascana (common name unknown) ts'ak kan.

sugar apple chäk 'oop.

sugar banana maasam (äh).

sugarcane chäk sukal; hach sukal; mehen sukal; sukal; ts'ibal sukal.

summer squash mumun sikil.

sunfruit chäk halol.

sunfruit halol.

sunfruit hach halol.

Swiss cheese plant hach 'ilon; 'ilon; kamis 'ilon (ah); säk 'ilon.

Syngonium chiapense (common name unknown) nukuch 'oochin$_2$.

T–t

taw taw.

temple tree nikte' (ah)$_2$.

Ternstroemia tepezapote (common name unknown) chäk pach.

thevetia, broadleaf (dog's tongue) hach tuch.

thevetia, broadleaf (dog's tongue) tuchil k'aax.

thevetia, broadleaf (dog's tongue) tuch.

three-nerved baccharis si'si' k'uuts.

ticktrefoil, Dixie tsaayentsaay (ah)$_2$.

tigerwood mehen k'än'ol.

timber sweet babah mehen 'oonte'; babah 'oonte'; kakache'.

tobacco hach k'uuts; k'uuts.

tomato p'ak.

tomato, husk beyanchan.

tomato, Mexican husk (Mayan husk tomato, ground cherry) p'akan (ah).

tourist tree chäklah.

ardisia (type of) ne nuk k'ik'ni' balum.

trumpet tree k'o'och (äh)$_2$; mehen k'o'och; nukuch k'o'och.

tuberose baknikte'.

Turkcap (Turk's turban) tämän che'.

turkey berry k'i'ix 'uk'unche'.

turkey berry nightshade (devil's fig, prickly solanum) 'uk'unche'$_2$.

twig-sedge siib; siibal.

V–v

vanilla buk'luch' (äh).

velvet bean bubuhchi' (ah)$_2$; wich ahyuk (ah).

velvet philodendron mehen 'oochin$_1$.

vine, virility mo'ak' (ah).

virola k'ik'che'.

W–w

water lily k'oben (äh)$_1$.

water vine hach ts'us; mehen ts'us.

water yam hach suli'.

watermelon sanyah.

wax mallow tämän che'.

West Indian birch chäklah.

West Indian jasmine nikte' $(ah)_2$; säk nikte'.

white leadtree tsäläm (ah).

white popinac (jumbay, white popinac, royal poinciana, flamboyant) tsäläm (ah).

wild hog plum huhup.

wild mahogany nukuch k'än'ol.

wild pear mehen koyoh.

wild pine ch'äm (äh).

wild tamarind wäch'.

Wimmeria bartlettii (common name unknown) k'än chululche'.

winged calabash k'än luuch.

wing-leaf soapberry habonsiyo; habón.

wormseed k'äxeex (äh).

X–x

xate hembra [Mex. Sp.] ch'ibix bo'oy (ah).

xate macho [Mex. Span.] (jade [Mex. Sp.]) hach $bo'oy_1$; säk $bo'oy_3$.

Y–y

yam suli' $(äh)_3$.

yam bean chi'kam.

yam, winged hach suli'.

yellow mombin k'inim.

yellow morning glory ts'ak k'ik'.

yellow rattle shaker its'in säk wawal.

yellow sapote o'ot' ya'.

yellow wood xoyok (ah).

Z–z

zingiberales muxan.

References

Allen, O. N., & Allen, E. K. (1981). *The Leguminosae: A source book of characteristics, uses, and nodulation*. Madison, WI: University of Wisconsin Press.

Anderson, W. R., Anderson, C., & Davis, C. C. (2006). Malpighiaceae. http://herbarium.lsa.umich.edu/malpigh. Retrieved January 2012.

Arango, J., Murgueitio, R. E., & Maass, B. L. (2004). Diversity and use of the calabash tree (Crescentia cujete) in Colombia. Poster presented at "Botanikertagung," Braunschweig, Germany, 5–10 September 2004.

Arriagada, J. E. (1995). Ethnobotany of *Clibadium* L. (Compositae, Heliantheae) in Latin America. *Economic Botany, 49*(3), 328–330.

Atran, S. (1999). Itzaj Maya folkbiological taxonomy. In D. L. Medin & S. Atran (Eds.), *Folkbiology* (pp. 119–203). Cambridge, MA: MIT Press.

Atran, S., Lois, X., and Ek', E. U. (2004). Plants of the Petén Itza' Maya: Plantas de los Maya Itza' del Petén. *Memoirs of the Museum of Anthropology 38*. Ann Arbor: University of Michigan.

Baer, M. & Baer, P. (1952). Materials on Lacandon culture of the pethá (pelhá) region. *Microfilm Collection of Manuscripts on Middle American Cultural Anthropology* 34. Chicago: University of Chicago Library.

Baer, P., & Merrifield, W. R. (1971). *Two studies on the Lacandones of Mexico*. Norman, OK: Summer Institute of Linguistics, University of Oklahoma.

Bansod, S., & Rai, M. (2008). Antifungal activity of essential oils from Indian medicinal plants against human pathogenic *Aspergillus fumigatus* and *A. niger. World Journal of Medical Sciences 3*(2), 81–88. http://idosi.org/wjms/3(2)08/7.pdf

Berg, C. C. (2007). Proposals for treating four species complexes in Ficus sub-genus Urostigma section Americanae (Moraceae). *Blumea, 52*, 295–312.

Beutelspacher-Baigts, C. R. (2009). Semillas utilizadas para elaborar artesanías Lacandones en Chiapas, México. *Lacandonia Revista Ciencias, 3*(2), 45–58.

Bishop Museum. (2013). *Hawaiian Kalo*. http://hbs.bishopmuseum.org/botany/taro/key/HawaiianKalo/Media/Html/whatistaro.html

Bolles, D. (2001). *Combined dictionary-concordance of the Yucatecan Mayan language*. http://www.famsi.org/reports/96072/#dictionary

Breedlove, D. E. (1981). *Flora of Chiapas Part 1: Introduction to the flora of Chiapas*. San Francisco, CA: California Academy of Sciences.

Breedlove, D. E. (1986). *Listados florísticos de México IV. Flora de Chiapas*. México: Instituto de Biología, UNAM.

Breedlove, D. E., & Laughlin, R. M. (1993). *The flowering of man: A Tzotzil botany of Zinacantan* (Smithsonian Contributions to Anthropology 35, Vol. 1). Washington DC: Smithsonian Institution Press.

Bricker, V., Po'ot, E., & de Po'ot, O. D. (1998). *Dictionary of the Maya language: As spoken in Hocaba Yucatan*. Salt Lake City: University of Utah Press.

Bruce, S. R. D. (1968). *Gramática del Lacandón*. Departamento de Investigaciones Antropológicas (Vol. 21). México: Instituto Nacional de Antropología e Historia.

Bruce, S. R. D. (1974). *El libro de Chan K'in*. Colección Cientifíca, Lingüística, 12. México: Instituto Nacional de Antropología e Historia.

Bruce, S. R. D. (1975). *Lacandon dream symbolism 1: Dream symbolism and interpretation*. México: Ediciones Euroamericanas Klaus Thiele.

Bruce, S. R. D. (1976). *Textos y dibujos Lacandones de Najá*. (Trilingual Edition: Lacandón-Spanish-English). Departamento de Lingüística

Colección Científica, Lingüística, Núm 45. México: Instituto Nacional de Antropología e Historia.

Bruce, S. R. D. (1979). *Lacandon dream Symbolism 2: Dream symbolism and interpretation among the Lacandon Mayas of Chiapas, Mexico*. México: Ediciones Euroamericanas Klaus Thiele.

Burger, W. C. (1962). Studies in New World Moraceae: Trophis, Clarisia, Acanthinophyllum. *Annals of the Missouri Botanical Garden, 49*(1/2), 1–34.

Bye, R. A., Jr., & Linares, E. (1990). Mexican market plants of 16th century. I Plants. Recorded in Historia Naturel de Nueva España. *Journal of Ethnobiology, 10*(2), 151–168.

California Rare Fruit Growers. (1996, 1997). *Banana: Musa species*. California Rare Fruit Growers, Inc. http://www.crfg.org/pubs/ff/banana.html

Canger, Una. (1970). *Lacandón de San Quintín Vocabulary of San Quintín*. Unpublished manuscript.

Cheng, K., Diemont, S. A. W., & Drew, A. P. (2011). Role of tao (*Belotia mexicana*) in the traditional Lacandon Maya shifting cultivation ecosystem. *Agroforestry Systems, 82*(3), 331–336. http://www.springerlink.com/content/j137243k1l4ux855/fulltext.pdf

Chudnoff, M. (1984). Tropical timbers of the world. *Agriculture Handbook, 607*. Washington, DC: US Department of Agriculture, Forest Service.

Clayton, W. D., & Renvoize, S. A. (1986). Genera Graminum: Grasses of the World. *Kew Bulletin Additional Series* 13, Royal Botanical Garden, Kew. Her Majesty's Stationery Office, London.

Clayton, W. D., Vorontsova, M. S., Harman, K. T., & Williamson, H. (2006 onwards). *GrassBase—The Online World Grass Flora*. The Board of Trustees, Royal Botanic Gardens, Kew. http://www.kew.org/data/grasses-db.html

Coe, F. G. (2008). Rama midwifery in eastern Nicaragua. *Journal of Ethnopharmacology, 117*, 136–157. http://www.bio-nica.info/biblioteca/coe%202008midwiferymedicinals.pdf

Comisión Nacional de Áreas Naturales Protegidas. (2006). *Programa de conservación y manejo área de flora y fauna Nahá*. 1a edición. Tlalpan C.P., México, DF: Dirección General de Manejo para la Conservación, CONANP. http://www2.ine.gob.mx/publicaciones/consultaPublicacion.html?id_pub=558&id_tema=12&dir=Consultas ISBN: 968-817-814-4

Cook, O. F. (1935). The Maya breadnut in southern Florida. *Science, 82*(2139), 615–616.

Cook, B. G., Pengelly, B. C., Brown, S. D., Donnelly, J. L., Eagles, D.A., & Franco, M.A. (2005). *Tropical forages: an interactive selection tool*. [CD-ROM], CSIRO, DPI&F(Qld), CIAT and ILRI, Brisbane, Australia. http://www.tropicalforages.info/key/Forages/Media/Html/Vigna_unguiculata.htm

Cortés-Rodríguez, G. R., & Inés-Agullar, R. (2002). Native species of bamboo in Mexico. In A. Kumar, I. V. Ramanuja Rao, & C. B. Sastry (Eds.), *Bamboo for sustainable development: Proceedings of the Vth International Bamboo Congress and the VIth International Bamboo Workshop, San José, Costa Rica, 2–6 November 1998* (pp. 89–95). Leiden, Netherlands: Brill Academic Publishers.

Croat, T. B. (1974). Case for selection for delayed fruit maturation in Spondias (Anacardiaceae). *Biotropica, 6*(2), 135–137.

Croat, T. B. (1978). *Flora of Barro Colorado Island*. Stanford: Stanford University Press.

D. R. Biblioteca Digital de la Medicina Tradicional Mexicana. 2009. *Atlas de las plantas de la medicina tradicional Mexicana*. Mexico: UNAM. http://www.medicinatradicionalmexicana.unam.mx

Dakina, K., & Wichmann, S. (2000). Cacao and chocolate: A Uto-Aztecan perspective. *Ancient Mesoamerica, 11*, 55–75.

Dallimore, W., Jackson, A. B., & Harrison, S. G. (1967). *A handbook of Coniferae and Ginkgoaceae* (4th ed.). NY: St. Martin's Press.

Davis, V. D. (1978). *Ritual of the Northern Lacandon Maya* (Unpublished doctoral dissertation). (PhD Dissertation), Tulane University, New Orleans, LA.

de Kroon, H., & Kalliola, R. (1995). Shoot dynamics of the giant grass *Gynerium sagittatum* in Peruvian Amazon floodplains, a clonal plant that does show self thinning. *Oecologia, 101*(1), 124–131.

Diemont, S. A. W. (2006). *Ecosystem management and restoration as practiced by the indigenous Lacandon Maya of Chiapas, Mexico* (Unpublished doctoral dissertation). Ohio State University, OH.

Diemont, S. A. W., Martin, J. F., Levy-Tacher, S. I., Nigh, R. B., et al. (2006). Lacandon Maya forest management: Restoration of soil fertility using native tree species. *Ecological Engineering, 28*(3), 205–212.

Domínguéz-Válzquez, G. and Islebe, G. A. 2008. Protracted drought during the late Holocene in the Lacandon rain forest, Mexico. *Vegetation History and Archaeobotany* 17:327–333.

Duke, J. A. (2003). *CRC handbook of medicinal spices*. Boca Raton, FL: CRC Press Ltd.

Duke, J. A. (2008). *Duke's handbook of medicinal plants of Latin America*. Boca Raton, FL: CRC Press.

Durán, F. A. (1999). *Estructura y etnobotánica de la selva alta perennifolia de Nahá Chiapas* (Unpublished master's thesis). Universidad Nacional Autónoma de México, Mexico.

Earle, C. J. (Ed.). (2013a). *Pinus pseudostrobus*. In *The gymnosperm database*. http://www.conifers.org/pi/Pinus_pseudostrobus.php

______. (2013b). *Podocarpus matudae*. In *The gymnosperm database*. http://www.conifers.org/po/Podocarpus_matudae.php

Elias, T. S. (1968). Flora of Panama. Part VI. Family 112. Vitaceae. *Annals of the Missouri Botanical Garden, 55*(2), 81–92.

Enquist, B. J., & Sullivan, J. J. (2001). *Vegetative key and descriptions of tree species of the tropical dry forests of upland sector Santa Rosa, area de conservación Guanacaste, Costa Rica*. http://www.acguanacaste.ac.cr/paginas_especie/plantae_online/EnquistSullivanTreeKey.pdf

Espinosa-García, F. J. (2000). *Malezas introducidas en México*. Universidad Nacional Autónoma de México. Centro de Investigaciones en Ecosistemas. Informe final SNIB-CONABIO proyecto No. U024. México DF. http://www.conabio.gob.mx/institucion/proyectos/resultados/InfU024.pdf

Gargiullo, M. B., Magnuson, B. L., & Kimball, L. D. (2008). *A field guide to plants of Costa Rica*. New York: Oxford University Press.

Garwood, N. C., & Bridgewater, S. (2005). Chamaedorea species. *Palms of Belize*. Rapid Color Guide 195. www.fmnh.org/plantguides/

Gentry, H. S. (1982). *Agaves of Continental North America*. Tucson: University of Arizona Press.

Gilman, E. F., & Watson, D. G. (1993). *Crescentia cujete; Calabash-Tree*. Fact Sheet ST-216, Environmental Horticulture Department, Florida Cooperative Extension Service, Institute of Food and Agricultural Sciences, University of Florida.

Gonzales, J. (2000). *Croton drago (Targuá)*. INBio. http://darnis.inbio.ac.cr/FMPro?-DB=UBIpub.fp3&-lay=WebAll&-Format=/ubi/detail.html&-Op=bw&id=308&-Find

Grandtner, M. M. (1997–2010). *World dictionary of trees*. http://www.wdt.qc.ca/

Grandtner, M. M. (2005). *Elsevier's dictionary of trees: With names in Latin, English, French, Spanish and other languages*. Amsterdam: Elsevier.

Grear, J. W. (1978). A revision of the New World species of *Rhynchosia* (Leguminosae-Faboideae). *Memoirs of the New York Botanical Garden, 31*, 1–168.

Gupta, Y. K., & Peshin, S. S. (2012). Do herbal medicines have potential for managing snake bite envenomation? *Toxicology International, 19*(2), 89–99. http://www.ncbi.nlm.nih.gov/pmc/articles/PMC3388772/

Gutiérrez, B. (1994). Icacinaceae. In V. Sosa, & A. Gómez-Pompa (Eds.), *Flora de Veracruz 8*. Instituto Ecología. A. C. and University of California, Xalapa, Veracruz, México.

Haeserijn, E. (1979). *Diccionario K'ekchi-Espanol*. Guatemala: Piedra Santa.

Hanan-Alipi, A. M. & Mondragón-Pichardo, J. 2009a. Bothriochloa laguroides. *Malezas de México*. http://www.conabio.gob.mx/malezasdemexico/poaceae/bothriochloa-laguroides/fichas/ficha.htm

Hanan-Alipi, A. M. & Mondragón-Pichardo, J. 2009c. Cnidoscolus multilobus. *Malezas de México*. http://www.conabio.gob.mx/malezasdemexico/euphorbiaceae/cnidoscolus-multilobus/fichas/ficha.htm

Hanan-Alipi, A. M. & Mondragón-Pichardo, J. 2009d. Digitaria setigera. *Malezas de México*. http://www.conabio.gob.mx/malezasdemexico/poaceae/digitaria-setigera/fichas/ficha.htm

Hanan-Alipi, A. M. & Mondragón-Pichardo, J. 2009e. Lantana trifolia. *Malezas de México*. http://www.conabio.gob.mx/malezasdemexico/verbenaceae/lantana-trifolia/fichas/ficha.htm

Hanan-Alipi, A. M. & Mondragón-Pichardo, J. 2009f. Solanum torvum. *Malezas de México*. http://www.conabio.gob.mx/malezasdemexico/solanaceae/solanum-torvum/fichas/ficha.htm

Hanelt, P., & Institute of Plant Genetics and Crop Plant Research. (2001). *Mansfeld's encyclopedia of agricultural and horticultural crops (except ornamentals)*. Berlin, New York: Springer.

Hellmuth, N. M. (2009). *Achiote as red dye and flavoring for cacao beverage for Maya and Aztec*. http://www.maya-archaeology.org/pre-Columbian_Mesoamerican_Mayan_ethnobotany_Mayan_iconography_archaeology_anthropology_research/achiote_Bixa_orellana_annatto_cacao_drink_red_dye_coloring.php

Hellmuth, N. M. (2010a). *Introduction to the pataxte form of cacao in Mayan ethnobotany*. http://www.maya-archaeology.org/Mayan_anthropology_ethnography_archaeology_art_history_iconography_epigraphy_ethnobotany/pataxte_pataste_pataschte_Belize_Mexico_Guatemala_Honduras_sacred_flowers_fruits_religion_cacao_chocolate_Theobroma_bicolor_balamte_jaguar_tree.php

Hellmuth, N. M. (2010b). *Theobroma bicolor: Guatemala, Mexico, and Belize*. http://www.maya-archaeology.org/Mayan_anthropology_ethnography_archaeology_art_history_iconography_epigraphy_ethnobotany/pataxte_pataste_pataschte_Belize_Mexico_Guatemala_Honduras_sacred_flowers_fruits_religion_cacao_chocolate_Theobroma_bicolor_balamte_jaguar_tree.php

Henderson, A., Galeano, G., & Bernal, R. (1995). *A field guide to the palms of the Americas*. Princeton, NJ: Princeton University Press.

Hernández-Nava, J. (2003). *Areas de protección de flora y fauna de Nahá y Metzabok*. Chiapas, México: Ficha Informativa de los Humedales de Ramsar.

Hernández-Ramos, M., Avila-Bello, C. H., & Morales-Mávil, J. E. (2007). Ethnobotánica y ecología de plantas utilizadas por tres curanderos contra la mordedura de serpiente en la región de Acayucan, Veracruz, México. *Boletín de la Sociedad Botánica de México, 81*, 89–100.

Hernandez-Solano, J. (1999). *Estudio integral de la selva Lacandona, Chiapas* (Unpublished tesis de licenciatura). Universidad Autónoma Chapingo, Texcoco, Estado de México.

Herrera, K., Lorence, D. H., Flynn, T., & Balick, M. J. (2010). Checklist of the vascular plants of Pohnpei, Federated States of Micronesia with local names and uses. *Allertonia, 10*, 1–204.

Higbee, E. C. (1947). Lonchocarpus: A fish-poison insecticide. *Economic Botany, 1*(4), 427–436.

Hodel, D. R. (1992). *Chamaedorea palms: The species and their cultivation*. Lawrence, Kansas: Allen Press.

Hofling, C. A., & Tesucún, F. F. (1997). *Itzaj Maya-Spanish-English dictionary-diccionario Maya Itzaj-Español-Ingles*. Salt Lake City, Utah: The University of Utah Press.

Houston, S. (1996). Symbolic sweatbaths of the Maya: Architectural meaning in the cross group at Palenque, Mexico. *Latin American Antiquity, 7*(2), 132–151.

Hurst, W. J. (2006). The determination of cacao in samples of archaeological interest. In C. McNeil (Ed.), *Chocolate of Mesoamerica: A cultural history of cacao* (pp. 105–113). Gainesville: University Press of Florida.

Irvine, M. S. (2005). *Passiflora online*. http://www.passionflow.co.uk/passiflora-hahnii.htm

Janzen, D. H. (1974). Swollen-thorn acacias of Central America. *Smithsonian Contributions to Botany Number 13*. Washington, DC: Smithsonian Institution Press.

Jiménez, Q. (1999). *Platymiscium pinnatum*. INBio. http://darnis.inbio.ac.cr/FMPro?-DB=UBIpub.fp3&-lay=WebAll&-Format=/ubi/detail.html&-Op=bw&id=2156&-Find

Kalegari, M. (2009). *Composição fitoquímica e atividades biológicas de Rourea induta Planch, Connaraceae. Curitiba* (Dissertação). Universidade Federal do Paraná.

Kalliola, R., & Renvoize, S. A. (1994). One or more species of Gynerium? (Poaceae). *Kew Bulletin, 49*(2), 305–320.

Kashanipour, R. A., & McGee, R. J. (2004). Northern Lacandon Maya medicinal plant use in the communities of Lacanja Chan Sayab and Naha', Chiapas, Mexico. *Journal of Ecological Anthropology, 8*, 47–66.

Kass, D. L. (n.d.) *Erythrina species—Pantropical multipurpose tree legumes*. www.fao.org/ag/AGP/AGPC/doc/Publicat/Gutt-shel/x5556e0b.htm

Kaufman, T., & Justeson, V. (2007). The history of the word for cacao in ancient Mesoamerica. *Ancient Mesoamerica, 18*(2), 193–237.

Kinsey, T. B. (2013). *Merremia tuberosa*—Woodrose. *Hawaiian plants and tropical flowers: A guide to the flowers and tropical plants of Hawaii*. http://wildlifeofhawaii.com

Kobayashi, K. D., McEwen, J., & Kaufman, A. J. (2007). *Ornamental ginger, red and pink*. Department of Tropical Plant and Soil Sciences. http://www.ctahr.hawaii.edu/oc/freepubs/pdf/OF-37.pdf

Kuijt, J. (2003). Monograph of Phoradendron (Viscaceae). *Systematic Botany Monographs, 66*, 1–643.

Landa, R., Madrid, S., Meli, P., Tolisano, J., & Yeager, C. (n.d.). Assessment of tropical forest and biodiversity conservation in Mexico. FAA sections 118–119 report. United States Agency for International Development.

Langeland, K. A. & Burks, C. (Eds.). 1998. *Identification and biology of non-native plants in Florida's natural areas*. Gainesville, FL: University of Florida.

Langeland, K. A., Ferrell, J. A., Sellers, B., MacDonald, G. E., Stocker, R. K. (1979). *Nephthytis, arrowhead vine*. Center for Aquatic and Invasive Plants. http://plants.ifas.ufl.edu/node/440

Laughlin, R. M. (1975). *The Great Tzotzil dictionary of San Lorenzo Zinacantán. Smithsonian Contributions to Anthropology 19*. Washington, DC: Smithsonian Institution Press.

Lay, K. K. (1949). A revision of the genus *Heliocarpus* L. *Annals of the Missouri Botanical Garden, 36*(4), 507–541.

Lee, H. J., Jeong, H. S., Kim, D. J., Noh, Y. H., et al. (2008). Inhibitory effect of citral on NO production by suppression of iNOS expression and NF-kappa B activation in RAW264.7 cells. *Archives of Pharmacal Research, 31*(3), 342–349.

Levy-Tacher, S. I. 2004. *El papel de Ochroma pyramidale Cav. en la agricultura tradicional Lacandona*. Report to Secretaria del Medio Ambiente y Recursos Naturales (Semarnat) of Mexico, San Cristóbal de las Casas, Mexico.

Levy-Tacher, S. I., Aguirre-Rivera, J. R., García-Perez, J. D., & Martínez-Romero, M. M. (2006). Aspectos florísticos de Lacanhá Chansayab, Selva Lacandona, Chiapas. *Acta Botanica Mexicana, 77*, 69–98.

Levy-Tacher, S. I., & Golicher, J. D. (2004). How predictive is traditional ecological knowledge: The case of the Lacandon Maya fallow enrichment system. *Interciencia, 29*, 496–503.

Lockhart, G. J. (1989). *The fighting herbs*. http://www.arthurleej.com/Fighting.pdf

Lois, X. (1998). Gender markers as "Rigid Determiners" of the Itzaj Maya world. *International Journal of American Linguistics, 64*(3), 224–282.

López-Toledo, L., Burslem, D., Martínez-Ramos, M., & García-Naranjo, A. (2008). Non-detriment findings report on *Guaiacum sanctum* in Mexico. *NDF Workshop Case Studies Wg 1- Trees Case Study 7 Guaiacum sanctum.* http://www.conabio.gob.mx/institucion/cooperacion_internacional/TallerNDF/Links-Documentos/Casos%20de%20Estudio/Trees/WG1%20CS7.pdf

MacDonald, M., Ferrell, J., Sellers, B., Langeland, K., et al. (2008). *Lantana species.* The University of Florida, IFAS Extension, Circular 1529, Invasive Species Management Plans for Florida. http://plants.ifas.ufl.edu/node/223

Madamba, R., Grubben, G. J. H., Asante, I. K., & Akromah, R. (2006). *Vigna unguiculata* (L.) Walp. In M. Brink & G. Belay (eds.), *PROTA 1: Cereals and pulses/Céréales et légumes secs.* Wageningen, Netherlands: PROTA. http://database.prota.org/PROTAhtml/Vigna%20unguiculata_En.htm

Mahdeem, H. (1994). Custard apples (Annona spp.). In J. E. Hernándo Bermejo & J. León (eds.), *Neglected Crops: 1492 from a different perspective* (pp. 85–92). Plant production and protection series No. 26. FAO, Rome, Italy. http://www.hort.purdue.edu/newcrop/1492/annonas.html

March, I. J. (1998). *Los Mayas Lacandones, Hach Winik: Problemas y potenciales para el desarrollo de un grupo indígena minoritario.* Proyecto perfiles indígenas de Chiapas. El Colegio de la Frontera Sur.

Marín, W. A. and Flores, E. M. (n.d.). *Astronium graveolens Jacq.* Part II—Species Descriptions, pp. 311–314. www.rngr.net/publications/ttsm/species/Astronium%20graveolens.../file

Marion, M. O. (1991). *Los hombres de la selva, un estudio de tecnología cultural en medio selvático.* México, DF: Inst. Nal. Antropol. Hist. (INAH).

Martin, S. (2006). Cacao in ancient Maya religion: First fruit from the maize tree and other tales from the underworld. In C. L. McNeil (Ed.), *Chocolate in Mesoamerica* (pp. 154–183). Gainesville: University Press of Florida.

Martínez, M. (1969). *Las plantas medicinales de México.* México, DF: Ediciones Botas.

Martínez, M. (1979). *Catálogo de nombres vulgares y científicos de plantas mexicanas.* México, DF: Fondo de Cultura Económica.

McGee, R. J. (2002). *Watching Lacandon Maya.* Boston: Allyn and Bacon.

McVaugh, R. (1983). Gramineae. In W. R. Anderson (ed.), *Flora Novo-Galiciana. A descriptive account of the vascular plants of Western Mexico* 14. Ann Arbor, Michigan: University of Michigan Press.

Miller, J. (2008). Vachellia collinsii. *Acacia systematics and ecological genomics.* Canberra, Australia: Center for Australian National Biodiversity. http://www.anbg.gov.au/jmiller/factsheets/Vachellia/collinsii.htm

Miranda, F. (1953). *La vegetación de Chiapas (segunada parte).* Tuxtla Gutiérrez, México: Gobierno del Estado de Chiapas.

Mirov, N. T. (1954). Composition of turpentines of Mexican pines. *Unasylva, 8*(4), 167–173. http://www.fao.org/docrep/x5373e/x5373e05.htm

Mondragón-Pichardo, J. & Vibrans, H. (2009). Phytolacca icosandra. *Malezas de México.* http://www.conabio.gob.mx/malezasdemexico/phytolaccaceae/phytolacca-icosandra/fichas/ficha.htm

Morton, J. F. (1981). *Atlas of medicinal plants of Middle America* (Vol. 2). Springfield, IL: Charles C. Thomas.

______. 1987. Avocado. In J. F. Morton (Ed.), *Fruits of warm climates* (pp. 91–102). Miami, FL: Julia J. F Morton. http://www.hort.purdue.edu/newcrop/morton/avocado_ars.html

______.______. Banana. In J. F. Morton (Ed.), *Fruits of warm climates* (pp. 29–46). Miami, FL.: Julia F. Morton. http://www.hort.purdue.edu/newcrop/morton/banana.html

______.______. Cherimoya. In J. F. Morton (Ed.), *Fruits of warm climates* (pp. 65–69). Miami, FL: Julia J. F. Morton http://www.hort.purdue.edu/newcrop/morton/cherimoya.html

______.______. Jamaica Cherry. In J. F. Morton (Ed.), *Fruits of warm climates* (pp. 65–69). Miami, FL: Julia F. Morton. https://hort.purdue.edu/newcrop/morton/jamaica_cherry.html

______.______. Loquat. In J. F. Morton (Ed.), *Fruits of warm climates* (pp. 103–108). Miami, FL: Julia F. Morton. http://www.hort.purdue.edu/newcrop/morton/loquat.html

______.______. Mexican Husk Tomato. In J. F. Morton (Ed.), *Fruits of warm climates* (pp. 434–437). Miami, FL: Julia F. Morton http://www.hort.purdue.edu/newcrop/morton/mexican_husk_tomato.html

______.______. Mexican Lime. In J. F. Morton (Ed.), *Fruits of warm climates* (pp.168–172). Miami, FL: Julia F. Morton. http://www.hort.purdue.edu/newcrop/morton/mexican_lime.html

______.______. Nance. In J. F. Morton (Ed.), *Fruits of warm climates* (pp.207–209). Miami, FL.: Julia F. Morton. http://www.hort.purdue.edu/newcrop/morton/nance.html

______.______. Purple Mombin. In J. F. Morton (Ed.), *Fruits of warm climates* (pp. 242–245). Miami, FL.: Julia F. Morton. http://www.hort.purdue.edu/newcrop/morton/purple_mombin.html#Varieties

______.______. Sansapote. In J. F. Morton (Ed.), *Fruits of warm climates* (pp.113–114). Miami: Julia F. Morton. http://www.hort.purdue.edu/newcrop/morton/sansapote.html

______.______. Sapodilla. In J. F. Morton (Ed.), *Fruits of warm climates* (pp.393–398). Miami, FL: Julia F. Morton. http://www.hort.purdue.edu/newcrop/morton/sapodilla.html

______.______. Sapote. In J. F. Morton (Ed.), *Fruits of warm climates* (pp. 398–402). Miami, FL.: Julia F. Morton. http://www.hort.purdue.edu/newcrop/morton/sapote_ars.html

______.______. Yellow Mombin. In J. F. Morton (Ed.), *Fruits of warm climates* (pp. 245–248). Miami, FL: Julia F. Morton. http://www.hort.purdue.edu/newcrop/morton/yellow_mombin_ars.html

Mudry, B. (n.d.). Quercus skinneri. *Woods of the world.* woodsoftheworld.org/displayspecies.php?&species_name=Quercus%20skinneri.

Murillo, R. M., Jakupovic, J., Rivera, J., & Castro, V. H. (2001). Diterpenes and other constituents from *Croton draco* (Euphorbiaceae). *Revista de Biología Tropical, 49*(1), 259–264.

National Peanut Board. (2006–2012). *History of peanuts.* http://www.nationalpeanutboard.org/classroom-history.php

National Tropical Botanical Garden. (2013). *Dioscorea bulbifera* (Dioscoreaceae). http://www.ntbg.org/plants/plant_details.php?plantid=4171#

Nations, J. D. (1989). Lacandon Maya bow and arrow: An ethnoarchaeological example of Postclassic lowland Maya weapon manufacture. In M. Gaxiola & J. E. Clark (Eds.), *La Obsidiana en Mesoamerica* (pp. 449–457). Mexico: Instituto Nacional de Antropología e Historia.

Nations, J. D. (1992). *Vocabulario Lacandón: plantas y animals.* Paper prepared for Conservation International, April 1992. Washington, DC.

Nations, J. D. (2006). *The Maya tropical forest: People, parks and ancient cities.* Austin: University of Austin Press.

Nations, J. D., & Nigh, R. B. (1980). The evolutionary potential of Lacandon Maya sustained-yield tropical forest agriculture. *Journal of Anthropological Research, 36*(1), 1–30.

Nee, M. (1993). Solanaceae II. In V. Sosa (Ed.), *Flora de Veracruz* (72nd ed., pp. 1–158). Xalapa, Veracruz, México: Instituto de Ecología.

Negi, J. S., Bisht, V. K., Bhandari, A. K., Singh, P., et al. (2011). Chemical constituents and biological activities of the genus *Zanthoxylum*: A eview. *African Journal of Pure and Applied Chemistry, 5*(12), 412–416. http://www.academicjournals.org/AJPAC

Nelson, C. (1998). Aegiphila monstrosa. IUCN Red List of threatened species. Version 2011.2. http://www.iucnredlist.org

Nigh, R. (2008). Trees, fire, and farmers: Making woods and soil in the Maya forest. *Journal of Ethnobiology, 28*(2), 231–243.

Ohashi, H., Tateishi, Y., Nemoto, T., & Endo, Y. (1988). Taxonomic studies on the Leguminosae of Taiwan III. *Science Reports of the Tohoku University Fourth Series (Biology), 39*, 191–248.

Olajuyigbe, O. O., & Afolayan, A. J. (2012). In vitro antibacterial and time-kill evaluation of the *Erythrina caffra* Thunb. Extract against bacteria associated with diarrhoea. *Scientific World Journal, 2012*, 1–8. doi:10.1100/2012/738314

Orwa, C., Mutua, A., Kindt, R., Jamnadass, R., & Anthony, S. (2009). *Agroforestree database: A tree reference and selection guide version 4.0*. http://www.worldagroforestry.org/sites/treedbs/treedatabases

Pennington, T. D., & Sarukhán, J. (1968). *Manual para la identificación de campo de los principales árboles tropicales de México*. México: INIF/FAO.

Pennington, T. D., & Sarukhán, J. (2005). *Árboles tropicales de México: Manual para la identification de las principales especies* (3rd ed.). México, DF: Universidad Nacional Autónoma de México.

Perales, H. R., Benz, B. F., & Brush, S. B. (2005). Maize diversity and ethnolinguistic diversity in Chiapas, Mexico. *PNAS, 102*(3), 949–954.

Peters, C. M., & Pardo-Tejeda, E. (2001). *Brosimum alicastrum* (Moraceae): Uses and potential in Mexico. *Economic Botany, 36*(2), 166–175.

PsychoTropia. (2011). *Ipomoea violacea linnaeus-morning glory vine*. http://psychotropia.co/?p=1215.

Raintree and Taylor, L. 1996. *CAMPECHE (Haematoxylon campechianum)*. RainTree tropical plant database. http://www.rain-tree.com/campeche.htm#.VmG6uL8-KFS

Rätsch, C. (1985a). *Das erlernen von zaubersprüchen ein beitrag zur ethnomedizin der Lacandonen von Naha'*. Berlin: Express-Edition.

Rätsch, C. (1985b). *Bilder aus der unsichtbaren Welt: Zaubersprüche und naturbeschreibung bei den Maya und den Lakandonen*. Munchen: Kindler.

Rätsch, C. (1994a). Ipomoea violacea linnaeus-morning glory vine. *PsychoTropia*. http://psychotropia.co/?p=1215

Rätsch, C. (1994b). Tsak: die heilpflanzen der Lakandonen von Naha'. *Jahrbuch für Ethnomedizin und Bewußsteinsforschung, 2*, 43–93.

Rieger, M. (2006). *Introduction to fruit crops*. Boca Raton, FL: CRC Press.

Rodríguez-Flores, C. I., & Stiles, G. F. (2005). Análisis ecomorfológico de una comunidad de colibríes ermitaños (Trochilidae, Phaetorninae) y sus flores en la Amazonia colombiana. *Ornitología Colombiana, 3*, 7–27.

Rodríguez-Sánchez, L. (2003–2012). *Terminalia amazonia*. Finca Leola. http://www.fincaleola.com/roble%20coral.htm

Rogers, D. J. (1964). Some botanical and ethnological considerations of *Manihot esculenta*. *Economic Botany, 17*(4), 369–377.

Rojas-Chávez, S. (2009). Mucuna argyrophylla. *Malezas de México*. http://www.conabio.gob.mx/malezasdemexico/fabaceae/mucuna-argyrophylla/fichas/pagina1.htm

Rosenblum, M. (2006). *Chocolate: A bittersweet saga of dark and light* (1st ed.). New York: North Point Press.

Roys, R. L. (1931). *The ethno-botany of the Maya*. New Orleans: Department of Middle American Research, University of Tulane.

Roys, R. L. (1933). *The book of Chilam Balam of Chumayel*. Washington DC: Carnegie Institution.

Ruan-Soto, F., Cifuentes, J., Mariaca, R., Limón, F., Pérez-Ramírez, L., & Sierra, S. (2009). Uso y manejo de hongos silvestres en dos comunidades de la selva Lacandona, Chiapas, México. *Revista Mexicana de Micología, 29*, 62–72. http://www.scielo.org.mx/pdf/rmm/v29/v29a9.pdf

Ruan-Soto, F., Garibay-Orijel, R., & Cifuentes, J. (2006). Process and dynamics of traditional selling wild edible mushrooms in tropical Mexico. *Journal of Ethnobiology and Ethnomedicine, 2*(3).

Rudd, V. E. (1965). The American species of *Ormosia* (Leguminosae). In V. E. Rudd (Ed.), *Leguminosae, Contributions from the United States National Herbarium* (32nd ed., pp. 279–384). Washington, DC: Smithsonian Institution Press.

Saade, R. L., & Montes-Hernández, S. (1994). Cucurbits. In J. E. Hernándo Bermejo and J. León (Eds.), *Neglected crops: 1492 from a different perspective* (pp. 63–77). Plant Production and Protection Series No. 26. FAO, Rome, Italy.

Sánchez-Vélez, A. (1988). El nanche (*Byrsonima crassifolia* l.) y otros elementos reforestadores no convencionales para los tropicos secas. *Revista Chapingo, 50*(51), 33–41.

Sauer, J., & Kaplan, L. (1969). Canavalia beans in American prehistory. *American Antiquity, 34*(4), 417–424.

Schoenhals, L. C. (1988). *A Spanish-English glossary of Mexican flora and fauna*. Mexico DF: Summer Institute of Linguistics.

Schultes, R. E. (n.d.). *De plantis toxicariis e Mundo Novo tropical commentationes IV*. http://wlbcenter.org/Schultes%20Publications/BotMusLeaf_22_133-164.pdf

Shumway, S. W., Letcher, S. L., Friberg, A., & DeMelo, D. (2009). *Rainforest plants*. http://wikis.wheatonma.edu/rainforest/index.php/Cyclanthaceae

Simpson, W. T. (1994). Resistance *moisture meter correction factors for four tropical wood species*. Res. Note FPL-RN-0260. Madison, WI: U.S. Department of Agriculture, Forest Service, Forest Products Laboratory. http://www.fpl.fs.fed.us/documnts/fplrn/fplrn260.pdf

Soustelle, G. (1966). *Collections Lacandons*. Paris: Museum National D'Histoire Naturelle.

Standley, P. C. (1930). Flora of Yucatan. *Fieldiana*: *Botany series* 3, III. Chicago: Field Museum of Natural History.

Standley, P. C., & Record, S. J. (1936). *The forests and flora of British Honduras* (Vol. XII). Chicago: Field Museum of Natural History.

Standley, P. C., & Steyermark, J. A. (1946). Flora of Guatemala. *Fieldiana*: *Botany series* 24, IV. Chicago: Field Museum of Natural History.

Standley, P. C., & Steyermark, J. A. (1949). Flora of Guatemala. *Fieldiana*: *Botany series* 24, V1. Chicago: Field Museum of Natural History.

Standley, P. C., & Steyermark, J. A. (1952). Flora of Guatemala. *Fieldiana*: *Botany series* 24, III. Chicago: Field Museum of Natural History.

Standley, P. C., & Steyermark, J. A. (1958). Flora of Guatemala. *Fieldiana*: *Botany series* 24, I. Chicago: Field Museum of Natural History.

Standley, P. C., & Williams, L. O. (1961). Flora of Guatemala. *Fieldiana*: *Botany series* 24, VII(1). Chicago: Field Museum of Natural History.

Standley, P. C., & Williams, L. O. (1966). Flora of Guatemala. *Fieldiana*: *Botany series* 24, III (1–2). Chicago: Field Museum of Natural History.

Standley P. C. & Williams, L. O. (1969). Flora of Guatemala. *Fieldiana: Botany series* 24, VIII (4). Chicago: Field Museum of Natural History.

Standley, P. C., & Williams, L. O. (1970). Flora of Guatemala. *Fieldiana*: *Botany series* 24, IX (1–2). Chicago: Field Museum of Natural History.

Standley, P. C., & Williams, L. O. (1971). Flora of Guatemala. *Fieldiana*: *Botany series*, 24, XI (1–3). Chicago: Field Museum of Natural History.

Stauffer, F. W., Rutishauser, R., & Endress, P. K. (2002). Morphology and development of the female flowers in *Geonoma interrupta* (Arecaceae). *American Journal of Botany, 89*, 220–229. http://www.amjbot.org/cgi/content/full/89/2/220

Stirton, C. H. (1987). *Ormosia schippii*. RBG Kew HerbCat database. http://www.kew.org/data/herb_digitisation.html

Stross, B. (1997). Mesoamerican copal resins. *U-Mut Maya, 6*, 177–186. http://www.utexas.edu/courses/stross/papers/copal.htm

Stross, B. (2008). K'U: The divine monkey. *Journal of Mesoamerican Languages and Linguistics* 1/1:1–34. http://www.utexas.edu/courses/stross/papers/StrossJMLL0001.pdf
Sullivan, G. (2008). *The drifting seed* 14(1).
Taylor, L. (2005). *The healing power of rainforest herbs: A guide to understanding and using herbal medicinals*. Garden City Park, NY: Square One Publishers.
Taylor, L. (2006). *Technical Data Report for matico (Piper aduncum, angustifolium).* http://www.rain-tree.com/reports/matico-tech-report.pdf
Tiwari, M., Dwivedi, U. N., & Kakkar, P. (2010). Suppression of oxidative stress and pro-inflammatory mediators by *Cymbopogon citratus* D. Stapf extract in lipopolysaccharide stimulated murine alveolar macrophages. *Food and Chemical Toxicology, 48*(10), 2913–9.
Tozzer, A. M. (1907). *A comparative study of the Mayas and the Lacandones*. London: Macmillan.
Trueblood, E. W. E. (1973). "Omixochitl" the tuberose (*Polianthes tuberosa*). *Economic Botany, 27*(2), 157–173.
University of Queensland. (2011). Giant reed *Arundo donax*. *Special edition of environmental weeds of Australia for biosecurity Queensland*. Queensland: University of Queensland. http://keyserver.lucidcentral.org/weeds/data/03030800-0b07-490a-8d04-0605030c0f01/media/Html/Arundo_donax.htm
Vargas-Soto, J. G., & Andrade, J. L. (2006). Clusia en México: La historia de un exilio. *Ciencias, 88*, 38–40.
Vargas-Soto, J. G., Andrade, J. L., & Winter, K. (2009). Carbon isotope composition and mode of photosynthesis in *Clusia* species from Mexico. *Photosynthetica, 47*(1), 33–40.
Vázquez-Yanes, C., Batis-Muñoz, A. I., Alcocer-Silva, M. I., Gual-Díaz, M., & Dirzo, S. (1999). *Alchornea latifolia. Árboles y arbustos potencialmente valiosos para la restauración ecológica y la reforestación*. Reporte técnico del proyecto J084. CONABIO - Instituto de Ecología,UNAM. http://www.conabio.gob.mx/institucion/proyectos/resultados/J084_Fichas%20de%20Especies.pdf
Vibrans, H. (2009). Piper umbellatum L. *Malezas de México*. http://www.conabio.gob.mx/malezasdemexico/piperaceae/piper-umbellatum/fichas/ficha.htm
Vibrans, H. (2011). Canavalia rosea (Sw.) DC. *Malezas de México*. http://www.conabio.gob.mx/malezasdemexico/fabaceae/canavalia-rosea/fichas/ficha.htm
Villaseñor, R., Espinosa-Garcia, J. L., & Espinosa-Garcia, F. J. (1998). *Catálogo de malezas de México*. México, DF: Universidad Nacional Autónoma de México. Consejo Nacional Consultivo Fitosanitario. Fondo de Cultura Económica.
Wellhausen, E. J., Fuentes, A. O., Hernandez-Corzo, A., & Mangelsdorf, P. C. (1957). *Races of maize in Central America. National Academy of Sciences*. National Research Council 511. Washington, DC.
World Conservation Monitoring Centre. (1998). *Guatteria anomala. IUCN 2012 Red List of Threatened Species.* Version 2. www.iucnredlist.org
Wunderlin, R. P. (1983). Revision of the arborescent bauhinias (Fabaceae: Caesalpinioideae: Cercideae) native to Middle America. *Annals of the Missouri Botanical Garden, 70*(1), 95–127.
Zamora, N. (1999). *Inga oerstediana* Benth. ex Seem. *Especies disponsibles*. INBio. http://darnis.inbio.ac.cr/

6 Ethnographic Inventory

6.1 Agriculture

Ch—ch

chäkäl hot. *Lit:* 'hot'. Refers to unfavourable soil conditions and plants that have a detrimental effect on the soil, neighbouring plants and crops. Plants include *si'si'k'uuts*, *'ak su'uk* (*Bothriochloa laguroides*), and *si'si'k'uuts* (*Erechtites hieracifolia*). Additionally, the presence of these plants indicates hard soil. *Ant:* **siis** 'cold'. *[Source:* BM*]*

chäkik cook v., to cook s.t. *Ant:* **sisik** 'cool something'. *Morph:* **chäk-ik**. *See:* **chäkäl**. *[Source:* BM*]*

chäk'an [chäk'an, chäk'än] field (savannah, clearing). *See:* **paakche' kol**. *[Note:* chäk'an [Itz.]. savannah (Hofling and Tesucún 1997); chakan [Yuc.]. savannah (Roys 1931); chak'an, chàak'an [Yuc.]. savanna (Bricker et al. 1998).*]* *[Source:* BM*]*

Ch'—ch'

ch'äk chop; fell with axe or machete. *See:* **halalch'äktik**; **ch'äkbil che'**. *[Note:* ch'ak.[Yuc.]. cut (with a blow), axe (Bricker et al. 1998); ch'ak [Itz.]. cut (Hofling and Tesucún 1997).*]*

ch'äkbil che' cleared land. *Lit:* 'cut trees'. The term refers to a freshly cleared forest or a very old cornfield that has lain fallow for 10 years or more. *Phase:* **el premier milpa** 'the first cornfield'. *See:* **yanchunil kol**. *[Source:* BM*]*

Electronic supplementary material The online version of this chapter (doi:10.1007/978-1-4614-9111-8_6) contains supplementary material, which is available to authorized users. Videos can also be accessed at http://link.springer.com/chapter/10.1007/978-1-4614-9111-8_6.

H—h

hal weed v; remove weeds. *Lit:* 'clear'. *See:* **halalch'äktik**. *[Source:* BM*]*

halalch'äktik [halalch'äktik, häläch'äktik] cut down brush with a machete. *Lit:* 'clear by chopping'. *Indicator:* **pet ak' (ah)** 'undetermined'; *Indicator:* **halol** (*Heliocarpus* spp.). *Morph:* **hal-al=ch'äk-t-ik**. *See main entry:* **hal**; **ch'äk**.

hortalisa vegetable garden. Relatively new to the culture, small vegetable gardens now occupy a focal point in the majority of Lacandon households. Introduced by Mexican NGOs, gardens consist of raised beds approximately 6×1.2 m (20′×4′). They are not particularly productive due to a number of factors, such as low soil fertility, hard soil, and improper plant spacing. The variety of crops grown is limited to radishes, carrots, green onions, cabbage, and lettuce. Peppers and tomatoes usually die at the seedling stage. *Variant:* **kolil nah**. *From:* hortaliza (Spanish).

K—k

kol cornfield, milpa (Sp.). (Fig. 4.11) *[Note:* kol [Yuc.]. to fell trees, to clear field; kòol. cornfield (Bricker et al. 1998: 132).*]*

kolil clear milpa; level brush with a machete. *See:* **nah kolil**. *[Note:* CXK and KP say this event happens in May; a similar phase called *nah kolil*, which entails clearing lands laid fallow for several years, occurs in March or April, according to BM. It is unclear which month this phase, and others, take place, since the Lacandon farmer does not schedule work around the Gregorian calendar. Moreover, weather patterns have changed, which affect the agricultural round. kòolil [Yuc.] felling season (Bricker et al. 1998).*]* *[Source:* BM; CXK; KP*]*

kolil nah house garden. *See:* **hortalisa**.

S. Cook, *The Forest of the Lacandon Maya*,
DOI 10.1007/978-1-4614-9111-8_6, © Springer Science+Business Media New York 2016

K'—k'

k'än k'up fungus, blight on corn. *Lit:* 'yellow cut'. **Man umuk'. P'iis chauistle.** 'The corn doesn't have any vigour. It's like *chauistle*.' (BM) According to BM, this corn disease begins at the top of the plant, attacking *uyi'ih* 'the corn tassel'. After this, it swiftly moves down the stalk, turning the leaves yellow and stunting the development of the ears. Within two weeks or so the plant is dead. *Chauitstli* (or, *chahuistle*) is a Nahuatl word for plague, skin fungus, or a disease caused by anything from an infectious agent to an evil spirit. The term is also applied to plants, and usually refers to a fungal infection. (Dr. Filiberto Malagón, pers. comm.). *Variant:* **k'än k'upul**. *From:* chauitstli Sp.<Nahuatl. *[Source:* BM*]*

L—l

lob [lob, lop'] bad; overgrown. *See:* **lo'obol**; **lo'obil**. (Appendix 2)

lu'uman päk'al orchard. *Lit:* 'cultivated land'. *Morph:* **lu'um-man päk'-al**. *[Source:* BM*]*

N—n

nah kolil felling season, main. *Lit:* 'main/great felling'. This is a phase in the agricultural cycle that refers to the preparation of the *milpa* for cultivation. The land is felled of brush and trees. The debris is left to dry out and then burned. After the ground has cooled, cultivation begins. *Time:* **nah ya'axk'in** 'main season'. *Morph:* **nah kol-il**. *See main entry:* **kol**. *Variant:* **kolil**. *[Source:* BM, CHX*]*

not'il stunted. *Morph:* **no'ot-il**. *[Note:* not' [Yuc.]. shrivel up, waste away (Bricker et al. 1998).*] [Source:* BM*]*

P—p

paak clearing; weeding at ground level. *See:* **paakche' kol**. *[Note:* paak [Itz.] N. clearing, paak. clearing (weeding with a machete) (Hofling and Tesucún 1997); páak [Yuc.] N. weeding (Bricker et al. 1998).*] [Source:* BM*]*

paakche' kol [pa(a)kche' kol, paakche' kor, pakche' kor] fallow. *Lit:* 'milpa cleared of trees'. More specifically, the term designates *milpas* that have been left fallow for a number of years. It is a transitional phase between annual cultivation and the reestablishment of woody vegetation and, ultimately, canopy closure. *Phase:* **acahual** 'fallow milpa'. *Morph:* **paak=che' kol**. *See:* **chäk'an**. *See main entry:* **paak** (Figs. 4.6 and 4.7). *[Note:* In published studies on Lacandon agroforesty, the term is written *pak che(') kol* but is variously glossed as 'planted tree milpa' (Nations and Nigh 1980; Kashanipour and McGee 2004), 'orchard-garden' (Nations and Nigh 1980), and 'tree milpa' (Diemont 2006). These translations suggest that the trees have been purposefully planted. There are two problems with this conclusion. First, the Lacandon verb to plant is *päk'*, whereas paak is a noun that means 'clearing', or a verb that means 'clear, weed'. Second, all the Lacandones that I interviewed on the matter said that the trees in the paakche' kol grow on their own, without human intervention.*] [Source:* BM; CKO; KM*]*

pak'al plant n., cultivated. *Morph:* **pak'-al**.

pak'alil garden. *Lit:* 'place of plants'. *Morph:* **pak'-al-il**.

päk' plant v., to plant something.

päybil näl maize, replanting of. *Lit:* 'maize is signalled'. This is a phase in the agricultural cycle that entails replanting the *milpa* with corn. This activity typically occurs in October, according to BM and KA. The *milpa* is not burned at this time. In the past, the Lacandones observed the blooming of the *ma'ax ak'* as a signal of when to commence this work (BM). *Indicator:* **ma'ax ak' (äh)** (*Paullinia* sp.). *Variant:* **päybil kol**. *[Source:* BM*]*

S—s

siis cold. Refers to favourable soil and beneficial plants, including *bo'oy* (Chamaedorea spp.) and *mäkäl* (*Xanthosoma* spp.). *Ant:* **chäkäl** 'hot'.

sisik cool v. to cool s.t. **Ne ts'oy ti' lu'umal. Ne hup ulu'umal.** '[Plants that are] good for the soil. They indicate the soil is soft.' (BM) Plants that are cool include: *hoben* (*Piper auritum*); *mäkuuläm* (Piper spp.); *mäkäl* (*Xanthosoma* spp.); *bo'oy* (*Chamaedorea* spp.); *ek' bahche'* (*Guatteria anomala*); and *chukuch le' xiiw* (*Heliconia* sp.). Additionally, the presence of these plants indicate soft, fertile soil. *Ant:* **chäkik** 'cook something'. *[Source:* BM*]*

T—t

täk'än kab fertile soil. *Lit:* 'ripe land'. *[Source:* AM; BM*]*

tumen näl maize, new. *Lit:* 'new maize'. Refers to the second crop of maize, which is usually sown in July. *[Source:* BM*]*

U—u

uchben kol milpa, old. The term refers to land that has been continuously cultivated for 4 years. *[Source:* BM*]*

W—w

wuts'ik näl fold-over maize. A term that refers to the act of bending over the maize stalks after the ears have ripened, to protect them against the rains while the maize dries. This is done in August or later in the fall. *Morph:* **wuts'-ik näl** (Fig. 4.8).

Y—y

yanchunil kol first milpa. This refers to a new cornfield that has been cut out of virgin forest. According to BM, when the northern region was decreed a reserve in 1995, the clearing of virgin forest for *milpas* ceased. *Morph:* **yan=chun-il kol**. *[Source:* BM*]*

6.2 Artefacts

6.2.1 Accessories

B—b

booch' umbrella. *Mat:* **ch'om mäkäl** 'vulture mäkäl (*Xanthosoma robustum*)'; *Part:* **le'** 'leaf'.

Ch'—ch'

ch'am [ch'am, ch'äm] grill rack. *Mat:* **mahaas (ah)** 'funeral tree (*Quararibea funebris*)'; *Mat:* **mäkuuläm** '*Piper* spp.'.

ch'uyu'$_2$ hanger, for clothes and meat. *Mat:* **mahaas (ah)** (*Quararibea funebris*); *Part:* **xa'ay** 'forked branches'. *[Source:* SK*]*

K—k

kib [kib,kip'] wax, candle, light. *Spec:* **pätbil kib** 'handmade candle (of beeswax)'; *Spec:* **säk kib** 'white wax (paraffin)'. *See:* **kibil kab**; **kab**.

kibil kab [kibil kab, kibil kap', kibih kab, kibih kab] beeswax. *Use:* **pätbil kib** 'handmade candles'; *Use:* **ti' banik k'an, häxan** 'for painting (sealing) cord, string'. *See:* **kab**. *See main entry:* **kib**.

P—p

pätbil kib candle, made of beeswax. *Lit:* 'formed wax'. These traditional candles were made from a long 1.8 m (6′) wick of twisted bark fibres and saturated with wild beeswax. *Mat:* **halol** 'halol (*Heliocarpus* spp.)'; *Part:* **usool** 'bark'; *Mat:* **kibil kab** 'beeswax'; *Sim:* **säkbil kib** 'paraffin wax candle'.

pixil chem canoe cover. *Mat:* **pahok (ah)** '*Geonoma oxycarpa*'.

S—s

säkbil kib candle, paraffin. *Lit:* 'whitened light'. *Sim:* **pätbil kib** 'handmade candle (of beeswax)'; *Gen:* **kib**. *Variant:* **säk kib**.

säk kib candle of paraffin wax. *Lit:* 'white light'. *Sim:* **pätbil kib** 'beeswax candle'; *Gen:* **kib**. *See:* **kibil kab**. *Variant:* **säkbil kib**.

W—w

waal$_2$ fan. *Mat:* **neeh uch'iich', ähkoox** 'tail feathers of birds, e.g., crested guan (*Penelope purpurascens*)'; *Mat:* **hach bo'oy** 'authentic bo'oy'.

6.2.2 Containers

B—b

box$_4$ cup, large. This is a large drinking vessel made from the hollowed out fruit of an inedible squash. *Mat:* **box** 'calabash (*Curcurbita* sp.)'. *[Source:* Bruce (1975)*]*

Ch'—ch'

ch'uyu'$_1$ [ch'uyu', ch'uyuk] basket, hanging. **ti' uch'uyik ba'al**. These hangers consist of a hand woven fibre net stretched over a circular ring made from sturdy vines, which is suspended by three cords of twisted fibre. *Mat:* **halol** '*Heliocarpus* sp.'; *Part:* **k'an** 'cord'; *Mat:* **nikte' ak'**

'undetermined'; *Part:* **me'et** 'foundation ring'; *Parts:* **yaal** 'ribs of the basket'; *Parts:* **hich'il, wich** 'knots, eyes'; *Parts:* **le'** 'leaf, net'. *Variant:* **ch'uch'uyil**. (Fig. 4.99) *[Source:* KP*]*

H—h

häyäk lek plate, metal. *Mat:* **maska'** 'metal'; *Sim:* **suulil lek**; *Sim:* **lekil wah** 'tortilla plate'; *Sim:* **lekil k'uh** 'god pot'; *Gen:* **lek** 'plate'.

K—k

kum pot. *Mat*: **k'ät** 'clay'.

kumil nahwah pot, cooking (ceremonial). This is a clay cooking cauldron used to cook *nahwah* 'ceremonial tamales'. *Mat:* **k'ät** 'clay'; *Sim:* **kumil puk'** 'mixing pot'; *Gen:* **kum**. *[Source:* Baer and Baer (1952)*]*

kumil puk' pot, cooking (ceremonial use). *Lit:* 'mixing pot'. A clay cauldron used for cooking ceremonial corn-based foods, i.e., *ch'ulha'* and *sa'*. *Mat:* **k'ät** 'clay'; *Sim:* **kumil nahwah** 'tamale pot'; *Gen:* **kum**. *[Source:* Baer and Baer (1952).*]*

K'—k'

k'äb luuch [k'äb luuch, k'ä' luuch] cup. *Lit:* 'hand cup'. A kind of ceremonial gourd drinking cup. *Loc:* **yatoch k'uh** 'god house'; *Use:* **ti' uk'ik balche'** 'for drinking balche''; *Mat:* **hach luuch** (*Crescentia cujete*); *Sim:* **hama'**. *[Source:* Baer and Baer (1952)*]*

k'ik' luuch cup, plastic. *Lit:* 'rubber cup'. *Mat:* **k'ik'** 'plastic'; *Gen:* **luuch** 'cup, bowl'. *[Source:* KP*]*

L—l

lek$_1$ [lek, läk] plate, bowl. *Mat:* **lek** (*Lagenaria siceraria*)'; *Spec:* **häyäk lek** 'metal or porcelain pot'; *Spec:* **suulil lek** (*Crescentia alata*); *Spec:* **lekil k'uh** 'god pot'; *Spec:* **lekil wah** 'tortilla plate'.

lekil wah plate, for tortillas. *Mat:* **lek** (*Lagenaria siceraria*); *Sim:* **lekil k'uh** 'god pot'; *Sim:* **suulil lek**; *Sim:* **häyäk lek** 'plate made of metal or porcelain'; *Gen:* **lek** 'plate'. *See main entry:* **lek**. *Variant:* **plate**.

luuch$_2$ cup, bowl. Likely derived from the verb 'to fold in the shape of a scoop, to scoop'. *Luuch* traditionally referred to the drinking cups made from the fruits of the tree by the same name, but today the word refers to cups made from any material. *Mat:* **hach luuch** (*Crescentia* spp.); *Sim:* **xikin luuch** 'tea cup'; *Sim:* **k'ik' luuch** 'plastic cup'; *Spec:* **k'äb luuch** 'hand cup'. *[Note:* luuch, lu'uch. drinking cup (Bruce 1975); luch [Itz.]. water gourd (Hofling and Tesucún 1997); lùuch [Yuc.]. vn. calabash. *Crescentia cujete* L. (Bricker et al. 1998).*]*

P—p

pak' bee hive, wasp nest. *Mat:* **chuh** (*Lagenaria siceraria*) (Figs. 4.44, 4.45, and 4.46).

S—s

suulil lek cup, small. Used in the preparation of *käkow* 'cocoa'. *Mat:* **sina'an luuch (äh)** (*Crescentia cujete, C. alata*); *Sim:* **häyäk lek** 'metal plate'; *Sim:* **lekil k'uh** 'god pot'; *Sim:* **lekil wah** 'tortilla plate'; *Gen:* **lek** 'plate'. *[Note:* sul [Yuc.]. soak, steep (Bricker et al. 1998); sul [Itz.]. soak (Hofling and Tesucún 1997).*]*

X—x

xaak basket. *Mat:* **'ilon** (*Monstera* sp.); *Mat:* **holop'** (*Philodendron* spp., *P. smithii*). (Fig. 4.95) *[Note:* xac (Baer and Merrifield 1971).*]*

xikin luuch teacup. *Lit:* 'ear cup'. *Mat:* **maska', k'ik'** 'metal or porcelain, plastic'; *Gen:* **luuch** 'cup, bowl'. *[Source:* Bruce (1975)*]*

6.2.3 Cordage

H—h

häxan string or cord that has been twisted. *Lit:* 'twisted'. *Mat:* **kih** '*Agave sisalana*'; *Mat:* **halol** (*Heliocarpus* spp.); *Prep:* **abälik uhol yok'ol awok** 'You roll the fibres down your leg.' *Morph:* **häx-an**. *See:* **k'an** (Figs. 4.52, 4.53, 4.54, and 4.55).

hit tie, braid, weave (loosely).

hitik ba'ay to weave or knot (nets, hammocks).

hol fibre, bark. The word refers to the processed fibre of *halol* (*Heliocarpus* spp.). Only young trees with trunks 30 cm (12″) in diameter are used, as the bark is easier to work with than that from older trees. According to KyP, *halol* is cut during the full moon; but he did not say why. The inner fibre is soaked in the river to bleach and soften it. When dry, it is shredded into long ribbons and then stored for future use. To form a long, thin cord, two or three ribbons are rolled together down the shin bone or thigh. *Mat:* **halol** 'cork (*Heliocarpus* spp.)'; *Part:* **uyoot'el** 'inner bark'; *Use:* **k'an** 'cord'; *Prep:* **ubähtik usool, pachil up'älik, pachil tsilik uyoot'el sool, ele ulähts'ämik yok'ol uk'um ti' hunts'it nah, pachil uch'uyik tu k'in ti' humbuhk'ä' k'in, pachil utsilik usool, pachil bälik tu yok'** 'One pounds the outer bark, then peels it off, then strips off the inner bark, then submerges it in the river for one month, then hangs them in the sun for five days, then shreds the bark, and then rolls it down the leg.' (Figs. 4.52, 4.53, 4.54, and 4.55) *[Source:* BM*]*

K'—k'

k'aax lashing, tie, rope.

k'an$_2$ cord. The terms refers to cords, ropes, hammocks, and spiders' silk. Several kinds of materials are used. *Mat:* **chäk taw** '*Trema micrantha*'; *Mat:* **kih** (*Agave sisalana*); *Mat:* **halol** '*Heliocarpus* spp.'; *Mat:* **holop'** (*Philodendron* spp.).

k'uch$_1$ thread, yarn. *Mat:* **tämän** (*Gossypium hirsutum*).

k'uch$_2$ spin thread. *See:* **pits'**.

P—p

pakche'il kih board for removing agave fibre. This is a narrow board about 30 cm (1′) longer than an agave leaf, used as a surface to scrape off the pulpy exterior of the leaf to expose the fibres within. This is a male task. *Mat:* **puuna' (äh)** (*Swietenia macrophylla*). *[Source:* Baer and Baer (1952)*]*

S—s

suum$_1$ rope, thick. *Mat:* **halol** (*Heliocarpus* spp.).

6.2.4 Craft

A—a

akpach warp (of weave). *Lit:* 'back face'. *[Source:* JK*]*

B—b

ba'ay bag (woven). Large, net bag used for carrying large loads. *Mat:* **halol** (*Heliocarpus* spp.); *Part:* **uyoot'el** 'inner bark'. *See:* **pooxal**; **ch'im** (Fig. 4.100).

but'u'o'och shuttle (weft bobbin). *Lit:* 'weft stuffer'. The shuttle stick is a slender dowel that holds the weft yarn. The shuttle is passed back and forth between the threads of the warp to weave in the weft. *Mat:* **chäk ya'** (*Manilkara achras*). *Morph:* **but'-u'=o'och**. *[Note:* butu'och (Baer and Baer 1952); 'ó'och [Yuc.] woof (Bricker et al. 1998).*]*

Ch—ch

che'il tu kuwätik warping board. *Lit:* 'the wooden board where one divides it'. This is a long, narrow board with wooden pegs for measuring and stretching warp yarn to the desired length and pattern before transferring the warp to the loom. The Lacandon version is laid horizontally on the ground, about 3 m (10′) long and 76 cm (2.5′) wide. *Mat:* **chäk ya'** (*Manilkara achras*). *Morph:* **che'-il tu k-u-wät-ik**. *[Source:* SK*]* *[*sd2 craft*]*

Ch'—ch'

ch'alol$_2$ spinning top. *Mat:* **hach ch'alol** (*Quercus corrugata*); *Part:* **wich** 'fruit'. *Variant:* **polbilich**. (Fig. 4.91) *[Note:* polbirich. spinning top Tozzer (1907).*]* *[Source:* AM*]*

ch'im$_1$ purse. A net shoulder bag. *Mat:* **kih** (*Agave sisalana*). (Fig. 4.102) *[Source:* Baer and Merrifield (1971)*]*

H—h

hälälte'$_1$ weaving sticks. *Lit:* 'arrow stick', including the shed bar and the beater. *Mat:* **chäk ya'** (*Manilkara achras*). *Variant:* **mamak**. *[Note:* hälälte' (Baer and Baer 1952); hala'te' [Yuc.] weaving sword (Bricker et al. 1998).*]* *[Source:* KM*]*

hälälte'$_2$ loom. *Mat:* **chäk ya'** (*Manilkara achras*). *Variant:* **säkäl**. (Fig. 4.93) *[Source:* AM*]*

M—m

mämäk heddle. Refers to the thread cords used to separate the warp threads and make a path for the shuttle. *Mat:* **k'an** 'cord' (*Heliocarpus* spp.). *[Note:* mämäk (Baer and Baer 1952); máamak, máamak [Yuc.]. heddle, arrange threads on a heddle (Bricker et al. 1998).*]*

me'et ring, trivet; woven rings. *Mat:* **'ilon** (*Monstera* spp.); *Part:* **ak'il** 'vines'; *Mat:* **hänan** (*Desmoncus orthacanthus*); *Part:* **che'il** 'stem'; *Mat:* **ya'ax hänan** (*Chamaedorea elatior*); *Part:* **che'il** 'stem'. (Fig. 4.51) *[Note:* me'et [Yuc.]. trivet, circular base, stand (Bricker et al. 1998).*]*

O—o

o'och$_1$ weft (woof); the horizontal threads woven into the warp, or the vertical threads. *Part:* **nook'** 'cloth (hand woven)'. *[Note:* 'o'och [Yuc.]. woof in weaving [cross threads] (Bricker et al. 1998).*]*

P—p

pecheech spindle. *Mat:* **mahaas (ah)** (*Quararibea funebris*); *Part:* **tukil** 'whorl'. *See:* **pits'**. *Variant:* **pacheech**. *[Note:* pachech (Baer and Baer 1952); pecheech [Itz.]. spindle (Hofling and Tesucún 1997); pechéech [Yuc.]. spindle (Bricker et al. 1998).*] [Source:* Baer and Baer (1952); KM; SK*]*

polbilich spinning top. *Mat:* **hach ch'alol** (*Quercus corrugata*). *Variant:* **ch'alol**. (Fig. 4.91) *[Source:* Tozzer (1907: 76)*]*

pooxal shoulder bag. Pooxal are small purses made from the hides of deer, jaguar, and ocelot. They are used by men for carrying arrow flints, steel and cotton for starting fires, cigars, and other personal items. The hides are tanned in a solution of water and the mashed inner bark of the puuna' (*Swietenia macrophylla*). *Mat:* **usoolel bäk'** 'leather'; *Mat:* **usoolel puuna'** (*Swietenia macrophylla*). *See:* **ch'im**. *From:* bolsa(?) (Spanish) (Figs. 4.108 and 4.109).

S—s

säkäl$_1$ cloth (traditional) hand-loomed cotton. *Mat:* **tämän** (*Gossypium hirsutum*). *See:* **säkälbil nook'**. *[Note:* säkäl. weave, loom (Bruce 1975); sakal [Yuc.]. length of woven cloth, loom (Bricker et al. 1998); säkäl [Itz.]. N. mending (Hofling and Tesucún 1997).*]*

säkäl$_2$ loom. *Mat:* **chäk ya'** (*Manilkara achras*). *Variant:* **hälälte'**. *[Note:* säkar, säkär [S. Lac.]. loom, weave (Canger 1970).*]*

säkälbil nook' [säkälbil nook', säkärbil nook', säkäbil nook', säkbil nook'] cloth, hand-loomed. *Lit:* 'woven cloth'. *[Source:* AM; Bruce (1975: 201)*]*

T—t

taabil ba'ay [taabil ba'ay, taabi(h) ba'ay, tabi(h) baay] tumpline. *Lit:* 'sack's cord'. *Mat:* **halol** (*Heliocarpus* spp.). *See:* **ba'ay**. *Variant:* **taabil kuch**. *[Source:* Baer and Baer (1952); Bruce (1975)*]*

tukil whorl (spindle). *Lit:* 'piler'. *Mat:* **mahaas (ah)** (*Quararibea funebris*); *Part:* **che'il** 'wood'; *Mat:* **pakay** (*Hymenaea courbaril*); *Part:* **nek'** 'seed'; *Mat:* **tuk'** (*Acrocomia mexicana*); *Part:* **usool uwich** 'fruit shell'; *Mat:* **aak** 'turtle'; *Part:* **box** 'shell'. *Morph:* **tuk-il**. *[Source:* AM; SK*]*

Ts'—ts'

ts'aab$_3$ shed. *Lit:* 'layer'. Refers to the opening created between the warp yarns by pulling some warp threads up and some down. *[Note:* ts'a'ats'a'ap [Yuc.]. wavy (interleaved, woven together) (Bricker et al. 1998).*] [Source:* KM*]*

U—u

uuh necklace. *Lit:* 'halo'. *Mat:* **nek'** 'seeds'; *Sim:* **uuhil k'äb** 'bracelet'. (Figs. 4.83, 4.84, 4.85, and 4.86) *[Note:* uuh. necklace (Bruce 1975); 'uh [Yuc.]. bead; moon (Bricker et al. 1998); 'uh [Itz.]. moon; bean (Hofling and Tesucún 1997).*]*

uuhil k'äb bracelet. *Lit:* 'halo around arm, hand'. *Mat:* **nek'** 'seeds'; *Sim:* **uuh** 'neclace'. *Morph:* **uuh-il k'äb**.

uyo'och tu' kubut'ik beater (batten). *Lit:* 'where one stuffs the weft'. The beater is used to force down the weft (horizontal threads) as they are woven. The Lacandon description might also refer to the weft bobbin. *Mat:* **chäk ya'** '*Manilkara achras*'. *See:* **but'u'o'och**. *Variant:* **hälälte'**. *[Source:* KM*]*

X—x

xuunche' warp end rods. *Lit:* 'end sticks'. Warp end bars are used to hold the warp (vertical threads) taut. One pole is secured to an upright post or tree and the other is attached to belt wrapped around the weaver's waist. *Mat:* **chäk ya'** 'chicle (*Manilkara achras*)'. *[Note:* xúun-che' [Yuc.]. bar at head or foot of backstrap loom (Bricker et al. 1998; Tozzer 1907) reports that bamboo sticks were used for the end rods.*] [Source:* Baer and Baer (1950), SK*]*

6.2.5 Furniture

B—b

bilich mattress. *Lit:* '(bed) roll'. *Mat:* **hach taw** (*Belotia mexicana*); *Mat:* **chäk taw** (*Trema micrantha*); *Part:* **uyoot'el** 'inner bark'. *Morph:* **bil-ich**. *See:* **taspach** 'back sheet/layer'.

Ch'—ch'

ch'aak bed. *Mat:* **puuna' (äh)** *Swietenia macrophylla.*

H—h

hach k'an hammock, traditional. *Mat:* **halol** (*Heliocarpus* spp.) (Fig. 4.104, Video 4.21)

K'—k'

k'an$_3$ hammock.

k'anche' chair, bench, stool. *Mat:* **puuna' (äh)** *Swietenia macrophylla. Morph:* **k'an=che'**.

P—p

pohche' table. *Mat:* **pukte'** (*Terminalia amazonia*); *Mat:* **ek' bahche'** (*Guatteria anomala*); *Mat:* **puuna' (äh)** (*Swietenia macrophylla*). *[Source:* BM; K'ayum Segundo*]*

pop mat. *Mat:* **säk bo'oy** *Chamaedorea* sp., *C. oblongata*; *Mat:* **hach bo'oy** 'authentic bo'oy'.

T—t

taspach mattress. *Lit:* 'back sheet/layer'. *Morph:* **tas-pach**. *See:* **bilich**. *[Source:* Bruce (1975)*]*

6.2.6 Hunting, Fishing, Trapping Items

B—b

baak antler hammer. *Lit:* 'bone'. *Use:* **bähik tok'** 'knap flint'.

bähik tok' knap flint.

Ch—ch

chulul$_3$ bow. *Mat:* **hach chulul** 'authentic chulul (*Tabebuia guayacan*)'; *Mat:* **säk chulul (äh)** 'white chulul (*Platymiscium dimorphandrum*)'; *Part:* **che'il** 'timber'.

chuste' [chuste', chuxte'] foreshaft. While my consultant listed two species used for the foreshaft, a number of different woods have been reported by researchers in the area: *säk ya' ek', k'ak'che', chäk ya', kekche', kukche', k'iis* (Tozzer 1907); *Cedrela odorata, Hirtella americana, H. triandra, sa kum che* (*Inga* sp.), *isa che* (*Licaria caudata*), *gik che, kiche* (*Virola guatemalensis*) (Durán 1999); *k'uche'* (*Cedrela mexicana*), *ya'* (*Manilkara zapota*), and *ek'* (*Haematoxylum campechianum*) (Nations 1989). *Mat:* **ya'** 'chicle (*Manilkara* spp.)'; *Mat:* **säk ya'** 'wild star apple (*Chrysophyllum mexicanum*)'. *[Note:* chuste. foreshaft (Tozzer 1907); chuiste (Durán 1999); chuste (Nations 1989).*] [Source:* BM*]*

H—h

hapak quiver. *Mat:* **halol** (Heliocarpus spp.); *Part:* **hach uyoot'el** 'outer bark'. *[Source:* Baer and Baer (1952: 45); KyP*]*

häläl$_3$ arrow. *Mat:* **hach 'ooh** (*Gynerium sagittatum*); *Part:* **uche'il utop'** 'flower stalks'; *Mat:* **hach häläl** (*Phragmites australis*); *Part:* **che'il** 'stalks'; *Use:* **yolil** 'shaft'; *Mat:* **xik'** 'wing feathers'; *Use:* **xik'** 'fletching'; *Mat:* **tok'** 'flint'; *Mat:* **tah** 'obsidian'; *Use:* **ni'** 'arrow point'; *Mat:* **kih** (*Agave sisalana*); *Mat:* **halol** (*Heliocarpus* spp.); *Use:* **k'äxal** 'binding'; *Mat:* **ek'** (*Haematoxylon campechianum*); *Mat:* **chäk 'oopche'** (*Clarisia biflora*); *Part:* **uyits** 'resin'; *Mat:* **kibil kab** 'wax'; *Use:* **ti' banik k'an** 'binding sealer'. (Figs. 4.78, 4.79, 4.80, and 4.81) *[Note:* herlerl. arrow (Tozzer 1907); halal [Yuc.]. reed. (Bricker et al. 1998); jäläl [Itz.]. carrizo. reed (Hofling and Tesucún 1997). SK says that wing feathers are called *xik'*, whereas the downy feathers (breast feathers) are called *k'uk'.*] *[Source:* BM;SK*]*

K—k

koh luts fish hook. *Lit:* 'tooth hook'. *Variant:* **luts käy**.

K'—k'

k'anil luts fishing line. The traditional hand fishing line was made of natural fibres twisted together to form a fine cord. The cord was then sealed with the plant resin. Today, nylon lines are used. *Mat:* **kih** (*Agave sisalana*); *Part:* **uyoot'el** 'inner fibre'; *Mat:* **ni' ts'ul** (*Mucuna argyrophylla*); *Part:* **yits** 'resin'. *Variant:* **lutsil**.

L—l

lama' spear, harpoon. *Mat:* **semet'** (*Rhipidocladum bartlettii*). *Variant:* **lomo'**. *[Source:* Bruce (1975)*]*

leh trap. *Mat:* **kih** 'agave (*Agave sisalana*)'; *Part:* **le'** 'leaves'; *Mat:* **babah** 'false mamery (*Calophyllum brasiliense*)'; *Part:* **k'äbche', mehen che'** 'branches, saplings'.

lomo' spear, harpoon. *Mat:* **semet'** (*Rhipidocladum bartlettii*). *Variant:* **lama'**. (Fig. 4.82) *[Note:* lama'. spear, harpoon (Bruce 1975: 186); lam [Yuc.]. penetrate, sink, push; lom [Yuc.]. puncture, stab; x lomob' [Yuc.]. knife (Bricker et al. 1998); lom [Itz.] stab; knife wound (Hofling and Tesucún 1997).*]*

luts hook.

lutsil line, fishing. *Mat:* **kih** (*Agave sisalana*). *Variant:* **k'anil luts**.

luts käy fish hook; to fish. *See main entry:* **luts**. *Variant:* **koh luts**. *[Source:* Bruce (1975)*]*

P—p

pak$_1$ notch. *Lit:* 'fold'. The notch where the arrowhead fits into the foreshaft. *[Source:* Tozzer (1907); Nations (1989).*]*

pakat bird bolt. This is a large, broad arrow point used for stunning birds. *Use:* **ti' kinsik ch'iich'** 'to kill birds'; *Mat:* **mahaas (ah)** (*Quararibea funebris*); *Mat:* **ton ku'uk** (*Cymbopetalum penduliflorum*). *[Note:* BM mentioned another kind of soft wood, like the *mahaas*, that is also used for the bird bolt. He did not indicate what kind of tree it was (*ton k'uk'*, perhaps). Baer and Baer (1952) say that the bird bolt and the foreshaft was made from the wood of the *subin. Subin* may be *subinte'*, which corresponds to *Acacia collinsii* (bull horn acacia). According to Nigh (2008: 238) *majas'akuche* corresponds to *Trichilia breviflora. Majas'akuche* is a name that I have not encountered in Naha'.*]*

S—s

siit blowgun. Blowguns were used throughout the Maya territories for killing small birds. They were made of wood or reeds. Information on Lacandon blowguns is scant, as they were replaced by the bow and arrow centuries ago. *[Source:* Bruce (1975).*]*

T—t

tok' flint arrowhead. When knapping the flint with a new hammer, the Lacandones invoke birds with straight, sturdy wing bones and hard skulls, such as eagles and woodpeckers, and specific plants with leaves that have straight, durable midribs. Infused with power to chip flint properly, the hammer will continue to perform from then on. *Loc:* **Boca Lacantun**; *Loc:* **Arroyo Santo Domingo** 'stream or river bed of Santo Domingo'. *[Note:* tok'. flint, arrowhead for killing large game (Bruce 1975).*]*

Ts—ts

tsuy nock. The term refers to the notch in the butt end of the arrow, which serves to keep the arrow in place on the bowstring as it drawn back. *Mat:* **puuna' (äh)** (*Swientenia macrophylla*). *[Note:* sui. nock (Nations 1989).*]*

X—x

xaakil chukkäy [xakil chuhkäy, xakil chukäy] fish trap. *Lit:* 'basket to catch fish'. *Mat:* **'ilon** (*Monstera* spp.). *Morph:* **xaak-il chuk-käy**. *See:* **xaak**.

Y—y

yolil shaft (of arrow). *Mat:* **'ooh** (*Gynerium sagittatum*); *Mat:* **hach häläl** (*Phragmites australis*). *Morph:* **y-ol-il**.

6.2.7 Musical Instruments

Ch—ch

chul flute. *Mat:* **'ämäy, chehew** 'undetermined'. *[Source:* AM; Bruce (1975)*]*

H—h

huub$_2$ horn, trumpet. If people were lost in the forest, they would roll up a large leaf into a horn and blow it to let their companions know where to find them. *Mat:* **ule' ähch'up** 'leaves of large orchids or bromeliads'; *Part:* **le'** 'leaf'; *Mat:* **huub** 'conch shell'. *[Source:* BM; Bruce (1975)*]*

K'—k'

k'ayum$_2$ drum. Named after the 'Lord of Music', *k'ayum* is a ceremonial drum made of clay and hide. It is played only in ritual contexts in the god house. *Mat:* **k'ät** 'clay'.

P—p

pax guitar. This is a flat, thin rectangular board with cross-pieces at either end over which five strings of henequen are strung (Tozzer 1907: 75). It looks like a zither. According to Soustelle (1966), this instrument was likely introduced from Africa. *[Note:* This instrument was unavailable to inspect, and the wood used is undetermined. Like many artefacts, it may have been made with *Manilkara achras*. pechakche'. (Tozzer 1907); pax (Soustelle 1966); pax [Yuc.]. play an instrument, celebrate (Bricker et al. 1998); pax [Itz.]. play marimba, play a musical instrument, music (Hofling and Tesucún 1997).*]*

S—s

so'ot rattle. This is a ceremonial rattle made from a gourd on a long handle wrapped with bark cloth and painted with annatto. According to AM, it is only played in accompaniment to *uk'ayil pom* 'Song of the Copal'. *Mat:* **chuh** (*Lagenaria siceraria*); *Mat:* **halol** (*Heliocarpus* spp.); *Mat:* **k'uxub** (*Bixa orellana*).

6.2.8 Religious Objects

B—b

bol$_1$ balche' urn. *Mat:* **k'ät** 'clay'. *Variant:* **pak, chan lak'**.

Ch—ch

chaachibil balche' strainer, cloth. This is a red, cotton bandana used to strain *balche'*. In the past, it was made from red bark cloth (Davis 1978). *Mat:* **nook'** 'cloth'.

chan lak' balche' urn. *Lit:* 'little wife'. Another name for pak. It is used in the drinking song, which is sung during a drinking ritual, which is performed to determine if a particular individual has lied. *Mat:* **k'ät** 'clay'. *See:* **bol**.

chäk hu'un ti' pak bark headband. *Lit:* 'bark cloth headband for pak'. This is a red bark cloth headband that is tied around the rim of the *balche'* urn. *Mat:* **ya'ax hänan** (*Chamaedorea elatior*); *Mat:* **hach hänan** (*Desmoncus orthocanthus*).

chemil balche' canoe for balche'. *Mat:* **hach taw** (*Belotia mexicana*); *Part:* **chunche'** 'trunk'; *Mat:* **pahok (ah)** (*Geonoma oxycarpa*); *Use:* **pix** 'cover'; *Part:* **le'** 'leaves'. (Fig. 4.60)

Ch'—ch'

ch'aak äkyum shelf, for god pots. *Lit:* 'bed of our lords'. This is a raised shelf located along west side of the god house used to store the god pots. *See:* **ch'aak**. *[Source:* Baer and Baer (1952)*]*

H—h

hama' cup. This is a large drinking cup incised with crude images of the sun, waves, and stick figures. It is used for ceremonial purposes only. *Loc:* **yatoch k'uh** 'god house'; *Use:* **ti' uk'ik ch'ulha', säkha'** 'for drinking ch'ulha' and säkha''; *Mat:* **hach luuch** (*Crescentia cujete*); *Mat:* **chuh** (*Lagenaria siceraria*); *Sim:* **k'ä' luuch** 'hand cup'. (Fig. 4.49)

K'—k'

k'anche'il pom shelf, for incense. *Lit:* 'bench for incense'. This is a shelf located in the god house used to store incense and other articles. *Morph:* **k'an=che'-il**. *[Source:* Baer and Baer (1952)*]*

k'ik'$_2$ rubber. *Lit:* 'blood'. *Use:* **tulis k'ik' (ah)** 'ceremonial figurines'; *Mat:* **k'ik'** 'Panama rubber tree (*Castilla elastica*)'; *Mat:* **hut'kih (ah)** (*Ipomoea* spp., *Ipomoea alba*); *Prep:* **Ahäts'ik uyak'il ahhut'kih, ayäts'ik uyits, pachil ahuytik yetel k'ik' yetel ha'** 'You pound the vines of the *hut'kih*, squeeze out the resin, and then stir it with the latex and water'.

L—l

lekil k'uh [lekil k'uh, lekih k'uh, läkih k'uh] ceremonial incensories. *Lit:* 'god's plate'. This is a large, clay bowl used for offerings to the gods. A stone or jade relic collected from the Mayan ruins at Yaxchilan is placed inside. It serves as a bench for the participating god to sit on. These stones are referred to as *k'anche' k'uh* 'god bench'. One side of the exterior of the bowl is decorated with an anthropomorphic face representing the god and vertical scratches around the rim to represent the god's headband, or *p'ookil k'uh*. The pots are white-washed with *k'ulta'an* 'quick lime' and painted with simple geometric designs in red and black pigments made from soot and *chäk lu'um* 'hematite'. Stripes indicate a male god pot and cross-hatching, a female god pot. These patterns may symbolize the woven designs on the ceremonial robe worn by men and the female skirt, respectively. *Mat:* **säk k'ät** 'white clay'; *Sim:* **lekil wah** 'tortilla plate'; *Sim:* **suulil lek**; *Sim:* **häyäk lek** 'plate made of metal or porcelain'; *Gen:* **lek** 'plate'. *Morph:* **lek-il k'uh**. *See:* **lek**. *See main entry:* **lek**. *Variant:* **plate**. (Fig. 4.63)

P—p

pak$_2$ balche' urn. *Mat:* **k'ät** 'clay'. *Variant:* **bol**. (Fig. 4.65)

päk'änil pom (äh)$_2$ incense; knobs of mashed pine resin or copal used to feed the gods. *Mat:* **tähte'** (*Pinus* spp.); *Mat:* **pom**; *Mat:* **hach pom** 'authentic pom'. *See main entry:* **pom**. *[Source:* AM*]*

pom incense. *Pom* is a gooey resin made from a number of species of *Bursera* and *Pinus*. Called *sil* 'gift' in Lacandon, lumps of the mashed resin are offered as food for the gods. *Pom* was and remains the central offering in Mayan ceremonies, a tradition handed down from the ancient Olmecs and from whom the word was borrowed. There are numerous sources of pom that are used by the Mayas in other regions. These include: *Bursera bipinnata; B. diversifolia; B. excelsa; B. tomentosa; B.jorullensis*; and, *B. penicillata* (Stross 1997). The Lacandones of Naha' extract resin almost exclusively from species of *Pinus*, but also from *hach pom* 'authentic pom' (unidentified), *Bursera simaruba, Protium copal*, and *P. schippii* (Durán 1999; Comisión Nacional de Áreas Naturales Protegidas 2006; Levy et al. 2006). *Loc:* **k'aax** 'forest'; *Use:* **päk'anil pom (äh)** 'nodules of incense'; *Mat:* **tähte'** 'pine resin (*Pinus* spp.)'; *Mat:* **hach pom** 'authentic pom'; *Mat:* **tsatsel pom** (*Protium copal*). *[Note:* pom. copal. *Protium copal* (Nations and Nigh 1980); pòom [Yuc.]. incense tree, (x) pomol-che'. *Jatropha gaumeri* (Bricker et al. 1998); pom [Itz.]. *Protium copal* (Atran et al. 2004). The chant to awaken the offerings of pom is provided in Tozzer (1907: 178–179).*]*

P'—p'

p'ätälche' board; hand-hewn board used to support the god pots during ceremonies. *Lit:* 'leave wood'. *Mat:* **puuna' (äh)** (*Swietenia macrophylla*). *Morph:* **p'ät-äl=che'**. *[Note:* pätähche' (McGee 2002).*]*

S—s

sil$_1$ gift. *Mat:* **pom** 'incense'; *Mat:* **k'ik'** 'rubber'. *[Note:* sil. gift, the name of the offering and the new braseritos (Tozzer 1907); síih, síil [Yuc.]. donate, make a gift (Bricker et al. 1998); siil [Itz.]. gift (Hofling and Tesucún 1997)*]*

sil$_2$ brazier (small); a container for fire. These are made along with the new god pots during the changing of the god pots ceremony. *Mat:* **k'ät** 'clay'.

T—t

tulis k'ik' (ah) figurines, ceremonial. *Lit:* 'whole rubber'. These are small, 5.08×3 cm (2×1.18″) rubber torsos of a man (*koho'*) and woman (*chel*) that are given to the gods to perform servitude. They are burned as offerings in certain ceremonies, such as *witsbil* and the changing of the god pots ceremony (k'ämkih?). When they are burned their souls rise up to the gods where they take up residence to be their servants. *Mat:* **k'ik'** 'rubber'. (Fig. 4.70) *[Note:* tulis k'ik'. rubber figurine (ceremonial) used as incense (Bruce 1975); tulis [Itz.] whole (Hofling and Tesucún 1997); tulis [Yuc.]. full, round, whole, complete (Bricker et al. 1998). In my research and that of Bruce (1975), k'ämkih is the name of an offering, whereas Davis (1978) says it is the name of the ceremony. For details on the structures of Lacandon ceremonies, see Davis (1978: 252–302).*] [Source:* AM*]*

X—x

xikal [xikal, xika] incense board. Refers to the sacred wooden board that holds *pom* 'incense'. *Mat:* **puuna' (äh)** (*Swietenia macrophylla*). (Fig. 4.67)

6.2.9 Tools

Ch—ch

chaachib [chaachib, chaachi'] colander. *Spec:* **hach chaachib** 'authentic colander'; *Spec:* **k'ik' chaachib** 'plastic colander'.

che'il häxbil k'ak' firedrill sticks. *Mat:* **k'uxub** (*Bixa orellana*); *Mat:* **ak' tsup** 'undetermined'; *Mat:* **häxal k'ak'** (*Hamelia calycosa*); *Mat:* **chulul** 'bow'. *See:* **häxal k'ak'**.

H—h

hach chaachib colander. *Lit:* 'authentic colander'. *Use:* **ti' chaachtik uyo'och, sa'** 'for straining food, like *atole* (a type of corn gruel)'; *Mat:* **hach luuch** (*Crescentia cujete*); *Part:* **pach** 'rind'; *Sim:* **k'ik' chaachib** 'plastic colander'; *Gen:*

chaachib. (Fig. 6.1) *[Note:* cha'achi'. sieve (Bruce 1975): chachib. strainer (Baer and Merrifield 1971); chaach [Itz.]. sifting (Hofling and Tesucún 1997); xcháachab'. sieve, colander (Bricker et al. 1998).*] [Source:* KP*]*

häxal k'ak'$_2$ firedrill. *Mat:* **häxal k'ak'** (*Hamelia calycosa*); *Mat:* **k'uxub** (*Bixa orellana*); *Part:* **uche'il** 'wood'. *See:* **che'il häxbil k'ak'**. *Variant:* **tikinche', che'il häxbil k'ak'**. (Fig. 6.2) *[Note:* tikinche'. taladro (firedrill) (Bruce 1974).*]* *[Source:* AM; BM*]*

huux whetstone, file. *Loc:* **Arroyo Santo Domingo** 'Santo Domingo stream (near Damasco on the Pénjamo-Bonampak road)'. *[Source:* Bruce (1975); Nations (1989)*]*

huyub [huyub, huyu'] ladle, large spoon. *Mat:* **k'uche' (äh)** (*Cedrela odorata*). *Morph:* **hu(u)y-ub**. *[Note:* huyu'. ladle, large spoon (Bruce 1975): huyub [S. Lac.]. spoon (Canger 1970); xhúuyub' [Yuc.]. instr. stirrer (Bricker et al. 1998); huyu' [Itz.]. batidor de madera, molinillo. stirring stick (Hofling and Tesucún 1997)*]*

Fig. 6.1 Hach chaachib 'authentic colander'. Photo Credit: Suzanne Cook

K—k

ka' millstone; traditional stone millstone, or *metate. Mat:* **tunich** 'stone'. (Fig. 4.72)

K'—k'

k'ek' che'$_2$ beater (cocoa). *Lit:* 'wood barbs'. *Mat:* **chäk ya'** (*Manilkara achras*); *Mat:* **mahaas (ah)** (*Quararibea funebris*). *Variant:* **mulinuh, mahs**. *[Note:* Compare k'ek' [Itz.]. barb. (Hofling and Tesucún 1997: 389)*]*

k'ik' chaachib colander. *Mat:* **k'ik'** 'plastic'; *Use:* **ti' chaachtik uyo'och** 'for straining foods'; *Sim:* **hach chaachib** 'authentic colander'; *Gen:* **chaachib**. *[Source:* KP*]*

M—m

mahs beater, cocoa. This is named after the wood used to make the object. *Mat:* **mahaas (ah)** (*Quararibea funebris*). *Variant:* **mulinuh, k'ek' che'**. *[Source:* Baer and Merrifield (1971)*]*

maska' machete (metal, iron). *[Note:* maska. machete, knife, iron (Bruce 1975); maskab' [Itz.]. machete (Hofling and Tesucún 1997); máaskab' [Yuc.]. machete (Bricker et al. 1998).*]*

mulinuh [mulinuh, mulinu, murinu(h)] beater (for cocoa), mill. *Mat:* **chäk ya'** (*Manilkara achras*); *Mat:* **mahaas (ah)**. *Variant:* **k'ek' che'**. *From:* molino, molinillo (Spanish). *[Source:* JK*]*

Fig. 6.2 Häxal k'ak' 'firedrill'. Photo Credit: Suzanne Cook

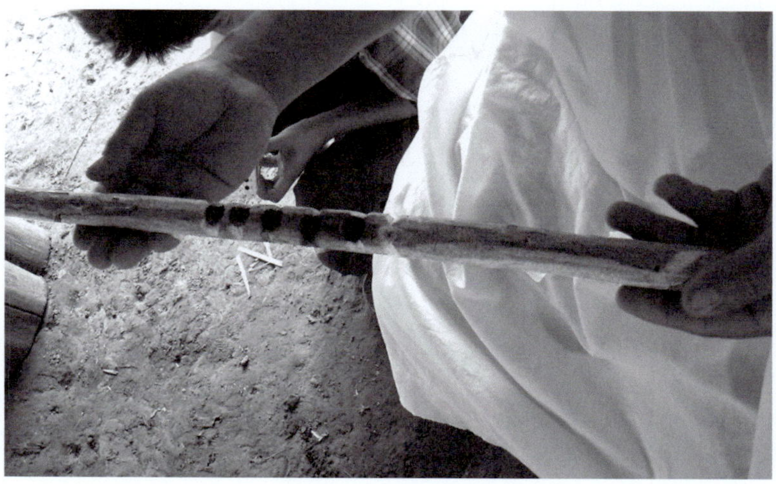

P—p

puuts' needle, awl.

X—x

xämäch griddle. This is a shallow, metal or earthenware pan used primarily for baking tortillas. *Mat:* **k'ät** 'clay'; *Mat:* **maska'** 'metal'. (Fig. 4.77) *[Note:* xämäch/xamach. griddle (Bruce 1975); xämäch [Itz.].comal. comal griddle (Hofling and Tesucún 1997); shamach, shàamach [Yuc.]. griddle (Bricker et al. 1998).*]*

xho'opil lu'um hoe. *Lit:* 'soil scraper'. *Morph:* **x-ho'op-il lu'um**. *[Source:* Bruce (1975)*]*

6.3 Clothing

Ch—ch

chäk hu'un bark cloth headbands. *Lit:* 'red bark cloth'. These ceremonial headbands are worn by participants and offered to the gods during certain rites, such as the changing of the god pots. They are painted with geometrical designs in annatto dye. *Mat:* **hach hu'un** '*Ficus* sp.'; *Mat:* **ak' hu'un** (*Poulsenia armata*); *Part:* **uyootel**. *Variant:* **p'okil hu'un**.

H—h

hach nook' bark cloth tunic. *Lit:* 'authentic clothes'. The original clothes were made of bark cloth. Although not very comfortable, the bark cloth tunic was worn during ceremonies. Children were also dressed in bark cloth or went naked. The preferred material was hand loomed cotton. *Mat:* **ak' hu'un** (*Poulsenia armata*); *Mat:* **bits'kal hu'un** '*Ficus* sp.'; *Mat:* **hach hu'un** (Ficus sp.). (Fig. 4.107)

K'—k'

k'äxnäk loincloth. *Lit:* 'belly tie'. This is a long, narrow length of material that is wound several times around the waist with ends hanging down in the front and at the back. They are no longer worn. *Mat:* **tämän** (*Gossypium hirsutum*). *[Source:* Tozzer (1907: 29)*]*

k'uk'man feather headdress worn by married only. It is made from the pelts of several kinds of birds, and tied onto the end of the woman's braid. *See:* **xik'**. (Fig. 4.111a) *[Source:* BM; KP; SK*]*

N—n

nook' [nook', nok'] cloth, clothes. *Mat:* **tämän** (*Gossypium hirsutum*); *Mat:* **bits'kal hu'un** '*Ficus* sp.'; *Mat:* **ak' hu'un** (*Poulsenia armata*); *Mat:* **hach hu'un** (*Ficus* sp.)

P—p

pänyo bandana. An article that appears to have fallen out of vogue, only elderly men wear this patterned kerchief around their necks. It is also used to strain *balche'* into the *balche'* urn. In the old days, bark cloth was used for this purpose. *Use:* **nook' ti' kaal** 'neck cloth'; *Use:* **ti' uchaachtik balche'** 'to strain the balche''; *Mat:* **tämän** (*Gossypium hirsutum*); *Mat:* **hu'un** 'bark cloth'. *Variant:* **chaachibil balche', unok'il balche'**. *From:* pañuelo (Spanish). *[Note:* chachib. gourd strainer (Baer and Merrifield 1971); chacheh. gourd sieve (Davis 1978); cháach (2) X; x cháachab', inst.[Yuc.]. sieve, colander (Bricker et al. 1998: 58)*]*

pik skirt. *Mat:* **tämän** 'cotton'. *Thes:* **che'**. (Fig 4.110)

pits' spin, cloth. Lacandones use the drop spindle, or supported spindle, to spin cotton fibres. The bottom tip of the spindle rests on the ground or in a bowl. The technique involves spinning the shaft between the fingers and then letting the shaft spin in the crook of the thumb and forefinger. The *utukil* 'whorl' at the bottom of the shaft maintains the momentum of the spinning shaft while the spinner feeds cotton onto the top of the shaft. She twists and drafts the cotton, pushing the spun fibres down the shaft as she goes. *See:* **k'uch$_2$**.

P'—p'

p'äyats k'uh headdress. Refers to the incised lines on the rims of god pots and anthropomorphic rubber figurines. *See:* **lekil k'uh**; **tulis k'ik' (ah)**. *[Source:* Bruce (1975)*]*

X—x

xikul tunic, poncho. This is a simple, white cotton tunic made from two rectangular pieces of cotton sown together up each side, with a v-neck gap cut out for the head and two gaps on either side for the arms. Southern Lacandon men and women wear ankle-length tunics, whereas northern Lacandon men wear a shorter, knee length version. Northern Lacandon women wear a blouse similar to the *xikul* over knee-length skirts. Although now made from commercially manufactured cotton, the *xikul* and traditional skirts were made of hand loomed cotton. Another version of the male *xikul* was sown up the front and back, rather than at the sides. Three stripes of blue and red thread were woven into the seams. These tunics were reserved for ceremonial use. Tozzer notes that the *xikul* was also made of agave fibre (1907: 29). *Mat:* **säkälbil nook'** 'woven cloth'. *See:* **hach nook'**. (Fig. 4.106)

6.4 Deities

Ä—ä

Äkyanto' Lord of Commerce, Travelers, and (western) Civilization. He is the brother of *Hachäkyum* and *Sukunkyum*. Like his brothers, he emerged from the tuberose and belongs to the *ma'ax* 'spider money' *onen. Äkyanto'* is the creator of western civilization and foreigners, commerce and money. He lives on "the other side of the sea", in the West. *Variant:* **kynatho', yantho', ho'**. *[Source:* Bruce 1968 (125–126)*]*

Äk'inchob [ak'inchob, ak'inchop'] Lord of Maize. The son-in-law of *Hachäkyum, Äk'inchob* is the most venerated of the gods who was not born of the tuberose. He is the one whom the Lacandones petition most often to carry their messages to *Hachäkyum*. He is also the protector of the *milpa*. *Variant:* **chob, ki'chäkchob**. *[Source:* Bruce (1968: 132).*]*

B—b

Bol$_2$ Lord of Wine. He is one of several deities who emerged from the tuberose. His face is depicted on the *balche'* urn, which is dedicated to him.

H—h

Hachäkyum Lord God. *Lit:* 'our true lord'. One of the three principal deities, he was the third to emerge from the tuberose, after his brothers, *Sukunkyum* and *Äkyanto'*. He is the Lacandones' principal deity: he gave form to the earth, creating the forests, rivers, and streams, and populated the land with the creatures of the earth and human beings. His *onen* is the *ma'ax* 'spider monkey'. *Morph:* **hach=ähk-yum**. *Variant:* **yumbilil ka'an, yolil ka'an, äkchanyum, äkyum**.

ha'nahk'uh [ha'nahk'uh, ha'nahk'uho', haha'nahk'uh] rain spirit(s). *Lit:* 'water house god(s)'. This is a class of servants who work under the direction of *Mensäbäk*, the Lord of Rain. They carry a fan of macaw feathers, which they use to sprinkle the soot on the clouds. Each one takes the name of the cardinal direction that they represent: *xämän* 'North (spirit)', *tseltsel xämän* 'Northeast (spirit)', *bulha' kilotalk'in* 'Flood of the East (spirit)', *tseltsel nohol* 'Southeast (spirit)', *nohol* 'South (spirit)', and *ch'ikink'uh* 'Lord of the West'. *Morph:* **ha'=nah=k'uh**. *Variant:* **ha'ha'nahk'uh, ha'nahwinik**. *[Source:* Bruce (1968: 127)*]*

I—i

Itsanal first assistant to the gods. He is the first assistant to the principal deities (Bruce 1975: 164) and corresponds to *Itzamná*, the main god in the mythology of Yucatecans during the Late Classic period (Bruce 1968: 130, 131; Schlesinger 2001: 69).

Itsanohk'uh Lord of Hail. *Lit:* 'Itzaj lord'. He is the sixth of the major gods who emerged from the tuberose. He belongs to the *k'ek'en* 'wild boar' *onen*. He is also the Lord of Cold, Lakes, and Alligators (Bruce 1975: 164). *Loc:* **Mensäbäk**. *[Source:* Bruce (1968, 1975)*]*

K—k

Kisin Devil. *Lit:* 'causer of death'. Known as the devil, Messenger of Death, and the Trembling, *Kisin* was created by *Hachäkyum* from the *'aak'älyoom* 'night blooming cestrum' (*Cestrum nocturnum*). He obeys the orders of *Sukunkyum*, Lord of the Underworld, who passes on the souls of degenerate people to him. Proud of his role as destructor of the world, he causes earthquakes by kicking the pillars holding up the earth. It is preordained that in the next and final destruction of the world, he will succeed in knocking the pillars down, collapsing the earth and releasing the jaguars from their caverns to devour all living beings. *Morph:* **kim-s-in**. *[Source:* Bruce (1968)*]*

K'—k'

K'akoch [K'akoch, Kak'och] Creator (the). *K'akoch* is the first god who created the world, the oceans, the first sun, and the tuberose from which the principal deities emerged. He is the god of the deities. Regarded by the Lacandones as a remote god, he plays an insignificant role in the religion.

K'ayum$_1$ [k'ayum, k'ayom] Lord of Music. *Lit:* 'song lord'. He is one of several deities who emerged from the tuberose following the birth of the principal deities. His face is depicted on the sacred drum, which is dedicated to him. *Morph:* **k'ay=yum**.

M—m

Mensäbäk Powder Maker, Lord of Rain. *Lit:* 'make soot'. He also created the Tzeltales. His *onen* is *k'ek'en* 'white-lipped peccary'. *Loc:* **Mensäbäk**. *Morph:* **men=säbäk**. *Variant:* **yum kanansäbäk**. *[Source:* Bruce (1975)*]*

S—s

Sukunkyum Eldest Lord. He is the eldest brother of the three principal deities, and was the first to emerge from the tuberose. He is the Lord of the Underworld, where he judges the souls of the dead and decides their fate. He also carries the sun through the Underworld at night and places it in the eastern horizon. He belongs to the *ma'ax* 'spider monkey' *onen*.

Ts'—ts'

Ts'ibatnah [Ts'ibanah] Painter of Houses. This deity features in Lacandon mytho-history as the one who was responsible for painting the gods' houses with the blood of people at the end of first and second world cycles. He is the sixth of the major gods who emerged from the tuberose. His *onen* is *k'ek'en* 'white-lipped peccary'. *Variant:* **nohk'uh, channah**. *[Source:* Bruce (1968)*]*

6.5 Food

A—a

axontik näl shell corn. *[Note:* ixo'om, ixo'omtik [Itz.]. shell corn (Hofling and Tesucún 1997).*]* *[Source:* Bruce (1975)*]*

B—b

balche'$_2$ fermented beverage. *Mat:* **hach balche'** (*Lonchocarpus* sp.); *Part:* **uyoot'el** 'inner bark'.

bu'ulil wah tamale, bean. Cooked beans are mashed and spread between two layers of corn dough. This is wrapped up in a banana or plantain leaf and then steamed. *Mat:* **bu'ul** 'beans'; *Mat:* **säkan** 'corn dough'; *Mat:* **le' pätan** 'plantain leaf'. *[Note:* buliwa. a flat tamale of corn and beans commonly offered to the gods (Tozzer 1907).*]*

Ch—ch

chäkbil wah tamale, meat. *Lit:* 'cooked tortilla'. This is made of cooked or raw meats stuffed between layers of corn dough, wrapped in a banana or plantain leaf, and then steamed. *Mat:* **bäk'** 'meat'; *Mat:* **säkan** 'corn dough'; *Mat:* **le' pätan** 'plantain leaf'. *[Source:* Bruce (1975)*]*

che'che' raw.

Ch'—ch'

ch'ooch' [ch'ooch', ch'och'] salt, commercial. *See:* **ta'anil kun**.

ch'ulha' gruel, ceremonial (a type of). A food offering similar to *ma'ats'* but made without quicklime and strained. *Mat:* **näl** 'maize'; *Sim:* **säkha'** 'white water, type of gruel'. *[Note:* ch'ula/ch'ulha'. ceremonial drink of corn (Bruce 1975).*]* *[Source:* Baer and Baer (1952)*]*

H—h

hach ch'uhuk sugar, unrefined. *Lit:* 'authentic sugar'. Sugarcane was squeezed in a press and primarily used to sweeten balche'. When processed sugar became available, this practice stopped. First introduced was brown sugar in the form of cones, called *piloncillo*, which are still sold in Mexican markets. Eventually white, refined sugar became available and has been used ever since. Honey was the original Mayan sweetener, gathered from wild honeybee hives *Use:* **ti' mentik balche', ti' tähkintik** 'for making (fermenting) balche' 'mead''; *Prep:* **mänbil** 'purchased'; *Mat:* **sukal** (*Saccharum officinarium*); *Sim:* **säk ch'uhuk** 'white sugar'. *[Note:* ch'uuk [Itz.]. sweet (Hofling and Tesucún 1997); ch'uhuk [Yuc.]. adj. sweet, n. candy (Bricker et al. 1998).*]* *[Source:* AM; Baer and Baer (1952)*]*

hach wah tamale; a 10–15 cm (4–6″) round, thick tamale, usually stuffed with beans. *Use:* **hanal** 'food'; *Mat:* **säkan, bu'ul** 'corn dough and beans'; *Prep:* **Umentik yetel bu'ul yetel säkan, pätbil yok'ol chukuch xib wa le' pätan wo säk wawal** 'One makes it with beans and corn dough, folded in a long leaf, such as a banana leaf or the leaf of the *Calathea lutea*'; *Gen:* **wah** 'tortilla'. *[Source:* KP*]*

hanal [hanal, hanan, hana] food. *Morph:* **han-al**. *See:* **o'och**. *[Note:* han(–an). eat, food (Bruce 1975); hanal [Yuc.]. food (Bricker et al. 1998); janal [Itz.]. food (Hofling and Tesucún 1997).*]*

huybil stirred. A method of cooking, e.g., *huybil ma'ats'* 'type of corn gruel', *huybil uye'el kax* 'scrambled eggs'. *Morph:* **huy-bil**. *See:* **huyub**.

huybil wah stirred tortilla. *Prep:* **Uhuytik wah ye'el kax yok'ol xämäch wa yok'ol ule' pätan, pachil utäpik yetel**

ule' mäkäl wa pätan, pachil umukik i uhayik yok'ol k'ak' 'One scrambles the tortilla with eggs on the griddle or a banana leaf, then covers it with an elephant ears leaf or a plantain leaf, then wraps it up, and then spreads it out on the fire'; *Gen:* **wah** 'tortilla'. *[Source:* KP*]*

I—i

'is wah tortilla (a type of). A sweet tortilla made from the dough of green maize. The dough is prepared without quicklime. *Prep:* **Amentik yetel kuxul näl, ahuch'ik** 'You make it with fresh, young ground maize'; *Gen:* **wah**. *[Source:* KM; KP*]*

K—k

kab [kab, kap'] honey, nectar. In the old days, honey was collected from the hives of wild bees. The hives would be cracked open, the honey scooped out by hand (bees and all), placed in a pot, and then boiled down. Honey was used for fermenting balche' and as a candy coating for roasted squash seeds. Before the introduction of sugarcane, honey was the only sweetener. Some Lacandones also raised wild honeybees in gourd containers hung up under the eaves of houses. *Use:* **hanal** 'food'; *Prep:* **ch'uhuk yetel sikil, yetel chäkbil k'um** 'candied squash seeds, with boiled squash'; *Use:* **tähkintik balche'** 'to ferment balche''. *[Note:* kab. honey, honeybee (Bruce 1975); kab' [Itz.]. honey, syrup, hive (Hofling and Tesucún 1997); kàab' [Yuc.]. honey, beehive (Bricker et al. 1998).*]*

K'—k'

k'ayem [k'ayem, k'ayen, k'eyem, k'eyen] gruel; a type of a cooked gruel made from *nixtamalized* corn. *Use:* **hanal** 'food'; *Mat:* **näl** 'maize'; *Prep:* **Achäkik näl yetel k'ulta'an, pachil achalha'tik axikik ukoh, pachil ahuch'ik, pachil apuk'ik säkan ich ha', pachil awuk'ik** 'You boil the maize with quicklime, then you rinse it and rub off skins off the kernels, then you grind it, then you dissolve it in water, and then you drink it.'. *[Note:* k'ayem. traditional corn drink of ground hominy (Bruce 1975); k'eyem [Yuc.]. posol (Bricker et al. 1998); k'eyem [Itz.]. atole (Hofling and Tesucún 1997).*]* *[Source:* KM*]*

k'äh gruel; a type of gruel made with ground, toasted maize and water. *Mat:* **näl** 'maize'; *Prep:* **Uk'elik yok'ol k'ak', uhuch'ik yetel ka', pachil uhuyik ich chäkäl ha'** 'One toasts it on the fire, grinds it with the *metate*, and then stirs it into hot water.' *[Source:* KM*]*

k'ool porridge, thick soup; made with corn dough or broken pieces of tortilla and augmented with pieces of meat, fungi, or vegetables. *Spec:* **k'oolbil kax wa 'ulum** 'thickened soup with chicken or turkey'; *Spec:* **k'oolbil k'ayoch'** 'thickened soup with a kind of fungus (*Pleurotus djamor*)'; *Spec:* **k'oolbil mäkäl** 'thickened soup made with *Xanthosoma mafaffa*'; *Spec:* **k'oolbil wah** 'thickened soup make with tortillas or corn dough'. *[Note:* k'óol [Yuc.]. thick sauce or gravy (Bricker et al. 1998: 156); k'ool [Itz.]. thickened soup (Hofling and Tesucún 1997: 398)*]* *[Source:* KP*]*

k'ulta'an quicklime. *Lit:* 'mashed ash?'; made from the burnt and crushed shells of *äht'unu'n* (river snails) or limestone. Nations (1979) reports that the limestone in the area is unsuitable for quicklime, while Tozzer (1907) claims that the limestone "pure enough for use as alkali" was unavailable in Chiapas (Tozzer 1907: 51); yet, one of my Lacandon consultants said that the Lacandones purchase quicklime from the Tzeltales, who retrieve it from dried riverbeds near the community of Lacandón, half an hour's drive from Naha'. Today, it is rare to find anyone who processes their own quicklime from snail shells. *Morph:* **k'ul-ta'an**.*]*

k'u'um [k'u'um, k'uum, k'um] *nixtamalized* corn (hulled corn), hominy. *Nixtamalization*, from the Nahautl word *nixtamalli*, entails soaking dry maize in quicklime to soften the kernels and loosen the hulls. The process also causes the grain to puff up to about twice its normal size, improves the flavour and aroma, and boosts the nutritional value of the grain. *K'u'um* is ground into dough and then used to make a variety of gruels and tortillas called *k'u'umbil wah*. The dough may be soured over night, which enhances the flavour. *Prep:* **Achäkik näl yetel uta'an, pachil achalha'tik axikik usool** 'You boil the corn in quicklime and then rinse it and remove the hulls.' *[Note:* k'ú'um [Yuc.]. hominy (Bricker et al. 1998: 159); k'u'um [Itz.]. nixtamal. leached corn (Hofling and Tesucún 1997: 401)*]* *[Source:* KM*]*

k'u'umbil wah [k'umbil wah, k'umil wah] tortilla. This is a large 25 cm (10″) tortilla made with *k'u'um*. *Mat:* **chich näl** 'hard maize'; *Prep:* **Achäkik näl yetel k'ulta'an, pachil axikik usoolil koh ich ha', pachil ahuch'ik, pachil apäk'ächtik** 'You boil the maize with quicklime, then you remove the corn hulls in water, then you grind it, and then you pat it out into tortillas.'; *Gen:* **wah**. *Morph:* **k'u'um-bil wah**. *See:* **k'u'um**.

k'uxubil wah tamale (of annatto). *Lit:* 'bread of annatto'. This is a ceremonial food offering. *Mat:* **k'uxub, säkan** 'annatto, corn dough'; *Gen:* **wah**. *Morph:* **k'uxub-bil wah**.

M—m

ma'ats' corn gruel (a type of). This is the basic corn gruel, prepared without boiling the maize in quicklime. Dough is pinched off and then mixed into a cup of water. *Mat:* **näl**

'maize'. *See:*. *[Note:* ma'ats'. drink of ground hominy mixed with water (Bruce 1975); maats. posol (Tozzer 1907); má'ats' [Yuc.]. hull (corn) (Bricker et al. 1998); maatz' [Itz.]. rough (of cooked corn) (Hofling and Tesucún 1997).*] [Source:* Tozzer (1907: 52).*]*

mukbil wah tortilla, buried. *Lit:* 'buried tortilla'. This is a large corn dough 'tamale, enclosed in large leaf and baked in coals. *Gen:* **wah**.

N—n

nahwah tamale, ceremonial. *Lit:* 'great tortilla'. This is a small, 10 cm (4″) ceremonial tamale that is filled with beans. Stacks of five are offered to the gods. *Prep:* **Amentik yetel wah yetel bu'ul; apätik yetel xiiw** 'You make it with tortilla and beans; you form it with a leaf.'; *Gen:* **wah**. *[Source:* KP*]*

O—o

o'och$_2$ meal, food. *See:* **hanal**.

'op' cracker, toast, tostada. *Mat:* **näl** 'maize'. *Variant:* **säk-pet**. *[Note:* 'òop' [Yuc.]. N. toast (Bricker et al. 1998); 'op' [Itz.]. toasted (Hofling and Tesucún 1997).*]*

P—p

patsto' [patsto', paato'] tamal; made from young maize, that has been ground and mixed with sugar, then patted into fresh corn husks, and then boiled or steamed. *Mat:* **säk näl** 'white maize'; *Use:* **ti' tap'ik wah** 'for tamales'; *Prep:* **Ahuch'ik näl, pachil apulik säk ch'uhuk yetel säkan, pachil apätik wah** 'You grind the maize, then you add sugar to the dough, and then you form a tamale with it.' *[Source:* Bruce (1975: 206); KP*]*

S—s

sa' gruel (a type of) repared with ground fresh corn, water, and sugar. *Mat:* **kuxul näl** 'fresh maize'; *Use:* **hanal** 'food'; *Prep:* **Ahäyk'intik ukoh, pachil ahuch'ik, pachil kusaastal pähil, pachil apuk'ik yok'ol ha', achaachtik, ch'uhuk apulik, pachil achäkik** 'You spread out the kernels in the sun to dry, then you grind them and let the dough sour until morning, then dissolve it in water, strain it, put sugar in, and then boil it.' *See:* **'ulisa'**; **säkha'**. *[Source:* KM*]*

säkan [säkan, säkän] dough. Dough is used in a variety of ways, in tortillas, tamales, gruels, and as a thickening agent for meat and vegetable porridges. While corn is the primary ingredient, dough may also be fortified with manioc, plantains, and when maize stocks are low, the seeds of the Maya breadnut (*Brosimum alicastrum*). *Mat:* **näl** 'maize (*Zea mays*)'; *Prep:* **wah, mukbil wah, chäkbil wah, bu'ulil wah** 'tortilla, tamale, meat tamale, bean tamale'; *Prep:* **ma'ats', sa'** 'corn gruels'; *Prep:* **k'oolbil** 'porridge'; *Mat:* **hach 'oox, k'än 'oox** (*Brosimum alicastrum* ssp. *alicastrum*); *Mat:* **ts'in** 'manioc (*Manihot esculenta*)'; *Prep:* **wah** 'tortilla'. *[Note:* säkän. dough, corn (Bruce 1975); säkan [Itz.]. corn dough (Hofling and Tesucún 1997); sakam [Yuc.]. dough (Bricker et al. 1998).*] [Source:* KM; KP*]*

säkha' [säkha', säka', saka'] gruel, ceremonial. *Lit:* 'white water'; a ceremonial food offering similar to sa'. *Loc:* **yatoch k'uh** 'god house'; *Mat:* **näl** 'maize'; *Sim:* **ch'ulha'** 'ceremonial type of gruel'. *See:* **sa'**. *[Source:* Bruce (1975)*]*

säkpet tortilla (crisp); a thin, crisp tortilla, approximately 15.24 cm (6″) in diameter made from the "face" of tortillas that have failed to puff up. *Lit:* 'white disc'. *Mat:* **näl** 'maize'; *Prep:* **Ak'ak'tik** 'You toast it over the fire.'; *Gen:* **wah**. *Variant:* **'op'**. *[Note:* säk-pet [Itz.]. fine fried tortilla (Hofling and Tesucún 1997).*] [Source:* KM; KP*]*

säk ch'uhuk sugar, white refined. *Lit:* 'white sweet'. Sugarcane was introduced into Mesoamerica by the Spanish. Ever since then, the Lacandones have been cultivating several varieties (or natural hybrids) and using the sugar in its raw, juiced form in place of the traditional sweetener, wild honey. After the introduction of processed brown and later, refined sugar, the cultivation of sugarcane fell out of favour and is now grown by only a few households for snacking on (BM). *Mat:* **sukal** (*Saccharum officinarium*); *Sim:* **hach ch'uhuk** 'brown sugar (*piloncillo*, Mexican brown sugar)'. *From:* azúcar (Spanish). *[Source:* BM*]*

T—t

ta'anil kun ashes, palm;. *Lit:* 'ashes of kun'. Extracted from the fan palm (give-and-take palm), this is the original salt Lacandones used before the introduction of salt. AM says it was the only salt he had while growing up and that it was not very tasty. *Mat:* **kun (ah)** (*Cryosophila stauracantha*); *Use:* **ch'ooch'** 'salt'; *Part:* **chuch ule'** 'leaf stem'; *Prep:* **Ak'ak'tik uchuch ule', pachil malik uta'anil i but'ik yok'ol kum wa luuch** 'You roast the leaf stems in the fires, then collect the ashes and put them in a pot or gourd container (for later use).'

Ts'—ts'

ts'ahik ti' ho'ol first fruits offering. *Lit:* 'give it to the head'. Lacandones give the first fruits of their crops to the gods before consuming them themselves.

ts'ak medicine; spice.

U—u

'ulisa' gruel made from fresh corn and left to sour. *Mat:* **näl** 'maize'. *Morph:* **'ul-'is-sa'**. *See:* **sa'**. *[Note:* ulisa'. gruel of green corn (Bruce 1975); u'ul 'soured atole' (Baer and Baer 1952); 'is-'ùul [Yuc.]. gruel made from green corn with a bit of salt (Bricker et al. 1998).*] [Source:* Bruce (1975)*]*

W—w

wah tortilla, tamale. This word also follows many food preparations made with corn dough, e.g., *k'oolbil wah* 'porridge thickened with corn dough'. *Mat:* **näl** 'maize'; *Spec:* **nahwah** 'great tortilla'; *Spec:* **hach wah** 'authentic tortilla'; *Spec:* **'is wah** 'sweet tortilla'; *Spec:* **mukbil wah** 'baked tamale enclosed in a large leaf'; *Spec:* **k'u'umbil wah** 'hulled corn tortilla'; *Spec:* **hoybil wah** 'spread tamale'; *Sim:* **säkpet** 'crisp tortilla'. *[Note:* Fewer women, especially the younger generation, make these traditional tortillas, because they require considerable time and energy to produce. Instead, they use the instant corn flour called Maseca. It is more convenient and requires little effort—just add water and mix. McGee (2002) explains that the switch from traditional means of subsistence to the consumption of pre-packaged foods is linked to several factors, particularly tourism, the abandonment of the traditional religion, and birth control. Faced with surviving in a modern world, smaller families devote the bulk of their time to activities that generate a cash income. If instant food frees up time, it also comes at a price: not only is it expensive, but it also lacks the nutrition of the traditional corn tortillas and thus lead to diminished overall health.*]*

6.6 Geography

A—a

'aktun cave.

'ak'alche' ['ak'alche', 'ak'älche', 'ak'ache', 'ak'äche'] swamp. *Lit:* 'treed water hole'. *Morph:* **ak'al-che'**. *[Note:* ak'ä(l) che'. swamp (Bruce 1975); ?àk'acYe?, y-àk'acYe?-Ir [S. Lac.]. swamp (Canger 1970); 'áak'al [Yuc.]. tank, reservoir (Bricker et al. 1998); 'ak'al [Itz.]. small lake, water hole in the forest (Hofling and Tesucún 1997).*]*

Ä—ä

äka'an level land. **Muy plano, mas kabal.** 'Very flat, low land.' Typically humid and swampy, it is the perfect environment for *hach pahok* (*Geonoma* sp.), *k'unche' pahok* (*Geonoma oxycarpa*), *miis* (*Carludovica labela*), and *hach ch'alol* (*Quercus corrugata*). *Morph:* **äk-a'an 'pooled, puddled'**. *[Source:* BM*]*

äk'ache' swamp.

Ch—ch

chäk 'aktun chäk 'aktun. *Lit:* 'red cave'. This is the name of a cave below ancient stone platforms, structures, and remnants of walls around Mensäbäk that may have been used for important ceremonies or as a look-out post, or territorial markers (Palka et al. 2008: 815). *Loc:* **Mensäbäk**. *Variant:* **El Mirador**. *[Source:* Palka et. al. (2008)*]*

chäk k'ät clay, red. Soil with this type of clay is the second choice medium for cultivation. Ek' lu'um 'black soil' is the first. *Sim:* **säk k'ät** 'white clay'; *Sim:* **k'än k'ät** 'yellow clay'; *Gen:* **k'ät** 'clay'. *Variant:* **chäk lu'um**. *[Source:* BM*]*

chun wits base of the hill. Plants that grow well here include: *ek' bahche'* (*Guatteria anomala*); *chäk ya'* (*Manilkara achras*); *ya'axche'* (*Ceiba pentandra*); and, *wäch'* (*Dialium guianense*). *[Source:* BM*]*

Ch'—ch'

ch'ach'anäk puddle(s). Areas with standing water. According to BM, *xiiw* (Zingiberales) prefer this environment. *[Note:* äka'an [Itz.]. puddled; äk'äknak. puddling. (Hofling and Tesucún 1997: 876). *[Source:* BM*]*

E—e

ek' lu'um soil (black). This is a rich and loamy, soil that is preferred cultivation. A number of trees that help nourish the soil and create this black earth, including *ek' bahche'* (*Guatteria anomala*), *halol* (*Heliocarpus donell-smithii*), *taw* (*Belotia mexicana*), *ponche'* (*Bernoullia flammea*), and *bits'* (*Inga* spp.). Nations and Nigh (1980) list other trees that the Lacandones believe benefit the soil for cultivation. These include: breadnut ramon (*Trophis mexicana*) and the ceiba tree (*Ceiba pentandra*). *See:* **ek' bahche'**; **halol**; **taw**; **ponche'**; **bits'**. *[Source:* AM; BM*]*

H—h

habän 'aktun cave. *Lit:* 'hollow rock/boulder'.

hach häläl petha' hach häläl petha'. This is the name of the place where *hach häläl* (*Phragmites australis*) grows. *Loc:* **Yaha'nah** 'La Cueva'. *[Source:* BM*]*

heek'el uk'um river fork. *Lit:* 'branch of the river'. *[Source:* Bruce (1975)*]*

ho'ol wits hilltop. *Lit:* 'head (of the) hill'. The soil is hard, with patches of clay and sandy areas. Plants that grow well here include, *bo'oy*, *chäk pahok*, and other palms, *chäk ya'*, *ya'*, and *'uuch*. Among the plants that do not grow well here is *ek' bahche'*. *See:* **bo'oy (äh)**; **chäk pahok**; **chäk ya'**; **ya'**; **'uuch (äh)**. *[Source:* BM*]*

ho' yahaha' ho' yahaha'. *Lit:* 'head of yahaha' (upstream of)'.

hup yielding, soft (soil). *Lit:* 'sink'. Ne hup ulu'uma'. *[Note:* hu'up lu'um. marshy ground, quicksand (Bruce 1975); hup [Yuc.]. sink, insert, húup-chá'at. step on something yielding (Bricker et al. 1998).*] [Source:* BM*]*

I—i

Itsanohk'uh Itsanok'uh. A sacred place named after the deity, Itsanohk'uh 'Lord of Hail'. It is situated due east of Naha' and south of Lake Mensäbäk. *[Source:* Bruce (1975)*]*

K—k

kab$_1$ land, world. *Syn:* **lu'um** 'land, earth, soil'. *[Note:* kàab' [Yuc.]. land, world (Bricker et al. 1998: 118); kab' [Itz.]. world (Hofling and Tesucún 1997: 332)*]*

kuyak kuyak. *Lit:* 'cave of the god(s)'. This is a cave located near the northern Lacandon community of Mensäbäk (Palka et al. 2008). *Morph:* **k'uh-il 'ak**. *[Source:* Palka et al. (2008)*]*

K'—k'

k'än k'ät clay, yellow. This type of clay is encountered throughout the forest. It is unsuitable for both wild and cultivated plants. *Sim:* **säk k'ät** 'white clay'; *Sim:* **chäk k'ät** 'red clay'; *Gen:* **k'ät** 'clay'. *[Source:* AM*]*

k'ät clay. *Use:* **Bol** 'balche' urn'; *Use:* **K'ayum** 'ceremonial drum'; *Use:* **lekil k'uh** 'god pots, incensories'; *Spec:* **chäk k'ät** 'red clay'; *Spec:* **k'än k'ät** 'yellow clay'; *Spec:* **säk k'ät** 'white clay'. *[Source:* BM*]*

L—l

lu'um [lu'um, luum] earth, soil, ground. Lacandones distinguish six types of soil, based on colour and friability. *Ek' lu'um* 'black earth' is preferred for *milpas*, because of the high levels of biomass. *Chäk lu'um* 'red earth' is second best, found on high, well-drained land (Nations and Nigh 1980: 9). Three other kinds of soil contain large amounts of clay and are thus unsuitable for cultivation. These include *säk' k'ät* 'white clay'; *k'än k'ät* 'yellow clay'; and *chäk k'ät* 'red clay'. The sixth kind of soil is *sa'am* 'sand'. *Spec:* **chäk lu'um** 'red soil'; *Spec:* **ek' lu'um** 'black soil'. *[Source:* AM; BM*]*

lu'umal [lu'umah] soil. *Lit:* 'of the earth'. *Morph:* **lu'um-al**. *[Source:* BM*]*

M—m

Mensäbäk Mensabak. Named after the Lacandon deity Mensäbäk 'The Powder Maker', it designates a lake and a northern Lacandon community about an hour's drive northwest of Naha'. Mensäbäk is a protected site and natural reserve. The area's main features are the number of large, aqua lakes connected by narrow waterways, and numerous archaeological sites and sacred caves dating as far back as the post-Classic period. Many of the sites are named after the Lacandon deities who dwell there, including *Itsanohk'uh*, *It'sanal*, *Mensäbäk*, and *K'ak'*. Other significant sites are *chäk aktun* (El Mirador), *kuyak* (/k'uhil ak/?), *paten*, Los Olores, and La Punta (Palka et. al. 2008: 812–815). *Loc:* **N** 17.07 O 91.37, 550 m elev. *[Source:* Palka et. al. (2008)*]*

N—n

Naha' Naha'. *Lit:* 'great water'. *Morph:* **nah=ha'**.

nohol south.

P—p

päk'änil pom$_1$ Päk'anil pom; a place where *pom* grows, possibly Tenosique. *Lit:* 'planted with pom'. *Morph:* **päk'-an-il=pom**. *[Source:* AM*]*

petha' lagoon.

S—s

sa'am$_1$ [sa'am, saam] sand.

Sa'am$_2$ Monte Líbano. *Lit:* 'sand'. Originally this was a northern Lacandon community, before it was colonized by Tzeltal immigrants. It is located about half an hour's drive south-east of Naha'.

säk k'ät clay (white); a clay that is mixed with sand, and used for making god pots. *Use:* **lekil k'uh** 'god pots, incensories'; *Sim:* **chäk k'ät** 'red clay'; *Sim:* **k'än k'ät** 'yellow clay'; *Gen:* **k'ät** 'clay'. *[Source:* BM*]*

Sibal Sibal. *See:* **siib**. On some maps it is spelled Sival.

sos gravel, calcium deposits. Refers to calcium deposits left in inundated areas, particularly in the town of Sos (Lacandón). The calcium is used for *nixtamalization*, a process that softens corn kernels and improves their nutritional value. The Spanish name for this material is *cal*. *Loc:* **Sos** 'Lacandón'.

T—t

tah (äh) obsidian. *Use:* **tok'** 'arrowhead'. *[Note:* tah (äh). obsidian, volcanic glass (Bruce 1975). Lacandones say they never use obsidian because it breaks easily (Bruce 1976: 62).*]*

tun rock, boulder. *See:* **tunich**. *[Note:* tun. limestone (in the process of hardening) (Tozzer 1907: 17).*]*

tunich stone. *Morph:* **tun-ich**. *See main entry:* **tun**. *[Note:* tätunich. hard, white stone (Tozzer 1907).*]*

tu' kubin k'in west. *Lit:* 'where the sun goes'.

tu' kutal k'in east. *Lit:* 'where the sun comes from'.

tu' tselel wits hillside, slope. *Lit:* 'where the hill slopes'. The soil here is stony and composed of clay and red soil. Plants that grow well here include, *mehen ch'alol* (*Quercus* sp.), *pixän k'ambul* (*Quercus Skinneri*), and *chechen* (*Metopium brownie*). (BM). *See:* **mehen ch'alol**; **pixan k'ambul (äh)**; **chechem (äh)**. *[Note:* tsel P, T. lie on one's side; skirt, sidle; tzel [Itz.]. slope (Hofling and Tesucún 1997: 624)*] [Source:* BM*]*

Ts—ts

tseltsel wits mountain side. *Morph:* **tsel-tsel=wits**.

Ts'—ts'

Ts'ibatnah Ts'ibatnah. *Lit:* 'inscribe house with paint'. This is a sacred site named after the Lacandon deity, "Painter of Houses". It is one of the largest of several lakes, located near the community of Mensäbäk. According to Lacandon beliefs about the afterlife, the souls of the dead pass through Ts'ibatnah on their way to or from the Underworld, where the Sukunkyum passes judgement on them: if they are good, he sends the souls to Ts'ibatnah or Mensäbäk, where they live in peace; if they are bad, he sends them to Metlan, where they remain with Kisin (BM). The lake is overhung by a precipitous cliff painted with murals. At the base there is a cave that contains remnants of human and animal bones, skulls, god pots, and pieces of obsidian. From a Lacandon perspective, these would belong to the good souls; from an historical perspective, the cave may have been a burial ground for highly esteemed people or an important ritual site or both (See Palka et al. 2008). *Loc:* **Mensäbäk**. *[Note:* Spelling follows Bruce (1968), however my Lacandon consultant BM says that there is no /t/ pronounced in this word.*] [Source:* BM*]*

ts'u' k'aax dense forest; primary forest. *Lit:* 'dense forest'.

W—w

wits hill.

X—x

xämän$_1$ north.

Y—y

ya'ax petha' green lagoon. *Lit:* 'green lagoon'. Located southwest of Naha'. Crocodiles are said to be here.

Yaha'nah La Cueva. *Lit:* 'great water house'. *Morph:* **yah=ha'=nah**. *From:* La Cueva (Spanish).

6.7 Housing

A—a

akam$_1$ [akam, äkam] foundation post. *Mat:* **ya'** 'chicle (*Manilkara* spp.)'; *Mat:* **hach ya'** 'sapodilla (*Manilkara zapote*)'; *Mat:* **säk ya'** (*Chrysophyllum mexicanum*); *Mat:* **säkche'** (*Bauhinia rubeleruziana*). *[Note:* akam. beam (of wood), rafter (Bruce 1975). Baer and Baer (1952: 3) say that *chäk ya'* (*Manilkara achras*) is also used for posts.*] [Source:* AM; BM*]*

akam tanil [akan tanih] pillars (front). The two house posts that support the girt. *Mat:* **sayok' (ah)** (*Aspidosperma megalocarpon*); *Mat:* **babah** (*Calophyllum brasiliense*). *See:* **taan**. *[Source:* SK*]*

atoch home. *See:* **yatoch**; **yatoch k'uh**. *Variant:* **nah**. *[Note:* NI*]*

B—b

bahbil wall (plank). *Lit:* 'nailed'. These are timber planks that are aligned horizontally, as opposed to vertically. This is a relatively new kind of wall, similar to the villagers' houses. *Mat:* **ya'** 'chicle'; *Mat:* **k'alok' che' (äh)** (*Billia colombiana*). *See:* **hi'che'**. *[Source:* BM; SK*]*

Ch—ch

chi' pohche' kitchen, ceremonial. *Lit:* 'table edge'. An open hut located beside the god house, where women prepare ceremonial foods. *Morph:* **chi'=pohche'**. *See main entry:* **pohche'**. *[Source:* Baer and Baer (1952); Davis (1978); KyP*]*

H—h

hach nah house, main part. *[Source:* BM*]*

hi'che' wall (of poles), fence. *Mat:* **ek' bahche'** (*Guatteria anomala*); *Mat:* **pukte'** (*Terminalia amazonia*); *Mat:* **hach taw** (*Belotia mexicana*); *Mat:* **puuna' (äh)** (*Swietenia macrophylla*); *Mat:* **ponche'** (*Bernoullia flammea*); *Mat:* **k'uche' (äh)** (*Cedrela odorata*); *Mat:* **halol** (*Heliocarpus* spp.); *Mat:* **k'ik'che'** (*Virola guatemalensis*). *[Note:* hi'che'. fence, pole wall (Bruce 1975); jil-che' [Itz.]. fence. (Hofling and Tesucún 1997); hil [Yuc.]. picket, wattle (Bricker et al. 1998).*]*

hi'che' ahpek' dog enclosure. *Mat:* **mahaas (ah)** (*Quararibea funebris*). *[Source:* SK*]*

hi'che' nah wall (of poles). A traditional Lacandon house wall is made of these vertical poles or narrow timber slats. This design has been replaced by *bahbil*, lit. 'nailed', which are wide timber planks that are aligned horizontally and nailed to upright supports. *See:* **hi'che'**. *[Note:* hi'che'. fence, pole wall (Bruce 1975);*] [Source:* BM; SK*]*

hiil battens. These are horizontal poles placed at intervals from the bottom of the roof to its peak. They are used for tying thatch to the roof. *Mat:* **babah** (*Calophyllum brasiliense*); *Part:* **mehen che', k'äbche'** 'saplings, branches'; *Mat:* **bahun che'** (*Cordia alliodora*); *Part:* **mehen che', k'äbche'** 'saplings, branches'; *Mat:* **haban uts'u' (ah)** (*Podachaenium eminens*); *Part:* **che'il** 'stalks'; *Mat:* **hach 'ooh** (*Gynerium sagittatum*); *Part:* **che'il** 'stalks'; *Mat:* **k'isis che'** 'undetermined'; *Part:* **mehen che', k'äbche'** 'saplings, branches'; *Mat:* **bamax (äh)** (*Pseudolmedia* spp.); *Part:* **mehen che'** 'saplings'; *Mat:* **che' chäk'an** 'savannah tree, e.g., *Vernonanthura patens*'; *Part:* **mehen che', k'äbche'** 'saplings, branches'; *Mat:* **pakah che'** 'undetermined'; *Part:* **mehen che', k'äbche'** 'saplings, branches'; *Mat:* **sayok' (ah)** (*Aspidosperma megalocarpon*); *Part:* **k'äbche'** 'branches'; *Mat:* **ek' bahche'** (*Guatteria anomala*); *Part:* **che'il** 'timber'. *[Source:* BM; SK*]*

hool door, window. *Lit:* 'hole'. *Mat:* **puuna' (äh)** (*Swietenia macrophylla*).

ho'olan che'il [ho'olan che'il, hooran che'il] ridgepole, ridge board. The ridgepole is comprised of three poles running the length of the apex of the roof. *Mat:* **säk ya'** (*Chrysophyllum mexicanum*); *Mat:* **chäk ya'** (*Manilkara achras*); *Mat:* **ponche'** (*Bernoullia flammea*); *Mat:* **säkyuuhche' (äh)** (*Blepharidium mexicanum*); *Mat:* **babah** (*Calophyllum brasiliense*); *Mat:* **bamax (äh)** (*Pseudolmedia* spp.). *Variant:* **ho'ol unah**. *[Source:* BM; Baer and Baer (1952)*]*

ho'ol unah [ho'ol unah, ho' unah, ho'nah, hoolnah, holnah] ridgepole, house peak. *Lit:* 'head of the house'. *Variant:* **ho'olan che'il**. *[Note:* jo'ol naj, jo'ol-nah [Itz.]. roof (lit. head of house) (Hofling and Tesucún 1997).*]*

K—k

ka'che' näl corncrib. This is a simple, three-sided hut, with raised floor and back-slanting roof used to store maize. The floor is raised to prevent rats and other four-legged pillagers from entering the hut. The walls are made of the same vertical poles used in traditional houses, but fitted closer together. The thatched roof extends over the front, providing shelter to dry tobacco. *Loc:* **kol** 'milpa'; *Mat:* **halol** (*Heliocarpus* spp.); *Part:* **che'il** 'wood'; *Part:* **k'äbche'** 'branches'; *Mat:* **hach taw** (*Belotia mexicana*); *Part:* **che'il** 'timber'; *Mat:* **haban uts'u'** (*Podachaenium eminens*); *Part:* **che'il** 'stalks'; *Mat:* **tuch** (*Thevetia ahouai*); *Part:* **che'il**; *Mat:* **kuti'** (*Talauma mexicana*); *Part:* **che'il** 'timber'. *[Note:* unähinäl. corncrib (Baer and Baer 1952).*]*

kelil nah roof support. *Lit:* 'shoulder of the house/roof'. Two tall posts, centred between the foundation posts at each end, that support the ridgepole. *[Note:* kalembah. shoulder (Bruce 1975); keléemb'al [Yuc.]. shoulder (Bricker et al. 1998); kelem [Itz.]. shoulder (Hofling and Tesucún 1997).*] [Source:* Baer and Baer (1952)*]*

kuch säkche'il [kuch säkche'il, ku säkche'il] beam (load bearing). *Lit:* 'carry load of the beam'. The two large beams that run the length of the hut and support the *säkche'il* 'rafters'. *[Source:* AM; BM*]*

K'—k'

k'oben hearth. Consists of three stones arranged in a circle on the ground with a metal grill placed on top. Fire brands are fed in between the stones. This kind of hearth is used primarily for boiling large pots of maize and tamales. A *xämäch* ‘grilling plate’ is placed on top of the stones or the grill for baking tortillas. *[Note:* k'oben. fireplace (Baer and Baer 1952); k'óob'en [Yuc.]. kitchen, hearthstone [refers to three hearthstones] (Bricker et al. 1998); k'ob'e'en [Itz.]. fireplace of three stones (Hofling and Tesucún 1997).*]*

k'unah [k'unah, k'unan] temple. *Variant:* **yatoch k'uh**. *[Note:* Likely, <k'uh=nah ‘god=house’*] [Source:* Bruce (1975)*]*

M—m

mooy overhang. This is a portion of the roof truss that extends beyond the exterior walls, like an awning, and held up by pillars. It is considered separate from the main house (*hach nah*). *[Note:* móoy [Yuc.]. rounded end of oval house (Bricker et al. 1998)*] [Source:* BM*]*

N—n

nah$_1$ house. *Mat:* **kuti'** (*Talauma mexicana*); *Sim:* **atoch'** ‘home’; *Spec:* **hach nah** ‘main house’. *See:* **nahil**.

nahil roof. *Mat:* **kun (ah)** (*Cryosophila stauracantha*); *Mat:* **xa'an** (*Sabal mexicana*); *Mat:* **pahok (ah)** (*Geonoma oxycarpa*). *See:* **pix nah**. *[Source:* BM*]*

O—o

okman house posts. *Lit:* ‘fabricated foot’. *Mat:* **ch'obenche'** ‘*Trichilia* sp.’; *Mat:* **chäk ya'** (*Manilkara achras*); *Mat:* **ek'** (*Haematoxylon campechianum*); *Mat:* **kakache'** (*Ocotea cernua*); *Mat:* **tuch** (*Thevetia ahouai*); *Mat:* **k'än chululche'** (*Wimmeria bartlettii*); *Spec:* **okman ukabal** ‘short pillar, corner posts’; *Spec:* **okman uka'anal** ‘tall pillar, interior posts’. *Morph:* **ok-man**. *See:* **akam**. *Variant:* **okben**. *[Source:* BM; SK*]*

okman uka'anal pillar; interior house posts. *Lit:* ‘high pillars’. *Gen:* **okman** ‘pillar’. *[Source:* BM*]*

okman ukabal pillar; corner post. *Lit:* ‘low pillars’. *Gen:* **okman** ‘pillar’. *[Source:* BM*]*

P—p

paasäl shed, lean-to. *[Note:* pasel, pàasel [Yuc.]. hut no more than two meters high in cornfield (Bricker et al. 1998).*] [Source:* Bruce (1975)*]*

pach unah back of house. *Morph:* **pach u-nah**.

pa'te' cage. *Lit:* ‘split tree/wood’. An enclosure for livestock, chickens, and fish; e.g., *pa'te' kax* ‘chicken coop’, *pa'te' käy* ‘farm raised fish’. *Morph:* **pa'-te'**. *[Note:* pa'ate'. cage, (chicken) coop. (Bruce 1975); pa' te' [S. Lac.] corral (Canger 1970).*]*

pa'te' kax chicken coop. *Mat:* **ya'ax balche'** (*Lonchocarpus* spp.); *Mat:* **mäkuuläm** (*Piper* spp.). *See main entry:* **pa'te'**.

päk'il nah wall, plaster or concrete. *Lit:* ‘spread of the wall’. *[Note:* päk'-bi:r nah [S. Lac.]. house wall (Canger 1970).*] [Source:* Bruce (1975)*]*

pib stove, oven. The main stove/oven, *pib* is raised thigh-high from the ground, three sides enclose the oven, while the fourth is left open to hold the firebrands. A sheet of metal coated with quicklime is placed over the gap. *Mat:* **puuna' (äh)** (*Swietenia macrophylla*).

pix nah thatch. *Lit:* ‘house cover’. *Mat:* **kun (ah)** (*Cryosophila stauracantha*); *Mat:* **pahok (ah)** (*Geonoma oxycarpa*); *Mat:* **xa'an** (*Sabal mexicana*). *Variant:* **na, nahil**. *[Source:* SK*]*

S—s

säkche'$_2$ rafters. *Lit:* ‘white tree’. *Mat:* **sa'yok' (ah)** ‘*Aspidosperma macrocarpon*’; *Mat:* **tso'ots bamax** ‘*Pseudolmedia oxyphyllaria*’; *Mat:* **babah** ‘*Calophyllum brasiliense*’; *Mat:* **hach bamax**; *Mat:* **yache' kab** ‘undetermined’; *Mat:* **k'isis che'** ‘undetermined’. *Variant:* **säkche' nah**. *[Source:* BM; K'ayum Segundo; SK*]*

T—t

taan$_2$ [taan, tan] girt, horizontal beams. *Lit:* ‘front’. *Mat:* **sa'yok' (ah)** ‘*Aspidosperma megalocarpon*’; *Mat:* **babah** (*Calophyllum brasiliense*); *Mat:* **yache' kab**. *Variant:* **tanah**. *[Source:* BM; SK*]*

tanah beam, girt. *See:* **taan**. *[Note:* u-tà:n nAh [S. Lac.]. viga (Canger 1970).*]*

W—w

waan$_2$ girt, side. *Lit:* ‘length’. *Mat:* **sa'yok' (ah)** (*Aspidosperma megalocarpon*); *Mat:* **babah** (*Calophyllum brasiliense*).

Y—y

yatoch home. *Morph:* **y-atoch**.

yatoch k'uh god house, temple. *Lit:* 'home of the gods'. *Morph:* **y-atoch k'uh**. *See main entry:* **atoch**. *Variant:* **k'unah**.

yokman pillar. *Morph:* **y-ok-man**. *See:* **okman**.

6.8 Medical Conditions

A—a

'aak'ä' ich evil eye. *Lit:* 'night eye'. The evil eye is a belief that a person or a malevolent force has the power to harm another person or their property simply by looking at them in an envious way or by complimenting them. Infants and children are particularly susceptible. Symptoms of someone who has been exposed to the evil eye include diarrhea, vomiting, fever, yawning, hiccoughs, and death. In Mexico, *mal de ojo* 'evil eye' is believed to be caused by witches and evil spirits, or virulent bacteria. The concept of the evil eye developed in the Old World, likely India, the Near East, and Europe (Roberts 1976, in Dundes 1981: 259). Conceivably, Portuguese and Spanish colonists introduced the concept to the Americas. *Mat:* **yo'och ähbaach'** (*Siparuna thecaphora*); *Part:* **le'** 'leaves'; *Prep:* **Ak'elik ule' ich k'ak', pachil pulik alkohol yok'ol ti', pachil pulik yok'ol uho'ol ti' mehen paalal** 'You sear the leaves over an open fire, then douse them with alcohol, and then place them on the head of the child.' *Variant:* **yak uwich winik**. *[Note:* aak'ä' ich. evil eye (Bruce 1975); ak'as-'ich [Itz.]. evil eye (Hofling and Tesucún 1997).*] [Source:* CNK; KyP; SK*]*

'ax wart. *Mat:* **ts'ak 'ax** 'wart medicine (undetermined)'; *Part:* **le', che'il** 'leaves and stems'; *Prep:* **Ayäts'ik uyits yok'ol a'ax** 'You squeeze the resin from the leaves onto your warts.' *[Source:* AM; KyP*]*

Ch—ch

chäkwilil fever. *Mat:* **mulix** (*Citrus aurantifolia*); *Part:* **uyits ule'** 'resin from the leaves'; *Prep:* **Apulik yok'ol awich, tsem, ho'ol** 'You put it on your face, chest, and head.'; *Mat:* **ma'ax ak' (äh)** (*Paullinia* sp.); *Part:* **che'il, le'** 'stems and leaves'; *Prep:* **te'** 'tea'; *Mat:* **puuna' (äh)** (*Swietenia macrophylla*); *Part:* **?** *[Note:* Remedies using *ma'ax ak'* and *puuna'* are provided by Kashanipour and McGee (2004).*] [Source:* AM*]*

chechem$_2$ *Metopium brownei*. This tree contains a caustic sap that can cause blindness, and even death if drunk. Plants that help reduce the symptoms include spiral ginger and gumbo limbo. *Mat:* **pahsa' ak' (äh)** (*Costus pulverulentus*); *Part:* **uyits che'il** 'resin from the stems'; *Prep:* **ha'li' ak'upik uche'il i ach'ahik uyits yok'ol uwich** 'just cut the stem and drip the resin into the eyes'; *Use:* **chäklah** (*Bursera simaruba*); *Part:* **uyits** 'resin'; *Prep:* **ha'li' apulik yok'ol uyahil** 'just put it on the wound.' *[Note:* chäklah remedy from Nations (2006).*] [Source:* AM; BM; KyP*]*

chibal bite, mosquito and other small insects. In general, *chibal* refers to any kind of bite, and in particular, mosquito bites. Some plant remedies include cedar, hairy beggarticks, and an undetermined vine. *Mat:* **chäk ak'** 'undetermined'; *Part:* **sool** 'bark'; *Prep:* **mäna'** 'none'; *Mat:* **k'uche' (ah)** (*Cedrela odorata*); *Part:* **le'** 'leaves'; *Prep:* **Acha'achtik ule' pachil apulik yok'ol chibal** 'You chew the leaves and then apply them to the bite.'; *Mat:* **ku'uchnook'** (*Bidens odorata*); *Part:* **le'** 'leaves'; *Prep:* **pixik uchibal yetel** 'wrap them around the bite'. *[Source:* Kashanipour and McGee (2004).*]*

chibal hämnen stomach ache. Plant remedies include a number of species. *Mat:* **hach mäkuuläm** (*Piper aduncum*); *Part:* **le'** 'leaves'; *Prep:* **Apets'ik ule' ich sisil ha' i uyuk'ik, 'oxwäts'** 'You crush the leaves in cold water and then drink it three times.'; *Mat:* **nikte' ak'**; *Part:* **uyits ak'** 'juice from the vine'; *Prep:* **uk'ul** 'beverage'; *Mat:* **bits'** (*Inga* spp.); *Part:* **le'** 'leaves'; *Prep:* **Mäna'. Ha'li' acha'achtik** 'None. You just chew the leaves.'; *Mat:* **chäklah** (*Bursera simaruba*); *Part:* **mots** 'roots'; *Prep:* **te'** 'tea'; *Mat:* **kakawat** '*Arachis hypogaea*'; *Part:* **le'** 'leaves'; *Prep:* **te'** 'tea'; *Mat:* **peesache' (äh)** (*Pimenta dioica*); *Part:* **le'** 'leaves'; *Prep:* **te'** 'tea'; *Mat:* **puhan** (*Muntingia calabura*); *Part:* **le'** 'leaves'; *Prep:* **Mäna'. Ha'li' acha'achtik** 'None. You just chew the leaves.' *[Note:* Remedy using *Inga* spp., *Pimenta dioica*, and *Muntingia calabura* is provided by Kashanipour and McGee (2004).*] [Source:* AM*]*

chibal kan snakebite. *Mat:* **mäkuuläm** '*Piper* spp., *P. hispidum*'; *Part:* **le'** 'leaves'; *Prep:* **te'** 'tea'; *Mat:* **ts'ak kan** (*Strychnos tabascana(?)*); *Part:* **mots** 'branches'; *Prep:* **Amuxik i apulik yok'ol; ahuytik yetel (otras plantas) ich ha' achäkik, pachil awuk'ik** 'You crush it and ... stir it with (other plants?) in the water, you boil it, then you drink it.'; *Mat:* **chäk pach** (*Ternstroemia tepezapote*); *Part:* **sool** 'bark'; *Prep:* **Achäkik ich ha', pachil ap'o'ik yetel ti'** 'You boil it in water and then you wash (the area) with it.' *[Note:* Medicinal preparation is provided by Durán (1999: 149). Translations in Lacandon and English are by the author.*]*

chibal koh toothache. *Mat:* **sensi'** (*Zingiber officinale*); *Part:* **wi'** 'rhizome'; *Prep:* **Che'che'. Ahupik ich akoh, wa**

apaats'tik yok'ol achi' 'Insert a raw piece in your tooth, or rub it on your mouth.'; *Mat:* **ya'ax 'oox** (*Brosimum alicastrum* var. *alicastrum*); *Part:* **uyits** 'latex'; *Prep:* **Axaak'tik uyits yetel ch'ooch', pachil asuulik tämän yetel chen but'ik yok'ol akoh** 'You mix together the latex and salt, then you soak a piece of cotton with it and stuff it in your tooth.'; *Mat:* **peesa'che' (äh)** (*Pimenta dioica*); *Part:* **le'** 'leaves'; *Prep:* **Ha'li' acha'achtik ule'** 'You just chew the leaves.'; *Mat:* **tuch** (*Thevetia ahouai*); *Mat:* **puuna' (äh)** (*Swietenia macrophylla*); *Part:* **wich** 'fruit'; *Prep:* **Achäkik uwich, awuk'ul, apaats'tik uwich ich achi'** 'You boil it and drink it and you rub the fruit in your gums.' *[Note: Tuch* remedy is from Nations and Nigh (1980).*] [Source:* AM; CKD; SK*]*

chiclero leishmaniasis. This is a disease caused by protozoan parasites (unicellular organisms with complex structures enclosed within a membrane) of the *Leishmania* genus and transmitted by certain species of sand fly. Chicleros 'chicle tappers' are susceptible to the disease, given that they work in the forest where the sand fly is found. *Mat:* **k'o'och (äh)** (*Cecropia obtusifolia, C. peltata*); *Part:* **muun ule', uyol** 'tender leaves and the growing tips'; *Prep:* **Uyächik ule', uyol i usuulik ich ha', pachil uts'ik ti'** 'One crushes the leaves and the shoots, steeps them in water, and then places them on (the affected area).' *From:* chiclero (Spanish). *[Source:* AM; BM; K'ayum segundo*]*

chok$_2$ chok. This is a serious condition that befalls dogs that have ingested the fruits of a plant by the same name. *Mat:* **yo'och ahpek'** (*Canavalia* spp.); *Part:* **nek'** 'seeds'; *Prep:* **k'elbil** 'toasted'. *[Source:* AM*]*

chuhul blister from burn, burn. *Mat:* **chäklah** (*Bursera simaruba*); *Part:* **soolche'** 'bark'; *Prep:* **Amuxik usool pachil apulik yok'ol uchuhul** 'You crush the bark and then put it on the burn.' *[Source:* Durán (1999). Lacandon translation by author.*]*

chupul swelling. *Mat:* **k'uxub** (*Bixa orellana*); *Part:* **Mäna'. Ha'li' kunyah** 'Only its name is invoked in a curing chant.'; *Mat:* **ya'** (*Manilkara achras*); *Part:* **Mäna'. Ha'li' kunyah** 'Only its name is invoked in a curing chant.'; *Mat:* **naba'** (*Myroxylon balsamum*); *Part:* **Mäna'. Ha'li' kunyah** 'Only its name is invoked in a curing.'; *Mat:* **ts'ak ahch'up** 'a plant medicine for the bite of a type of caterpillar (*Cissus biformafolia*)'; *Part:* **uyits** 'resin'; *Prep:* **Apaats'tik uyits yok'ol** 'You rub the resin on (the area).'; *Mat:* **baba'** (*Sapindus saponaria*); *Part:* **uyits** 'sap'; *Prep:* **Ha'li' apulik yok'ol uyah** 'You just put it on the sore area.' *[Note:* Remedy using ts'ak ahch'up is provided by Kashanipour and McGee (2004). Lacandon translation by the author.*] [Source:* AM*]*

H—h

hubul hämnen diarrhea. *Lit:* 'loose stomach'. *Mat:* **kuti'** (*Talauma mexicana*); *Part:* **sool** 'bark'; *Prep:* **te'** 'tea'; *Mat:* **tsäk'ats** (*Licania platypus*); *Part:* **unek'** 'the seed'; *Prep:* **Achäkik uwich ich ha' pachil awuk'ik. Ne k'ah.** 'You boil the seed in water and then you drink it. It is very bitter.'; *Mat:* **sa'yok' (ah)** (*Aspidosperma megalocarpon*); *Part:* **sool** 'bark'; *Prep:* **te'** 'tea'; *Mat:* **k'ante' ak'** (*Erythrina coralloides*); *Part:* **sool, uyits** 'bark and resin'; *Prep:* **Che'che'. Acha'achtik usool** 'You just chew the bark.'; *Mat:* **hach mäkuuläm** (*Piper aduncum*); *Part:* **le'** 'leaves'; *Prep:* **Upets'ik ule' ich sisil ha' i uyuk'ik, 'oxwäts'** 'One crushes leaves in cold water and then drinks it three times.'; *Mat:* **sensi' (ah)** (*Zingiber officinale*); *Part:* **wi'** 'rhizome'; *Prep:* **che'che'** 'raw'; *Mat:* **lo'k'in** (*Parathesis* sp.); *Part:* **sool** 'bark'; *Prep:* **te'** 'tea'; *Mat:* **peesache' (äh)** (*Pimenta dioica*); *Part:* **le'** 'leaves'; *Prep:* **te'** 'tea'; *Mat:* **axux** (*Allium sativa*); *Prep:* **te'** 'tea'. *[Note:* According to "lol k'in", "xu'u" 'garlic' and "kanti ak'" are used to treat hubul hämnen, and *Pimenta dioica* is used to treat colic (Durán 1999).*] [Source:* AM; K'ayum segundo*]*

K'—k'

k'ak'il boils, skin eruptions. *Mat:* **ts'ak k'ak'il** 'undetermined'; *Mat:* **babah** (*Calophyllum brasiliense*); *Part:* **uyits usool** 'resin from the bark'; *Prep:* **Ha'li' ats'ahik uyits yok'ol k'ak'il** 'You just apply the resin to the boils.' *[Source:* AM*]*

k'ik' ha' perspiration. *Lit:* 'blood water'. *Mat:* **mulix** (*Citrus aurantifolia*); *Part:* **uyits ule'** 'resin from the leaves'; *Prep:* **Apulik yok'ol awich, atsem, aho'ol** 'You put it on your face, your chest, and your head.' *[Source:* AM*]*

k'ik' uchi' bleeding mouth; could be caused by any number of conditions. *Lit:* 'bleeding mouth'. *Mat:* **chukum che'** (*Croton draco*); *Part:* **uyits usool** 'resin from the bark'; *Prep:* **Mäna'. Ha'li' apaats'tik ti'** 'You just rub it on.' *[Source:* AM*]*

M—m

mal de orín en niños bed-wetting, children. *Mat:* **k'o'och (äh)** (*Cecropia obtusifolia*); *Part:* **yol** 'growing tips'; *Prep:* **te'** 'tea (infusion)'; *Mat:* **chäkchop'** (*Lantana* spp.); *Part:* **le'** 'leaves'; *Prep:* **Ha'li' ahantik che'che'** 'You just eat the leaves raw.' *[Note: chäk chok* remedy provided by Kashanipour and McGee (2004); *k'o'och* remedy is from Durán (1999).*]*

mehen k'ak'il measles, chicken pox, small pox. *Lit:* 'small boils'. In Lacandon dream interpretation, puhun (*Muntingia calabura*) foretells small pox, chicken pox, and measles (Bruce 1975: 143–144). *See:* **k'ak'il**. *[Source:* Bruce (1975)*]*

N—n

nok'ol worms, intestinal. *Mat:* **k'äxeex (äh)** (*Chenopodium ambrosioides*); *Use:* **ts'ak ti' nok'ol** 'medicine for worms'; *Part:* **mots** 'roots'; *Prep:* **Achäkik umots ich ha', pachil awuk'ik** 'You boil the roots in water and then drink it.' *[Source:* AM; BM: KM; KP*]*

S—s

sak'il itchiness. *Mat:* **pahsa' ak' (äh)** (*Costus pulverulentus*); *Part:* **uyits** 'resin'; *Prep:* **Mäna'** 'None.'

sarna scabies. *Mat:* **'ukunte' (äh)** (*Sapium lateriflorum*); *Part:* **uyits** 'resin'; *Prep:* **Ha'li' apulik yok'ol** 'You just put it on.' *From:* sarna (Spanish). *[Source:* AM; KA; KP*]*

säkpähk'ak' oral disease (a type of). Possibly stomatitis, which is an inflammation of the oral mucosa. In Lacandon dream interpretation, to dream of someone suffering from *säkpähk'ak'* foretells seeing an animal with the mucosa of its mouth peeling (Bruce 1975: 215). *Mat:* **säkpähk'ak' che'** (*Psychotria poeppigiana*); *Prep:* **Ha'li' kunyah** 'Only the name is invoked in a curing chant.'; *Mat:* **säk yolte' ik (äh)** 'white walking-stick pepper'; *Part:* **wich** 'fruit'; *Prep:* **Mäna'** 'None.' *[Note:* säkpäk'ak. a certain mortal illness, possibly trichinosis (Bruce 1975: 215).*] [Source:* AM*]*

säk se'em catarrh, cough. *Mat:* **bak'che'** 'epiphytes, e.g., *Tillandsia festucoided*'; *Part:* **le'** 'leaves'; *Prep:* **te'** 'tea'; *Mat:* **sensi' (ah)** (*Zingiber officinale*); *Part:* **wi'** 'rhizome'; *Prep:* **te'** 'tea'; *Mat:* **ts'us** (*Vitis* spp.); *Part:* **mots** 'roots'; *Prep:* **te'** 'teas'. *See:* **se'em**. *[Note:* The remedy, *Tillandsia festucoided*, is provided by Durán, and *Zingiber officinale* and *Vitis* spp., by Kashanipour and McGee.*] [Source:* Durán (1999); Kashanipour and McGee (2004).*]*

se'em common cold. *Mat:* **hach 'ak (äh)** (*Cymbopogon citratus*); *Part:* **le'** 'leaves'; *Prep:* **Achäkik seis ule', pachil awuk'ik** 'You boil six leaves in water and then drink it.'; *Mat:* **san lorenso** (*Kalanchoe pinnata*); *Part:* **le'** 'leaves'; *Prep:* **Apokik ule', pachil abut'ik ich chakal ha', pachil awuk'ik** 'You sear the leaves over an open fire, then put them in hot water, and then drink it.'; *Mat:* **mulix** (*Citrus aurantifolia*); *Part:* **k'ab** 'juice'; *Prep:* **chäkbil** 'boiled'. *[Source:* AM; KP*]*

siim [siim, siin] mucus. *[Note:* síin [Yuc.]. mucus (Bricker et al. 1998); siim [Itz.]. mucus (Hofling and Tesucún 1997).*]*

Ts'—ts'

ts'ikin nok'ol (ah) larvae of the botfly. **Ek'. Batak 5 cm (2″). Yan uyaach, chäk, uhupik ubah ti' awoot'el ak'äb, aho'ol, apach, y awok, ti' kuch'ihil. Säk unok'ol, yan utso'ots. Ne sak awoot'el.Yan a pulik uyits uk'uuts y pachil apaats'tik (AM)**. This is described as a white worm with black hair running in bands around the body. It gestates under the skin of its host, gnawing away at the flesh. According to AM, the adult is similar to *ya'ax käch*, a large black fly, approximately 5 cm (2″) long, with a red stinger, which it uses to deposit eggs under the skin of your hands, your head, your back, and your legs. The affected area becomes swollen and very itchy. To draw it out, Lacandones apply tobacco juice to the affected area and then massage the larva out. Although it is unclear what kind of insect lays the eggs, it could be *Dermatobia hominis*. Other species of fly implicated in myiasis are members of the Calliphoridae family, e.g., blowflies and screwworm flies; some of these deposit their eggs on mosquitoes or houseflies, which in turn, release the eggs into the bodies of mammals. *Mat:* **hach k'uuts** (*Nicotiana tabacum*); *Part:* **uyits ule'** 'resin from the leaves'; *Prep:* **Ucha'achtik uk'uuts wa ule' uk'uuts, pachil utubik uyits yok'ol ti', pachil kuhook'ol ahnok'ol kuhook'sik** 'One chews the cigar or some tobacco leaves then spits the juice on it, then, when the worm emerges, one pulls it out.'

ts'uk uchi' oral disease. *Lit:* 'mouth rot'. Possibly hand-foot-mouth disease, this is a contagious virus that is usually seen in babies and small children. Symptoms include white and red pustules inside the lips and throat, accompanied by fever and lack of appetite. *Mat:* **chukum che'** (*Croton draco*); *Part:* **uyits usool** 'resin from the bark'; *Prep:* **Mäna'. Ha'li' apaats'tik ti'** 'You just rub it on.'; *Mat:* **ik** 'chile pepper'; *Part:* **wich** 'fruit'; *Prep:* **Ha'li' acha'achtik** 'You just chew it.' *[Note:* According to AM, it is ingested in the context of a curing ritual.*] [Source:* AM; NI*]*

U—u

uyoot'el ahmehen paalal baby rash. *Mat:* **yo'och ik mehen** (*Psychotria* sp.); *Part:* **le'** 'leaves'; *Prep:* **Achäkik ule', pachil ahixtik uyoot'el yetel** 'You boil the leaves and then rub them on the skin.' *[Source:* AM*]*

uyoot'el ahwinik adult skin condition (rash). *Mat:* **yo'och ik nukuch** (*Psychotria panamensis* var. *panamensis*); *Part:* **le'** 'leaves'; *Prep:* **Achäkik ule' ich ha', pachil ap'o'ik uyoot'el yetel ha'** 'You boil the leaves in water and then wash the skin with the water.' *[Source:* AM*]*

X—x

xeh vomiting, nausea.

Y—y

yah [yah, yaah] painful. *[Note:* yaah [S. Lac.]. painful, sickly (Canger 1970); yah [Yuc.]. painful (Bricker et al. 1998); yaj [Itz.]. painful, pain (Hofling and Tesucún 1997).*]*

yahil pain, wound. *Mat:* **tuch** (*Thevia ahouai*); *Part:* **uyits, ule'** 'resin and leaves'; *Prep:* **Amuxik i pulik yok'ol ti'** 'You crush it and put it on (the affected area).' *Morph:* **yah-il**. *See main entry:* **yah**. *[Note:* ya'hil. illness (Bruce 1975); yajil [Itz.]. pain, wound, sore (Hofling and Tesucún 1997). Remedy provided by Durán 1999). Translation by author.*]* *[Source:* Durán (1999).*]*

yah ubaakel bone pain. *Mat:* **ch'obenche'** (*Trichilia* sp.); *Part:* **le'** 'leaves'; *Prep:* **?**; *Mat:* **maasan kitan** (*Equisetum hyemale* var. *affine*); *Part:* **che'il** 'stalks'; *Prep:* **?** 'unknown'; *Mat:* **k'o'och** (*Cecropia peltata*); *Part:* **le'** 'leaves'; *Prep:* **te'** 'tea'; *Mat:* **kopo', tsoy** (*Clusia salvinii*); *Part:* **k'ä'che'** 'branches'; *Prep:* **te'** 'tea'. *[Note:* The remedies using *Cecropia peltata* and *Clusia salvinii* are provided by Durán.*]* *[Source:* AM; Durán (1999)*]*

yah uchi' ulceration. *Mat:* **tuch** (*Thevetia ahouai*); *Part:* **uyits, ule'** 'resin and leaves'; *Prep:* **Asuulik ich chäkal ha', pachil achalik achi' yetel** 'You steep it in hot water, then you rinse your mouth with it.' *[Note:* Lacandon translation by author.*]* *[Source:* Durán (1999).*]*

yah uch'aay sore kidneys. *Mat:* **chäklah** (*Bursera simaruba*); *Use:* **ts'ak ti' ch'aay, chen ma' hach winik, ts'ak ti' kah** 'It is medicine for kidneys, but it is not traditional; it is the medicine of the villagers (Tzeltales).'; *Part:* **soolche'** 'bark'; *Prep:* **Chäkbil. Awuk'ik 'oxwats' k'in ti' hunts'it na'** 'Boiled. You drink it three times a day for one month.' *[Source:* AM*]*

yah upix knee pain. *Mat:* **chäk wawal** (*Calathea* sp.); *Part:* **le'** 'leaves'; *Prep:* **Apulik ule' yok'ol upix** 'You apply the leaves to the knee.'; *Mat:* **hach k'uuts** 'tobacco'; *Part:* **le'** 'leaves'; *Prep:* **Apokik yok'ol k'ak', atap'ik yok'ol pix** 'You roast (10 leaves) in the fire then cover the knee with them.' *[Source:* AM*]*

References

Atran, S., Lois, X., and Ek', E. U. (2004). Plants of the Petén Itza' Maya: Plantas de los Maya Itza' del Petén. *Memoirs of the Museum of Anthropology 38*. Ann Arbor: University of Michigan.

Baer, M. & Baer, P. (1952). Materials on Lacandon culture of the pethá (pelhá) region. *Microfilm Collection of Manuscripts on Middle American Cultural Anthropology* 34. Chicago: University of Chicago Library.

Baer, P., & Merrifield, W. R. (1971). *Two studies on the Lacandones of Mexico*. Norman, OK: Summer Institute of Linguistics, University of Oklahoma.

Bricker, V., Po'ot, E., & de Po'ot, O. D. (1998). *Dictionary of the Maya language: As spoken in Hocaba Yucatan*. Salt Lake City: University of Utah Press.

Bruce, S. R. D. (1968). *Gramática del Lacandón*. Departamento de Investigaciones Antropológicas (Vol. 21). México: Instituto Nacional de Antropología e Historia.

Bruce, S. R. D. (1974). *El libro de Chan K'in*. Colección Cientifíca, Lingüística, 12. México: Instituto Nacional de Antropología e Historia.

Bruce, S. R. D. (1975). *Lacandon dream symbolism 1: Dream symbolism and interpretation*. México: Ediciones Euroamericanas Klaus Thiele.

Bruce, S. R. D. (1976). *Textos y dibujos Lacandones de Najá*. (Trilingual Edition: Lacandón-Spanish-English). Departamento de Lingüística Colección Científica, Lingüística, Núm 45. México: Instituto Nacional de Antropología e Historia.

Canger, Una. (1970). *Lacandón de San Quintín Vocabulary of San Quintín*. Unpublished manuscript.

Comisión Nacional de Áreas Naturales Protegidas. (2006). *Programa de conservación y manejo área de flora y fauna Nahá*. 1a edición. Tlalpan C.P., México, DF: Dirección General de Manejo para la Conservación, CONANP. http://www2.ine.gob.mx/publicaciones/consultaPublicacion.html?id_pub=558&id_tema=12&dir=Consultas ISBN: 968-817-814-4

Davis, V. D. (1978). *Ritual of the Northern Lacandon Maya* (Unpublished doctoral dissertation). Tulane University, New Orleans, LA.

Diemont, S. A. W. (2006). *Ecosystem management and restoration as practiced by the indigenous Lacandon Maya of Chiapas, Mexico* (Unpublished doctoral dissertation). Ohio State University, OH.

Dundes, A. (1981). Wet and dry, the evil eye: An essay in Indo-European and Semitic worldview. In A. Dundes (Ed.), *The evil eye: A casebook* (pp. 257–309). Madison, WI: The University of Wisconsin Press.

Durán, F. A. (1999). *Estructura y etnobotánica de la selva alta perennifolia de Nahá Chiapas* (Unpublished master's thesis). Universidad Nacional Autónoma de México, Mexico.

Hofling, C. A., & Tesucún, F. F. (1997). *Itzaj Maya-Spanish-English dictionary-diccionario Maya Itzaj-Español-Ingles*. Salt Lake City, Utah: The University of Utah Press.

Kashanipour, R. A., & McGee, R. J. (2004). Northern Lacandon Maya medicinal plant use in the communities of Lacanja Chan Sayab and Naha', Chiapas, Mexico. *Journal of Ecological Anthropology, 8*, 47–66.

Levy-Tacher, S. I., Aguirre-Rivera, J. R., García-Perez, J. D., & Martínez-Romero, M. M. (2006). Aspectos florísticos de Lacanhá Chansayab, selva Lacandona, Chiapas. *Acta Botanica Mexicana, 77*, 69–98.

McGee, R. J. (2002). *Watching Lacandon Maya*. Boston: Allyn and Bacon.

Nations, J. D. (1979). Snail shells and maize preparation: A Lacandon Maya analogy. *American Antiquity, 44*(3), 568–571.

Nations, J. D. (1989). Lacandon Maya bow and arrow: An ethnoarchaeological example of Postclassic lowland Maya weapon manufacture. In M. Gaxiola & J. E. Clark (Eds.), *La obsidiana en Mesoamerica* (pp. 449–457). Mexico: Instituto Nacional de Antropología e Historia.

Nations, J. D. (1992). *Vocabulario Lacandón: plantas y animals*. Paper prepared for Conservation International, April 1992. Washington, DC.

Nations, J. D. (2006). *The Maya tropical forest: People, parks and ancient cities*. Austin: University of Austin Press.

Nations, J. D., & Nigh, R. B. (1980). The evolutionary potential of Lacandon Maya sustained-yield tropical forest agriculture. *Journal of Anthropological Research, 36*(1), 1–30.

Nigh, R. (2008). Trees, fire, and farmers: Making woods and soil in the Maya Forest. *Journal of Ethnobiology, 28*(2), 231–243.

Palka, J. W., Balderas, F. S., Hollingshead, I., Deeb, R. et al. (2008). recorrido arqueológico en Mensabak, Chiapas y los Mayas Postclásicos e históricos en las tierras bajas. In J. P. Laporte, B. Arroyo, and H. Mejía (Eds.), *XXI Simposio de Investigaciones Arqueológicas en Guatemala, 2007* (pp. 808–835). Guatemala, Museo Nacional de Arqueología y Etnología. http://www.asociaciontikal.com/pdf/54_-_Palka.07.pdf

Roberts, J. M. (1976). Belief in the evil eye in world perspective. In C. Maloney (Ed.), *The evil eye* (pp. 223–278). New York: Colombia University Press.

Roys, R. L. (1931). *The ethno-botany of the Maya*. New Orleans: Department of Middle American Research, University of Tulane.

Schlesinger, V. (2001). *Animals and plants of the ancient Maya: A guide*. Austin, TX: University of Austin Press.

Soustelle, G. (1966). *Collections Lacandons*. Paris: Museum National D'Histoire Naturelle.

Stross, B. (1997). Mesoamerican Copal Resins. *U-Mut Maya, 6*, 177–186. http://www.utexas.edu/courses/stross/papers/copal.htm

Tozzer, A. M. (1907). *A comparative study of the Mayas and the Lacandones*. London: Macmillan.

Appendix 1
Lacandon Plants Unidentified Botanically

A–a

ak' tsup *Lit:* 'tsup vine'. The species looks like a large, thick, jungle vine, approximately 7.6 cm (3″) thick, descending from the canopy to the forest floor. The Lacandones cut off pieces for use as a firedrill. These vines may actually be the aerial roots of *Dendropanax arboreus*, an epiphytic tree. Durán's Lacandon consultants say that *Hamelia calycosa* is used for the same purpose. *Use:* **che'il häxbil k'ak'** 'firedrill'; *Part:* **ak'** 'vine'. *SD:* Plants. *Thes:* **ak'**. *[Note:* jaxa kak. *Hamelia calycosa* (Durán 1999); tzup [Itz.]. lion's paw tree. *Dendropanax arboreus* (Atran et al. 2004; Hofling and Tesucún 1997).*] [Source:* AM; BM*]* [\sd2 fuel*]*

Ä–ä

'ämäy reed (a large type of). ***Yan ubäho', semet' i bambu. Chen ma' 'ooh. Es chich, haban uts'u', yan umo'okol.*** 'Its brothers are semet' (*Rhipidocladum bartlettii*) and bamboo (*Bambusa vulgaris*) but not *'ooh* (*Gynerium sagittatum*). It has a hard stalk, a hollow pith, and jointed nodes.' (BM) According to AM, it once grew in abundance in Naha' but now it is only found in Yahapetha'. Sections of the stalks were used as flutes. Although the species has not yet been identified, it could be a species of *Guadua,* perhaps *Guadua amplexifolia*. *Loc:* **Yahapetha'** 'Guineo'; *Use:* **chul** 'flute'; *Part:* **che'il** 'stalk'; *Sim:* **semet'** '*Rhipidocladum bartlettii*'; *Sim:* **bambu** 'bamboo (*Bambusa vulgaris* 'Vittata')'. *SD:* Plants. *[Note:* The name was unfamiliar to BMjr or other young Lacandones whom I interviewed.*] [Source:* AM; BM*] [* \sd2 musical_instruments*]*

B–b

bahche' bahche'. *Lit:* 'nail tree'. *Loc:* **k'aax** 'forest'; *Spec:* **ek' bahche'** 'black bahche' (*Guatteria anomala*)'; *Spec:* **säk bahche'** 'white bahche' (*Guarea glabra*)'. *SD:* Plants.

Ch–ch

chäkhun che' *Lit:* 'red bark cloth tree'. A tree with a straight trunk, approximately 30 cm (11″) in diameter, with somewhat smooth, exterior bark and bright orange inner bark. It is unclear whether or not the tree provides fibre for barkcloth. *See:* **chäk hu'un**. *SD:* Plants. *Thes:* **che'**. *[Source:* AM; BM*]*

chäk 'akte' *Lit:* 'red 'akte''. A spiny palm variety of hach 'akte' (*Astrocaryum mexicanum*). According to BM, it is distinguished by its reddish leaf sheath. No uses were reported. *Loc:* **pach wits** 'behind the hills'; *Sim:* **ya'ax 'akte'** 'green 'akte''; *Gen:* **hach 'akte'** 'authentic 'akte' (*Astrocaryum mexicanum*)'. *SD:* Plants. *[Source:* AM; BM*]*

chäk hach chulul *Lit:* 'red authentic bow'. *Use:* **chulul** 'bows'; *Sim:* **ek' hach chulul** 'black chulul'; *Gen:* **hach chulul** 'authentic chulul'. *SD:* Plants. *Thes:* **che'**. *[Source:* AM*] [* \sd2 hunting_and_fishing*]*

chäk nikte' ak' *Lit:* 'red blossom-tree vine'. *Use:* **uyoomtik käkow yetel** 'One froths cacao with it'; *Part:* **yits uyol** 'resin from the growing tip'. *Variant:* **chäk ak'**. *SD:* Plants. *Thes:* **ak'**. *[Source:* AM*] [* \sd2 food*]*

chäk 'oox *Lit:* 'red breadnut'. ***Mas mihin que hach 'oox. As chäkchäk. Yuul usool. Utahal ti' marso.*** 'Its (fruit) is much smaller than that of hach 'oox. It is sort of red. The skin is smooth. It ripens in March.' (BM). *Use:* **hanal** 'food'; *Part:* **wich** 'fruit'; *Prep:* **che'che'** 'raw'; *Sim:* **hach 'oox** 'authentic breadnut (*Brosimum alicastrum*)'; *Sim:* **ya'ax 'oox**; *Sim:* **k'än 'oox** 'yellow breadnut (*Brosimum alicastrum*)'; *Gen:* **'oox**. *SD:* Plants. *Thes:* **che'**. *[Note:* Likely a tree in the Moraceae family. I was unable to locate the tree, and so without plant materials and photos it is difficult to determine the taxon. The only species that comes closest to BM's description is *Trophis racemosa*. Few references are made to *chäk 'oox* in the Mayan botanical literature. Burger (1962: 31) identifies cha cox as *Trophis racemosa* ssp. *ramon* (Schlecht. & Cham.) W. Burger. He describes the species as being adapted to drier conditions

S. Cook, *The Forest of the Lacandon Maya*,
DOI 10.1007/978-1-4614-9111-8, © Springer Science+Business Media New York 2016

than those tolerated by other members of *Trophis*. It grows on dry hillsides, in ravines, river valleys, and forests from sea level to 2500 m (Burger 1962: 10). The only other reference to *chäk 'oox* is provided in Atran et al. (2004), where "*chäk ox*" [Itz.] is associated with *Brosimum alicastrum.] [Source:* AM; BM*]* [\sd2 food*]*

chäk pahok This is a small palm that is approximately 1.8 m (6′) tall. The name refers to the red inflorescence and growing tips. *Loc:* **wits** 'hills'; *Spec:* **mehen chäk pahok** 'small red pahok'; *Sim:* **hach pahok** 'authentic pahok'; *Sim:* **k'unche' pahok** 'k'unche' pahok (*Geonoma oxycarpa*)'; *Gen:* **pahok**. *SD:* Plants. *[Source:* AM; BM*]*

chäk tso'ots bamax *Lit:* 'red hairy bamax'. *Use:* **hanal** 'food'; *Part:* **wich** 'fruit'; *Prep:* **che'che'** 'raw'; *Sim:* **k'än tso'ots bamax** 'yellow hairy bamax'; *Gen:* **tso'ots bamax** '*Pseudolmedia oxyphyllaria*'. *SD:* Plants. *Thes:* **che'**. [\sd2 food*]*

chehew bamboo (a type of). ***Nuk uk'i'ixel, tak ba'ik sukal utsa'apil [<ts'ap?].*** 'It has large thorns and sections like those of sugarcane' (CK, in Bruce 1975: 142). This word was unfamiliar to all of my consultants, except for AM. From CK's description, *chehew* could be a species of Guadua. *G. amplexifolia* is the only species listed in the most recent Lacandon botanical literature (Comisión Nacional de Áreas Naturales Protegidas 2006), although there may be others in the area, namely *G. aculeata, G. paniculata* (syn. *Chusquea spinosa*) and *G. longfolia* (Gibb Cooper, pers. comm.). All are thorny and clumping. *Use:* **chul** 'flute'. *See:* **'ämäy**. *SD:* Plants. *[Note:* ché'eh [Yuc.]. be exposed (points, tips); cheh 'stick out', expose (Bricker et al. 1998).*] [Source:* CK, in Bruce (1975).*]* [\sd2 musical_instruments*]*

chixtun che' *Lit:* 'gravel tree'. *Loc:* **wits** 'the hills'; *Use:* **hanal** 'food'; *Part:* **nek'** 'seeds'; *Prep:* **ak'elik unek' ich xämäch, pachil amäk'ik, p'iis kakawat** 'You toast the seeds on the griddle and then you eat them like peanuts'. *Morph:* **chich=tun che'**. *SD:* Plants. *Thes:* **che'**. *[Note:* BM was unfamiliar with this name.*] [Source:* AM*]* [\sd2 food*]*

Ch'–ch'

ch'ox ***Nukuch che'. Uyäch'ik ule', ya'ax.*** 'This is a big tree. You crush the leaves to get blue (dye).' (AM). *Loc:* **Mensäbäk** 'Mensäbäk'; *Use:* **ti' ubanik ba'al** 'for dying things'; *Part:* **le'** 'leaves'; *Prep:* **ayäch'ik ule'** 'You crush the leaves'. *SD:* Plants. *Thes:* **che'**. *[Note:* ch'ox. Indigofera suffruticosa (Baer and Merrifield 1971); ch'òoh [Yuc.]. Indigofera suffruticosa Miller, Indigofera mucronata. An herbaceous plant used to treat varicose veins, flowers or leaves used for dye, branches used for brooms (Bricker et al. 1998: 86); ch'oj [Itz.] tinto de añil. indigo dye. Cf tzitz (Hofling and Tesucún 1997).*] [Source:* AM*]* [\sd2 craft*]*

ch'ulkeh (ah) *Lit:* 'the deer urine'. A tall shrub that grows along roadsides and disturbed areas. It bears large, yellow sunflowers. The Lacandones consider it *lo'obil* 'a weed'. *Use:* **ts'ak ti' uyoot'el ahpek'** 'medicine for dog's skin, mange'; *Part:* **le'** 'leaves'; *Prep:* **uch'ulik ule' ich ha', pachil uhixtik uyoot'el ahpek' yetel, 'oxwäts'** 'one soaks the leaves in water and then scrubs the dog's skin with them, three times'. *Morph:* **ch'ul-keh**. *SD:* Plants. *[Note:* chul-ceh [Yuc.]. Lit. deer-chul. A decoction is employed as a bath for aching bones, convulsions, giddiness, and an unidentified disease (Roys 1931: 239). The word was unfamiliar to BM.*] [Source:* AM*]* [\sd2 medicinal*]*

E–e

ek' hach chulul *Lit:* 'black authentic bow'. *Use:* **chulul** 'bows'; *Sim:* **chäk hach chulul** 'red authentic chulul'; *Gen:* **hach chulul**. *SD:* Plants. *Thes:* **che'**. *[Source:* AM*]* [\sd2 hunting_and_fishing*]*

H–h

hach koyoh coyo persea (wild pear). A cultivated relative of *Persea*. *Use:* **hanal** 'food'; *Part:* **wich** 'fruit'; *Prep:* **che'che'** 'raw'; *Sim:* **nukuch koyoh** 'large koyoh' (*Persea schiedeana*); *Gen:* **koyoh**. *Variant:* **mehen koyoh**. *SD:* Plants. *Thes:* **che'**. *[Source:* AM; BM*]* [\sd2 food*]*

hach nikte' *Lit:* 'authentic nikte''. **Hach nikte' yan uche'il, batak 30 cm (12″). P'iis che'. Unuktal p'iis uka'anil ubakel uho'ol nah. Unuk utan ule' utop'. Latanup ule' utop'. Yan ubok, ne ki'.** 'Hach nikte' has a trunk about 30 cm (12″) in diameter, the same as a tree. It grows as high as the roof of a house. The petals are large. There are four petals. They are very fragrant.' (BM) BM said the plant is invoked in a prayer to consecrate the copal. Cultivated. *Use:* **tsoy uwich** 'ornamental'. *SD:* Plants. *Thes:* **ak'**. *[Source:* BM*]* [\sd2 ornamental \sd2 religion*]*

hach pahok *Lit:* 'authentic pahok'. According to BM, this tree is similar to *chäk pahok*, though much taller and with larger leaves. It prefers a low, level terrain with rich soil. It does not grow well in the hills. *Loc:* **äka'an** 'low, level ground'; *Use:* **che'il k'ak'** 'firewood'; *Use:* **pixnah** 'thatch'; *Part:* **le'** 'leaves'; *Sim:* **chäk pahok** 'red pahok'; *Sim:* **k'unche' pahok** '*Geonoma oxycarpa*'; *Gen:* **pahok (ah)**. *Variant:* **pahok**. *SD:* Plants. *[Source:* BM*]* [\sd2 construction \sd2 fuel*]*

hach pom *Lit:* 'authentic incense'. A tree whose resin is collected and mashed into *pom* 'incense'. According to AM, the tree does not grow in Naha'. He recalls it being abundant in a place called *Päk'änil Pom* and thinks it might still grow there and in Chancalá. BM says it also grows in Mensäbäk. *Loc:* **Mensäbäk** 'Mensäbäk'; *Loc:* **Päk'änil Pom** 'Tenosique (?)'; *Loc:* **chankalá** 'Chancala'; *Use:* **pom** 'incense'; *Part:* **uyits** 'resin'; *Sim:* **tsatsel pom** '*Protium copal*'; *Gen:* **pom**. *See:* **tähte'**. *SD:* Plants. *Thes:* **che'**. *[Note:* tasi pom. *Bursera simaruba* (Nations and Nigh 1980); tasi' pom. *Bursera simaruba* (Baer and Merrifield 1971).*] [Source:* AM; BM*] [*\sd2 other \sd2 religion*]*

ho'te' che' A species whose stalks were used to make arrows. According to AM, it was preferred over *'ooh* (*Gynerium sagittatum*), which has now supplanted it. It could be *Phragmites australis*, which AM said was used before they switched to *'ooh*. *Use:* **häläl** 'arrows'; *Part:* **che'il** 'stalk'. *See:* **hach häläl**. *SD:* Plants. *[Source:* AM*] [*\sd2 hunting_and_fishing*]*

K–k

koyoh [koyoh, koyok] coyo persea (wild pear). The species is a wild relative of *'oon* (*Persea americana*). The Lacandones distinguish two kinds: *nukuch koyoh* 'large koyoh' and *hach koyoh* 'authentic koyoh'. The former is cultivated and the latter is wild. Both produce elongated avocado-like fruit. According to my Lacandon consultants, the fruit of *hach koyoh* is similar to that of *nukuch koyoh* only smaller. *Nukuch koyoh* is *Persea schiedeana*. *Hach koyoh* was unavailable to inspect, but it is probably *Beilschmiedia anay*. Both occur in the region. *Persea schiedeana* grows in mountain forests from southern Mexico to Panama in temperate, lowland and montane wet forests at altitudes between 400 and 1600 m (1312–5248′) (Gutiérrez-Carvajal and Dorantes-López 2003–2004). One of the common names for *Persea schiedeana* is "coyo" (González-Rosas et al. 1985; Hellmuth 2013: 25). The fruit is large, thick-skinned, and pendulous. The scant, soft, stringy pulp is considered by some people to be inedible (Schoenhals 1988: 13). *Beilschmiedia anay* bears obovate, green- or black-skinned fruit. Its pulp is higher in oil than avocados and sweeter (Borys et al. 1993). It grows wild in forests in northeast Chiapas at altitudes between 400 and 700 m (1312–2296′) (Pennington and Sarukhán 2005: 176). It likely occurs in the northern Lacandon settlement of Mensäbäk. *Use:* **hanal** 'food'; *Part:* **wich** 'fruit'; *Prep:* **che'che'** 'raw'; *Spec:* **nukuch koyoh** 'large koyoh (*Persea schiedeana*)'; *Spec:* **hach koyoh** 'authentic koyoh'. *SD:* Plants. *Thes:* **che'**. *[Note:* koyoh. wild avocado (Bruce 1975); koyok. unidentified wild avocado-like fruit (Nations 2006); koyok. *Beilschmiedia anay* (Nations and Nigh 1980).*] [Source:* BM*] [*\sd2 food*]*

K'–k'

k'an₁ *Lit:* 'cord'. **Ne chich usool. Yan uk'an uwich, p'iis kakawat. Ne ki'. Ahk'ek'en kuhantik.** 'The bark is very hard. The fruits have cords, the same as peanuts. They're very tasty. The white-lipped peccary eats them.' (BM). *Use:* **hanal** 'food'; *Part:* **nek'** 'seeds'; *Prep:* **k'elbil** 'toasted'. *See:* **pakah che'**. *SD:* Plants. *Thes:* **che'**. *[Source:* BM*] [*\sd2 food*]*

k'änse ak' Pach uwich. Yan uyits ne tsoy ti' koh, chi' (AM). 'The fruit is surrounded by a hard shell. The resin is very good for treating the teeth and mouth.' (AM) According to BM, this is an unarmed, woody vine that climbs high up into the canopy. The vines are as thick as tree branches. When cut, they exude a yellowish resin. The species could be Securidaca diverifolia. Compare "kän sehak" (Kashanipour and McGee 2004) and "kach che ak" (Durán 1999). Both sources report that the vine is used to treat diarrhea. It is also used to treat toothaches in Malaysia (Phyknome, http://mapping.fbb.utm.my/phyknome/node/6). *Use:* **ts'ak ti' chi', koh** 'medicine for the mouth (cankers) and teeth'; *Part:* **uyits** 'resin'; *Prep:* **mäna'** 'none'. *SD:* Plants. *Thes:* **che'**. *[Note:* kän sehak. unidentified (Kashanipour and McGee 2004); kach che ak, a kan ak. Securidaca diversifolia (Durán 1999).*] [Source:* AM; BM*] [*\sd2 medicinal*]*

k'än suum (ah) *Lit:* 'yellow rope'. **Ak'. Yan utop'. Ne k'än utop'. Febrero utop'ol. Ahhach t'eech ich utop'. (AM).** 'This is a vine that bears very yellow flowers. The flowers open in February. The 'real t'eech' (a kind of horsefly) is inside the blossoms.' (AM) KP provided a similar description for säktun. There is still another plant called säk suum that bears white flowers. *Sim:* **säk suum (ah)** 'white suum'; *Gen:* **suum**. *See:* **säktun (äh)**. *SD:* Plants. *Thes:* **ak'**. *[Note:* kan sum. Sinclairia deppeana (Durán 1999).*] [Source:* AM*]*

k'än tso'ots bamax *Lit:* 'yellow hairy bamax'. *Use:* **hanal** 'food'; *Part:* **wich** 'fruit'; *Prep:* **che'che** 'raw'; *Sim:* **chäk tso'ots bamax** 'red bamax'; *Gen:* **tso'ots bamax** '*Pseudolmedia oxyphyllaria*'. *SD:* Plants. *Thes:* **che'**. *[Source:* SK*] [*\sd2 food*]*

k'i'ix (ah) *Lit:* 'the thorn'. BM identified a thorny stem as *ahk'i'ix* and AM identified a long, flattened pod, resembling the fruit of a *Mimosa* species as *ahk'i'ix*. These identifications were provided at separate times, thus it is uncertain whether they belong to the same plant. *SD:* Plants. *[Note:* jarochkiix. *Mimosa ervendbergii* (Nigh 2008).*] [Source:* AM; BM*]*

k'isis che' A large, tropical tree bearing clusters of small, round drupes that turn black and sweet when ripe. The wood contains a resin that is *ne ts'ak* 'very strong' (harmful). (BM) The only reference to the Lacandon word is found in Bruce, who describes *"k'isiseh"* as a tropical species of black haw (1976: 90). Black haw is the English common name for *Viburnum prunifolium* L. Bruce was probably only familiar with the Black haw from his native Oklahoma, since *Viburnum prunifolium* is prevalent in North America and was used by the indigenous groups there for medicinal purposes. I was unable to locate the plant, so its botanical identity is uncertain. Neither the Lacandon name nor *Viburnum* species appear in the extant Lacandon ethnobotanical literature or botanical inventories. Nevertheless, Morton (1933) lists five species from (southern) Chiapas: (1) *Viburnum guatemalensis*, (2) *V. lautum*, (3) *V. blandum*, (4) *V. jucundum*, and (5) *V. elatum*. Breedlove (1986) adds to this list: *V. acutifolium*; *V. actifolium* [sic?]; *V. discolor* (also a Guatemalan species); *V. hartwegii*; *V. mendax*; and, *V. obtusatum*. *Use:* **hiil** 'battens'; *Part:* **mehen che', k'äbche'** 'saplings, branches'; *Use:* **säkche'** 'rafters'. *Variant:* **k'isis**. *SD:* Plants. *Thes:* **che'**. *[Source:* BM*]* *[*\sd2 construction*]*

k'uyuch ak' *Lit:* 'twisted vine'. **Chan mihin uyak'il. Yan uwich, ne nuk uwich, batak 15.25–20.32 cm (6–8″) wolis, t'elt'el. Ya'ax. Yan uyits, säk uyits. Yan unek'. Säk unek'. Ne ma' chich. Uhantik pokbil. Ne hup. P'iis ahp'ix. Unek' ahantik tambien.** 'It has tiny vines. Its fruits are very large, about 15.25–20.32 cm (6–8″) round, and ridged. They're green. There is white resin. The seeds are white and not very hard. One eats (the fruit) roasted. They're very soft, the same as chayote (*Sechium edule*). You can eat the seed too.' (AM) According to BM, the flavour is similar to potatoes. *Use:* **hanal** 'food'; *Part:* **wich** 'fruit'; *Prep:* **pokbil** 'roasted'. *Morph:* **k'uy-uch ak'**. *SD:* Plants. *Thes:* **ak'**. *[Source:* AM; BM*]* *[*\sd2 food*]*

M–m

mehen chäk pahok *Lit:* 'small red pahok'. *Gen:* **chäk pahok**. *SD:* Plants. *[Source:* AM*]*

mehen 'oop che' *Lit:* 'small custard apple tree'. **P'iis 'oop. K'än uwich. Chan mihin uwich, batak 5 cm (2″). Ch'uhuk.** 'It's the same as *'oop*. The fruit is yellow and small, about 5 cm (2″) in diameter. It's sweet.' (AM). *Loc:* **Yahapetha'** 'Guineo'; *Use:* **hanal** 'food'; *Part:* **wich** 'fruit'; *Prep:* **che'che'** 'raw'; *Sim:* **hach 'oop che'** 'authentic custard apple tree (*Annona reticulata, A. cherimoya*)'; *Sim:* **chäk 'oop che'** '*Clarisia biflora*'; *Gen:* **'oop che'**. *SD:* Plants. *[Source:* AM*]* *[*\sd2 food*]*

mehen säk bahche' *Lit:* 'small white bahche''. *Loc:* **k'aax** 'forest'; *Sim:* **nukuch säk bahche'** '*Guarea* sp.'; *Gen:* **säk bahche'** '*Guarea glabra*'. *SD:* Plants. *Thes:* **che'**. *[Source:* BM*]*

mehen tu'xikin *Lit:* 'small rotten ear'. *Sim:* **nukuch tu'xikin** 'large rotten ear'; *Gen:* **tu'xikin**. *SD:* Plants. *[Note:* mehen tu'xikin. *Aristolochia foetida* (Rätsch 1994a).*]* *[Source:* Rätsch (1994a)*]*

N–n

nikte' ak' *Lit:* 'blossom-tree vine'. **Es ak'. Uche'il uyak'il. Ne chukuch uyak'il. As k'änk'än utop', batak 6.35 mm (1/4″). Usool yan ubok, p'iis ubok peesache'. Ha'li' uche'il yan ubok. Kuk'anik ti' metik nah, ti' ukäxik uyokman nah. Uch'ihil ha'li' ich k'aax, naach. Ahya'ax ak' uboho'.** 'It's a vine. The stems of the vines are very long. The blossoms are yellow and about 6.35 mm (1/4″) in diameter. The bark has an odor similar to allspice (*Pimenta dioica*). Only the stems have an odor. One uses them to make houses, for tying the house pillars. It only grows in the forest, far away. The 'green vine' is its companion.' (BM) This species was unavailable to inspect, but if it is a companion of *ya'ax ak'*, then it may be a species of Bignoniaceae. Kashanipour and McGee (2004) and Nations (2006) surmise that it corresponds to a liana called 'Santo Palo, bejuco de pimienta' (Spanish). In Itzaj, *ajpimijeentaj ak'* translates as 'bejuco de pimienta' or 'bejuco de agua'; it smells like *pimienta gorda* 'allspice' (Hofling and Tesucún 2000: 110). *Bejuco pimienta* corresponds to *Arrabidaea floribunda*, according to Lindsay (2011); however, the flowers of this species are pinkish-purple, whereas those of *nikte' ak'* are yellow. *Callichlamys latifolia* has yellow flowers, but they are larger than 6.35 mm. According to Kashanipour and McGee, the bark is boiled and the liquid drunk to relieve bone and muscle aches (2004: 63). My Lacandon consultant, AM, said that he drinks the juice of the vine to relieve a stomach ache. KP said it made a tasty beverage. *Loc:* **k'aax** 'forest'; *Use:* **ti' uhich'ik uyokman ti' utaan** 'for tying house posts to the girt'; *Part:* **uyak'il** 'its vines'; *Use:* **ch'uyu'** 'the frame for hanging baskets'; *Part:* **uyak'il** 'its vines'; *Use:* **uk'ul** 'beverage'; *Part:* **uyits ak'** 'resin from the vine'; *Use:* **chibal hämnen** 'stomach ache'; *Part:* **uyits ak'** 'juice from the vine'; *Use:* **yah ubaakel** 'bone ache'; *Sim:* **ya'ax ak' (ah)** 'blue/green vine (*Arrabidaea* sp.)'; *Sim:* **säk ak'** 'white vine (*Arrabidaea verrucosa*)'. *Variant:* **nikte'**. *SD:* Plants. *Thes:* **ak'**. (Appendix 1.1) *[Note:* nikte' ak'. *Clematis* spp. (Kashanipour and McGee 2004); nikte' ak. palo santo. *Clematis* spp. (Nations 2006).*]* *[Source:* AM; BM; KM; KP*]* *[*\sd2 craft \sd2 cordage \sd2 medicinal \sd2 beverage*]*

Appendix 1.1 (a) Nikte' ak' (b) Nikte' ak'

nukuch kopo' Es nukuch che'. Uche'il batak 1 m (3′). Mäna' uyak'il. Chan mihin utop', säk. Uwich batak 2.5 cm (1″) wolis. K'än usool uwich cuando täk'än. Uyo'och ahba'ats'. 'It's a large tree. The trunk is about 1 m (3′) around. The flowers are tiny and white. The fruit is about 2.5 cm (1″) round. When they're ripe they turn yellow. It's the food of the howler monkey.' (BM) The specimen showed to me was a tree with a rangy habit, the stems were light grey and flexible. Leaves were large and oblong, about 18 cm (7″) long by 8 cm (3″) wide, slightly cordate at the base with numerous lateral veins running from nearly horizontal to 45 degrees from the main vein, which was thick, light green, and raised above the lamina. Secondary veins were prominent and raised. Leaves were dark green, thick, stiff, and leathery. Petioles were depressed along the centre like a shallow channel. Neither fruit nor flowers were present at the time, making a positive identification difficult. The leaves, however, resemble those of *Ficus lapathifolia* (Liebm.) Miq., an endangered species (World Conservation Monitoring Centre 1998). *Loc:* **k'aax** 'forest'; *Use:* **mukbil wah** 'large baked tamales'; *Part:* **le'** 'leaves'; *Sim:* **mehen kopo'** 'small kopo' (*Ficus pertusa*)'; *Sim:* **kopo' wits** 'kopo' of the hills'; *Gen:* **kopo' (ah)**. *SD:* Plants. *Thes:* **che'**. *[Note:* nukuch kopo. *Coussapoa oligocephala* (Durán 1999).*] [Source:* AM; BM; KM*] [*\sd2 food*]*

nukuch k'ik'ni' balum *Lit:* 'large bleeding-nosed jaguar'. *Use:* **hanal** 'food'; *Part:* **wich** 'fruit'; *Prep:* **che'che'** 'raw'; *Sim:* **mehen k'ik'ni' balum** 'small k'ik'ni' balum (*Eugenia* sp., *Ardisia compressa*)'; *Sim:* **hach k'ik'ni' balum** '*Ardisia compressa*'; *Gen:* **k'ik'ni' balum (ah)**. *Variant:* **ne nuk k'ik'ni' balum**. *SD:* Plants. *[Source:* AM*] [*\sd2 food*]*

nukuch p'ap'ax *Lit:* 'large nettle'. *Gen:* **p'ap'ax**. *SD:* Plants. *Thes:* **che'**. *[Source:* BM*]*

nukuch tu'xikin *Lit:* 'large rotten ear'. AM described it as a vigorous, invasive, climbing plant with thick vines 10–15.24 cm (4–6″) in diameter. He said it has large, tubular flowers that look like *tu'xikin*. *Use:* **ti' ukinsik k'ik'** 'to kill the latex of *Castilla elastica*'; *Part:* **uyits uche'il** 'resin from the stems'; *Sim:* **mehen tu'xikin** 'small rotten ear'; *Gen:* **tu'xikin**. *SD:* Plants. *Thes:* **ak'**. *[Note:* nukuch tu'xikin. *Aristolochia grandiflora* (Rätsch 1994a).*] [Source:* AM*] [*\sd2 other*]*

nukuch ts'ulha' (äh) *Sim:* **chäk ts'ulha' (äh)** '*Crinum amabile*'; *Sim:* **ts'ulha'il petha' (äh)** '*Hymenocallis littoralis*'; *Gen:* **ts'ulha' (äh)**. *SD:* Plants. *[Source:* AM*]*

O–o

'oop ma'ax *Lit:* 'monkey annona'. *Sim:* **hach 'oop** 'authentic custard apple'; *Sim:* **'oopil k'aax** 'custard apple of the forest, wild custard apple'; *Gen:* **'oop** '*Annona* spp.'. *SD:* Plants. *[Source:* BM*]*

'op' ak' *Lit:* 'burst vine'. A thin, forest vine with woody stems armed with straight thorns. BM describes it as "a tree with long spines and fruit so hard that even birds can't eat them". *Sim:* **tsuk'in ak' (ah)** '*Smilax domingensis*'; *Sim:* **koke' ak' (ah)** '*Smilax lanceolata*'. *See:* **k'ul ak'**. *SD:* Plants. *Thes:* **che'**. *[Note:* op ak'. *Smilax* sp. (Nations and Nigh 1980); op ak'. *Smilax* spp. (Kashanipour and McGee 2004); o pak. *Strychnos brachistantha* (Durán 1999); op ak' [Lac.]. *Dioscorea bartlettii* (in Atran et al. 2004); ix kokolmeekaj [Itz.]. *Dioscorea bartlettii* (Atran et al. 2004).*] [Source:* BM*]*

Appendix 1.2 Pakah che'

P–p

pakah che' [pakahche', pakache', pahache'] (Flacourtiaceae). ***Ne chich usool. Yan uk'an uwich. P'is kakawat. Ne ki'. Ähk'ek'en kumäk'ik.*** 'The skin of fruit is very tough. The fruits have cords. They're like peanuts. They're really tasty. The *hach k'ek'en* 'white-lipped peccary' eats them all the time.' (BM) AM adds that the fruits are round and about 7.6 cm (3″) in diameter. The shells are hard and black, and the nutmeat is white and soft. The *ts'u'ts'u'* 'coatimundi' relishes the fruits. Neither fruit nor flowers were present, making it difficult to determine the taxon. Durán (1999) reports that his Lacandon consultants eat the seeds of *Casearia aculeata*, which they call *"pa chac che"*. The tree that my consultants identified does not, however, resemble this species. The leaves are not serrated and the petioles are longer. *Use:* **hanal** 'food'; *Part:* **wich** 'fruit'; *Prep:* **k'elbil** 'toasted'; *Use:* **hiil** 'battens'; *Part:* **mehen che', k'äbche'** 'saplings, branches'. *See:* **k'an**. *SD:* Plants. *Thes:* **che'**. (Appendix 1.2) *[Note:* pa chac che. *Casearia* aff. *aculeata* Jacq. (Durán 1999).*] [Source:* AM; BM*] [*\sd2 food \sd2 construction*]*

payok che' (ah)$_1$ A tall forest tree with glossy, lance-shaped leaves approximately 13 cm (4″) by 4.5 cm (1.7″). They are simple, alternate, smooth, entire, slightly bullate, and arranged horizontally in two tiers along thin, zig-zagging branchlets. According to BM the fruit is a small capsule, similar to that of the *k'ik'che'* tree (*Virola guatemalensis*), which turns red at maturity.When ripe, in June, the fruits draw a variety of birds including *pän* (keel-billed toucan), *t'ut'* (white-crowned parrot), and *pichik'* (collarded araçari toucan). *SD:* Plants. *Thes:* **che'**. *[Note:* a pa yok che. *Casearia bartlettii* (Durán 1999).*] [Source:* BM*]*

pet ak' (ah) [pet ak', pet' ak'] **Ne k'än utop'. Chan mihin utop', pechepech, ne yaap'. Uyahchunil, ahpet ak' yetel halol uwaakal kuluk'ul utop', ahäläch'äktik lo'obil i che'.** 'The flowers are very yellow. The flowers are very small and numerous and overlapping. *Pet ak'* and the *halol* (*Heliocarpus* spp.) are the first to bloom in the year. When they do, it is time to clear the underbrush and small trees [in the *milpa.*] After the flowers fall to the ground, you can plant corn.' (AM) From the description the species may be *Senna racemosa* (Mill.) H. S. Irwin & Barneby. *Senna racemosa* is a large shrub or small/medium deciduous tree that is covered with sweetly fragrant, lemon-yellow blossoms, which attract bees and butterflies. Fruit is a brown seed pod. Native to the Yucatan Peninsula, Cuba and Central America as far as Costa Rica, it flourishes on limestone. *Indicator*: *Part:* **utop'** 'its flowers'; *Activity:* **halalch'äktik** 'cut down underbrush with a machete'. *SD:* Plants. *Thes:* **ak'**. *[Note:* jai patan ak. *Senna racemosa* (Durán 1999).*] [Source:* AM*] [*\sd2 agriculture*]*

P'–p'

p'ap'ax nettle. A kind of stinging nettle with a vine-like habit. *Loc:* **k'aax** 'forest'; *Spec:* **nukuch p'ap'ax** 'large p'ap'ax'; *Sim:* **tsaah (äh)** '*Cnidoscolus multilobus*'. *[Note:* ppoppox [Yuc.]. *Tragia nepetaefolia, T. Guameri, T. yucatanensis*: "This plant takes its name from the complaint for which it is the remedy, an ache in the limbs or gout. Ppoppox. Nettles of this and which contain some poison Zac-ppoppox (white ppoppox) is one kind; chac-ppoppox (red ppoppox) is another. …The yax-(green-) ppoppox is still another kind." (Motul.) (Roys 1931: 278).*] [Source:* BM]

S–s

säktun (äh) *Lit:* 'white rock'. A large, thick liana with stems that sag and loop every which way, winding up trees and forming a thick twisted trunk at its base. The leaves are simple, elliptic, acuminate, medium-green and glossy, approximately 16 cm (6.5″) long and half as wide, on long petioles.

According to KP, it bears fragrant, yellow blossoms that are home to the *t'eech* (a type of horsefly). *See:* **säk suum (ah)**. *SD:* Plants. *Thes:* **ak'**. *[Note:* kan sum. *Sinclairia deppeana* (Durán 1999).*] [Source:* KP*]*

säk bobo ule' ***Nuk uche'. K'än uwich. Uyo'och ahkachok'. Octubre yan uwich. Ule', ya'ax, säk tuyalam.*** 'It is a large tree. The fruit is yellow. It is food for the *kachok'* 'mealy blue-crowned parrot' (*Amazona farinosa*). It fruits in October. The leaves are green [above] and white below.' (AM) The observed species was found in secondary forest close to the road outside of Naha'. *Loc:* **k'aax**. *SD:* Plants. *Thes:* **che'**. *[Source:* AM*] [*\sd2 fuel*]*

säk luuch *Lit:* 'white gourd'. According to AM, säk luuch is an old variety of calabasa which is not cultivated anymore. *Use:* **luuch** 'cups'; *Sim:* **hach luuch** 'authentic luuch (*Crescentia cujete*)'; *Sim:* **sina'an luuch** 'stretched out luuch (*Crescentia cujete, C. alata*)'; *Gen:* **luuch**. *SD:* Plants. *[Source:* AM*] [*\sd2 containers*]*

säk pukte' *Lit:* 'white pukte''. A large tree with a trunk approximately 1 m (3′) in diameter. The wood is very hard and heavy. The pith of mature trees contains stones. I was unable to locate the tree, however Diemont et al. (2006) identify the species as *Bucida buceras*, and Durán (1999) identifies it as *Ulmus mexicana*. The timber of both species is hard and heavy, and contains high levels of silica. The silica could be the "stones" that BM refers to. In Itzaj, *pukte'* corresponds to *Bucida buceras*. There are a number of folk specifics, named according to the colour of the heartwood, i.e., *b'ox pukte'* 'black pukte'', *k'än pukte'* 'yellow pukte'', and *säk pukte'* 'white pukte'' (Atran et al. 2004: 100; Hofling and Tesucún 1997: 525), and they all correspond to *Budica buceras*. The tree my Lacandon consultants called *pukte'* corresponds to *Terminalia amazonia*, a species that belongs to the same family as *Bucida buceras*. *Terminalia amazonia* and *Bucida buceras* are similar with respect to their leaves, but *T. amazonia* is a much larger tree, growing to 50 m (164′), whereas *B. buceras* tends to be bushy with thorny branches (Morton 1993). The fruits of *T. amazonia* are winged achenes, similar to the keys of maple trees, whereas the fruit of *B. buceras* are urn-shaped, reddish-brown berries borne in drooping clusters. *T. amazonia* lacks the high silica content of *B. buceras*. *Ulmus mexicana* may be mistaken for *Terminalia amazonia*, as it also produces winged achenes. It appears that the conceptual range of *pukte'* includes members of Combretaceae and Ulmaceae, but more research is needed to confirm this. Although no uses for *säk pukte'* were mentioned by my consultants, Diemont (2006) reports that the timber of *Bucida buceras* is used in construction and as firewood in the southern Lacandon community. According to Nations and Nigh (1980: 12), the blossoms of *Bucida buceras* signal the time to sow corn. *Gen:* **pukte'** 'bullywood, canshan (*Terminalia amazonia*)'. *SD:* Plants. *Thes:* **che'**. *[Note:* sa puk te. *Bucida buceras*. Timber used in construction and as fuel (Diemont et al. 2006); pokte'. *Bucida buceras* (Nations and Nigh 1980). sac pucte. *Ulmus mexicana* (Durán 1999); puk-te' [Itz.]. *Bucida buceras*. bullet tree (Hofling and Tesucún 1997).*] [Source:* BM*]*

Appendix 1.3 Säk suum flowers

säk suum (ah) Mehen che'. Yan utop'. Säk utop'. Chan mihin, batak .63 cm (.25″). Febrero kutop'ol ahek' t'eech ich utop'. 'A small tree with tiny, white flowers about .63 cm (.25″) in diameter. When it blooms in February the *ek' t'eech* 'black horsefly' is inside of the flowers.' (AM). *Sim:* **k'än suum (ah)** 'yellow suum'; *Gen:* **suum**. *Variant:* **säktun (äh)**. *SD:* Plants. *Thes:* **che'**. (Appendix 1.3) *[Source:* AM*]*

suum₂ *Lit:* 'rope'. *Spec:* **k'än suum (ah)** 'yellow rope'; *Spec:* **säk suum (ah)** 'white rope'. *SD:* Plants.

Ts'–ts'

ts'ak 'ax *Lit:* 'wart medicine'. ***Es lo'obil. Chukuch ule'.*** 'It's a weed. It has long leaves.' (AM) The plant is further described as a two-foot herbaceous bush with 30.48–45.7 cm (12–18″) long, smooth leaves that sting like nettles. It bears small white flowers followed by 5.08 cm (2″) blue fruit. *Loc:* **k'aax** 'forest'; *Use:* **ts'ak ti' 'ax** 'medicine for warts'; *Part:* **le', che'il** 'leaves and stems'; *Prep:* **Ayäts'ik uyits yok'ol 'ax** 'You squeeze the resin from the leaves onto the warts'. *SD:* Plants. *[Note:* AM says that this never worked him, although Mateo Viejo swore by its effectiveness.*] [Source:* AM; KyP*] [*\sd2 medicinal*]*

ts'ak kan ak'₂ *Lit:* 'snake medicine vine'. This is a vigorous, twining, thin-stemmed vine with alternate, light- to medium-green, heart-shaped leaves, 7.6–10 cm (3–4″) in diameter. Primary veins are palmate. It grows rampant in disturbed areas. Leaves contain a resin with a strong chlorine odour and bitter taste that produces a slight numbing sensation on the tongue. Although its flowers were unavailable, SK said they they were blue and looked similar to morning glories. According to SK, this species is considered one of best snake bite remedies. *Use:* **chibal kan** 'snakebite'; *Part:* **le'** 'leaves'; *Prep:* **Ayäch'ik ule', pachil apulik yok'ol ti'** 'You crush the leaves and then you apply them to [the area]'; *Part:* **mots** 'roots'; *Prep:* **Apech'ik, achäkik yok'ol ha', pachil awuk'ik** 'You crush the roots, boil them in water and then you drink it'. *SD:* Plants. *Thes:* **ak'**. *[Note:* SK says that there are several kinds of plants that are used to treat snakebite and that they are all called *ts'ak kan*. He described a species with a bushy habit and bluish leaves. It produces nodules at its base similar to potatoes (perhaps a species of *Dioscorea*). He says it is the best snakebite medicine of them all. He says this plant is good but not as powerful as the aforementioned one. It's been reported that the northern Lacandones use *Strychnos tabascana* for snakebite. They apply crushed roots to the affected area and drink an infusion of the roots mixed with other plants (Durán 1999: 145). Op ak'. *Smilax* spp. is still another snakebite plant, according to Kashanipour and McGee (2004).*]* *[Source:* AM; AT; KyP; SK*]* *[*\sd2 medicinal*]*

ts'ak k'ak'il [ts'ak k'ak'il, ts'ahk'ak'ih, ts'ak'ak'ih] *Use:* **ts'ak ti' k'ak'il** 'medicine for skin eruptions'; *Part:* **uyits** 'resin'; *Prep:* **Ha'li' apulik uyits yok'ol k'ak'il** 'You just put the resin on the skin eruption'. *SD:* Plants. *[Source:* KyP*]* *[*\sd2 medicinal*]*

Y–y

ya'ax 'akte' *Lit:* 'green akte''. A kind of spiny palm and variety of *hach 'akte'* (*Astrocaryum mexicanum*). According to BM, it is distinguished by its large heart and dull green, and somewhat reddish leaves. Possibly *Bactris mexicana*. *Loc:* **Yaha'nah** 'La Cueva'; *Sim:* **chäk 'akte'** 'red 'akte''; *Gen:* **hach 'akte'**. *SD:* Plants. *[Source:* BM*]*

ya'ax nukuch 'oochin P'iis nukuch 'ooch'in, chen ya'ax uwich. 'It's the same as *nukuch 'oochin*, only its fruit stays green.' (AM). *Gen:* **nukuch 'oochin** 'large 'oochin'. *SD:* Plants. *Thes:* **ak'**. *[Source:* AM*]*

ya'ax tuch *Sim:* **hach tuch** 'authentic tuch (*Thevetia ahouai*)'; *Sim:* **mehen tuch** 'small tuch (*Thevetia ahouai*)'; *Gen:* **tuch**. *SD:* Plants. *Thes:* **che'**. *[Source:* BM*]*

yache' kab *Lit:* 'large honey(bee) tree'. *Use:* **tanah, taan** 'girt'; *Use:* **säkche'** 'rafters'; *Part:* **che'il** 'timber'. *SD:* Plants. *[Note:* ya che kap. *Siparuna andina* (Durán 1999).*]* *[*\sd2 construction*]*

Appendix 2
General Botanical Terms in Lacandon

A–a

ak' vine. **Yan uwich. Kulubul, kuhook'ol ukaanil. Pachil, kubin, kuchuktik ubah yok'ol che'. Kubak'ik ubah. Yan uchan me'ex. Kubak'ik ubah, yan ume'ex, kubin uyak'il. Tan ubin uyak'il kuna'akal, kubin. Tan uk'axik ubah. Yan ume'ex, tan ulubul, tan uhutul tulahyanchunil. Tan ubin, tan ubin, tan ubin, tan uk'axik ubah. K'uchul tan nukuch che', nen ka'anan, ya kuhook'ol ule', kuhook'ol, yaab, ule'. Pachil yan utop'. Kuyantal top'. Kuwaalik utop'. Pachil kulahhutul utop'. Yan uwich yan uwich yan uwich. Ele lati' ak'.** 'It has seeds. When they fall to the ground they send out runners. The runners go along and grab onto the trees and wind themselves around them. They have small roots from which smaller vines develop and go along. They climb up as they go and tie themselves together. They have small whiskers (aerial roots), which fall off and drop to the base of the vine. They go, and go, and go, tying themselves as they go. When they get to the face of a large tree, way up high, their leaves emerge. A lot of leaves come out. Then, their flowers bloom. The flowers burst open. Then, the flowers all drop off and spread their seeds everywhere. That is what a vine is.' (BM). *Spec:* **ya'ax ak'** 'green vine'; *Spec:* **säk ak'** 'white vine'; *Spec:* **ma'ax ak'** 'monkey vine'; *Spec:* **nikte' ak'** 'flowering vine'. *Thes:* **ak'**. *[Note:* When asked to provide an example of a vine (either a name or a drawing), Lacandon consultants invariably refer to, or draw, the large vines that are wrapped around and hang down from tall trees; ak. vine, something fresh and green, tongue (Roys 1931: 209); àak' (2) N. vine (Bricker et al. 1998: 3); ak' (1). N. bejuco. vine (Hofling and Tesucún 1997: 146).*] [Source:* BM*]*

ak'il uk'äb [ak'il uk'äb, ak'ir uk'äp', ak'ir uk'ä'] vine, guide vine. *Lit:* 'vine of the branch'. Refers to vines growing from branches, and aerial roots. *Morph:* **ak'-il u-k'äb**. *Variant:* **uyak'il k'äb, uyak'il**. *[Source:* AM; BM*]*

B–b

baat ahp'ix seed of the chayote squash. *Lit:* 'axe of the chayote squash'. *Use:* **hanal** 'food'; *Prep:* **chäkbil** 'boiled'. *See:* **p'ix (ah)**. *[Source:* KM*]*

bakal [bakal, bakar, bäkäl, bäkär] corncob. The top kernels of the cob (*ho'ol ubakalel*) are reserved for making *posol*. The rest of the kernels are ground into flour, and the best are saved for seed. *Posol* that is made from the tops of corncobs is offered to the gods. *[Source:* BM; SK*]*

bak'che' epiphytes. *Lit:* 'encircle tree'. **Bak'che' p'iis ak' pero ma' ak'. Bak'che' yan ak', yan ak' bak'che', yan, pero yan hach bak'che' mäna' uyak'il, mana' uyaak'il. Yan uchan wi', yan umots'. Kukaxik ubah yok'ol uk'äbche'. Lati' ubak'che'. Kuk'axik ubah. Yan uk'uuk'tikil uwi'. Uk'uuk'tikil es una bola en la raiz, hay salen en la raiz. Lehih umots tu kuk'axik ubah yok'ol uk'äbche'. Ule', yan chukuch ule', yan chukuch ule'. Uyool ma' neh ka'anan. Ki'ki'ts'abil, ki'ki'ts'abil; ma' seeb, ma' seebseeb, hump'el hunts'itna' yan uhel ule'. Hunts'itna' yan uhel ule'. Mix uch'ihil una'akal ubakel, mäna'. Ne ki'ts'abil, uchben uhutik ule', uhutik uka'hook'ol tumen. Lati' ubak'che'. Yan utop'. Kutop'änkäl ti' bix bähe' ti' yan ti' ya'axk'in. Yan uchan top'; yan yan säk utop'. Yan k'aaxa'an utop'. Yan k'än. Es el lehih bak'che', yan utop'. Yan uwich, yan uwich yan umehen ch'iich' yan to'an ne täk'än, ne yi'h, kuhantik mehen ch'iich'. Kupulik uhel mas ubak'che' ti' uhel che' yan uwich.** '*Bak'che'* is like a vine, but it isn't a vine. *Bak'che'* has vines, but a true *bak'che'* doesn't have any vines. *Bak'che'* has a small bulb and roots. The roots attach themselves on the tree branches. This is *bak'che'*. It ties itself around. It has *k'uuk'tikil. K'uuk'tikil* is a ball in the root from which the roots emerge. These roots tie themselves in the branches. It has long leaves. The growth tip doesn't grow

S. Cook, *The Forest of the Lacandon Maya*,
DOI 10.1007/978-1-4614-9111-8, © Springer Science+Business Media New York 2016

very high. It grows very slowly; it's not fast. In one month there is a leaf. The next month there is another leaf. It doesn't climb, either. Gradually the old leaves drop off, one falls off and a new one emerges. That is *bak'che'*. It has flowers. They bloom at this time (September). There are small flowers. There are white flowers. The flowers are different. There are yellow ones. These are *bak'che'*. They have flowers. They have fruit. When they are very ripe, very hard, the little birds flock to eat them. They drop the seeds and where they land a new *bak'che'* will grow.' (BM). *Loc:* **k'aax** 'forest'. *See:* **kuxum che'**. *Platymiscium dimorphandrum [Source:* AM; BM*]*

box$_2$ shell, peel, rind. *[Note:* box [Yuc.]. shell, the thick skin of a fruit (Roys 1931); box, 'u-bo:x-e:r che' [S. Lac]. cáscara muy gruesa de árbol (Canger 1970).*]*

Ch–ch

chan le' sepal. *Lit:* 'little leaf'. *See:* **le' uchuch top'**. *[Source:* KM*]*

chan tso'otsil filament, includes anther. *Lit:* 'little hairs'. *[Source:* KM*]*

chan utop' stigma. *Lit:* 'little flower'. *See main entry:* **top'**. *[Source:* AM; KM*]*

che' tree. **Che' lati' päytan uwich ulubul pachil uwaak'al. En hach t'an uyalik uwaak'al uwich, brotar, uwaak'al uwich. Päytan uwaak'al utichil uchan, uk'aba', uyool. Uyool nen chan mihin, ne ki'ts'aabil tuhook'ol, waak'al uyool. Pachil tuhook'ol uchan le' uyool, waak'al ule'. Pachil uyool tan utichil uyool ule'. Lati' che'. Tan uch'ihil ma' chan mehen ule'. Tan ubutik uhuutul mehen le', uhel tumen uhook'ol uyool uhel ule'. E lati' che'. Tan uch'ihil. Ah, lati' tuwolol yan nukuch che' wa mehen che', puro bä'ik waak'al. Ese che'. Ma' ak', che'. Wolol nukuch—puk'te', ek' bahche', puuna'. Tuwolol päytan uwich lubul yok'ol lu'um, uwaak'al uwich, uhook'ol uyool, tichil uyool.** 'Trees, they begin as seeds, which fall to the ground and then sprout. In the Lacandon language they say *"uwaak'al uwich"*. First they sprout and emerge. This is called a sprout. The sprout is small, it comes up very slowly. Then the small leaves of the shoot emerge. That is a tree. As it grows the leaves large leaves fill it, the smaller leaves fall off, and new leaves emerge. So that is a tree. It keeps growing. Um, all of the trees, large and small, start out as sprouts. This is a tree. It's not a vine, but a tree. All of the large ones—*pukte', ek' bahche', puuna'*—were first a seed that fell to the ground and sprouted.' (BM) The general term for tree, bush, or plant with woody stalks and branches. Small trees, saplings, and plants with branches and a stem are all called *mehen che'* 'small tree'. In its most restricted sense, che' refers to trees. Excluded from the category are palms, reeds, and plants with herbaceous stalks instead of trunks, e.g., zingerbales. *[Source:* BM*]*

che'il trunk, stalk, main stem. *Lit:* 'stem/stalk of'. *See:* **chun**. *[Source:* BM*]*

che'il uchuch pedicel; stem of a sucker. *See:* **chuch**. *[Source:* BM (2010)*]*

che'il uk'äb peduncle. *Lit:* 'stem of its branch'. *[Source:* AM*]*

che'il ule' leaf stalk, leaf stem. *Lit:* 'stalk of leaf'. *[Source:* BM*]*

che'il umots lateral roots.

che'il utuch'il rachis (banana); the stalk of the inflorescence of bananas or plantains. *[Source:* SK*]*

chich näl maize, hard. In reference to ears of maize that are completely dried. *Use:* **k'u'umbil wah, säkän** 'nixtamalized corn tortilla, dough'; *Ant:* **kuxul näl** 'young maize'.

chuch stem, tender stalk, sucker. *[Note:* chúuch [Yuc.]. stem, stalk; esophagus; vein, artery (Bricker et al. 1998).*]* *[Source:* BM*]*

chuch uchan top' style. *Lit:* 'stem of a small blossom'. *See:* **chuch utop'**. *[Source:* AM*]*

chuch uche'il ule' leaf base, particularly the leaf that wraps around the false stem of the banana plant. *[Source:* BM*]*

chuch ule' petiole; leaf stem. *Lit:* 'stem of the leaf'. *[Source:* AM; BM*]*

chuch utop' receptacle.

chulul$_1$ heart (pith). *See:* **ts'u'**. *[Source:* AM*]*

chulul$_2$ [chulul, chulu', chuluk] bow (tree). *Lit:* 'bow'. Lacandones distinguish two kinds of *chulul* that they use for making bows—*hach chulul* 'authentic chulul' and *säk chulul* 'white/false chulul'. The best is *hach chulul* 'authentic chulul'. There are two kinds—*ek' chulul* 'black chulul' and *chäk chulul* 'red chulul'. Previous researchers in the area have identified *hach chulul* as *Guaiacum sanctum* (Nations and Nigh 1980; Tozzer 1907), but according to Toledo et al. (2008), Naha' is outside of the distributional range for this species. Miranda (1953) surmises that the Lacandones may have made their traditional bows out of *Tabebuia guayacan* (Durán 1999). *Use:* **chulul** 'bow'; *Part:* **che'il** 'wood'; *Use:* **puustah** 'shavings'; *Spec:* **hach chulul** 'authentic chulul'; *Spec:* **säk chulul (äh)** 'white chulul (*Platymiscium dimorphandrum*)'. *Variant:* **hach chulul**. *Thes:* **che'**. *[Note:* chulul. Platymiscium yucatanum (March 1998). chulul

[Yuc.]. Apoplanesia paniculata. A tree from which they make bows (Roys 1931: 239); chulúul [Yuc.]. Apoplanesia paniculata Presl. (Bricker et al. 1998).*] [Source:* BM*]*

chun base, trunk; corm. *See:* **chunche'**. *[Note:* chùun [Yuc.]. base, origin; trunk [tree] (Bricker et al. 1998: 75).*]*

chunche' tree trunk. *Lit:* 'base of tree'. *See:* **che'il**. *[Note:* chunche'il [Itz.]. tree trunk (Hofling and Tesucún 1997: 217)*]*

chun ule' stem base. *[Source:* BM*]*

chun uwi' basal plate. *Lit:* 'base of tuber or bulb'. *[Source:* BM*]*

Ch'–ch'

ch'ib [ch'ip', ch'i'] mid rib, main leaf vein. *Variant:* **xich'el ule'**. *[Source:* BM; SK*]*

ch'ibil ule' vein (of leaf), mid rib. *See:* **ch'ib**. *[Source:* BM*]*

ch'ot nodes (of a corm). *Lit:* 'twist, turn'. *[Source:* SK*]*

H–h

hachil näl corn ear. *[Source:* Bruce (1975)*]*

hach lo'obil weeds, underbrush. *Lit:* 'authentic weeds'. The term refers to a variety of uncultivated plants, bushes, and trees that appear in a milpa after it has been cleared for the first time. The main ones include: *ch'ayok'* (*Solanum* spp.); *habän uts'u'* (*Podachaenium eminens*); *ki'bok* (*Clibadium arboreum* and other members of the Asteraceae family); *ku'uchnook'* (*Bidens odorata*); *hach hoben* (*Piper auritum*); *mäkuuläm* (*Piper* spp.); *si'si'k'uuts* (*Erechtites hieracifolia*); and, *tekoox* (*Phytolacca* spp.). Some species are eaten, others are used for medicinal and cleansing purposes. Some are considered good for the soil, e.g., *ch'ayok', mäkuuläm,* and *tekoox. [Source:* BM*]*

hach uyoot'el [hach uyoot'el, hach uyoot'er, hach uyoot'eh] bark (outer). *Lit:* 'its true skin'. *See main entry:* **oot'**. *[Source:* BM*]*

halo'och [halo'och, haro'ch] corn shucks. *Use:* **ti' tap'ik wah, 'is wah** 'for wrapping tamales of green corn'. *See:* **halo'ochil näl**. *[Note:* halo'ochi(l) 'shucks', halo'ochi(l) näl 'corn shucks' (Bruce 1975); holoch [Yuc.]. corn husk (Roys 1931); holó'och [Yuc.]. corn husk (Bricker et al. 1998); jolo'och [Itz.]. corn husk (Hofling and Tesucún 1997). According to BM, *halo'och* refers to the ear of corn not the husk.*] [Source:* BM; KP*]*

halo'ochil näl corn shucks. *Morph:* **halo'och-il näl**. *See main entry:* **halo'och**.

heekel näl corn, ear. *Morph:* **heek-el**. *Variant:* **heek**. *[Note:* héek [Yuc.]. young corn ear (Bricker et al. 1998).*]*

heek'el$_1$ [heek'el, heek'er, heek'eh] branch (small); leaf stem. *See:* **heek'el uk'um**. *[Note:* héek' [Yuc.] break; branch (Bricker et al. 1998).*]*

heek'el$_2$ lobes of a leaf. *[Source:* AM*]*

ho'ol ubakalel corncob (top part). *Lit:* 'head of it corncob'. This is the only part of the corn used to make *posol. [Source:* SK*]*

I–i

its resin, sap latex. *See:* **yits**.

K–k

kanil [kanil, kanih] sub-aerial stems (runner, stolon, off-set, sucker). *Lit:* 'its snake'. *Variant:* **k'anil**. *[Source:* AM; BM; KM; CNK*]*

kanil uyak'il stolon. *Lit:* 'snake of its vine'. *Morph:* **kan-il u-y-ak'-il**. *See:* **kanil**. *[Source:* BM*]*

koh kernel. *Lit:* 'tooth'.

kuxul näl maize, fresh ear. *Lit:* 'new maize'. *Use:* **hanal** 'food'; *Prep:* **chäkbil** 'boiled (in husks)'; *Prep:* **pokbil** 'roasted'; *Prep:* **sa'** 'gruel (atole)'; *Prep:* **'is wah** 'tortillas'; *Use:* **ti' bo'otik k'uh** 'offering'; *Prep:* **p'iis sa'** 'like atole'; *Ant:* **chich näl** 'hard maize'. *Morph:* **kux-ul**. *See main entry:* **näl**. *[Note:* kuxul-näl [Itz.]. ear of new corn (Hofling and Tesucún 1997).*] [Source:* BM; SK*]*

kuxum [kuxum, kuxun] mould. According to BM, *kuxum* means not living and rotting. The word designates anything that grows on something else and causes it to decay. According to AM, *kuxum* encompasses a variety of organisms that live on something else, i.e., mould, rust, scum, algae, lichen, and fungi. *Kuxum* is divided into two main groups: (1) *kuxum che'* 'mould that grows on trees or wood and (2) *kuxum lu'um* 'mould that grows on the ground'. It is not a general term for fungi. All forms of *kuxum* are considered inedible, even poisonous, and so the Lacandones avoid them. However, some kinds are used for other purposes. For example, *Cookeina stricholoma* and *C. tricholoma* are used as whistles, and *Clavulinopsis* sp. is used to remove warts (Ruan-Soto et al. 2009: 68). *Spec:* **kuxum che'** 'tree mould'; *Spec:* **kuxum lu'um** 'earth mould'. *See:* **kib lu'um**. *[Note:* kuxum [Itz.]. mold. (Hofling and Tesucún 1997); kùuxum [Yuc.]. moss, mold; algae; mushoom (Bricker et al. 1998).*] [Source:* AM; BM; CKD; KP*]*

kuxum che'₁ [kuxumche', kuxunche', kuxuche'] mould on tree. **Ukuxul wolol che' luum. Es lo'obil. Laba' ti', wolol lab.** 'It lives on all decomposing wood. It's a weed. It's belongs to decomposed things, all decomposed things. Strictly speaking, *kuxum che'* refers to inedible wood fungi, e.g., bracket fungi, coral fungi, and *Cookeina* spp. But it also includes anything that grows on trees, such as tree-mosses, epiphytic ferns, and lichens. *Bak'che'*, flowering epiphytes, are an exception. Sim: kuxum lu'um 'earth mould'. *Sim:* **kuxum lu'um** 'earth mould'. *[Note:* kuxumche'. flat mushrooms which grow on dead and rotting wood (Bruce 1975); cuxum-che [Yuc.]. tree mold, tree rust. Very red and round (Roys 1931); kùuxum [Yuc.]. moss, old, algea, mushroom; kùuxum che'. tree's mushroom (Bricker et al. 1998); kuxun [Itz.]. mold (Hofling and Tesucún 1997)*] [Source:* AM; BM; KP; SK*]*

kuxum lu'um [kuxum lu'um, kuxun lu'um] mould on earth. Refers to inedible fungi, small ferns, mosses, and lichens that grow on the ground. *Sim:* **kuxum che'** 'tree mould'. *See:* **k'u' kan**. *See main entry:* **kuxum**. *[Source:* AM*]*

K'–k'

k'anil sub-aerial stems; i.e., runner, stolon, off-set, sucker. *Lit:* 'snake like'. *Variant:* **kanil**.

k'äb [k'äb, k'äp', k'ä'] branch, limb. *Lit:* 'hand, arm'. *Variant:* **k'äb'che'**.

k'äbche' [k'äbche', k'äp'che', k'ä'che'] branch, tree. *Morph:* **k'äb=che'**.

k'äb ule' [k'äb ule', k'äp' ule', k'ä' ule'] branch, of leaves. *[Source:* AM; BM; SK*]*

k'ek' branches, lateral. *Lit:* 'barb'. *[Note:* k'ek' [Itz.]. barb (Hofling and Tesucún 1997).*] [Source:* BM*]*

k'i'ilix thorns (blunt), lenticels; the small protuberances along the stems of vines that may contain an irritating resin. *See:* **tuch'**. *[Source:* BM*]*

k'i'ix thorn, spine, prickle.

k'ik'i'ix ak' bramble. *Lit:* 'very thorny vine'. *See:* **halo'och k'i'ix**. *Thes:* **ak'**. *[Source:* BM; K'ayum Segundo*]*

k'uuk' sprout, shoot, sucker. *See:* **k'uuk' che'**. *[Note:* k'uuk' [Itz.]. retoño. sprout (Hofling and Tesucún 1997); k'ú'uk' [Yuc.]. sprout, shoot (Bricker et al. 1998).*] [Source:* BM*]*

k'uuk' che' sprout, sucker; includes plants and saplings that sprout from nurse trees, but not *bak'che'* 'ephipytes'. *Lit:* 'tree sprout'. *Morph:* **k'uuk'**. *[Note:* k'uuk' [Itz.] retoño. sprout (Hofling and Tesucún 1997: 401); k'ú'uk' [Yuc.]. N. sprout, shoot (Bricker et al. 1998: 158).*] [Source:* AM; BM*]*

L–l

laak' cupule. *Lit:* 'spouse'. *[Source:* KP*]*

le' leaf. *See:* **xiiw**.

le' uchuch top' sepal. *See:* **chuch**. *[Source:* BM (2010)*]*

le' usool leaf scale; i.e., the fleshy scales of an onion. *Lit:* 'leaves of skin'. *[Source:* SK*]*

le' utop' petal. *Lit:* 'leaf of a flower'. *[Source:* AM; BM*]*

le' uyool sheath (leaf), pseudo-stem sheath. *Lit:* 'leaf of a shoot'. *[Source:* BM*]*

le' xiiw leaves, large; typically those of plants in the order Zingiberales. *[Source:* AM; BM; KM; KP; SK*]*

lo'obil [lo'obil, loobil, loobih] weeds, underbrush. **Ahlo'obil, ooo, ne seeb. Bähe' ulo'obil wa yan ne yi'h uwich, uhutul uwich. Bähe', hutih yok'ol lu'um, ka'p'ek'in, 'oxp'ek'in, yan ule'. Ne seeb usaastal uhel ule', usaastal uhel ule', usaastal uhel ule'. Seebseeb uch'ihil lo'obil. Seeb uman ki' k'uchul hunts'itna', yan utop' tuka'ten uhel äkna'.Yan uwich uka'hutul uwich. Ne yan ich paakche' kol. Yan su'uk, yan 'ak su'uk, winik su'uk yan, yan hach lo'obil, yan hach su'uk, yan hoben. Tuwolol lati' lo'obil, porke ne seeb uch'ihil pero ma' ne nuktal. Lati' paakche' kol, lo'obil. Yan hump'el ne tsoy. Yan huntul ne tsoy—tekoox. Ne tsoy ti' kol, porque tekoox ne muun, ne siis. Ma' chäkal. Lehih ne ma' tsoy, 'ak. Ahak' ne ma' tsoy. Ahwaalka'an, ne ma' tsoy. Lehih ma' tsoy ti' kol. Ah'ak, chäkal. Chäkal es calor, porque ukanil kukinsik näl. Chich umuk'; kuhupik, kuhupik näl; uchäkik, uchäkik. Ahwaalka'an uchäkik näl. Ahhach su'uk tambien uchäkik näl, porque ne chich umuk'. Ne seeb ulobtal, ne seeb ulobtal. Wolol hach lo'obil, ki'ki'ts'abil, ma' ne seeb, pero hach su'uk, ne seeb. Lati' ne ma' tsoy.** 'Lo'obil, oooh, they grow quickly. Today, if the seeds are very ripe and they drop on the ground, in two or three days there are leaves. The next day there are leaves and the next day there are more leaves; in a month there are flowers, in another month, they have seeds again. There are a lot of kinds in the fallow cornfields. There is *su'uk* (Poaceae), *'ak su'uk* (*Bothriochloa* sp.), winik su'uk (*Andropogon bicornis*), *hach lo'obil* 'authentic brush', *hach su'uk* (*Digitaria* sp.), and *hoben* (*Piper auritum*). All of these are called *lo'obil*, because they grow rapidly but they don't grow tall. There is one kind that is very good—*tekoox* (*Phytolacca icosandra*). It is good for the cornfield because it is tender and cold. It's not hot. The one that's very bad is *'ak*

(*Bothriochloa* sp.). *Waalka'an* (*Pteridium* spp.) is very bad. *'Ak* is hot, because kills the corn. They're very strong; they pierce the corn; they cook it. *Waalka'an* cooks the corn. *Hach su'uk* cooks the corn, because it's very strong. Quickly they become dense, overgrown. *Hach lo'obil* grows slowly and steadily, except for *hach su'uk*; it grows very quickly. That one is very bad.' (BM) Other herbaceous plants referred to as *lo'obil* include: *si'si'k'uuts* (*Erechtites hieracifolia*); *tsaayent-saay* (*Desmodium* spp.); *ku'uchnook'* (*Bidens odorata*); and, *ch'ayok'* (*Solanum* spp., esp. *S. nigrum*). *Morph:* **lo'ob-il**. *See main entry:* **lob**. *[Note:* lob. bad, overgrown (i.e. road or trail); loobil. weeds, underbrush (Bruce 1975); sak robir [S. Lac] (Canger 1970); lö'ob' [Yuc.]. be hurt, harmed, damaged; covered (with weeds, brambles, etc.) (Bricker et al. 1998); lob [Yuc.]. weed, bad; loobol. weedy (Roys 1931).*] [Source:* BM*]*

lo'obol [looboh] weedy. *Morph:* **lo'ob-ol**. *See:* **lo'obil**. *See main entry:* **lob**. *[Note:* lob. bad, overgrown (i.e., road or trail) (Bruce 1975: 187); loobil. weeds (Bruce 1975: 187); lo'ob'ol [Yuc.]. weedy (Bricker et al. 1998: 170)*] [Source:* BM*]*

M–m

me'ex tendril, aerial roots. *Lit:* 'whiskers'. *[Source:* AM; BM*]*

mehen che' small tree, sapling. *See main entry:* **che'**. *[Source:* BM*]*

mehen k'uuk' axillary bud. *See:* **k'uuk'**. *[Source:* BM*]*

mehen le' leaflet. *[Source:* BM*]*

mehen mots root hairs, secondary roots. *[Source:* BM*]*

mehen tsayal form fruit; a term applied to young fruit forming on certain plants, such as *näl, k'um, p'ix,* and *kubul. Morph:* **tsay-al**. *[Note:* tsay [Yuc.]. T. form ears(corn) (Bricker et al. 1998: 41).*] [Source:* AM*]*

mook [mo'ok, mook] knot, node; hard ridges encircling a palm stalk or cane between internodes. *[Note:* Each knot indicates one month's growth (BM). mo'ok [Yuc.]. be knotted; mòok vn. knot (Bricker et al. 1998); mo'ok kaanyaj [Itz.]. section of cane (from one knot to another) (Hofling and Tesucún 1997).*] [Source:* BM*]*

mots root. *[Note:* motz. root (Roys 1931); mòots. root (Bricker et al. 1998).*]*

muun tender (green, young); with reference to leaves, plants, and squashes. *[Note:* mum [Itz.] (1) A. tierno. tender. (Hofling and Tesucún 1997: 456); mun_2 mùn [S.Lac]\ A blando, tierno (como pan Bimbo) (Canger 1970); munil. green, tender (Roys 1931).*]*

N–n

nah che'il stem, main. *[Source:* BM*]*

nap' bark markings. The patches of different colours and textures on bark. *[Source:* AM*]*

nek' [nek', näk'] seed, nut. *Nek'* also refers to testicles, and for that reason people will use *wich* 'fruit' instead. *[Source:* BMjr*]*

nikte'$_1$ *Lit:* 'flower-tree'.

ni' uyool growing tip, apex. *Lit:* 'nose of the shoot'. *Variant:* **yool$_1$**.

noy pulp. *[Note:* noy [Yuc.]. the pulp of a fruit or the starchy interior of certain plants (Roys 1931); noy N. carnaza. flesh (of fruit); marrow, fine part (Hofling and Tesucún 1997).*]*

O–o

ok sprout; the incipient phase of a seedling. *Lit:* 'foot'. *See:* **yokman**. *[Source:* SK*]*

ool [ool, oor, or] apex; heart. *See:* **yool**.

oot' skin. *[Source:* BM*]*

'op' ch'uplal *Lit:* 'burst female'. Refers to the seeds of *ahk'ante'* (*Erythrina berteroana*) and *'äm ch'uplal* (*Ormosia isthmensis*). Both are solid red and similar in shape and size. *Use:* **uuh** 'necklaces'; *Part:* **nek'** 'seed'; *Male:* **'op' ton**. *See:* **'äm**.

'op' ton *Lit:* 'burst male'. Refers to the black and red seeds of *Ormosia schippii. Use:* **uuh** 'necklaces'; *Part:* **nek'** 'seeds'; *Female:* **'op' ch'uplal** '*Ormosia isthmensis*'. *See:* **'äm**.

P–p

pach$_1$ top of the leaf. *Lit:* 'back'. *Ant:* **taan** 'front'. *[Source:* AM*]*

pach$_2$ skin; bark.

pach (ah)$_3$ seed pod. *[Note:* BM said this was not the name for a pine cone. A pine cone is simply called *nek'* 'seed'.*] [Source:* BM; BMjr*]*

pats'kih soft, with reference to seeds that are immature. *[Source:* BM*]*

peek' bunch, hand, or tier of bananas and plantains; not the whole stalk. Also a nominal classifier to count bunches, e.g., *hun-peek' pätan* 'one bunch of plantains'. *See:* **kucho'**.

[Note: peek' [Itz.]. bunch (of plantains) (Hofling and Tesucún 1997).*] [Source:* SK*]*

peek'el hand of bananas. *Lit:* 'its bunch (of bananas)'. *Morph:* **peek'-el**. *See main entry:* **peek'**. *[Source:* SK*]*

S–s

säkba' näl corn stalk. *[Note:* sak'ab', n2. corn stalk [Yuc.] (Bricker et al. 1998); säk'ä' [Itz.]. corn stalk (Hofling and Tesucún 1997).*] [Source:* BM*]*

soobil näl smut (corn fungas). *Ustilago maydis* (Persoon) Roussel. This is a black fungus that grows on green maize. Although edible, the Lacandones do not report eating it. Known in Mexico as *huitlacoche* (cuitlacoche), from Nahuatl, *cuitlacotzin* (Bruce 1975: 219), it is a delicacy eaten throughout Mexico. *See:* **näl**. *[Note:* Term not recognized by BM. sop' P. pile up [Yuc.] (Bricker et al. 1998: 249)*] [Source:* Bruce (1975: 219)*]*

sool shell, bark, peel. *See:* **box**. *[Note:* sor (?U-)so:r [S. Lac.]. cáscara (de palo, frutas, etc.) (Canger 1970); zol [Yuc.]. husk, shell (Roys 1931); sool [Itz.]. cáscara. shell, skin (Hofling 1997); sóol [Yuc]. rind, peel, shell, husk, bark, hull (Bricker 1998).*] [Source:* BM*]*

sool che' [sool che', soor che'] bark, exterior. *Lit:* 'shell/peel of tree'. *See main entry:* **sool**.

sool uwich lemma (the lower bract of a grass floret). *See main entry:* **sool**. *[Source:* BM*]*

suli'$_1$ [suli', suri'] tuber/root, edible. *Lit:* 'contact with the ground'. This is a general term for edible tubers, rhizomes, and roots, i.e., *Discorea alata, Ipomoea batatas, Pachyrhizus erosus, Solanum tuberosum*, and *Xanthosoma yucatanense. Use:* **hanal** 'food'; *Part:* **wi'** 'tuber'. *Morph:* **suul-i'**. *[Note:* Possibly derived from sùul [Yuc.]. contact with the ground (Bricker 1998: 251). suri, surí, chäk is. Dioscorea spp.; chikam, surí. Pachyrhizus erosus (Nations and Nigh 1980); suri, makäl. Xanthosoma yucatanense (Nations 1992, in Atran et al. 2004); ix pä'yak' [Itz.]. Dioscorea bulbifera, D. alata (Atran et al. 2004).*]*

su'uk (ah) [su'uk, suuk] grass. Poaceae *Panicoideae* e.g., Andropogoneae Dumort.(in part), Paniceae Haller. **Uwich primero, päytan uwich. Ne yi'h uwich, uhutul. Chen suuk mas ne seeb. Bähe' kuhutul uwich, oxp'elk'in uwaak'al uwich. Chen lati' su'uk uwaak'al uwich ule', yan ule', yan ule', seeb. Pachil, mas seebutichil uyool ule', uyol ule'. Ele lati' su'uk. Pero su'uk mas ne seeb, ma' xantal. Mas ne seeb uhook'ol ule'. Mas ne seeb ahche'. Ma' nuk. Ma' ka'anan ubakel. Mihin, p'iis uyak'il. P'iis uyak'ilil yan ule' kubin uyak'ileh chichin. Yan ule'. Ese lati' su'uk. Seeb uwich, ma' ne xantal uwich. Lati' su'uk.** 'First there are seeds. When they are very ripe they fall. Grass grows very rapidly, after that. Today the seeds fall and in three days they burst their skins (sprout). Then the leaves open up, a leaf here, then a leaf there, it's quick. After this, the sprouts shoot up, the leaves open on the shoots. And so that is grass. But grass grows very rapidly. It grows much faster than trees. Grass isn't as big. It is a small plant with (runners) that have leaves. That is what grass is. The seeds form quickly, it doesn't take long. That is grass.' (BM). The word *su'uk* derives from *so'k* (Mix-Zoq). The Lacandones distinguish a number of kinds of *su'uk*, undoubtedly because they interfere with cultivation. According to Nigh (2008: 235), Old World pasture grasses (e.g., *Pennisetum*) and the universal bracken (*Pteridium aquinilum*) impede the management of transitional phases between annual cultivation. *Spec:* **wakäx su'uk** '*Pennisetum purpureum*'; *Spec:* **winik su'uk** 'Andropogon glomeratus'; *Spec:* **hach su'uk** '*Digitaria setigera*'; *Spec:* **'ak su'uk** '*Bothriochloa laguroides*'. *[Note:* suk. grass (Nigh 2008); sú'uk [Yuc.]. grass. Brachiaria fasciculata (Bricker et al. 1998: 250); zuuc [Yuc.]. grass (Roys 1931); su'ukil chäk'an [Itz.]. Leptocoryphium lanatum (Atran et al. 2004); su'ukil b'o'oy [Itz.]. Paspalum caespitosum (Atran et al. 2004).*] [Source:* AM; BM*]*

T–t

taan$_1$ [taan, tan] underside of leaf. *Lit:* 'front'. *Ant:* **pach** 'back'. *[Source:* AM*]*

ta'an ash, dust; pollen. *See:* **ta'anil utop'**. *[Note:* ta'an. ashes (Bruce 1975); ta'an [Yuc.]. lime; ash, pollen (Bricker et al. 1998); ta'an [Itz.] ash, lime (Hofling and Tesucún 1997: 580).*]*

ta'anil utop' pollen. *Lit:* 'dust of the flower'. *See main entry:* **ta'an**. *[Source:* BM*]*

tahal mature, ripen. *Morph:* **tah-al**. *See:* **täk'än**. *[Note:* tah [Yuc.]. mature, ripen (Bricker et al. 1998).*]*

tak'lan che'$_1$ [tak'lan che', tak'lah che'] mistletoes. *Lit:* 'sticking all over trees'. Santalales. The Lacandon forest abounds in vigorous, "obligate hemi-parasitic plants" known as mistletoes. Yet, only three species have been reported in Naha' (Comisión Nacional de Áreas Naturales Protegidas 2006; Durán 1999): *Phoradendron nervosum, Phoradendron piperoides*, and *Phoradendron uspanatum*. One species, *Phoradendron* aff. *flavens*, has been recorded from the southern Lacandon territory (Levy-Tacher et al. 2006). All are referred to in Lacandon as *tak'lah che'*. Despite the myriad leaf shapes, flowers, and fruits, Lacandon consultants differentiated the species by size—they are either *nukuch* 'large' or *mehen* 'small'. The Lacandones say that these plants harm fruit trees. *Loc:* **k'aax, paakche' kol, mehen kol** 'forest, fallow milpas, home gardens'; *Spec:* **mehen tak'lan che'** 'small tak'lah che' (*Phoradendron crassifolium*)'; *Spec:*

nukuch tak'lan che' 'large tak'lah che' (*Struthanthus orbicularis*)'. *Morph:* **tak'-l-a'an**. *Variant:* **iich'ak tolok**. *Thes:* **ak'**. *[Note:* sak jara che. *Phoradendron nervosum* (Durán 1999); a kukin che. *Phoradendron piperoides* (Durán 1999); a takan de ak. Begonia glabra (Durán 1999).*] [Source:* AM; BM; KP*]*

täk'än ripe; cooked. *Morph:* **täk'-än**. *See:* **tahal**; **yi'h**. *[Note:* täk'-a'n [S. Lac.]. ripe (Canger 1970); tak'an [Yuc.]. aj. mature, ripe (Bricker et al. 1998); tak' [Itz.]. A, N. ripening (Hofling and Tesucún 1997).*]*

te' tree. Usually a bound morpheme signifying tree or bush. A number of bushes and trees take the—te' suffix, including: *akte'; balumte'; itsänte'; akunte'; ukunte'; k'änte'; k'änte' ak'; k'unk'unte'; läkte'; läkte'; nikte' (ak'); 'oonte'; pukte'; subinte';* and, *tähte'*. One exception is *te' usin* 'buzzard tree'. Additionally, *te'* sometimes alternates with *che'*, e.g., *k'unk'unte'* and *k'unk'unche'* (*Oecopetalum mexicanum*). See: **che'**. *[Note:* -te' [Itz.]. wood, tree (Hofling and Tesucún 1997); -te' [Yuc.]. tree (Bricker et al. 1998).*]*

tichil sprout, from the ground. Lit: 'emergence'. Morph: **tich-il**.

top' flower.

top' ak' vine flower, flowering vine.

top' che' flower on woody shrubs and trees and includes flowering epiphytes growing in the crooks of tree branches. Lit: 'tree flower'.

top' che'il flowering shrub or tree. **P'iis mehen lo'obil. Uch'ihil as nuk. Yan uche'il.** 'It's the same as a small weed. It grows somewhat large. It has a woody stem.' (BM).

top' lo'obil flower of a weed.

tuch' lenticel; bud, protusion. Lit: 'point'. *[Source:* BM*]*

tuch'il pätan bell, the male bud of a plantain or banana plant. Morph: **tuch'-il**. See: **chuch utop'**. *[Source:* KM*]*

tuch'tuch' blistered, bumpy surface. Ant: **yuul** 'smooth'. See: **tuch'**. See main entry: **tuch'**. *[Note:* tuch' [Yuc.]. P, T. raise, become erect (Bricker et al. 1998); tuch'- [Itz.]. point (Hofling and Tesucún 1997).*] [Source:* BM*]*

Ts–ts

tsaayal uwich cauliflory; the manner in which fruits grow directly on the branches of a tree, rather than on peduncles. Lit: 'spliced on fruit'. *[Source:* BM*]*

tso'ots$_2$ tree moss. Lit: 'hair'. See: **kuxum che'**. *[Source:* AM*]*

tsuk näl corn silk. *[Note:* tsùuk [Yuc.]. hair, silk [corn] (Bricker et al. 1998); tzuk [Itz.]. corn silk (Hofling and Tesucún 1997).*] [Source:* Bruce (1975)*]*

Ts'–ts'

ts'u' pulp, pith, heartwood, core. See: **chulul$_1$**. *[Note:* ts'u'. pulp (Bruce 1968); ts'u.[Yuc.]. soft inner part of a plant or fruit (Roys 1931); ts'u' [Yuc.]. middle, centre (Bricker et al. 1998); tz'u' [Itz.]. centre (Hofling and Tesucún 1997).*]*

ts'u' uyool utop' axillary bud. Lit: 'pith of the flower bud'. *[Source:* BM*]*

W–w

waak' burst open; refers to the blooming of flowers, the splitting open of seed pods, and the sprouting of a seeds. *[Source:* KM*]*

waal [waal, wal, war] frond; leaf of a palm or fern. Lit: 'fan'. See: **waalka'an**. *[Note:* waal N. palm frond, fan of feathers for fire. (Hofling and Tesucún 1997: 661); wáal [Yuc.]. page; leaf (Bricker et al. 1998).*]*

waalka'an [waalka'an, waarka'an] fern. Lit: 'tall fan'. Polypodiales. A variety of ferns occur in the area. Lacandones distinguish the plants according to where they grow, i.e., *waalka'an k'aax* 'forest fern' and *waalka'an kol* 'fern of milpas'. Another is *Alsophila* sp. *[Note:* wàrka'n [S.Lac.]. helecho (Canger 1970); x-ual-canil [Yuc.]. *Pteridium caudatum* (L.) Maxon. (Millsp. III, 8) (Roys 1931); wáal [Yuc.]. page, leaf (Bricker et al. 1998); waal [Itz.]. wild palm, palm frond (Hofling and Tesucún 1997).*] [Source:* BM; SK*]*

wi' tuber, edible root and rhizome.

wich$_1$ fruit. See: **nek'**. *[Source:* BM; SK*]*

wich$_2$ pollen sacks. *[Source:* AM*]*

winikil [winikil, winikih, winkih] people; refers to the strands of male flowers of *Chamaedorea tepijilote* and *C. alternans*, and the rows of kernels on an ear of maize. Lit: 'its people'. *[Source:* AM; BM; SK*]*

wolol uwich spikelet, seed head of grass. Lit: 'seeds/fruits all over'. *[Source:* BM*]*

X–x

xa'ay [xa'ay, xaay] fork (of branch or pole). See: **k'ek'**. *[Note:* xay [Yuc.]. the fork of a branch (Roys 1931); xá'ay [Yuc.]. forked pole (Bricker et al. 1998: 255); xa'ay [Itz.].

fork, forked stick (Hofling and Tesucún 1997: 669)*]* *[Source:* BM; KM*]*

xib [xib,xip'] leaf. See: **xiiw**. Variant: **xibil**.

xich'el ule' vein. Lit: 'nerve of a leaf'. See: **ch'ib**. *[Note:* xìich' [Yuc.]. tendon, nerve, sinew (Bricker et al. 1998).*]* *[Source:* BM*]*

xiiw$_1$ leaf, large paddle-shaped leaf. One Lacandon consultant, SK, said that xiiw refers to the large leaves of Zingerbales plants, such as ginger, heliconia, canna lily, and prayer plants. Use: **ti' utap'ik ba'al, säkan, k'ät, bu'ulil wah, nahwah, mukbil wah** 'to cover dough, clay, bean tamales, ceremonial tamales, buried (baked) tamales'. Variant: **xib, xibil**. *[Source:* BM*]*

xiiw$_2$ Zingiberales. **Ahxiiw, yan uwi', xiiw yan uwi'. Yan umots. Umots mana' uximbal; ma' uximbal. Mana' ukaanil, ha'li' yan uwi'. Yan uhook'ol uyool. Uyool, lati' uk'aba' uyool ule'. Lati' xiiw. Chukuch uwaan ule'. Kooch utan ule', chukuch ule'. Lati' uk'aba' xiiw. Mäna', ma' ka'anan ubakel. P'iis komo hump'el metro uka'anan ubakel, o ka'p'el metro, ha'li'. Mana', ma' ba'il che'. Lati' uk'aba' xiiw. Mana' uche'il. Ha'li' yan uchuch ule', nada mas, ha'li'. Uchuch ule' es komo, komensa a formar su le'. Xiiw uk'anik ti' tap'ik nahwah, ti' utap'ik k'eyem, wah, säkan, bak' wa ti' k'ak'tik käy, ti' mukik wah xiiw. Hay tipos de le' xiiw: mäkäl, nukuch komo' xiiw, säk muxan xiiw, säk wowo xiiw, chukuch le' xiiw. Ule' pätan sirve para mukbil wah, pero se dice le' xiiw, le' pätan. Le' pätan se dice.** 'The *xiiw* has a tuber. It has roots. The roots don't travel; they don't have runners. They only have tubers. There is a shoot that goes out (sprouts). The shoot is called the leaf shoot. That's *xiiw*. The leaves are long and wide. That's called a *xiiw*. They are not very tall. They're only about one or two meters tall. They don't grow taller, they're not like a tree. They don't have trunks. They only have leaf stems. That's all. The leaf stem is like the beginning of a leaf. You use *xiiw* for covering ceremonial tamales, *k'eyem* (a type of a cooked gruel made from nixtamalized corn), tortillas, corn dough, meat, or for barbequing fish and baking stuffed tortillas. There are kinds of *xiiw* leaves: *mäkäl* (*Xanthosoma* sp.); *nukuch komo'* (*Renealmia* sp.); *säk muxän xiiw* (*Heliconia librata*); *säk wawal* (*Calathea lutea*); and, *chukuch le' xiiw* (*Heliconia* sp.). The leaves of bananas serve as wrappers for the large, baked tamale; but they are called "*xiiw*". (BM) From BM's description it appears that the word designates the long, broad, paddle-shaped leaves of plants in Zingiberales order. Plant families within this order and which occur in Naha' include: Cannaceae (cannas); Costaceae (costus); Heliconiaceae (heliconias); Marantaceae (prayer plants); Musaceae (bananas and plantains); *Xanthosoma* (taro, elephant ears); and, Zingiberaceae (gingers). The word *xiiw* is derived from the Nahuatl word *xihuitl* 'turquoise, grass', in Yucatec, it means herb, grass, weed and in Itzaj, it refers to herb, grass, and medicinal plant. Although this is suggestive of a botanical category in these languages, it is also likely that *xiiw*, in Lacandon at least, simply refers to the distinctive leaves. *[Note:* xiu [Yuc.]. plant, herb (in general) (Roys 1931); xíiw [Yuc.]. herb, weed, grass (Bricker et al. 1998); xiw [Itz.]. medicinal plant, herb, weed, grass (Hofling and Tesucún 1997).*]*

xuul$_2$ [xuul, xuur, xur] tip of the plant. Lit: 'end, terminus'. See: **yool**$_1$. *[Source:* BM*]*

Y–y

yaal shoot, sucker; the male inflorescences of palms. Lit: 'its offspring'. Morph: **y-aal**. See: **yool**$_1$. *[Note:* Lacandones regard plants that produce male inflorescences as the female plants, even though they are aware that they are really male plants. Lacandon is not the only language which regards plant suckers as offspring: basal suckers, particularly of bananas, are referred to as "children" in other cultures. For example, in Hawaiian they are called *keiki* 'children'.*]*

yaalal usool scales (of a bulb), phloem; vascular cambium. Lit: 'layers of its skin'. Morph: **y-aal-al**. Variant: **yaal sool**. *[Source:* BM*]*

yaalal uts'u' growth rings of a tree; the layers of bark surrounding the core. Lit: 'layers of/around the pith'. See: **ts'u'**. Variant: **yaalal, yaalal uts'u'**. *[Source:* BM*]*

yaal sool vascular cambium (phloem). Lit: 'layer of skin/bark'. See: **sool**. Variant: **yaalal uts'u'**. *[Source:* BM*]*

yi'h [yi'h, yih] hard; in reference to very mature seeds, fruit, stems, vines, and stalks. See: **täk'än**; **tähäl**. *[Source:* BM; NukKM*]*

yi'ih [yi'ih, yiih, yiik] spikelet, sprout, with reference to grasses (Poaceae). *[Note:* yi'ih. espiga (Bruce 1968); yi'ij [Itz.]. espiga. sprout (Hofling and Tesucún 1997); 'í'ih [Yuc.]. staminate inflorescence of corn plant (Bricker et al. 1998).*]* *[Source:* BM*]*

yiik' sprout, green bud. See: **yi'ih**. *[Source:* Bruce (1975)*]*

yits [yiits, yits] resin, sap, latex. Morph: **y-its**. See: **kab**. See main entry: **its**.

yits ya' latex, gum. Latex from the *Manilkara* genus of trees. Use: **ti' cha'ach** 'for chewing'. See main entry: **its**. *[Source:* KyP*]*

yool$_1$ [yool, yoor, yor] bud. Morph: **y-ool**. See: **ool**; **yaal**. Variant: **ni' uyool**. *[Source:* BM*]*

yool$_2$ heart (of plant). See: **ool**. *[Note:* yol tuk, yol tuts. palm hearts, palm cabbage (Bruce 1975).*] [Source:* AM; BM*]*

yool umots root cap. Lit: 'apex of root'. Morph: **y-ool u-mots**. See: **yool$_1$**. *[Source:* BM*]*

yool utop' stigma. Lit: 'the bud of the flower'. See: **yool$_1$**. *[Source:* BM*]*

yoot'el bark, inner. Lit: 'its skin'. Morph: **y-oot'-el**. See main entry: **oot'**. *[Source:* SK*]*

Appendix 3
Seasons

Ä–ä

äxp'äli' solstice (winter). **Ahk'in umaan tsel ka'an.** 'The the sun passes over the side of the sky' (AM). To dream of this event foretells your death before the next winter solstice (Bruce 1975: 246). Morph: **äx-p'äl-i'**. SD: SEASONS. *[Note:* äxp'äli', äh xp'äli'. winter solstice (Bruce 1975); äxpäli. winter solstice (McGee 2002); ixp'äl [Itz.]. agn. peeler (female); p'äl. to shell beans (Hofling and Tesucún 1997).*] [Source:* AM; BM; Bruce (1975)*] [* \sd2 astronomical_event*]*

Ch–ch

chukpah ya'axk'in year end; November. Lit: 'the year is completed'. SD: SEASONS. *[Source:* BM*] [*\sd2 season_terms*]*

chumuk ya'axk'in the midde of the year; May, June and July. SD: SEASONS. *[*\sd2 season_terms*]*

H–h

ha' rainy season. Lit: 'water'. There are two rainy seasons. One begins in November and is marked by successive northers. The other one begins in April and continues until the end of the summer, punctuated by brief period of dry weather in May. See: **yok ha' (ah)**. SD: SEASONS. *[Source:* BM*] [* \sd2 season_terms*]*

M–m

man tu p'iis end of the season. Lit: 'the time has passed'. SD: SEASONS. *[Source:* BM*] [* \sd2 season_terms*]*

N–n

nah ya'axk'in main season. Lit: 'full/great year'. **Tan chumukil ya'axk'in. Mana' ha'. Hutul ule' uchben. Uhook'ol uhel tumen ule'. Hasta nah ya'axk'in apäk'al.** 'It's in the middle of the year. There is no rain. The old leaves drop and new leaves come out. In the main season you plant.' According to BM, this is period in the agricultural cycle called nah kolil 'main felling'. Traditionally, this event fell was begun in April or May. But today changes in weather patterns compel farmers to delay it. Activity: **nah kolil** 'main/great felling'; Gen: **ya'axk'in** 'year'. See: **chumuk ya'axk'in**. Variant: **tan chumukil ya'axk'in**. SD: SEASONS. *[Note:* Lacandon seasons may not correspond consistently with Gregorian calendar, especially in light of climate change.*] [Source:* BM*] [*\sd2 season_terms \sd2*]*

O–o

okolbel ya ya'axk'in growing season. Lit: 'the year has commenced'. SD: SEASONS. *[Source:* BM*] [*\sd2 season_terms*]*

T–t

tan uxupul ya'axk'in autumn; August, September, October. Lit: 'the year is ending; August-October'. This is a very wet time of year, with torrential rains and flooding. SD: SEASONS. *[Note:* bulha' kil utal k'in (McGee 2002)*] [*\sd2 season_terms*]*

Y–y

ya'axk'in$_1$ dry season. A time when the sky is clear and blue. See: **ya'axk'in**. SD: SEASONS. *[Note:* ya'axk'in. spring, year (Bruce 1975).*] [Source:* BM*] [*\sd2 season_terms*]*

ya'axk'in$_2$ year. Spec: **nah ya'axk'in** 'full, main year'; Spec: **chumuk ya'axk'in** 'middle of the year'; Spec: **yahchunil ya'axk'in** 'beginning of the year'; Spec: **mäna'**

S. Cook, *The Forest of the Lacandon Maya*,
DOI 10.1007/978-1-4614-9111-8, © Springer Science+Business Media New York 2016

ya'axk'in 'end of the year'. SD: SEASONS. *[Source:* BM*]* *[*\sd2 season_terms*]*

yanchunil ya'axk'in spring. Lit: 'the beginning of the year'. In Chiapas and other tropical places that experience distinct wet and dry seasons, spring is the beginning of the dry season. In Naha', this occurs in January. Morph: **yan=chun-il ya'ax=k'in**. SD: SEASONS. *[Note:* a season in eastern Chiapas covers the so-called dry period, January through May (Amram 1937: 30).*]* *[*\sd2 season_terms*]*

yok ha' (ah) rainy season (the beginning of). Lit: 'foot of rain'. SD: SEASONS. *[* \sd2 season_terms*]*

Appendix 4
Weather

A–a

as k'a'am utosol ha' drizzle. Lit: 'the drizzle is somewhat strong'. SD: WEATHER. *[Source:* BM*]* *[* \sd2 rain*]*

Ch–ch

chäkalbil weather. SD: WEATHER. *[Source:* BM*]* *[* \sd2 term/state*]*

chäkal k'in midday. Lit: 'hot sun'. SD: WEATHER. *[* \sd2 term/state*]*

K'–k'

k'axa'an uchäkalbil climate change. **Antes, mas que hubuhk'ä' winik, mas ne su' utal ha', ma' chäkal, mas sisinäk. Tal ha', pero ma' ne k'am. Päk'tik näl ti' marzo. Ahora, k'axa' uchäkalbil. Ma' uhach tal ha', mas chäkal. Bähe' ti' noviebre ne k'am talil ha'. Ne bulul. Bu'ul lah tsukil ule'. Päk'tik näl ti' mayo wo junio—mas pachil. Ma' hach tak uch'ihil. Umots ma' ne chukuch. K'än k'up', yaab.** 'About twenty years ago, the rains were more consistent, it wasn't so hot, it was cooler.It rained, but it wasn't so strong. We planted corn in March. Now, the weather has changed. It doesn't rain much, (and) it is much warmer. These days the rains are very strong in November. Now, we have to wait until May or June to plant corn, much later than in the past. The corn doesn't grow as strong. The roots aren't very long. The plants get this disease called k'än k'up', more frequently. This year, there was severe flooding. All the leaves of the beans rotted.' (BM). SD: WEATHER. *[Source:* BM*]* *[* \sd2 weather_patterns*]*

N–n

nahil ka'an cloudy. Lit: 'full sky'. SD: WEATHER. *[* \sd2 weather_patterns*]*

näklil ka'an overcast. Lit: 'lean against the sky'. Morph: **näk-l-ik**. Variant: **nahil ka'an**. SD: WEATHER. *[*\sd2 weather_patterns*]*

ne ch'ik yum rain (downpour). See: **ne k'a'am ha'**. SD: WEATHER. *[Source:* AM*]* *[*\sd2 rain \sd2 Ethnographic*]*

ne heben clear sky. Lit: 'very open'. SD: WEATHER. *[* \sd2 weather_patterns*]*

ne k'a'am ha' strong rain. See: **ne ch'ik yum**. SD: WEATHER. *[* \sd2 rain*]*

S–s

sisinak cool, refreshing. SD: WEATHER. *[Source:* BM*]* *[*\sd2 term/state*]*

T–t

tal ha' rain. Lit: 'come water'. SD: WEATHER. *[* \sd2 rain*]*

titih ha' rain subsides. Lit: 'dry up water'. Morph: **CV(C)-tih=ha'**. SD: WEATHER. *[Source:* BM*]* *[* \sd2 rain*]*

tosol ha' drizzle. SD: WEATHER. [Note: x tóos-ha' [Yuc.]. drizzle (Bricker et al. 1998); tos [Itz.]. Af, T. sprinkle, drizzle (Hofling and Tesucún 1997).] *[Source:* BM*]* *[* \sd2 rain*]*

S. Cook, *The Forest of the Lacandon Maya*,
DOI 10.1007/978-1-4614-9111-8,

Ts–ts

tsitsil ha' [tsitsil ha', tsitsir, tsitsi ha'] fine mist. Lit: 'sprinkle of water'. Said in reference to rain that is abating. Morph: **ts'its' 'sprinkle -il POSS ha' 'water'**. See: **yo'oxol**. SD: WEATHER. *[Note:* tsíits-ha'at [Yuc.]. tv. asperge, sprinkle (Bricker et al. 1998: 43); tzitzitz-ja' [Itz.]. sprinkle water; mist, fine mist (Hofling and Tesucún 1997: 628).*] [Source:* BM*] [* \sd2 rain \sd2 Ethnographic*]*

X–x

xämän$_2$ norther. A period of constant rain that is brought on the winds from the north. Rainy periods last from a few days to a week, subside for a few days, and then resume. SD: WEATHER. *[* \sd2 weather_patterns \sd2 Ethnographic*]*

Y–y

yo'oxol [yo'oxol, yo'oxoh] mist. Lit: 'it mists/steams'. Morph: **y-oox-ol**. See: **ts'its'il ha'**. SD: WEATHER. *[Note:* 'óoxoh [Yuc.]. hot, sultry (Bricker et al. 1998); oox [Itz.]. steam, make vapor (Hofling and Tesucún 1997).*] [Source:* BM; KM*] [* \sd2 rain\sd2 Ethnographic*]*

Appendix 5 Lacandon Texts

Take Care of the Forest, by CK, Naha', 1991

Source: [Video_p_1.mp4 – Take Care of the Forest]
The author asked CK to talk to the children about what was most important to him and what he thought the children needed to hear. This is what he said.

Take care of it, take care of the forest.
Take care of it.

When you grow up,
when you grow up,
you take care of the forest.
When I die, you take care of it.
Take care of the gods.
You take care of them.

Eh, if your father is dead,
your father is dead,
Bol is dead,
K'ayum is dead,
K'in is dead,
take care of it.

Take of your father.
Take care of your mother.
And, take care of the forest.

Take care of the forest,
because the forest,
because the forest exists for us.
Don't burn it.
During the dry season if you start a fire,
put it out!
Or if you are sleeping in the forest,
put it out!

And so there isn't any rain,
like now, there isn't any water (current) flowing,
there isn't any,
so one doesn't burn the forest.

Weed your milpa.
Take good care of it.
Plant your maize.
And then,
and then,
when it is finished,
you have your maize.

You keep praying to the gods.
You offer the first fruits.
You offer maize.
You give it to the gods, first.

Okay…

You offer it to the gods.
You make a little bit of *posol*.
After you offer it to the gods, then you can eat it.
After you finish your offering, you eat it.

Your mother,
take care of your mother.
It is your mother's work,
so don't annoy your mother,
your father,
anyone.
Be calm.

Take care of the forest,
like before, like the ancestors did.

In the past,
there wasn't so much pain in the forest.
There was game.
This wasn't a problem.
But now there isn't any.
So where can we go?
There are many villagers,
many Tzeltales.
We are surrounded.
That is why there aren't any animals.
Well, there are, but you have to really look to find some.

S. Cook, *The Forest of the Lacandon Maya*,
DOI 10.1007/978-1-4614-9111-8, © Springer Science+Business Media New York 2016

You heed the gods.
You take care of them.
Take care of the gods.
Take care of them.

Make the temple, whoever wants to,
whether it is you or someone else, your younger brother perhaps,
whoever wants to,
whether it is you or your older brother or whoever.
Together, you take care of the gods.
If you want to drink *balche'*, you will be cheerful. Sing the songs well.
Learn the songs of the gods.
Learn all the songs of the ancestors.
Like the ancestors did before, you learn them all.
Learn them all by heart.

Well, okay, you (say) you learn those songs of the Tzeltales.
You go to learn them there (in their church).
Perhaps this is so,
just don't abandon the gods completely. Don't.
They have been with you up until now,
just like in the past.

Don't annoy your mother.
Don't annoy your father.
Take care of your mother.
If your mother has fallen ill,
you talk to the gods.
(Do it) for whomever falls ill,
whether it is you or your children, if there are any.

You say you are buying medicine now, because there are no curing prayers for all the illnesses.
Only medicine.
For a cold,
there is medicine.
For a stomach ache,
there is medicine.
There is medicine for everything.

Well, there is the "true stomach ache", but that is different.
For that you talk to the gods.
When you pray to the gods, it gets better.
They help (cure) it, like they did before.

Before, (when) there weren't any villagers.
(there were just) the gods.
All the gods were prayed to about everything.
Everyone did this.

They offered the *balche'*...
Bark cloth was given.
Annatto was given,
given to the gods.
When it was finished,
there were tortillas for the people.
There were meat tamales.
There was *balche'*.
There was *posol*.
There was *posol* for the gods.
These were the offerings on which the ancestors were raised.

Where I was raised,
where I grew up with my father,
there was meat.
there was *posol*.
there were songs.
there was copal.
there was rubber.
These were the offerings to the gods.
The singing was wonderful,
(and) the dancing. Ah, this is marvellous for (the gods).
You take care of them.
Take care of the forest.
Take care of it.
Take good care of the forest.
Take good care of it.

Secret of the *Balche'*, by AM, Naha', 2003[1]

Source: [Video 4.3 Ut'anil *Balche'* - Secret of the *Balche'*]
AM recites the secret of the balche', to assist the fermentation of the beverage. He does this after he has filled the canoe with water and sugar. After he finishes the chant, he takes the temperature with a thermometer and then fills the canoe with balche' bark. He repeated the chant a second time, but only the first part is transcribed below. He said one needs to repeat the chant many times, up to four times, to ensure that the balche' ferments properly.

White light in my hand,
I am the true Bol.
I am the true *etsel* cup.

White light in my hand,
I am the true Bol.
I am the true *etsel* cup.
I am the true Bol.

When the foam of my pot of food rises,
you immediately watch the pot of our solution for us,
the pot of my food,
so that the foam (rises).

[1]See Tozzer's version, in Chant 17 (1907:177).

When it rises, come move it for me *chäk xok*.
The *chäk xok* is standing on the right side,
she is suddenly submerging
on the right side of our canoe,
on the left side of our canoe.

When the foam of my pot of food rises,
come move it for me (so) my *balche'* dissolves.
All come throw in (the venom) for me.

The *t'int'a* turtles[2] and the yellow turtles are suddenly submerging.
All come,
… striped mojarra, red mojarra, *sohom*,[3]
come move it for us,
the pot of our solution.
Come.

The *hach 'aak*[4] is rolling about.
T'int'a turtles,
snapping turtles,
come move my pot of food,
the pot of our solution.

This makes a churning noise.
… The rolling about of the yellow turtles makes a noise along the length of our canoe.

The *chäk xok* is standing on the right side of our canoe.
She is standing on the left side of our canoe.

When the foam of my pot of food rises.
I kindle the fire
with the tail of the penelopina.
It is going to fan the fire.
I am fanning it on the right side of our canoe.

When the foam of the pot of my food rises,
come rattle snakes,
come bite the hand cups.
Bite our small cups.
Bite my hand cup.
Bite our small cups.
Bite our *balche'* urn.
Bite my hand cup.
Bite my hand cup.
Bite our large cups.
Bite them for us.

Come, throw in the venom from your fangs,
rattle snakes,
feathered serpents.
Black snakes,
soot snakes,
bite the right of my *balche'*.
Bite the base of my *balche'*.

When the foam of my pot of food rises,
kindle its fire.
When the foam rises,
the *balche'* roils.
This makes noise on the right side of our canoe.

I am the true Bol.

When the foam rises, the rattle snakes are the ties of our canoe,
the cover of our canoe.[5]
The cover of our canoe are nettles,
the devil nettles,
(which) I used as ties for our canoe.
The branch snakes are the ties on the right side of our canoe,
the ties of our canoe.

Branch snakes,
soot snakes,
feathered serpents,
black snakes,
soot snakes,
come throw the venom from your fangs into our canoe.
Bite the left side of our canoe.
Bite the base of our canoe.

Come.
Come make my *balche'* dissolve.
Come bite the base of my *balche'*,
black snakes,
soot snakes,
feathered serpents.

Black snakes,
rattle snakes,
come.
Bite the right side of our canoe.
Bite my *balche'*.
Bite the left side of my *balche'*.
Bite the base of my *balche'*.

When the foam of my pot of food rises, kindle the fire.
…My canoe is frothing.

On the right side of our canoe, the *chäk xok* is rolling about.
The garfish
they are suddenly submerging.

---

[2]Likely, this is the name of a kind of turtle.

[3]According to my Lacandon consultants, this is a type of black fish, bigger than a mojarra. Cf. sohom. (Tozzer 1907:54); tsau. "large fresh-water mojarras" (Motul., in Roys 1931:340).

[4]Three species and one subspecies were identified as hach 'aak by two Lacandon consultants: (1) Central American river turtle (*Dermatemys mawii*); (2) white-cheeked mud turtle (*Kinosternon leucostomum*); and, (3) red-cheeked mud turtle (*Kinosternon scorpioides*).

[5]The *balche'* canoe is covered with palm leaves and then tied down with vines. These vines are what AM is referring to.

Soot snakes,
come throw the venom from your fangs into my hand cups,
into my large cups,
the large hand cups.
Bite our *balche'* urn.
Move the pot of our solution for me.
Come,
rattlesnakes,
black snakes,
nettle ties of our canoe.

The white walking stick chilli peppers,
when I gathered them up,
when I dumped them into the right side of our canoe,
the foam rises,
the *balche'* roils.

This makes a noise.
On the right side of our canoe,
the frog makes the noise.

Red turtles, yellow turtles, soot snakes, black snakes,
I am the true Bol.

Song of the Cacao, by JK, Naha', 2003

Source: [Video 4.6 Uk'aay Käkow - Song of the Cacao]
This work song is sung while preparing ceremonial cacao. It is sung to invoke the tsuk'in vine, which is used to froth the cacao. The cacao beans are roasted, ground, and then mixed with water. The tender tips of the tsuk'in vine are crushed and then added to the beverage, and then the beverage is beaten with a wooden beater, to raise the foam. Most of the preparation takes place in the ceremonial kitchen located beside the god house. Women seldom enter the god house, except for when they froth the cacao, which they do in front of the god pots that will receive the offering.

I toast my cacao.
I push the fire together for my cacao.
I fetch the fire for my cacao.
I fetch it.

I walk around to gather firewood for my cacao.
I push the fire-brands together.
I fetch the pieces (of) the pot of my cacao.[6]
I fetch the wood.
I stir my cacao.
I toast it.
I break it up.

[6]Unclear. JK may be referring to the number of small bowls that are used in the process of making cacao.

I break up my cacao.
I move my cacao around.

I wash the table for my cacao.
I go fetch the water for my cacao.
I go carry its water.

I fetch the cup.
I wash the cup for my cacao.
I wash the table.

I fetch the beater.
I beat my cacao.
I pass my cacao through.
I grind my cacao.

The beater,
it is the beater for my cacao.

I move my cacao around.
I fetch the small cup for my cacao,
for the food of our Lord.

When you come, you see my offering to our Lord.

I move it around.
I toast my cacao.
I move it around.

I fetch the *tsuk'in* for my cacao.
I snap off the *tsuk'in* for my cacao
I pass the *tsuk'in* (through).
I grind my cacao.
I re-grind my cacao.

I fetch the beater.
I fetch the cup for my cacao.
I re-grind it.
I dissolve it.
I dissolve my cacao.

I move my cacao around.
I beat my cacao.
I mash my cacao.
I crush it.

I fetch the small cup of our Lord.
I offer it to our Lord.
I fetch the beater (for) my cacao.
I am going to go fetch the beater.
I chip it into pieces.

If it pleases you, sarsaparilla,
come see me offer it to our Lord.
You must throw down your juice,
twisted vine, sarsaparilla.
Here it is.
You put in your juice there ....

I peel my *tsuk'in*.
Do you see it?

I re-grind my *tsuk'in*.
I put it there for when you come see it.

I crush it.
I re-grind it.
I re-grind my cacao.
I re-grind it.
I dissolve my cacao.
I toast my cacao.
I offer it to our Lord

You must come see it (when) I go beat it in the god house,
(when) I go to beat it at the mouth of our Lord's plate.[7]

I will go toast my cacao.
I am walking through.
I will dissolve it.
I empty the bowl.
I will go with the small cup.
I go set the small cup at the mouth of our Lord's plate.
I beat my cacao.

You come put in your juice for me, oh sarsaparilla.
Where are you, oh sarsaparilla?
Sit down and put in your juice for me …
You are watching me snap it.

I am snapping you off.
I re-grind my cacao.
I am putting you in the foam of my cacao.
I am re-grinding (the cacao) with it.

I grind it smooth on the millstone.
I re-grind it on my table.
I am crushing it.
I am kneading the *tsuk'in* for my cacao.
I am re-grinding my cacao.

You see this is my offering to our Lord.
You come to see it,
oh *tsuk'in* vine,
oh sarsaparilla.

You gather together here,
twisted vine,
twisted vine of Jesus.

How to Build a House, by BM, Naha', 2011

Source: [Video 4.10 How to Build a House]
This text was recorded in an interview with BM, who describes how to build a traditional, thatched house.

First,
I look for the pillars.
The *(y)okman* 'pillars' are first.
I have to look for four pillars.
Two for the tall pillars',
to support the front beam.

Then I go look for the *utaan* 'front beams'.
I bring all four front beams.

Then I go look for
the *uwaan* 'side beams'.
They're called *kuchsäkche'il* 'load bearing beams'.
O.K. I bring all the load bearing beams.

Then,
I look for and chop down
the *usäkche'il* 'rafters'.
O.K. I bring all the rafters.

I go look for the *uhiil* 'battens'.
The battens have to be cut.
I bring all the battens.

Then,
I bring all *ho'olan che'il*[8] 'ridgepoles'.
I bring them all.

Then, I will go and look for the vines.
I look for the monkey vine (*Paullinia* sp., *Serjania* sp.).
No, not the monkey vine, not the monkey vine--maybe a few--
nikte', the *säk ak'* 'white vine' (*Arrabidaea verrucosa*), and the *ya'ax ak'* 'green vine' (*Arrabidaea* sp.), to tie the pillars to the front beams and to tie the rafters.
All the rafters have to be tied.

When I finish tying the rafters, I tie the battens.
I tie the battens.

(When) all the battens are finished,
I will go look for *ukunil* 'the thatch'.
I go look for the smooth *holop'* (*Philodendron smithii, P. tripartitum*),
to weave its thatch.
You have to tie it (with) either *Philodendron* spp., or *Hampea stipitata*.

O.K.
I weave the thatch with these.
(When) I finish weaving the thatch,
I have my home.

BM talks about how long it takes to build his house.

It takes one day for me to drive in the pillars.
To raise the beams, one day.

[7]The god pots have anthropomorphic faces moulded on one side of the pot. The mouth has a large, protruding bottom lip on which food offerings are placed.

[8]Cf. *hó'ol-nah-che'* [Yuc.]. ridgepole (Bricker 1998:110).

Then another day to raise and tie its rafters (to) its battens. That's two days, and a third to weave its thatch. Three days.
In three days I have my house.
It's not big.

BM describes his selection of species for parts of the house.

O.K. First --it's like I said-- first the pillars.
For the pillars, I look for the hardest (wood).
Ch'obenche' is very strong, it's very hard. Chicle (*Manilkara achras*), that's the hardest. Chicle lasts in my house for two or three years.

O.K. (As for) the front beams, the best are *Aspidosperma megalocarpon* and *Calophyllum brasiliense*. They are very straight. Their trunks are very straight. They're very strong. For the side beams, *Aspidosperma megalocarpon* is equally good. *Aspidosperma megalocarpon* is also good for the ridgepoles, or *Calophyllum brasiliense* is just as good. It's very straight.

For the load bearing beams, the names (of the trees) that are very good are *ahtso'ots bamax* (*Pseudolmedia oxyphyllaria*) or *hach bamax* (*P. oxyphyllaria*).[9] These are very good for the load bearing beams. If there is an old fallow milpa, a good tree is the savannah tree. The savannah tree is very good for the load bearing beams. The same goes for the battens. It's very good for battens. The savannah tree is very straight. It's very good. It's not heavy. It lasts.

O.K. For the battens, the savannah tree is good or, if there are any, *Aspidosperma megalocarpon*, *Calophyllum brasiliense* or *Pseudolmedia oxyphyllaria* saplings. All of them, (and) *tso'ots bamax*, are very good for battens.

O.K. For the roof, *Cryosophila stauracantha* is very good. *Sabal mexicana* is very good too. *Cryosophila stauracantha* (is good), because the leaves are wide (so) the rain doesn't come in. They're the best. They're smooth, so not much rain (gets in). But they are denser and thicker to weave. *Geonoma oxycarpa* is just as good and faster to (thatch). It's faster, because the leaves are long. *Sabal mexicana* is like that too. The leaves are long and very wide. The leaf blades are wide, so the work goes faster. One can finish (the roof) quickly. *Sabal mexicana* lasts longer, *Sabal mexicana* and *Cryosophila stauracantha*. They last if there is smoke or fire. They will last ten years. Smoke is very strong. It's like I said, it has strength for the entire house. Smoke is very strong. It dries (the thatch) quickly. When it rains, it dries quickly. If there isn't any smoke, the thatch rots quickly because it's cold.

So those are all very good. There are only three: *Geonoma oxycarpa, Sabal mexicana*, and *Cryosophila stauracantha*.

[9]Obviously, these are two different species of trees to BM.

Flint Song, by AM, Naha', 2003

Source: [Video 4.15 Uk'aay Tok' – Song of the Flint]
This chant is recited when a new antler punch is broken in, to give strength both to the antler and the stone hammer. According to BM, it is sung so that the antler does not split.

Here when I sat down at my flint,
is head is the white-breasted sparrow hawk,
the head of the sparrow hawk,
when I sat down.

When I sat down to chip I spread out the veins,
my great antler lying across the head removes the flakes.[10]

When I sat down to chip the cortex,
I sat down to chip,
the hammer is the white-breasted hawk,[11]
it is my hammer, the vein of the harpy eagle,
its great thigh(bone) to its great shin bone,
are the bones of my antler,
the hardness of my antler,
the hardness of my flint,
when I sat down to chip the cortex.
The great antler leaning across the head removes the tongues.

When I sat down to chip the cortex,
I spread out the sea,[12]
I spread out the veins of the white chamaedorea,
I spread out the veins of the sea.
The parlour palm is all spread out.
I spread out the veins of the white chamaedorea veins,
when the great antler removes the tongues.
I spread them out on the cortex, when I sat down to chip.

The red-capped woodpecker,[13]
(it is) the strength of the wing of the red-capped woodpecker.

Eh, when I sat down to flake my flint, they are the talons of an eagle,
the talons of the golden eagle, the harpy eagle,
these are their talons.

[10]Half of the stone is the striking surface, called "the head", or in knapping terminology, the platform. The flakes are referred to as *ch'ibix bo'oy* and *u-yak'-al* 'its tongues' *Ch'ibix bo'oy* also refers to the parlour palm (*Chamaedorea elegans* Mart.).

[11]To make the striking surface smoother, the white breasted hawk is mentioned.

[12]Unclear reference to the sea. It may refer to the ridges of waves. This reference occurs in a similar version sung by another Lacandon man, recorded by the author.

[13]It is unclear why the wings would be the focus, instead of its strong beak. According to one Lacandon consultant, the red-capped woodpecker is significant because the knocking sound it makes—bap, bap, bap, SLAP—is similar to the sound made when knapping flint.

When I sat down to chip, I spread out the veins.
When I sat down to spread out my flint well,
it removes the tongues.[14]
All the veins are spread out.

When I sat down to chip, my flint is the white-breasted sparrow hawk.
It is my hammer.
Its head is secured across my hand.
The head of my quartz is the white-breasted sparrow hawk.
It is my hammer.

My great antler removes the tongues.
They are scattered, when I sat down at my antler.

I lean it against the head.
It is the hardness of my flint.
When I sat down to chip, (it is) the red-capped woodpecker.

I spread out the midrib of the plantain (leaf), the (leaf of) the white calathea.
The midribs of the white heliconia are all spread out.
I spread out the fish-tail palm. (The mid ribs) are the veins (of my flint).
They all are extended, the leaves of the parlour palm, the spines of the chocho palm.

The young corn ears are rubbed between the hands, the kernels of corn,[15]
when my flint blades are born,
when I sat down to chip.

When I sat down to split the talons of the black hawk-eagle,
the talons of the harpy eagle,
my hammer is the white-breasted hawk.
That is my quartz, when I tied on his head.

When I separated the veins of my flint,
the antler leans across the head…
It removes the tongues.

The hardness of my antler (is) the chocho palm,
it is the centre of my stone.
This is my antler.
The hardness of the *pahok* palm is the hardness of my antler.
The hardness of the shin bone of the harpy eagle,
its upper arm, its shoulder bone,
that is the hardness of my flint.

When I sat down to chip, I lean it across the head.
The antler removes the tongues.
When I sat down to chip my flint, the cortex
I removed, the entire cortex.
The great antler removes the tongues.

When I sat down to chip my flint,
the red-capped woodpecker, the strength of its wing is the hardness of my flint when I burnished it.
The red-capped woodpecker is my flint.

[14]Although literally 'tongue, vine', the word is also used metaphorically to refer to flint blades.

[15]Unclear reference.

Song for Spinning Thread, by JK, Naha', 2003

Source: [Video 4.19 Uk'aay K'uch – Song for Spinning Thread]

The following work song[16] is akin to a magical charm that is recited to help speed up the process of spinning thread. The singer calls upon spiders and some bees, to help her with her task by providing their own silk to expand the volume of thread it will take to make a tunic. Throughout the song the singer alludes to the different phases involved in spinning with a drop spindle and weaving on a backstrap loom. JK was the only woman whom I approached that was willing to sing this song. Only the older women have had experience spinning and weaving, which may be one reason why younger women claimed they did not know the song. The other elderly women I approached claimed not to know it well.

Oh, where are you, orb spiders?
How many you are, little orb spiders,
orb spiders of the branches.
You all gather together here to spin it for me.
You come to help me.
You put in your thread in my thread for me,
so that my thread will expand here.
You come to see it on my spindle here.
You help me spin.

How many little orb spiders there are.
How many large orb spiders there are.
How many trapdoor spiders there are.
All of them,
the tarantulas all around,
you all gather together here
for my spinning here,
for my cloth.

You have to pile (your silk) on top for me,
so that my spinning (thread) increases,
so that it increases.
You have to come help me spin,
where you are here, below.

[16]The author follows Davis (1978), by calling the songs sung by women "work songs".

How many are all below.
You help me.
Sit down and stuff in your brood for me here.
You have to come fill it.
You have to come and stack it on my spinning here.
You have to come and stack it up for me here.

How many *chi'* there are.[17]
How many yellow *säk k'oho'*,[18]
how many cave bees,
how many there are.
How many *yuus* are all here.[19]
You all gather together here to stack my spinning for me,
to put in your young here.
Here, come stack them up for me here, so that it increases,
so that my thread increases,
so that it increases,
so that it increases faster.

You are here,
oh little orb spiders.
How many orb spiders there are that increase my thread,
that come to see it on my spindle here.

You have to come and stuff in your thread in my thread where I spread it out,
where I wind it up.
I wind up my thread,
so that when you come you stuff your silk in here.
You have to help me spin it.
You have to help me to spin it.
You hear me spin.
I spin it

When I finish spinning it,
I will wind it up.
When you come help me, I enlarge my (ball of) thread.
You come help me.
I fetch the end rods.
You will come help me stretch the warp on the end rods here.
You come help me.
You are here.

How many you are,
orb spiders,
how many you are.
How many you all are here.

When you come and gather together you will pile your silk on my thread for me,
so that my thread expands for my cloth,
so that it expands.
You have to come to help me weave it here on the end rods,
where the weaving sword is.
You are helping me.

How many orb spiders there are;
little orb spiders,
everywhere.
You are far up in the sky.
You are on the branches of the big trees.
All of you, come down here.

You see that I am spinning here.
You help me with all of my spinning here.
You help me spin,
you come help me to spin my cloth,
so that it may expand faster when I weave it on my shuttle,
when I put the cloth where it lies,
where it lies face back.[20]
You come help me.

You have to come help me with my spinning here.
You have to come to help me spin it so that it expands faster on my spindle,
so that it expands faster.
You have to come help me with my cloth.
You have to roll it for me here,
where I am rolling it,
so that my thread may expand faster.
You help me.

Come.
You will spin my thread.
You have to come help me here spin thread for my cloth.
You have to come help me so that I may spin here on the spindle,
so that it may expand faster for my cloth.

I am spinning my cloth.
You come see it on the end rods.
You will come and stuff in your silk for me.

Little orb spiders, orb spiders,
everywhere on the bowl of my thread there,[21]
you will come see it.
You must come see it,
when you come.

You stuff in (your silk) for me,
(while) I am weaving my cloth.

[17]This is a very small variety of stingless honeybee (AM).

[18]This is a large and somewhat reddish, stingless bee with brown and white wings. It makes its hive in hollow logs. It produces a lot of honey. (AT) The golden honey is mild, thin, and very sweet. For this reason, these bees are also raised in gourd hives.

[19]This is a small species of grey, stingless, honeybee. It makes it hive in fallen logs. The hives yield up to two litres of honey. (BM)

[20]Conceivably, JK is describing her weaving the weft into the warp.

[21]JK is referring to the small bowl in which the tip of the spindle is supported.

I am spinning my cloth.
I am making my cloth.

You come see it on the end rods.
You will come see it there on the warp
where I weave it.
You will come see it.

When you come see my cloth, you stuff in your silk in for me,
so that it goes faster,
so that I spin it,
so that I make it swell,
so that I can swell my cloth.
This is my cloth for my end-rods.

I take the beater,
I smooth down (the weft) again.
I smooth it down again, to make it snug.

Come, stuff in silk.
Night is falling.
Come, stuff in your silk.
It is nightfall when I swell it,
when I cover it up and lie down.

You have to help me weave it.
You bring a great deal of your silk, which you stuff in.
You all help me spin thread,
so that I can weave my cloth,
my husband's tunic is what I weave,
so that it expands,
so that it goes faster.

Although it is very long,
may it shrink more quickly.
You have a great deal of silk, which you stuff inside.

You help me so that my work goes faster.
Although I get up from there,
I get up and I go on my way,
you help me.
(Although) I leave you,
you come help me by stuffing your silk into my thread,
so that it expands quickly.
You are the ones that help me with my husband's cloth,
so that it finishes faster.

After three days, it is finished.
I will weave another one.

You are very strong.
You help me.
You are the spiders,
the dear little orb spiders.
How many you are,
orb spiders.

You have to help me with my work.
You have to help me spin.

When you come put your silk on my spindle,
you are watching my spindle,
you are coming to stuff it in where I raise (it) up high in the rafters.[22]
I set it up above,
so that you will come and put your silk on my spindle. You are very strong, when you come and put it in.
You help me spin it.
You help me.
You are the ones that pile up your silk (and) stuff it in for me.
You come and fill the woof of my woven cloth
where it is laying,
where I expand it.
You fill up the woof of my woven cloth.
You come help me.

Song for Grinding, by JK, Naha', 2003

Source: [Video 4.7 Uk'aay Huch' - Song for Grinding]
In this women's work song, JK calls upon the birds to help her make the nahwah, the ceremonial tamales that will be offered in a ch'ulha'kih ceremony.

These are my shells.
You must come help me.
You help me with my work.
I grind my shells.
You must come here.

How many speckled hawks,
how many quail there are,
how many are they all.
You all come together for me,
to my work here for our Lord Jesus,
here for our Lord,
here for my work here.

I grind here.
You have to come grind it for me here.
I am grinding it here.
You have to come see my work, for my offering to our Lord.

How many quail there are.
How many partridges,
how many tinamous,[23]
how many partridges there are.
You come here to help me.

[22]The rafters is where various articles are typically stored.

[23]My consultant, AM, identified several birds with this name, including: buffy-crowned (highland) wood partridge; long-tailed wood partridge; and great tinamou.

You stick your work in here,
when you help me with my grinding here.
You come to help me here,
with my work here.

You have to come help grind it for me here.
You have to grind here.
You have come help me,
large parrot,
mealy blue-crowned parrot.

How many macaws there are.
How many there are.
Large parrots,
how many there are.
You come to help me.
You help me with my work here.
I am paying the Lord here.
You have to come help me. I am working for our Lord.

You have to come crush for me.
You have to come chip it.
You have to come crush it all here.

Mealy blue-crowned parrots, quails, …everywhere,
you come all at once here,
Ani,
all of you come here for my work for our Lord here.
Come help me here with my offering to our Lord here.
You have to come help me here with my offering, to make it go faster.
You gather together here for your work here,
for our work,
for our grinding
You have to crush it for me.
You have to chip it all for me here.
You have to crush it all for me here.

How many chestnut-capped brush-finches,
ani,
itsanal,[24]
you all gather together here,
for our work to our Lord here.
You make it finish more quickly,
so that it goes faster,
so that it quickly piles up.

We work for our Lord here,
so that it finishes faster here.
You help me here.
You have to help me regrind it.
I re-grind my work.[25]

You have to help me.
You come quickly, so it finishes faster.

How many are the large parrots, how many mealy-blue-crowned parrots.
How many (you are). *Itsanal*, everywhere.
Here, you all gather together here.
You are here to work for our Lord here,
so our re-grinding goes faster,
so that it goes faster,
so that it piles up faster,
so that it goes faster.

How many there are of you who come, how many (you are).
How many there are of you that remain in the branches.
How many there are of you.
Quails all around,
how many there are of you that stopped by.
All of you gather together here.
You have to see me work for our Lord here.
So many of you stopped here.
You are here, you are all here to help me here.

You see that the re-grinding finishes faster,
so we can fold the tamales,
so we can make them,
so that they pile up faster.

Don't let the sun rise.
Don't let the sun rise.
Stop the dawn.
Don't let the sun come up, for our work here.
Don't let the sun come up, for our work here,
so that our folding goes faster.

You bring the plate for us to fill,
so that our folding them piles up faster,
so that our making them here piles up faster.
The plate overflows here,
so that it goes faster.

How many you are here, you all gather your companions together.
How many there are.
Curassow, how many there are.
Crested guan, how many there are here.
Here you all gather together here,
so we can pulverize it here. All kinds of birds everywhere.
You all gather together here,
for our work for our Lord here,
so it goes faster,
so our folding them goes faster.

Don' t let the sun rise.
Don't let it rise,
so it goes faster,

[24]Undetermined species. According to AM, the name refers to a number of species of small, (grass) seed-eating birds.

[25]Dough.

so it piles up faster,
so it goes faster.

Here you all gather together here, for our work to re-grind it.
When it finishes, we knead it.
When it finishes, we knead it on the table,
We will fold quickly.
You bring the plate and we will fold them.
You sweep the plate.
You sweep the plate well, for us to fill it.

You all gather together here.
How many you are here.
You all gather together.

We grind it all on the millstone, to crush it.
Come, let's all crush it,
so that it goes faster.

You are very strong.
You crush it here.
You are very strong.
You crush all the kernels.
You fetch the water to knead it.
You finish kneading it,
so that it can be re-ground.

I am regrinding it.
I re-grind my ground corn.
I regrind it.
I knead it,
I knead my offering for our Lord.

You are coming to help me.
How many there are.
You are many.
From far away, (now) here you all are.
You are all here gathered together.
Here, you come here to work.

How many kinds of birds there are.
Small birds everywhere, one by one, they all gather together here.
The ani,
the *ts'ah k'ak'*,
all of them,
you all gather together to help me with our work,
so that it finishes faster.
You come to crush my dough for me,
so that it finishes faster.

My dough piles up faster.
It piles up.
Quickly it piles up.
I grind it so it piles up faster.

How many there are when I fold (the tamales).
They are piling up when I fold them.
Quickly the plate fills up.
Another and another (plate) you bring,
so that I can fill it again,
so it piles up.
The folding goes very well.
It hurries along.

How many quail there are, how many speckled hawks there are.
How many different kinds of birds there are.
Yellow birds,
everywhere,
one by one, you gather together here
so we can crush it,
so we can fold (the tamales),
so that it finishes faster,
for our Lord,
for my offering to our Lord,
so I can pay our Lord,
so it goes faster.

You pinch (a piece) off the edge and stick it on the edge (of the tamale),[26]
for my offering to our Lord.
It finishes faster.
You place the edge on the leaf of the tamale.

When my re-grinding is finished,
you will come to help me fold (the tamales).
You come to help me,
when you all gather together here.
You all gather together here,
so we can fold them on the table here.

I wash the pot for the tamales.
I wash the pot.
I cook them,
I cook the tamales.
When the folding finishes,
they will cook.

Quickly, we gather together to fold them,
to make it go faster,
so that we can cook them.

I am finished re-grinding (the dough).
Let's fold them now.
We have to fold them on the plate,
put them on the plate.
I'm going to knead (the dough), if it is not well ground.
If it is well ground, then it will be fine.

Quickly,
all of you come to our work here.

[26]Before closing up the tamale, a piece of the leaf wrapper is torn off and placed on the dough.

Go fetch the water for me.
Let's go carry the water for the kneading.
You have to wash the pot.

The tamales need to cook.
There is the grill rack.
There is the grill rack,
the *Quararibea funebris* grill rack.

You have to come help me with my work.
You have to come help me.
It's ready.
I am washing the table.
The table is all washed.
You have to come help me with our grinding on the millstone.
You have to come help me with my work here on the millstone here.
You have to help me.

How many birds there are.
Here you all gather together for me here,
for my work here,
ani,
ts'ah k'ak',
--everyone;
mealy blue-crowned parrots,
little parrots,
brown-hooded parrots,
everyone is here.
You gather together;
Parakeets,
you all gather together for me.
You all gather together here to work here, for our work for our Lord,
for our offering to our Lord here,
so that the it piles up faster,
so that it all piles up faster.

All the large parrots,
all of them,
all the mealy blue-crowned parrots,
how many there are.
The brown-hooded parrots,
all of them are here.
You gather together here for our work here.
You help me with all my work here.
You have to come help me with my work,
so that finishes faster.

Don't let the sun rise.
Don't let it.
Don't let the fire of dawn arrive.
Don't let it dawn.
Grab the sun, please.
Stop it,
so we can fold them,
so we can fold them for our Lord,
so we can work for our Lord.

You help me work here.
You fetch the plate.
When you fill it,
you come out to fold them.
You fill the plate with them,
fill the plate with …our work for our Lord.

All you large parrots everywhere.
Mealy blue-crowned parrots everywhere.
How many macaws there are everywhere.
You all gather together for me.
How many *itsanal* there are here.
You gather together for my work to our Lord here,
so we help crush (it) for our Lord here,
for my work,
for our work for our Lord here,
so that it piles up faster,
so the dough will pile up faster--
the dough will quickly pile up.

I am working for our Lord.
Quickly, my tamales will pile up.
Quickly they will pile up,
my work for our Lord.

The plate is filling up.
You help us make them on the plate.
You have to dust off the plate well.
You dust off the plate well.
Fill it well.

You have to look at it.
Wash the table.
You have to wash the table completely,
for us to grind our work.

Go fetch the water.
Fetch the water,
for our work,
for our dough.

You have to wash the cooked corn.
You have to wash it.
You fetch the basket.
When you wash the cooked corn,
fetch its basket to wash my cooked corn.
Let's wash it.

You have to help me carry the water.
You have to come help me with our work here,
to grind on the millstone here,
our work here on the table,
so that it piles it faster, the piled up dough of my work.

Quickly, my folding is piling up.
Quickly it piles up.
Quickly, my grinding it finishes.
Quickly, it finishes.
You are very strong.

How many there are.
How many curassows there are.
How many crested guans there are.

My grinding piles up.
Quickly, it piles up. Quickly, the partridges,
the mealy blue-crowned parrots,
all of them come.
The parakeets,
all of them,
you all gather together here,
for my work for the Lord here.

I scrub the cooked corn.
I scrub the cooked corn.
I wash it so that it goes faster.

I am filling it faster.
I am cooking it.
I am taking it to our Lord.
I go cook it in the god house,
(where) I finish it (when) my husband's prayers finish.

Quickly, my cooking it finishes.
The *atole* (ceremony) finishes.
I cook the *atole*.
My making the *atole* goes faster,
my offering to our Lord.
I cook the *atole*.
I boil the *atole*.
I cook it.

When it is finished,
I am going to grind the *atole*,
grind the *atole* for my offering to our Lord,
for my offering to our Lord.

...You wash the cups... Quickly, you wash the cups,
so I can fill them,
so I can fill them.
(When) I finish dissolving (the *atole*), I will fill them.

You are many, when you gather together where I crush (the corn) for our Lord.
You crush it all.
When you help me crush the kernels,
when you help me crush them.
you have to help me grind it.
You have to help me with my work here,
for my offering to our Lord,
for my offering at the end of the nahwah ritual,
the end of the *balche'* ritual,
for our Lord,

It is late afternoon.
It is cooling off.
The praying in the late afternoon finishes.
It is cooling off; the praying is finished.
The offerings of the first fruits finishes,
the offerings of tamales.

When the afternoon comes,
he gives the first fruits.
When he pinches off the head (of the offering), then I may eat it.
We will eat it.
You will eat it.

The *balche'* is later, at the end of the afternoon.
When the *balche'* finishes in the late afternoon,
you have to help me give the offerings to the Lord.

You have to come see it.
When you come,
you see that I am finishing it so that my husband can finish his prayers to our Lord.
You have to come see our crushing.
When it is finished,
we will rest.
We will rest.

When we finish folding the tamales,
we will wash our hands.
You will wash your hands.
We will go wash them in the water.

I pass through to wash my hands,
wash my hands.
Wash them well.
Wash your hands well.
Your hands are dirty.
You wash them well.
Wash all your fingernails well.
You wash them well.

We knead the dough.
I am making it.
We grind it to fold them,
to fold the tamales,
to fold them.

The grinding is finished.
The folding it finished.
Eh, the folding is finished.

Bibliography

Amram, D. W. Jr. (1937). Eastern Chiapas. *Geographical Review, 27* (1), 19–36.

Atran, S., Lois, X., and Ek', E. U. (2004). Plants of the Petén Itza' Maya: Plantas de los Maya Itza' del Petén. *Memoirs of the Museum of Anthropology 38*. Ann Arbor: University of Michigan.

Baer, P., & Merrifield, W. R. (1971). *Two studies on the Lacandones of Mexico*. Norman, OK: Summer Institute of Linguistics, University of Oklahoma.

Borys, M. W., Leszczyñska-Borys, H., Ramirez-Marañen, S., Castro, L., & del Campo, F. (1993). An avocado relative: Beilschmiedia anay (Blake) Kosterm: A fruit source. *California Avocado Society Yearbook, 77*, 125–136.

Breedlove, D. E. (1986). *Listados Florísticos de México IV. Flora de Chiapas*. México: Instituto de Biología, UNAM.

Bricker, V., Po'ot, E., & de Po'ot, O. D. (1998). *Dictionary of the Maya Language: As spoken in Hocaba Yucatan*. Salt Lake City: University of Utah Press.

Bruce, S. R. D. (1968). *Gramática del Lacandón. Departamento de Investigaciones Antropológicas* (Vol. 21). México: Instituto Nacional de Antropología e Historia.

Bruce, S. R. D. (1975). *Lacandon Dream Symbolism 1: Dream symbolism and interpretation*. México: Ediciones Euroamericanas Klaus Thiele.

Bruce, S. R. D. (1976). *Textos y Dibujos Lacandones de Najá*. (Trilingual Edition: Lacandón-Spanish-English). Departamento de Lingüística Colección Científica, Lingüística, Núm 45. México: Instituto Nacional de Antropología e Historia.

Burger, W. C. (1962). Studies in New World Moraceae: Trophis, Clarisia, Acanthinophyllum. *Annals of the Missouri Botanical Garden, 49*(1/2), 1–34.

Canger, Una. (1970, Unpublished MS). Lacandón de San Quintín Vocabulary. of San Quintín.

Comisión Nacional de Áreas Naturales Protegidas. (2006). *Programa de Conservación y Manejo Área de Flora y Fauna Nahá*. 1a edición. Tlalpan C.P., México, DF: Dirección General de Manejo para la Conservación, CONANP. http://www2.ine.gob.mx/publicaciones/consultaPublicacion.html?id_pub=558&id_tema=12&dir=Consultas ISBN: 968-817-814-4

Davis, V. D. (1978). Ritual of the Northern Lacandon Maya (PhD Dissertation), Tulane University, New Orleans, LA.

Diemont, S. A. W. (2006). Ecosystem management and restoration as practiced by the indigenous Lacandon Maya of Chiapas, Mexico Ph.D dissertation, Ohio State University, OH.

Diemont, S. A. W., Martin, J. F., Levy-Tacher, S. I., Nigh, R. B., et al. (2006). Lacandon Maya forest management: Restoration of soil fertility using native tree species. *Ecological Engineering, 28*(3), 205–212.

Durán, F. A. (1999). *Estructura y etnobotánica de la selva alta perennifolia de Nahá Chiapas* M.Sc. Thesis. Universidad Nacional Autónoma de México, Mexico.

González-Rosas, H., Salazar-García, S., & Vázquez-Valdivia, V. (1985). Propagation in vitro of Chinini (Persea schiedeana Nees). *California Avocado Society Yearbook, 69*, 125–132.

Gutiérrez-Carvajal, L., & Dorantes-López, J. (2003–2004). Aguacatillo. *Especies forestales de uso tradicional del estado de Veracruz Potencialidades de especies con uso tradicional del estado de Veracruz, como opción para establecer Plantaciones Forestales Comerciales*. CONAFOR–CONACYT–UV. http://www.verarboles.com/Aguacatillo/aguacatillo.html

Hellmuth, N. M. (2013). *Maya Ethobotany complete inventory: Fruits, nuts, root crops, grains, construction materials, utilitarian uses, sacred plants, sacred flowers*. Asociacion FLAAR Mesoamerica. http://www.maya-ethnobotany.org

Hofling, C. A., & Tesucún, F. F. (1997). *Itzaj Maya-Spanish-English Dictionary-Diccionario Maya Itzaj-Español-Ingles*. Salt Lake City, Utah: The University of Utah Press.

Hofling, C. A., & Tesucún, F. F. (2000). *Tojt'an: Diccionario Maya Itzaj-Castellano*. Iximulew, Guatemala: Cholsamaj Fundacion.

Kashanipour, R. A., & McGee, R. J. (2004). Northern Lacandon Maya medicinal plant use in the communities of Lacanja Chan Sayab and Naha', Chiapas, Mexico. *Journal of Ecological Anthropology, 8*, 47–66.

Levy-Tacher, S. I., Aguirre-Rivera, J. R., García-Perez, J. D., & Martínez-Romero, M. M. (2006). Aspectos florísticos de Lacanhá Chansayab, Selva Lacandona, Chiapas. *Acta Botanica Mexicana, 77*, 69–98.

Lindsay, C. E. (2011). *Culturally modified landscapes from past to present: Yalbac, Belize*. MA thesis, University of Illinois Urbana-Champaign.

McGee, R. J. (2002). *Watching Lacandon Maya*. Boston: Allyn and Bacon.

Miranda, F. (1953). *La Vegetación de Chiapas (segunada parte)*. Tuxtla Gutiérrez, México: Gobierno del Estado de Chiapas.

Morton, C.V. (1933). The Mexican and Central American Species of Viburnum. *Contributions from the National Herbarium, 26*, 339–366. Washington DC: Smithsonian Institution Press. http://www.sil.si.edu/smithsoniancontributions/USNatHerb/pdf_hi/0026.07.pdf

Morton, J. F. (1993). The Black Olive (Bucida Buceras L.), a tropical timber tree has many faults as an ornamental. *Proceedings of the Florida State Horticultural Society, 106*, 338–343.

Nations, J. D. (1992). *Vocabulario Lacandón: plantas y animals*. Paper prepared for Conservation International, April 1992. Washington, DC.

Nations, J. D. (2006). *The Maya Tropical Forest: People, parks and ancient cities*. Austin: University of Austin Press.

Nations, J. D., & Nigh, R. B. (1980). The evolutionary potential of Lacandon Maya sustained-yield tropical forest agriculture. *Journal of Anthropological Research, 36*(1), 1–30.

Nigh, R. (2008). Trees, fire, and farmers: Making woods and soil in the Maya Forest. *Journal of Ethnobiology, 28*(2), 231–243.

Pennington, T. D., & Sarukhán, J. (2005). *Arboles Tropicales de México: Manual para la Identification de las Principales Especies* (3rd ed.). México, DF: Universidad Nacional Autónoma de México.

Rätsch, C. (1994a). Ipomoea violacea linnaeus-morning glory vine. *PsychoTropia*. http://psychotropia.co/?p=1215

Rätsch, C. (1994b). Tsak: die heilpflanzen der Lakandonen von Naha'. *Jahrbuch für Ethnomedizin und Bewußsteinsforschung, 2*, 43–93.

Roys, R. L. (1931). *The Ethno-Botany of the Maya*. New Orleans: Department of Middle American Research, University of Tulane.

Ruan-Soto, F., Cifuentes, J., Mariaca, R., Limón, F., Pérez-Ramírez, L., & Sierra, S. (2009). Uso y manejo de hongos silvestres en dos comunidades de la Selva Lacandona, Chiapas, México. *Revista Mexicana de Micología, 29*, 62–72. http://www.scielo.org.mx/pdf/rmm/v29/v29a9.pdf

Schoenhals, L. C. (1988). *A Spanish-English glossary of Mexican Flora and Fauna*. Mexico DF: Summer Institute of Linguistics.

Toledo, L. L., Burslem, D. F., Martínez-Ramos, M., & García-Naranjo, A. (2008). Non-Detriment Findings Report on Guaiacum Sanctum in Mexico. *Ndf Workshop Case Studies WG 1—trees, Case study 7 Guaiacum sanctum*. Mexico. http://www.conabio.gob.mx/institucion/cooperacion_internacional/TallerNDF/Links-Documentos/Casos%20de%20Estudio/Trees/WG1%20CS7.pdf

Tozzer, A. M. (1907). *A comparative study of the Mayas and the Lacandones*. London: Macmillan.

World Conservation Monitoring Centre (1998). *Ficus lapathifolia*. In IUCN 2012. IUCN Red List of Threatened Species. Version 2, 2012. http://www.iucnredlist.org

The videos for this book can be accessed at:
http://www.springer.com/us/book/9781461491101

Index

A
Agriculture
  cattle ranching, 44
  chop, 309
  cleared land, 309
  clearing, weeding, 310
  climate change, 44
  cold, 310
  commercial enterprises, 44
  cornfield, milpa, 309
  fallow *milpa,* 38
  farming, 41–44
  felling season, 310
  fertile soil, 310
  first milpa, 311
  flooded *milpa,* 45
  fold-over maize, 311
  fungus, 310
  'hot', 309
  jurup che, 37
  Kolil nah 'house garden', 39
  Kol '*milpa*', 37, 39
  Lacandon household, 37
  maize, replanting of, 310
  milpa, old, 310
  monocropping, 44
  natural regeneration, 37
  overgrown, 310
  Paakche' kol, 39–41
  vegetable garden, 309
  weed, 309
Artefacts
  accessories, 311
  ceremonial kitchen utensils
    chocolate, 69, 78
    cooking, 73, 76
    Ka' grindstone, 69, 77
    net full of gourd cups, 73, 75
    table, and bowls, 73, 75
    xämäch clay griddle, 73, 76
  containers, 311–312
    AT and gourd beehive, 61
    gourd, 62–64
    gourd beehive hangs, house rafters, 61, 62
    gourd beehive interior, 61, 62
    pottery, 64
    sources, 61
  cordage, 312–313
  craft, 313–314
  fibre
    bark, 65
    Cord made from hol, 65, 66
    Hol drying on line, 65
    pounded bark fibre, 65, 68
    rafters in bedroom, 65, 67
    rolling fibres of hol, 65–67
    wooden fibre beater, 65, 68
  fishing, 77–79
  furniture, 61, 314–315
  god house
    balche' chem in foreground, 65, 68
    god pot, 68, 70
    K'ayum drum, 68, 70
    Kiln, 69, 71
    Luuch 'cup', 68
    Mashing pom in chemil pom, 65, 69
    Men's pak (Lt), 68, 72
    Pak, 68, 71
    women's pak (Rt), 68, 72
    xikal, 69, 72
  hunting
    arrows, 74–75
    binding, 78
    bows, 74
    fletching, 75
    income, 74
    quiver, 77
  hunting, fishing, trapping items, 315–316
  inedible offerings
    cigars, 70, 73
    rubber, 70, 72
    rubber figurines, 70, 73
    rubber several weeks, 70, 73
  religion and ritual, 65
  religious objects, 317–318
  tools, 318–320
  trapping, 77
  vines, 65

B
Bilingual Lacandon, 9
Botanical classification
  cultivated plants, 97
  folk and botanical taxa, 98
  forest plants, 97
  generics, 98
  intermediate generics, 99
  life forms, 97–98
  specifics, 98
  unique beginner (UB), 97
  varietals, 98

S. Cook, *The Forest of the Lacandon Maya*,
DOI 10.1007/978-1-4614-9111-8, © Springer Science+Business Media New York 2016

Botanical inventory
Abelmoschus moschatus, 107
abnormal guatteria (black cork), 141
abraza palo, 148
acacia, bull horn, 245
Acacia mayana, 105
Acoelorraphe wrightii, 117
Aegiphila monstrosa, 177–179
Agave sisalana, 170
Alchornea latifolia, 207
alligatorwood (muskwood), 235
allspice (Jamaican pepper), 224–226
Alpinia purpurata, 250
Amaryllidaceae *Allium sativa L.,* 107
amaryllis and lilies, 264
American burnweed (fireweed, pilewort), 213
American muskwood, 214–215
Andropogon glomeratus, 268, 269
angelica tree (potatowood, sweetwood, white gumbo limbo), 232
annatto (lipstick tree), 187, 189
Annona cherimoya, 240
Annona muricata, 222, 223
Annona scleroderma, 217
Anthurium huixtlense, 210
Anthurium schlechtendalii (bird's nest type), 209
Anthurium sp., 210
apple, custard, 129, 156, 216–217
apple, monkey, 168
apple, wild star, 241
ardisia, 209, 228–229
Ardisia compressa, 187, 189, 221
Ardisia paschalis, 222
arrabidaea (type of), 273
Arrabidaea verrucosa, 235
arrowhead vine (goosefoot plant), 202, 213
artist's bracket (artist's conch, flacher lackporling), 177
arum, 215
Asclepiadaceae, 170–171
Astrocaryum mexicanum, 143–144
Astronium graveolens, 200–202
authentic bean, 148
avocado, American (alligator pear), 128, 215–216, 277
avocado tree, 216
balsam of Peru, 208
bamboo, golden (striped bamboo, painted bamboo), 113
Bambusa vulgaris, 114
banana, 117
banana, Cuban red, 117
banana 'Lady Finger' (date banana, fig banana, sugar banana), 197
banana, Orinoco (tall), 222
banana, type of, 145
barbasco, 191
bauhinia, rio rubeleruz, 210
bayal (basket palm, liana palm), 150–151, 162
bead tree (horse-eye beans), 108
bean, 212–213
bean, castor, 189, 200, 213
bean, castor (green), 277
bean, castor (red), 126
bean, Jack (Chickasaw lima bean), 119
bean, lima, 126, 165
bean, Little Marble, 165
bean, rice (small red bean), 252
bean, Saint Mary's, 142
bean, Saint Mary's (crucifixion bean, elephant creeper, woolly morning glory), 163
beans, bush, 130
beans, climbing (pole beans), 105
bean, shelling, 118, 119
beans, Jack (Chickasaw lima bean), 278
bean, twining snout, 133
beardgrass, silver. *Lit:* grass geass, 105
beardgrass, silver. *Lit:* grass weed, 105
Belotia mexicana, 159
Bidens odorata, 177, 178
billia, 178
Billia colombiana, 178, 180
bitterwood, 139–140
Bixa orellana, 193–194
black bean, 141
blackberry; brambles, 161
black bush beans, 119
black lima bean, 141–142
bloody croton, 134
bone blossom tree, 110–111
Bothriochloa laguroides, 105
bottle gourd, 133–134
Bouvardia longiflora, 210
bouvardia, scented, 210
Brasil blanco, 130
Brazil beauty-leaf, Santa Maria, 109–110
Brazilian fern tree (Brazilian fire tree, Mexican fern tree), 225
breadnut, Mayan, 183–184, 217–218, 277
Bromelia pinguin, 136, 137
Brosimum alicastrum ssp. *Alicastrum,* 277
Brosimum sp. leaves, 156
bullhoof, 197
bull horn, 104
bullock's heart, 128
bullywood, 181–182, 227–228
bunchosia (peanut butter tree), 243
Bursera simaruba, 123
bush beans, 125
Byrsonima crassifolia, 182–184
cabbage, 174
cacao, 169–170
cacao, mocambo, 112
cacao, wild, 112
cactus, climbing (orchid cactus, leaf cactus), 195
Caesalpinia major, 220
calabasa (squash, pumpkin), 191–192, 277
calabash, 118, 154, 195, 197, 244
calabash, bottle gourd, 133–134, 161–162, 171
calabash, Mexican (winged calabash), 183
calabash tree, 197, 201
calathea, 267
calathea Cigar, 235, 240
Calathea crotalifera, 168
Calathea lutea, 240–241
Calathea macrosepala, 177, 275
caliandra, 126–127
Calliandra calothyrsus, 127
Canavalia ensiformis, 120
candle tree, 107, 169
cane, giant (common reed), 215
canistel (yellow sapote, egg fruit), 217
canna, 119–120
Canna indica, 119–121
Capsicum frutescens, 121
Caribbean grape, 280
Carica papaya, 228, 229
Carludovica labela, 205

Castilla elastica, 185, 187, 188
catalina, 123–124
Cecropia spp., 189, 190
cedar, 190–191
Cedrela odorata, 191, 192
ceiba (cotton tree, floss silk), 273–275
Ceiba pentandra, 273, 274
Cestrum nocturnum, 103, 104
chamaedorea, 117–118, 121, 138–139, 235–236, 249–250, 275
Chamaedorea arenbergiana, 236
Chamaedorea cataractarum, 137
Chamaedorea elatior, 275, 276
Chamaedorea elegans, 136–137
Chamaedorea ernesti-augusti, 185
Chamaedorea neurochlamys, 182, 183
Chamaedorea pinnatifrons, 145, 148
Chamaedorea sp., 138–139
Chamaedorea tepejilote, 219–220
Chamaeorea pinnatifrons, 236
Chamaesyce hirta, 270
cherimoya (custard apple), 240
cherry, 255–256
chestnut, Guiana, 174
chicle, 130, 160–161, 273
chile (type of), 275–276
chile pepper, 140, 236
chile pepper, red, 126
chile pepper, yellow, 183
chile, Tabasco, 152
Chione chiapasensis, 200, 221
chocolate pudding fruit, 265
choco palm, spiny palm, 103–104
Chrysophyllum mexicanum, 241, 242
cilantro (coriander, Chinese parsley), 225
Cissus biformifolia, 257–258
Cladium jamaicense, 242, 243
Clarisia biflora, 128
Clethra sp., 175, 176
Clibadium arboreum, 123, 124, 171
clusia (pitch apple), 134, 262
Clusia flava, 169
Clusia lundellii, 135, 262–263
Cnidoscolus multilobus, 252, 253
coccoloba, 113–114
Coffea, 169
coffee, 169
coffee-shade tree, 247
coffee, wild (limoncillo), 278
coffee, wild (typical Panama balsamo), 212
Coix lacryma-jobi, 234
cojoba, 119, 200
Cookeina tricholoma, 177, 178
copal, 252
coralberry (marlberry), 153, 187, 202
coral fungi, 177
coral tree, 180
coral tree (coral bean tree), 178, 180
Cordia alliodora, 110
Cordia dodecandra, 270
coriander, long, 174
cork, 159
Costus pulverulentus, 219
cotton, 249
cow pea (black eyed pea), 118–119
crabgrass (East Indian crabgrass, bristly crabgrass), 158
Crescentia cujete, 154
Crinum amabile, 130
Croton draco, 134
Cryosophila stauracantha, 175, 176
cucurbit (Guadeloupe cucumber, meloncito, speckled gourd), 231
Cucurbita moschata, 191, 193
cushaw pumpkin (pipian), 243
custard apple tree, 156, 216
Cydista aequinoctilis, 107, 108
daisy family, 123, 171
daisy tree, giant, 142, 171
deer eye (ox eye), 210
Dendropanax arboreus, 232, 233
Desmodium tortuosum, 254
Desmoncus orthocanthus, 151
devil nettle (bull nettle), 252
Dieffenbachia sp., 131
Digitaria setigera, 158
Dioscorea alata, 158
Dioscorea bartlettii burl, 191, 192
Dioscorea bulbifera, 115
Diospyros digyna, 265
Dipholis sp., 246
Disciphania calocarpa, 222, 223
dog-killer, 133
dogwood (christmasbush), 207
Doliocarpus dentatus, 233
dumbcane, 131
ear flower, 250
elephant ears, 126, 199, 246
elephant ears, capote, 148, 159
elephant ears, malanga (Caribbean cabbage), 154, 238
Entada gigas, 174
Epiphyllum sp., 196
Eriobotrya japonica, 255, 259
Eryngium foetidum, 174
Erythrina berteroana, 180
Esquisetum hyemale, 198
eugenia (type of), 202
fabaceae *Phaseolus vulgaris,* 119
false jade (monkey-tail), 182
false mamery, 109
fat pork, 221
fern, bracken, 266–267
fern, shield, 267
fern tree, 266
ficus, 152, 173
Ficus pertusa, 200, 201
Ficus sp., 153, 165, 173
fig, rock (strangler fig), 173
fig, Sonoran strangler (perforated fig, plumleaved fig tree, strangler fig), 200
fish-tail palm (Ernest August's palm), 185
flame tree, 226
Florida tibisee (smallcane), 244–245
frangipani, 127, 209–210, 239
funeral tree (funeral swizzlestick tree), 198
fungi (types of), 142
fungi, bonnets, 206
fungi, white bonnets, 206
fungus, jelly, 196
fungus, red cup, 170, 177
fungus, split-gill, 270, 271
garlic, 107
garlic vine, 107
Geonoma oxycarpa, 192, 194
giant cane, 162

Botanical inventory (*cont.*)
giant groundsel, 206
ginger, 172, 220
ginger, ostrich plume (pine cone ginger), 172
ginger, red, 249
ginger, spiral, 219
glassy wood, 200
glorybower (bagflower, bleeding-heart), 167
goat's milk tree, 112
gonzalagunia, thyrsoid, 206
Gonzalagunia thyrsoidea, 207
grape, Caribbean (water vine), 159–160, 204
grape, wild, 264
grass, bushy bluestem (bushy beardgrass), 268
grass, napier (merker grass, cane grass, elephant grass), 240, 267
grass, saw (fen-sedge, twig-sedge), 242
ground cherry, 204
guaba_inga, 116
Guarea glabra, 235
Guatemala blepharidium, 234
Guatteria anomala, 141
guava, 129, 226, 239
guayacan trumpet tree, 148
gumbo limbo, 123
Gynerium sagittatum, 215, 216
hairy beggarticks (Spanish needle), 177
Hamelia calycosa, 162
Hampea stipitata, 263
Hawaiian wood rose (Spanish arborvine, Spanish woodbine, yellow morning glory), 260
heliconia (type of), 134, 239
Heliconia librata, 134, 154–155, 239
heliocarpus (sunfruit), 125, 150
Heliocarpus appendiculatus, 125
Heliocarpus donnell-smithii, 150, 238
hibiscus, 249
hicatee fig, 131
hubu, 254
huevos de caballo, 250
Hymenocallis littoralis, 264
icacina, 174–175
inga (shimbillo), 115–116
inga (type of), 200
inga, Belize, 116
Inga oerstediana, 247
Inga pavoniana, 252
Inga punctata, 211
Ipomoea alba, 152
Ipomoea batatas, 167
Ipomoea sp., 251
ironwood, black (leadwood), 132
Isthmus ormosia, 108
Jaltomata repandidentata, 135
Jamaica cherry, 227
Jamaican nettletree, 129
Jatropha curcas, 243
jicama (yam bean, Mexican potato), 131
Job's tears, 233
jungleplum (bustic), 245
Justicia sp., 258
Kalanchoe pinnata, 232
Lagenaria siceraria, 195, 196
lancepod, 111–112, 144
Lantana trifolia, 122
lemongrass, 103, 144
lemon pepper (joint wood, buttonwood), 258
Leucaena leucocephala, 257
Licania platypus, 256
Licaria alata, 239
lily, cahaba (shoals spider lily), 263
lily, giant spider, 129
lime, Mexican (key lime), 205
limoncillo, 278
locustberries (serrets), 275
logwood (bloodwood tree, inkwood), 140
Lonchocarpus rugosus, 273, 274
Lonchocarpus sp., 144, 273, 274
loquat (Japanese medlar, Japanese nispero), 255
louteridium, 191
Louteridium donnell-smithii, 190, 191
lycianthes, 135, 191
Lycianthes heteroclite, 135
macawood (granadillo), 236
magnolia, heartflower tree, 176–177
mahogany, 229
mahogany, bastard (wild mahogany, mosquito wood), 212
maiden plum, 181
maize, 142, 181, 208–209, 239
majagua (sun-fruit), 150
majagua, white, 237
Malvaviscus arboreus, 249
mammee sapote, 142
mango, 199
Manihot esculenta, 129, 160, 261–262, 277–278
manioc, 129, 159, 240, 261–262, 278
Marantaceae *Calathea* sp., 129–130
marmalade plum, 142
mastate, 105
mata rama (God bush), 165, 215
Melothria pendula, 231, 232
Metopium brownei, 130, 131
Mexican pepperleaf (eared piper, rootbeer plant), 151
milady, my lady, 232–233
mint, 205, 258
mistletoe, 204
mombin (hog plum), 124
mombin (wild hog plum), 163
mombin (yellow mombin, Spanish plum, gully plum), 187
mombin, green (hog plum), 273
mombin, red (hog plum), 273
mombin, yellow (Spanish plum, hog plum, scarlet plum), 182
monkeysoap (guanacaste), 225–226
monkeytail (shadow palm), 218
Monstera sp., 152–153
moonflower (moon vine), 152–153
Moraceae *Ficus* spp., 173
Moraceae *Pseudolmedia spp.*, 112
Morinda yucatanensis, 272
morning glory, 125–126, 163, 251
Mosquitoxylum jamaicense, 212
mountain bribri, 252
mountain moho, 263
Mucuna argyrophylla, 211
Mucuna sloanei hilum, 269
Muntingia calabura, 228
Musa acuminata ×*Musa balbisiana*, 125, 145, 147, 157, 163, 165, 168, 186, 222, 224
mushroom, pink oyster, 180–181
musk mallow, 107
nance, 132, 182
nectandra (type of), 202, 213
nectandra, globose, 142

nickernut, 220
night-blooming cestrum, 103
night-blooming cestrum (night-blooming jasmine, night jessamine), 265
nightshade, 135
nightshade, American black, 135–136
nightshade, rovirosa, 135
nikte' ak', 235
nukuch koyoh, 212
nut, physic, 343
oak, 135, 200
oak, chicarro (Mexican red oak), 226
oak, corrugated, 149, 212
obtuse leaf oreopanax, 181
Ocotea cernua, 110
ocotea, drooping, 110, 168–169
Oecopetalum mexicanum, 174–175
onion, 258–259
onion cordia (Spanish elm), 110
onion, white, 240
onion, wild (scallion), 258
orange, mandarin, 125
orange, sweet, 132
orchid, 190
orchid tree, 233
Oreopanax obtusifolius, 181
Ormosia isthmensis, 108, 109
Ormosia schippii, 79
Oudemansiella aff. *steffendii*, 142
Oxyrhynchus trinervius, 165
Pachira aquatica, 174
palm (fan palm, give-and-take palm, rootspine palm), 175
palm, cat (cascade palm), 137
palm, choco, 143
palm, corozo, 251
palm, coyol (grugru palm, macaw palm, macauba palm), 251
palm, Everglades, 117
palm, fan (sabal, Mexican palmetto, Texas palmetto), 268–270
palm, monkeytail (shadow palm), 218
palm, pacaya, 138, 219
palm, Panama hat, 205
palm, parlor, 136, 138
palm, woodsonia, 137–138
palo blanco, 130–131
Panama balsamo, 279–280
Panama rubber tree, 185
papaya, 228
parathesis (type of), 196, 271
Parathesis sp., 197, 271
Passiflora aff. ambigua, 141
Passiflora biflora, 192, 193
Passiflora cookii, 138
Passiflora serratifolia, 227
passion flower (bat-leaved type), 192
passion fruit vine, 140
passion fruit, wild, 227
peanut, 169, 243–244
pear, vegetable (pear squash), 185, 230–231, 240, 277
Pennisetum purpureum, 269
pepper, 202
pepper, bird (chiltepe, chile tepin, bird's eye), 247–248
pepperbush, 175
pepper, chile, 121, 126, 140, 166, 210, 241
pepper, Jamaican (cordoncillo), 202
petunia, Mexican (Mexican bluebell), 275
philodendron, 162–163
Philodendron donnell-smithii, 156
philodendron, heart leaf (velvet philodendron), 202
Philodendron hederaceum, 202, 203
Philodendron inaequilaterum, 148, 149
Philodendron radiatum, 185, 186
Philodendron sagittifolium, 213, 214
Philodendron smithii, 163, 280
philodendron, split leaf, 192
Philodendron tripartitum, 280
Phoradendron crassifolium, 204
Physalis pruinosa, 230
pillpod sand-mat, 270
Pimenta dioica, 225
pineapple, 218
pinguin (wild pine), 136
Pinus sp., 248
piper, 162
Piper aduncum, 154
Piper amalago, 258, 261
Piper auritum, 152
piper, matico (bamboo piper), 154
piper, monkey hand, 237–238
Piper sp. plant, 199
pitch pine, 248–249
plantain, 22–223
plantain (type of), 157, 168, 199, 224
plantain (white), 240
Platymiscium dimorphandrum, 237
Pleuranthodendron lindenii, 168
Plumeria rubra f. rubra, 127
Pluteus harrisii, 142, 206
poaceae, 103
Podachaenium eminens, 143
podocarp, 136
podocarp, Jalisco, 114
Podocarpus matudai, 115
poinsettia (Christmas star, Christmas flower), 122
poisonwood tree, 130
pokeweed, tropical (red inkplant), 249
Polianthes tuberosa, 110, 111
Polyporus tenuiculus, 271
potato, 220
potato, air, 115
potato, sweet, 126, 129, 167, 238
potato tree (mullein nightshade, tobacco tree, big eggplant), 265
potato, white, 247
Pouteria mammosa, 142, 143
prickly-ash (Hercules' club), 194–195
Protium copal, 255
Pseudolmedia oxyphyllaria, 145, 260
Pseudolmedia sp., 113
Psidium guajava, 226
psychotria, 278
Psychotria panamensis var. panamensis, 280
Psychotria poeppigiana, 234
Psychotria sp., 279
psychotria, type of, 278
pumpwood (trumpet tree), 189, 202, 213
purslane, 181
Quararibea funebris, 198, 199
Quercus skinneri, 226
Quercus sp., 200, 201
red bean, 125
red chinah, 132
red vine, 124
reed, 150, 162

Botanical inventory (*cont.*)
reed, common (giant cane), 127, 155, 239
Renealmia sp., 172, 173
Rhipidocladum bartlettii, 241, 242
Rhynchosia erythrinoides, 133
Rhynchosia pyramidalis, 205, 206
Ricinus communis, 126
royal fig (lateralflower sapium), 264
Rubus sp., 161, 162
Sabal mexicana, 268, 270
Saccharum officinarum, 157, 204
safflower (false saffron), 244
säk suum flowers, 280
San Lorenzo (air plant), 232
sansapote, 254
Sapindus saponaria, 143, 245
Sapium lateriflorum, 266
sapote (type of), 150
sapote, white, 268
sarsaparilla, 257, 259
sarsaparilla (greenbriar), 171–172
Saurauia yasicae, 206, 207
Schipp ormosia, 108
Schistocarpha eupatorioides, 123, 124
Schizolobium parahybum, 225
Schizophyllum commune, 271
scouringrush horsetail, 197
seagrape, Barbados, 210
seagrape, Honduras, 199
sea heart, 173
Sechium edule, 231
Selaginella silvestris, 194, 195
shaving brush tree, 140
Siparuna thecaphora, 279
sisal (henequen), 170
slipplejack, 198
smilax, 129, 277
Smilax luculenta, 172
soapberry, 108–109, 198, 245
Solanum americanum, 135, 136
Solanum erianthum, 265, 267
Solanum rovirosanum, 135, 136
Solanum torvum, 185, 187, 265, 267
sore-mouth bush (hot lips), 234
soursop, 222
Spathiphyllum phryniifolium, 221–222
spikemoss, 194
spiny pod (milkweed), 170
split gill, 271
Spondias purpurea forma lutea, 182
Spondias sp., 163, 164
squash seeds summer squash, 206
stinktoe (sausage tree, cuapinol), 219
Struthanthus orbicularis, 166
Strychnos tabascana, 258
sugarcane, 129, 157, 204, 246, 261
Swietenia macrophylla, 229, 230
Swiss cheese plant, 152, 166, 169, 238
Syngonium chiapense, 213, 214
Syngonium podophyllum, 203, 213, 214
Talauma mexicana, 176
taw, 247
Telanthophora grandifolia, 206, 207
Terminalia amazonia, 227, 228
Ternstroemia tepezapote, 128
Theobroma bicolor, 112
Thevetia ahouai, 251
Thevetia, broadleaf (cascabel, dog's tongue), 159, 204, 250–251
threeleaf (lavender popcorn), 122
three-nerved baccharis, 244
ticktrefoil, Dixie, 252
timber sweet, cayenne rosewood, 110
tobacco, 153, 193
tomato, 229
tomato, husk, 114
tomato, Mexican husk (Mayan husk tomato, ground cherry), 230
Trema micrantha, 247, 248
Trichilia sp., 140
Turkcap (Turk's turban, wax mallow, ladies teardrop, Scotchman's purse), 249
turkey berry (devil's fig, prickly nightshade), 185
turkeyberry nightshade (devil's fig, prickly solanum), 265
vanilla, 119
Vanilla planifolia, 119, 121
velvet bean (cowhage, cowitch), 119
velvet bean (horse-eye bean, deer eye), 267
Vigna umbellata, 252
Vigna unguiculata, 118
vine, virility (snout bean), 205
virola, 187, 188
Virola guatemalensis, 188
Vitis sp., 265
Vitis tiliifolia, 204, 205
water lily, 189
watermelon, 231
white leadtree (jumbay, white popinac), 254
wild tamarind, 267
Wimmeria bartlettii, 183
wing-leaf soapberry, 143
wormseed (epazote, Mexican tea), 184
Xanthosoma mafaffa, 126, 238
Xanthosoma robustum, 140
xate macho (jade), 145, 161, 236
ya'ax'oon, 277
yam (barbasco), 246
yam, winged (greater yam, water yam, purple yam), 158
yellow rattle shaker (rattlesnake calathea), 168
yellow wood, 271
Zanthoxylum sp., 195
zingiberales, 207
Botanical-Lacandon Index, 281–292
Botanical terms, Lacandon, 343–351

C

Chiapas, 1, 2
Clothing
bandana, 320
bark cloth
headbands, 320
tunic, 320
children garment, 90, 93
headdress, 320
loincloth, 320
men
bark cloth xikul, 88, 91
cotton xikul, 88, 91
leather purse-back, 88, 92
leather purse-front, 88, 91
poncho, 320
women
Clusia flava blossom, 89, 93

feather hair ornament, 89, 93
traditional women's dress, 89, 92
Containers
AT and gourd beehive, 61
basket, 311, 312
bee hive, 312
cup, large, 311
cup, small, 312
gourd, 62–64
gourd beehive hangs, house rafters, 61, 62
gourd beehive interior, 61, 62
'hand cup', 312
plate, bowl, 312
plate, metal, 312
pot, cooking, 312
pottery, 64
sources, 61
Cordage
bag (woven), 313
board for removing agave fibre, 313
lashing, tie, rope, 313
rope, thick, 313
string/cord, twisted, 312
'Swietenia macrophylla', 313
Craft
cloth (traditional) hand-loomed cotton, 314
heddle, 313
necklace, 314
'sack's cord', 314
shed, 314
spindle, 314
spinning top, 313
warp (of weave), 313
warp end rods, 314
warping board, 313
weaving sticks, 313
weft, 313
Crafts
artistic skill and attention, 79
bows and arrows, 74, 80
charming effect, 79
dyes and stains, 86, 90
jewellery
necklace with hamburger beans, 79, 80
seed necklace, 77
toucan beak, 79, 80
musical instruments, 81–82
pottery, 80–81
spinning, 82–83
weaving techniques, 83–86
Ba'ay 'net bag, 84–85, 88
backstrap loom, 84
basket, 84, 85
basket bottom, 84, 86
basket rim, 84, 85
hammock, knots, 86, 90
hand spun cotton xikul, 84
hanging basket, 86, 87
traditional hammock, 86, 89
woven bag, loop, 86, 88
woven fabric, 83
woven purse, 86, 88, 89

D

Deities, 321–322

E

English-Lacandon Index, 218–292
Ergative, 9
Ethnobotanies, 5
Ethnographic inventory
accessories, 311
agriculture (*see* Agriculture)
clothing, 320–321
containers, 311–312
cordage, 312–314
craft, 313–314
deities, 321–322
food
ashes, palm, 324
corn gruel, 323
dough, 324
fermented beverage, 322
first fruits offering, 325
gruel, 323, 324
honey, nectar, 323
meal, food, 324
nixtamalized corn, 323
porridge, 323
quicklime, 323
salt, commercial, 322
shell corn, 322
sugar, 324
sugar, unrefined, 322
tamal, 324
tamale, ceremonial, 323
tamale, meat, 322
tortilla, 322–323
tortilla, tamale, 324
fuel (*see* Fuel)
furniture, 314–315
geography (*see* Geography)
housing, 327–330
hunting, fishing, trapping items, 315–316
medical conditions (*see* Medical conditions)
musical instruments, 316–317
religious objects, 317–318
tools, 318–320

F

Farming
crops, 42–43
fallow *milpas*, 41
harvest, 43–44
nah kolil, 41
preservation, 44
SK's description, agricultural round, 42
squash, 44
Food. *See also* Ethnographic inventory
beans, 26
fruits, 27
fungi, 27
maize, 25–26
gruels, 26
porridges, 26
tamales, 26
tortilla dough, 26
ritual offerings
edible offerings, 28
first fruit rituals, 29
vegetables, 27

Fuel
'authentic pahok', 336
bead tree, 108
bitterwood, 139
bullywood, 181–182
Clarisia biflora, 128
coffee, 278
Cojoba, 212
Guatemala blepharidium, 234–235
logwood, 140
Ormosia isthmensis, 347
palm, 251
pitch pine, 248–249
'tsup vine', 335
wild tamarind, 267
Furniture, 314–315

G
Geography
areas with standing water, 325
cave, 325
clay, yellow, 325
earth, soil, ground, 326
green lagoon, 354
Itsanok'uh, 326
kuyak, 326
land, world, 326
level land, 325
Mensabak, 326
mountain side, 327
Naha', 326
obsidian, 327
Päk'anil pom, 326
'red cave', 325
sand, 326
soil (black), 325
Ts'ibatnah, 327
wits hill, 327

H
Housing
akam, 329
chicken coops, 45, 50
construction
battens and shelf, god pots, 51, 56
battens in god house, 51, 58
ceremonial kitchen, 53, 59
concrete house, 52, 60
god house, 51
god house interior, 51, 58
hi'che', 51
hool door, 328
jailhouse, 52, 60
Kun thatch, 51, 55
lashing, 51, 57
mooy, 50
parts, god house, 51, 56
roof, 49
rot-resistant timber, 46
smoke vent, 51, 57
vertical pole house walls, 52, 59
cooking
hearth, 45, 51
range, 45, 52
traditional stove, 45, 55
corn crib, 355
extended families' compounds, 46
garden, compost, 45, 47
girt, 329
hearth, 328–329
home, 330
homestead, 46, 48
house, main part, 328
house posts, 329
kitchen, 45, 328
Koh's backyard, 45, 48
Kohs' kitchen, 2000, 45, 51
Lacandon house, 45, 50
new Kohs' kitchen, 2011, 47, 52
old Kohs' kitchen, 45, 52
outdoor sink, 45, 49
overhang, 329
rafters, 329
shed, 329
storage shed, 45, 49
stove, 45, 54
vegetable gardens, 45
wall, 329
woman on hammock, 45, 51
Hunting, fishing, trapping items
blowgun, 316
bow, 315
fish hook, 316
fishing line, 316
fish trap, 316
flint arrowhead, 316
harpoon, 315
nock, 316
notch, 316
quiver, 315
shaft, 316

I
Inventory format
botanical, 5–6
enthnographics, 6

L
Lacandones
botanical inventory, 2
collection methods, 2
consultants, 2–5
history, 1
homesteads, 1
jungle settlements, 1
Lacandon forest. *See* Lacandones
Lacandon orthographic conventions, 10
Lacandon plants botanically unidentified, 335–342
Lacandon rainforest
climate
Chiapas physical map, 14, 15
Lacandon area, 14, 17
Lacandon forest, 14, 16
Northern Lacandon territory, 14, 18
deforestation and human activities, 13
Flooded Thorn Woodland, 17
Maya Lowlands, 13
moist soils and humidity, 13

Naha' and Mensäbäk", 13–14
secondary vegetation
black soil, 18, 19
Canoe by rock cliff, 18, 20
Cattle in pasture, 18, 22
fallow milpa, 18
forest, 10 year fallow, 18, 20
Lacandon homestead, 18, 23
Lagoon, 18, 21
Naha', 2010, 18, 22
Schoolyard, 18, 23
second-growth and successional forest, 17–18
soil structure, 18, 19
waterfall in southern Lacandon territory, 18, 20
topography and climate, 13
vegetation
Flooded Thorn Woodland, 17
Lower Montane Rain Forest, 14–15
Montane Rain Forest, 15
Pine forest, 16–17
Pine-Oak-Liquidambar, 15–16
secondary vegetation, 17–23
Lacandon Texts, 357–369
Lacandon traditional culture
agriculture (*see* Agriculture)
artefacts (*see* Artefacts)
clothing (*see* Clothing)
Crafts (*see* Crafts)
food (*see* Food)
food plants, 25, 29–31
housing (*see* Housing)
material and nutritional requirements, 25
material plants, 25, 31–34
subsistence and material culture, 25
Language system, 9

M
Maya. *See* Lacandones
Medical conditions
baby rash, 332
bed-wetting, children, 331
blister from burn, 331
boils, skin eruptions, 331
chibal, 330–331
diarrhea, 331
evil eye, 330
fever, 330
itchiness, 332
larvae of botfly, 332
leishmaniasis, 331
painful, 333
snakebite, 330
stomach ache, 330
swelling, 331
toothache, 330–331
vomiting, nausea, 333
worms, intestinal, 332
Medicines
air and wind, 96
and curing, 90, 92, 93, 96, 97
plant medicines, 90, 94–96
pregnancy, 90, 93
Monolingual Lacandon, 9
Musical instruments, 316–317

N
Naha, 1–3

P
Polysynthetic language, Lacandon, 10
Pronunciation guide, 10

R
Religious objects
balche' urn, 318
board, 318
ceremonial incensories, 317
cup, 317
figurines, 318
incense board, 317
shelf, for god pots, 317
shelf, for incense, 317
strainer, cloth, 317

S
Seasons, 353, 354
Spanish households, 9

T
Tools, 318–320

V
Vine bark cloth, 105

W
Weather
clear sky, 355
climate change, 355
cloudy, 355
cool, refreshing, 355
drizzle, 355
mid-day, 355
mist, 356
overcast, 355
rain, 356

Y
Yucatecan, 9

MIX
Papier aus verantwortungsvollen Quellen
Paper from responsible sources
FSC® C105338

If you have any concerns about our products,
you can contact us on
ProductSafety@springernature.com

In case Publisher is established outside the EU,
the EU authorized representative is:
Springer Nature Customer Service Center GmbH
Europaplatz 3, 69115 Heidelberg, Germany

Printed by Libri Plureos GmbH
in Hamburg, Germany